Multimedia Signal Processing

Multimedia Signal Processing

Theory and Applications in Speech, Music and Communications

Saeed V. Vaseghi
Professor of Communications and Signal Processing
Department of Electronics, School of Engineering and Design
Brunel University, UK

John Wiley & Sons, Ltd

Copyright © 2007 John Wiley & Sons Ltd, The Atrium, Southern Gate, Chichester,
West Sussex PO19 8SQ, England

Telephone (+44) 1243 779777

Email (for orders and customer service enquiries): cs-books@wiley.co.uk
Visit our Home Page on www.wiley.com

Other Wiley Editorial Offices

John Wiley & Sons Inc., 111 River Street, Hoboken, NJ 07030, USA

Jossey-Bass, 989 Market Street, San Francisco, CA 94103-1741, USA

Wiley-VCH Verlag GmbH, Boschstr. 12, D-69469 Weinheim, Germany

John Wiley & Sons Australia Ltd, 42 McDougall Street, Milton, Queensland 4064, Australia

John Wiley & Sons (Asia) Pte Ltd, 2 Clementi Loop #02-01, Jin Xing Distripark, Singapore 129809

John Wiley & Sons Canada Ltd, 22 Worcester Road, Etobicoke, Ontario, Canada M9W 1L1

Wiley also publishes its books in a variety of electronic formats. Some content that appears in print may not be available in electronic books.

Anniversary Logo Design: Richard J. Pacifico

Library of Congress Cataloging in Publication Data

Vaseghi, Saeed V.
Multimedia signal processing: theory and applications in speech, music and communications / Saeed V. Vaseghi.
p. cm. Includes index.
ISBN 978-0-470-06201-2 (alk. paper)
1. Signal processing. 2. Multimedia systems. I. Title.
TK5102.9.V38 2007
621.382′2—dc22

2007014614

British Library Cataloguing in Publication Data

A catalogue record for this book is available from the British Library

ISBN 978-0-470-06201-2 (HB)

Typeset in 9/11pt Times by Integra Software Services Pvt. Ltd, Pondicherry, India

To my dears Luke, Jalil, Iran, Geraldine, Simin, Sima, Mona, Malisha

Contents

Preface

The applications of digital signal processing (DSP) in information and communication technology are pervasive and include multimedia technology, cellular mobile communication, adaptive network management, radar, pattern recognition, medical signal processing, financial data forecasting, artificial intelligence, decision making systems, control systems and search engines.

The aim of this book is to provide an accessible text to the theory and applications of DSP. This is an ambitious task as signal processing covers such a wide range of topics that it would take several volumes to cover the entire subject area. Nevertheless, I have tried to cover a set of core topics and applications that would enable interested readers to develop a practical understanding of the design and working of modern application of signal processing.

For example, the popular MP3 music coder is based on a combination of a number of signal processing tools including filter banks, discrete Fourier transform, discrete cosine transform, auditory masking thresholds, Huffman code and rate-distortion theory. All these topics are covered in this book as is their combined application in a single system such as MP3.

This book is arranged in three parts and eighteen chapters.

Part I Basic Digital Signal Processing in five chapters introduces the Fourier and z transforms, digital filters, sampling and quantisation.

Chapter 1 begins with an introduction to digital signal processing, and provides a brief review of signal processing methodologies and applications.

Chapter 2 presents the theory and application of Fourier analysis and synthesis of signals. This chapter covers discrete Fourier transform (DFT), fast Fourier transform (FFT) and discrete cosine transform (DCT). Important engineering issues such as time-frequency resolutions, the effect of windowing and spectral leakage are discussed. Several applications of Fourier transform, such as the design of a spectrogram, are presented.

Chapter 3 provides an introduction to z-transform. The relationship between z-transform, Laplace transform and Fourier transform are explored and z-transform is derived from the Laplace transform. The concept of z-transfer function and the associated poles and zeros of the transfer function are introduced.

Chapter 4 introduces the theory and design of digital filters. This chapter starts from a basic explanations of the working of filters and introduces different popular methods for filter design. This chapter also covers the design of quadrature mirror filters which have applications in coding and wavelet analysis.

Chapter 5 introduces the theory and practice of sampling for conversion of analogue signals to discrete-time signals, the process of quantisation of analogue samples of a discrete-time signal to digital values and the process of resampling of a discrete-time signal.

Part II Model-based Signal Processing covers the theory and applications of probability and information models, Bayesian inference, Wiener filters, Kalman filters, adaptive filters, linear prediction models, eigen analysis, principal component analysis (PCA) and independent component analysis (ICA).

Chapter 6 presents probability and information models. This chapter begins with an introduction to random signals, stochastic processes, probabilistic models and statistical measures. The concepts of entropy, stationary, non-stationary and ergodic processes are introduced and some important classes of random processes are considered. The effects of transformation of a signal on its statistical distribution are considered.

Chapter 7 is on Bayesian inference. In this chapter the estimation problem is formulated within the general framework of Bayesian inference. The chapter includes Bayesian theory, classical estimators, the estimate–maximise method, the Cramér–Rao bound on the minimum-variance estimate, Bayesian classification, and the k-means method of modelling of the space of a random signal. This chapter provides a number of examples on Bayesian estimation of signals observed in noise.

Chapter 8 considers Wiener filters. The least square error filter is first formulated through minimisation of the expectation of the squared error function over the space of the error signal. Then a block-signal formulation of Wiener filters and a vector space interpretation of Wiener filters are considered. The frequency response of the Wiener filter is derived through minimisation of the mean square error in the frequency domain. Some applications of the Wiener filter are considered. A case study of the Wiener filter for removal of additive noise provides particularly useful insight into the operation of the filter.

Chapter 9 considers adaptive filters. The chapter begins with the state-space equation for Kalman filters. The recursive least squared (RLS) filter, which is an exact sample-adaptive implementation of the Wiener filter, is derived in this chapter. Then the steepest-descent search method for the optimal filter is introduced. The chapter concludes with a study of the least mean squared (LMS) adaptive filters.

Chapter 10 considers linear prediction. Forward prediction, backward prediction, lattice predictors and sub-band predictors are studied. This chapter introduces a modified predictor for the modelling of the short-term and the pitch period correlation structures. A maximum a posteriori (MAP) estimate of a predictor model that includes the prior probability density function of the predictor is introduced. This chapter concludes with the application of linear prediction in signal restoration.

Chapter 11 considers hidden Markov models (HMMs) for non-stationary signals. The chapter begins with an introduction to the modelling of non-stationary signals and then concentrates on the theory and applications of hidden Markov models. The hidden Markov model is introduced as a Bayesian model, and methods of training HMMs and using them for decoding and classification are considered. The chapter also includes the application of HMMs in noise reduction.

Chapter 12 covers the immensely important and related subjects of eigen analysis, principal component analysis and independent component analysis. Several illustrative example of applications of each method are provided.

Part III Applications of Digital Signal Processing in Speech, Music and Telecommunications covers speech processing, music processing, speech enhancement, echo cancellation and communication signal processing.

Chapter 13 on music signal processing begins with an introduction to musical notes, musical intervals and scales. The physics of some string and pipe instruments namely guitar, violin and trumpet are studied. This chapter includes a study of the anatomy of the ear and the psychoacoustics of hearing. The chapter ends with a study of music coders and MP3.

Chapter 14 on speech processing starts with an introduction to the physiology of speech production. The linear prediction model of speech and the harmonic plus noise model of speech are presented as is a model that combines the two. Speech coding for mobile phones and speech recognition for voice-dialling are considered in this chapter.

Chapter 15 on speech enhancement considers noise reduction, packet loss concealment and bandwidth extension. On noise reduction we consider methods such as spectral subtraction, Wiener filter and Kalman filter. A particular feature of this chapter is the use of a combination of linear prediction model and harmonic plus noise model as a framework for speech enhancement.

Chapter 16 covers echo cancellation. The chapter begins with an introduction to telephone line echoes, and considers line echo suppression and adaptive line echo cancellation. Then the problems of acoustic echoes and acoustic coupling between loudspeaker and microphone systems are considered. The chapter concludes with a study of a sub-band echo cancellation system

Chapter 17 is on blind deconvolution and channel equalisation. This chapter begins with an introduction to channel distortion models and the ideal channel equaliser. Then the Wiener equaliser, blind equalisation using the channel input power spectrum, blind deconvolution based on linear predictive models, Bayesian channel equalisation, and blind equalisation for digital communication channels are considered. The chapter concludes with equalisation of maximum phase channels using higher-order statistics.

Chapter 18 covers wireless communication. Noise, fading and limited radio spectrum are the main factors that constrain the capacity and the speed of communication. For improved efficiency modern mobile communication systems rely on signal processing methods at almost every stage from source coding to the allocation of time bandwidth and space resources. In this chapter we consider how communication signal processing methods are employed for improving the speed and capacity of communication systems.

The following companion website provides Matlab programs and audio–visual animations related to the subjects covered in this book http://dea.brunel.ac.uk/cmsp/mmsp.

Saeed V. Vaseghi

Acknowledgement

I wish to thank Saeed Ayat for his technical proofreading of this book and his suggestions and contributions. Many thanks to the publishing team at John Wiley, Sarah Hinton, Mark Hammond, Jen Beal, Brett Wells, Katharine Unwin and Lyn Imeson for all their support and contributions. Thanks also to Ales Prochazka, Helena Jetelova, Esi Zavarehei, Emir Tourajlic, Ben Milner, Qin Yan, Dimitrios Rentzos, Charles Ho, Ali Ghorshi, Phuay Low and Aimin Chen.

Symbols

$\boldsymbol{A}$	Matrix of predictor coefficients
a_k	Linear predictor coefficients
$\boldsymbol{a}$	Linear predictor coefficients vector
$\boldsymbol{a}^{inv}$	Inverse linear predictor coefficients vector
a_{ij}	Probability of transition from state i to state j in a Markov model
$\alpha_i(t)$	Forward probability in an HMM
$b(m)$	Backward prediction error, Binary state signal
$\beta_i(t)$	Backward probability in an HMM
$\boldsymbol{C}_{xx}$	Covariance matrix of $\boldsymbol{x}$
$c_{xx}(m)$	autocovariance of signal at lag m
$c_{XX}(k_1, k_2, \cdots, k_N)$	k^{th} order cumulant of $x(m)$
$C_{XX}(\omega_1, \omega_2, \cdots, \omega_{k-1})$	k^{th} order cumulant spectra of $x(m)$
$\boldsymbol{D}$	Diagonal matrix
Δf	Frequency resolution
$\delta(t)$	Dirac delta function
$e(m)$	Estimation error or prediction error
$\mathcal{E}[x]$	Expectation of x
f	Frequency variable
f_c	Filter cutoff frequency
F_0	Fundamental frequency
F_s	Sampling frequency
$f_X(\boldsymbol{x})$	Probability density function for process $\boldsymbol{X}$
$f_{X,Y}(\boldsymbol{x}, \boldsymbol{y})$	Joint probability density function of $\boldsymbol{X}$ and $\boldsymbol{Y}$
$f_{X\|Y}(\boldsymbol{x} \| \boldsymbol{y})$	Probability density function of $\boldsymbol{X}$ conditioned on $\boldsymbol{Y}$
$f_{X;\Theta}(\boldsymbol{x}; \theta)$	Probability density function of $\boldsymbol{X}$ with θ as a parameter
$f_{X\|S,\mathcal{M}}(\boldsymbol{x} \| \boldsymbol{s}, \mathcal{M})$	Probability density function of $\boldsymbol{X}$ given a state sequence $\boldsymbol{s}$ of an HMM $\mathcal{M}$ of the process $\boldsymbol{X}$
$\Phi(m, m-1)$	State transition matrix in Kalman filter
G	Filter gain factor
$\boldsymbol{h}$	Filter coefficient vector, Channel response
$\boldsymbol{h}_{\max}$	Maximum-phase channel response
$\boldsymbol{h}_{\min}$	Minimum-phase channel response
$\boldsymbol{h}^{\text{inv}}$	Inverse channel response

$H(f)$	Channel frequency response
$H^{\text{inv}}(f)$	Inverse channel frequency response
$H(z)$	z-transfer function
$\boldsymbol{H}$	Observation matrix, Distortion matrix
$\boldsymbol{I}$	Identity matrix
$\boldsymbol{J}$	Fisher's information matrix
$\lvert\boldsymbol{J}\rvert$	Jacobian of a transformation
JND	Just noticeable distortion level
$\boldsymbol{K}(m)$	Kalman gain matrix
$k(x)$	Kurtosis
λ	Eigenvalue
Λ	Diagonal matrix of eigenvalues
m	Discrete time index
m_k	k^{th} order moment
$\mathcal{M}$	A model, e.g. an HMM
μ	Adaptation step size
μ_x	Expected mean of vector $\boldsymbol{x}$
$n(m)$	Noise
$\boldsymbol{n}(m)$	A noise vector of N samples
$n_i(m)$	Impulsive noise
$N(f)$	Noise spectrum
$N^*(f)$	Complex conjugate of $N(f)$
$\overline{N(f)}$	Time-averaged noise spectrum
$\boldsymbol{N}(\boldsymbol{x}, \mu_{\boldsymbol{xx}}, \Sigma_{\boldsymbol{xx}})$	A Gaussian pdf with mean vector μ_{xx} and covariance matrix Σ_{xx}
$O(\cdot)$	In the order of $(\cdot)$
P	Filter order (length)
$P_X(\boldsymbol{x}_i)$	Probability mass function of $\boldsymbol{x}_i$
$P_{X,Y}(\boldsymbol{x}_i, \boldsymbol{y}_j)$	Joint probability mass function of $\boldsymbol{x}_i$ and $\boldsymbol{y}_j$
$P_{X\mid Y}(\boldsymbol{x}_i \mid \boldsymbol{y}_j)$	Conditional probability mass function of $\boldsymbol{x}_i$ given $\boldsymbol{y}_j$
$P_{NN}(f)$	Power spectrum of noise $n(m)$
$P_{XX}(f)$	Power spectrum of the signal $x(m)$
$P_{XY}(f)$	Cross–power spectrum of signals $x(m)$ and $y(m)$
$Q(x_1, x_2, x_3, \ldots)$	Cumulant
θ	Parameter vector
$\hat{\theta}$	Estimate of the parameter vector θ
r_k	Reflection coefficients
$r_{xx}(m)$	Autocorrelation function
$\boldsymbol{r}_{xx}(m)$	Autocorrelation vector
$\boldsymbol{R}_{\boldsymbol{xx}}$	Autocorrelation matrix of signal $\boldsymbol{x}(m)$
$\boldsymbol{R}_{\boldsymbol{xy}}$	Cross-correlation matrix
T_s	Sampling period
$\boldsymbol{s}$	State sequence
$\boldsymbol{s}^{ML}$	Maximum-likelihood state sequence
σ_n^2	Variance of noise $n(m)$
$\Sigma_{\boldsymbol{nn}}$	Covariance matrix of noise $\boldsymbol{n}(m)$
$\Sigma_{\boldsymbol{xx}}$	Covariance matrix of signal $\boldsymbol{x}(m)$
σ_x^2	Variance of signal $x(m)$
σ_n^2	Variance of noise $n(m)$

$x(m)$	Clean signal
$\hat{x}(m)$	Estimate of clean signal
$\boldsymbol{x}(m)$	Clean signal vector
$X(f)$	Frequency spectrum of signal $x(m)$
$X^*(f)$	Complex conjugate of $X(f)$
$\overline{X(f)}$	Time-averaged frequency spectrum of the signal $x(m)$
$X(f, t)$	Time-frequency spectrum of the signal $x(m)$
$\boldsymbol{X}$	Clean signal matrix
$\boldsymbol{X}^{\mathrm{H}}$	Hermitian transpose of $\boldsymbol{X}$
$y(m)$	Noisy signal
$\boldsymbol{y}(m)$	Noisy signal vector
$\hat{\boldsymbol{y}}(m \mid m-i)$	Prediction of $\boldsymbol{y}(m)$ based on observations up to time $m-i$
$\boldsymbol{Y}$	Noisy signal matrix
$\boldsymbol{Y}^{\mathrm{H}}$	Hermitian transpose of $\boldsymbol{Y}$
Var	Variance
ω	Angular frequency in radian/sec
ω_c	Cutoff angular frequency in radian/sec
ω_0	Fundamental angular frequency
ω_s	Angular sampling frequency
w_k	Wiener filter coefficients
$\boldsymbol{w}(m)$	Wiener filter coefficients vector
$W(f)$	Wiener filter frequency response
z	z-transform variable

Abbreviations

ADC	Analogue to digital converter
AR	Autoregressive process
ARMA	Autoregressive moving average process
ATH	Absolute threshold of hearing
ATRAC	Adaptive Transform Acoustic Coding
AWGN	Additive white Gaussian noise
bps	Bits per second
BSS	Blind signal separation
CD	Compact disc
cdf	Cumulative density function
CELP	Code Excited Linear Prediction
Companding	**Comp**ressing Exp**anding**
DAC	Digital to analogue converter
dB	Decibels: 10log10(power ratio) or 10log10(amplitude ratio)
DCT	Discrete cosine transform
Det()	determinant
DFT	Discrete Fourier transform
DNA	Deoxyribonucleic acid
DoA	Direction of arrival
DSP	Digital signal processing
DTW	Dynamic time warping
EM	Estimate-maximise
EM	Electro-magnetic
FFT	Fast Fourier transform
FIR	Finite impulse response
GMM	Gaussian mixture model
GSM	Global system for mobile
HMM	Hidden Markov model
HNM	Harmonic plus noise model
Hz	Unit of frequency in cycles per second
ICA	Independent component analysis
IDCT	Inverse discrete cosine transform
IDFT	Inverse discrete Fourier transform

IFFT	Inverse fast Fourier transform
IID	Independent identically distributed
IIR	Infinite impulse response
ISD	Itakura–Saito distance
ISI	Inter symbol interference
ITU	International Telecommunication Union
JND	Just noticeable distortion
KLT	Karhunen–Loève transform
LF	Liljencrants–Fant
LMS	Least mean squared error
LP	Linear prediction model or Lowpass filter
LPC	Linear prediction coding
LPSS	Spectral subtraction based on linear prediction model
LS	Least square
LSAR	Least square AR interpolation
LSE	Least square error
LSF	Line spectral frequency
LSP	Line spectral pair
LTI	Linear time invariant
MA	Moving average process
MAP	Maximum a posterior estimate
M-ary	Multi-level signalling
MAVE	Minimum absolute value of error estimate
MFCC	Mel frequency cepstral coefficients
MIMO	Multiple input multiple output
ML	Maximum likelihood estimate
MMSE	Minimum mean squared error estimate
MOS	Mean Opinion Score
MP3	MPEG-1 Audio Layer 3
MPEG	Moving Picture Experts Group
ms	Milliseconds
NLMS	Normalised least mean squared error
PCA	Principal component analysis
pdf	Probability density function
PLC	Packet loss concealment
pmf	Probability mass function
PRNG	Pseudo random number generators
psd	Power spectral density
QMF	Quadrature mirror filter
QR	Q is an orthogonal matrix and R is an upper triangular matrix
QRD	Orthogonal matrix decomposition
RF	Radio frequency
RLS	Recursive least square
ROC	Region of convergence
SH or S/H	Sample and hold
SINR	Signal to impulsive noise ratio
SNR	Signal to noise ratio
SPL	Sound pressure level
SQNR	Signal to quantisation noise ratio

std	Standard deviation
STFT	Short time Fourier transform
SVD	Singular value decomposition
ToA	Time of arrival
VAD	Voice activity detector
Var	Variance
ZI	Zero insertion

Part I

Basic Digital Signal Processing

1 Introduction

Signal processing provides the theory, the methods and the tools for such purposes as the analysis and modelling of signals, classification and recognition of patterns, extraction of information from signals, synthesis and morphing of signals – morphing is creating a new voice or image out of the existing samples. Signal processing is concerned with the modelling, extraction, communication and utilisation of information patterns and structures in a signal process.

Applications of signal processing methods are very wide and include audio hi-fi, TV and radio, cellular mobile phones, voice recognition, vision, antenna arrays, radar, sonar, geophysical exploration, medical electronics, bio-medical signal processing, physics and in general any system that is concerned with the communication or processing and retrieval of information. Signal processing plays a central role in the development of the new generations of mobile telecommunication and intelligent automation systems and in the efficient transmission, reception, decoding, organisation and retrieval of information content in search engines.

This chapter begins with a definition of signals, and a brief introduction to various signal processing methodologies. We consider several key applications of digital signal processing in biomedical signal processing, adaptive noise reduction, channel equalisation, pattern classification/recognition, audio signal coding, signal detection, spatial processing for directional reception of signals, Dolby noise reduction, radar and watermarking.

1.1 Signals and Information

A signal is the variation of a quantity such as air pressure waves of sounds, colours of an image, depths of a surface, temperature of a body, current/voltage in a conductor or biological system, light, electromagnetic radio waves, commodity prices or volume and mass of an object. A signal conveys information regarding one or more attributes of the source such as the state, the characteristics, the composition, the trajectory, the evolution or the intention of the source. Hence, a signal is a *means to convey information* regarding the past, the current or the future states of a variable.

For example, astrophysicists analyse the spectrum of signals, the light and other electromagnetic waves, emitted from distant stars or galaxies to deduce information about their movements, origins

Multimedia Signal Processing: Theory and Applications in Speech, Music and Communications Saeed V. Vaseghi

and evolution. Imaging radars calculate the round trip delay of reflected light or radio waves bouncing from the surface of the earth to produce maps of the earth.

A signal can be a function of one dimension, that is a function of one variable, such as speech or music whose amplitude fluctuations are a function of the time variable, or a signal can be multidimensional such as an image (i.e. reflected light intensity) which is a function of two-dimensional plane or video which is a function of two-dimensional plane and time. Note that a photograph effectively projects a view of objects in three-dimensional space onto a two-dimensional image plane where depth information can be deduced from the shadows and gradients of colours.

The information conveyed in a signal may be used by humans or machines (e.g. computers or robots) for communication, forecasting, decision-making, control, geophysical exploration, medical diagnosis, forensics etc.

The types of signals that signal processing systems deal with include text, image, audio, video, ultrasonic, subsonic, electromagnetic waves, medical, biological, thermal, financial or seismic signals.

Figure 1.1 illustrates a simplified overview of a communication system composed of an information source $I(t)$ followed by a signalling system $T[\cdot]$ for transformation of the information into variation of a signal $x(t)$ that carries the information, a communication channel $h[\cdot]$ for modelling the propagation of the signal from the transmitter to the receiver, additive channel and background noise $n(t)$ that exists in every real-life system and a signal processing unit at the receiver for extraction of the information from the received signal.

In general, there is a mapping operation (e.g. modulation) that maps the output $I(t)$of an information source to the physical variations of a signal $x(t)$ that carries the information, this mapping operator may be denoted as $T[\cdot]$ and expressed as

$$x(t) = T[I(t)] \tag{1.1}$$

The information source $I(t)$ is normally discrete-valued whereas the signal $x(t)$ that carries the information to a receiver may be continuous or discrete. For example, in multimedia communication the information from a computer, or any other digital communication device, is in the form of a sequence of binary numbers (ones and zeros) which would need to be transformed into a physical quantity such as voltage or current and modulated to the appropriate form for transmission in a communication channel such as a radio channel, telephone line or cable.

As a further example, in human speech communication the voice-generating mechanism provides a means for the speaker to map each discrete word into a distinct pattern of modulation of the acoustic vibrations of air that can propagate to the listener. To communicate a word w, the speaker generates an acoustic signal realisation of the word $x(t)$; this acoustic signal may be contaminated by ambient noise and/or distorted by a communication channel or room reverberations, or impaired by the speaking abnormalities of the talker, and received as the noisy, distorted and/or incomplete signal $y(t)$ modelled as

$$y(t) = h[x(t)] + n(t) \tag{1.2}$$

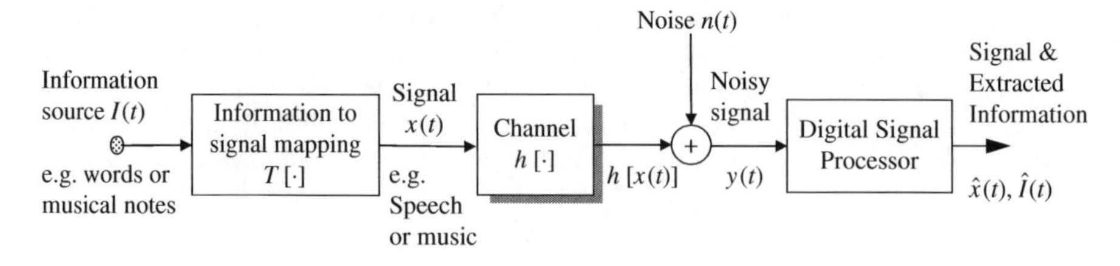

Figure 1.1 Illustration of a communication and signal processing system.

In addition to conveying the spoken word, the acoustic speech signal conveys information on the prosody (i.e. pitch intonation and stress patterns) of speech and the speaking characteristic, accent and the emotional state of the talker. The listener extracts this information by processing the signal $y(t)$.

In the past few decades, the theory and applications of digital signal processing have evolved to play a central role in the development of modern telecommunication and information technology systems.

Signal processing methods are central to efficient mobile communication, and to the development of intelligent man/machine interfaces in such areas as speech and visual pattern recognition for multimedia systems. In general, digital signal processing is concerned with two broad areas of information theory:

(a) Efficient and reliable coding, transmission, reception, storage and representation of signals in communication systems such as mobile phones, radio and TV.
(b) The extraction of information from noisy and/or incomplete signals for pattern recognition, detection, forecasting, decision-making, signal enhancement, control, automation and search engines.

In the next section we consider four broad approaches to signal processing.

1.2 Signal Processing Methods

Signal processing methods provide a variety of tools for modelling, analysis, coding, synthesis and recognition of signals. Signal processing methods have evolved in algorithmic complexity aiming for optimal utilisation of the available information in order to achieve the best performance. In general the computational requirement of signal processing methods increases, often exponentially, with the algorithmic complexity. However, the implementation costs of advanced signal processing methods have been offset and made affordable by the consistent trend in recent years of a continuing increase in the performance, coupled with a simultaneous decrease in the cost, of signal processing hardware.

Depending on the method used, digital signal processing algorithms can be categorised into one or a combination of four broad categories. These are transform-based signal processing, model-based signal processing, Bayesian statistical signal processing and neural networks, as illustrated in Figure 1.2. These methods are briefly described in the following.

1.2.1 Transform-Based Signal Processing

The purpose of a transform is to express a signal or a system in terms of a combination of a set of elementary simple signals (such as sinusoidal signals, eigenvectors or wavelets) that lend themselves to relatively easy analysis, interpretation and manipulation. Transform-based signal processing methods include Fourier transform, Laplace transform, z-transform, and wavelet transforms.

The most widely applied signal transform is the Fourier transform (introduced in Chapter 2) which is effectively a form of vibration analysis; a signal is expressed in terms of a combination of the sinusoidal vibrations that make up the signal. Fourier transform is employed in a wide range of applications including popular music coders, noise reduction and feature extraction for pattern recognition. The Laplace transform, and its discrete-time version the z-transform (introduced in Chapter 3), are generalisations of the Fourier transform and describe a signal or a system in terms of a set of transient sinusoids with exponential amplitude envelops.

In Fourier, Laplace and z-transform, the different sinusoidal basis functions of each transform all have the same duration and differ in terms of their frequency of vibrations and the amplitude envelopes.

In contrast wavelets are multi-resolution transforms in which a signal is described in terms of a combination of elementary waves of different dilations. The set of basis functions in a wavelet

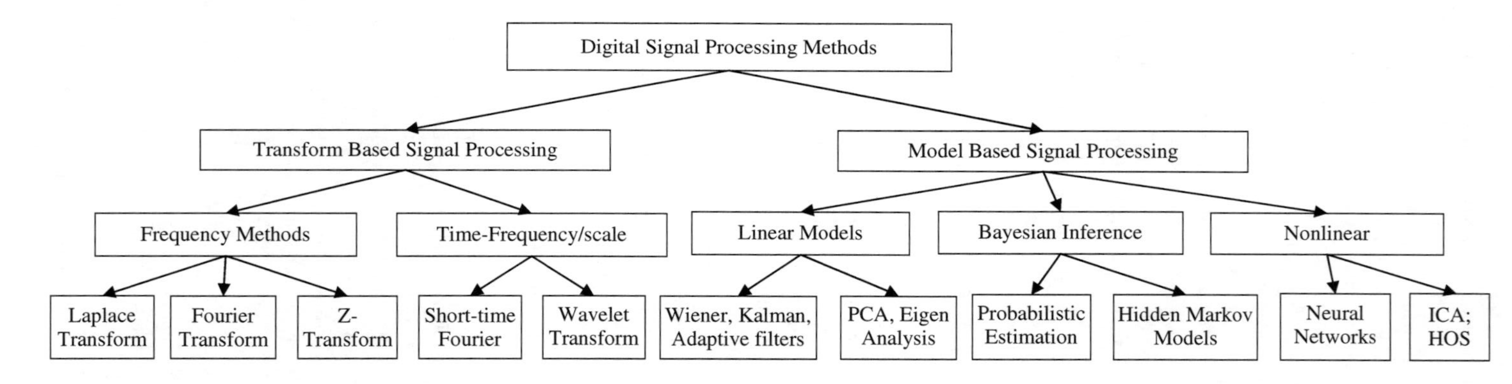

Figure 1.2 A broad categorisation of some of the most commonly used signal processing methods. ICA = Independent Component Analysis, HOS = Higher order statistics. Note that there may be overlap between different methods and also various methods can be combined.

is composed of contractions and dilations of a single elementary wave. This allows non-stationary events of various durations in a signal to be identified and analysed. Wavelet analysis is effectively a tree-structured filter bank analysis in which a set of high pass and low filters are used repeatedly in a binary-tree structure to split the signal progressively into sub-bands as explained in Chapter 4 on digital filters.

1.2.2 Source-Filter Model-Based Signal Processing

Model-based signal processing methods utilise a parametric model of the signal generation process. The parametric model normally describes the predictable structures and the expected patterns in the signal process, and can be used to forecast the future values of a signal from its past trajectory.

Model-based methods normally outperform non-parametric methods, since they utilise more information in the form of a model of the signal process. However, they can be sensitive to the deviations of a signal from the class of signals characterised by the model.

The most widely used parametric model is the linear prediction model, described in Chapter 10. Linear prediction models have facilitated the development of advanced signal processing methods for a wide range of applications such as low-bit-rate speech coding in cellular mobile telephony, digital video coding, high-resolution spectral analysis, radar signal processing and speech recognition.

1.2.3 Bayesian Statistical Model-Based Signal Processing

Statistical signal processing deals with random processes; this includes all information-bearing signals and noise. The fluctuations of a random signal, or the distribution of a class of random signals in the signal space, cannot be entirely modelled by a predictive equation, but it can be described in terms of the statistical average values, and modelled by a probability distribution function in a multi-dimensional signal space. For example, as described in Chapter 10, a linear prediction model driven by a random signal can provide a source-filter model of the acoustic realisation of a spoken word. However, the random input signal of the linear prediction model, or the variations in the characteristics of different acoustic realisations of the same word across the speaking population, can only be described in statistical terms and in terms of probability functions.

Bayesian inference theory provides a generalised framework for statistical processing of random signals, and for formulating and solving estimation and decision-making problems. Bayesian methods are used for pattern recognition and signal estimation problems in applications such as speech processing, communication, data management and artificial intelligence. In recognising a pattern or estimating a signal, from noisy and/or incomplete observations, Bayesian methods combine the evidence contained in the incomplete signal observation with the prior information regarding the distributions of the signals and/or the distributions of the parameters associated with the signals. Chapter 7 describes Bayesian inference methodology and the estimation of random processes observed in noise.

1.2.4 Neural Networks

Neural networks are combinations of relatively simple non-linear adaptive processing units, arranged to have a structural resemblance to the transmission and processing of signals in biological neurons. In a neural network several layers of parallel processing elements are interconnected with a hierarchically structured connection network. The connection weights are trained to 'memorise patterns' and perform a signal processing function such as prediction or classification.

Neural networks are particularly useful in non-linear partitioning of a signal space, in feature extraction and pattern recognition, and in decision-making systems. In some hybrid pattern recognition systems neural networks are used to complement Bayesian inference methods. Since the main objective of this book is to provide a coherent presentation of the theory and applications of statistical signal processing, neural networks are not discussed here.

1.3 Applications of Digital Signal Processing

In recent years, the development and commercial availability of increasingly powerful and affordable digital computers has been accompanied by the development of advanced digital signal processing algorithms for a wide variety of applications such as noise reduction, telecommunication, radar, sonar, video and audio signal processing, pattern recognition, geophysics explorations, data forecasting, and the processing of large databases for the identification, extraction and organisation of unknown underlying structures and patterns. Figure 1.3 shows a broad categorisation of some DSP applications. This section provides a review of several key applications of digital signal processing methods.

Part III of this book covers the applications of DSP to speech processing, music processing and communications. In the following an overview of some applications of DSP is provided. Note that these applications are by no means exhaustive but they represent a useful introduction.

1.3.1 Digital Watermarking

Digital watermarking is the embedding of a signature signal, i.e. the digital watermark, underneath a host image, video or audio signal. Although watermarking may be visible or invisible, the main challenge in digital watermarking is to make the watermark secret and imperceptible (meaning invisible or inaudible). Watermarking takes its name from the watermarking of paper or money for security and authentication purposes.

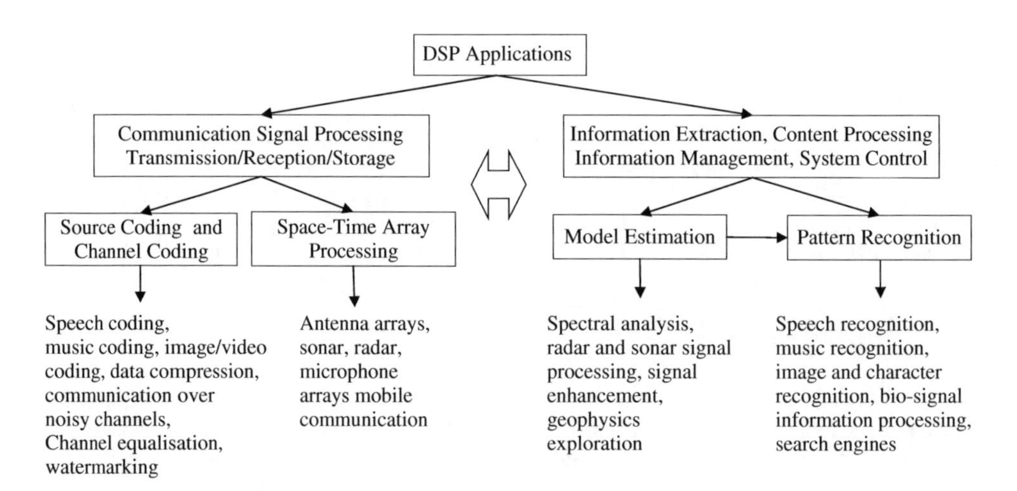

Figure 1.3 A classification of the applications of digital signal processing.

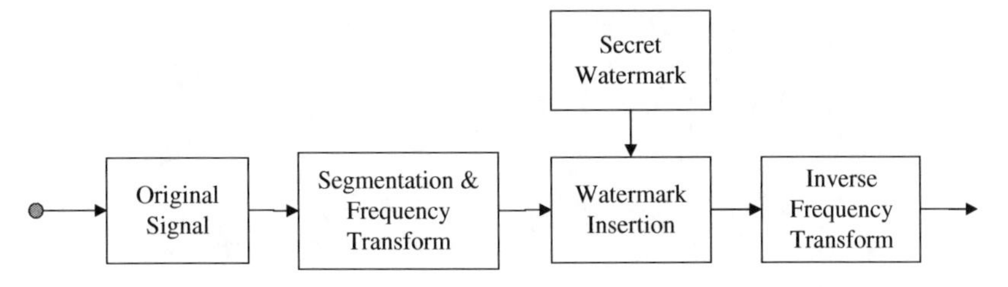

Figure 1.4 A simplified illustration of frequency domain watermarking.

Watermarking is used in digital media for the following purposes:

(1) Authentication of digital image and audio signals. The watermark may also include owner information, a serial number and other useful information.
(2) Protection of copyright/ownership of image and audio signals from unauthorised copying, use or trade.
(3) Embedding of audio or text signals into image/video signals for subsequent retrieval.
(4) Embedding a secret message into an image or audio signal.

Watermarking has to be robust to intended or unintended degradations and resistant to attempts at rendering it ineffective. In particular watermarking needs to survive the following processes:

(1) Changes in the sampling rate, resolution and format of the signal.
(2) Changes in the orientation of images or phase of the signals.
(3) Noise and channel distortion.
(4) Non-linear imperceptible changes of time/space scales. For example non-linear time-warping of audio or non-linear warping of the dimensions of an image.
(5) Segmentation and cropping of the signals.

The simplest forms of watermarking methods, Figure 1.4, exploit the time–frequency structure of the signal together with the audio-visual perceptual characteristics of humans. The watermark signal is hidden in the parts of the host signal spectrum, where it is invisible or inaudible. For example, a simple way to embed a watermark into an audio signal is to transform the watermark such that it closely follows the envelope of the time-varying spectrum of the audio signal. The transformed watermark is then added to the audio signal.

An example of invisible watermarking is shown in Figure 1.5. The figure shows a host image and another image acting as the watermark together with the watermarked image and the retrieved watermark.

1.3.2 Bio-medical Signal Processing

Bio-medical signal processing is concerned with the analysis, denoising, synthesis and classification of bio-signals such as magnetic resonance images (MRI) of brain or electrocardiograph (ECG) signals of heart or electroencephalogram (EEG) signals of brain neurons.

An electrocardiograph signal is produced by recording the electrical voltage signals of the heart. It is the main tool in cardiac electrophysiology, and has a prime function in the screening and diagnosis of cardiovascular diseases.

Figure 1.5 Illustration of invisible watermarking of an image, clockwise from top-left: a picture of my son, the watermark, watermarked image and retrieved watermark. The watermark may be damaged due to modifications such as a change of image coding format.

Electroencephalography is the neurophysiologic measurement of the electrical activity of the neurons in brain picked up by electrodes placed on the scalp or, in special cases, on the cortex. The resulting signals are known as an electroencephalograph and represent a mix of electrical signals and noise from a large number of neurons.

The observations of ECG or EEG signals are often a noisy mixture of electrical signals generated from the activities of several different sources from different parts of the body. The main issues in the processing of bio-signals, such as EEG or ECG, are the denoising, separation and identification of the signals from different sources.

An important bio-signal analysis tool, considered in Chapter 12, is known as independent componentanalysis (ICA). ICA is primarily used for separation of mixed signals in multi-source multi-sensor applications such as in ECG and EEG. ICA is also used for beam forming in multiple-input multiple-output (MIMO) telecommunication.

The ICA problem is formulated as follows. The observed signal vector $\boldsymbol{x}$ is assumed to be a linear mixture of M independent source signals $\boldsymbol{s}$. In a linear matrix form the mixing operation is expressed as

$$\boldsymbol{x} = \boldsymbol{A}\boldsymbol{s} \tag{1.3}$$

The matrix $\boldsymbol{A}$ is known as the *mixing matrix* or the observation matrix. In many practical cases of interest all we have is the sequence of observation vectors $[\boldsymbol{x}(0), \boldsymbol{x}(1), \ldots, \boldsymbol{x}(N-1)]$. The mixing matrix $\boldsymbol{A}$ is unknown and we wish to estimate a demixing matrix $\boldsymbol{W}$ to obtain an estimate of the original signal $\boldsymbol{s}$.

This problem is known as *blind source separation* (BSS); the term blind refers to the fact that we have no other information than the observation $\boldsymbol{x}$ and an assumption that the source signals are independent of each other. The demixing problem is the estimation of a matrix $\boldsymbol{W}$ such that

$$\hat{s} = \boldsymbol{W}\boldsymbol{x} \tag{1.4}$$

The details of the derivation of the demixing matrix are discussed in Chapter 12. Figure 1.6 shows an example of ECG signal mixture of the hearts of a pregnant mother and foetus plus other noise and interference. Note that application of ICA results in separation of the mother and foetus heartbeats. Also note that the foetus heartbeat rate is about 25% faster than the mother's heartbeat rate.

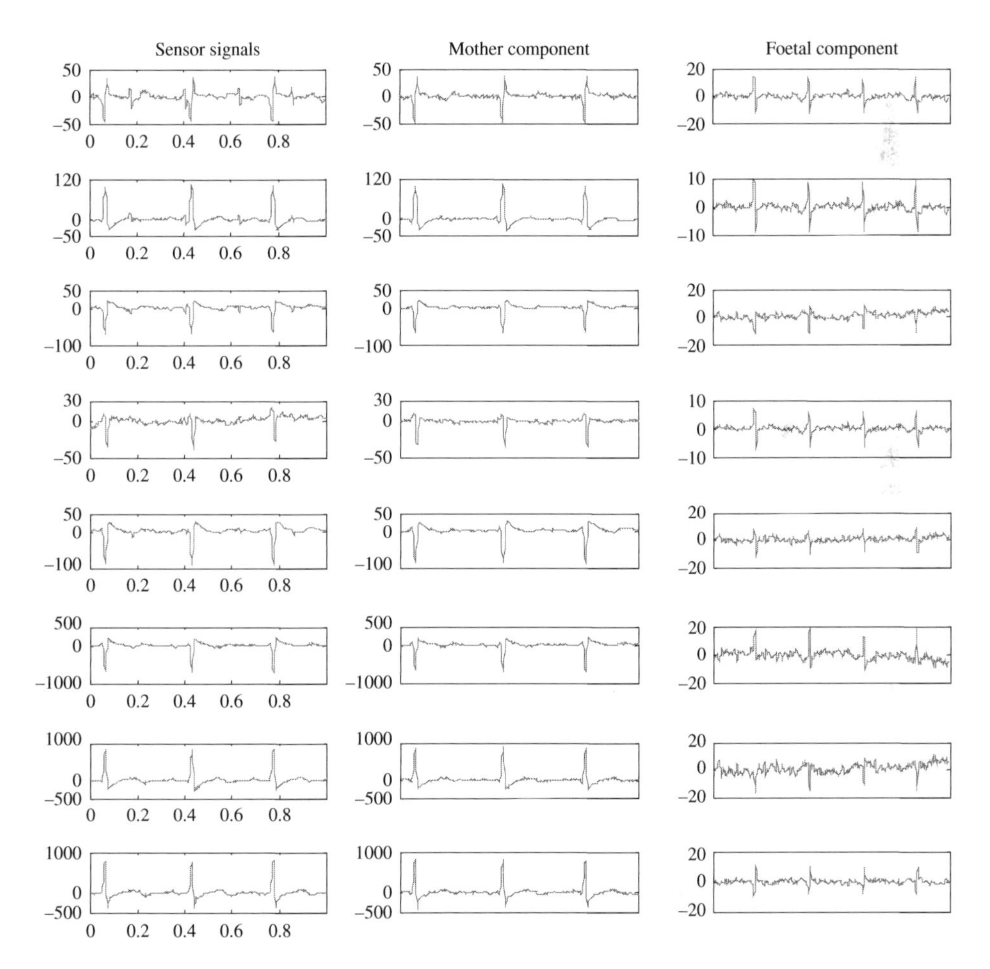

Figure 1.6 Application of ICA to separation of mother and foetus ECG. Note that signals from eight sensors are used in this example (see Chapter 12).

1.3.3 Adaptive Noise Cancellation

In speech communication from a noisy acoustic environment such as a moving car or train, or over a noisy telephone channel, the speech signal is observed in an additive random noise. In signal measurement systems the information-bearing signal is often contaminated by noise from its surrounding environment. The noisy observation $y(m)$ can be modelled as

$$y(m) = x(m) + n(m) \tag{1.5}$$

where $x(m)$ and $n(m)$ are the signal and the noise, and m is the discrete-time index. In some situations, for example when using a mobile telephone in a moving car, or when using a radio communication device in an aircraft cockpit, it may be possible to measure and estimate the instantaneous amplitude of the ambient noise using a directional microphone. The signal $x(m)$ may then be recovered by subtraction of an estimate of the noise from the noisy signal.

Figure 1.7 shows a two-input adaptive noise cancellation system for enhancement of noisy speech. In this system a directional microphone takes as input the noisy signal $x(m) + n(m)$, and a second directional microphone, positioned some distance away, measures the noise $\alpha n(m+\tau)$. The attenuation factor α and the time delay τ provide a rather over-simplified model of the effects of propagation of the noise to different positions in the space where the microphones are placed. The noise from the second microphone is processed by an adaptive digital filter to make it equal to the noise contaminating the speech signal, and then subtracted from the noisy signal to cancel out the noise. The adaptive noise canceller is more effective in cancelling out the low-frequency part of the noise, but generally suffers from the non-stationary character of the signals, and from the over-simplified assumption that a linear filter can model the diffusion and propagation of the noise sound in the space.

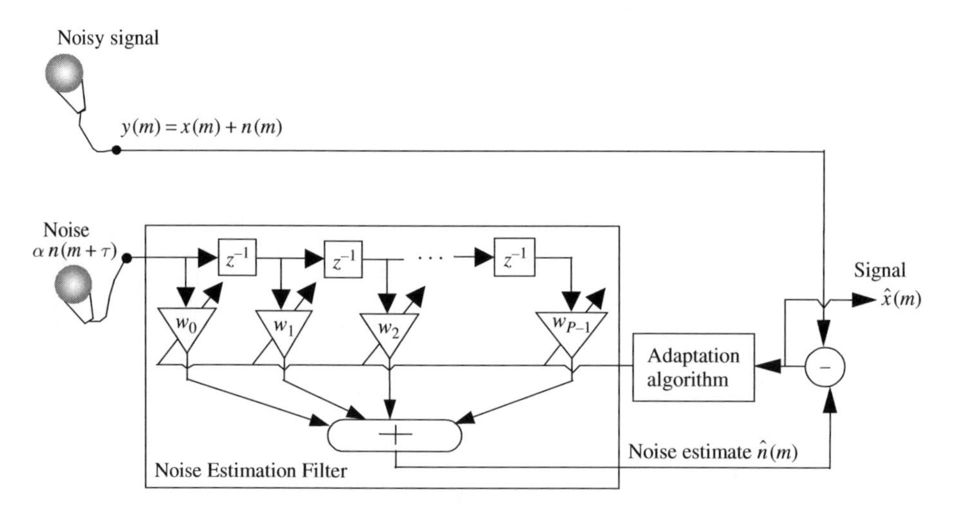

Figure 1.7 Configuration of a two-microphone adaptive noise canceller. The adaptive filter delay elements (z^{-1}) and weights wi model the delay and attenuation that signals undergo while propagating in a medium.

1.3.4 Adaptive Noise Reduction

In many applications, for example at the receiver of a telecommunication system, there is no access to the instantaneous value of the contaminating noise, and only the noisy signal is available. In such cases the noise cannot be cancelled out, but it may be reduced, in an average sense, using the statistics of the signal and the noise process. Figure 1.8 shows a bank of Wiener filters for reducing additive noise when only the noisy signal is available. The filter bank coefficients attenuate each noisy signal frequency in inverse proportion to the signal-to-noise ratio at that frequency. The Wiener filter bank coefficients, derived in Chapter 8, are calculated from estimates of the power spectra of the signal and the noise processes.

1.3.5 Blind Channel Equalisation

Channel equalisation is the recovery of a signal distorted in transmission through a communication channel with a non-flat magnitude or a non-linear phase response. When the channel response is unknown the process of signal recovery is called blind equalisation. Blind equalisation has a wide range of applications, for example in digital telecommunications for removal of inter-symbol interference due to non-ideal channel and multi-path propagation, in speech recognition for removal of the effects of the microphones and the communication channels, in correction of distorted images, analysis of seismic data, de-reverberation of acoustic gramophone recordings etc.

In practice, blind equalisation is feasible only if some useful statistics of the channel input are available. The success of a blind equalisation method depends on how much is known about the characteristics of the input signal and how useful this knowledge can be in the channel identification

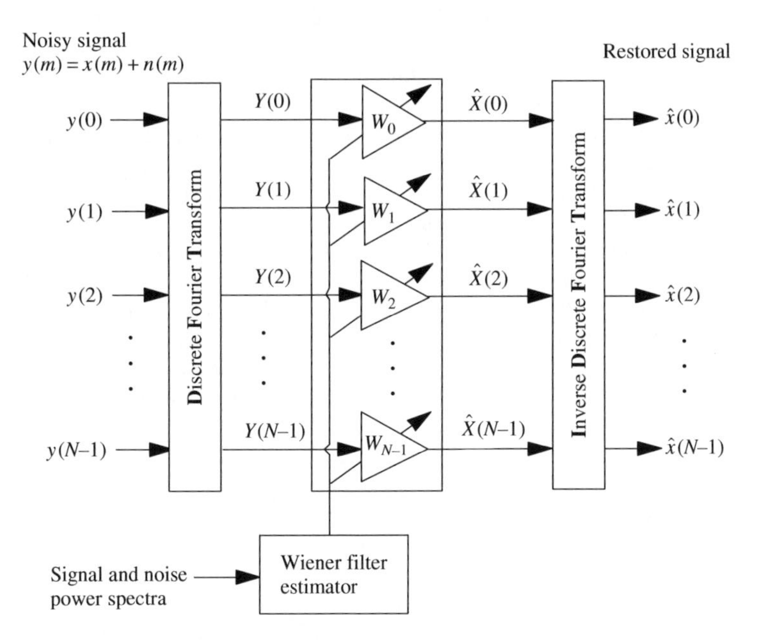

Figure 1.8 A frequency-domain Wiener filter for reducing additive noise.

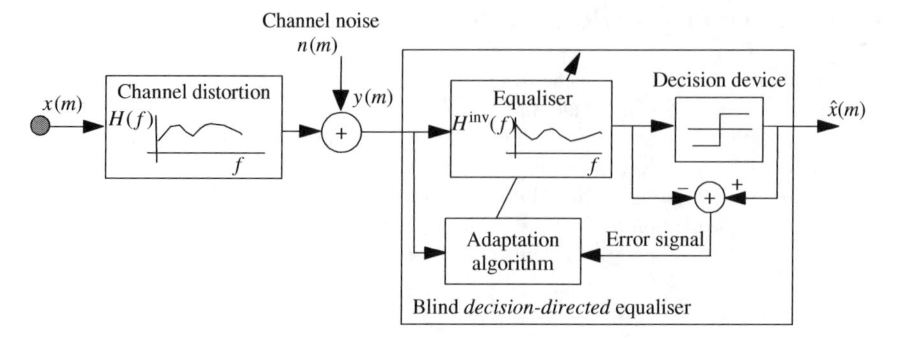

Figure 1.9 Configuration of a decision-directed blind channel equaliser.

and equalisation process. Figure 1.9 illustrates the configuration of a decision-directed equaliser. This blind channel equaliser is composed of two distinct sections: an adaptive equaliser that removes a large part of the channel distortion, followed by a non-linear decision device for an improved estimate of the channel input. The output of the decision device is the final estimate of the channel input, and it is used as the desired signal *to direct* the equaliser adaptation process. Blind equalisation is covered in detail in Chapter 17.

1.3.6 Signal Classification and Pattern Recognition

Signal classification is used in detection, pattern recognition and decision-making systems. For example, a simple binary-state classifier can act as the detector of the presence, or the absence, of a known waveform in noise. In signal classification, the aim is to design a minimum-error system for *labelling* a signal with one of a number of likely classes of signal.

To design a classifier, a set of models are trained for the classes of signals that are of interest in the application. The simplest form that the models can assume is a bank, or codebook, of waveforms, each representing the prototype for one class of signals. A more complete model for each class of signals takes the form of a probability distribution function. In the classification phase, a signal is labelled with the nearest or the most likely class. For example, in communication of a binary bit stream over a band-pass channel, the binary phase-shift keying (BPSK) scheme signals the bit '1' using the waveform $A_c \sin \omega_c t$ and the bit '0' using $-A_c \sin \omega_c t$.

At the receiver, the decoder has the task of classifying and labelling the received noisy signal as a '1' or a '0'. Figure 1.10 illustrates a correlation receiver for a BPSK signalling scheme. The receiver has two correlators, each programmed with one of the two symbols representing the binary states for the bit '1' and the bit '0'. The decoder correlates the unlabelled input signal with each of the two candidate symbols and selects the candidate that has a higher correlation with the input.

Figure 1.11 illustrates the use of a classifier in a limited-vocabulary, isolated-word speech recognition system. Assume there are V words in the vocabulary. For each word a model is trained, on many different examples of the spoken word, to capture the average characteristics and the statistical variations of the word. The classifier has access to a bank of $V+1$ models, one for each word in the vocabulary and an additional model for the silence periods. In the speech recognition phase, the task is to decode and label an acoustic speech feature sequence, representing an unlabelled spoken word, as one of the V likely words or silence. For each candidate word the classifier calculates a probability score and selects the word with the highest score.

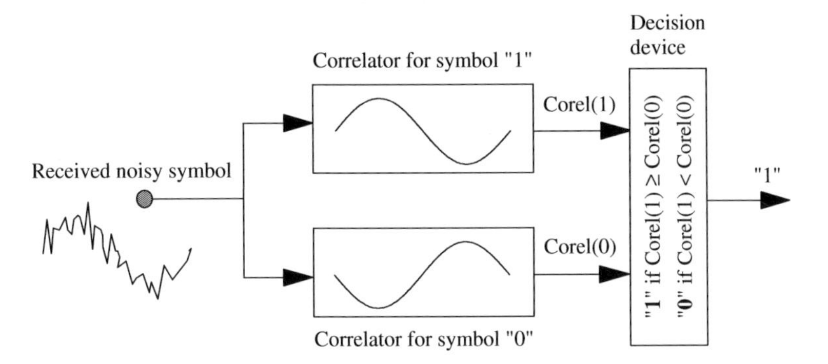

Figure 1.10 Block diagram illustration of the classifier in a binary phase-shift keying demodulation.

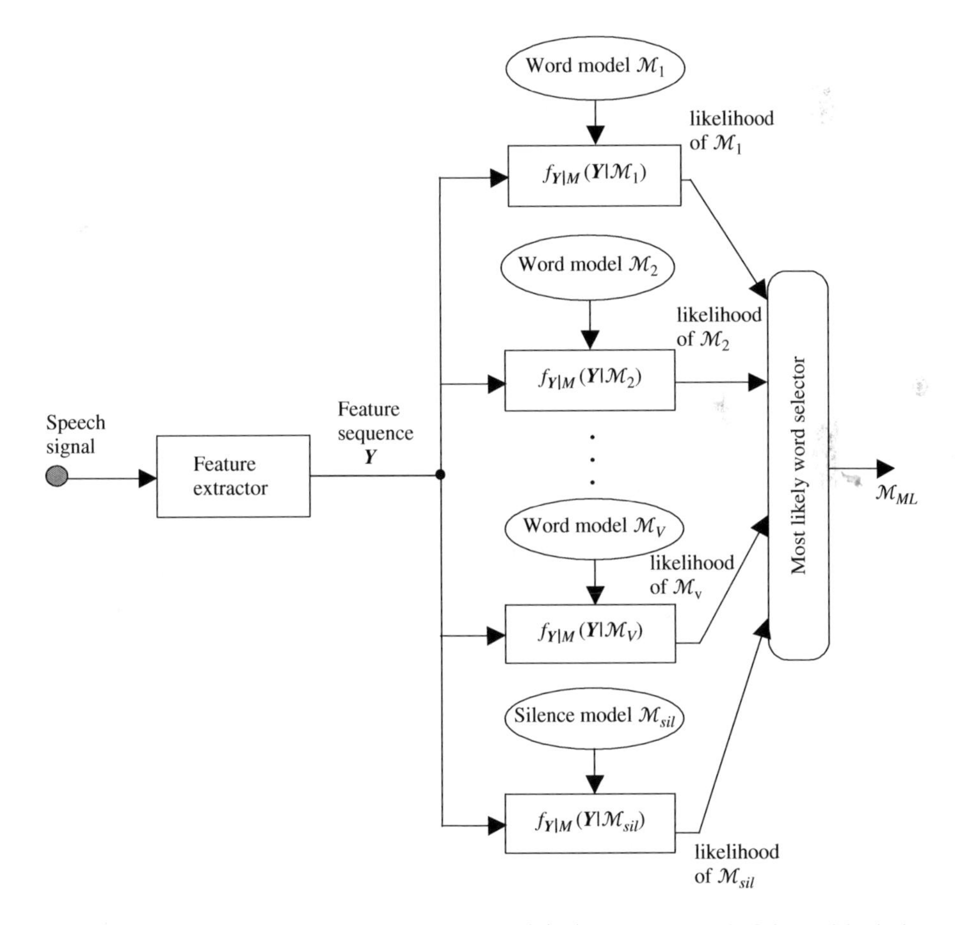

Figure 1.11 Configuration of speech recognition system, $f(\boldsymbol{Y}|\mathcal{M}_i)$ is the likelihood of the model $\mathcal{M}_i$ given an observation sequence $\boldsymbol{Y}$.

1.3.7 Linear Prediction Modelling of Speech

Linear predictive models (introduced in Chapter 12) are widely used in speech processing applications such as low-bit-rate speech coding in cellular telephony, speech enhancement and speech recognition. Speech is generated by inhaling air into the lungs, and then exhaling it through the vibrating glottis cords and the vocal tract. The random, noise-like, air flow from the lungs is spectrally shaped and amplified by the vibrations of the glottal cords and the resonance of the vocal tract. The effect of the vibrations of the glottal cords and the resonance of the vocal tract is to shape the frequency spectrum of speech and introduce a measure of correlation and predictability on the random variations of the air from the lungs. Figure 1.12 illustrates a source-filter model for speech production. The source models the lungs and emits a random excitation signal which is filtered, first by a pitch filter model of the glottal cords and then by a model of the vocal tract.

The main source of correlation in speech is the vocal tract modelled by a linear predictor. A linear predictor is an adaptive filter that forecasts the amplitude of the signal at time m, $x(m)$, using a linear combination of P previous samples $[x(m-1), \mathrm{L}, x(m-P)]$ as

$$\hat{x}(m) = \sum_{k=1}^{P} a_k x(m-k) \tag{1.6}$$

where $\hat{x}(m)$ is the prediction of the signal $x(m)$, and the vector $\boldsymbol{a}^{\mathrm{T}} = [a_1, \ldots, a_P]$ is the coefficients vector of a predictor of order P. The prediction error $e(m)$, i.e. the difference between the actual sample $x(m)$ and its predicted value $\hat{x}(m)$, is defined as

$$e(m) = x(m) - \sum_{k=1}^{P} a_k x(m-k) \tag{1.7}$$

In speech processing, the prediction error $e(m)$ may also be interpreted as the random excitation or the so-called innovation content of $x(m)$. From Equation ((1.7)) a signal generated by a linear predictor can be synthesised as

$$x(m) = \sum_{k=1}^{P} a_k x(m-k) + e(m) \tag{1.8}$$

Linear prediction models can also be used in a wide range of applications to model the correlation or the movements of a signal such as the movements of scenes in successive frames of video.

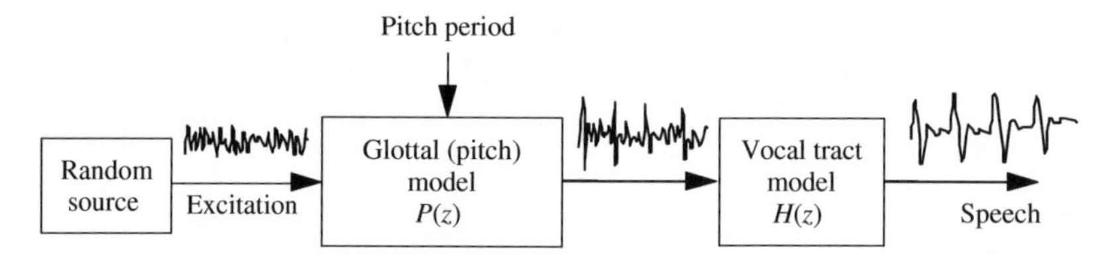

Figure 1.12 Linear predictive model of speech.

1.3.8 Digital Coding of Audio Signals

In digital audio, the memory required to record a signal, the bandwidth and power required for signal transmission and the signal-to-quantisation-noise ratio are all directly proportional to the number of bits per sample. The objective in the design of a coder is to achieve high fidelity with as few bits per sample as possible, at an affordable implementation cost.

Audio signal coding schemes utilise the statistical structures of the signal, and a model of the signal generation, together with information on the psychoacoustics and the masking effects of hearing. In general, there are two main categories of audio coders: model-based coders, used for low-bit-rate speech coding in applications such as cellular telephony; and transform-based coders used in high-quality coding of speech and digital hi-fi audio. Figure 1.13 shows a simplified block diagram configuration of a speech coder–decoder of the type used in digital cellular telephones. The speech signal is modelled as the output of a filter excited by a random signal. The random excitation models the air exhaled through the lungs, and the filter models the vibrations of the glottal cords and the vocal tract. At the transmitter, speech is segmented into blocks of about 20 ms long during which speech parameters can be assumed to be stationary. Each block of speech samples is analysed to extract and transmit a set of excitation and filter parameters that can be used to synthesise the speech. At the receiver, the model parameters and the excitation are used to reconstruct the speech.

A transform-based coder is shown in Figure 1.14. The aim of transformation is to convert the signal into a form where it lends itself to a more convenient and useful interpretation and manipulation. In Figure 1.14 the input signal may be transformed to the frequency domain using a discrete Fourier transform or a discrete cosine transform or a filter bank. Three main advantages of coding a signal in the frequency domain are:

(a) The frequency spectrum of a signal has a relatively well-defined structure, for example most of the signal power is usually concentrated in the lower regions of the spectrum.
(b) A relatively low-amplitude frequency would be masked in the near vicinity of a large-amplitude frequency and can therefore be coarsely encoded without any audible degradation.
(c) The frequency samples are orthogonal and can be coded independently with different precisions.

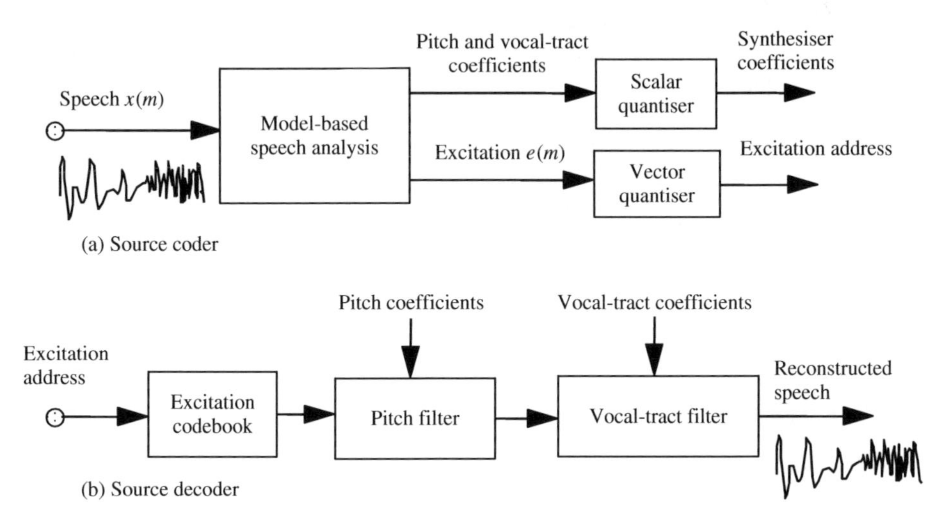

Figure 1.13 Block diagram configuration of a model-based speech (a) coder and (b) decoder.

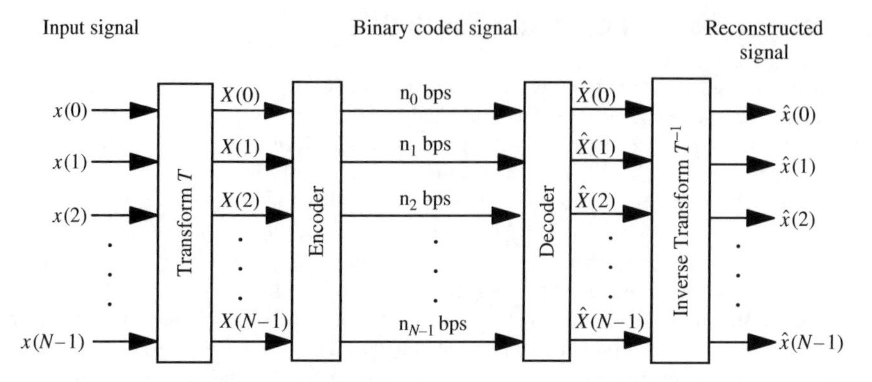

Figure 1.14 Illustration of a transform-based coder.

The number of bits assigned to each frequency of a signal is a variable that reflects the contribution of that frequency to the reproduction of a perceptually high-quality signal. In an adaptive coder, the allocation of bits to different frequencies is made to vary with the time variations of the power spectrum of the signal.

1.3.9 Detection of Signals in Noise

In the detection of signals in noise, the aim is to determine if the observation consists of noise alone, or if it contains a signal. The noisy observation $y(m)$ can be modelled as

$$y(m) = b(m)x(m) + n(m) \tag{1.9}$$

where $x(m)$ is the signal to be detected, $n(m)$ is the noise and $b(m)$ is a binary-valued state indicator sequence such that $b(m) = 1$ indicates the presence of the signal $x(m)$ and $b(m) = 0$ indicates that the signal is absent. If the signal $x(m)$ has a known shape, then a correlator or a matched filter can be used to detect the signal as shown in Figure 1.15. The impulse response $h(m)$ of the matched filter for detection of a signal $x(m)$ is the time-reversed version of $x(m)$ given by

$$h(m) = x(N-1-m) \quad 0 \le m \le N-1 \tag{1.10}$$

where N is the length of $x(m)$. The output of the matched filter is given by

$$z(m) = \sum_{k=0}^{N-1} h(k)y(m-k) \tag{1.11}$$

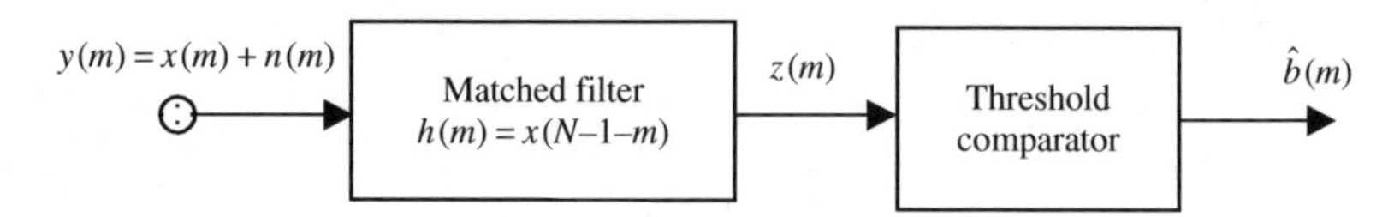

Figure 1.15 Configuration of a matched filter followed by a threshold comparator for detection of signals in noise.

Table 1.1 Four possible outcomes in a signal detection problem.

$\hat{b}(m)$	$b(m)$	Detector decision
0	0	Signal absent (*Correct*)
0	1	Signal absent (*Missed*)
1	0	Signal present (*False alarm*)
1	1	Signal present (*Correct*)

The matched filter output is compared with a threshold and a binary decision is made as

$$\hat{b}(m) = \begin{cases} 1 & \text{if abs}\,(z(m)) \geq \text{ threshold} \\ 0 & \text{otherwise} \end{cases} \tag{1.12}$$

where $\hat{b}(m)$ is an estimate of the binary state indicator sequence $b(m)$, and it may be erroneous in particular if the signal-to-noise ratio is low. Table 1.1 lists four possible outcomes that together $b(m)$ and its estimate $\hat{b}(m)$ can assume. The choice of the threshold level affects the sensitivity of the detector. The higher the threshold, the less the likelihood that noise would be classified as signal, so the false alarm rate falls, but the probability of misclassification of signal as noise increases. The risk in choosing a threshold value θ can be expressed as

$$\mathcal{R}(\text{Threshold} = \theta) = P_{\text{False Alarm}}(\theta) + P_{\text{Miss}}(\theta) \tag{1.13}$$

The choice of the threshold reflects a trade-off between the misclassification rate $P_{\text{Miss}}(\theta)$ and the false alarm rate $P_{\text{FalseAlarm}}(\theta)$.

1.3.10 Directional Reception of Waves: Beam-forming

Beam-forming is the spatial processing of plane waves received by an array of sensors such that the waves' incidents at a particular spatial angle are passed through, whereas those arriving from other directions are attenuated. Beam-forming is used in radar and sonar signal processing (Figure 1.14) to steer the reception of signals towards a desired direction, and in speech processing for reducing the effects of ambient noise.

To explain the process of beam-forming, consider a uniform linear array of sensors as illustrated in Figure 1.16. The term *linear array* implies that the array of sensors is spatially arranged in a straight

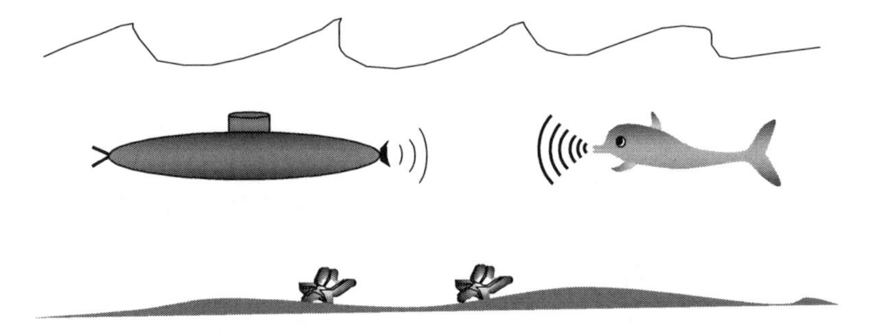

Figure 1.16 Sonar: detection of objects using the intensity and time delay of reflected sound waves.

line and with equal spacing d between the sensors. Consider a sinusoidal far-field plane wave with a frequency F_0 propagating towards the sensors at an incidence angle of θ as illustrated in Figure 1.16. The array of sensors samples the incoming wave as it propagates in space. The time delay for the wave to travel a distance of d between two adjacent sensors is given by

$$\tau = \frac{d \sin\theta}{c} \tag{1.14}$$

where c is the speed of propagation of the wave in the medium. The phase difference corresponding to a delay of τ is given by

$$\varphi = 2\pi\frac{\tau}{T_0} = 2\pi F_0\frac{d \sin\theta}{c} \tag{1.15}$$

where T_0 is the period of the sine wave. By inserting appropriate corrective time delays in the path of the samples at each sensor, and then averaging the outputs of the sensors, the signals arriving from the direction θ will be time-aligned and coherently combined, whereas those arriving from other directions will suffer cancellations and attenuations. Figure 1.17 illustrates a beam-former as an array of digital filters arranged in space. The filter array acts as a two-dimensional space-time signal processing system. The space filtering allows the beam-former to be steered towards a desired direction, for example

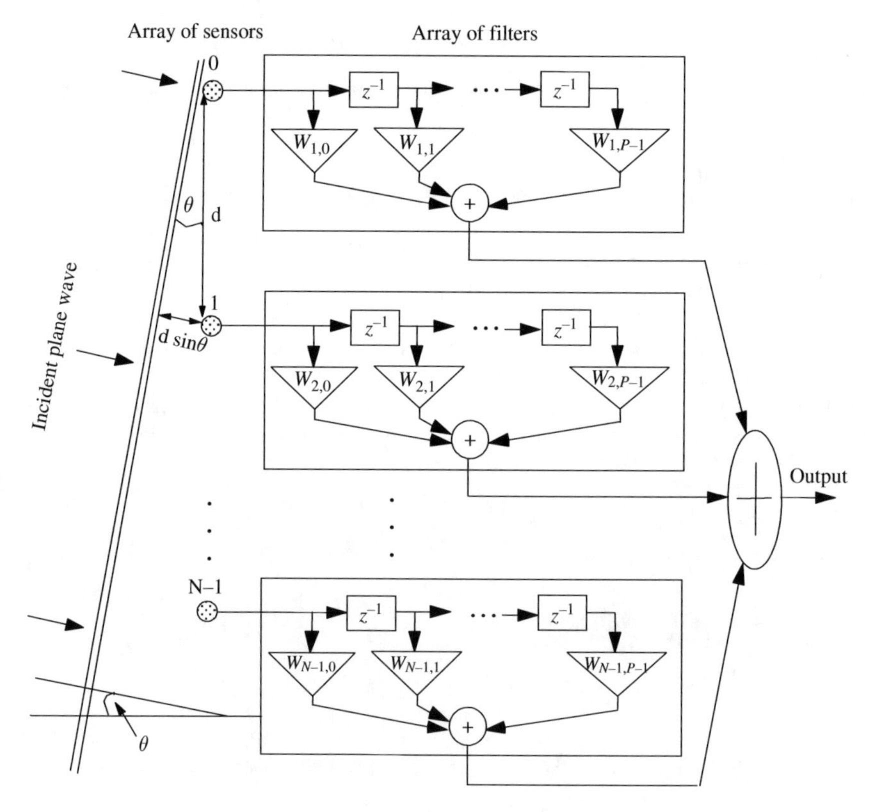

Figure 1.17 Illustration of a beam-former, for directional reception of signals.

towards the direction along which the incoming signal has the maximum intensity. The phase of each filter controls the time delay, and can be adjusted to coherently combine the signals. The magnitude frequency response of each filter can be used to remove the out-of-band noise.

1.3.11 Space-Time Signal Processing

Conventionally transmission resources are shared among subscribers of communication systems through the division of time and frequency leading to such resource-sharing schemes as time division multiple access or frequency division multiple access. Space provides a valuable additional resource that can be used to improve both the communication capacity and quality for wireless communication systems.

Space-time signal processing refers to signal processing methods that utilise simultaneous transmission and reception of signals through multiple spatial routes. The signals may arrive at the destinations at different times or may use different time slots. Space-time signal processing, and in particular the division of space among different users, is an important area of research and development for improving the system capacity in the new generations of high-speed broadband multimedia mobile communication systems.

For example, in mobile communication the multi-path effect, where a radio signal propagates from the transmitter to the receiver via a number of different paths, can be used to advantage in space-time signal processing. The multiple noisy versions of a signal, arriving via different routes with different noise and distortions, are processed and combined such that the signals add up constructively and become stronger compared with the random uncorrelated noise. The uncorrelated fading that the signals suffer in their propagation through different routes can also be mitigated.

The use of transmitter/receiver antenna arrays for beam-forming allows the division of the space into narrow sectors such that the same frequencies, in different narrow spatial sectors, can be used for simultaneous communication by different subscribers and/or different spatial sectors can be used to transmit the same information in order to achieve robustness to fading and interference. In fact combination of space and time can provide a myriad of possibilities, as discussed in Chapter 18 on mobile communication signal processing. Note that the ICA method, described in Section 1.3.2 and Chapter 12, is often used in space-time signal processing for separation of multiple signals at the receiver.

1.3.12 Dolby Noise Reduction

Dolby noise reduction systems work by boosting the energy and the signal-to-noise ratio of the high-frequency spectrum of audio signals. The energy of audio signals is mostly concentrated in the low-frequency part of the spectrum (below 2 kHz). The higher frequencies that convey quality and sensation have relatively low energy, and can be degraded even by a low amount of noise. For example when a signal is recorded on a magnetic tape, the tape 'hiss' noise affects the quality of the recorded signal. On playback, the higher-frequency parts of an audio signal recorded on a tape have smaller signal-to-noise ratio than the low-frequency parts. Therefore noise at high frequencies is more audible and less masked by the signal energy. Dolby noise reduction systems broadly work on the principle of emphasising and boosting the low energy of the high-frequency signal components prior to recording the signal. When a signal is recorded it is processed and encoded using a combination of a pre-emphasis filter and dynamic range compression. At playback, the signal is recovered using a decoder based on a combination of a de-emphasis filter and a decompression circuit. The encoder and decoder must be well matched and cancel each other out in order to avoid processing distortion.

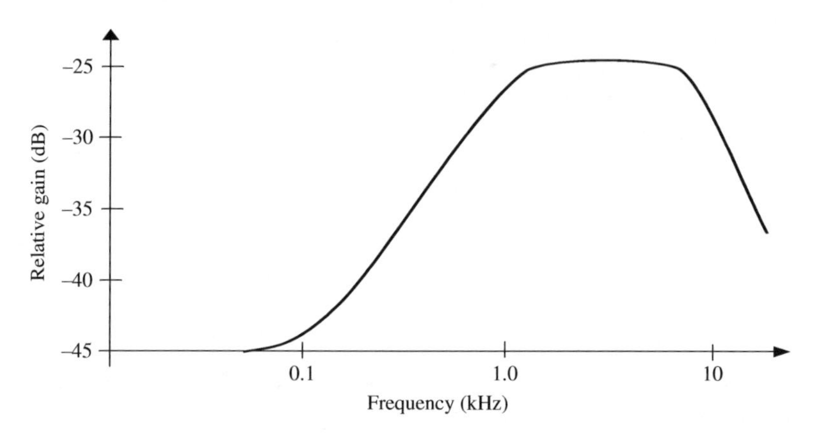

Figure 1.18 Illustration of the pre-emphasis response of Dolby C: up to 20 dB boost is provided when the signal falls 45 dB below maximum recording level.

Dolby developed a number of noise reduction systems designated Dolby A, Dolby B and Dolby C. These differ mainly in the number of bands and the pre-emphasis strategy that that they employ. Dolby A, developed for professional use, divides the signal spectrum into four frequency bands: band 1 is low-pass and covers 0 Hz to 80 Hz; band 2 is band-pass and covers 80 Hz to 3 kHz; band 3 is high-pass and covers above 3 kHz; and band 4 is also high-pass and covers above 9 kHz. At the encoder the gain of each band is adaptively adjusted to boost low-energy signal components. Dolby A provides a maximum gain of 10 to 15 dB in each band if the signal level falls 45 dB below the maximum recording level. The Dolby B and Dolby C systems are designed for consumer audio systems, and use two bands instead of the four bands used in Dolby A. Dolby B provides a boost of up to 10 dB when the signal level is low (less than 45 dB than the maximum reference) and Dolby C provides a boost of up to 20 dB as illustrated in Figure 1.18.

1.3.13 Radar Signal Processing: Doppler Frequency Shift

Figure 1.19 shows a simple diagram of a radar system that can be used to estimate the range and speed of an object such as a moving car or a flying aeroplane. A radar system consists of a transceiver

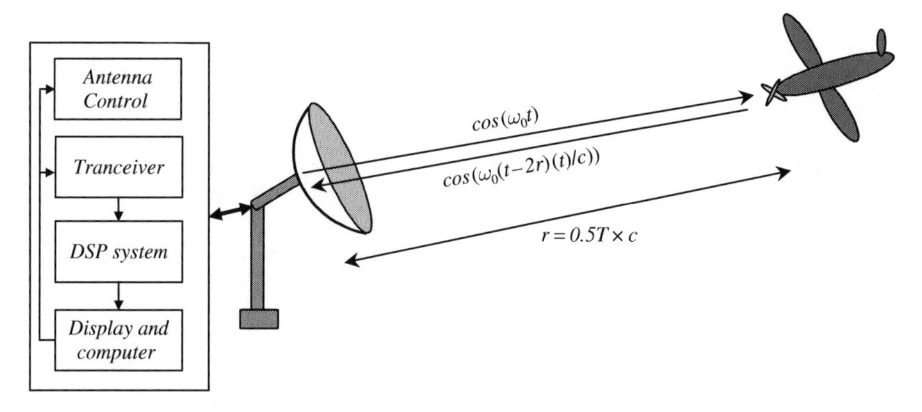

Figure 1.19 Illustration of a radar system.

(transmitter/receiver) that generates and transmits sinusoidal pulses at microwave frequencies. The signal travels with the speed of light and is reflected back from any object in its path. The analysis of the received echo provides such information as range, speed and acceleration. The received signal has the form

$$x(t) = A(t)\cos\{\omega_0[t - 2r(t)/c]\} \tag{1.16}$$

where $A(t)$, the time-varying amplitude of the reflected wave, depends on the position and the characteristics of the target, $r(t)$ is the time-varying distance of the object from the radar and c is the velocity of light. The time-varying distance of the object can be expanded in a Taylor series as

$$r(t) = r_0 + \dot{r}t + \frac{1}{2!}\ddot{r}t^2 + \frac{1}{3!}\dddot{r}\,t^3 + \cdots \tag{1.17}$$

where r_0 is the distance, $\dot{r}$ is the velocity, $\ddot{r}$ is the acceleration etc. Approximating $r(t)$ with the first two terms of the Taylor series expansion we have

$$r(t) \approx r_0 + \dot{r}t \tag{1.18}$$

Substituting Equation (1.18) in Equation (1.16) yields

$$x(t) = A(t)\cos[(\omega_0 - 2\dot{r}\omega_0/c)t - 2\omega_0 r_0/c] \tag{1.19}$$

Note that the frequency of reflected wave is shifted by an amount

$$\omega_d = 2\dot{r}\omega_0/c \tag{1.20}$$

This shift in frequency is known as the Doppler frequency. If the object is moving towards the radar then the distance $r(t)$ is decreasing with time, $\dot{r}$ is negative, and an increase in the frequency is observed. Conversely if the object is moving away from the radar then the distance $r(t)$ is increasing, $\dot{r}$ is positive, and a decrease in the frequency is observed. Thus the frequency analysis of the reflected signal can reveal information on the direction and speed of the object. The distance r_0 is given by

$$r_0 = 0.5T \times c \tag{1.21}$$

where T is the round-trip time for the signal to hit the object and arrive back at the radar and c is the velocity of light.

1.4 Summary

This chapter began with a definition of signal and information and provided a qualitative explanation of their relationship. A broad categorisation of the various signal processing methodologies was provided. We considered several key applications of digital signal processing in biomedical signal processing, adaptive noise reduction, channel equalisation, pattern classification/recognition, audio signal coding, signal detection, spatial processing for directional reception of signals, Dolby noise reduction, radar and watermarking.

Further Reading

Alexander S.T. (1986) Adaptive Signal Processing Theory and Applications. Springer-Verlag, New York.
Davenport W.B. and Root W.L. (1958) An Introduction to the Theory of Random Signals and Noise. McGraw-Hill, New York.

Ephraim Y. (1992) Statistical Model Based Speech Enhancement Systems. Proc. IEEE, 80 (10), pp. 1526–1555.
Gallager R.G. (1968) Information Theory and Reliable Communication. Wiley, New York.
Gauss K.G. (1963) Theory of Motion of Heavenly Bodies. Dover, New York.
Haykin S. (1985) Array Signal Processing. Prentice-Hall, Englewood Cliffs, NJ.
Haykin S. (1991) Adaptive Filter Theory. Prentice-Hall, Englewood Cliffs, NJ.
Kailath T. (1980) Linear Systems. Prentice Hall, Englewood Cliffs, NJ.
Kalman R.E. (1960) A New Approach to Linear Filtering and Prediction Problems. Trans. of the ASME, Series D, Journal of Basic Engineering, 82, pp. 35–45.
Kay S.M. (1993) Fundamentals of Statistical Signal Processing, Estimation Theory. Prentice-Hall, Englewood Cliffs, NJ.
Kung S.Y. (1993) Digital Neural Networks. Prentice-Hall, Englewood Cliffs, NJ.
Lim J.S. (1983) Speech Enhancement. Prentice Hall, Englewood Cliffs, NJ.
Lucky R.W., Salz J. and Weldon E.J. (1968) Principles of Data Communications. McGraw-Hill, New York.
Marple S.L. (1987) Digital Spectral Analysis with Applications. Prentice-Hall, Englewood Cliffs, NJ.
Oppenheim A.V. and Schafer R.W. (1989) Discrete-Time Signal Processing. Prentice-Hall, Englewood Cliffs, NJ.
Proakis J.G., Rader C.M., Ling F. and Nikias C.L. (1992) Advanced Signal Processing. Macmillan, New York.
Rabiner L.R. and Gold B. (1975) Theory and Applications of Digital Processing. Prentice-Hall, Englewood Cliffs, NJ.
Rabiner L.R. and Schafer R.W. (1978) Digital Processing of Speech Signals. Prentice-Hall, Englewood Cliffs, NJ.
Scharf L.L. (1991) Statistical Signal Processing: Detection, Estimation, and Time Series Analysis. Addison Wesley, Reading, MA.
Shannon C.E. (1948) A Mathematical Theory of Communication. Bell Systems Tech. J., 27, pp. 379–423, 623–656.
Therrien C.W. (1992) Discrete Random Signals and Statistical Signal Processing. Prentice-Hall, Englewood Cliffs, NJ.
Van-Trees H.L. (1971) Detection, Estimation and Modulation Theory. Parts I, II and III. Wiley, New York.
Vaseghi S. (2006) Advanced Digital Signal Processing and Noise Reduction, Wiley, New York.
Widrow B. (1975) Adaptive Noise Cancelling: Principles and Applications. Proc. IEEE, 63, pp. 1692–1716.
Wiener N. (1948) Extrapolation, Interpolation and Smoothing of Stationary Time Series. MIT Press, Cambridge, MA.
Wiener N. (1949) Cybernetics. MIT Press, Cambridge, MA.
Wilsky A.S. (1979) Digital Signal Processing, Control and Estimation Theory: Points of Tangency, Areas of Intersection and Parallel Directions. MIT Press, Cambridge, MA.

2 Fourier Analysis and Synthesis

Fourier's method of description of a signal in terms of a combination of elementary trigonometric functions had a profound effect on the way signals are viewed, analysed and processed. In communication and signal processing such fundamental concepts as frequency spectrum and bandwidth result from the Fourier representation of signals.

The Fourier method is the most extensively applied signal analysis and synthesis tool. The applications of the Fourier transform in telecommunication and signal processing include filtering, correlation, music processing, signal coding, signal synthesis, feature extraction for pattern identification as in speech or image recognition, spectral analysis and radar signal processing.

The Fourier transform of a signal lends itself to easy interpretation and manipulation, and leads to the concept of frequency analysis and synthesis. Furthermore, even some biological systems, such as the human auditory system, perform some form of frequency analysis of the input signals.

This chapter begins with an introduction to Fourier himself, the complex Fourier series and Fourier transform, and then considers the discrete Fourier transform, the fast Fourier transform, the 2-D Fourier transform and the discrete cosine transform. Important engineering issues such as the trade-off between the time and frequency resolutions, problems with finite data length, windowing and spectral leakage are considered.

2.1 Introduction

Jean Baptiste Joseph Fourier (1768–1830), introduced the method of expansion of a function in terms of cosine and sine waves, now known as the Fourier series, in his major work on the mathematical theory of heat conduction, *The Analytic Theory of Heat*. He established the partial differential equation governing heat diffusion and solved it using an infinite series of trigonometric (sine and cosine) functions.

It may be noted that before Fourier introduced the concept of mathematical analysis and synthesis of signals using sinusoidal functions, musical instrument makers had already created string musical instruments which can be considered as a form of mechanical Fourier synthesisers.

Multimedia Signal Processing: Theory and Applications in Speech, Music and Communications Saeed V. Vaseghi

Fourier was born in Auxerre, France. By the early age of 13 Fourier's main interest was mathematics. He attended the École Royale Militaire and in 1790 became a teacher there. Fourier became involved in the French Revolution and served two prison terms. He continued his studies at the École Normale in Paris, having as his teachers great mathematicians such as Lagrange, Laplace and Monge. He taught at the Collège de France and later moved to a position at the École Polytechnique where in 1797 he succeeded Lagrange as the Professor of analysis and mechanics.

Fourier, together with Monge and Malus, joined Napoleon as scientific advisors in his invasion of Egypt. There Fourier organised the Cairo Institute. After Napoleon returned to Paris, he asked Fourier to take an administrative post in Grenoble. It was during his time in Grenoble between 1804 and 1807 that Fourier completed his important work and memoir *On the Propagation of Heat in Solid Bodies*. Fourier's work on expansion of functions as trigonometric series was refereed by a committee including Lagrange, Laplace and Monge; however initially they were not persuaded by his work and considered it insufficiently vigorous. Fourier later completed his famous book *Théorie Analytique de la Chaleur* where he showed how functions can be represented by a trigonometric series.

Objectives of a Transform

The primary objective of signal transformation is to express a signal in terms of a combination of a set of simple *elementary* signals, known as the *basis functions*. The transform's output should lend itself to convenient analysis, interpretation, modification and synthesis.

In Fourier transform the basic elementary signals are a set of sinusoidal signals (sines and cosines) with various periods of repetition giving rise to the concept of frequency defined as the number of cycles per second in units of Hz. Many indispensable concepts in communication and signal processing theory, such as the concepts of bandwidth, power spectrum and filtering, result from the Fourier description of signals.

In Fourier analysis a signal is decomposed into its constituent sinusoidal vibrations. The amplitudes and phases of the sinusoids of various frequencies form the frequency spectrum of the signal. In inverse Fourier transform a signal can be synthesised by adding up its constituent frequencies.

It turns out that many signals that we encounter in daily life – such as speech, car engine noise and music – are generated by some form of vibrations and have a periodic or quasi-periodic structure. Furthermore, the cochlea in the human hearing system performs a kind of vibration analysis of the input audio signals. Therefore the concept of frequency analysis is not a purely mathematical abstraction in that some biological systems including humans and many species of animals have evolved sensory mechanisms that make use of the frequency analysis concept. Note that stringed musical instruments, such as guitars, may be considered as a form of mechanical Fourier synthesisers, since they create signals from sinusoidal waves produced by strings.

The power of the Fourier transform in signal analysis and pattern recognition lies in its simplicity and its ability to reveal spectral structures that can be used to characterise a signal. This is illustrated in Figure 2.1 for the two extreme cases of a sine wave and a purely random signal. For a periodic signal, such as a sine wave or a train of pulses, the signal power is concentrated in extremely narrow band(s) of frequencies indicating the existence of a periodic structure and the predictable character of the signal. In the case of a pure sine wave as shown in Figure 2.1(a) the signal power is concentrated in just one frequency. For a purely random signal as shown in Figure 2.1(b) the average signal power is spread equally in the frequency domain indicating the lack of a predictable structure in the signal.

A Note on Comparison of Fourier Series and Taylor Series

The Taylor series expansion describes a function $f(x)$ *locally* in terms of the differentials of the function at point $f(x = a)$. In contrast, as shown next, Fourier series describes a function *globally*

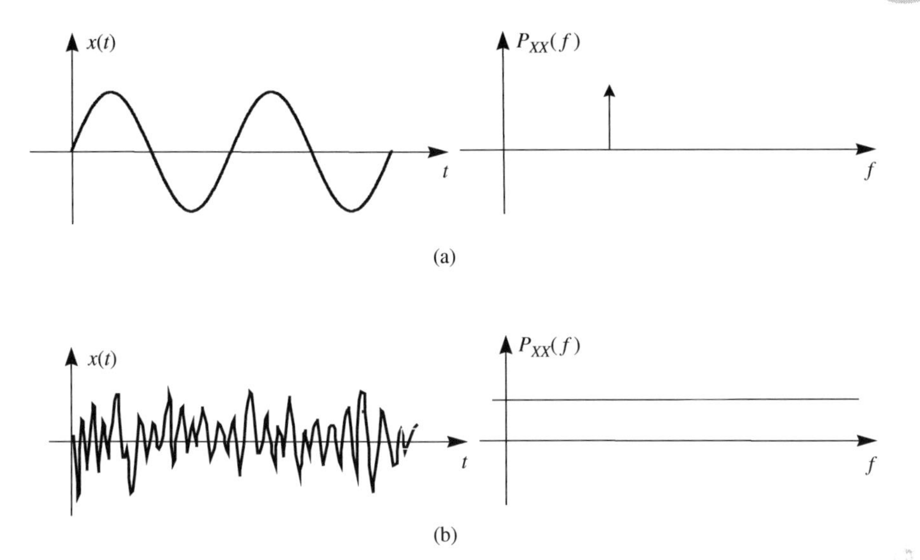

Figure 2.1 Fourier analysis reveals the signal structure. The concentration or spread of the power of a signal in frequency indicates the correlated or random character of a signal: (a) a predictable signal, such as a sine wave is concentrated in extremely narrow band(s) of frequencies, (b) a white noise is spread across the frequency.

in terms of sine waves and cosine waves whose weighting coefficients are calculated over the entire length of the signal.

Notation: In this chapter the symbols t and m denote continuous and discrete time variables, and f and k denote continuous and discrete frequency variables respectively. The variable $\omega = 2\pi f$ denotes the angular frequency in units of rad/s, it is used interchangeably (within a scaling of factor of 2π) with the frequency variable f in units of Hz.

2.2 Fourier Series: Representation of Periodic Signals

A periodic signal can be described in terms of a series of harmonically related (i.e. integer multiples of a fundamental frequency) sine and cosine waves.

The following three sinusoidal functions form the *basis functions* for Fourier analysis.

$$x_1(t) = \cos \omega_0 t \tag{2.1}$$

$$x_2(t) = \sin \omega_0 t \tag{2.2}$$

$$x_3(t) = \cos \omega_0 t + \mathrm{j} \sin \omega_0 t = e^{\mathrm{j}\omega_0 t} \tag{2.3}$$

A cosine function is an even function with respect to the vertical axis (amplitude) at time $t = 0$ and a sine function is an odd function. A weighted combination of a sine and a cosine at angular frequency ω_0 can model any phase of a sinusoidal signal component of $x(t)$ at that frequency.

Figure 2.2(a) shows the sine and the cosine components of the complex exponential (cisoidal) signal of Equation (2.3), and Figure 2.2(b) shows a vector representation of the complex exponential in a complex plane with real (Re) and imaginary (Im) dimensions. The Fourier basis functions are periodic with a period of $T_0 = 1/F_0$ and an angular frequency of $\omega_0 = 2\pi F_0$ radians/second, where F_0 is the frequency in units of Hz.

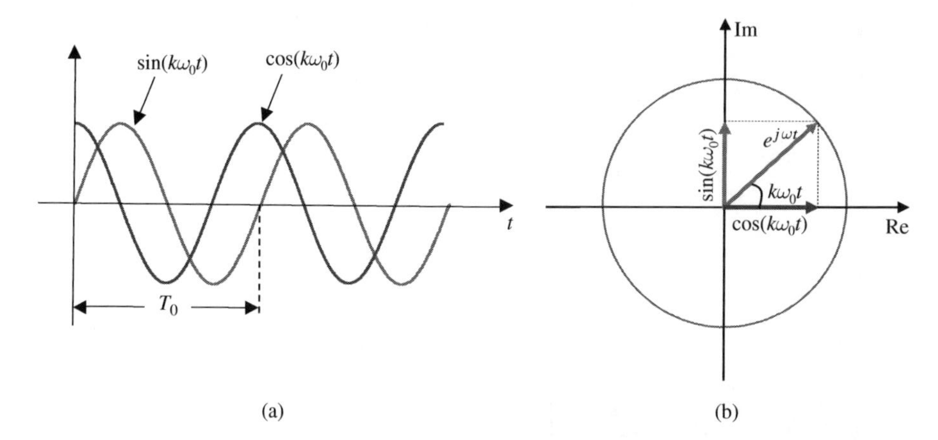

Figure 2.2 Fourier basis functions: (a) real and imaginary components of a complex sinusoid, (b) vector representation of a complex exponential. If the cosine is considered as the in-phase component then the sine is the quadrature component.

> **Matlab function locus_sin_vs_cos( )**
> Plots the locus of a sine wave versus a cosine wave for varying time and phase. The frequency (the number of cycles per second) of the sinusoidal signal can also be varied. Note that the locus of a sine vs. a cosine is a circle. As shown, the starting point of the circle depends on the phase.

The Properties of Fourier's Sinusoidal Basis Functions

The following properties make the sinusoids an ideal choice as the elementary building block basis functions for signal analysis and synthesis.

(i) Orthogonality; two sinusoidal functions of *different* frequencies have the following orthogonal property:

$$\int_{-\infty}^{\infty} \sin(\omega_1 t)\sin(\omega_2 t)dt = -\frac{1}{2}\int_{-\infty}^{\infty} \cos(\omega_1+\omega_2)t\; dt + \frac{1}{2}\int_{-\infty}^{\infty} \cos(\omega_1-\omega_2)t\; dt = 0 \tag{2.4}$$

For sinusoids the integration interval can be taken over one period (i.e. $T = 2\pi/(\omega_1+\omega_2)$ and $T = 2\pi/(\omega_1-\omega_2)$. Similar equations can be derived for the product of cosines, or sine and cosine, of different frequencies. Orthogonality implies that the sinusoidal basis functions are 'independent' and can be processed independently. For example in a graphic equaliser we can change the relative amplitudes of one set of frequencies, such as the audio bass, without affecting other frequencies, and in music coding the signals in different frequency bands are coded independently and allocated different numbers of bits.

(ii) The sine and cosine components of $e^{j\omega t}$ have only a relative phase difference of $\pi/2$ or equivalently a relative time delay of a quarter of one period i.e. $T_0/4$. This allows the decomposition of a signal in terms of orthogonal cosine (in-phase) and sine (quadrature) components.

(iii) Sinusoidal functions are infinitely differentiable. This is a useful property, as most signal analysis and synthesis methods require the signals to be differentiable.

(iv) A useful consequence of transforms the Fourier and Laplace transforms is that relatively difficult differential analysis on the time domain signal become simple algebraic operations on the transformed signal.

The Basis Functions of Fourier Series

Associated with the complex exponential function $e^{j\omega_0 t}$ is a set of harmonically related complex exponentials of the form

$$[1, e^{\pm j\omega_0 t}, e^{\pm j2\omega_0 t}, e^{\pm j3\omega_0 t}, \dots] \tag{2.5}$$

The set of exponential signals in Equation (2.5) are periodic with a fundamental frequency $\omega_0 = 2\pi/T_0 = 2\pi F_0$ where T_0 is the period and F_0 is the fundamental frequency. These signals form the set of *basis functions* for the Fourier series analysis. Any linear combination of these signals of the form

$$x(t) = \sum_{k=-\infty}^{\infty} c_k e^{jk\omega_0 t} \tag{2.6}$$

is also a periodic signal with a period of T_0. Conversely any periodic signal $x(t)$ can be synthesised from a linear combination of harmonically related exponentials.

The Fourier series representation of a periodic signal, with a period of T_0 and angular frequency $\omega_0 = 2\pi/T_0 = 2\pi F_0$, is given by the following synthesis and analysis equations:

$$x(t) = \sum_{k=-\infty}^{\infty} c_k e^{jk\omega_0 t} \quad k = \cdots, -1, 0, 1, \cdots \quad \textbf{Synthesis equation} \tag{2.7}$$

$$c_k = \frac{1}{T_0} \int_{-T_0/2}^{T_0/2} x(t) e^{-jk\omega_0 t} dt \quad k = \cdots, -1, 0, 1, \cdots \quad \textbf{Analysis equation} \tag{2.8}$$

Fourier Series Coefficients

The complex-valued Fourier series coefficient c_k conveys the amplitude (a measure of the strength) and the phase (or time delay) of the frequency content of the signal at frequency $k\omega_0$ Hz. Note from the Fourier analysis Equation (2.8), that the coefficient c_k *may be interpreted as a measure of the correlation of the signal* $x(t)$ *and the complex exponential* $e^{-jk\omega_0 t}$.

The representation of a signal in the form of Equation (2.7) as the sum of its constituent harmonics is referred to as the *complex Fourier series* representation. The set of complex coefficients $\dots, c_{-1}, c_0, c_1, \dots$ is known as the *frequency spectrum* of the signal.

Equation (2.7) can be used as a synthesiser (as in a music synthesiser) to generate a signal as a weighted combination of its elementary frequencies. Note from Equations (2.7) and (2.8) that the complex exponentials that form a periodic signal occur only at discrete frequencies which are integer multiple harmonics of the fundamental frequency ω_0. Therefore the spectrum of a periodic signal, with a period of T_0, is *discrete* in frequency with *discrete spectral lines* spaced at integer multiples of $\omega_0 = 2\pi/T_0$.

Positive and Negative Frequencies

Note from the Fourier series representation equation (2.7) that for a *real-valued* signal $x(t)$ for each positive frequency $e^{jk\omega_0}$ there needs to be its complex conjugate $e^{-jk\omega_0}$; so that in the Fourier synthesis equation the imaginary parts cancel out. This gives rise to the concept of negative frequencies.

Example 2.1 *Derivation of Fourier series coefficients*
Given the Fourier series synthesis equation (2.7), obtain the Fourier analysis equation (2.8) i.e. the coefficients of the Fourier series.

Solution: To obtain c_k, the coefficient of the kth harmonic, multiply both sides of Equation (2.7) by $e^{-jn\omega_0 t}$ and integrate over one period to obtain

$$\int_{-T_0/2}^{T_0/2} x(t)e^{-jn\omega_0 t}dt = \int_{-T_0/2}^{T_0/2} \sum_{k=-\infty}^{\infty} c_k e^{jk\omega_0 t} e^{-jn\omega_0 t} dt = \sum_{k=-\infty}^{\infty} c_k \int_{-T_0/2}^{T_0/2} e^{j(k-n)\omega_0 t} dt \tag{2.9}$$

where n is an integer. From the orthogonality principle the integral of the product of two complex sinusoids of different frequency is zero. Hence, for $n \neq k$ the integral over one period of $e^{j(k-n)\omega_0 t}$ in the r.h.s. of Equation (2.9) is zero. For $n = k$, $e^{j(k-n)\omega_0 t} = e^0 = 1$ and its integral over one period is equal to T_0 and the r.h.s. of Equation (2.9) is equal to $c_k T_0$. Hence for $n = k$ we have

$$c_k = \frac{1}{T_0} \int_{-T_0/2}^{T_0/2} x(t)e^{-jk\omega_0 t} dt \tag{2.10}$$

Example 2.2 Find the frequency spectrum of a 1 kHz ($F_0 = 1000$ cycles/second, $\omega_0 = 2000\pi$) sine wave, shown in Figure 2.3(a).

$$x(t) = \sin(2000\pi t) \qquad -\infty < t < \infty \tag{2.11}$$

Solution A: The Fourier synthesis equation (2.7) can be written as

$$x(t) = \sum_{k=-\infty}^{\infty} c_k e^{jk2000\pi t} = \cdots + c_{-1}e^{-j2000\pi t} + c_0 + c_1 e^{j2000\pi t} + \cdots \tag{2.12}$$

Now the sine wave can be expressed as

$$x(t) = \sin(2000\pi t) = \frac{1}{2j} e^{j2000\pi t} - \frac{1}{2j} e^{-j2000\pi t} \tag{2.13}$$

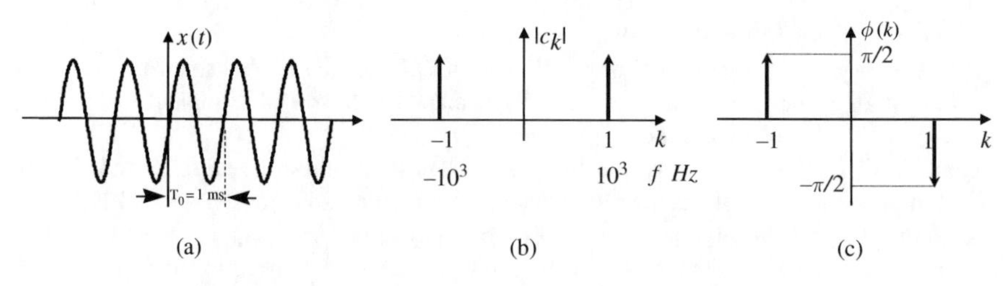

Figure 2.3 (a) A sine wave, (b) its magnitude spectrum, (c) its phase spectrum. Note that negative frequency is a consequence of the requirement to represent a real-valued signal as a combination of complex conjugate parts.

Equating the coefficients of Equations (2.12) and (2.13) yields

$$c_1 = \frac{1}{2}\mathrm{j},\ c_{-1} = -\frac{1}{2}\mathrm{j} \text{ and } c_{k\neq\pm1} = 0 \tag{2.14}$$

Figure 2.3(b) shows the magnitude and phase spectrum of the sine wave, where the spectral lines c_1 and c_{-1} correspond to the 1 kHz and -1 kHz frequencies respectively.

Solution B: Substituting $\sin(2000\pi t) = \frac{1}{2\mathrm{j}}e^{\mathrm{j}2000\pi t} - \frac{1}{2\mathrm{j}}e^{-\mathrm{j}2000\pi t}$ in the Fourier analysis equation (2.8) yields

$$c_k = \frac{1}{T_0}\int_{-T_0/2}^{T_0/2} \underbrace{\left(\frac{1}{2\mathrm{j}}e^{\mathrm{j}2000\pi t} - \frac{1}{2\mathrm{j}}e^{-\mathrm{j}2000\pi t}\right)}_{\sin(2000\pi t)} e^{-\mathrm{j}k2000\pi t}dt$$

$$= \frac{1}{2\mathrm{j}T_0}\int_{-T_0/2}^{T_0/2} e^{\mathrm{j}(1-k)2000\pi t}dt - \frac{1}{2\mathrm{j}T_0}\int_{-T_0/2}^{T_0/2} e^{-\mathrm{j}(1+k)2000\pi t}dt \tag{2.15}$$

Since sine and cosine functions are positive-valued over one half of a period and odd symmetric (equal and negative) over the other half, it follows that Equation (2.15) is zero unless $k = 1$ or $k = -1$. Hence

$$c_1 = \frac{1}{2\mathrm{j}} \quad \text{and } c_{-1} = -\frac{1}{2\mathrm{j}} \quad \text{and } c_{k\neq\pm1} = 0 \tag{2.16}$$

Example 2.3 Find the frequency spectrum of a periodic train of pulses, shown in Figure 2.4, with an amplitude of 1.0, a fundamental frequency of 100 Hz and a pulse 'on' duration of 3 milliseconds (0.003 seconds).

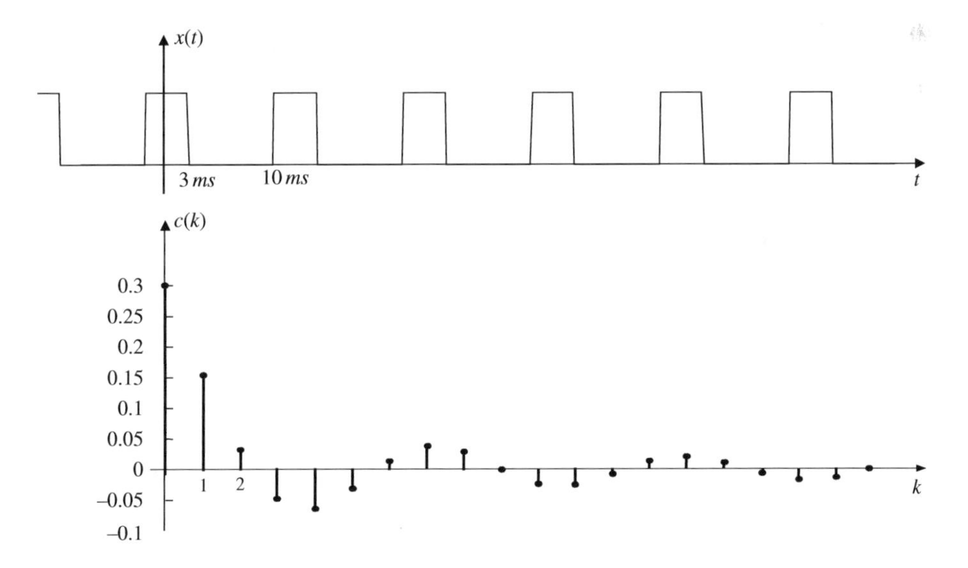

Figure 2.4 A rectangular pulse train and its discrete frequency 'line' spectrum (only positive frequencies shown).

Solution: The pulse period $T_0 = 1/F_0 = 0.01$ s, and the angular frequency $\omega_0 = 2\pi F_0 = 200\pi$ rad/s. Substituting the pulse signal in the Fourier analysis equation (2.8) gives

$$\begin{aligned} c_k &= \frac{1}{T_0}\int_{-T_0/2}^{T_0/2} x(t)e^{-jk\omega_0 t}\,dt = \frac{1}{0.01}\int_{-0.0015}^{0.0015} e^{-jk200\pi t}\,dt \\ &= \left.\frac{e^{-jk200\pi t}}{-j2\pi k}\right|_{t=-0.0015}^{t=0.0015} = \frac{e^{j0.3\pi k} - e^{-j0.3\pi k}}{j2\pi k} \\ &= \frac{\sin(0.3\pi k)}{\pi k} = 0.3\frac{\sin(0.3\pi k)}{0.3\pi k} = 0.3\mathrm{sinc}(0.3k) \end{aligned} \tag{2.17}$$

where $\mathrm{sinc}(x) = \sin(\pi x)/\pi x$. For $k = 0$ as $c_0 = \sin(0)/0$ is undefined, differentiate the numerator and denominator of Equation (2.17) w.r.t. the variable k to obtain

$$c_0 = \frac{0.3\pi\cos(0.3\pi \times 0)}{\pi} = 0.3 \tag{2.18}$$

Example 2.4 In Example 2.3, write the formula for synthesising the signal up to the Nth harmonic, and plot a few examples for increasing number of harmonics.

Solution: The equation for the synthesis of a signal up to the Nth harmonic content is given by

$$\begin{aligned} x(t) &= \sum_{k=-N}^{N} c_k e^{jk\omega_0 t} = c_0 + \sum_{k=1}^{N} c_k e^{jk\omega_0 t} + \sum_{k=1}^{N} c_{-k} e^{-jk\omega_0 t} \\ &= c_0 + \sum_{k=1}^{N} [\mathrm{Re}(c_k) + j\ \mathrm{Im}(c_k)][\cos(k\omega_0 t) + j\sin(k\omega_0 t)] \\ &\quad + \sum_{k=1}^{N} [\mathrm{Re}(c_k) - j\ \mathrm{Im}(c_k)][\cos(k\omega_0 t) - j\sin(k\omega_0 t)] \\ &= c_0 + \sum_{k=1}^{N} [2\mathrm{Re}(c_k)\cos(k\omega_0 t) - 2\mathrm{Im}(c_k)\sin(k\omega_0 t)] \end{aligned} \tag{2.19}$$

The signals in Figure 2.5 were obtained using a MatLab code to generate a synthesised pulse train composed of N harmonics. In this example there are five cycles in an array of 1000 samples. Figure 2.5 shows the waveform for the number of harmonics equal to: 1, 3, 6 and 100.

Matlab function FourierHarmonicSynthesis( )

Generates and displays the harmonics of a periodic train of pulses. Starting with the fundamental frequency, at each iteration of a loop an additional harmonic is added to the signal.

The signal is displayed in time, frequency and spectrogram and also played back through speakers.

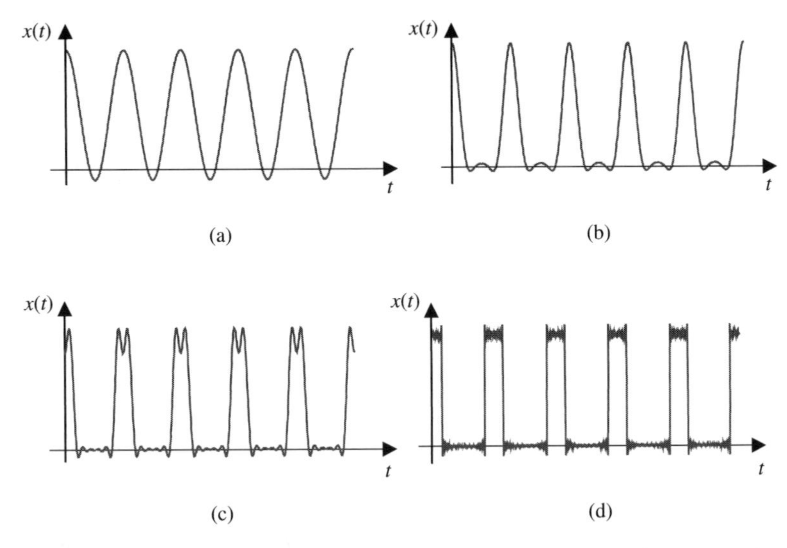

Figure 2.5 Illustration of the Fourier synthesis of a periodic pulse train, and the Gibbs phenomenon, with the increasing number of harmonics in the Fourier synthesis: (a) $N = 1$, (b) $N = 3$, (c) $N = 6$, and (d) $N = 100$. Note that the over-shoot in (d) exhibits the Gibbs phenomenon.

Fourier Synthesis at Discontinuity: Gibbs Phenomenon

The sinusoidal basis functions of the Fourier transform are smooth and infinitely differentiable. In the vicinity of a discontinuity the Fourier synthesis of a signal exhibits ripples as shown in Figure 2.5(d). The peak amplitude of the ripples does not decrease as the number of harmonics used in the signal synthesis increases. This behaviour is known as the *Gibbs phenomenon.* For a discontinuity of unit height, the partial sum of the harmonics exhibits a maximum value of 1.09 (i.e. an overshoot of 9%) irrespective of the number of harmonics used in the Fourier series. As the number of harmonics used in the signal synthesis increases, the ripples become compressed toward the discontinuity but the peak amplitude of the ripples remains constant.

2.3 Fourier Transform: Representation of Nonperiodic Signals

The Fourier representation of non-periodic signals can be obtained by considering a non-periodic signal as a special case of a periodic signal with an infinite period. If the period of a signal is infinite, then the signal does not repeat itself and hence it is non-periodic.

The Fourier series representation of periodic signals consist of harmonically related sinusoidal signals with a discrete spectra, where the spectral lines are spaced at integer multiples of the fundamental frequency F_0. Now consider the discrete spectra of a periodic signal with a period of T_0, as shown in Figure 2.6(a). As the period T_0 increases, the fundamental frequency $F_0 = 1/T_0$ decreases, and successive spectral lines become more closely spaced. In the limit, as the period tends to infinity (i.e. as the signal becomes non-periodic) the discrete spectral lines merge and form a continuous spectrum.

Therefore, the Fourier equations for a non-periodic signal (known as the Fourier transform), must reflect the fact that the frequency spectrum of a non-periodic signal is a continuous function of frequency. Hence, to obtain the Fourier transform relations the discrete-frequency variables and the

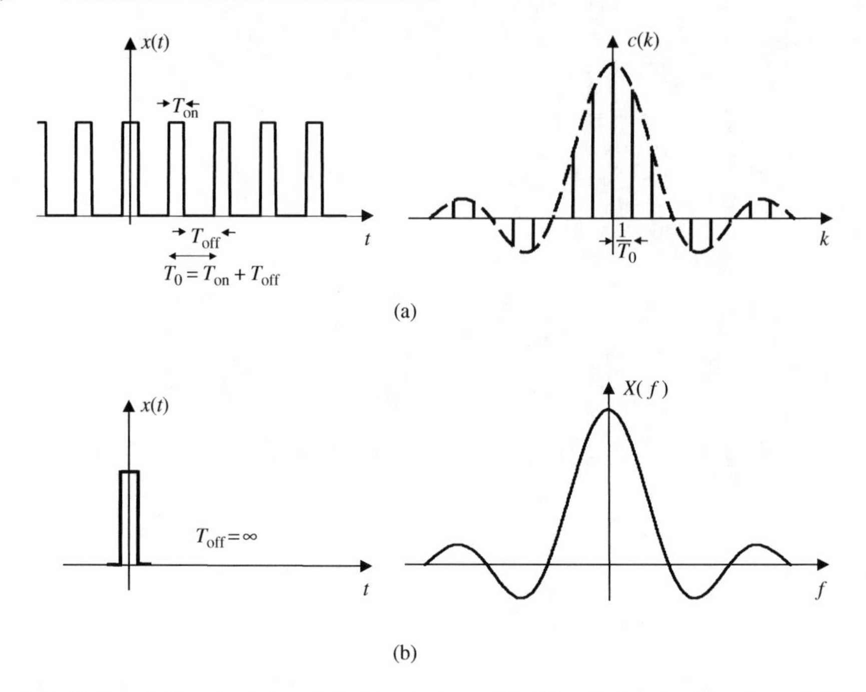

Figure 2.6 (a) A periodic pulse train and its line spectrum $c(k)$, (b) a single pulse from the periodic train in (a) with an imagined 'off' duration of infinity; its spectrum $X(f)$ is the envelope of the spectrum of the periodic signal in (a).

discrete summation operations in the Fourier series equations (2.7) and (2.8) are replaced by their continuous-frequency counterparts. That is the discrete summation sign Σ is replaced by the continuous summation integral sign $\int$, the discrete harmonics of the fundamental frequency kF_0 is replaced by the continuous frequency variable f, and the discrete frequency spectrum c_k is replaced by a continuous frequency spectrum, say $X(f)$.

The Fourier analysis and synthesis equations for non-periodic signals, known as the *Fourier transform pair*, are given by Fourier Transform (Analysis) Equation

$$X(f) = \int_{-\infty}^{\infty} x(t) e^{-j2\pi ft} dt \tag{2.20}$$

Inverse Fourier Transform (Synthesis) Equation

$$x(t) = \int_{-\infty}^{\infty} X(f) e^{j2\pi ft} df \tag{2.21}$$

Note from Equation (2.20), that $X(f)$ *may be interpreted as a measure of the correlation of the signal* $x(t)$ *and the complex sinusoid* $e^{-j2\pi ft}$.

The condition for the existence (i.e. computability) of the Fourier transform integral of a signal $x(t)$ is that the signal must have finite energy, that is

$$\text{Signal Energy} = \int_{-\infty}^{\infty} |x(t)|^2 \, dt < \infty \tag{2.22}$$

Example 2.5 *Derivation of the inverse Fourier transform*
Given the Fourier transform equation (2.20) derive the inverse Fourier transform equation (2.21).

Solution: Similar to derivation of the Fourier transform from the Fourier series analysis equation, the inverse Fourier transform can be derived from the Fourier series synthesis equation by considering the case of an aperiodic signal as a single period of a periodic signal which does not repeat itself (the period imagined to be infinity padded with zeros).

Consider the Fourier series analysis equation (2.8) for a periodic signal with a period of T_0 (and a fundamental frequency of $F_0 = 1/T_0$) and the Fourier transform equation (2.20) for its non-periodic version (i.e. consisting of only one period of the signal with a duration of T_0). Comparing these equations reproduced below

$$\underbrace{c_k = \frac{1}{T_0}\int_{-T_0/2}^{T_0/2} x(t)e^{-j2\pi kF_0 t}\,dt}_{\textit{Fourier Series}}, \quad \underbrace{X(f) = \int_{-\infty}^{\infty} x(t)e^{-j2\pi ft}\,dt = \int_{-T_0/2}^{T_0/2} x(t)e^{-j2\pi ft}\,dt}_{\textit{Fourier Transform}} \tag{2.23}$$

where for the Fourier transform we have substituted the limits of $\pm\infty$ with $\pm T_0/2$ as the non-periodic signal in Fourier transform is assumed to consist of only one period of the periodic signal. Note that as $T_0 \to \infty$ the periodic and aperiodic signals of the Fourier series and transform become identical.

Now from Equation (2.23) we observe that the Fourier transform at frequency $f = kF_0$, $X(f = kF_0)$ has the same integral as the Fourier series, with the only difference being a multiplicative factor of $1/T_0$, hence we have

$$c_k = \frac{1}{T_0} X(kF_0) \quad \text{as } T_0 \to \infty \tag{2.24}$$

Using Equation (2.24), the Fourier series synthesis equation (2.7) for a periodic signal can be rewritten as

$$x(t) = \sum_{k=-\infty}^{\infty} \frac{1}{T_0} X(kF_0)e^{j2\pi kF_0 t} = \sum_{k=-\infty}^{\infty} X(kF_0)e^{j2\pi kF_0 t}\Delta F \tag{2.25}$$

where $\Delta F = 1/T_0 = F_0$ is the frequency spacing between successive spectral lines of the spectrum of a periodic signal as shown in Figure 2.6.

Now as the period T_0 tends to infinity, that is as the signal becomes aperiodic, $\Delta F = 1/T_0$ tends towards zero, then the discrete frequency variables kF_0 and the discrete frequency interval ΔF should be replaced by continuous frequency variables f and df and the discrete summation sign is replaced by the continuous integral sign. Thus as $T_0 \to \infty$ (i.e. $\Delta F \to 0$) Equation (2.25) becomes the inverse Fourier transform:

$$x(t) \underset{T_0 \to \infty}{\Rightarrow} \int_{-\infty}^{\infty} X(f)e^{j2\pi ft}\,df \tag{2.26}$$

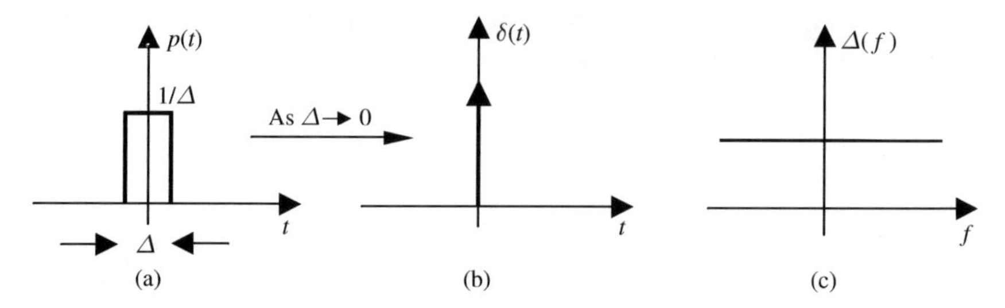

Figure 2.7 (a) A unit-area pulse, (b) the pulse becomes an impulse as $\Delta \to 0$ and amplitude tends to infinity but area remains equal to unity, (c) the spectrum of the impulse function.

Example 2.6 *The spectrum of an impulse function*
The impulse function, also known as delta function or Dirac function (after Paul Dirac), is a theoretical function with immense practical use in system analysis in such concepts as the description of a system by its impulse response.

Consider the unit-area pulse $p(t)$ shown in Figure 2.7(a). As the pulse width Δ tends to zero the pulse tends to an impulse. The impulse function shown in Figure 2.7(b) is defined as a unit-area pulse with an infinitesimal time width as

$$\delta(t) = \underset{\Delta \to 0}{limit}\ p(t) = \begin{cases} 1/\Delta & |t| \le \Delta/2 \\ 0 & |t| > \Delta/2 \end{cases} \tag{2.27}$$

Note that the delta function has a width of zero, $\Delta = 0$, an amplitude of infinity $1/\Delta = \infty$ and an area of unity; it is easy to see that the integral of the impulse function is given by

$$\int_{-\infty}^{\infty} \delta(t)dt = \Delta \times \frac{1}{\Delta} = 1 \tag{2.28}$$

All signals can be considered as a combination of a finite or infinite number of weighted and shifted impulse functions as can be deduced from the following sampling property of impulse function.

The important *sampling property of the impulse function* is defined as

$$\int_{-\infty}^{\infty} x(t)\delta(t-T)dt = x(T) \tag{2.29}$$

Hence Equation (2.29) outputs a sample of a signal $x(t)$ at time instance T, i.e. $x(T)$. Equation (2.29) follows from the property of $\delta(t-T)$ which has a value of zero at any point other than at $t = T$ where the integral of the function is equal to one as expressed in Equation (2.28).

The Fourier transform of the impulse function is obtained as

$$\Delta(f) = \int_{-\infty}^{\infty} \delta(t)e^{-j2\pi ft}dt = e^{0}\int_{-\infty}^{\infty} \delta(t)dt = 1 \tag{2.30}$$

The impulse function is used to describe the response of systems (impulse response) and it is also used as a *test function* again to obtain the impulse response of a system. This is because as expressed

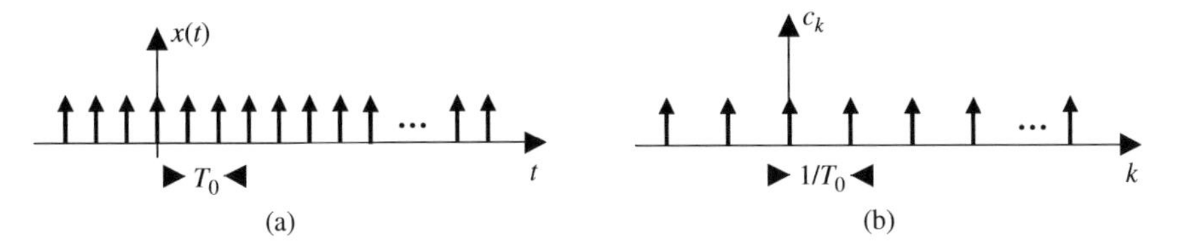

Figure 2.8 A periodic impulse train $x(t)$ and its Fourier series spectrum c_k.

by Equation (2.30) and shown in Figure 2.7(c) *an impulse is a spectrally rich signal containing all vibrations or frequencies in equal amounts.* Furthermore from the sampling equation (2.29) any signal can be constructed as a combination of shifted and weighted impulses.

Example 2.7 *The spectrum of a (sampling) train of impulses*
Find the spectrum of a 10 kHz periodic impulse train $x(t)$ shown in Figure 2.8(a) with an amplitude of $A = 1$ mV and expressed as

$$x(t) = \sum_{m=-\infty}^{\infty} A \times \underset{\Delta \to 0}{\text{limit}}\ p(t - mT_0) \times \Delta = \sum_{m=-\infty}^{\infty} A \int_{-\infty}^{\infty} \delta(t - mT_0) dt \tag{2.31}$$

where $p(t)$ is a unit-area pulse with width Δ as shown in Figure 2.7(a)

Solution: The period $T_0 = 1/F_0 = 0.1$ ms and the amplitude $A = 1$ mV. For the time interval of one period $-T_0/2 < t < T_0/2 x(t) = A\delta(t)$, hence we have

$$c_k = \frac{1}{T_0} \int_{-T_0/2}^{T_0/2} x(t) e^{-jk\omega_0 t} dt = \frac{1}{T_0} \int_{-T_0/2}^{T_0/2} A\delta(t) e^{-jk\omega_0 t} dt = \frac{A}{T_0} e^0 \int_{-T_0/2}^{T_0/2} \delta(t) dt = \frac{A}{T_0} = 10 \text{ V} \tag{2.32}$$

As shown in Figure 2.8(b) the spectrum of a periodic impulse train in time with a period of T_0 is a periodic impulse train in frequency with a period of $F_0 = 1/T_0$

Example 2.8 *The spectrum of rectangular pulse: sinc function*
The rectangular pulse is a particularly important signal in signal analysis and communication. A rectangular pulse is inherent whenever a signal is segmented and processed frame by fame since each frame can be viewed as the result of multiplication of the signal and a rectangular window. Furthermore, the spectrum of a rectangular pulse can be used to calculate the bandwidth required by pulse radar systems or by digital communication systems that transmit modulated pulses to represent binary numbers.
The Fourier transform of a rectangular pulse, in Figure 2.9(a), of duration T seconds is obtained as

$$\begin{aligned} R(f) &= \int_{-\infty}^{\infty} r(t) e^{-j2\pi ft} dt = \int_{-T/2}^{T/2} 1.e^{-j2\pi ft} dt \\ &= \frac{e^{j2\pi fT/2} - e^{-j2\pi fT/2}}{j2\pi f} = T\frac{\sin(\pi fT)}{\pi fT} = T\text{sinc}(fT) \end{aligned} \tag{2.33}$$

where $\text{sinc}(x) = \sin(\pi x)/\pi x$.

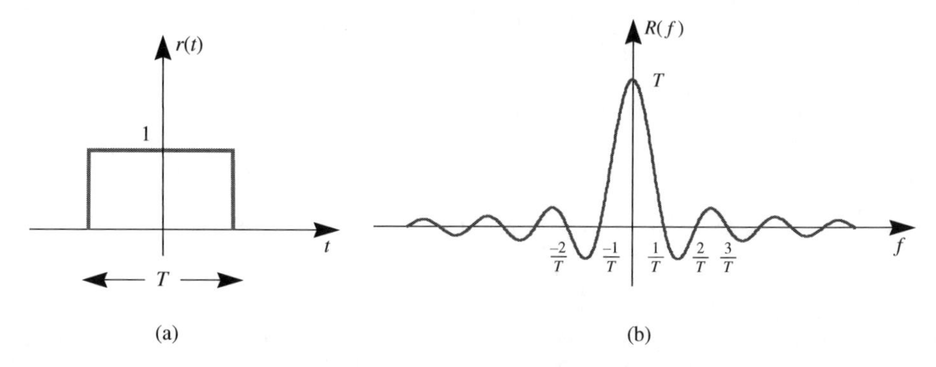

Figure 2.9 (a) A rectangular pulse, (b) its spectrum is a sinc function. Note the bandwidth of the main lobe (between $\pm 1/T$) is $2/T$.

Figure 2.9(b) shows the spectrum of the rectangular pulse. Note that most of the pulse energy is concentrated in the *main lobe* within a bandwidth of 2/Pulse_Duration, i.e. $BW = 2/T$. However, there is pulse energy in the side lobes that may interfere with other electronic devices operating at the side lobe frequencies.

MatLab function Rect_Sinc()
Generates a rectangular pulse and displays the pulse, its magnitude spectrum and its spectrogram for the varying pulse width. The program illustrates the inverse relationship between the pulse width and the bandwidth.

2.3.1 Comparison of Laplace and Fourier Transforms

As mentioned at the beginning of this chapter, Laplace was Fourier's teacher and initially, as part of a review committee, he was not impressed with Fourier's work. Perhaps this makes the relation between Fourier and Laplace transforms more interesting.

The Fourier transform assumes that the signal is *stationary*, this implies that the frequency content of the signal (the number of frequency components, their frequencies, magnitude and phase) does not change over time. Hence the signal is transformed into a combination of stationary sine and cosine waves of time-invariant frequencies, magnitudes and phases. In contrast Laplace transform can model a *non-stationary* signal as a combination of rising, steady and decaying sine waves.

The Laplace transform of $x(t)$ is given by the integral

$$X(s) = \int_{0}^{\infty} x(t)e^{-st}\,dt \tag{2.34}$$

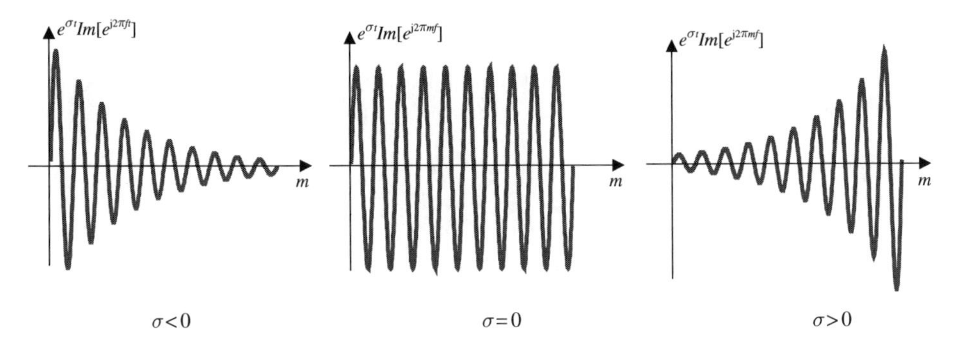

Figure 2.10 The basis functions of Laplace transform.

where the complex variable $s = \sigma + j\omega$. Note that $e^{-st} = e^{-\sigma t}e^{-j\omega t}$ is a complex sinusoidal signal $e^{-j\omega t}$ with an envelope of $e^{-\sigma t}$. The inverse Laplace transform is defined by

$$x(t) = \frac{1}{2\pi j}\int_{\sigma_1 - j\infty}^{\sigma_1 + j\infty} X(s)e^{st}ds \tag{2.35}$$

where σ_1 is a vertical contour in the complex plane chosen so that all singularities of $X(s)$ are to the left of it. The elementary functions (kernels) for the Laplace transform are damped or growing sinusoids of the form $e^{-st} = e^{-\sigma t}e^{-j\omega t}$ as shown in Figure 2.10. These basis functions are particularly suitable for transient signal analysis. In contrast, the Fourier basis functions are steady complex exponential, $e^{-j\omega t}$, of time-invariant amplitudes and phase, suitable for steady state or time-invariant signal analysis.

Laplace transform is particularly useful in solving linear ordinary differential equations as it can transform relatively difficult differential equations into relatively simple algebraic equations. Specifically differentiation in time becomes multiplication by the frequency s as $de^{st}/dt = se^{st}$ and integration in time becomes division by s as $\int e^{st} = e^{st}/s$.

The Laplace transform is a one-sided transform with the lower limit of integration at $t = 0$, whereas the Fourier transform equation (2.21) is a two-sided transform with the lower limit of integration at $t = -\infty$. However, for a one-sided signal, which is zero-valued for $t < 0$, the limits of integration for the Laplace and the Fourier transforms are identical. In that case if the variable s in the Laplace transform is replaced with the frequency variable $j2\pi f$ then the Laplace integral becomes the Fourier integral. Hence for a one-sided signal, *the Fourier transform is a special case of the Laplace transform* corresponding to $s = j2\pi f$ and $\sigma = 0$. Also for a two-sided signal the bilateral Laplace transform (with limits of integration between $\pm\infty$) becomes Fourier transform when the variable s is replaced by $j2\pi f$. The relation between the Fourier and the Laplace transforms are discussed further in Chapter 3 on Z-transform. Note that Laplace transform can accommodate the initial state of a signal whereas Fourier transform assumes that the signal is in a stationary steady state.

2.3.2 Properties of the Fourier Transform

There are a number of Fourier transform properties that provide further insight into the transform and are useful in reducing the complexity of the application of Fourier transform and its inverse transform.

Linearity and Superposition

The Fourier transform is a linear operation, this mean the principle of superposition applies. Hence the Fourier transform of the weighted sum of two signals $x(t)$ and $y(t)$

$$z(t) = ax(t) + by(t) \tag{2.36}$$

is the weighted sum of their Fourier transforms $X(f)$ and $Y(f)$, i.e.

$$\begin{aligned} Z(f) &= \int_{-\infty}^{\infty} z(t)e^{-j2\pi ft}dt = \int_{-\infty}^{\infty} [ax(t) + by(t)]e^{-j2\pi ft}dt = a\int_{-\infty}^{\infty} x(t)e^{-j2\pi ft}dt + b\int_{-\infty}^{\infty} y(t)e^{-j2\pi ft}dt \\ &= aX(f) + bY(f) \end{aligned} \tag{2.37}$$

Conjugate Symmetry

This property states that if the time domain signal $x(t)$ is real-valued (in practice this is often the case) then

$$X(f) = X^*(-f) \tag{2.38}$$

where the superscript asterisk * denotes the complex conjugate operation. Equation (2.38) follows from the fact that if in the complex variable $e^{-j2\pi f}$ the frequency variable f is replaced by $-f$ and the result is conjugated we will have $(e^{-j2\pi-f)})^* = e^{-j2\pi f}$.

From Equation (2.38) it follows that $\mathrm{Re}\{X(f)\}$ is an even function of f and $\mathrm{Im}\{X(f)\}$ is an odd function of f. Similarly the magnitude of $X(f)$ is an even function and the phase angle is an odd function.

Time Reversal

Time reversing a signal (from time variable t to $-t$) is equivalent to a complex conjugate operation in the frequency domain.

$$\int_{-\infty}^{\infty} x(-t)e^{-j2\pi ft}dt = \int_{-\infty}^{\infty} x(t)e^{j2\pi ft}dt = X^*(f) \tag{2.39}$$

In terms of the Fourier analysis integral of Equation (2.20) reversing the time variable of the input signal has the same effect as reversing the time variable of the complex basis function from $e^{-j2\pi ft}$ to $e^{+j2\pi ft}$. For example, as shown in Example 2.2, $\sin(t) = e^{+j2\pi ft}/2j - e^{-j2\pi ft}/2j$ has coefficients $c_1 = 1/2j$ and $c_{-1} = -1/2j$ whereas its time-reversed version $\sin(-t) = e^{-j2\pi ft}/2j - e^{+j2\pi ft}/2j$ has coefficients $c_1 = -1/2j$ and $c_{-1} = 1/2j$.

Matlab function TimeReversal( )

Demonstrates that taking the FFT of a signal followed by complex conjugating and inverse FFT results in the time reversal of the signal.

Time Shifting and Modulation (Frequency Translation)

Let $X(f) = \mathcal{F}[x(t)]$ be the Fourier transform of $x(t)$. If the time domain signal $x(t)$ is delayed by an amount T_0, the effect on its spectrum $X(f)$ is a phase change of $e^{-j2\pi T_0 f}$ as

$$\mathcal{F}[x(t-T_0)] = \int_{-\infty}^{\infty} x(t-T_0)e^{-j2\pi ft}dt = e^{-j2\pi fT_0}\int_{-\infty}^{\infty} x(t-T_0)e^{-j2\pi f(t-T_0)}dt$$

$$= e^{-j2\pi fT_0}\underbrace{\int_{-\infty}^{\infty} x(\tau)e^{-j2\pi f\tau}d\tau}_{X(f)\ \text{Fourier transform of}\ x(\tau)} = e^{-j2\pi fT_0}X(f) \tag{2.40}$$

where the time variable $\tau = t - T_0$.

Conversely, if $X(f)$ is shifted by an amount F_0, the effect on the inverse Fourier transform is

$$\mathcal{F}^{-1}[X(f-F_0)] = \int_{-\infty}^{\infty} X(f-F_0)e^{j2\pi ft}df = e^{j2\pi F_0 t}\int_{-\infty}^{\infty} X(f-F_0)e^{j2\pi(f-F_0)t}df$$

$$= e^{j2\pi F_0 t}\underbrace{\int_{-\infty}^{\infty} X(\nu)e^{j2\pi\nu t}d\nu}_{\text{x(t) = Inverse Fourier transform of X}(\nu)} = e^{j2\pi F_0 t}x(t) \tag{2.41}$$

where the frequency variable $\nu = f - F_0$.

Matlab function Amplitude_Modulation( )
Demonstrates the modulation property of the Fourier transform. A narrowband noise is multiplied by a carrier sine wave; the result is that the narrowband noise is modulated and shifted to the frequency of the sine wave. This process is also known as amplitude modulation.
Matlab function Phase_and_Delay( )
Demonstrates the relationship between phase and delay. The spectrum of a signal is multiplied by a phase term exp(- i*2*pi*f*delay). The result is converted to time domain via inverse FFT. The displays of the original signal and the phase-modified signal shows a relative delay.

Note that the modulation property Equation (2.41) states that multiplying a signal $x(t)$ by $e^{j2\pi F_0 t}$ translates the spectrum of $x(t)$ onto the frequency channel F_0, this is the frequency translation principle of amplitude modulation in telecommunication.

Differentiation and Integration

As with the Laplace transform, through Fourier transform the mathematical operations of differentiation and integration can be transformed into simple algebraic operations. Let $x(t)$ be a continuous time signal with Fourier transform $X(f)$.

$$x(t) = \int_{-\infty}^{\infty} X(f) e^{j2\pi ft} df \tag{2.42}$$

Differentiating both sides of the Fourier transform equation (2.42) we obtain

$$\frac{dx(t)}{dt} = \int_{-\infty}^{\infty} X(f) \frac{de^{j2\pi ft}}{dt} df = \int_{-\infty}^{\infty} \underbrace{j2\pi f X(f)}_{\substack{\text{Fourier Transform} \\ \text{of } dx(t)/dt}} e^{j2\pi ft} df \tag{2.43}$$

That is *multiplication* of $X(f)$ by the factor $j2\pi f$ is equivalent to differentiation of $x(t)$ in time. Similarly division of $X(f)$ by $j2\pi f$ is equivalent to integration of the function of time $x(t)$.

$$\int_{-\infty}^{t} x(\tau)d\tau = \underbrace{\int_{-\infty}^{\infty} x(\tau)u(t-\tau)d\tau}_{\text{Convolution of x(t) and a step fuction u(t)}} = \underbrace{\int_{-\infty}^{\infty} X(f)U(f)e^{j2\pi ft} df}_{\substack{\text{Convolution of x(t) and u(t) is the inverse Fourier transform of the} \\ \text{product of their spectra } X(f)U(f)}}$$

$$= \int_{-\infty}^{\infty} X(f) \underbrace{\left(\frac{1}{j2\pi f} + \pi\delta(f) \right)}_{\substack{\text{U(f); Fourier transform of} \\ \text{a step function u(t)}}} e^{j2\pi ft} df$$

$$= \underbrace{\int_{-\infty}^{\infty} \frac{1}{j2\pi f} X(f) + \pi X(0)\delta(f) e^{j2\pi ft} df}_{\text{Fourier transform of } \int_{-\infty}^{t} x(\tau)d\tau} \tag{2.44}$$

where $u(t)$ is a step function, $U(f)$ is the Fourier transform of $u(t)$ and the impulse term $\pi X(0)\delta(f)$ reflects the average value that can result from the integration.

Matlab function Differentiation_Integration_Demo( )

Demonstrates that multiplication of the spectrum of a signal by i*2*pi*f followed by IFFT results in differentiation of the signal. Division of the spectrum of a signal by i*2*pi*f followed by IFFT results in integration of the signal.

Time and Frequency Scaling

If $x(t)$ and $X(f)$ are Fourier transform pairs then compression of time scale results in expansion of bandwidth and vice versa. This inverse relationship between time scale and bandwidth may be expressed as

$$x(\alpha t) \stackrel{\mathcal{F}}{\longleftrightarrow} \frac{1}{\alpha} X\left(\frac{f}{\alpha}\right) \tag{2.45}$$

Equation (2.44) can be proved as

$$\int_{-\infty}^{\infty} x(\alpha t)e^{-j2\pi ft}dt = \int_{-\infty}^{\infty} x(\alpha t)e^{-j2\pi \frac{f}{\alpha}\alpha t}dt = \frac{1}{\alpha}\int_{-\infty}^{\infty} x(t')e^{-j2\pi \frac{f}{\alpha}t'}dt' = \frac{1}{\alpha}X\left(\frac{f}{\alpha}\right) \tag{2.46}$$

where $t' = \alpha t$. For example try to say something very slowly, then $\alpha > 1$, your voice spectrum will be compressed and you may sound like a slowed-down tape or compact disc; you can do the reverse and the spectrum would be expanded and your voice shifts to higher frequencies. This property is further illustrated in Section 2.9.5.

Convolution

The convolution integral of two signals $x(t)$ and $h(t)$ is defined as

$$y(t) = \int_{-\infty}^{\infty} x(\tau)h(t-\tau)d\tau \tag{2.47}$$

The convolution integral is also written as

$$y(t) = x(t) * h(t) \tag{2.48}$$

where asterisk * denotes the convolution operation. The convolution integral is used to obtain the time-domain response of linear systems to arbitrary inputs, as will be discussed in later sections.

The Convolution Property of the Fourier Transform

It can be shown that convolution of two signals in the time domain corresponds to multiplication of the signals in the frequency domain, and conversely multiplication in the time domain corresponds to convolution in the frequency domain. To derive the convolutional property of the Fourier transform take the Fourier transform of the convolution of the signals $x(t)$ and $h(t)$ as

$$\int_{-\infty}^{\infty} \underbrace{\left(\int_{-\infty}^{\infty} x(\tau)h(t-\tau)d\tau\right)}_{y(t)=x(t)*h(t)} e^{-j2\pi ft}dt = \int_{-\infty}^{\infty} \underbrace{\left(\int_{-\infty}^{\infty} h(t-\tau)e^{-j2\pi f(t-\tau)}dt\right)}_{H(f)} x(\tau)e^{-j2\pi f\tau}d\tau$$
$$= X(f)H(f) \tag{2.49}$$

Principle of Duality

Comparing the Fourier transform and the inverse Fourier transform relations we observe a symmetric relation between them. In fact the main difference between Equations (2.20) and (2.21) is a negative sign in the exponent of $e^{-j2\pi ft}$ in Equation (2.21). This symmetry leads to a property of Fourier transform known as the duality principle and stated as

$$x(t) \overset{\mathcal{F}}{\longleftrightarrow} X(f)$$
$$X(t) \overset{\mathcal{F}}{\longleftrightarrow} x(f) \tag{2.50}$$

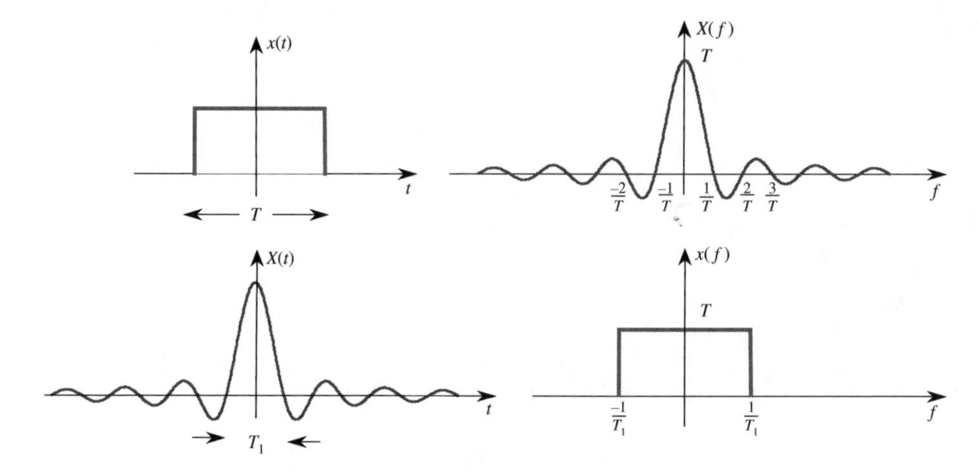

Figure 2.11 Illustration of the principle of duality.

As illustrated in Figure 2.11 the Fourier transform of a rectangular function of time $r(t)$ has the form of a sinc pulse function of frequency sinc(f). From the duality principle the Fourier transform of a sinc function of time sinc(t) is a rectangular function of frequency $R(f)$.

Parseval's Theorem: Energy Relations in Time and Frequency

Parseval's relation states that the energy of a signal can be computed by integrating the squared magnitude of the signal either over the time domain or over the frequency domain. If $x(t)$ and $X(f)$ are a Fourier transform pair, then

$$\text{Energy} = \int_{-\infty}^{\infty} |x(t)|^2 dt = \int_{-\infty}^{\infty} |X(f)|^2 df \tag{2.51}$$

This expression, referred to as Parseval's relation, follows from a direct application of the Fourier transform. Note that every characteristic value of a signal can be computed either in time or in frequency.

Example 2.9 *The spectrum of sine wave pulse*
Sine wave pulses are used by radars and pulse amplitude modulation and other communication systems. In this example we obtain and sketch the frequency spectrum of the following sine wave pulse.

$$x(t) = \sin(2\pi F_0 t) \quad -NT_0/2 \le t \le NT_0/2 \tag{2.52}$$

where $T_0 = 1/F_0$ is the period and $\omega_0 = 2\pi/T_0$.

Solution: Substitute for $x(t)$ and its limits in the Fourier transform equation (2.21)

$$X(f) = \int_{-NT_0/2}^{NT_0/2} \sin(2\pi F_0 t) e^{-j2\pi ft} dt \tag{2.53}$$

Substituting $\sin(2\pi F_0 t) = (e^{j2\pi F_0 t} - e^{-j2\pi F_0 t})/2j$ in (2.53) gives

$$X(f) = \int_{-NT_0/2}^{NT_0/2} \frac{e^{j2\pi F_0 t} - e^{-j2\pi F_0 t}}{2j} e^{-j2\pi f t} dt$$

$$= \int_{-NT_0/2}^{NT_0/2} \frac{e^{-j2\pi(f-F_0)t}}{2j} dt - \int_{-NT_0/2}^{NT_0/2} \frac{e^{-j2\pi(f+F_0)t}}{2j} dt \tag{2.54}$$

Evaluating the integrals yields

$$X(f) = \frac{e^{-j\pi(f-F_0)NT_0} - e^{j\pi(f-F_0)NT_0}}{4\pi(f-F_0)} - \frac{e^{-j\pi(f+F_0)NT_0} - e^{j\pi(f+F_0)NT_0}}{4\pi(f+F_0)}$$

$$= \frac{-j}{2\pi(f-F_0)} \sin(\pi(f-F_0)NT_0) + \frac{j}{2\pi(f+F_0)} \sin(\pi(f+F_0)NT_0) \tag{2.55}$$

The signal spectrum can be expressed as the sum of two shifted (frequency-modulated) sinc functions as

$$X(f) = -0.5\, jNT_0 \text{sinc}((f-F_0)NT_0) + 0.5j\, NT_0 \text{sinc}((f+F_0)NT_0) \tag{2.56}$$

Note that energy of a finite duration sine wave pulse is spread in frequency across the main lobe and the side lobes of the sinc function. Figure 2.12 demonstrates that as the window length increases the energy becomes more concentrated in the main lobe and in the limit for an infinite duration window the spectrum tends to an impulse positioned at the frequency F_0.

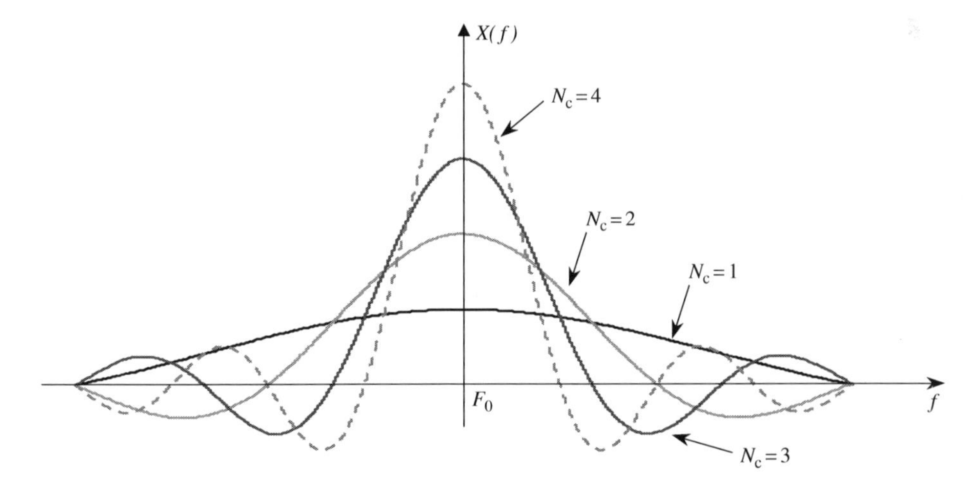

Figure 2.12 The spectrum of a finite duration sine wave with the increasing length of observation. N_c is the number of cycles in the observation window. Note that the width of the main lobe depends inversely on the duration of the signal.

Example 2.10 *Calculation of the bandwidth required for transmission of digital data at a rate of r_b bits per second*

In its simplest form binary data can be represented as a sequence of amplitude modulated pulses as

$$x(t) = \sum_{m=0}^{N-1} A(m)p(t - mT_b) \tag{2.57}$$

where $A(m)$ may be $+1$ or -1 and

$$p(t) = \begin{cases} 1 & |t| \leq T_b/2 \\ 0 & |t| > T_b/2 \end{cases} \tag{2.58}$$

where $T_b = 1/r_b$ is pulse duration. The Fourier transform of $x(t)$ is given by

$$X(f) = \sum_{m=0}^{N-1} A(m)P(f)e^{-j2\pi mT_b f} = P(f)\sum_{m=0}^{N-1} A(m)e^{-j2\pi mT_b f} \tag{2.59}$$

The power spectrum of this signal is obtained as

$$\begin{aligned} E[X(f)X^*(f)] &= E\left[\sum_{m=0}^{N-1} A(m)P(f)e^{-j2\pi mT_b f}\sum_{n=0}^{N-1} A(n)P^*(f)e^{j2\pi mT_b f}\right] \\ &= |P(f)|^2 \sum_{m=0}^{N-1}\sum_{n=0}^{N-1} E[A(m)A(n)]e^{-j2\pi(m-n)T_b f} \end{aligned} \tag{2.60}$$

Now assuming that the data is uncorrelated we have

$$E[A(m)A(n)] = \begin{cases} 1 & m = n \\ 0 & m \neq n \end{cases} \tag{2.61}$$

Substituting Equation (2.61) in (2.60) we have

$$E[X(f)X^*(f)] = N\,|P(f)|^2 \tag{2.62}$$

From Equation (2.61) the bandwidth required from a sequence of pulses is basically the same as the bandwidth of a single pulse. From Figure 2.11 the bandwidth of the main lobe of spectrum of a rectangular pulse is $2r_b$ Hz.

$$B_{\mathrm{Rec}} = \frac{2}{T_p} = 2r_p \tag{2.63}$$

Note that the bandwidth required is twice the pulse rate r_p.

Example 2.11 *Calculation of the bandwidth required for transmission of digital data at a rate of 1 mega pulses per second*

A data transmission rate of 1 mega pulses per second, with a pulse duration of 1 microsecond, requires a bandwidth of 2 mega Hz. Note that in M-array pulse modulation method of data transmission each pulse can carry $\log_2(M)$ bits. For example in 64-array modulation each pulse has 64 different states (such as 64 different amplitudes or 64 different combinations of phase and amplitude) and it can signal $\log_2(64) = 6$ bits. Hence with a 64-array modulation method, 1 mega pulses per second carry 6 mega bits per second.

2.3.3 Fourier Transform of a Sampled (Discrete-Time) Signal

A sampled signal $x(m)$ can be modelled as a sequence of time shifted and amplitude scaled discrete-time unit-impulse functions as

$$x(m) = \sum_{m=-\infty}^{\infty} x(t)\delta(t - mT_s) \tag{2.64}$$

where m is the discrete time variable and T_s is the sampling period. The Fourier transform of $x(m)$, a sampled version of a continuous signal $x(t)$, can be obtained from Equation (2.21) as

$$X(f) = \int_{-\infty}^{\infty} \sum_{m=-\infty}^{\infty} x(t)\delta(t - mT_s)e^{-j2\pi ft}dt = \sum_{m=-\infty}^{\infty} \int_{-\infty}^{\infty} x(t)\delta(t - mT_s)e^{-j2\pi ft}dt$$

$$= \sum_{m=-\infty}^{\infty} x(mT_s)e^{-j2\pi mfT_s} = \sum_{m=-\infty}^{\infty} x(mT_s)e^{-j2\pi mf/F_s} \tag{2.65}$$

For convenience of notation, without loss of generality, it is often assumed that sampling frequency $T_s = 1/F_s = 1$, hence

$$X_s(f) = \sum_{m=-\infty}^{\infty} x(m)e^{-j2\pi mf} \tag{2.66}$$

The inverse Fourier transform of a sampled signal is defined as

$$x(m) = \int_{-1/2}^{1/2} X_s(f)e^{j2\pi fm}df \tag{2.67}$$

where the period of integration is $-F_s/2 \le f \le F_s/2$ and it is assumed that the sampling frequency $F_s = 1$. Note that $x(m)$ and $X_s(f)$ are equivalent in that they contain the same information in different domains. In particular, as expressed by Parseval's theorem, the energy of the signal may be computed either in the time or in the frequency domain as

$$\text{Signal Energy} = \sum_{m=-\infty}^{\infty} x^2(m) = \int_{-1/2}^{1/2} |X_s(f)|^2 df \tag{2.68}$$

Example 2.12 Show that the spectrum of a sampled signal is periodic with a period equal to the sampling frequency F_s.

Solution: Substitute $f + kF_s$ for the frequency variable f in Equation (2.65)

$$X(f + kF_s) = \sum_{m=-\infty}^{\infty} x(mT_s)e^{-j2\pi m\frac{(f+kF_s)}{F_s}} = \sum_{m=-\infty}^{\infty} x(mT_s)e^{-j2\pi m\frac{f}{F_s}} \underbrace{e^{-j2\pi m\frac{kF_s}{F_s}}}_{=1} = X(f) \tag{2.69}$$

Figure 2.13(a) shows the spectrum of a band-limited continuous-time signal. As shown in Figure 2.13(b) after the signal is sampled its spectrum becomes periodic.

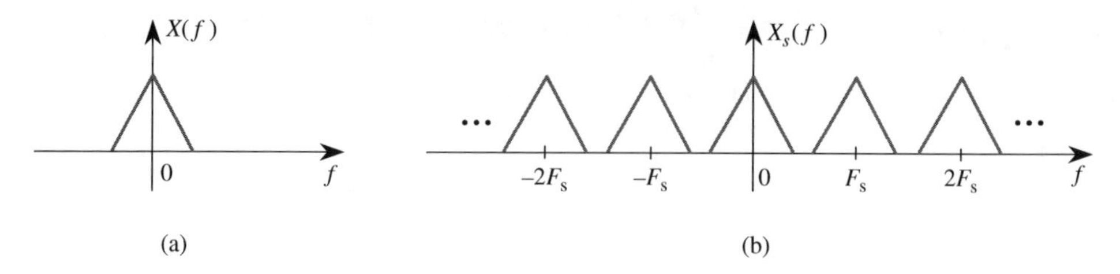

Figure 2.13 The spectrum of (a) a continuous-time signal, and (b) its discrete-time sampled version.

2.4 Discrete Fourier Transform

Discrete Fourier transform (DFT) is the Fourier transform adapted for digital signal processing. The DFT of a continuous-time signal can be derived by three operations of: (1) sampling in time, (2) segmenting the sampled signal into segments of length N samples and (3) sampling the discrete-time signal segment in frequency domain.

Note that just as sampling a signal in time (or space) renders its spectrum a *periodic* function of frequency, it follows that sampling a signal in frequency renders it periodic in time (or space). This is also an aspect of the principle of duality introduced earlier in this chapter.

The first two steps yields the Fourier transform of a sampled and windowed signal as

$$X(f) = \int_{-\infty}^{\infty} x(t) \underbrace{\sum_{m=0}^{N-1} \delta(t - mT_s)}_{N-\text{point sampling}} e^{-j2\pi ft} dt = \sum_{m=0}^{N-1} x(mT_s) e^{-j2\pi f m T_s} \tag{2.70}$$

where $\delta(.)$ is the Kronecker delta function. Without loss of generality it is usually assumed that the sampling period T_s and hence the sampling frequency F_s is equal to 1, i.e. $T_s = 1/F_s = 1$.

Note that in general $X(f)$ is a continuous and periodic function of the frequency variable f. The periodicity in $X(f)$ is introduced as a result of sampling as shown in Example 2.12. The final step is sampling in frequency domain that is evaluating $X(f)$ at discrete frequencies $f = kF_s/N$ where k is an integer and F_s is the sampling rate

$$X(k) = X(f)\delta\left(f - \frac{k}{N}F_s\right) \tag{2.71}$$

Note that $X(k)$ is shorthand for $X(kFs/N)$, that is the kth discrete frequency corresponds to an actual frequency of kF_s/N.

As illustrated in Figure 2.14 the input to DFT is N samples of a discrete signal $[x(0), \ldots, x(N-1)]$ and the output consists of N uniformly-spaced samples $[X(0), \ldots, X(N-1)]$ of the frequency spectrum of the input. When a non-periodic continuous-time signal is sampled, its Fourier transform becomes a periodic but *continuous* function of frequency, as shown in Equation (2.69). As shown above the discrete Fourier transform is derived from sampling the Fourier transform of a discrete-time signal at N discrete-frequencies corresponding to integer multiples of the frequency sampling interval $2\pi/N$. *The DFT is effectively the Fourier series of a sampled signal.*

For a finite duration discrete-time signal $x(m)$ of length N samples, the discrete Fourier transform (DFT) and its inverse (IDFT) are defined as

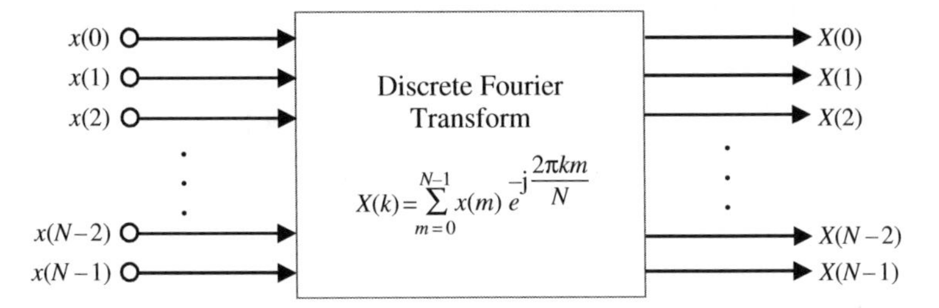

Figure 2.14 Illustration of the DFT as a parallel-input parallel-output signal processor.

Discrete Fourier transform (DFT) analysis equation

$$X(k)=\sum_{m=0}^{N-1} x(m)e^{-j\frac{2\pi}{N}mk} \qquad k=0,\ldots,N-1 \tag{2.72}$$

Inverse discrete Fourier transform (IDFT) synthesis equation

$$x(m)=\frac{1}{N}\sum_{k=0}^{N-1} X(k)e^{j\frac{2\pi}{N}mk} \qquad m=0,\ldots,N-1 \tag{2.73}$$

Note that the basis functions of a DFT are: 1, $e^{-j\frac{2\pi}{N}}, e^{-j\frac{4\pi}{N}}, \ldots, e^{-j\frac{2(N-1)\pi}{N}}$.

The DFT equation can be written in the form of a linear system transformation, $\boldsymbol{X}=\mathbf{W}\boldsymbol{x}$, as the transformation of an input vector $\boldsymbol{x}=[x(0)x(1)\ldots x(N-1)]$ to an output vector $\boldsymbol{X}=[X(0)X(1)\ldots X(N-1)]$ as

$$\begin{bmatrix} X(0)\\ X(1)\\ X(2)\\ \vdots\\ X(N-2)\\ X(N-1)\end{bmatrix} = \underbrace{\begin{bmatrix} 1 & 1 & 1 & \cdots & 1 & 1\\ 1 & e^{-j\frac{2\pi}{N}} & e^{-j\frac{4\pi}{N}} & \cdots & e^{-j\frac{2(N-2)\pi}{N}} & e^{-j\frac{2(N-1)\pi}{N}}\\ 1 & e^{-j\frac{4\pi}{N}} & e^{-j\frac{8\pi}{N}} & \cdots & e^{-j\frac{4(N-2)\pi}{N}} & e^{-j\frac{4(N-2)\pi}{N}}\\ \vdots & \vdots & \vdots & \ddots & \vdots & \vdots\\ 1 & e^{-j\frac{2(N-2)\pi}{N}} & e^{-j\frac{4(N-2)\pi}{N}} & \cdots & e^{-j\frac{2(N-2)^2\pi}{N}} & e^{-j\frac{2(N-2)(N-1)\pi}{N}}\\ 1 & e^{-j\frac{2(N-1)\pi}{N}} & e^{-j\frac{4(N-1)\pi}{N}} & \cdots & e^{-j\frac{2(N-1)(N-2)\pi}{N}} & e^{-j\frac{(N-1)^2\pi}{N}} \end{bmatrix}}_{\text{Fourier Transform Matrix}} \begin{bmatrix} x(0)\\ x(1)\\ x(2)\\ \vdots\\ x(N-2)\\ x(N-1)\end{bmatrix} \tag{2.74}$$

where the elements of the transformation matrix are $w_{km}=e^{-j\frac{2\pi}{N}km}$ where m is the discrete-time index and k is the discrete-frequency index. The form of Equation (2.74) is used later in Equation (2.92) for development of the fast Fourier transform (FFT).

The DFT spectrum consists of N uniformly spaced samples taken from one period (2π) of the continuous spectrum of the discrete time signal $x(m)$. At a sampling rate of F_s the discrete-frequency index k corresponds to kF_s/N Hz.

A periodic signal has a discrete spectrum. Conversely any discrete frequency spectrum belongs to a periodic signal. Hence *the implicit assumption in the DFT theory, is that the input signal* $[x(0),\ldots,x(N-1)]$ *is periodic with a period equal to the observation window length of N samples.* Figure 2.15 illustrates the derivation of the Fourier transform and the DFT from the Fourier series as discussed in the preceding sections. Note that DFT is equivalent to the Fourier series. Table 2.1 shows the discrete/continuous, periodic/non-periodic form of the inputs and outputs of the Fourier transform methods.

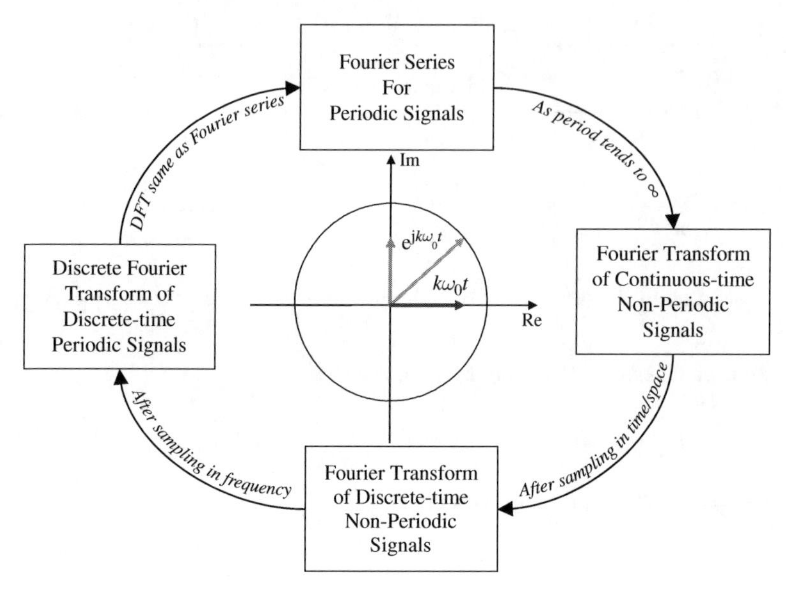

Figure 2.15 Illustration of the 'conceptual evolution' from Fourier series of a periodic signal to DFT of a sampled signal. The circle inside shows the in-phase and quadrature components of the basis function of Fourier transform.

Table 2.1 The form of input and outputs of different Fourier methods

	Input (time or space)		Output (frequency)	
	Periodic or non-periodic	Continuous or discrete	Periodic or non-periodic	Continuous or discrete
Fourier series	*Periodic*	*Either*	*Either*	*Discrete*
Fourier transform	*Non-periodic*	*Continuous*	*Non-periodic*	*Continuous*
Fourier transform of discrete-time signals	*Non-periodic*	*Discrete-time*	*Periodic*	*Continuous*
Discrete Fourier transform	*Either*	*Discrete-time*	*Periodic*	*Discrete*

Example 2.13 *Comparison of Fourier series and DFT* The discrete Fourier transform and the Fourier series are given by

$$X(k) = \sum_{m=0}^{N-1} x(m) e^{-j\frac{2\pi}{N}km} \qquad k = 0, \ldots, N-1 \tag{2.75}$$

$$c_k = \frac{1}{T_0} \int_{-T_0/2}^{T_0/2} x(t) e^{j2\pi k F_0 t}\, dt \qquad k = 0, \ldots, N-1 \tag{2.76}$$

Note that the kth DFT coefficient $X(k)$ equals the kth Fourier series coefficient c_k if:

$x(t)$ is sampled and replaced with $x(m)$ and the integral sign $\int$ is replaced with the discrete summation sign $\sum$

$x(m)$ is assumed to be periodic with a period of N samples.

It follows that the DFT of N samples of a signal $[x(0), \ldots, x(N-1)]$ is identical to Fourier series of periodic signal $\{\ldots [x(0), \ldots, x(N-1)], [x(0), \ldots, x(N-1)], [x(0), \ldots, x(N-1)], \ldots\}$ with a period of N samples.

Example 2.14 *Derivation of inverse discrete Fourier transform*
The discrete Fourier transform is given by

$$X(k) = \sum_{m=0}^{N-1} x(m) e^{-j\frac{2\pi}{N}km} k = 0, \ldots, N-1 \tag{2.77}$$

Multiply both sides of the DFT equation by $e^{j2\pi kn/N}$ and take the summation as

$$\sum_{k=0}^{N-1} X(k) e^{+j\frac{2\pi}{N}kn} = \sum_{k=0}^{N-1}\sum_{m=0}^{N-1} x(m) e^{-j\frac{2\pi}{N}km} e^{+j\frac{2\pi}{N}kn} = \sum_{m=0}^{N-1} x(m) \underbrace{\sum_{k=0}^{N-1} e^{-j\frac{2\pi}{N}k(m-n)}}_{\begin{array}{ll} N & \text{if } m=n \\ 0 & \text{otherwise} \end{array}} \tag{2.78}$$

Using the orthogonality principle, the inverse DFT can be derived as

$$x(m) = \frac{1}{N} \sum_{k=0}^{N-1} X(k) e^{-j\frac{2\pi}{N}km} \qquad m = 0, \cdots, N-1 \tag{2.79}$$

A Note on DFT Channels of Real-Valued Signals

For N samples of a real-valued signal (e.g. speech or image), the DFT will consist of $N/2+1$ unique samples, comprised of a real-valued sample at frequency zero, $X(0)$, followed by $N/2-1$ complex-valued samples at $X(1), \ldots, X(N/2-1)$, followed by a real-valued sample at frequency $k = N/2$ (a total of N values since each complex number is composed of two values: a real and an imaginary part). The DFT samples at $k = N/2+1, \ldots, N-1$ are complex conjugates of those as $k = N/2-1 : -1 : 2$.

2.4.1 Frequency-Time Resolutions: The Uncertainty Principle

An important practical issue in applications of DFT is the issue of resolution, that is frequency versus time (or space) resolutions. Resolution is the ability to resolve the details of a signal in time or frequency or space. For example a satellite imaging system may have a resolution of 1 metre which means that it cannot capture details or objects smaller than 1 metre. Similarly in this section we will show that the frequency resolution of DFT is the width of each DFT bin between two successive discrete frequencies. The ability to resolve two closely spaced frequencies is inversely proportional to the DFT length, that is the length of the input signal.

The DFT length and hence the resolution of DFT is limited by the non-stationary character of most signals and also by the allowable delay in communication systems. Signals such as speech, music or image are generated by non-stationary – i.e. time-varying and/or space varying – systems or processes.

For example speech is composed of a sequence of short-duration sounds called phonemes each of which has a different spectral-temporal composition; music is composed of notes and sounds whose frequency compositions vary with time and an image is composed of various objects. As Fourier transform assumes that the signals within the DFT window of N samples are stationarity, the spectra of non-stationary signal events would be averaged over the duration of the window and would not be resolved and shown as separate signal events.

When using the DFT it is desirable to have a high resolution in time or space (that means small window size or number of input samples N) in order to obtain the details of the temporal/spacial characteristics of each individual elementary event, sound or object in the input signal. However, there is a fundamental trade-off between the length, of the input signal (i.e. the time or space resolution) and the frequency resolution of the DFT output.

The DFT takes as the input a window of N uniformly spaced discrete-time samples $[x(0), x(1), \ldots, x(N-1)]$ with a total duration of $\Delta T = N.T_s$, and outputs N spectral samples $[X(0), X(1), \ldots, X(N-1)]$ spaced uniformly between 0 Hz and the sampling frequency $F_s = 1/T_s$ Hz. Hence the frequency resolution of the DFT spectrum Δf, i.e. the frequency space between successive frequency samples, is given by

$$\Delta f = \frac{1}{\Delta T} = \frac{1}{NT_s} = \frac{F_s}{N} \tag{2.80}$$

Note that the frequency resolution Δf and the time resolution ΔT are inversely proportional in that they cannot both be simultaneously increased, in fact $\Delta T \Delta f = 1$. This is known as the uncertainty principle.

Example 2.15 A DFT is to be used in a DSP system for the analysis of an analogue signal with a frequency content of up to 10 kHz. Calculate: (i) the minimum sampling rate F_s required, (ii) the number of samples required for the DFT to achieve a frequency resolution of 10 Hz at the minimum sampling rate; and (iii) the time resolution.

Solution:
Sampling rate $> 2 \times 10$ kHz, say 22 kHz

$$\Delta f = \frac{F_s}{N} \qquad 10 = \frac{22000}{N} \qquad N \geq 2200$$

$$\Delta T = \frac{1}{\Delta f} = 100\,\text{ms}$$

Example 2.16 Write a Matlab program to explore the spectral resolution of a signal consisting of two sine waves, closely spaced in frequency, with the varying length of the observation window.

Solution: In the Matlab program the two sine waves have frequencies of 100 Hz and 110 Hz, and the sampling rate is 1 kHz. We experiment with two time windows of length $N_1 = 1024$ with a theoretical frequency resolution of $\Delta f = 1000/1024 = 0.98$ Hz, and $N_2 = 64$ with a theoretical frequency resolution $\Delta f = 1000/64 = 15.62$ Hz.

Figure 2.16 shows a plot of the spectrum of two closely spaced sine waves for different DFT window lengths and hence resolutions.

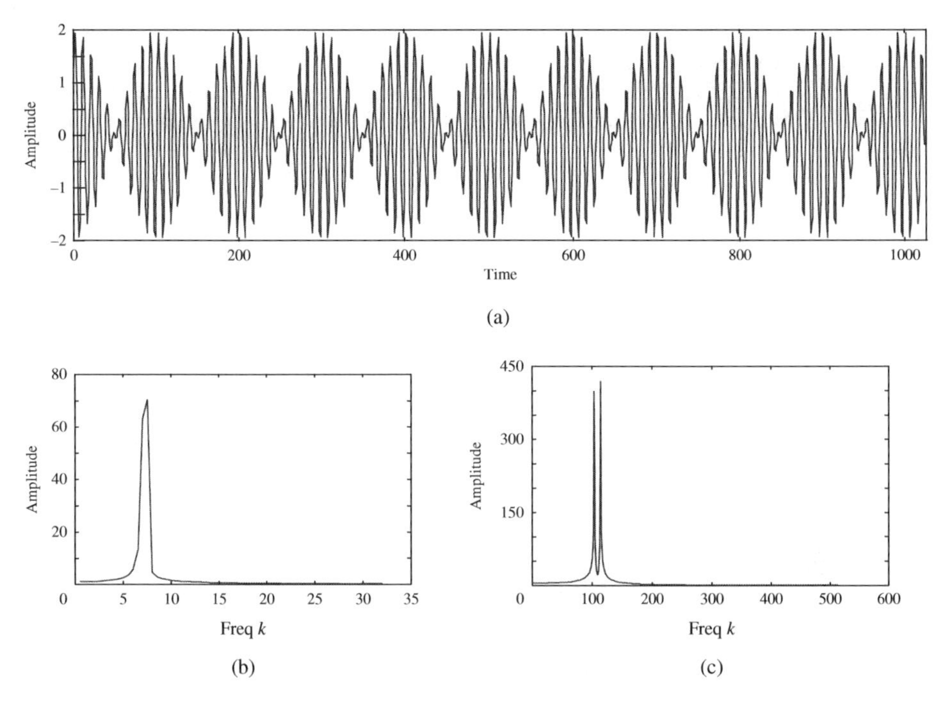

Figure 2.16 Illustration of time and frequency resolutions: (a) sum of two sine waves with 10 Hz difference in their frequencies, note a beat of 10 Hz (b) the spectrum of a segment of 64 samples from demonstrating insufficient frequency resolution to separate the sine waves, (c) the spectrum of a segment of 1024 samples has sufficient resolution to show the two sine waves.

2.4.2 The Effect of Finite Length Data on DFT (Windowing)

Matlab function TimeFreqResolSineWaves( )
Demonstrates the increase in frequency resolution with the increasing length of a signal containing two closely spaced sine waves. Initially the frequency resolution is low and the sine waves cannot be resolved. As the length of window increases the frequency resolution increases and the sine waves become distinguishable.

Matlab function TimeFreqResolMusic( )
Displays the spectrograms of music (trumpet and piano) to demonstrate the inverse relationship between time and frequency resolutions. Three cases are demonstrated: high time resolution and low frequency resolution, medium time and frequency resolutions, low time resolution and high frequency resolution.

In signal processing the DFT is usually applied to a relatively short segment of a signal of the order of a few tens of ms. This is because usually either we have a short length signal or when we have a long signal, due to various considerations such as stationarity of a signal or the allowable system delay, the DFT can only handle one segment at a time. A short segment of N samples of a signal, or a slice of N samples from a signal, is equivalent to multiplying the signal by a unit-amplitude rectangular pulse window of N samples. Therefore an N-sample segment of a signal $x(m)$ is equivalent to

$$x_w(m) = w(m)x(m) \tag{2.81}$$

where $w(m)$ is a rectangular pulse of N samples duration given as

$$w(m) = \begin{cases} 1 & 0 \le m \le N-1 \\ 0 & \text{otherwise} \end{cases} \tag{2.82}$$

Multiplying two signals in time is equivalent to the convolution of their frequency spectra. Thus the spectrum of a short segment of a signal is convolved with the spectrum of a rectangular pulse as

$$X_w(k) = W(k) * X(k) \tag{2.83}$$

The result of this convolution is some spreading of the signal energy in the frequency domain as illustrated in the next example.

Example 2.17 Find the DFT of a rectangular window given by

$$w(m) = \begin{cases} 1 & 0 \le m \le N-1 \\ 0 & \text{otherwise} \end{cases} \tag{2.84}$$

Solution: Taking the DFT of $w(m)$, and using the convergence formula for the partial sum of a geometric series, $S_N = 1 + \alpha + \cdots + \alpha^{N-1} = \frac{1-\alpha^N}{1-\alpha}$, we have

$$W(k) = \sum_{m=0}^{N-1} w(m) e^{-j\frac{2\pi}{N}mk} = \frac{1 - e^{-j2\pi k}}{1 - e^{-j\frac{2\pi}{N}k}} = e^{-j\frac{(N-1)}{N}\pi k} \frac{\sin(\pi k)}{\sin(\pi k/N)} \tag{2.85}$$

Note that for the integer values of k, $W(k)$ is zero except for $k = 0$. There are $N-1$ zero-crossings uniformly spread along 0 to F_s the sampling frequency. The discrete-time rectangular window has a sinc-shaped spectrum with a main-lobe bandwidth of $2F_s/N$.

Example 2.18 Find the spectrum of an N-sample segment of a complex sine wave with a fundamental frequency $F_0 = 1/T_0$.

Solution: Taking the DFT of $x(m) = e^{-j2\pi F_0 m}$ we have

$$\begin{aligned} X(k) &= \sum_{m=0}^{N-1} e^{-j2\pi F_0 m} e^{-j\frac{2\pi}{N}mk} = \sum_{m=0}^{N-1} e^{-j2\pi(F_0 - \frac{k}{N})m} \\ &= \frac{1 - e^{-j2\pi(NF_0 - k)}}{1 - e^{-j2\pi(NF_0 - k)/N}} = e^{-j\frac{(N-1)}{N}\pi(NF_0 - k)} \frac{\sin(\pi(NF_0 - k))}{\sin(\pi(NF_0 - k)/N)} \end{aligned} \tag{2.86}$$

Note that for integer values of k, $X(k)$ is zero at all samples but one, that is $k = 0$.

2.4.3 End-Point Effects in DFT; Spectral Energy Leakage and Windowing

A window that rolls down smoothly is used to mitigate the effect of spectral leakage due to the discontinuities at the end of a segment of a signal. In DFT the input signal is assumed to be periodic, with a period equal to the length of the observation window of N samples. For a sinusoidal signal when there is an integer number of cycles within the observation window, as in Figure 2.17(a), the assumed periodic waveform is the same as an infinite-duration pure sinusoid. However, if the observation window contains a non-integer number of cycles of a sinusoid then the assumed periodic waveform will not be a pure sine wave and will have end-point discontinuities. The spectrum of the signal then differs from the spectrum of a sine wave as illustrated in Figure 2.17(b). The overall effects of finite length window and end-point discontinuities are:

(1) The spectral energy which could have been concentrated at a single point is spread over a large band of frequencies.
(2) A smaller amplitude signal located in frequency near a larger amplitude signal, may be interfered with, or even obscured, by one of the larger spectral side-lobes.

The end-point problems may be alleviated using a window that gently drops to zero. One such window is a raised cosine window of the form

$$w(m) = \begin{cases} \alpha-(1-\alpha)\cos\frac{2\pi m}{N} & 0 \le m \le N-1 \\ 0 & \text{otherwise} \end{cases} \tag{2.87}$$

For $\alpha = 0.5$ we have the Hanning window also known as the raised cosine window

$$w_{\text{Hanning}}(m) = 0.5 - 0.5\cos\frac{2\pi m}{N} \quad 0 \le m \le N-1 \tag{2.88}$$

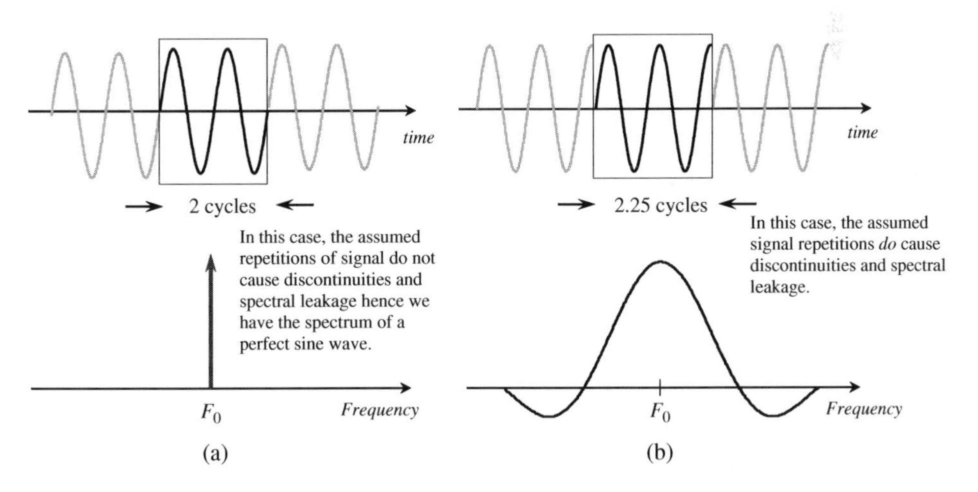

Figure 2.17 The DFT spectrum of $\exp(j2\pi fm)$: (a) an integer number of cycles within the N-sample analysis window, (b) a non-integer number of cycles in the window. Note that DFT theory assumes the signal is periodic hence outside the window the signal is assumed to repeat itself with a periodicity equal to the window length.

For $\alpha = 0.54$ we have the Hamming window

$$w_{\text{Hamming}}(m) = 0.54 - 0.46\cos\frac{2\pi m}{N} \quad 0 \leq m \leq N-1 \tag{2.89}$$

The Blackman window is given by

$$w_{\text{Blackman}}(m) = 0.42 - 0.5\cos\frac{2\pi m}{N+1} - 0.08\cos\frac{4\pi m}{N+1} \quad 0 \leq m \leq N \tag{2.90}$$

The main difference between various windows is the trade-off between bandwidth of the main lobe and the amplitude of side-lobes. Figure 2.18 shows the rectangular window, Hamming window, Hanning window and Blackman window and their respective spectra.

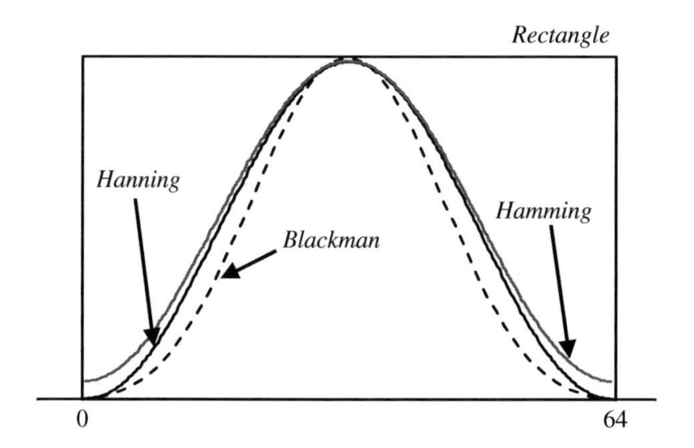

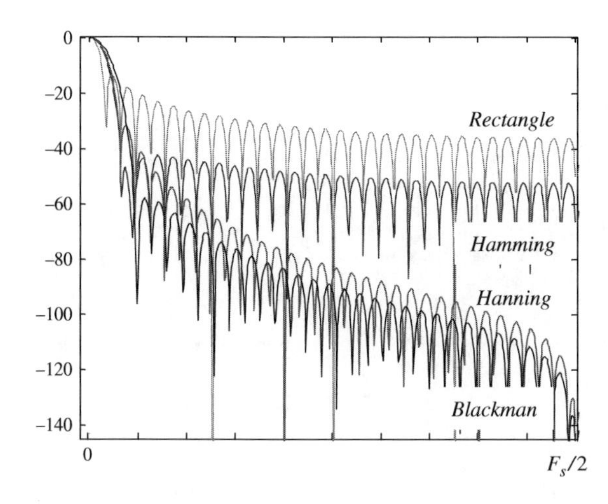

Figure 2.18 Four windows and their frequency spectra.

2.4.4 Spectral Smoothing: Interpolation Across Frequency

The spectrum of a short length signal can be interpolated to obtain a smoother looking spectrum. Interpolation of the frequency spectrum $X(k)$ is achieved by *zero-padding* of the time domain signal $x(m)$. Consider a signal of length N samples $[x(0), ..., x(N-1)]$. Increase the signal length from N to $2N$ samples by padding N zeros to obtain the padded sequence $[x(0), ..., x(N-1), 0, ..., 0]$. The DFT of the padded signal is given by

$$X(k) = \sum_{m=0}^{2N-1} x(m)e^{-j\frac{2\pi}{2N}mk}$$
$$= \sum_{m=0}^{N-1} x(m)e^{-j\frac{\pi}{N}mk} \qquad k = 0, ..., 2N-1 \tag{2.91}$$

The spectrum of the zero-padded signal, Equation (2.91), is composed of $2N$ spectral samples, N of which, $[X(0), X(2), X(4), X(6), ..., X(2N-2)]$ are the same as those that would be obtained from a DFT of the original N samples, and the other N samples $[X(1), X(3), X(5), X(6), \ldots, X(2N-1)]$ are interpolated spectral lines that result from zero-padding. Note that zero padding does not increase the spectral resolution, it merely has an *interpolating or smoothing* effect in the frequency domain, as illustrated in Figures 2.19 and 2.20.

Matlab function zeropadding( )
Demonstrates the increase in the apparent frequency resolution resulting from zero-padding of the sum of two sine waves. See Figure 2.20.

2.5 Short-Time Fourier Transform

In real-life and real-time applications signals have finite durations and before the application of Fourier transform the signals are often divided into relatively short length segments for two main reasons:

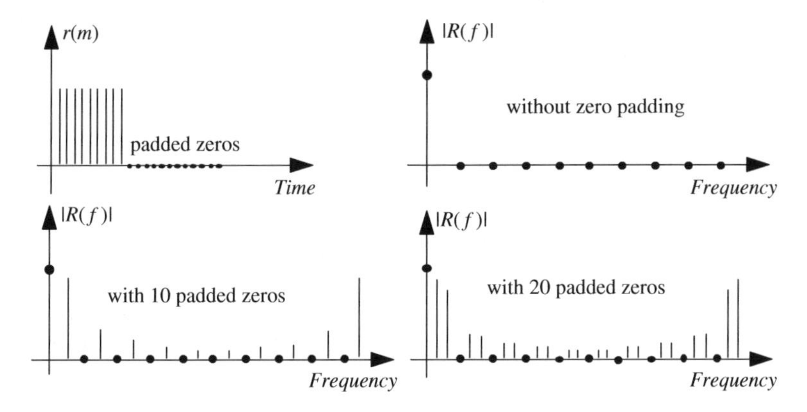

Figure 2.19 Illustration of the interpolating effect, in the frequency domain, of zero padding a signal in the time domain.

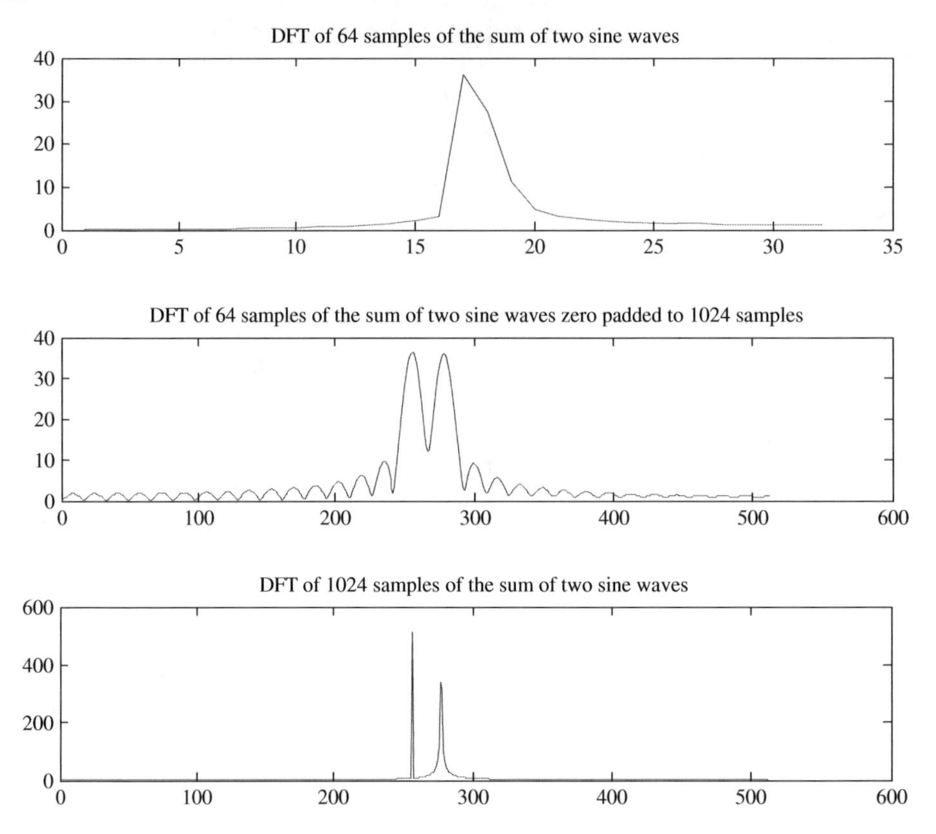

Figure 2.20 Spectra of the sum of two sine waves at frequencies of 250 Hz and 270 Hz sampled at 1000 Hz. Top panel: magnitude DFT of 64 samples of the signal, middle panel: magnitude DFT of 64 samples zero-padded to 1024 samples, bottom panel: magnitude DFT of 1024 samples of the signal.

(i) The existing limits on the tolerable time delay in communication systems, and the requirements to reduce the computational complexity as much as possible, imply that signals need to be segmented into relatively short duration segments.

(ii) Fourier theory assumes that the signals are stationary; this means the signal statistics, such as the mean, the power, and the power spectrum, are time-invariant. Most real-life signals such as speech, music, image and noise are non-stationary in that their amplitude, power, spectral composition and other features changes continuously with time.

In using the Fourier transform on non-stationary signals, the signal is divided into appropriately short-time windows, such that within each window the signal may be assumed to be time-invariant. The Fourier transform applied to the short signal segment within each window is known as the short-time Fourier transform (STFT). Figure 2.21 illustrates the segmentation of a speech signal into a sequence of overlapping, Hamming windowed, short segments. The choice of window length is a compromise between the time resolution and the frequency resolution. For audio signals a time window of about 25 ms, corresponding to a frequency resolution of 40 Hz, is normally adopted.

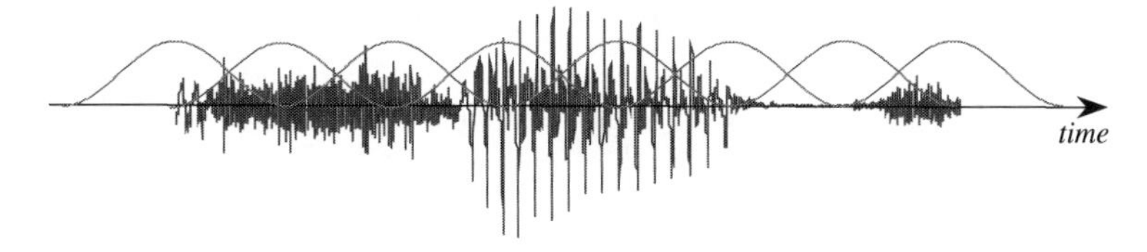

Figure 2.21 Segmentation of speech using Hamming window for STFT; in this case successive windows overlap by half window length.

2.6 Fast Fourier Transform (FFT)

In this section we consider computationally fast implementation of the discrete Fourier transform. The discrete Fourier transform equation (2.60) can be rewritten as

$$X(k) = \sum_{m=0}^{N-1} x(m) w_N^{mk} \quad k = 0, \ldots, N-1 \tag{2.92}$$

where $w_N = e^{-j2\pi/N}$. The DFT equation (2.92) can be expressed in a matrix transformation form as

$$\begin{bmatrix} X(0) \\ X(1) \\ X(2) \\ \vdots \\ X(N-1) \end{bmatrix} = \begin{bmatrix} 1 & 1 & 1 & \cdots & 1 \\ 1 & w & w^2 & \cdots & w^{(N-1)} \\ 1 & w^2 & w^4 & \cdots & w^{2(N-1)} \\ \vdots & \vdots & \vdots & \ddots & \vdots \\ 1 & w^{(N-1)} & w^{2(N-1)} & \cdots & w^{(N-1)^2} \end{bmatrix} \begin{bmatrix} x(0) \\ x(1) \\ x(2) \\ \vdots \\ x(N-1) \end{bmatrix} \tag{2.93}$$

In a compact form the DFT equation (2.92) can be written as

$$\boldsymbol{X} = \boldsymbol{W}_N \boldsymbol{x} \tag{2.94}$$

where the output vector $\boldsymbol{X}$ is the Fourier transform of the input sequence $\boldsymbol{x}$. The inverse DFT in matrix notation can be written as

$$\boldsymbol{x} = \boldsymbol{W}_N^{-1} \boldsymbol{X} \tag{2.95}$$

Note that the individual elements of the inverse DFT matrix W_N^{-1} are the inverse of the individual elements of the DFT matrix $\boldsymbol{W}_N$. From Equation 2.91 the direct calculation of the Fourier transform requires $N(N-1)$ multiplications and a similar number of additions.

Algorithms that reduce the computational complexity of the discrete Fourier transform are known as fast Fourier transform (FFT) methods. FFT methods utilise the periodic and symmetric properties of the Fourier basis function $w_N = e^{-j2\pi/N}$ to avoid redundant calculations. Specifically, FFT methods utilise the following:

1. The periodicity property of $w_N^{mk} = w_N^{(m+N)k} = w_N^{m(k+N)}$.
2. The complex conjugate symmetry property $w_N^{-mk} = \left(w_N^{mk}\right)^* = w_N^{(N-m)k}$.

In the following we consider two basic forms of FFT algorithms: decimation-in-time FFT and decimation-in-frequency FFT.

2.6.1 Decimation-in-Time FFT

Decimation-in-time FFT divides a sequence of input samples into a set of smaller sequences, and consequently smaller DFTs. Consider the DFT of the sequence $[x(0), x(1), x(2), x(3), \ldots, x(N-2), x(N-1)]$

$$X(k) = \sum_{m=0}^{N-1} x(m) w_N^{mk} \quad k = 0, \ldots, N-1 \tag{2.96}$$

Divide the input sample sequence into two sub-sequences with even and odd numbered discrete-time indices such as $[x(0), x(2), x(4), \ldots, x(N-2)]$ and $[x(1), x(3), x(5), \ldots, x(N-1)]$. The DFT equation (2.96) can be rearranged and rewritten as

$$X(k) = \sum_{m \in \text{even}} x(m) w_N^{mk} + \sum_{m \in \text{odd}} x(m) w_N^{mk} \tag{2.97}$$

$$X(k) = \sum_{m=0}^{N/2-1} x(2m) w_N^{2mk} + \sum_{m=0}^{N/2-1} x(2m+1) w_N^{(2m+1)k} \tag{2.98}$$

$$X(k) = \sum_{m=0}^{N/2-1} x(2m) \left(w_N^2\right)^{mk} + w_N^k \sum_{m=0}^{N/2-1} x(2m+1) \left(w_N^2\right)^{mk} \tag{2.99}$$

Now $w_N = e^{-j2\pi/N}$ hence

$$w_N^2 = \left(e^{-j2\pi/N}\right)^2 = e^{-j2\pi/(N/2)} = w_{N/2} \tag{2.100}$$

Using Equation (2.100), the DFT equation (2.100) can be written as

$$X(k) = \underbrace{\sum_{m=0}^{N/2-1} x(2m) w_{N/2}^{mk}}_{N/2 \text{ point DFT}} + \underbrace{w_N^k \sum_{m=0}^{N/2-1} x(2m+1) w_{N/2}^{mk}}_{N/2 \text{ point DFT}} \tag{2.101}$$

It can be seen that the N-sample DFT algorithm of Equation (2.101) is composed of two DFTs of size $N/2$. Thus the computational requirement has been halved from $N(N-1)$ operations to approximately $2(N/2)^2$ operations. For any N, a power of 2, this process of division of a DFT into two smaller-sized DFTs can be iterated $\log_2 N - 1 = \log_2(N/2)$ times to arrive at DFTs of size 2, with $\mathbf{W}_2$ defined as

$$\mathbf{W}_2 = \begin{bmatrix} 1 & 1 \\ 1 & -1 \end{bmatrix} \tag{2.102}$$

From Equation (2.102) the DFT of size 2 can be obtained by a single addition and a single subtraction and requires no multiplication. Figure 2.22 shows the flow graph for the computation of an elementary 2-point DFT called a butterfly.

Each stage of the decomposition requires $N/2$ multiplications; therefore the total number of complex multiplications required is $(N/2)\log_2(N/2)$. The total number of complex additions is $N \log_2 N$.

For example for $N = 256$ the computation load for the direct DFT is

$$256 \times 255 = 65,280 \quad \text{complex multiplications}$$

$$256 \times 255 = 65,280 \quad \text{complex additions}$$

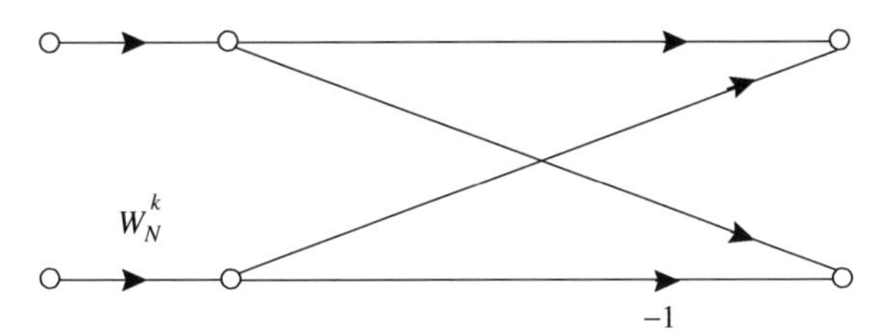

Figure 2.22 Flow graph of a basic 2-point DFT butterfly.

Using the FFT method the computational load is reduced considerably to

$$128 \times \log_2 128 = 896 \quad \text{complex multiplications}$$

$$256 \times \log_2 256 = 2048 \quad \text{complex additions}$$

Example 2.19 Consider the implementation of decimation-in-time FFT of an 8-point sequence $[x(0), x(1), x(2), \ldots, x(8)]$.

In the signal flow diagram of Figure 2.23(a) the 8-point DFT is divided into two 4-point DFTs. In Figure 2.23(b) each 4-point DFT is further divided and represented as the combination of two 2-point DFTs. Figure 2.23(c) shows the signal flow diagram for the computation of a 2-point DFT. The FFT diagram of Figure 2.23(d) is obtained by substituting back the basic butterfly unit in the 2-point DFT of Figure 2.23(b) and then substituting the result in the 4-point DFT blocks of Figure 2.23(a).

2.6.2 Bit Reversal

The FFT requires the grouping together of the even-indexed samples and odd-indexed samples at each stage. The result of this process for an 8-point sequence is shown in Figure 2.23 and Table 2.2. Consider the binary-index representation of the input and output sequences of the 8-point DFT of Figure 2.23(d).

It can be seen that the time index of the input signal $x(m)$ in the binary format is a bit-reversed version of the naturally ordered sequence. For example $x(100)$, i.e. $x(4)$, takes the place of $x(001)$, i.e. $x(1)$, and $x(110)$, i.e. $x(5)$, takes the place of $x(011)$, i.e. $x(3)$, and so on.

2.6.3 Inverse Fast Fourier Transform (IFFT)

The main difference between the Fourier transform and the inverse Fourier transform is the sign of the exponent of the complex exponential basis function: i.e. a minus sign for the Fourier transform $e^{-j2\pi/N}$, and a plus sign for the inverse Fourier transform $e^{j2\pi/N}$. Therefore the IFFT system can be obtained by replacing the variable w by w^{-1} and scaling the input sequence by a factor of N as expressed in Equation (2.70).

2.6.4 Decimation-in-Frequency FFT

Decimation-in-time FFT divides a sequence of samples into a set of smaller sequences, and consequently smaller discrete Fourier transformations. Instead of dividing the time sequence, we can divide

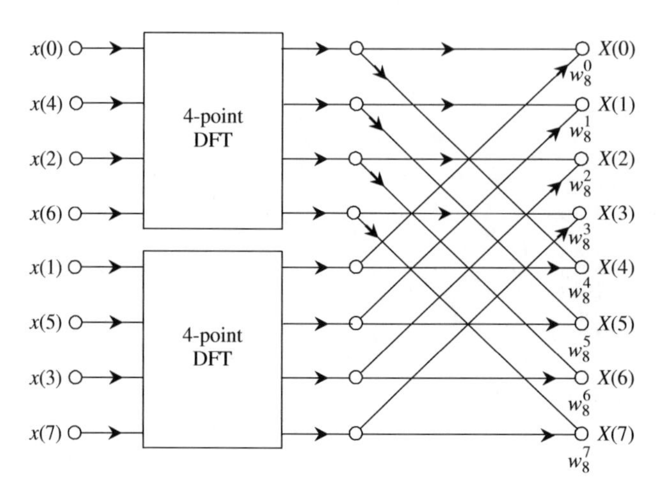

(a) Two 4-point DFTs combined into an 8-point DFT.

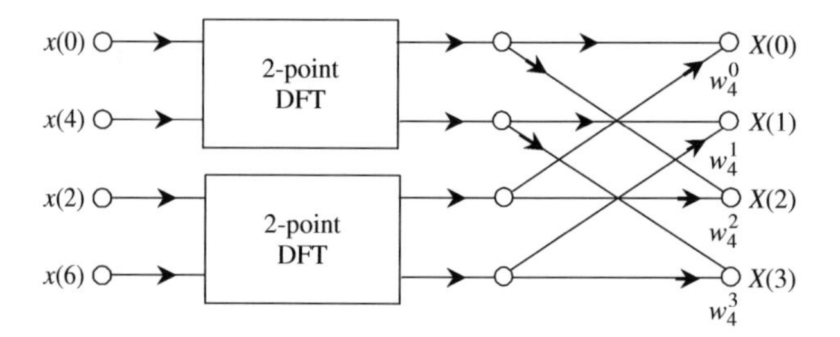

(b) Two 2-point DFTs combined into a 4-point DFT.

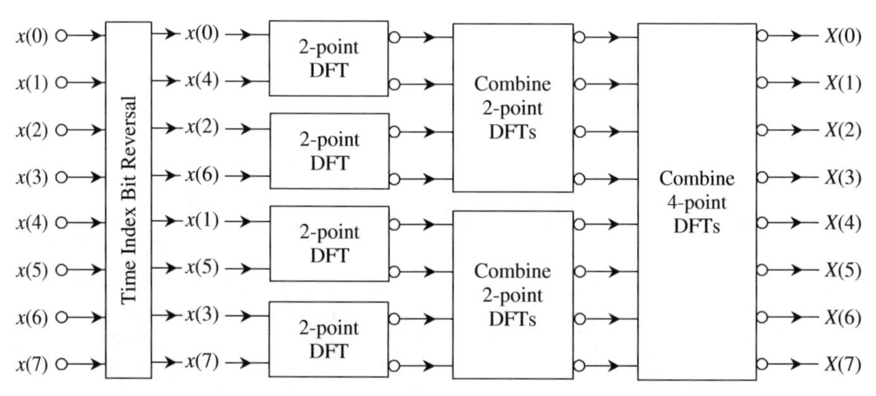

(c) An 8-point DFT implemented as a combination of 2-point DFTs followed by a combination of 4-point DFTs. Note the input sequence is rearranged through bit reversal of its time index.

Figure 2.23 Illustration of implementation of decimation-in-time FFT for an 8-point sequence.

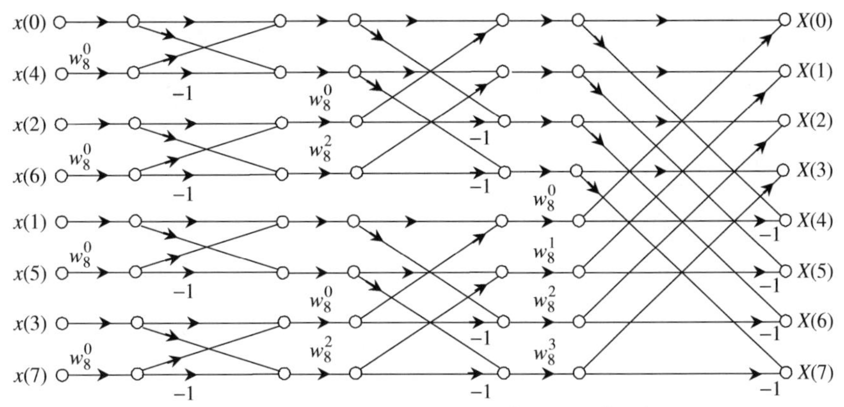

(d) The detailed implementation of the 8-point FFT. Note $w_8^0 = 1$ and $w_8^2 = w_4^1$.

Figure 2.23 (Continued)

Table 2.2 Illustration of bit reversal process in FFT.

Sample number	Bit address	Time reversed address	Input sample	Output sample
0	000	000	$x(000)$	$X(000)$
1	001	100	$x(100)$	$X(001)$
2	010	010	$x(010)$	$X(010)$
3	011	110	$x(110)$	$X(011)$
4	100	001	$x(001)$	$X(100)$
5	101	101	$x(101)$	$X(101)$
6	110	011	$x(011)$	$X(110)$
7	111	111	$x(111)$	$X(111)$

the output sequence leading to decimation-in-frequency FFT methods. Consider the DFT of the sequence $[x(0), x(1), x(2), x(3), \ldots, x(N-2), x(N-1)]$

$$X(k) = \sum_{m=0}^{N-1} x(m)w^{mk} \qquad k = 0, \ldots, N-1 \tag{2.103}$$

Divide the output sample sequence into two sample sequences with even and odd numbered discrete-time indices as $[X(0), X(2), \ldots, X(N-2)]$ and $[X(1), X(3), \ldots, X(N-1)]$. The even-number frequency samples of the DFT equation (2.103) can be rewritten as

$$X(2k) = \sum_{m=0}^{N/2-1} x(m)w_N^{2mk} + \sum_{m=N/2}^{N-1} x(m)w_N^{2mk} \tag{2.104}$$

$$X(2k) = \sum_{m=0}^{N/2-1} x(m)w_N^{2mk} + \sum_{m=0}^{N/2-1} x(m+N/2)w_N^{2(m+N/2)k} \tag{2.105}$$

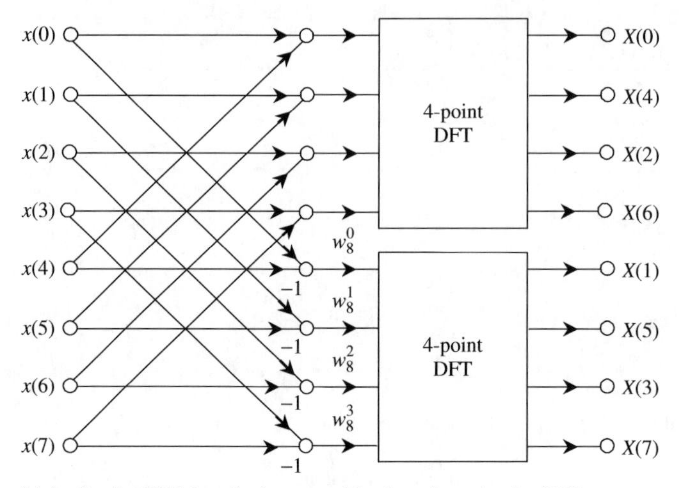

(a) An 8-point DFT described as a combination of two 4-point DFTs.

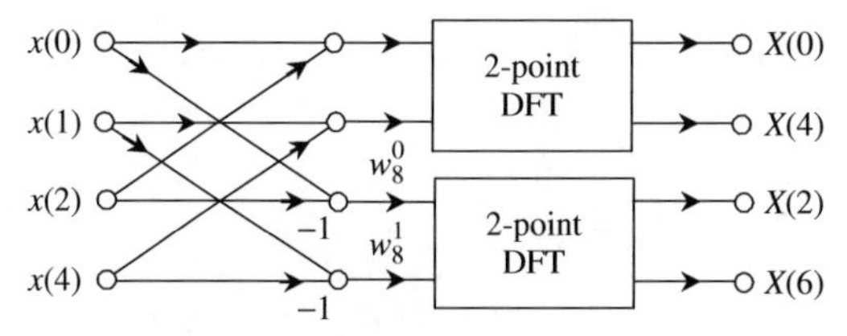

(b) A 4-point DFT described in terms of two 2-point DFTs.

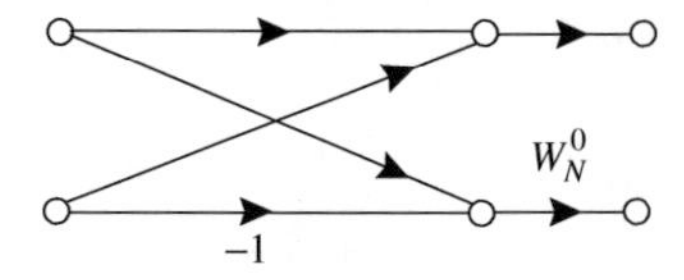

(c) The basic butterfly for computation of a 2-point DFT.

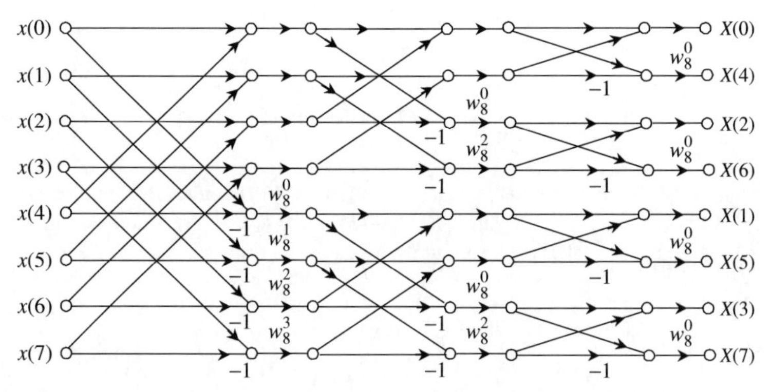

(d) The detailed implementation of the 8-point FFT. Note $w_8^0 = 1$ and $w_8^2 = w_4^1$.

Figure 2.24 Illustration of decimation-in-frequency FFT for an 8-point sequence.

Since w_N^{2mk} is periodic with a period of N we have

$$w_N^{2(m+N/2)k} = w_N^{2mk} w_N^{Nk} = w_N^{2mk} \tag{2.106}$$

and $w_N^2 = w_{N/2}$, hence Equation (2.105) can be written as

$$X(2k) = \sum_{m=0}^{N/2-1} [x(m) + x(m+N/2)] w_{N/2}^{mk} \tag{2.107}$$

Equation (2.107) is an $N/2$-point DFT. Similarly, for the odd-number frequency samples the DFT equation (2.103) can be rewritten as

$$X(2k+1) = \sum_{m=0}^{N/2-1} x(m) w_N^{m(2k+1)} + \sum_{m=N/2}^{N-1} x(m) w_N^{m(2k+1)} \tag{2.108}$$

$$X(2k+1) = \sum_{m=0}^{N/2-1} x(m) w_N^{m(2k+1)} + \sum_{m=0}^{N/2-1} x(m+N/2) w_N^{(m+N/2)(2k+1)} \tag{2.109}$$

Since w_N^{2km} is periodic with a period of N we have

$$w_N^{(m+N/2)(2k+1)} = w_N^{m(2k+1)} \underbrace{w_N^{kN}}_{=1} \underbrace{w_N^{N/2}}_{=-1} = -w_N^{m(2k+1)} \tag{2.110}$$

Therefore

$$X(2k+1) = \sum_{m=0}^{N/2-1} [x(m) - x(m+N/2)] w_N^{m(2k+1)} \tag{2.111}$$

or

$$X(2k+1) = \sum_{m=0}^{N/2-1} [x(m) - x(m+N/2)] w_{N/2}^{km} w_N^m \tag{2.112}$$

Equation (2.112) is an $N/2$-point DFT of the sequence $[x(m) - x(m+N/2)]w_N^m$. Thus as before the N-point DFT can be implemented as the sum of 2 $N/2$-point DFTs.

Example 2.20 Consider the decimation-in-frequency FFT of an 8-point sequence $[x(0), x(1), x(2), \ldots, x(8)]$.
In the signal flow diagram of Figure 2.24(a) the 8-point DFT is divided into two 4-point DFTs. In Figure 2.24(b) each 4-point DFT is further divided and represented as the combination of two 2-point DFTs. Figure 2.24(c) shows the signal flow diagram for the computation of a 2-point DFT. The FFT diagram of Figure 2.24(d) is obtained by substituting back the basic butterfly unit in the 2-point DFT of Figure 2.24(b) and then substituting the result in the 4-point DFT blocks of Figure 2.24(a).

2.7 2-D Discrete Fourier Transform (2-D DFT)

For a two-dimensional signal such as an image, the discrete Fourier transform is essentially two one-dimensional DFTs, consisting of the DFT of the columns of the input signal matrix followed by the DFT of the rows of the matrix or vice versa. The 2-D DFT is defined as

$$X(k_1, k_2) = \sum_{m_1=0}^{N_1-1} \sum_{m_2=0}^{N_2-1} x(m_1, m_2) e^{-j\frac{2\pi}{N_1} m_1 k_1} e^{-j\frac{2\pi}{N_2} m_2 k_2} \quad k_1 = 0, \ldots, N_1 - 1$$

$$k_2 = 0, \ldots, N_2 - 1 \tag{2.113}$$

and the 2-D inverse discrete Fourier transform (2-D IDFT) is given by

$$x(m_1, m_2) = \frac{1}{N_1 N_2} \sum_{k_1=0}^{N_1-1} \sum_{k_2=0}^{N_2-1} X(k_1, k_2) e^{j\frac{2\pi}{N_1}m_1 k_1} e^{j\frac{2\pi}{N_2}m_2 k_2} \quad m_1 = 0, \ldots, N_1 - 1$$

$$m_2 = 0, \ldots, N_2 - 1 \tag{2.114}$$

The 2-D DFT equation (2.113) requires $N_1(N_1 - 1)N_2(N_2 - 1)$ multiplications and a similar number of additions. For example the 2-D DFT for an image block of size 8×8 pixels involves 3136 multiplications. Using the fast Fourier transform of the rows and columns, the number of multiplications required is reduced considerably to $(N_1/2)\log_2(N_1/2)(N_2/2)\log_2(N_2/2) = 64$ multiplications.

2.8 Discrete Cosine Transform (DCT)

The Fourier transform expresses a signal in terms of a combination of complex exponentials composed of cosine (real part) and sine (imaginary part) waveforms. A signal can be expressed purely in terms of cosine or sine basis functions leading to cosine transform and sine transform. For many processes, the cosine transform is a good approximation to the data-dependent optimal Karhunen–Loève transforms. The cosine transform is extensively used in speech and image compression and for feature extraction. The discrete cosine transform is given by

$$X(k) = 2\sum_{m=0}^{N-1} x(m)\cos\frac{\pi k(2m+1)}{2N} \qquad 0 \le k \le N-1 \tag{2.115}$$

$$x(m) = \frac{1}{N}\sum_{k=0}^{N-1} w(k)X(k)\cos\frac{\pi k(2m+1)}{2N} \qquad 0 \le k \le N-1 \tag{2.116}$$

where the coefficient $w(k)$ is given by

$$w(k) = \begin{cases} \frac{1}{2} & k = 0 \\ 1 & k = 1, \cdots, N-1 \end{cases} \tag{2.117}$$

Figure 2.25 shows the first six basis functions of the discrete cosine transform for a DCT of length 128. A useful property of DCT is its ability to compress and concentrate most of the energy of a signal into a relatively few low-frequency coefficients, as shown in Examples 2.21 and 2.22 and Section 2.9.7. DCT is a good approximation to principal component analysis, also known as Karhunen–Loève transform (KLT), described in Chapter 12.

A modified form of the DCT, which is a unitary orthonormal transform, is given by

$$X(k) = \sqrt{\frac{2}{N}}w(k)\sum_{n=0}^{N-1} x(m)\cos\frac{\pi k(2m+1)}{2N} \quad 0 \le k \le N-1 \tag{2.118}$$

$$x(m) = \sqrt{\frac{2}{N}}\sum_{k=0}^{N-1} w(k)X(k)\cos\frac{\pi k(2m+1)}{2N} \quad 0 \le k \le N-1 \tag{2.119}$$

where the coefficient $w(k)$ is given by

$$w(k) = \begin{cases} \frac{1}{\sqrt{2}} & k = 0 \\ 1 & k = 1, \cdots N-1 \end{cases} \tag{2.120}$$

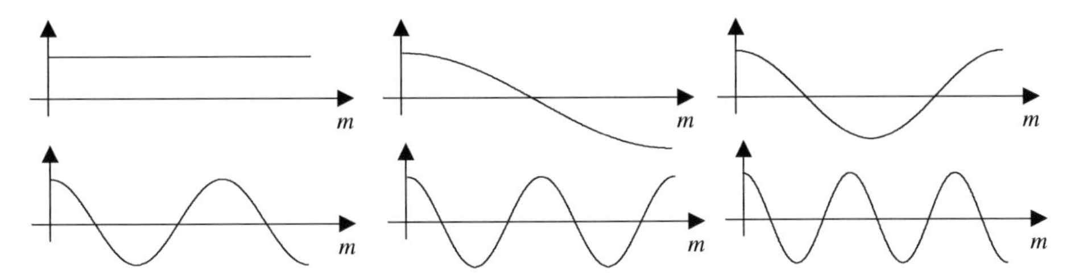

Figure 2.25 The first six basis functions of the discrete cosine transform.

Example 2.21 Compare the DCT and the DFT of a discrete-time impulse function.

Solution: The discrete-time impulse function is given by

$$\delta(m) = \begin{cases} 1 & m = 0 \\ 0 & m \neq 0 \end{cases} \tag{2.121}$$

$$\Delta^{DCT}(k) = 2\sum_{n=0}^{N-1} \delta(m) \cos \frac{\pi k(2m+1)}{2N} = 2\cos \frac{\pi k}{2N} \quad 0 \leq k \leq N-1 \tag{2.122}$$

The Fourier transform of impulse function

$$\Delta^{DFT}(k) = \sum_{m=0}^{N-1} \delta(m) e^{-j\frac{2\pi}{N}km} = 1 \quad 0 \leq k \leq 1 \tag{2.123}$$

Figure 2.25 shows the DFT spectrum and the DCT spectrum of an impulse function.

Example 2.22 *Comparison of DFT and DCT of an impulse function*

Solution: The Matlab program for this exercise is

```
x = zeros(1, 1000);
x(1) = 1;
xdct = dct(x);
xfft = abs(fft(x));
```

The plots in Figure 2.26 show that the DCT of a discrete-time impulse is more concentrated towards low frequency compared with the DFT of a discrete-time impulse.

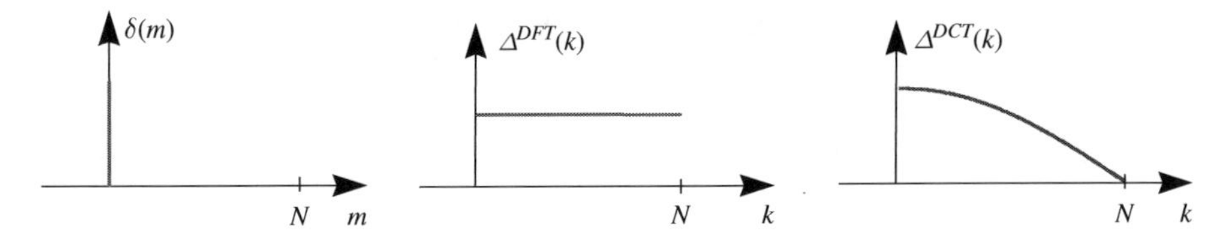

Figure 2.26 A discrete-time impulse signal and its DFT and DCT spectra.

2.9 Some Applications of the Fourier Transform

The Fourier signal analysis/synthesis method is used in numerous applications in science, engineering and data analysis. In this section we consider several application of the Fourier transform.

2.9.1 Spectrogram

The spectrogram is a plot of the variation of the short time magnitude (or power) spectrum of a signal with time. A shown in Figure 2.27, the signal is divided into overlapping windowed segments of appropriately short duration (about 25 ms for audio signals), each segment is transformed with an FFT, and the magnitude frequency vectors are stacked and plotted with the vertical axis representing the frequency and the horizontal axis the time. The magnitude values are colour coded with black colour representing the lowest value and a light colour representing the largest value.

The spectrogram provides a powerful description of the time variation of the spectrum of a time-varying signal. This is demonstrated in Figure 2.28, which shows the spectrogram of voice and music. In the music-only periods the quasi-periodic instances of hitting of the drum are clearly visible. Also the periods where voice is present can be clearly distinguished from the pitch structure of the vocal cord excitation. In fact looking at the histogram, and relying on the pitch structure of speech, it is possible to manually segment the signal into music-only and voice-plus-music periods with ease and with a high degree of accuracy.

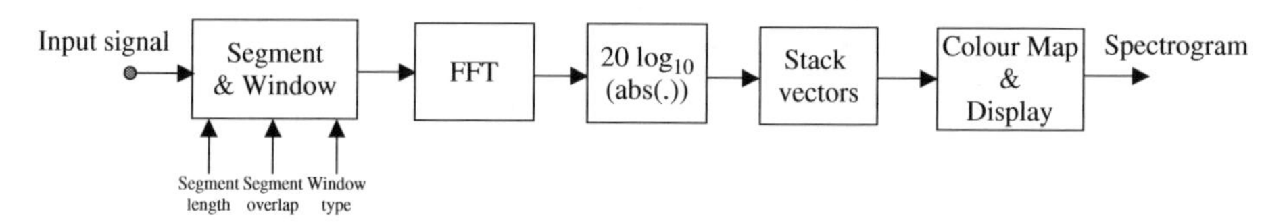

Figure 2.27 Block diagram illustration of spectrogram.

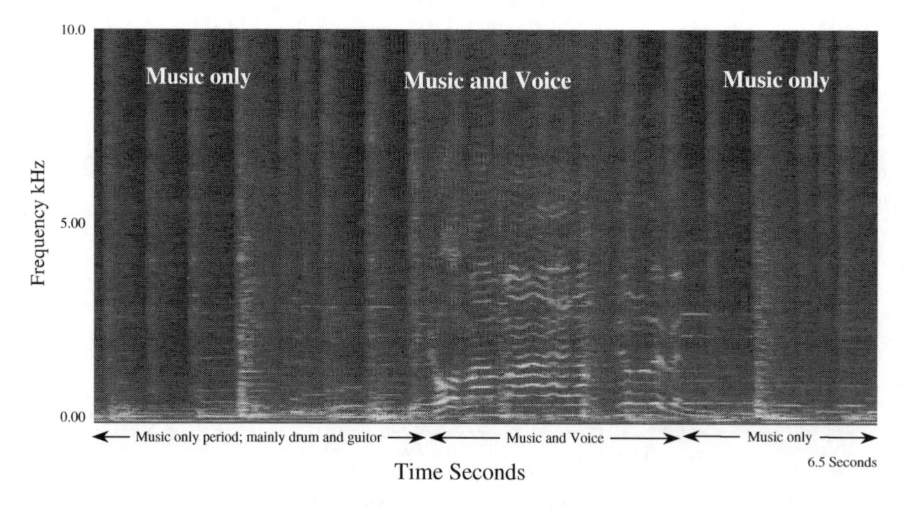

Figure 2.28 Spectrogram of music and voice.

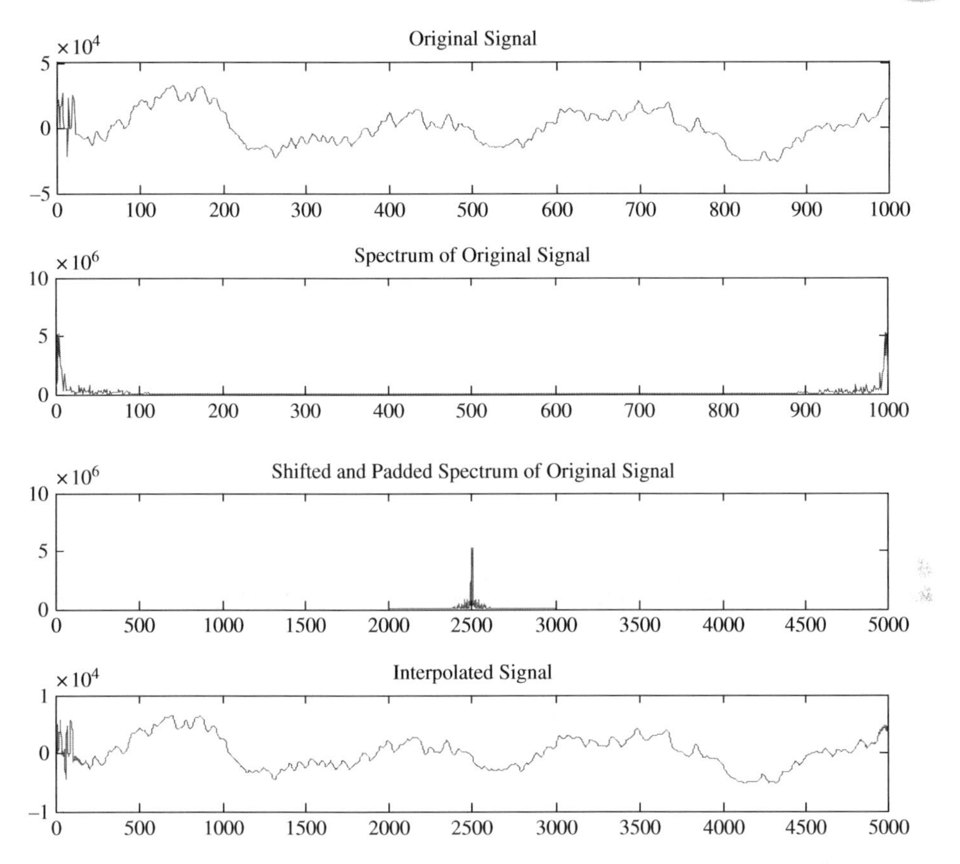

Figure 2.29 Illustration of time-domain interpolation using the DFT. Note the spectrum of the signal is shifted and padded and then transformed to time domain via IDFT. The interpolation factor in this case is 5.

2.9.2 Signal Interpolation

From the time – frequency duality principle, just as zero-padding of a signal in time (or space) leads to interpolation of its frequency spectrum (see Section 2.4.4), zero-padding in frequency followed by inverse transform provides interpolation of a signal in time (or space) as illustrated in Figure 2.29.

Consider a signal of length N samples $[x(0), x(1)\ldots, x(N-1)]$ together with its DFT $[X(0), X(1), \ldots, X(N-1)]$. Increase the length of signal spectrum from N to $2N$ samples by padding N zeros to obtain the padded sequence $X_{\text{padded}} = [X(0), \ldots, X((N+1)/2), 0, 0, \ldots, 0, X((N+1)/2), X(1), \ldots, X(N-1)]$. Note the padding is done in a manner to preserve the symmetric property of the Fourier transform of a real-valued signal. The inverse DFT of the padded signal gives an interpolated signal $x_i(m)$

$$x_i(m) = \frac{1}{2N} \sum_{k=0}^{2N-1} X_{\text{padded}}(k) e^{\mathrm{j}\frac{2\pi}{2N}mk} \quad m = 0, \ldots, 2N-1 \tag{2.124}$$

Noting that the padded zero-valued spectral components have no effect on the sum product of Equation (2.124), hence we may write Equation (2.124) as

$$x_i(m) = \frac{1}{2N} \sum_{k=0}^{N-1} X(k) e^{\mathrm{j}\frac{2\pi}{2N}mk} \quad m = 0, \ldots, 2N-1 \tag{2.125}$$

The $2N$-point signals given by Equation (2.125) consist of the N original samples plus N interpolated samples. For interpolation by a factor of M Equation (2.125) may be modified as

$$x(m) = \frac{1}{MN} \sum_{k=0}^{N-1} X(k) e^{\mathrm{j}\frac{2\pi}{MN}mk} \quad m = 0, \ldots, MN-1 \tag{2.126}$$

Matlab function DFT_TD_Interpolation( )
Demonstrates the time domain (TD) interpolation of a signal through zero-padding of its frequency spectrum followed by inverse Fourier transform. Before zero-padding the fftshift function is used to place the zero-frequency (dc) component at the centre of spectrum and to swap the left and right halves of the spectrum. The two ends of the spectrum are then zero-padded. The ifftshift function is used to swap back the spectrum and perform an inverse fft operation to yield the interpolated time domain signal. The function displays the original signal, its spectrum, zero-padded spectrum and the time domain interpolated signal.

2.9.3 Digital Filter Design

Given the required frequency response of a digital filter $H_{\mathrm{d}}(f)$, the inverse Fourier transform can be applied to obtain the impulse response $h_{\mathrm{d}}(m)$, and hence the coefficients, of an FIR filter implementation as

$$h_{\mathrm{d}}(m) = \int_{-F_s/2}^{F_s/2} H_{\mathrm{d}}(f) e^{\mathrm{j}2\pi fm}\, df \tag{2.127}$$

where F_s is the sampling rate. In Chapter 4 we consider the methods for digital filter design in some detail. The following example, and the MatLab program, illustrates the design of a digital audio bass filter using the inverse discrete Fourier transform equation.

2.9.4 Digital Bass for Audio Systems

A digital bass filter boosts the low-frequency part of the signal spectrum. The required frequency response of a typical bass booster is shown in Figure 2.30(a). The inverse DFT is used to obtain the impulse response corresponding to the specified frequency response. In theory the inverse frequency response has an infinite duration. A truncated version of the impulse response and its frequency spectrum are shown in Figure 2.30(a) and (b). Note that the frequency response of the filter obtained through IDFT is a good approximation of the specified frequency response.

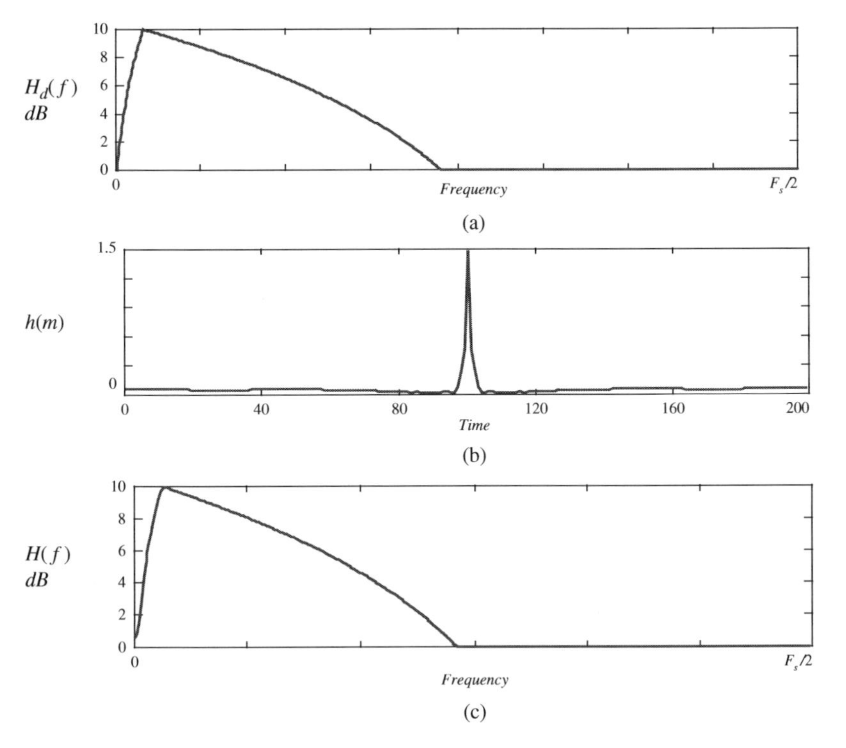

Figure 2.30 (a) The specified frequency response of a digital bass filter, (b) truncated inverse DFT of the frequency response specified in (a), and (c) the frequency response of the truncated impulse response.

2.9.5 Voice and Pitch Conversion

Another example of the application of Fourier transform is in pitch conversion. A simple method to change the pitch and voice characteristics is frequency warping. This follows time–frequency scaling relationship of the Fourier transform pair

$$x(\alpha t) \overset{\mathcal{F}}{\longleftrightarrow} \frac{1}{\alpha} X\left(\frac{f}{\alpha}\right) \tag{2.128}$$

Figure 2.31 shows a segment of a speech signal and its spectrum for three values of α : $\alpha = 1$, i.e. the original signal; $\alpha = 2$ corresponds to a compression in the time domain by a factor of 2:1 or equivalently an expansion in the frequency domain by a factor of 1:2; and $\alpha = 0.5$ corresponds to an expansion in the time domain by a factor of 1:2 or equivalently a compression in the frequency domain by a factor of 2:1. A familiar example of Equation (2.128) is when an audio signal is played back at a speed slower or faster than the recording speed. For $\alpha < 1$ the speech spectrum moves to the lower frequencies and the voice assumes a more male and gravel sounding quality, whereas for $\alpha > 1$ the voice shifts to the higher frequencies and becomes more feminine and higher pitched.

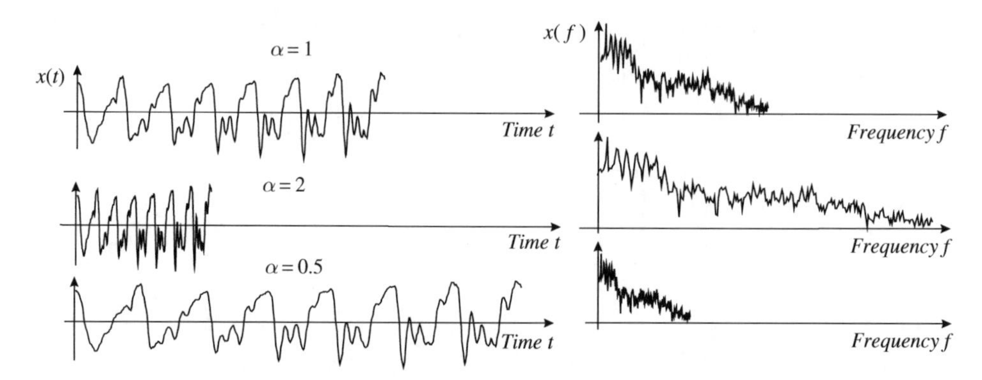

Figure 2.31 Illustration of a signal and its spectrum played back at twice and half the recording rate.

2.9.6 Radar Signal Processing: Doppler Frequency Shift

Figure 2.32 shows a simple sketch of a radar system which can be used to estimate the range and speed of an object such as a moving car or a flying aeroplane. Radar consists of a transceiver (transmitter/receiver) that generates and transmits sinusoidal pulses at microwave frequencies. The signal travels with the speed of light and is reflected from any object in its path. The analysis of the received echo provides such information as range, speed and acceleration. The received signal has the form

$$x(t) = A(t)\cos[\omega_0(t - 2r(t)/c)] \tag{2.129}$$

where $A(t)$ the time-varying amplitude of the reflected wave depends on the position and characteristics of the target, $r(t)$ is the time-varying distance of the object from the radar, $2r(t)/c$ is the round trip delay and the constant c is the velocity of light.

The time-varying distance of the object can be expanded in a Taylor's series form as

$$r(t) = r_0 + \dot{r}t + \frac{1}{2!}\ddot{r}t^2 + \frac{1}{3!}\dddot{r}t^3 + \cdots \tag{2.130}$$

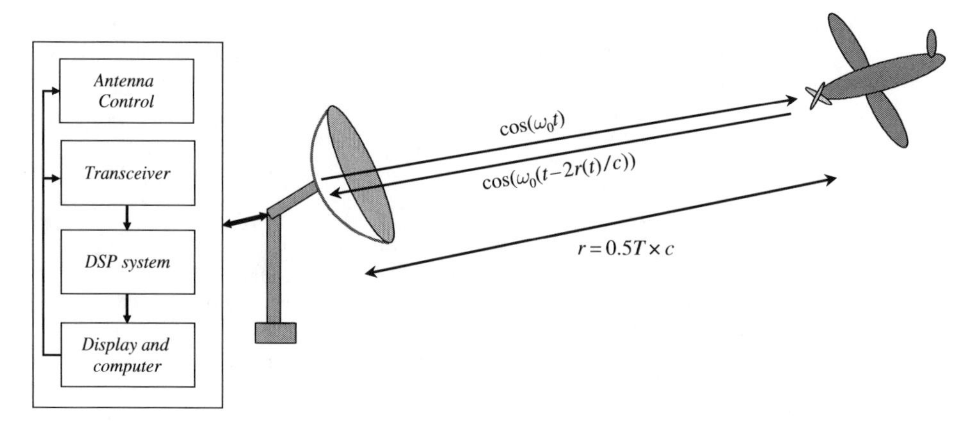

Figure 2.32 Illustration of a radar system.

where r_0 is the distance, $\dot{r}$ is the velocity, $\ddot{r}$ is the acceleration etc.

Approximating $r(t)$ with the first two terms of the Taylor's series expansion we have

$$r(t) \approx r_0 + \dot{r}t \tag{2.131}$$

Substituting Equation (2.131) in Equation (2.129) in yields

$$x(t) = A(t)\cos[(\omega_0 - 2\dot{r}\omega_0/c)t - 2\omega_0 r_0/c] \tag{2.132}$$

Note that the frequency of reflected wave is shifted by an amount of

$$\omega_d = 2\dot{r}\omega_0/c \tag{2.133}$$

This shift in frequency is known as the Doppler frequency. If the object is moving towards the radar, then the distance $r(t)$ is decreasing with time, $\dot{r}$ is negative and an increase in the frequency is observed. Conversely if the object is moving away from the radar then the distance $r(t)$ is increasing, $\dot{r}$ is positive and a decrease in the frequency is observed. Thus the frequency analysis of the reflected signal can reveal information on the direction and speed of the object. The distance r_o is given by

$$r_0 = 0.5T \times c \tag{2.134}$$

where T is the round-trip time for the signal to hit the object and arrive back at the radar and c is the velocity of light.

(a) (b)

633	59	40	31	25	19	13	8
51	39	32	28	23	18	12	8
36	32	29	26	23	18	13	9
29	27	26	25	23	18	14	9
23	22	22	21	19	17	12	9
17	17	17	17	17	14	11	7
11	11	12	12	11	10	8	6
5	5	5	6	6	6	4	4

(c)

Figure 2.33 (a) A picture of my son, (b) the picture was divided into 8×8 pixels blocks and reconstructed from the four lowest frequency DCT coefficients of each block, (c) the r.m.s. value of DCT coefficients of (8×8 pixels blocks) to the nearest integer averaged over all the blocks in the picture.

2.9.7 Image Compression in Frequency Domain

The discrete cosine transform is widely used in compression of still pictures and video. The DCT has the property that it compresses most of the signal energy into a relatively few low-frequency coefficients. The higher frequency coefficients can be discarded with little visible effect on the perceived quality of an image. This is illustrated in the following example. The image in Figure 2.33(a) was segmented into sub-blocks of $8 \times 8 = 64$ pixels (picture elements). Each block was transformed to frequency using DCT. For each block the lowest $5 \times 5 = 25$ coefficients were retained and other coefficients were simply discarded; this corresponds to a compression ratio of 64:25 or 2.6:1 or some 61% compression. The compressed image reconstructed via inverse DCT is shown in Figure 2.33(b). The error image Figure 2.33(c) is obtained by subtracting the original and the compressed image.

Figure 2.33(c) shows the root mean squared value of the DCT coefficients. This was obtained by taking the DTC of sub-blocks of $8 \times 8 = 64$ pixels of the image and then averaging the square of the 8×8 DCT values. This figure illustrates that in the DCT domain the energy is mostly concentrated in the lower indexed coefficients (i.e. in the top left corner of the table). Note particularly that the four DCT coefficients in the top left corner have the highest values.

2.10 Summary

This chapter began with a brief introduction to Joseph Fourier and his work. Starting with the complex Fourier series representation of a periodic signal, the Fourier signal analysis and synthesis methods were introduced. The Fourier transform of a finite energy aperiodic signal was derived as the Fourier series of a periodic version of the same signal but with an infinite 'off' duration. It was shown that sampling a signal, in time or space, renders its spectrum periodic. Similarly, sampling a signal in frequency renders a discrete spectrum which may be interpreted as the Fourier series of a periodic signal in time or space.

The discrete Fourier transform (DFT) was derived by sampling the Fourier transform in the frequency domain (which makes the signal periodic in time or space domain) and it was argued that the DFT is the same as the complex Fourier series. The important issues of time – frequency resolutions, spectral leakage, windowing and zero-padding for spectral interpolation were considered. The fast Fourier transform (FFT) was presented. The discrete cosine transform (DCT), widely used in speech and image processing, was introduced. The chapter concluded with several applications of Fourier transform.

Further Reading

Bracewell, R.N. (1965) The Fourier Transform and Its Applications, Mcgraw-Hill, New York.

Brault J.W. and White O.R. (1971) The Analysis and Restoration of Astronomical Data via the Fast Fourier Transform, Astron. & Astrophys., 13, pp. 169–189.

Brigham E.O. (1988) The Fast Fourier Transform and its Applications, Prentice-Hall, Englewood Cliffs, NJ.

Cooley J.W. and Tukey J.W. (1965) An Algorithm for the Machine Calculation of Complex Fourier Series, Mathematics of Computation, 19 (90), pp. 297–301.

Fourier J.B.J. (1878) Théorie Analytique de la Chaleur, Trans. A. Freeman; Repr. Dover Publications, 1952.

Grattam-Guiness I. (1972) Joseph Fourier (1768–1830): A Survey of His Life and Work, MIT Press.

Exercises

2.1 (i) Write the basis functions for complex Fourier series, (ii) state three useful properties of the Fourier basis functions and (iii) using the orthogonality principle derive the inverse DFT.

2.2 Explain why the Fourier transform has become the most applied tool in signal analysis and state three applications of the Fourier transform.

2.3 Draw a block diagram illustration of a spectrogram and label the main signal processing stage in each block. State the main parameters and choices regarding time – frequency resolution and windowing in spectrograms. State your choice of sampling rate, DFT time window length and the resulting time and frequency resolution for analysis of

(i) a speech signal with a bandwidth of 4 kHz
(ii) a music signal with a bandwidth of 20 kHz.

2.4 (i) A segment of N samples of a signal is padded with $3N$ zeros. Derive the DFT equation for the zero-padded signal. Explain how the frequency resolution of the DFT of a signal changes by zero-padding.
(ii) Assuming that a segment of 1000 samples of a signal is zero padded with 3000 extra zeros. Calculate the time resolution, the actual frequency resolution and the interpolated apparent frequency resolution. Assume the signal is sampled at 44,100 Hz.
(iii) Write a mathematical description of an impulse function and derive its Fourier transform and hence sketch its frequency spectrum.

2.5 Explain why the rectangular pulse and its frequency spectrum are so important in communication signal processing. Obtain and sketch the spectrum of the rectangular pulse given by

$$x(t) = \begin{cases} 1 & |t| \leq T/2 \\ 0 & \text{otherwise} \end{cases}$$

Find an expression relating the bandwidth of the main lobe of the spectrum to the pulse duration. Calculate the percentage of the pulse power concentrated in the main lobe of the pulse spectrum. Obtain the Fourier transform of a rectangular pulse of duration 1 microsecond and hence the bandwidth required to support data transmission at a rate of 1 megabit per second.

Find the main bandwidth of the spectrum of the rectangular pulse in terms of the pulse duration T.

2.6 State the minimum number of samples per cycle required to convert a continuous-time sine wave to a discrete-time sine wave.

Hence state the Nyquist–Shannon sampling theorem for sampling a continuous-time signal with a maximum frequency content of B Hz.

Calculate the bit rate per second and the bandwidth required to transmit a digital music signal sampled at a rate of 44,100 samples per second and with each sample represented with a 16-bit word.

2.7 Express the DFT equation in terms of a combination of a cosine transform and a sine transform. State the relation between DFT and complex Fourier series.

(i) What is the physical interpretation of the magnitude and phase of $X(k)$? Assuming a sampling rate of $F_s = 10\,\text{kHz}$, and a DFT length of $N = 256$, what is the actual frequency value corresponding to the discrete frequency k?
(ii) Obtain the length N of the input signal of the DFT to yield a frequency resolution of 40 Hz at a sampling rate of 8000 samples per second. What is the time resolution of the DFT?

2.8 Obtain the DFT of the following sequences

$$\boldsymbol{x} = [3, 1], \text{ and}$$

$$\boldsymbol{x}_p = [3, 1, 0, 0].$$

Quantify the improvements in the actual frequency resolution and the apparent frequency resolution of a DFT if N signal samples are padded by N zeros.

2.9 Find an expression for and sketch the frequency spectrum of a sine wave of angular frequency $\omega_0 = 2\pi/T_0$. Obtain an expression for the energy and power of this sine wave.

Obtain and sketch the frequency spectrum of the following signal

$$x(t) = 10\sin(1000\pi t) + 5\sin(3000\pi t) + 2\sin(4000\pi t)$$

Calculate the total signal power and the power at each individual frequency, and hence deduce Parseval's theorem.

2.10 Find the frequency spectrum of a periodic train of pulses with a period of 10 kHz and a pulse 'on' duration of 0.04 milliseconds. Write the formula for synthesising the signal up to the 10th harmonic, and plot a few examples for the increasing number of harmonics.

2.11 Find the spectrum of a shifted impulse function defined as

$$\delta(t - T) = \underset{\Delta \to 0}{\text{limit}}\, p(t) = \begin{cases} 1/\Delta & |t - T| \le \Delta/2 \\ 0 & |t - T| > \Delta/2 \end{cases}$$

and hence deduce the relationship between the time-delay and the phase of a signal.

2.12 Find the Fourier transform of a burst of N cycles of a sine wave of period T_0 seconds. You can model a burst of sine wave as an infinite duration signal multiplied by a rectangular window, and then employ the convolutional property of the Fourier transform for the product of two signals. Sketch the spectrum of the signal and show the effect of increasing the length of the signal in time on its frequency spectrum.

2.13 Find and sketch the frequency spectrum of the following signals.

(a) $x(t) = \delta(t)$

(b) $x(t) = \delta(t - k) + \delta(t + k)$

(c) $x(t) = \sum_{k=-\infty}^{\infty} \delta(t - kT_s)$

2.14 Assume that a signal $x(t)$ has a band limited spectrum $X(f)$ with the highest frequency content of B Hz, and that $x(t)$ is sampled periodically with a sampling interval of T_s seconds.

(a) Show that the frequency spectrum of the sampled signal $x_s(t)$ is periodic with a period of $F_s = 1/T_s$ Hz, and

(b) hence deduce, and state, the Nyquist theorem.

2.15 Show that a time-delayed signal $x(m - m_0)$ can be modelled as the convolution of $x(m)$ and a delayed delta function $\delta(m - m_0)$. Hence, using the Fourier transform, show that a time delay of m_0 is equivalent to a phase shift of $e_0^{-j2\pi fm}$.

2.16 Derive the discrete Fourier transform (DFT) and the inverse discrete Fourier transform (IDFT) equations. State the fundamental assumption in derivation of the DFT equations.

Define the terms frequency resolution and time resolution and write the expression for the frequency resolution of the DFT.

2.17 Write the DFTs of the sequence $\{x(0), x(1), x(2), x(3)\}$ and its zero-padded version $\{x(0), x(1), x(2), x(3), 0, 0, 0, 0\}$. Discuss the frequency-interpolating property of the zero-padding process as indicated by the DFT of the zero-padded sequence.

2.18 The Fourier transform of a sampled signal is given by

$$X(f) = \sum_{m=0}^{N-1} x(m)e^{-j2\pi fm}$$

Using the above equation, prove that the spectrum of a sampled signal is periodic, and hence state the sampling theorem.
A sinusoidal signal of frequency 10 kHz, $\sin(2\pi 10^4 t)$, is sampled, incorrectly, at a rate of 15 kHz. Sketch the frequency components of the sampled signal in the range 0 to 30 kHz.

2.19 An analogue signal $x(t)$ with the spectrum $X(f)$ sampled originally at a rate of 10 kHz is to be converted to a sampling rate of 40 kHz. In preparation for digital interpolation, the sampled signal is zero-inserted by a factor of three zeros in-between every two samples.

Derive an expression for the spectrum of the zero-inserted signal in terms of the original spectrum $X(f)$.

Sketch the spectrum of the zero-inserted signal, and hence state the cut-off frequency of the interpolation filter.

2.20 (i) The discrete Fourier transform (DFT) equation is given by

$$X(k) = \sum_{m=0}^{N-1} x(m)e^{-j\frac{2\pi}{N}mk}$$

Explain the effects of finite length data on the DFT spectrum of the signal, and suggest a solution.
(ii) Calculate the DFT of the sequence $[1, 1]$ and its zero-padded version $[1, 1, 0, 0]$. Discuss the interpolating property of the zero-padding operation as indicated by the DFT of the zero-padded sequence.

2.21 A signal is sampled at a rate of 20,000 samples per second. Calculate the frequency resolution of the DFT, in units of Hz, when

(i) a signal window of 500 samples is used as input to the DFT
(ii) the number of signal samples is increased from 500 to 1000
(iii) the number of signal samples 500 is augmented (padded) with 500 zeros.

2.22 Write the equations for, and explain the relations between, the Fourier and Laplace transforms.

Using the Fourier transform integral, and the superposition and time delay properties, obtain and sketch the spectrum of the pulse, shown in the figure below, of duration $T = 10^{-6}$ seconds.

Calculate the bandwidth of the main lobe of the spectrum of the pulse where most of the pulse energy is concentrated.

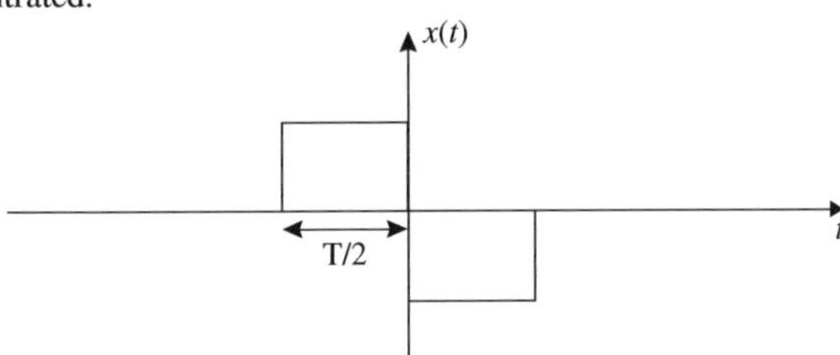

2.23 A DFT is used as part of a digital signal processing system for the analysis of an analogue signal with significant frequency content of up to 1 MHz. Calculate

(i) the minimum sampling rate F_s required, and
(ii) the number of samples in the DFT window required to achieve a frequency resolution of 2 kHz at the minimum sampling rate. Explain the effects of finite-length data on the DFT spectrum of a signal, and suggest a solution.

3 z-Transform

z-Transform is a frequency domain tool for the analysis of the transient and the steady-state response of discrete-time filters and systems. The z-transform is the discrete-time counterpart of the Laplace transform and hence a generalisation of the Fourier transform of a discrete-time signal. Like Laplace transform for continuous-time systems, the z-transform allows insight into the transient behaviour, the steady state behaviour, and the stability of discrete-time models and systems. A working knowledge of the z-transform is essential for the design and analysis of digital filters described in Chapter 4.

This chapter begins with the definition of Laplace transform and derives the z-transform equation from the Laplace transform of a discrete-time signal. A useful aspect of the Laplace and the z-transforms are the representation of a system in a complex plane (s-plane, z-plane) in terms of the locations of the poles and the zeros (i.e. the roots of the polynomial) of the system transfer function. In this chapter the z-plane, and its associated unit circle, are derived from sampling the s-plane of the Laplace transform. We study the representation of a system in terms of the system transfer function. The roots of the transfer function equation, known as the poles and the zeros of the transfer function, provide useful insight into the response and stability of the system. Several examples illustrating the physical significance of poles and zeros and their effect on the impulse and frequency response of a system are considered.

3.1 Introduction

The z-transform is a frequency domain method for representation, analysis and synthesis of signals and systems. The z-transform for discrete-time signals is the counterpart of the Laplace transform for continuous-time signals. The relationships between the z-transform, the Laplace and the Fourier transforms are illustrated in Figure 3.1 and explained in the following sections. These transforms are indispensable mathematical tools in the design and analysis of signal processing systems, and for prediction and monitoring of the stability of a system.

A working knowledge of the z-transform is essential for understanding the behaviour of discrete-time filters and systems. It is partly through the use of such transforms that we can formulate a closed-form mathematical description of the relationship between the input and the output of a

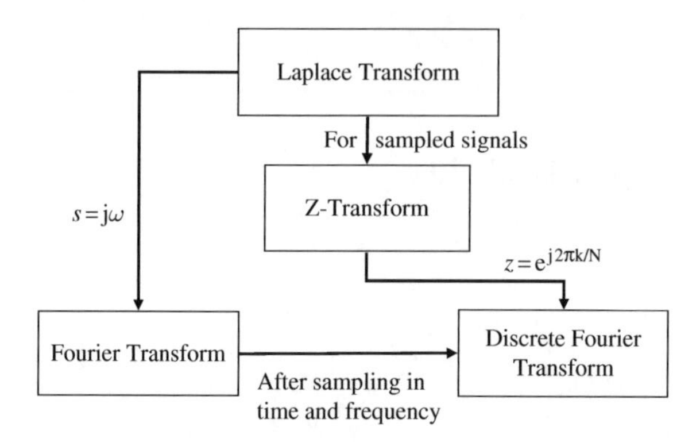

Figure 3.1 Illustration of the relationships between Laplace transform, z-transform and Fourier transform. For discrete-time signals Laplace transform becomes z-transform.

system in the frequency domain; use the mathematical description to design the system; and then analyse and predict the stability, the transient response and the steady state characteristics of the system.

The z-transform describes a signal in terms of a power series of a complex exponential variable, z. The z-transform was first introduced in 1958 by E.I. Jury. The use of the letter z to name a transform is similar to the use of the letter s in s-transform which is another name for the Laplace transform. In fact z-transform is based on Laurent series which is an alternative to Taylor series. In Taylor series a signal is expressed in terms of a combination of its differentials. In contrast, in frequency analysis methods such as Fourier, Laplace and z-transform a signal is modelled in terms of a combination of complex exponentials as explained in this chapter.

A mathematical description of the relationship between the input and the output of a system can be formulated either in the time domain or in the frequency domain using Laplace transform, Fourier transform or z-transform. Time-domain and frequency domain representation methods offer alternative insights into a system, and depending on the application at hand it may be more convenient to use one method in preference to another.

Time domain system analysis methods are based on differential equations which describe the system output as a weighted combination of the differentials (i.e. the rates of change with time) of the system input and output signals. In contrast the frequency domain methods, mainly the Laplace transform, the Fourier transform and the z-transform, describe a system in terms of its response to the individual frequency constituents of the input signal.

In Section 3.2.1 we explore the close relationships between the Laplace, the Fourier and the z-transforms, and observe that all these transforms employ various forms of complex exponentials as their basis functions. The description of a system in the frequency domain can reveal valuable insight into the behaviour and stability of the system in response to different input frequencies.

Frequency domain signal processing and system analysis offers some particular advantages as through transformation to frequency domain differential equations in time domain are reduced to relatively simple algebraic equations in frequency domain. Differentiation and integration operations in time domain are equivalent, in the frequency domain, to multiplication and division by the complex frequency variable respectively. Furthermore, the frequency components of the system can be treated independently and the transient and the steady state characteristics of a system can be predicted by analysing the roots of the Laplace transform or the z-transform equations, the so-called poles and zeros of a system.

3.2 Derivation of the z-Transform

The z-transform is the discrete-time form of the Laplace transform. In this section the z-transform is derived simply as the Laplace transform of a discrete-time signal. The Laplace transform $X(s)$, of a continuous-time signal $x(t)$, is given by the integral

$$X(s) = \int_{0}^{\infty} x(t)e^{-st}dt \tag{3.1}$$

where the complex variable $s = \sigma + j\omega$. Note that the Laplace basis functions $e^{st} = e^{\sigma t}e^{j\omega t}$ are composed of sinusoids with exponentially rising/decaying envelopes.

The inverse Laplace transform is defined by a *line integral*, also called the Bromwich integral or the Fourier–Mellin integral, defined as

$$x(t) = \frac{1}{2\pi j}\int_{\sigma_1 - j\infty}^{\sigma_1 + j\infty} X(s)e^{st}ds \tag{3.2}$$

where σ_1 is selected so that $X(s)$ is analytic (no singularities) for $s > \sigma_1$.

The z-transform can be derived from Equation (3.1) by sampling the continuous-time signal $x(t)$. For a sampled signal $x(mT_s)$, normally denoted as $x(m)$ by assuming the sampling period $T_s = 1$, the Laplace transform becomes

$$X(e^s) = \sum_{m=0}^{\infty} x(m)e^{-sm} \tag{3.3}$$

Note that we have replaced $X(s)$ with $X(e^s)$ and we can now substitute the variable e^s in Equation (3.3) with the variable z, $z = e^s$, to obtain the one-sided z-transform equation as

$$X(z) = \sum_{m=0}^{\infty} x(m)z^{-m} \tag{3.4}$$

The two-sided z-transform is defined as

$$X(z) = \sum_{m=-\infty}^{\infty} x(m)z^{-m} \tag{3.5}$$

Note that for a one-sided signal, i.e. $x(m) = 0$ for $m < 0$, Equations (3.4) and (3.5) are equivalent.

3.2.1 Relationships Between Laplace, Fourier, and *z*-Transforms

The Laplace transform, the Fourier transform and the z-transform are closely related *frequency analysis* methods that employ complex exponentials as their basis functions. As shown in Figure 3.1, the Laplace transform is a generalisation of the Fourier transform of a continuous-time signal, and the z-transform is a generalisation of the Fourier transform of a discrete-time signal.

In the previous section it is shown that the z-transform can be derived as the Laplace transform of a discrete-time signal. In the following the relation between the z-transform and the Fourier transform is explored. Using the equation

$$z = e^s = e^{\sigma}e^{j\omega} = re^{j2\pi f} \tag{3.6}$$

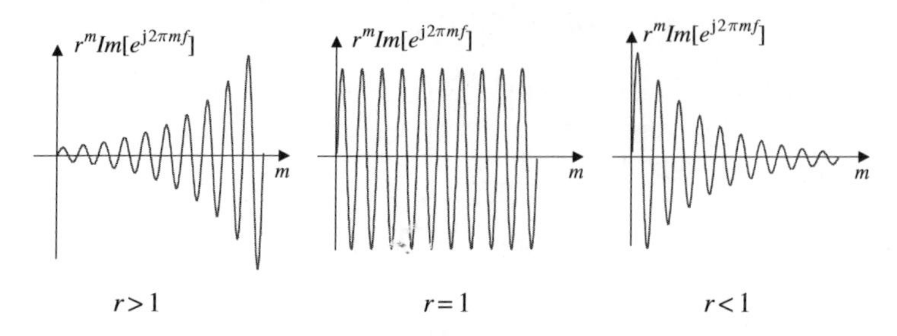

Figure 3.2 The z-transform basis functions are sinusoids with exponential envelope.

where $s=\sigma+j\omega$ and $\omega=2\pi f$, we can rewrite the z-transform equation (3.4) in the form

$$X(z)=\sum_{m=0}^{\infty} x(m)r^{-m}e^{-j2\pi mf} \tag{3.7}$$

Note that when $r=e^{\sigma}=1$ (i.e. $\sigma=0$ on $j\omega$ axis) the z-transform becomes the Fourier transform of a sampled signal given by

$$X(z=e^{-j2\pi f})=\sum_{m=0}^{\infty} x(m)e^{-j2\pi fm} \tag{3.8}$$

Therefore, the z-transform is a generalisation of the Fourier transform of a sampled signal. Like Laplace transform, the basis functions for the z-transform are damped, or growing, sinusoids of the form $z^m=e^{sm}=e^{\sigma m}e^{j\omega m}=r^m e^{j2\pi fm}$ as shown in Figure 3.2. These signals are particularly suitable for transient signal analysis. In contrast, the Fourier basis functions are steady complex sinusoids, $e^{j2\pi fm}$, of time-invariant amplitudes, suitable for steady state or time-invariant signal analysis.

A similar relationship exists between the Laplace transform and the Fourier transform of a continuous time signal. The Laplace transform is a one-sided transform with the lower limit of integration at $t=0^-$, whereas the Fourier transform equation (3.21) is a two-sided transform with the lower limit of integration at $t=-\infty$. However for a one-sided signal, which is zero-valued for $t<0^-$, the limits of integration for the Laplace and the Fourier transforms are identical. In that case if the variable s in the Laplace transform is replaced with the frequency variable $j2\pi f$ then the Laplace integral becomes the Fourier integral. Hence for a one-sided signal, the Fourier transform is a special case of the Laplace transform corresponding to $s=j2\pi f$ and $\sigma=0$.

Example 3.1 Show that the Laplace transform of a sampled signal is periodic with respect to the frequency axis $j\omega=j2\pi f$ of the complex variable $s=\sigma+j\omega$.

Solution: In Equation (3.3) substitute $s+jk2\pi$ where k is an integer variable, for the frequency variable s to obtain

$$\begin{aligned} X(e^{s+jk2\pi}) &= \sum_{m=0}^{\infty} x(m)e^{-(s+jk2\pi)m} = \sum_{m=0}^{\infty} x(m)e^{-sm}\underbrace{e^{-jk2\pi m}}_{=1} \\ &= \sum_{m=0}^{\infty} x(m)e^{-sm} = X(e^{s}) \end{aligned} \tag{3.9}$$

Hence the Laplace transform of a sample signal is periodic with a period of 2π as shown in Figure 3.3(a).

3.3 The z-Plane and the Unit Circle

The frequency variables $s = \sigma + j\omega$ of the Laplace transform and $z = re^{j\omega}$ of the z-transform are complex variables with real and imaginary parts and can be visualised in a two-dimensional plane. Figure 3.3(a) and (b) shows the s-plane of Laplace transform and the z-plane of z-transform. In the s-plane the vertical $j\omega$ axis is the frequency axis (this axis is also the location of Fourier transform) and the horizontal σ axis gives the exponential rate of decay or growth of the amplitude of the complex sinusoid as also shown in Figure 3.2.

As shown in Example 3.1 when a signal is sampled its Laplace transform, and hence the s-plane, becomes periodic with respect to the $j\tilde{\omega}$ axis. This is illustrated in Figure 3.3(a) by the periodic horizontal dashed lines. Periodic processes repeat with a period of 2π and hence can be conveniently represented using a circular polar diagram such as the z-plane and its associated unit circle.

Now imagine bending the $j\omega$ axis of the s-plane of the sampled signal of Figure 3.3(a) in the direction of the left-hand half of the s-plane to form a circle such that the points π and $-\pi$ meet. The resulting circle is called the *unit circle*, and the resulting diagram is called the z-plane. The area to the left of the s-plane, i.e. for $\sigma < 0$ or $r = e^{\sigma} < 1$, is mapped into the area inside the unit circle; this is the region of stable signals and systems which can be represented by a combination of damped exponential sinusoids. The area to the right of the s-plane, $\sigma > 0$ or $r = e^{\sigma} > 1$, is mapped onto the outside of the unit circle; this is the region of unstable signals and systems. The $j\tilde{\omega}$ axis, with $\sigma = 0$ or $r = e^{\sigma} = 1$, is itself mapped onto the unit circle line. Hence the Cartesian co-ordinates used in the s-plane for continuous time signals, Figure 3.3(a), are mapped into a polar co-ordinates in the z-plane for discrete-time signals, Figure 3.3(b). Figure 3.4 illustrates that an angular distance of 2π, i.e. once round the unit circle, corresponds to a frequency of F_s Hz where F_s is the sampling frequency. Hence a frequency of f Hz corresponds to an angle ϕ given by

$$\phi = \frac{2\pi}{F_s} f \qquad \text{radians} \tag{3.10}$$

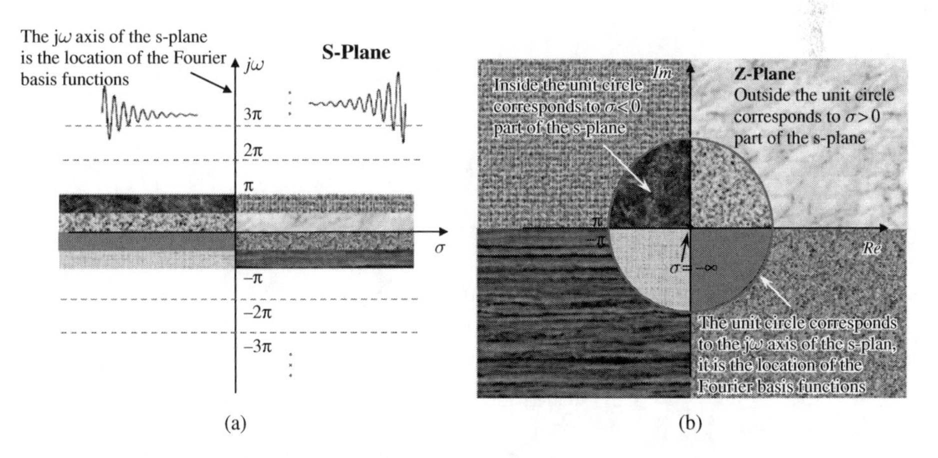

Figure 3.3 Illustration of (a) the s-plane of a discrete-time signal, note that consequent to sampling the s-plane becomes periodic, and (b) the periodicity of s-plane is modelled by the z-plane. Note that the $j\omega$ axis of s-plane (the locus of Fourier transform) is mapped onto the unit circle of the z-plane at the radius of 1 and the areas to its left and right are mapped into and outside the unit circle respectively.

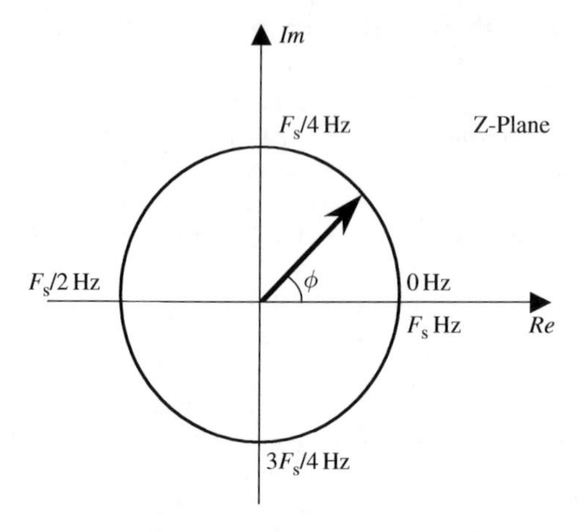

Figure 3.4 Illustration of mapping a frequency of f Hz to an angle of ϕ radians.

For example at a sampling rate of $F_s = 40\,\text{kHz}$, a frequency of $5\,\text{kHz}$ corresponds to an angle of $2\pi \times 5/40 = 0.25\pi$ radians or 45 degrees, a signal

3.3.1 The Region of Convergence (ROC)

Since the z-transform is an infinite power series, it exists (here the word *exists* means it is *computable*) only for those values of the variable z for which the series converges to a finite sum. *The region of convergence (ROC) of X(z) is the set of all the values of z for which X(z) attains a finite computable value.*

Example 3.2 Determine the z-transform, the region of convergence, and the Fourier transform of the following signal.

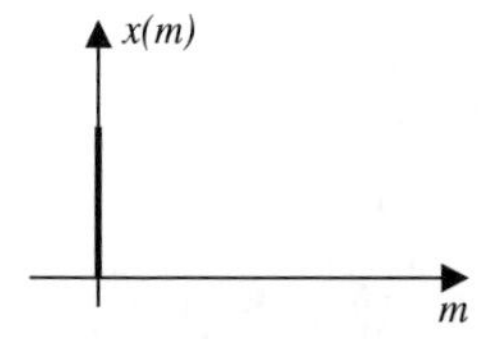

$$x(m) = \delta(m) = \begin{cases} 1 & m = 0 \\ 0 & m \neq 0 \end{cases} \tag{3.11}$$

Solution: Substituting for $x(m)$ in the z-transform equation (3.4) we obtain

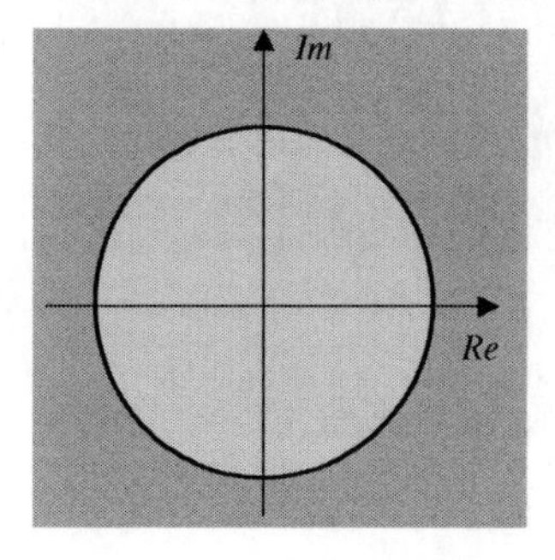

$$X(z) = \sum_{m=-\infty}^{\infty} x(m)z^{-m} = \delta(0)z^0 = 1 \tag{3.12}$$

For all values of the variable z we have $X(z) = 1$, hence, as shown by the shaded area of the figure on the left, the region of convergence is the entire z-plane. The Fourier transform of $x(m)$ may be obtained by evaluating $X(z)$ in Equation (3.12) at $z = e^{j\omega}$ as

$$X(e^{j\omega}) = 1 \tag{3.13}$$

Example 3.3 Determine the z-transform, the region of convergence, and the Fourier transform of the following signal.

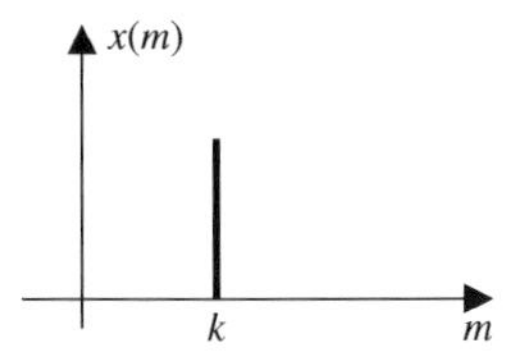

$$x(m) = \delta(m-k) = \begin{cases} 1 & m = k \\ 0 & m \neq k \end{cases} \tag{3.14}$$

Solution: Substituting for $x(m)$ in the z-transform equation (3.4) we obtain

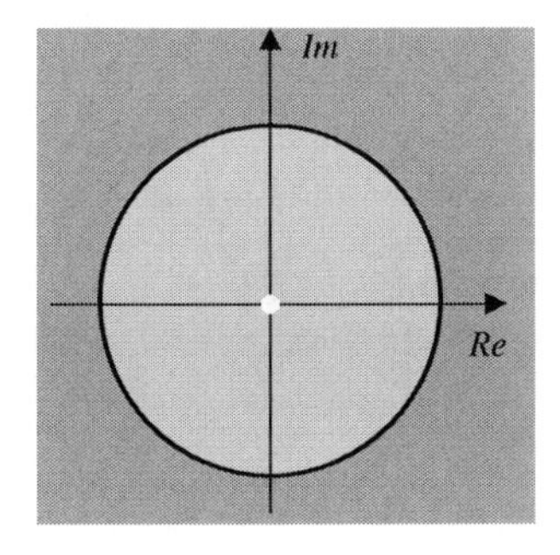

$$X(z) = \sum_{m=-\infty}^{\infty} \delta(m-k)z^{-m} = z^{-k} \tag{3.15}$$

The z-transform is $X(z) = z^{-k} = 1/z^k$. Hence $X(z)$ is finite-valued for all the values of z except for $z = 0$. As shown by the shaded area in the figure on the left, the region of convergence is the entire z-plane except the point $z = 0$. The Fourier transform is obtained by evaluating $X(z)$ in Equation (3.15) at $z = e^{j\omega}$ as

$$X(e^{j\omega}) = e^{-j\omega k} \tag{3.16}$$

Example 3.4 Determine the z-transform, the region of convergence, and the Fourier transform of the following signal.

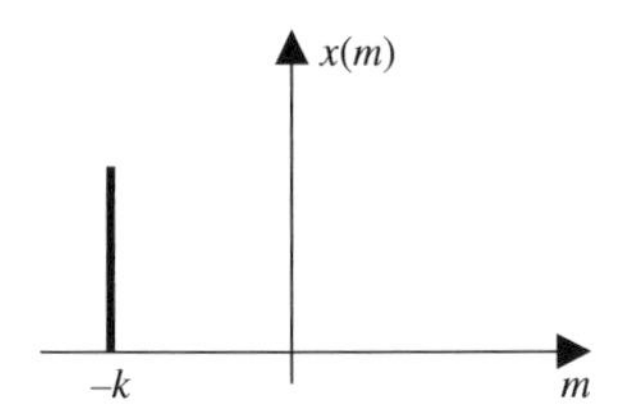

$$x(m) = \delta(m+k) = \begin{cases} 1 & m = -k \\ 0 & m \neq -k \end{cases} \tag{3.17}$$

Solution: Substituting for $x(m)$ in the z-transform equation (3.4) we obtain

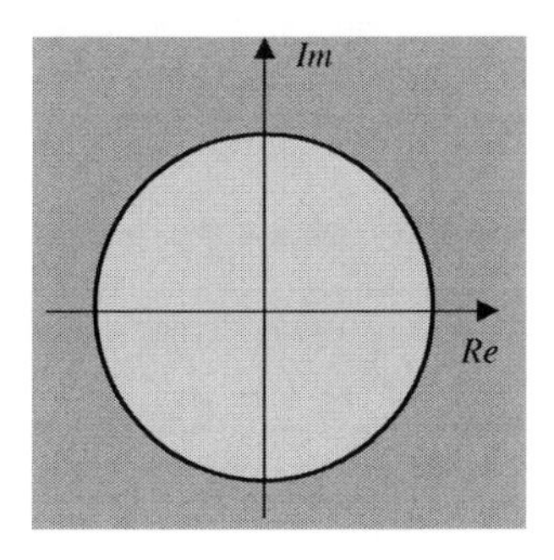

$$X(z) = \sum_{m=-\infty}^{\infty} x(m)z^{-m} = \sum_{m=-\infty}^{\infty} \delta(m+k)z^{-m} = z^{k} \tag{3.18}$$

The z-transform is $X(z) = z^k$. Hence $X(z)$ is finite-valued for all the values of z except for $z = \infty$. As shown by the shaded area in the figure on the left, the region of convergence is the entire z-plane except the point $z = \infty$ which is not shown! The Fourier transform is obtained by evaluating $X(z)$ in Equation (3.18) at $z = e^{j\omega}$ as

$$X(e^{j\omega}) = e^{j\omega k} \tag{3.19}$$

Example 3.5 Determine the z-transform, the region of convergence, and the Fourier transform of the following signal.

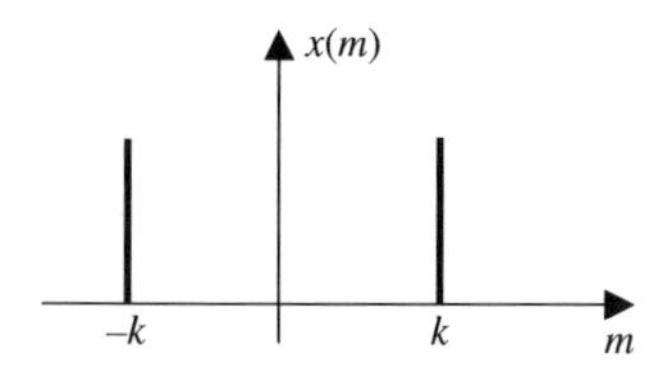

$$x(m) = \delta(m+k) + \delta(m-k) = \begin{cases} 1 & m = \pm k \\ 0 & m \neq \pm k \end{cases} \tag{3.20}$$

Solution: Substituting for $x(m)$ in the z-transform equation (3.4) we obtain

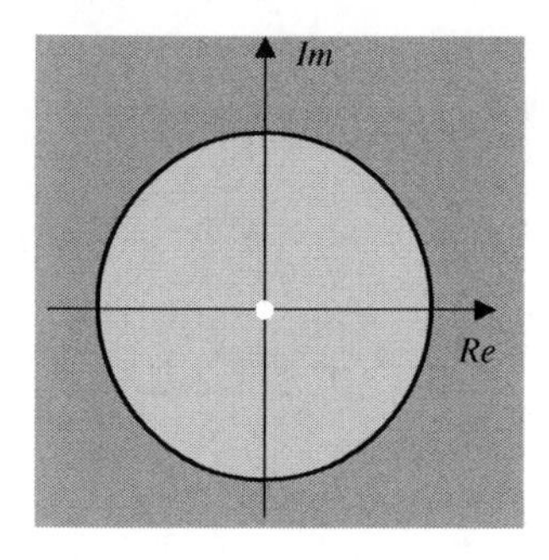

$$X(z) = \sum_{m=-\infty}^{\infty} (\delta(m+k) + \delta(m-k))\, z^{-m} = z^{k} + z^{-k} \tag{3.21}$$

Hence $X(z)$ is finite-valued for all the values of z except for $z = 0$ and $z = \infty$. As shown by the shaded area in the figure on the left, the region of convergence is the entire z-plane except the points $z = 0$ and $z = \infty$. The Fourier transform is obtained by evaluating Equation (3.21) at $z = e^{j\omega}$ as

$$X(e^{j\omega}) = e^{-j\omega k} + e^{+j\omega k} = 2\cos(\omega k) \tag{3.22}$$

Example 3.6 Determine the z-transform and the region of convergence of the following exponentially decaying transient signal.

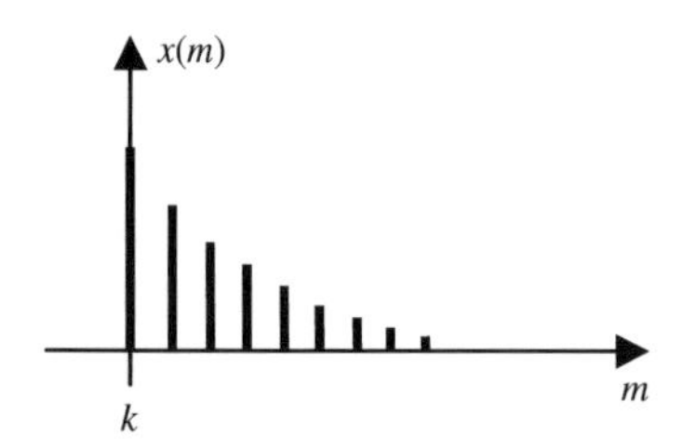

$$x(m) = \begin{cases} \alpha^{m} & m \geq 0 \\ 0 & m < 0 \end{cases} \tag{3.23}$$

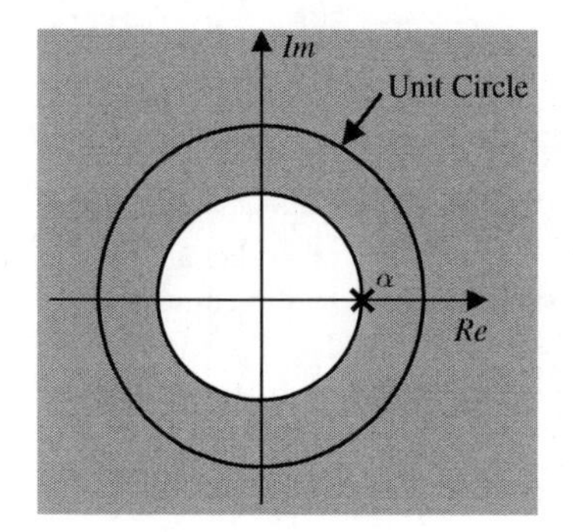

Solution: Note that as the signal is exponentially decaying $\alpha < 1$. Substituting for $x(m)$ in the z-transform equation (3.4) we obtain

$$X(z) = \sum_{m=-\infty}^{\infty} x(m) z^{-m} = \sum_{m=0}^{\infty} \alpha^{m} z^{-m} = \sum_{m=0}^{\infty} \left(\alpha z^{-1}\right)^{m} \tag{3.24}$$

This infinite power series converges only if $|\alpha z^{-1}| < 1$. Therefore the ROC is $|z| > |\alpha|$. As shown by the shaded area in the figure on the left, the region of convergence excludes a disc of radius α.

Example 3.7 Determine the z-transform and region of convergence of the following left-sided sequence.

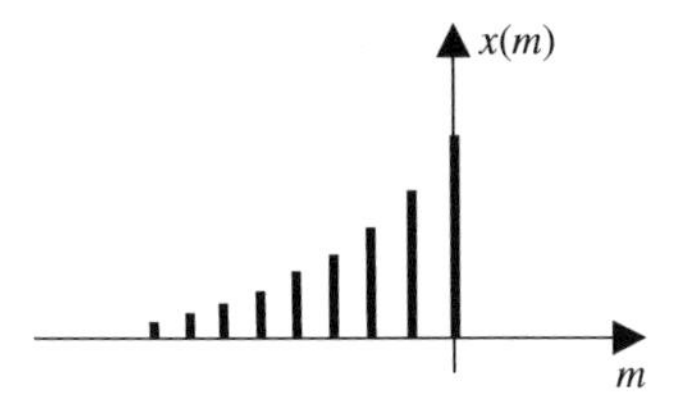

$$x(m) = \begin{cases} 0 & m > 0 \\ \alpha^m & m \le 0 \end{cases} \quad (3.25)$$

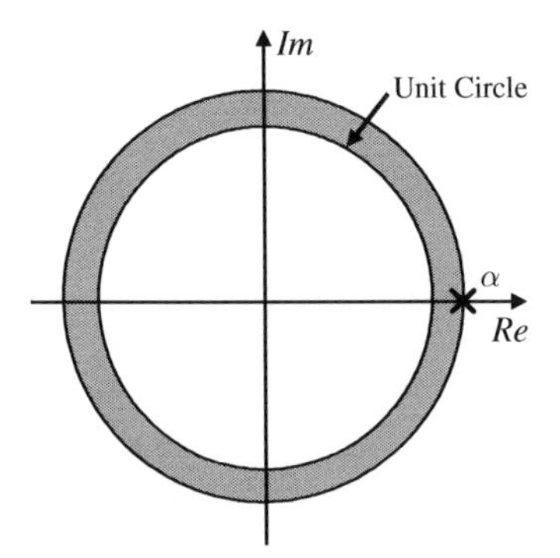

Solution: Substituting for $x(m)$ in the z-transform equation (3.4) we obtain

$$X(z) = \sum_{m=-\infty}^{\infty} x(m)z^{-m} = \sum_{m=-\infty}^{0} \alpha^m z^{-m} = \sum_{m=0}^{\infty} \left(\alpha^{-1} z\right)^m \quad (3.26)$$

Note that for $\left(\alpha^{-1}\right)^m$ to be a decaying sequence $|\alpha| > 1$. This infinite power series converges only if $\left|\alpha^{-1}z\right| < 1$. Therefore the ROC is $|z| < |\alpha|$. As shown by the shaded area of the figure, the region of convergence is confined to a disc of radius α.

Example 3.8 Determine the z-transform and region of convergence of the following left-sided sequence.

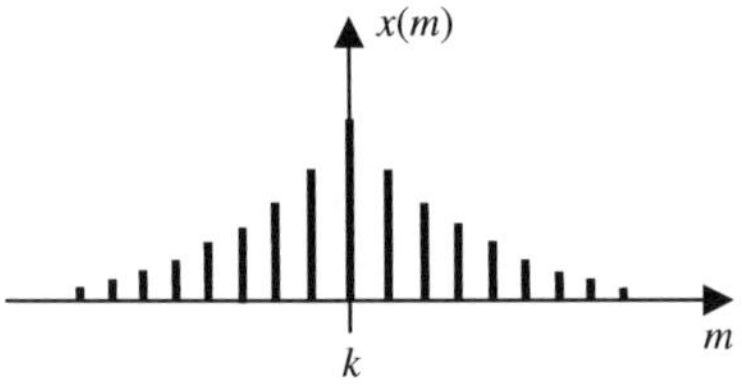

$$x(m) = \begin{cases} \alpha^m & m \ge 0 \\ \beta^m & m < 0 \end{cases} \quad (3.27)$$

Solution: Substituting for $x(m)$ in the z-transform equation (3.4) we obtain

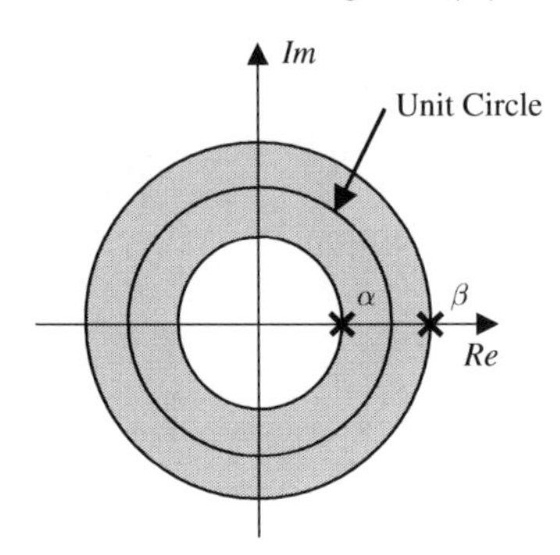

$$X(z) = \sum_{m=-\infty}^{\infty} x(m)z^{-m} = \sum_{m=0}^{\infty} \alpha^m z^{-m} + \sum_{m=-\infty}^{-1} \beta^m z^{-m} \quad (3.28)$$

$$X(z) = \underbrace{\sum_{m=0}^{\infty} \left(\alpha z^{-1}\right)^m}_{\text{Causal}} + \underbrace{\sum_{m=1}^{\infty} \left(\beta^{-1} z\right)^m}_{\text{Anti-causal}} \quad (3.29)$$

This infinite power series converges only if $\left|\alpha z^{-1}\right| < 1$ and $\left|\beta^{-1}z\right| < 1$. As shown by the shaded area on the left-hand figure, the region of convergence corresponds to the area of $\alpha < z < \beta$. Note that in this example a causal signal is a decaying exponential and an anti-causal signal is a rising exponential.

3.4 Properties of z-Transform

As z-transform is a generalisation of the Fourier transform of a sampled signal it has similar properties to the Fourier transform as described in the following.

Linearity

Given two signals $x_1(m)$ and $x_2(m)$ and their z-transforms $X_1(z)$ and $X_2(z)$ respectively

$$x_1(m) \overset{z}{\Leftrightarrow} X_1(z) \tag{3.30}$$

and

$$x_2(m) \overset{z}{\Leftrightarrow} X_2(z) \tag{3.31}$$

The linearity implies that for a linear combination of $x_1(m)$ and $x_2(m)$ we have

$$a_1 x_1(m) + a_2 x_2(m) \overset{z}{\Leftrightarrow} a_1 X_1(z) + a_2 X_2(z) \tag{3.32}$$

Equation (3.32) is known as the *superposition principle.*

Example 3.9 Given the following two signals

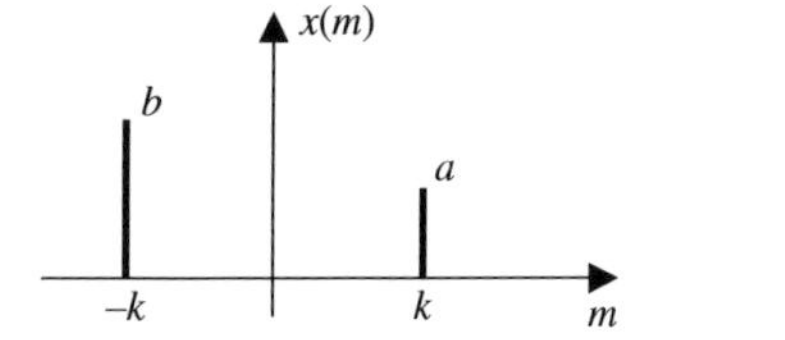

$$x_1(m) = \delta(m-k) \tag{3.33}$$

$$x_2(m) = \delta(m+k) \tag{3.34}$$

Determine the z-transform of

$$x(m) = a x_1(m-k) + b x_2(m+k) \tag{3.35}$$

Solution: Substituting for $x(m)$ in the z-transform equation (3.4) we obtain

$$\begin{aligned} X(z) &= \sum_{m=-\infty}^{\infty} (a\delta(m-k) + b\delta(m+k))\, z^{-m} \\ &= \sum_{m=-\infty}^{\infty} a\delta(m-k) z^{-m} + \sum_{m=-\infty}^{\infty} b\delta(m+k) z^{-m} = a z^{-k} + b z^{+k} \end{aligned} \tag{3.36}$$

It is clear from the second line of the above solution that the z-transform of the combination of two time domain signals $x(m) = x_1(m) + x_2(m)$ can be written as the sum of the z-transforms of the individual signals $x_1(m)$ and $x_2(m)$.

Time Shifting or Delay

The variable z can also be interpreted as a time delay operator (see Figure 3.5) or equivalently a phase changer; z^{-k} corresponds to a time delay of k sampling interval or phase change of $e^{-j\omega kT}$. Given a

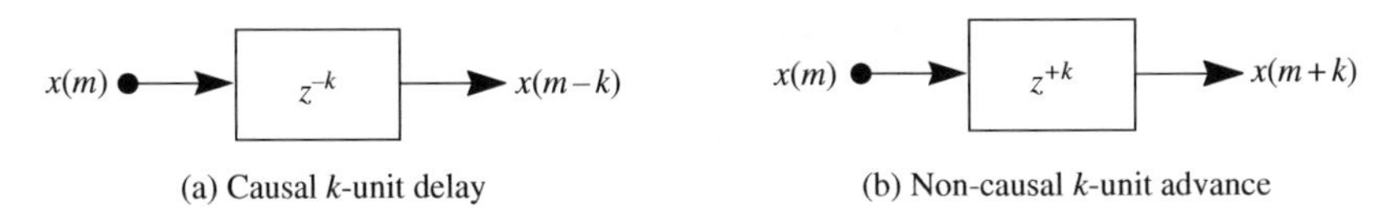

(a) Causal k-unit delay (b) Non-causal k-unit advance

Figure 3.5 Illustration of the variable z^{-k} as a k-unit delay operator.

signal $x(m)$ and its z-transform $X(z)$

$$x(m) \overset{z}{\Leftrightarrow} X(z)$$

for $x(m-k)$ we have

$$x(m-k) \overset{z}{\Leftrightarrow} z^{-k}X(z) \tag{3.37}$$

This property can be proved by taking the z-transform of $x(m-k)$

$$X(z) = \sum_{m=-\infty}^{\infty} x(m-k)z^{-m} = \sum_{n=-\infty}^{\infty} x(n)z^{-(n+k)} = z^{-k}\sum_{n=-\infty}^{\infty} x(n)z^{-n} = z^{-k}X(z) \tag{3.38}$$

where we have made a variable substitution $n = m - k$. Hence, the effect of a time shift by k sampling-interval units is equivalent to multiplication of the z-transform by z^{-k}. *Note that z^{-1} delays the signal by 1 unit and z^{-k} by k units, and z^{+1} is a non-causal time advance unit and z^{+k} advances a signal in time by k units.*

Example 3.10 Determine the z-transform and region of convergence of a time-delayed version of Example 3.6 given as

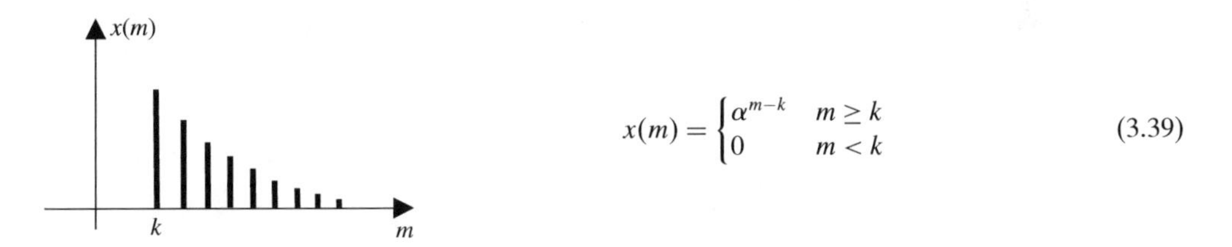

$$x(m) = \begin{cases} \alpha^{m-k} & m \geq k \\ 0 & m < k \end{cases} \tag{3.39}$$

Solution: Substituting for $x(m)$ in the z-transform equation (3.4) we obtain

$$\begin{aligned} X(z) &= \sum_{m=-\infty}^{\infty} x(m)z^{-m} = \sum_{m=k}^{\infty} \alpha^{m-k}z^{-m} = z^{-k}\sum_{m=k}^{\infty} \alpha^{m-k}z^{-(m-k)} \\ &= z^{-k}\sum_{m=0}^{\infty} \alpha^{m}z^{-m} = z^{-k}\sum_{m=0}^{\infty} \left(\alpha z^{-1}\right)^{m} \end{aligned} \tag{3.40}$$

This infinite power series converges only if $|\alpha z^{-1}| < 1$ and $z \neq 0$. Therefore the ROC is $|z| > |\alpha|$ which excludes $z = 0$ as required.

Multiplication by an Exponential Sequence (Frequency Translation)

The z-transform relation for the product of a signal $x(m)$ and the exponential sequence z_0^m is given by

$$z_0^m x(m) \overset{z}{\Leftrightarrow} X(z/z_0) \tag{3.41}$$

This prope:ty can be shown by substituting $z_0^m x(m) \overset{z}{\Leftrightarrow} X(z/z_0)$ in the z-transform equation

$$\sum_{m=-\infty}^{\infty} x(m) z_0^m z^{-m} = \sum_{m=k}^{\infty} x(m)(z/z_0)^{-m} = X(z/z_0) \tag{3.42}$$

Note that for the case when $z = e^{j\omega}$ and $z_0 = e^{j\omega_0}$, we have the frequency modulation equation

$$e^{j\omega_0 m} x(m) \overset{F}{\Leftrightarrow} X(e^{j(\omega-\omega_0)}) \tag{3.43}$$

Convolution

Given two signals $x_1(m)$ and $x_2(m)$ and their z-transforms $X_1(z)$ and $X_2(z)$ respectively

$$x_1(m) \overset{z}{\Leftrightarrow} X_1(z)$$

$$x_2(m) \overset{z}{\Leftrightarrow} X_2(z)$$

the convolutional property states that

$$x_1(m)_* x_2(m) \overset{z}{\Leftrightarrow} X_1(z) X_2(z) \tag{3.44}$$

where the asterisk * denotes the convolution operation. That is the convolution of two signals in the time domain is equivalent to multiplication of their z-transforms and vice versa.

Differentiation in the z-Domain

Given the z-transform pair

$$x(m) \overset{z}{\Leftrightarrow} X(z)$$

then

$$mx(m) \overset{z}{\Leftrightarrow} -z dX(z)/dz \tag{3.45}$$

This property can be proved by taking the derivative of the z-transform equation w.r.t. the variable z as

$$\begin{aligned} \frac{dX(z)}{dz} &= \frac{d}{dz} \sum_{m=-\infty}^{\infty} x(m) z^{-m} = -\sum_{m=\infty}^{\infty} mx(m) z^{-m-1} \\ &= -z^{-1} \sum_{m=\infty}^{\infty} mx(m) z^{-m} \end{aligned} \tag{3.46}$$

3.5 z-Transfer Function, Poles (Resonance) and Zeros (Anti-resonance)

The z-transfer function of a system gives the relationship between the z-transforms of the input and the output of the system. The frequency response of a system can be obtained from its z-transfer function by substituting for $z = e^{j\omega}$.

Consider the general linear time-invariant difference equation describing the input–output relationship of a discrete-time linear system (e.g. a digital filter)

$$y(m) = \sum_{k=1}^{N} a_k y(m-k) + \sum_{k=0}^{M} b_k x(m-k) \tag{3.47}$$

Note that a difference equation is the discrete-time counterpart of a differential equation, it describes a signal in terms of the rates of change (differential or differences) of input and output signals.

In Equation (3.47) the signal $x(m)$ is the filter input, $y(m)$ is the filter output, and a_k and b_k are the filter coefficients. Taking the z-transform of Equation (3.47) and noting that the z-transforms of $x(m-k)$ and $y(m-k)$ are $z^{-k}X(z)$ and $z^{-k}Y(z)$, we obtain

$$Y(z) = \sum_{k=1}^{N} a_k z^{-k} Y(z) + \sum_{k=0}^{M} b_k z^{-k} X(z) \tag{3.48}$$

Equation (3.48) can be rearranged and expressed in terms of the ratio of a numerator polynomial $Y(z)$ and a denominator polynomial $X(z)$ as

$$H(z) = \frac{Y(z)}{X(z)} = \frac{b_0 + b_1 z^{-1} + \cdots + b_M z^{-M}}{1 - a_1 z^{-1} - \cdots - a_N z^{-N}} = \frac{\sum\limits_{k=0}^{M} b_k z^{-k}}{1 - \sum\limits_{k=1}^{N} a_k z^{-k}} \tag{3.49}$$

$H(z)$ is known as the system transfer function or z-transfer function as it transfer the input to the output. The frequency response of a system $H(\omega)$ may be obtained by substituting $z = e^{j\omega}$ in Equation (3.49).

3.5.1 Poles and Zeros

One of the most useful aspects of the z-transform analysis is the description of a system in terms of the poles and zeros of the system. The zeros of a transfer function $H(z)$ are the values of the variable z for which the transfer function (or equivalently its numerator) is zero. Therefore the zeros are the roots of the numerator polynomial in Equation (3.49). The poles of $H(z)$ are the values of the variable z for which $H(z)$ is infinite. This happens when the denominator of $H(z)$ is zero. Therefore the poles of $H(z)$ are the roots of the denominator polynomial of Equation (3.49).

To obtain the poles and zeros of $H(z)$ rewrite the numerator and denominator polynomials to avoid negative powers of the variable z as

$$H(z) = \frac{b_0 z^{-M}}{z^{-N}} \times \frac{z^M + (b_1/b_0)\, z^{M-1} + \cdots + (b_M/b_0)}{z^N - a_1 z^{N-1} - \cdots - a_N} \tag{3.50}$$

Now the numerator and denominator polynomials of $H(z)$ may be factorised and expressed as

$$H(z) = b_0 z^{-M+N} \times \frac{(z-z_1)(z-z_2)\cdots(z-z_M)}{(z-p_1)(z-p_2)\cdots(z-p_N)} \tag{3.51}$$

$$\text{or } H(z) = Gz^{-M+N} \times \frac{\prod_{k=1}^{M}(z - z_k)}{\prod_{k=1}^{N}(z - p_k)} \tag{3.52}$$

where the gain $G = b_0$, and z_k are the zeros and p_k the poles of $H(z)$ respectively. Thus the transfer function $H(z)$ has M finite zeros i.e. the roots of the numerator polynomial at $z = z_1, z_2, \ldots, z_M$, N finite poles i.e. the roots of the denominator polynomial at $z = p_1, p_2, \ldots, p_N$, and there are also $|N - M|$ zeros (if $N > M$), or poles (if $N < M$), at the origin $z = 0$. Poles or zeros may also occur at $z = \infty$. A zero exists at infinity if $H(z = \infty) = 0$ and a pole exists at infinity if $H(z = \infty) = \infty$. If we count the numbers of poles and zeros at zero and infinity we find that $H(z)$ has exactly the same number of poles as zeros. A further important point to note is that for a system with real-valued coefficients a_k and b_k, complex-valued poles or zeros always occur in complex conjugate pairs.

Interpretation of the Physical Function of Poles and Zeros

The description of a system in terms of its poles and zeros is an extremely useful tool for analysis of the system behaviour.

The poles model the resonances of the system i.e. they model the peak energy concentrations of the frequency spectrum. The poles represent the roots of the feedback part of the transfer function of a system. For a stable system each pole should have a magnitude of less than one and must lie inside the unit circle as shown in the following examples. When the poles are on the unit circle they act as oscillators and produce sine waves in response to an input impulse function. When the poles are outside unit circle the system is unstable.

The zeros model the anti-resonances of the system i.e. they model the troughs (deeps) of the frequency spectrum. The zeros are the roots of the feed-forward part of the transfer function of a system. There is no restriction on the values of zeros other than that required to obtain a desired frequency or impulse response.

A useful graphical abstraction of the transfer function of a discrete-time system $H(z)$ is the *pole–zero plot* in a complex polar plane. The locations of the poles are shown by crosses ($\times$) and the locations of the zeros by circles (**O**) as described in the following examples.

Example 3.11 Find the z-transform and plot the pole–zero diagram of the following right-sided discrete-time signal (Figure 3.6(a))

$$x(m) = \begin{cases} \alpha^m & m \geq 0 \\ 0 & m < 0 \end{cases} \tag{3.53}$$

Solution: Substituting for $x(m)$ in the z-transform equation (3.4) we obtain

$$X(z) = \sum_{m=0}^{\infty} \alpha^m z^{-m} = \sum_{m=0}^{\infty} \left(\alpha z^{-1}\right)^m \tag{3.54}$$

For $|a| < 1$, using the convergence formula for the partial sum of the N first terms of a geometric series, $S_N = 1 + \alpha + \cdots + \alpha^{N-1} = \frac{1-\alpha^N}{1-\alpha}$, this power series converges to

$$X(z) = \frac{1}{1 - \alpha z^{-1}} = \frac{z}{z - \alpha} \tag{3.55}$$

Therefore $X(z)$ in this case has a single zero at $z = 0$ and a single pole at $z = \alpha$. The pole–zero plot is shown in Figure 3.6(b).

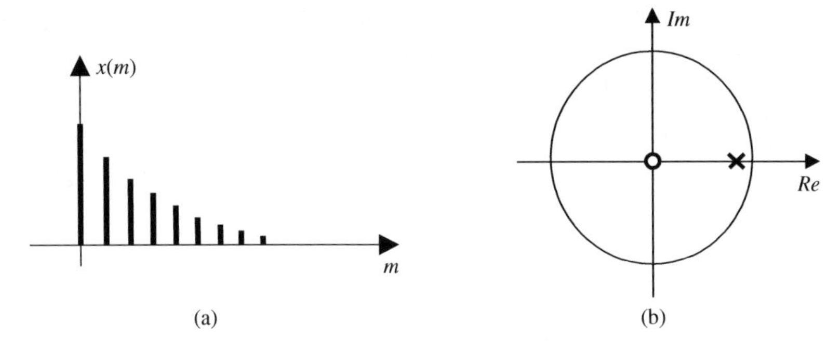

Figure 3.6 (a) An exponentially decaying signal, (b) its pole–zero representation.

3.5.2 The Response of a Single (First-Order) Zero or Pole

As shown in the following examples, the effect of a first-order zero is to introduce a deep or trough in the frequency spectrum of the signal at the frequencies 0 or π radians, where π corresponds to half the sampling frequency $F_s/2$. Conversely, the effect of a first-order pole is to introduce a peak in the frequency spectrum of the signal at frequencies 0 or π. These are illustrated in the following examples.

Example 3.12 Consider the first-order feed-forward filter of Figure 3.7 given by

$$y(m) = \alpha x(m-1) + x(m) \tag{3.56}$$

Taking the z-transform of Equation (3.56) yields

$$Y(z) = \alpha z^{-1} X(z) + X(z) = X(z)(1 + \alpha z^{-1}) \tag{3.57}$$

From Equation (3.57) the z-transfer function is given by

$$H(z) = \frac{Y(z)}{X(z)} = 1 + \alpha z^{-1} \tag{3.58}$$

Equating $H(z) = 0$ yields a zero at $z = -\alpha$. Substituting $z = e^{j\omega}$ in Equation (3.58) yields the frequency response

$$H(e^{j\omega}) = 1 + \alpha e^{-j\omega} \tag{3.59}$$

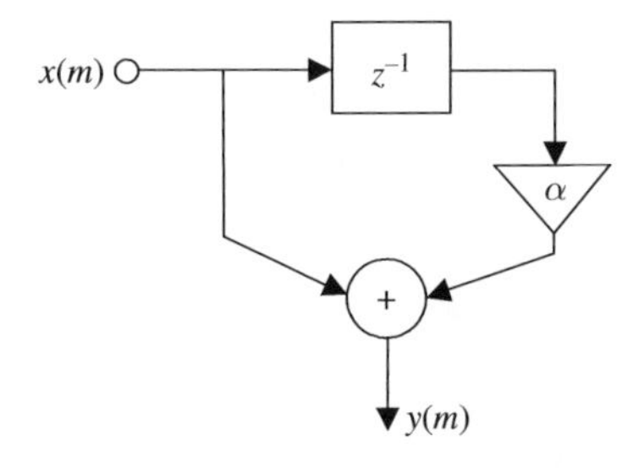

Figure 3.7 A first-order feed-forward discrete-time system.

Now at an angular frequency $\omega = 0$, $H(0) = 1 + \alpha$ and at $\omega = \pi$, $H(\pi) = 1 - \alpha$. Hence when $\alpha = 1$, $H(\pi) = 0$ and when $\alpha = -1$, $H(0) = 0$. Figure 3.8 shows the variation of the frequency response of the first-order single-zero system with the radius of the zero α. On the frequency axis the angular frequency $\omega = \pi$ corresponds to a frequency of $F_s/2$ Hz where F_s is the sampling rate. Note that for the positive values of α the zero is on the left half of the z-plane and the system has a low-pass frequency response, and conversely for the negative values of α the zero is on the right half of the z-plane and the system has a high-pass frequency response.

Example 3.13 Consider the first-order feedback system of Figure 3.9 given by

$$y(m) = \alpha y(m-1) + x(m) \tag{3.60}$$

Taking the z-transform of Equation (3.60) yields

$$Y(z) = \alpha z^{-1} Y(z) + X(z) \tag{3.61}$$

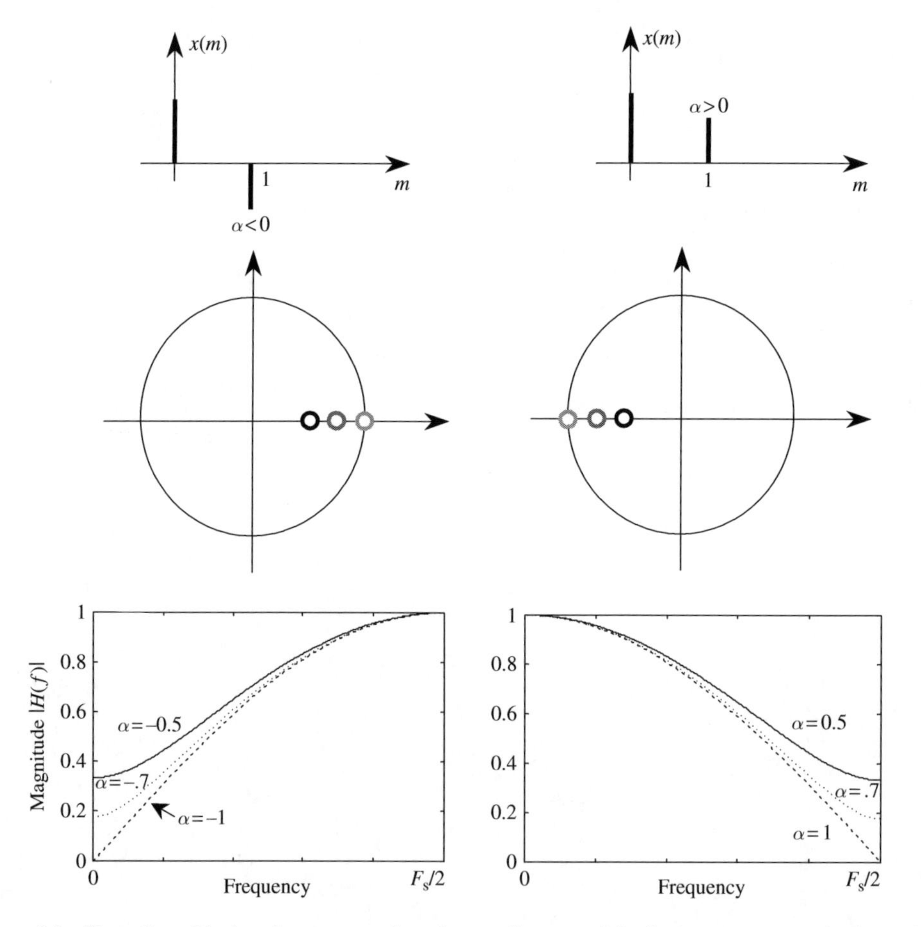

Figure 3.8 Illustration of the impulse response, the pole–zero diagram and the frequency response of a first order system with a single zero, for the varying values of the zero α.

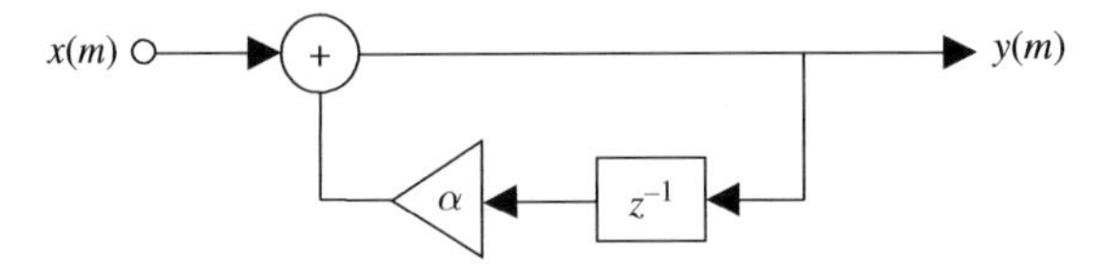

Figure 3.9 A first order feedback discrete-time system.

From Equation (3.61) the transfer function is given by

$$H(z) = \frac{Y(z)}{X(z)} = \frac{1}{1 - \alpha z^{-1}} = \frac{z}{z - \alpha} \tag{3.62}$$

The transfer function $H(z)$ has a pole at $z = \alpha$ and a zero at the origin. Substituting $z = e^{j\omega}$ in Equation (3.58) gives the frequency response

$$H(e^{j\omega}) = \frac{1}{1 - \alpha e^{-j\omega}} \tag{3.63}$$

Now at $\omega = 0$; $H(0) = 1/(1 - \alpha)$ and at $\omega = \pi$, $H(\pi) = 1/(1 + \alpha)$. Figure 3.10 shows the variation of the frequency response of the first-order single-pole system with the pole radius α. On the frequency axis the angular frequency $\omega = \pi$ corresponds to a frequency of $F_s/2$ Hz where F_s is the sampling rate. Note that for the positive values of α the pole is on the right half of the z-plane and the system has a low-pass frequency response, Figure 3.10(a), and conversely for the negative values of α the pole is on the left half of the z-plane and the system has a high-pass frequency response, Figure 3.10(b).

Matlab function FirstOrderPoleZeroDemo( )
This animation program plots the pole-zero, impulse response and frequency response of a single real pole and zero as a function of their position and radius. The angular frequency of the poles and zeros is either 0 or pi.

3.5.3 The Response of a Second-order Pair of Zeros or Poles

As shown in the following examples, the effect of a second-order pair of zeros is to introduce a deep or trough in the frequency spectrum of the signal at a frequency that depends on the angular position of the zeros. The effect of a second-order pair of poles is to introduce a peak in the frequency spectrum of the signal at a frequency that depends on the angular position of the poles in the unit circle.

Example 3.14 Consider the second-order feed-forward system of Figure 3.11 given by

$$y(m) = b_0 x(m) + b_1 x(m-1) + b_2 x(m-2) \tag{3.64}$$

Taking the z-transform of Equation (3.64) yields

$$Y(z) = b_0 X(z) + b_1 z^{-1} X(z) + b_2 z^{-2} X(z) \tag{3.65}$$

From Equation (3.65) the transfer function is given by

$$H(z) = \frac{Y(z)}{X(z)} = b_0 + b_1 z^{-1} + b_2 z^{-2} \tag{3.66}$$

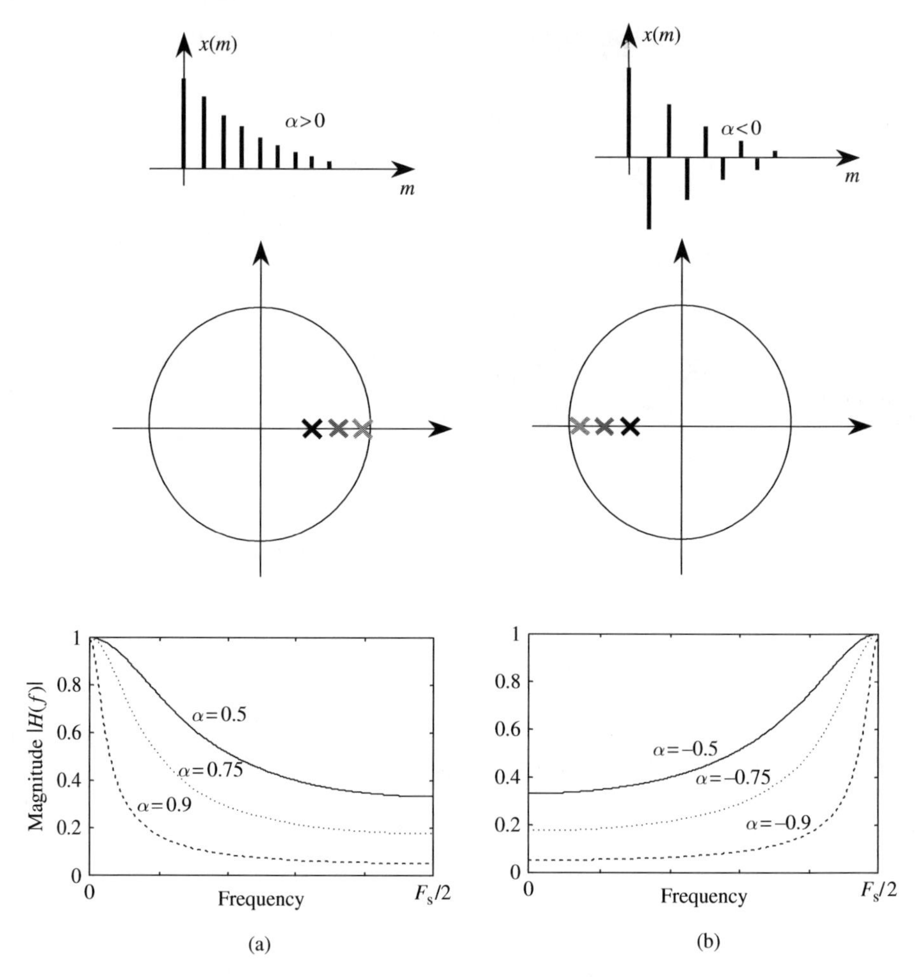

Figure 3.10 Illustration of the impulse response, the pole–zero diagram and the frequency response of a first order system with a single pole, for the varying values of the pole radius α.

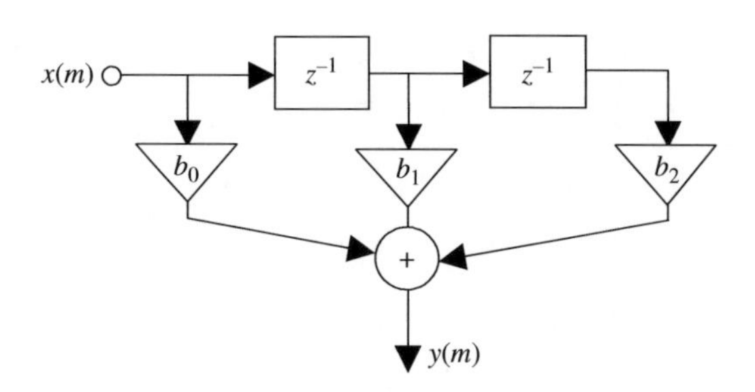

Figure 3.11 A second-order feed-forward discrete-time system.

Equation (3.66) can be factorised and expressed in terms of the zeros of the transfer function as

$$H(z) = G\left(1 - z_1 z^{-1}\right)\left(1 - z_1^* z^{-1}\right) \tag{3.67}$$

where the gain factor $G = b_0$. Note that since the coefficients of the transfer function polynomial in Equation (3.62) are real-valued, the roots of the polynomial have to be either complex conjugates or real-valued.

For a pair of complex conjugate poles $z_1 = re^{j\phi}$ and $z_1^* = re^{-j\phi}$ Equation (3.67) can be written in a polar form in terms of the angular frequency ϕ and the radius of the poles r as

$$\begin{aligned} H(z) &= G\left(1 - re^{j\phi} z^{-1}\right)\left(1 - re^{-j\phi} z^{-1}\right) \\ &= G\left(1 - 2r\cos(\phi) z^{-1} + r^2 z^{-2}\right) \end{aligned} \tag{3.68}$$

Comparing Equations (3.66) and (3.68) we have $b_0 = G$, $b_1 = -2Gr\cos(\phi)$ and $b_2 = Gr^2$. Figure 3.12(a) illustrates the variation of the zero frequency with the angular position of the zeros for a

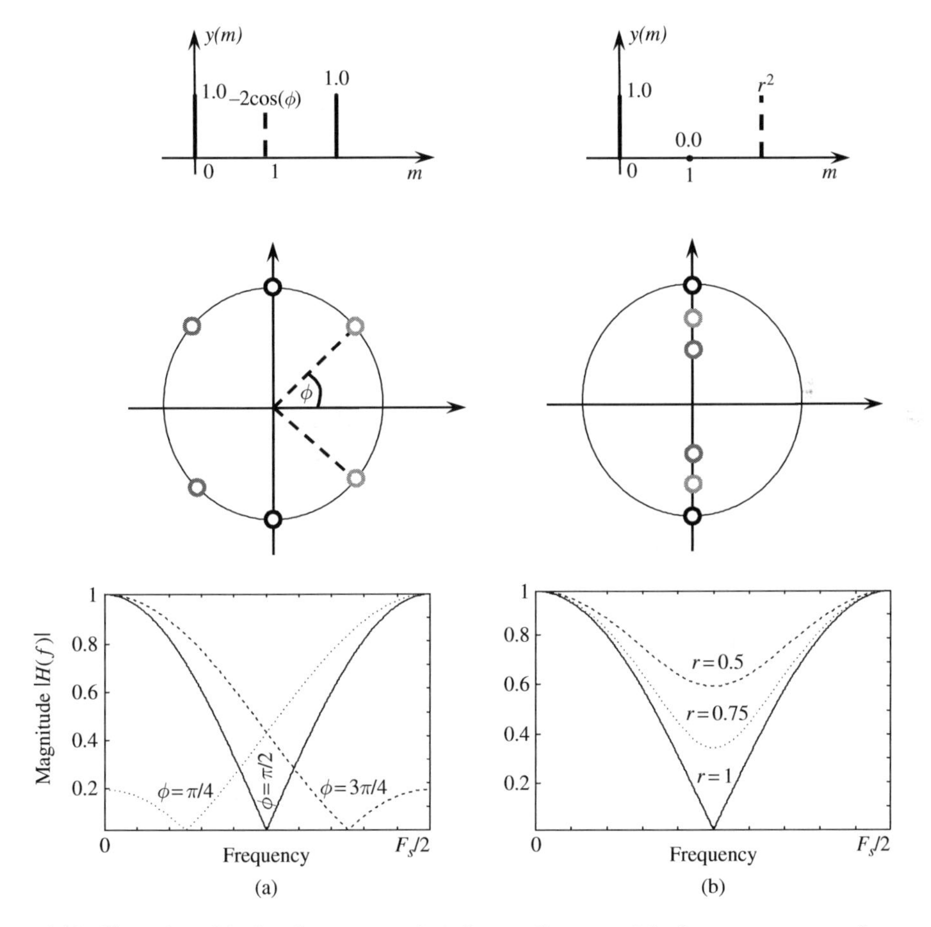

Figure 3.12 Illustration of the impulse response, the pole–zero diagram and the frequency response of a second order system with a pair of complex conjugate zeros, for (a) the varying values of the angular position ϕ at a fixed radius of 1 and (b) the varying radius r at a fixed angle of $\pi/2$.

complex conjugate pair of zeros. Figure 3.12(b) illustrates the variation of the depth and the bandwidth of the trough introduced by a complex conjugate pair of zeros with the radius of the zeros.

Example 3.15 Consider the second-order feedback filter of Figure 3.13 given by

$$y(m) = a_2 y(m-2) + a_1 y(m-1) + gx(m) \tag{3.69}$$

Taking the z-transform of Equation (3.69) we have

$$Y(z) = a_2 z^{-2} Y(z) + a_1 z^{-1} Y(z) + gX(z) \tag{3.70}$$

Rearranging Equation (3.70) we obtain the z-transfer function

$$H(z) = \frac{Y(z)}{X(z)} = \frac{g}{1 - a_1 z^{-1} - a_2 z^{-2}} \tag{3.71}$$

Equation (3.71) can be expressed in terms of the poles of the z-transfer function as

$$H(z) = \frac{g}{(1 - z_1 z^{-1})(1 - z_1^* z^{-1})} \tag{3.72}$$

Note that since the coefficients of the polynomial $H(z)$ are real-valued, the roots of this polynomial have to be complex conjugate valued or real valued. For a pair of complex conjugate poles $z_1 = re^{j\phi}$ and $z_1^* = re^{-j\phi}$ Equation (3.72) can be rewritten in a polar form in terms of the angular frequency and the radius of the poles as

$$\begin{aligned} H(z) &= \frac{g}{(1 - re^{j\phi} z^{-1})(1 - re^{-j\phi} z^{-1})} \\ &= \frac{g}{1 - 2r\cos(\phi) z^{-1} + r^2 z^{-2}} \end{aligned} \tag{3.73}$$

Figure 3.14(a) illustrates the variation of the resonance frequency with the angular position of the poles for a complex conjugate pair of poles. Figure 3.14(b) illustrates the variation of the bandwidth of the resonance of a complex conjugate pair of poles with the radius of the poles.

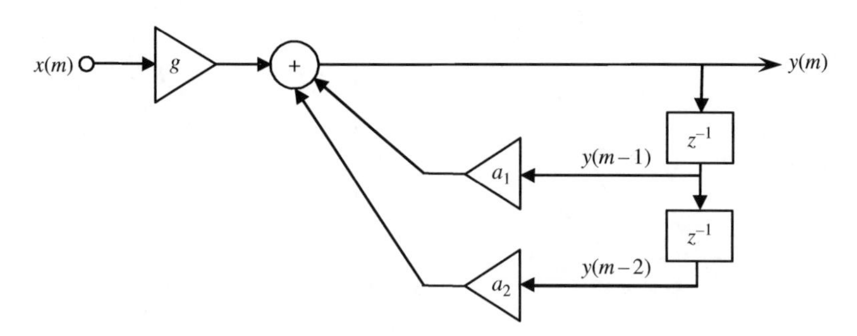

Figure 3.13 A second-order discrete-time feedback system.

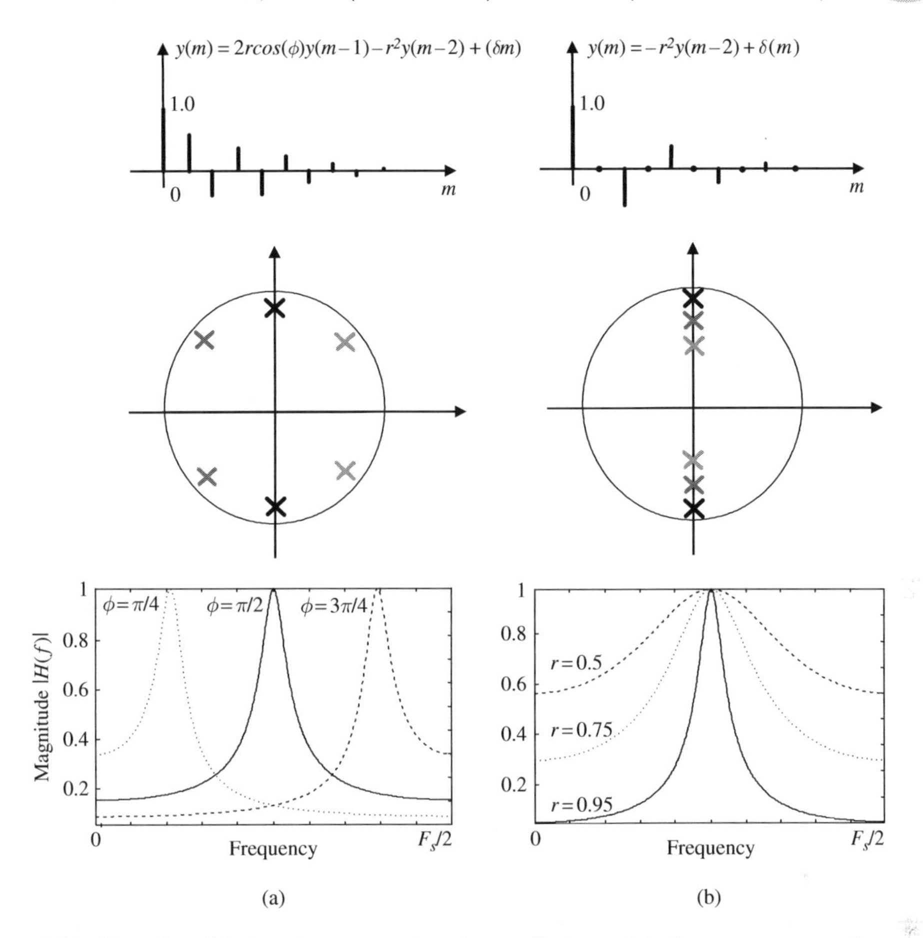

Figure 3.14 Illustration of the impulse response, the pole–zero diagram and the frequency response of a second order system with a pair of complex conjugate poles, for (a) the varying values of the angular position ϕ at a fixed radius of 0.95 and (b) the varying radius r at a fixed angle of $\pi/2$.

Matlab function Pole_Zero_demo( )

This animation program demonstrates the frequency response of a pair of complex poles and a pair of complex zeros as a function of their position in terms of the angular frequency and radius. In the first experiment the radius is constant and the angular frequency of the poles and zeros varies, in the second experiment the angular frequency is constant and the radius varies.

Matlab function Sound_of_Poles( )
This animation program plays the sound of a the impulse response of

(i) a complex pair of poles with varying frequency
(ii) a set of harmonically related poles with varying fundamental frequency.

The impulse response, the pole–zero diagram and the spectrogram are displayed.

3.5.4 Poles, Resonance and Oscillations

Resonance is the oscillatory exchange of energy between two interchangeable forms: for example in a vibrating spring–mass system, the energy conversion is between kinetic energy and the potential energy stored in a stretched/compressed spring; in a swinging pendulum, the energy is converted between kinetic energy and gravitational potential energy; and in an inductor–capacitor (LC) system, the energy is converted from electrical energy to magnetic energy and back. The main parameters of resonance are the frequency of oscillations, and the damping or energy loss per cycle related to bandwidth. Poles model resonances of a system.

Each resonance can be modelled by a pair of complex poles, where the angle of the pole determines the frequency of the oscillation of the resonance and the radius of the poles determines the bandwidth of the resonance. The bandwidth is related to energy loss i.e. the damping property of oscillations; the lower the bandwidth the less the loss of energy in each cycle of oscillation. For example, a pair of poles on the unit circle (i.e. with a pole radius of one) have a bandwidth of zero and can model the lossless oscillations of a sinusoid.

3.6 z-Transform of Analysis of Exponential Transient Signals

Although z-transforms are mostly used for system analysis and design, they can also be use to analyse and synthesise signals. After all, as explained earlier z-transform is a generalisation of the Fourier transform.

Example 3.16 *z-Transform of an exponentially decaying/rising transient signal*
Consider the z-transform of following exponential transient signal.

$$x(m) = \begin{cases} \alpha^m & 0 \leq m < N \\ 0 & m < 0, m \geq N \end{cases} \tag{3.74}$$

Substituting for $x(m)$ in the z-transform equation (3.4) we obtain

$$X(z) = \sum_{m=0}^{N-1} x(m)z^{-m} = \sum_{m=0}^{N-1} \alpha^m z^{-m} = \sum_{m=0}^{N-1} \left(\alpha z^{-1}\right)^m \tag{3.75}$$

Applying the formula for partial sum of the N first terms of a geometric series, $S_N = 1 + \alpha + \cdots + \alpha^{N-1} = \frac{1-\alpha^N}{1-\alpha}$, we have

$$X(z) = \frac{1 - \left(\alpha z^{-1}\right)^N}{1 - \alpha z^{-1}} \tag{3.76}$$

For $\alpha < 1$ and $\alpha^N << 0$ Equation (3.76) becomes

$$X(z) = \frac{1}{1 - \alpha z^{-1}} \tag{3.77}$$

For the general case when α is not constrained to be less than one, Equation (3.76) can also be written as

$$X(z) = \frac{\prod_{k=0}^{N-1} \left(1 - \alpha e^{-j2\pi k/N} z^{-1}\right)}{1 - \alpha z^{-1}} = \prod_{k=1}^{N-1} \left(1 - \alpha e^{-j2\pi k/N} z^{-1}\right) \tag{3.78}$$

To obtain Equation (3.78) from Equation (3.75) we have used the relationship $(1 - x^N) = \prod_{k=0}^{N-1} (1 - e^{-j2\pi k/N} x)$ with $x = \alpha z^{-1}$. The zeros of Equation (3.78) are given by

$$z = \alpha e^{j2\pi k/N} \quad k = 1, \cdots, N-1 \tag{3.79}$$

Figure 3.15 shows the zeros of z-transform for an exponentially rising signal with zeros outside the unit circle and an exponentially decaying signal with zeros inside the unit circle. Also note that this exponential signal can be represented either by a single pole as in Equation (3.77) or by $N - 1$ zeros as in Equation (3.78).

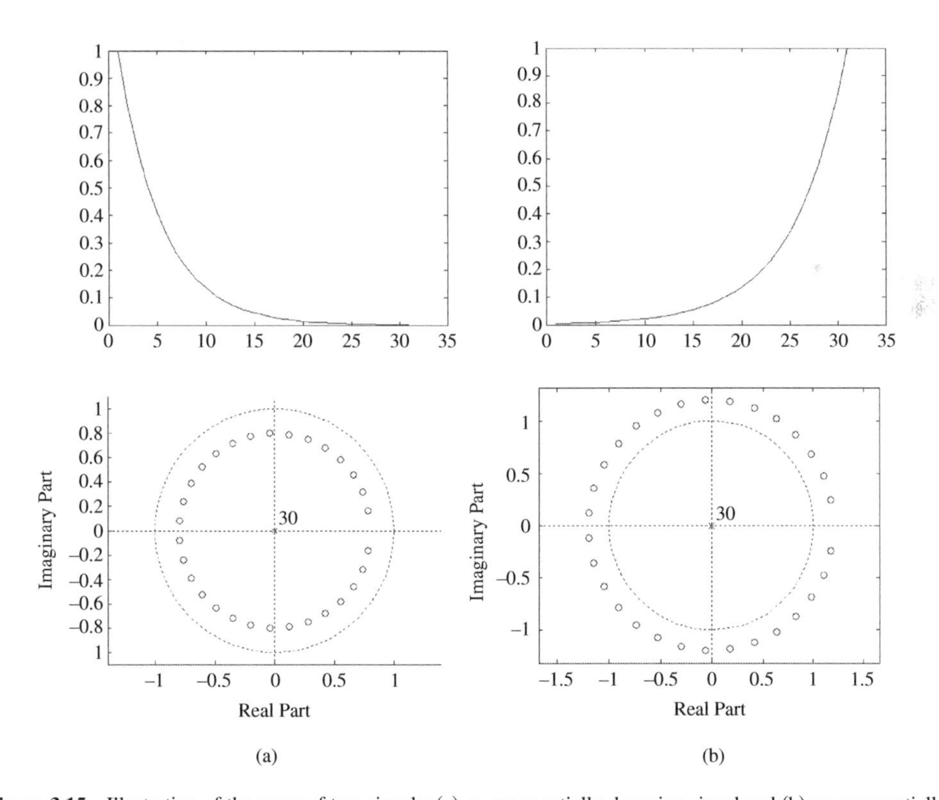

Figure 3.15 Illustration of the zeros of two signals: (a) an exponentially decaying signal and (b) an exponentially rising signal.

Example 3.17 Determine the z-transform of the following exponentially rising and decaying transient signal shown in Figure 3.16(a)

$$x(m) = \begin{cases} e^{\beta m} & 0 \le m < N \\ e^{\beta N - \beta(m-N)} & N \le m < 2N \\ 0 & m < 0, m \ge 2N \end{cases} \tag{3.80}$$

where $\beta > 0$.

Solution: The solution is the same as that in Equations (3.75)–(3.78) except substitute the exponentials $e^{\beta m}$ or $e^{-\beta m}$ for α.

$$X(z) = \sum_{m=-\infty}^{\infty} x(m) z^{-m} = \sum_{m=0}^{N-1} e^{\beta m} z^{-m} + \sum_{m=N}^{2N-1} e^{\beta N - \beta(m-N)} z^{-m} \tag{3.81}$$

$$X(z) = \underbrace{\sum_{m=-(N-1)}^{0} e^{-\beta m} z^{m}}_{\text{Anti-causal}} + z^{-N} e^{\beta N} \underbrace{\sum_{m=0}^{N-1} e^{-\beta m} z^{-m}}_{\text{Causal}} \tag{3.82}$$

$$X(z) = \prod_{k=1}^{N-1} \left(1 - e^{-\beta} e^{-j2\pi k/N} z\right) + z^{-N} e^{\beta N} \prod_{k=1}^{N-1} \left(1 - e^{-\beta} e^{-j2\pi k/N} z^{-1}\right) \tag{3.83}$$

As shown in Figure 3.16(b), the roots of the two parts of Equation (3.83) form two sets of uniformly spaced zeros. One set of zeros, corresponding to the decaying part of the exponential signal, are inside the unit circle and are given by

$$z_k = e^{-\beta} e^{-j2\pi k/N} \quad k = 1, \cdots, N-1 \tag{3.84}$$

The other set of zeros which are outside the unit circle correspond to the rising exponential part of the signal.

$$z_k = e^{\beta} e^{j2\pi k/N} \quad k = 1, \cdots, N-1 \tag{3.85}$$

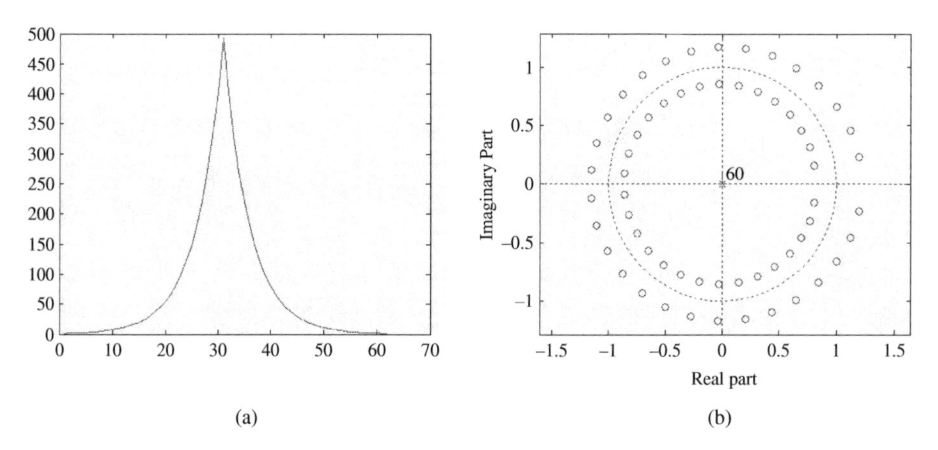

Figure 3.16 Illustration of the zeros of a signal composed of two parts: exponentially rising signal (the zeros outside the circle) and exponentially decaying (the zeros inside the circle).

Note that the different positions of zeros in effect decompose the rising exponentials (outside the unit circle) from the decaying exponentials (inside the unit circle).

Example 3.18 *z-Transform of a sine wave*

> **Matlab function Zeros_of_Exponential( )**
> Displays the zeros of the z-transform of rising and decaying exponential signals. The zeros of a rising exponential are outside the unit circle whereas the zeros of a decaying exponential are inside the unit circle.

Consider the z-transform of N samples of a cosine waveform with an exponential envelope $e^{\alpha m}\cos(2\pi fm)$:

$$\begin{aligned} X(z) &= \sum_{m=0}^{N-1} e^{\alpha m}\cos(2\pi fm)z^{-m} = 0.5\sum_{m=0}^{N-1} e^{\alpha m}\left(e^{j2\pi fm}+e^{-j2\pi fm}\right)z^{-m} \\ &= 0.5\sum_{m=0}^{N-1}\left(e^{\alpha}\left(e^{j2\pi f}+e^{-j2\pi f}\right)z^{-1}\right)^m = 0.5\sum_{m=0}^{N-1}\left(e^{\alpha}\cos(2\pi f)z^{-1}\right)^m \end{aligned} \tag{3.86}$$

Using a similar procedure as in the previous example, the z-transform of the cosine wave can be written as

$$X(z) = \prod_{n=1}^{N-1}\left(1 - e^{\alpha}\cos(2\pi f)e^{-j2\pi k/N}z^{-1}\right) \tag{3.87}$$

The zeros of the z-transform are given by

$$z_k = e^{\alpha}\cos(2\pi f)e^{-j2\pi k/N} \quad k = 1,\ldots,N \tag{3.88}$$

Figure 3.17 shows the zeros of the z-transfer function of 31 samples of an exponentially rising sine wave.

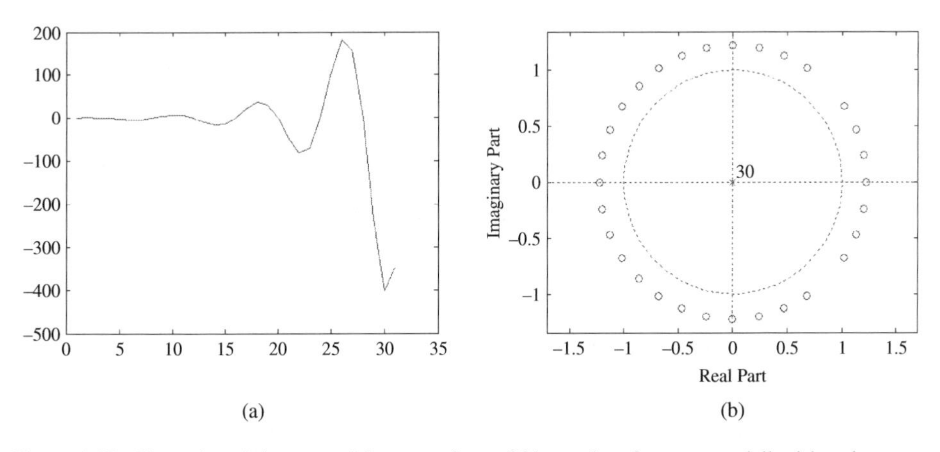

Figure 3.17 Illustration of the zeros of the z-transform of 31 samples of an exponentially rising sine wave.

3.7 Inverse z-Transform

The inverse z-transform can be obtained using one of the following four methods:

(1) Inspection method for relatively simple functions.
(2) Partial fraction method.
(3) Long division method.
(4) Contour integration.

In the inspection method each simple term of a polynomial $H(z)$, is substituted by its time-domain equivalent. For the more complicated functions of z, the partial fraction method is used to describe the polynomial in terms of simpler terms, and then each simple term is substituted by its time-domain equivalent term. The long division method is also relatively simple. The most complicated method is the contour integral method which is often avoided in favour of the other simpler methods.

3.7.1 Inverse z-Transform by Inspection

In this method the discrete-time equation for a signal or a system is obtained from its z-transform by recognising simple z-transform pairs and substituting the time-domain terms for their corresponding z-domain terms.

Example 3.19 Find the inverse z-transform of

$$H(z) = \frac{Y(z)}{X(z)} = \frac{1}{1-\alpha z^{-1}} \tag{3.89}$$

Solution: From Equation (3.85) we have

$$Y(z) = \alpha z^{-1} Y(z) + X(z) \tag{3.90}$$

By inspection and through the substitution of $z^{-k}Y(z)$ for $y(m-k)$ and $z^{-k}X(z)$ for $x(m-k)$ we obtain the discrete-time equivalent of Equation (3.90) as

$$y(m) = \alpha y(m-k) + x(m) \tag{3.91}$$

Now if the input $x(m)$ is a discrete-time impulse $\delta(m)$ given by

$$\delta(m) = \begin{cases} 1 & m = 0 \\ 0 & m \neq 0 \end{cases} \tag{3.92}$$

then the output of the feedback system of Equation (3.91) will be

$$y(m) = \begin{cases} \alpha^m & m \geq 0 \\ 0 & m < 0 \end{cases} \tag{3.93}$$

3.7.2 Inverse z-Transform by Partial Fraction

For the more complicated functions of the variable z, partial fraction methods may be used to describe the z domain polynomial in terms of simpler elementary terms. For this method we need to know that the elementary z-transform of both $1/(1-\alpha z^{-1})$ and $z/(z-\alpha)$ transforms to $\alpha^n u(n)$ or to $-\alpha^n u(-n-1)$ depending on whether $|z| > .|\alpha|$ or $|z| < |\alpha$.

The discrete-time equation for a signal or a system is obtained from its z-transform by recognising z-transform pairs and substituting the time domain functions for the z-domain functions.

Consider the polynomial

$$H(z) = \frac{\sum_{k=0}^{M} b_k z^{-k}}{1 - \sum_{k=1}^{N} a_k z^{-k}} = G z^{-M+N} \times \frac{\prod_{k=1}^{M} (z - z_k)}{\prod_{k=1}^{N} (z - p_k)} \tag{3.94}$$

Using the partial fraction technique $H(z)$ can described as

$$H(z) = \sum_{k=1}^{N} \frac{g_k}{1 - \alpha_k z^{-k}} \tag{3.95}$$

Now the inverse z-transform of each term in Equation (3.95) is given by

$$Z^{-1}\left(\frac{g_k}{1 - \alpha_k z^{-k}}\right) = \begin{cases} g_k \alpha^m & m \geq 0 \\ 0 & m < 0 \end{cases} \tag{3.96}$$

Example 3.20 Find the inverse z-transform of

$$H(z) = \frac{1}{(1 - 0.7z^{-1})(1 - 0.9z^{-1})} \tag{3.97}$$

Solution: Using the partial fraction method $H(z)$ can be written as

$$H(z) = \frac{a_1}{(1 - 0.7z^{-1})} + \frac{a_2}{(1 - 0.9z^{-1})} \tag{3.98}$$

where the coefficients a_1 and a_2 are given by

$$a_1 = H(z)\left(1 - 0.7z^{-1}\right)\big|_{z=0.7} = \frac{1}{(1 - 0.9/0.7)} = -3.5 \tag{3.99}$$

and

$$a_2 = H(z)\left(1 - 0.9z^{-1}\right)\big|_{z=0.9} = \frac{1}{(1 - 0.7/0.9)} = 4.5 \tag{3.100}$$

Substituting Equations (3.99) and (3.100) in Equation (3.98) yields

$$H(z) = \frac{-3.5}{(1 - 0.7z^{-1})} + \frac{4.5}{(1 - 0.9z^{-1})} \tag{3.101}$$

Now using Equation (3.101) the inverse z-transform of the system is obtained as

$$y(m) = -3.5 \times 0.7^m + 4.5 \times 0.9^m \quad m \geq 0 \tag{3.102}$$

This is the impulse response of the system, that is the response of the system with zero-initial condition to an impulse. Note that Equation (3.101) is a parallel combination of two first-order feedback systems

$$y(m) = y_1(m) + y_2(m) = -3.5x_1(m) + 0.7y_1(m-1) + 4.5x_2(m) + 0.7y_2(m-1) \tag{3.103}$$

The time-domain difference equation for Equation (3.97) can be obtained directly by inspection (after multiplying the terms in the two brackets) as

$$y(m) = 1.6y(m-1) - 0.63y(m-2) + x(m) \tag{3.104}$$

3.7.3 Inverse z-Transform by Long Division

For an example of the long division method, consider the simple z-transfer function

$$H(z) = \frac{1}{1 - az^{-1}} \tag{3.105}$$

The long division of this function is

$$\begin{array}{r l} & 1 + az^{-1} + a^2z^{-2} + \cdots \\ 1 - az^{-1} \,) & 1 \\ & \underline{1 - az^{-1}} \\ & \quad az^{-1} \\ & \quad \underline{az^{-1} - a^2z^{-2}} \\ & \qquad a^2z^{-2} \\ & \qquad \cdots \end{array} \tag{3.106}$$

The result is $H(z) = 1 + az^{-1} + a^2z^{-2} + \cdots$ whose inverse z-transform can be seen by inspection to be the sequence $h(m) = a^m u(m)$ where $u(m)$ is a step function that ensures the sequence exists for non-negative time index.

3.7.4 Inverse z-Transform by Contour Integral

This is the most difficult method of the inverse z-transform and it is usually avoided in favour of the partial fraction or other methods described above. However, it is useful to show the equation here as there is a similarity in the form of the contour integral and the form of the inverse Fourier or Laplace transforms. The inverse z-transform integral is defined as

$$x(m) = \frac{1}{j2\pi} \oint_C X(z) z^{m-1} dz \tag{3.107}$$

where C is a closed contour in the ROC in the complex plane around the origin.

Application of z-Transform to Filter Design and System Models

The z-transform is an indispensable tool in the design, implementation and stability analysis and control of digital filters as shown in detail in the next chapter. The z-transform is used throughout this book in the modelling of the input–output relationship of systems such as linear prediction models of speech in Chapters 13 and 14, Wiener and adaptive filters in Chapters 8 and 9, linear prediction models in Chapter 10 and in channel equalisation in Chapter 17.

3.8 Summary

This chapter provided an introduction to the z-transform method. The z-transform is a frequency domain method for analysis and synthesis of signals and systems. z-Transform provides an alternative to time-domain representation of signals and systems The z-transform equation was derived as the Laplace transform of a discrete-time signal. Like Laplace transform, z-transform is a generalisation of the Fourier transform; whereas the complex Fourier transform describes a signal or system in terms of a set of stationary complex sine waves, the z-transform describes a signal or system in terms of a set of complex sine waves with amplitude envelopes which may be exponentially rising or decaying or steady. The concept of z-transfer function as a means of description of the relationship between the z-transforms of the input and output of a system was introduced and the z-plane was derived from the s-plane of Laplace transform. The roots of the z-transfer function are the zeros and poles of the system or signal. The effects of poles and zeros on the spectrum of a signal were explored. Finally the inverse z-transform was considered.

Further Reading

Jury E.I. (1964) Theory and Application of the Z-Transform Method, John Wiley.
Oppenheim A.V., Willsky A.S., and Hamid S. (1996) Signals and Systems, Prentice Hall.

Exercises

3.1 Write the equations for, and explain the relationships between, the Fourier, the Laplace and the z-transforms.

3.2 Determine the z-transform, the region of convergence of the z-transform, and the Fourier transform of the following signals. For each signal sketch the time-domain signal $x(m)$, its magnitude spectrum $|X(f)|$ and the pole–zero diagram.

(a) $x(m) = \delta(m)$
(b) $x(m) = \delta(m-k)$
(c) $x(m) = \delta(m-k) + \delta(m+k)$, $x(m) = \sum_{k=-M}^{M} \delta(m-kT)$
(d) $x(m) = 0.9^{|m|}$

3.3 Find the poles and zeros of the following transfer functions and plot their pole–zero diagram and their frequency response.

(a) $H_1(z) = 1 - 0.5z^{-1}$
(b) $H_2(z) = 1 - 0.9z^{-1} + 0.81z^{-2}$
(c) $H_3(z) = 1 + z^{-4}$
(d) $H_4(z) = \dfrac{1}{1 - 0.5z^{-1}}$
(e) $H_5(z) = \dfrac{1}{1 - 0.9z^{-1} + 0.81z^{-2}}$
(f) $H_6(z) = \dfrac{1 + z^{-4}}{1 + 0.6561z^{-4}}$

3.4 Using the inspection method or the partial fraction method, find the inverse z-transform of each transfer function in exercise 3.3.

3.5 Describe the effects of a complex conjugate pair of poles and zeros on the frequency response of a filter. For each system whose poles and zeros are depicted in the z-plan diagrams below find:

(i) The corresponding transfer function.
(ii) The time-domain equation from an inverse z-transform.
(iii) The frequency response.

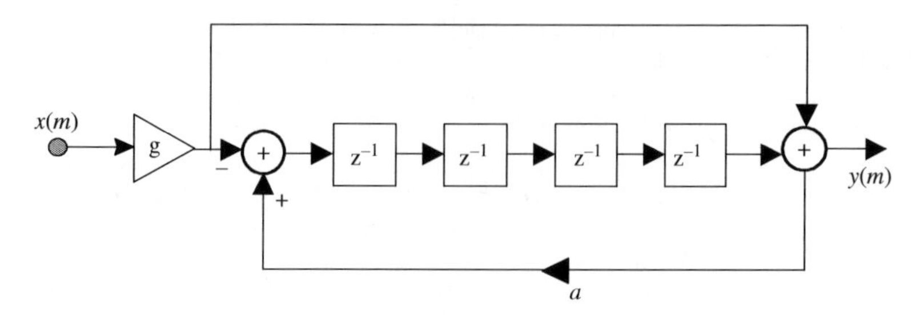

Figure 3.18 Illustration of a 4th order comb filter.

3.6 Consider the following second-order all-pole system operating at a sampling rate of 20 KHz and with its transfer function given by

$$H(z) = \frac{G}{1 + a_1 z^{-1} + a_2 z^{-2}}$$

Rewrite the transfer function $H(z)$ in terms of the radius and the angular frequency of the poles. Calculate the values of the coefficients a_1 and a_2 for a pole radius of 0.9 and a pole frequency of 2.5 kHz.

3.7 Using the inverse z-transform find the discrete-time difference equation corresponding to the second-order system in Example 3.5.

3.8 The input–output relation of a finite impulse response filter is given as

$$y(m) = x(m) - 2.5x(m-1) + 5.25x(m-2) - 2.5x(m-3) + x(m-4)$$

(a) Obtain the z-transfer function of this filter.
(b) Obtain and plot the zeros of this filter.

3.9 The difference equation relating the input and output of the filter shown in Figure 3.18 is

$$y(m) = ay(m-4) + gx(m) - gx(m-4)$$

(i) Taking the z-transform of the difference equation, find the transfer function of the filter.
(i) Describe the transfer function in polar form, and for a value of $a = 0.6561$ obtain the pole and zeros of the filter and sketch its pole–zero diagram.

3.10 The z-transfer function of a finite impulse response filter $H(z)$ is given by

$$H(z) = \left(1 - 0.5z^{-1} + 0.25z^{-2}\right)\left(1 - 2z^{-1} + 4z^{-2}\right)$$

Plot the zeros of this filter on a pole–zero diagram. Obtain the equation for the frequency response $H(\omega)$ of this filter.

3.11 Consider the following second-order FIR filters

$$H_1(z) = 1 - 0.7071z^{-1} + 0.25z^{-2}$$
$$H_2(z) = 1 - 2.8284z^{-1} + 4z^{-2}$$

(i) Write down an expression, and derive the zeros of, the z-transfer function of

$$H(z) = H_1(z)H_2(z)$$

(ii) Using the inverse z-transform obtain the discrete-time difference equation for $H(z)$.
(iii) Sketch the frequency response of $H(z)$ and suggest an application for this filter. Discuss the effect of varying the value of a on the frequency response of the filter.

3.12 A signal can be characterised with a fourth order linear prediction model with two resonances modelled with four poles at

$$z_1 = 0.98e^{j\pi/4}, \quad z_1^* = 0.98e^{-j\pi/4}, \quad z_2 = 0.9e^{j3\pi/4}, \quad z_2^* = 0.9e^{-j3\pi/4}$$

(i) Write the z-transfer function of a linear prediction model of this signal in terms of the product of the terms containing the individual poles and then arrange the z-transfer function in terms of the cascade of two second-order all-pole models.
(ii) Write the time-domain difference equations describing the input–output relation of each second-order linear prediction model in part (i).

4 Digital Filters

Filters are a basic component of all signal processing and telecommunication systems. The primary functions of a filter are one or more of the following: (a) to confine a signal into a prescribed frequency band or channel, for example as in anti-aliasing filter or a radio/TV channel selector; (b) to decompose a signal into two or more sub-band signals for sub-band signal processing, for example in music or image coding or wavelet decomposition; (c) to modify the frequency spectrum of a signal, for example in channel equalisation or audio graphic equalisers; and (d) to model the input–output relation of a system such as a mobile communication channel, voice production, musical instruments, telephone line echo, and room acoustic response.

In this chapter we introduce the general form of the equation for a linear time-invariant filter and consider the various methods of description of a filter in time and frequency domains. We study different filter forms and structures and the design of low-pass filters, band-pass filters, band-stop filters and filter banks. We consider several applications of filters such as in audio graphic equalisers, noise reduction filters in Dolby systems, image deblurring, and image edge emphasis.

4.1 Introduction

Filters are widely employed in signal processing and communication systems in applications such as channel equalisation, noise reduction, radar, audio processing, video processing, biomedical signal processing, and analysis of economic and financial data. For example, in a radio receiver band-pass filters, or tuners, are used to extract the signals from a radio channel. In an audio graphic equaliser the input signal is filtered into a number of sub-band signals and the gain for each sub-band can be varied manually with a set of controls to change the perceived audio sensation. In a Dolby system, pre-filtering and post-filtering are used to minimise the effect of noise. In hi-fi audio a compensating filter may be included in the preamplifier to compensate for the non-ideal frequency-response characteristics of the speakers. Filters are also used to create perceptual audio-visual effects for music, films and in broadcast studios.

Multimedia Signal Processing: Theory and Applications in Speech, Music and Communications Saeed V. Vaseghi

The Primary Functions of Filters

The main functions of a filter can be listed as one of the following:

(a) To confine a signal into a prescribed frequency band as in low-pass, high-pass, and band-pass filters.
(b) To decompose a signal into two or more sub-bands as in filter-banks, graphic equalisers, sub-band coders, frequency multiplexers.
(c) To modify the frequency spectrum of a signal as in telephone channel equalisation and audio graphic equalisers.
(d) To model the input–output relationship of a system such as telecommunication channels, human vocal tract, and music synthesisers.

Different Classes of Filters

Depending on the form of the filter equation and the structure of implementation, filters may be broadly classified into the following classes:

(a) Linear filters versus non-linear filters.
(b) Time-invariant filters versus time-varying filters.
(c) Adaptive filters versus non-adaptive filters.
(c) Recursive filters versus non-recursive filters.
(d) Direct-form, cascade-form, parallel-form and lattice structures.

In this chapter we are mainly concerned with linear time-invariant (LTI) filters. These are a class of filters whose output is a linear combination of the input and whose coefficients do not vary with time. Time-varying and adaptive filters are considered in later chapters.

4.1.1 Alternative Methods for Description of Filters

Filters can be described using the following time or frequency domain methods:

(a) *Time domain input–output relationship.* As described in Section 4.2 a difference equation is used to describe the output of a discrete-time filter in terms of a weighted combination of the input and previous output samples. For example a first-order filter may have the difference equation

$$y(m) = a\, y(m-1) + x(m) \tag{4.1}$$

where $x(m)$ is the filter input, $y(m)$ is the filter output and a is the filter coefficient.

(b) Impulse response. A filter can be described in terms of its response to an impulse input. For example the response of the filter of Equation (4.1) to a discrete-time impulse input at $m = 0$ is

$$y(m) = a^m \quad m = 0, 1, 2, \ldots \tag{4.2}$$

$y(m) = a^m = 1, a, a^2, a^3, a^4, \ldots$ for $m = 0, 1, 2, 3, 4 \ldots$ and it is assumed that $y(m) = 0$ for $m < 0$.

Impulse response is useful because: (i) any signal can be viewed as the sum of a number of shifted and scaled impulses, hence the response of a linear filter to a signal is the sum of the responses to all the impulses that constitute the signal; (ii) an impulse input contains all frequencies with equal energy, and hence it excites a filter at all frequencies; and (iii) impulse response and frequency response are Fourier transform pairs.

(c) *Transfer function, poles and zeros.* The transfer function of a digital filter transfers the input frequency content to output frequency content. The transfer function $H(z)$ is the ratio of the z-transforms of the filter output and input given by

$$H(z) = \frac{Y(z)}{X(z)} \tag{4.3}$$

For example the transfer function of the filter of Equation (4.1) is given by

$$H(z) = \frac{1}{1 - az^{-1}} \tag{4.4}$$

A useful method of gaining insight into the behaviour of a filter is the pole–zero description of a filter described in Chapter 3. As described in Section 3.5, poles and zeros are the roots of the denominator and numerator of the transfer function respectively.

(d) *Frequency response.* Viewed from the frequency domain, a linear filter simply multiplies the amplitudes of each input frequency component by a frequency-dependent factor and adds a frequency-dependent value to the phase of each frequency component of the input signal. The addition of a phase value is equivalent to a time delay of the input frequency. The phase change usually models the delay at each frequency as it propagates through the filter.

The frequency response of a filter describes how the filter alters the magnitude and the phase of the input signal frequencies. The frequency response of a filter can be obtained by taking the Fourier transform of the impulse response of the filter, or by simple substitution of the frequency variable $e^{j\omega}$ for the z variable $z = e^{j\omega}$ in the z-transfer function as

$$\begin{aligned} H(z = e^{j\omega}) &= \frac{Y(e^{j\omega})}{X(e^{j\omega})} \\ &= \frac{|Y(e^{j\omega})|\, e^{j\varphi_Y(\omega)}}{|X(e^{j\omega})|\, e^{j\varphi_X(\omega)}} \\ &= \underbrace{|H(e^{j\omega})|}_{\text{Magnitude response}}\, \underbrace{e^{j\varphi_H(\omega)}}_{\text{Phase response}} \end{aligned} \tag{4.5}$$

where $\varphi_Y(\omega) = \tan^{-1}(\mathrm{Im}(Y(\omega))/\mathrm{Re}(Y(\omega)))$ denotes the phase of $Y(e^{j\omega})$. The frequency response of a filter is a complex variable and can be described in terms of the filter magnitude response $|H(e^{j\omega})|$ and the phase response $e^{j\varphi_H(\omega)} = e^{j\varphi_Y(\omega)}/e^{j\varphi_X(\omega)} = e^{j\varphi_Y(\omega) - j\varphi_X(\omega)}$ of the filter.

4.2 Linear Time-Invariant Digital Filters

Linear time-invariant (LTI) filters are a class of filters whose output is a linear combination of the input signal samples and whose coefficients do not vary with time. The *linear* property entails that the filter response to a weighted sum of a number of signals, is the weighted sum of the filter responses to the individual signals. This is the *principle of superposition.* The term *time-invariant* implies that the filter coefficients and hence its frequency response is fixed and does not vary with time.

In the time domain the input–output relationship of a discrete-time linear filter is given by the linear difference equation

$$y(m) = \underbrace{\sum_{k=1}^{N} a_k y(m-k)}_{\text{Feedback of previous output samples}} + \underbrace{\sum_{k=0}^{M} b_k x(m-k)}_{\text{Feed forward of input samples}} \tag{4.6}$$

where, as shown in Figure 4.1, $\{a_k, b_k\}$ are the filter coefficients, and the output $y(m)$ is a linear combination of the previous N output samples $[y(m-1), \ldots, y(m-N)]$, the present input sample $x(m)$ and the previous M input samples $[x(m-1), \ldots, x(m-M)]$. The characteristic of a filter is completely determined by its coefficients $\{a_k, b_k\}$. For a time-invariant filter the coefficients $\{a_k, b_k\}$ are constants calculated to obtain a specified frequency response.

The filter transfer function, obtained by taking the z-transform of the difference equation (4.6), is given by

$$H(z) = \frac{\sum_{k=0}^{M} b_k z^{-k}}{1 - \sum_{k=1}^{N} a_k z^{-k}} \tag{4.7}$$

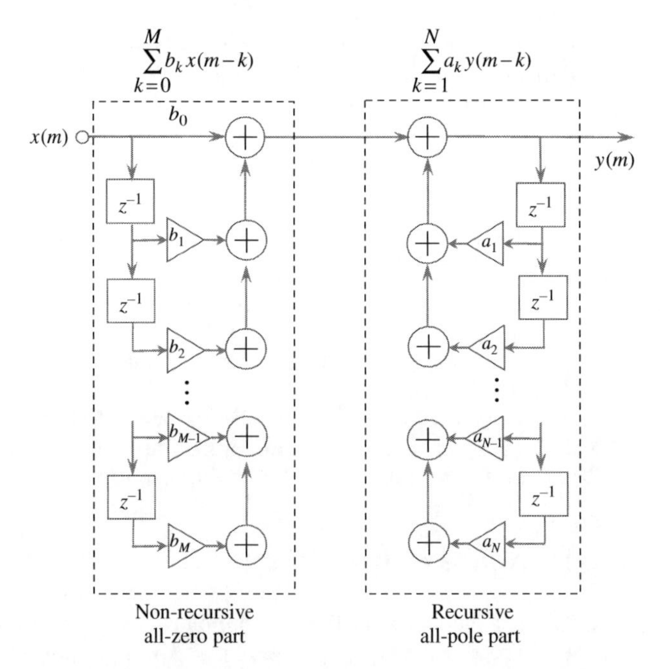

Figure 4.1 Illustration of a direct-form pole-zero filter showing the output is composed of the sum of two vector products: a weighted combination of the input samples $[b_0, \ldots, b_M][x(m), \ldots, x(M-1)]^T$ plus a weighted combination of the output feedback $[a_1, \ldots, a_N][y(m-1), \ldots, y(m-N)]^T$. T denotes transpose.

The frequency response of this filter can be obtained from Equation 4.7 by substituting the frequency variable $e^{j\omega}$ for the z variable, $z = e^{j\omega}$, as

$$H(e^{j\omega}) = \frac{\sum_{k=0}^{M} b_k e^{-j\omega k}}{1 - \sum_{k=1}^{N} a_k e^{-j\omega k}} \tag{4.8}$$

Since from Fourier transform a signal is a weighted combination of a number of sine waves, it follows, from superposition principle, that in frequency domain linear filtering can be viewed as a linear combination of the frequency constituents of the input multiplied by the frequency response of the filter.

Filter Order

The order of a discrete-time filter is the highest discrete-time *delay* used in the input–output equation of the filter. For example, in Equations (4.6) and 4.7 the filter order is the larger of the values of N or M. For continuous-time filters the filter order is the order of the highest differential term used in the input–output equation of the filter.

4.3 Recursive and Non-Recursive Filters

Figure 4.1 shows a block diagram implementation of the linear time-invariant filter in Equation (4.1). The transfer function of the filter in Equation 4.7 is the ratio of two polynomials in the variable z and may be written in a cascade form as

$$H(z) = H_1(z)H_2(z) \tag{4.9}$$

where $H_1(z)$ is the transfer function of a feed-forward, *all-zero*, filter given by

$$H_1(z) = \sum_{k=0}^{M} b_k z^{-k} \tag{4.10}$$

and $H_2(z)$ is the transfer function of a feedback, *all-pole*, recursive filter given by

$$H_2(z) = \frac{1}{1 - \sum_{k=1}^{N} a_k z^{-k}} \tag{4.11}$$

4.3.1 Non-Recursive or Finite Impulse Response (FIR) Filters

A non-recursive filter has no feedback and its input–output relation is given by

$$y(m) = \sum_{k=0}^{M} b_k x(m-k) \tag{4.12}$$

As shown in Figure 4.2 the output $y(m)$ of a non-recursive filter is a function only of the input signal $x(m)$. The response of such a filter to an impulse consists of a finite sequence of $M+1$ samples, where M is the filter order. Hence, the filter is known as a *Finite-Duration Impulse Response* (FIR) filter. Other names for a non-recursive filter include all-zero filter, feed-forward filter or moving average (MA) filter, a term usually used in statistical signal processing literature.

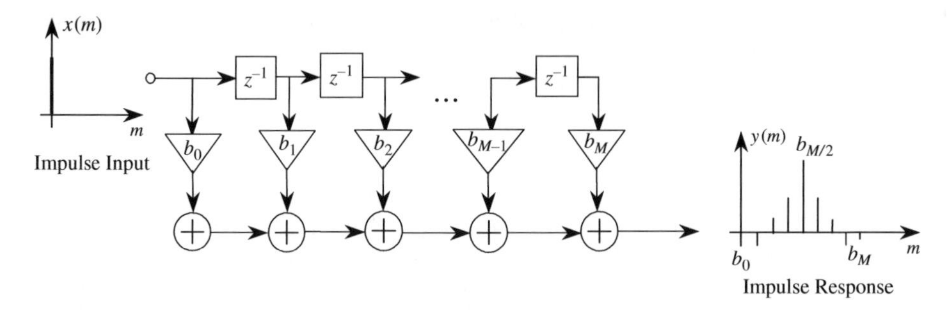

Figure 4.2 Direct-form FIR filter. Note that each filter output is the sum of the product of $[b_0, b_1, \ldots b_{M-1}]$ and $[x(m), x(m-1)], \ldots, x(m-M+1)$.

4.3.2 Recursive or Infinite Impulse Response (IIR) Filters

A recursive filter has feedback from output to input, and in general its output is a function of the previous output samples and the present and past input samples as described by the equation

$$y(m) = \sum_{k=1}^{N} a_k y(m-k) + \sum_{k=0}^{M} b_k x(m-k) \tag{4.13}$$

Figure 4.1 shows a direct form implementation of Equation (4.13). In theory, when a recursive filter is excited by an impulse, the output persists forever. Thus a recursive filter is also known as an *Infinite-Duration Impulse Response* (IIR) filter. Other names for an IIR filter include feedback filter, pole–zero filter and auto-regressive-moving-average (ARMA) filter, a term usually used in statistical signal processing literature.

A discrete-time IIR filter has a z-domain transfer function that is the ratio of two z-transform polynomials as expressed in Equation 4.7; it has a number of poles corresponding to the roots of the denominator polynomial and it may also have a number of zeros corresponding to the roots of the numerator polynomial.

The main difference between IIR filters and FIR filters is that an IIR filter is more compact in that it can usually achieve a prescribed frequency response with a smaller number of coefficients than an FIR filter. A smaller number of filter coefficients implies less storage requirements and faster calculation and a higher throughput. Therefore, generally IIR filters are more efficient in memory and computational requirements than FIR filters. However, it must be noted that an FIR filter is always stable, whereas an IIR filter can become unstable (for example if the poles of the IIR filter are outside the unit circle) and care must be taken in design of IIR filters to ensure stability.

Figure 4.3 shows a particular case of an IIR filter when the output is a function of N previous output samples and the present input sample given by

$$y(m) = \sum_{k=1}^{N} a_k y(m-k) + gx(m) \tag{4.14}$$

The transfer function of this filter is given by

$$H(z) = \frac{g}{1 - \sum_{k=1}^{N} a_k z^{-k}} \tag{4.15}$$

This filter is called an all-pole filter as it only has poles.

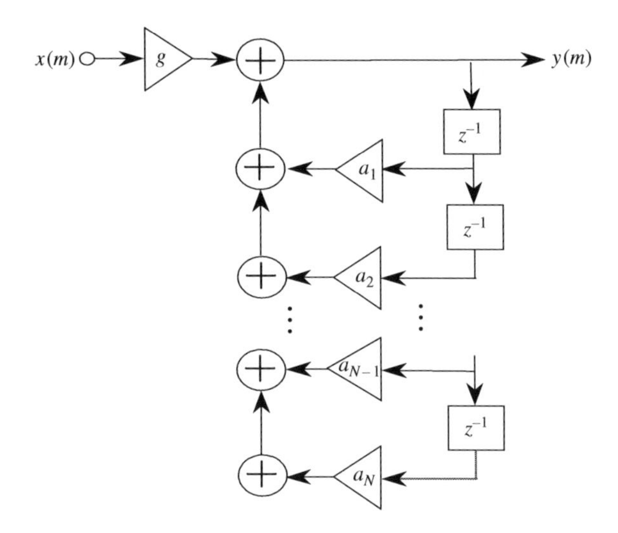

Figure 4.3 Direct-form all-pole IIR filter.

4.4 Filtering Operation: Sum of Vector Products, A Comparison of Convolution and Correlation

The filtering operation, as expressed in Equations (4.6) and (4.13) and illustrated in Figures 4.1 and 4.2, involves the summation of the result of multiplication of the filter coefficient vectors with the input and output signal vectors. Figure 4.1 shows that the filter output is obtained as the sum of two vector products: a weighted combination of the input samples, $[b_0, \ldots, b_M][x(m), \ldots, x(M-1)]^T$, plus a weighted combination of the output feedback $[a_1, \ldots, a_N][y(m-1), \ldots, y(m-N)]^T$.

As explained earlier, any signal can be expressed as a combination of shifted and scaled impulses. Hence, it follows that the operation of filtering of a signal $x(m)$ can be mathematically expressed as the convolution of the input signal and the impulse response of the filter $h(m)$ as

$$y(m) = \sum_{k=0}^{N} h(k)x(m-k) \tag{4.16}$$

The filtering, or convolution operation, of Equation (4.16) illustrated in Figure 4.4, is composed of the following four sub-operations:

(1) ***Fold*** the signal $x(k)$ to yield $x(-k)$, this is done because the samples with the earliest-time index (i.e. most distance past) go into filter first.
(2) ***Shift*** the folded input signal $x(-k)$ to obtain $x(m-k)$.
(3) ***Multiply*** $x(m-k)$ by the impulse response of the filter $h(k)$.
(4) ***Sum*** the results of the vector product $h(k)x(m-k)$ to obtain the filter output $y(m)$.

As shown in Section 2.3.2 in the frequency domain the convolution operation becomes a multiplication operation, hence Equation (4.16) becomes

$$Y(f) = H(f)X(f) \tag{4.17}$$

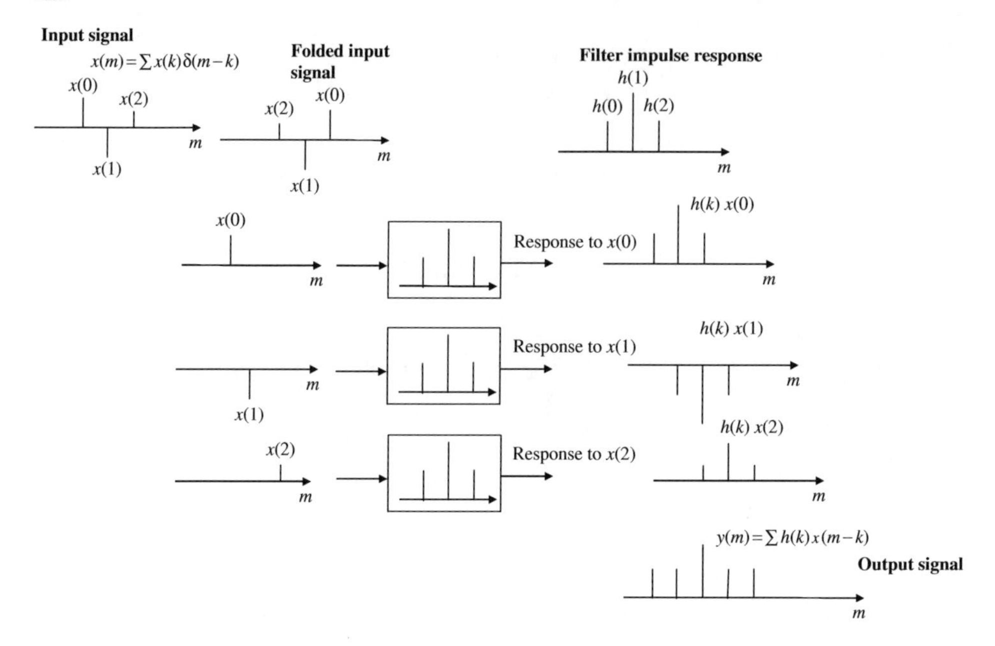

Figure 4.4 Illustration of the output of a filter in response to an input as the sum of the impulse responses of the filter to individual input samples. This is the convolution of input and the impulse response.

where $X(f)$, $Y(f)$ and $H(f)$ are the input, the output and the filter response at frequency f respectively.

Now, to explore the relationship between convolution, correlation and filtering, consider the (cross-)correlation of the two sequences $x(k)$ and $h(k)$ defined as

$$r(m) = \sum_{k=0}^{N} h(k)x(k-m) \tag{4.18}$$

Note in Equation (4.18) that the discrete-time index m is the *correlation lag*. From Equations (4.16) and (4.18) the main difference between convolution and correlation is that in convolution of two signals, one of the two signals, say $x(k)$, is folded in time to become $x(-k)$, as illustrated in Figure 4.5. Hence, the relation between correlation and convolution can be expressed as

$$\text{Conv}\,(h(k)x(k)) = \text{Corr}\,(h(k)x(-k)) \tag{4.19}$$

The Fourier transform of Equation (4.16), of convolution of two sequences $h(k)$ and $x(k)$, gives Equation (4.17) in the frequency domain, repeated here

$$Y(f) = H(f)X(f) \tag{4.20}$$

whereas the Fourier transform of Equation (4.18), of correlation of two sequences $h(k)$ and $x(k)$, gives

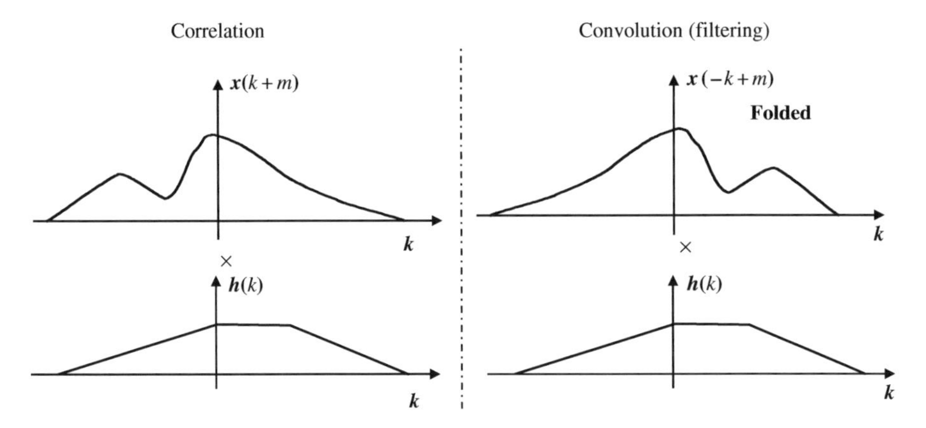

Figure 4.5 A comparative illustration of convolution and correlation operations. Both operations are the sum product of two signals. The main difference is that convolution involves folding one of the two signals.

$$R(f) = H(f)X^*(f) \tag{4.21}$$

where the asterisk * denotes complex conjugate. Note that the folding of a signal in time domain (time reversal) is equivalent to a complex conjugate operation in the frequency domain.

Matlab function CorrConvDemo( )
Demonstration of the equivalence of convolution and correlation operations when one of the two signals is flipped. The program demonstrates that correlation of a signal with its flipped version is equal to convolution of the signal with itself and that convolution of a signal with its flipped version is equal to autocorrelation.

4.5 Filter Structures: Direct, Cascade and Parallel Forms

In this section we consider different structures for realisation of a digital filter. These structures offer various trade-offs between complexity, cost of implementation, computational efficiency and stability.

4.5.1 Direct Filter Structure

The direct-form realisation of a filter is a direct implementation of Equation (4.6). Figures 4.1, 4.2 and 4.3 show the direct-form implementation of an IIR filter with poles and zeros, an all-zero (or FIR) filter and all-pole IIR filter respectively. Figure 4.6 shows a more efficient form of the direct implementation of an IIR filter. The direct-form structure provides a convenient method for the implementation of FIR filters. However, for IIR filters the direct-form implementation is not normally used due to problems with the design and operational stability of direct-form IIR filters.

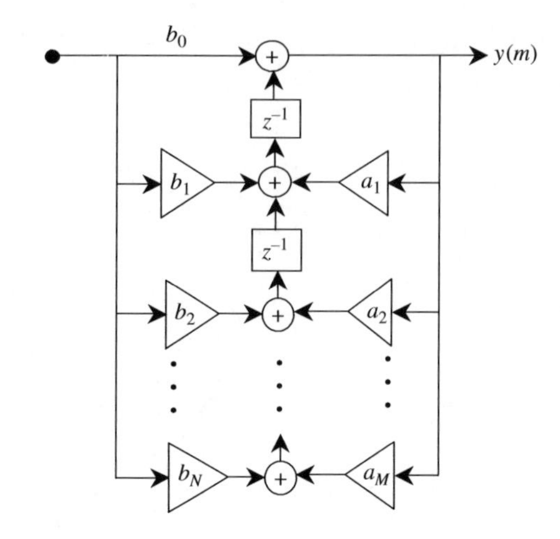

Figure 4.6 An efficient realisation of an IIR Filter.

4.5.2 Cascade Filter Structure

The cascade implementation of a filter is obtained by expressing the filter transfer function $H(z)$ in a factorised form as

$$H(z)=\frac{\sum\limits_{k=0}^{M} b_k z^{-k}}{1-\sum\limits_{k=1}^{N} a_k z^{-k}}=G\frac{(1-z_1z^{-1})(1-z_1^*z^{-1})(1-z_2z^{-1})(1-z_2^*z^{-1})\cdots(1-z_{M/2}z^{-1})(1-z_{M/2}^*z^{-1})}{(1-p_1z^{-1})(1-p_1^*z^{-1})(1-p_2z^{-1})(1-p_2^*z^{-1})\cdots(1-p_{N/2}z^{-1})(1-p_{N/2}^*z^{-1})} \tag{4.22}$$

where G is the filter gain and the poles p_k and zeros z_k are either complex conjugate pairs or real-valued. The factorised terms in Equation (4.22) can be grouped in terms of the complex conjugate pairs and expressed as cascades of second-order terms as

$$\begin{aligned}H(z)=G&\left(\frac{(1-z_1z^{-1})(1-z_1^*z^{-1})}{(1-p_1z^{-1})(1-p_1^*z^{-1})}\right)\times\left(\frac{(1-z_2z^{-1})(1-z_2^*z^{-1})}{(1-p_2z^{-1})(1-p_2^*z^{-1})}\right)\\&\times\cdots\times\left(\frac{(1-z_{M/2}z^{-1})(1-z_{M/2}^*z^{-1})}{(1-p_{N/2}z^{-1})(1-p_{N/2}^*z^{-1})}\right)\end{aligned} \tag{4.23}$$

Each bracketed term in Equation (4.23) is the z-transfer function of a second-order IIR filter as shown in Figures 4.7 and 4.8(a). Equation (4.23) can be expressed in compact notation as

$$H(z)=G\prod_{k=1}^{K}H_k(z) \tag{4.24}$$

where for an IIR filter, assuming that $N>M$, the variable K is the integer part of $(N+1)/2$ and for an FIR filter K is the integer part of $(M+1)/2$. Note that a cascade filter may also have one or several first-order sections with real-valued zeros and/or real-valued poles and that when N or M are odd numbers there will be at least one first-order zero or first-order pole in the cascade expression.

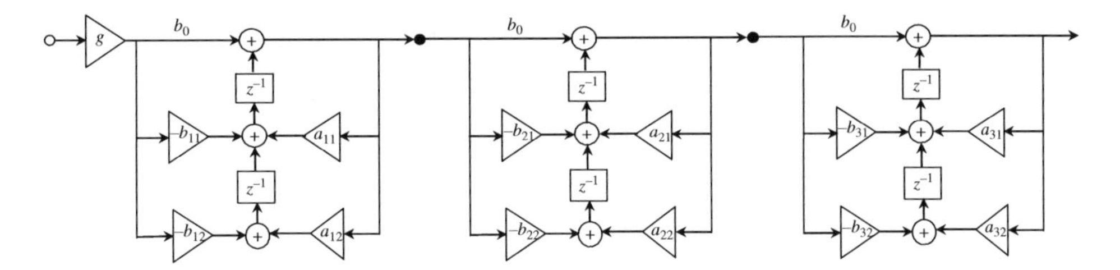

Figure 4.7 Realisation of an IIR cascade structure from second order sections.

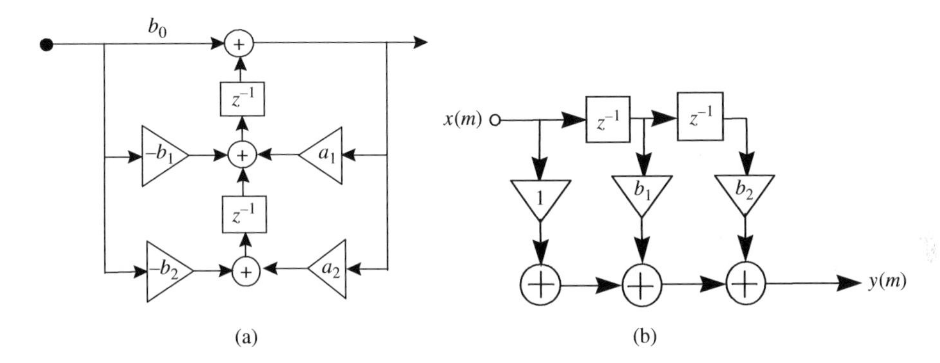

Figure 4.8 Realisation of a second-order section of (a) an IIR filter, (b) an FIR filter.

For an IIR filter each second-order cascade section has the form

$$H_k(z) = \frac{1 + b_{k1}z^{-1} + b_{k2}z^{-2}}{1 + a_{k1}z^{-1} + a_{k2}z^{-2}} \tag{4.25}$$

For an FIR filter, as shown in Figure 4.8(b), each second-order cascade section has the form

$$H_k(z) = 1 + b_{k1}z^{-1} + b_{k2}z^{-2} \tag{4.26}$$

4.5.3 Parallel Filter Structure

An alternative to the cascade implementation described in the previous section is to express the filter transfer function $H(z)$, using the partial fraction method, in a parallel form as the parallel sum of a number of second-order and first-order terms as

$$H(z) = K + \sum_{k=1}^{N_1} \frac{e_{0k} + e_{1k}z^{-1}}{1 - a_{1k}z^{-1} - a_{2k}z^{-2}} + \sum_{k=1}^{N_2} \frac{e_k}{1 - a_k z^{-1}} \tag{4.27}$$

or

$$H(z) = K + \sum_{k=1}^{N_1+N_2} H_k(z) \tag{4.28}$$

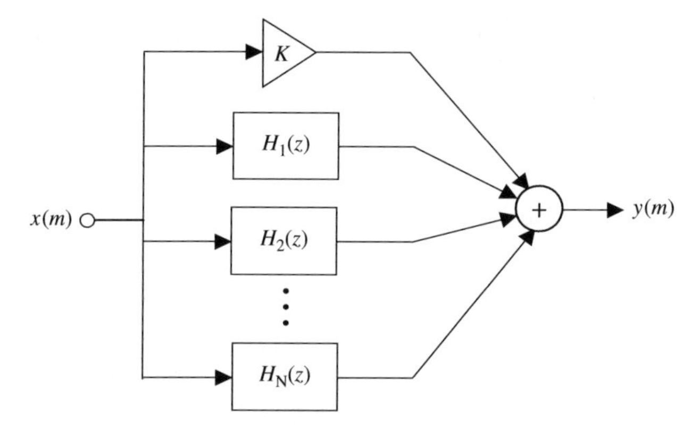

Figure 4.9 A parallel-form filter structure.

where in general the filter is assumed to have N_1 complex conjugate poles, N_1 real zeros and N_2 real poles and K is a constant. Figure 4.9 shows a parallel filter structure.

MatLab provides a function (called *residue*) for conversion of a direct form polynomial to a parallel form polynomial.

4.6 Linear Phase FIR Filters

Linearity of phase is a desirable property in many filtering situations. Phase is essentially the frequency equivalent of delay, and linearity implies that the output frequencies, harmonics etc. will have the same timing relationships with respect to each other as the input frequencies. For example, in the case of musical signals the timing relationships of the output frequencies (or musical notes and harmonics) of a linear phase system would be the same as those of the input frequencies.

It takes a finite amount of time before the input signal to a filter propagates through the filter and appears at the output. A delay of T seconds in time domain appears as a phase change of $e^{-j2\pi Tf}$ in frequency domain. The time delay, or phase change, caused by a filter depends on the type of the filter (e.g. linear/non-linear, FIR/IIR) and its impulse response. An FIR filter of order M, with $M+1$ coefficients, has a time delay of $M/2$ samples.

A linear phase filter is one whose phase $\phi(f)$ is a linear function of the frequency variable f. In the following it is shown that an FIR filter with a symmetric impulse response $h(k)$, or equivalently symmetric coefficients, has a linear phase response. Consider an FIR filter with a symmetric impulse response as

$$h(k) = h(M-k) \qquad k = 0, 1, \ldots, M \tag{4.29}$$

The frequency response of the filter is the Fourier transform of its impulse response, given by

$$H(f) = \sum_{k=0}^{M} h(k)e^{-j2\pi fk} \tag{4.30}$$

Expanding Equation (4.30) and using the assumed symmetry of the filter's impulse response – i.e. $h(0) = h(M)$, $h(1) = h(M-1)$ and so on – we have

$$H(f) = h(0) + h(1)e^{-j2\pi f} + h(2)e^{-j4\pi f} + \cdots + h(M-1)e^{-j2(M-1)\pi f} + h(M)e^{-j2M\pi f}$$

$$= e^{-jM\pi f}\left[h(M/2) + \underbrace{h(0)e^{jM\pi f} + h(M)e^{-jM\pi f}}_{2h(0)\cos(M\pi f)} + \underbrace{h(1)e^{j(M-2)\pi f} + h(M-1)e^{-j(M-2)\pi f}}_{2h(1)\cos((M-2)\pi f)} + \cdots\right] \tag{4.31}$$

where it is assumed that the filter length $M+1$ is an odd integer. Hence the frequency response of the filter $H(f)$ can be expressed as

$$H(f) = e^{-jM\pi f}\left[h(M/2) + \sum_{m=0}^{(M-1)/2} 2h(m)\cos((M-2m)\pi f)\right] \tag{4.32}$$

The term in the bracket is real-valued. The filter phase is the term $e^{-jM\pi f}$, and the filter phase response is given by

$$\phi(f) = \begin{cases} -M\pi f & \text{if } H_a(f) \geq 0 \\ -M\pi f + \pi & \text{if } H_a(f) < 0 \end{cases} \tag{4.33}$$

where $H_a(f)$ is the real-valued term within the bracket of Equation (4.32). Thus the phase response of an FIR filter with a symmetric impulse response is a linear function of the frequency variable f. When the amplitude of the frequency response equation (4.32) goes negative the phase undergoes a change of π radians.

4.6.1 Location of the Zeros of a Linear Phase FIR Filter

The transfer function of an FIR filter with a symmetric impulse response $[h(0) = h(M), h(1) = h(M-1), \ldots]$ can be rearranged and written as

$$\begin{aligned} H(z) &= \sum_{m=0}^{M} h(m)z^{-m} = h(0) + h(1)z^{-1} + \cdots + h(M)z^{-M} \\ &= h(M) + h(M-1)z^{-1} + \cdots + h(0)z^{-M} \\ &= z^{-M}\sum_{m=0}^{M} h(m)z^{m} = z^{-M}H(z^{-1}) \end{aligned} \tag{4.34}$$

Form Equation (4.34) we have the relationship $H(z) = z^{-M}H(z^{-1})$ which implies that if $H(z)$ has a zero at $z = z_1$ then it must also have a zero at $z = 1/z_1$. Therefore, in general the z-transfer function of a linear phase FIR filter can be factorised as

$$H(z) = G\underbrace{(1 - r_1e^{-j\varphi_1}z^{-1})(1 - r_1e^{j\varphi_1}z^{-1})}_{\text{Complex conjugate zero pair}}\underbrace{(1-(1/r_1)e^{-j\varphi_1}z^{-1})(1-(1/r_1)e^{j\varphi_1}z^{-1})}_{\text{Reciprocal zero pair}}\cdots \tag{4.35}$$

where r_k and φ_k are the radius and angular frequency of the kth complex pair of zeros. Thus the zeros of a linear phase filter occur in reciprocal pairs mirrored w.r.t. the unit circle, the exception being when the zeros are on the unit circle.

Example 4.1 The input–output relationship of a finite impulse response filter is given as

$$y(m) = x(m) - 2.5x(m-1) + 5.25x(m-2) - 2.5x(m-3) + x(m-4) \tag{4.36}$$

(i) Write the z-transfer function of this filter.
(ii) Show that the filter has a linear phase response.
(iii) Find and plot the zeros of this filter and explain the constraints on the position of the zeros of a linear phase filter.

Solution: The z-transfer function of the finite impulse response filter $H(z)$ is given by

$$H(z) = 1 - 2.5z^{-1} + 5.25z^{-2} - 2.5z^{-3} + z^{-4} \tag{4.37}$$

$$\begin{aligned} H(f) &= 1 - 2.5e^{-j2\pi f} + 5.25e^{-j4\pi f} - 2.5e^{-j6\pi f} + e^{-j8\pi f} \\ &= e^{-j4\pi f}\left[5.25 + \left(e^{j4\pi f} + e^{-j4\pi f}\right) - \left(2.5e^{j2\pi f} + 2.5e^{-j2\pi f}\right)\right] \end{aligned} \tag{4.38}$$

Hence the frequency response of the filter $H(f)$ can be expressed as

$$H(f) = e^{-j4\pi f}\left[5.25 + 2\cos(4\pi f) - 5\cos(2\pi f)\right] \tag{4.39}$$

The filter's phase response is given by

$$\phi(f) = \begin{cases} -4\pi f & \text{if } H_a(f) \geq 0 \\ -4\pi f + \pi & \text{if } H_a(f) < 0 \end{cases} \tag{4.40}$$

where $H_a(f)$ is the real-valued term within the bracket of Equation (4.45). The transfer function $H(z)$ can be factorised as

$$\begin{aligned} H(z) &= \left(1 - 0.5z^{-1} + 0.25z^{-2}\right)\left(1 - 2z^{-1} + 4z^{-2}\right) \\ &= \left(1 - 0.5e^{-j\pi/3}z^{-1}\right)\left(1 - 0.5e^{j\pi/3}z^{-1}\right)\left(1 - 2e^{-j\pi/3}z^{-1}\right)\left(1 - 2e^{j\pi/3}z^{-1}\right) \end{aligned} \tag{4.41}$$

Note that the factorisation in the first line of Equation (4.41) can be achieved by writing $H(z) = \left(1 + az^{-1} + bz^{-2}\right)\left(1 + cz^{-1} + dz^{-2}\right)$ and comparing the product of the coefficients with the coefficient

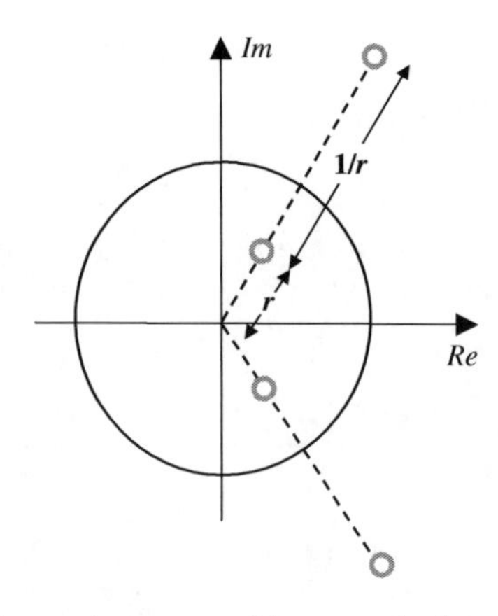

Figure 4.10 Symmetric structure of the zeros of a linear phase FIR filter.

values in Equation (4.37). The factorisation in the second line of (4.41) can be obtained from the roots of the quadratic equations in the first line. The zeros of the filter are positioned at radii of 0.5 and 2 and angles of ±60° as shown in Figure 4.10.

Example 4.2 Consider a unit delay filter with the following impulse response.

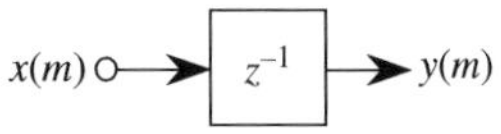

$$h(m) = \delta(m-1) = \begin{cases} 1 & m=1 \\ 0 & \text{otherwise} \end{cases} \qquad (4.42)$$

The frequency response of the filter is obtained as the Fourier transform of the impulse response as

$$H(f) = \sum_{m=0}^{M} h(m)e^{-\mathrm{j}2\pi mf} = e^{-\mathrm{j}2\pi f} \qquad (4.43)$$

The filter magnitude frequency response is

$$|H(f)| = |e^{-\mathrm{j}2\pi f}| = 1 \qquad (4.44)$$

and the filter phase response is

$$\phi(f) = -2\pi f \qquad (4.45)$$

This filter has a unity magnitude response and a linear phase response shown in Figure 4.11. This is called a pure delay filter as its only effect is to delay the input.

Example 4.3 Consider a filter with the following impulse response

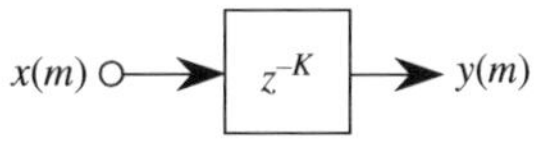

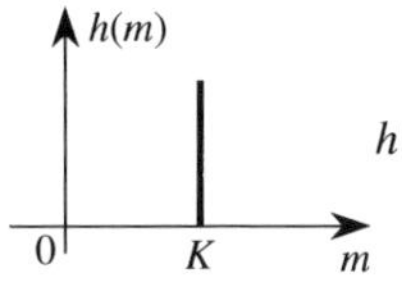

$$h(m) = \delta(m-K) = \begin{cases} 1 & m=K \\ 0 & \text{otherwise} \end{cases} \qquad (4.46)$$

The frequency response of the filter is obtained as the Fourier transform of the impulse response as

$$H(f) = \sum_{m=0}^{M} h(m)e^{-\mathrm{j}2\pi mf} = e^{-\mathrm{j}2K\pi f} \qquad (4.47)$$

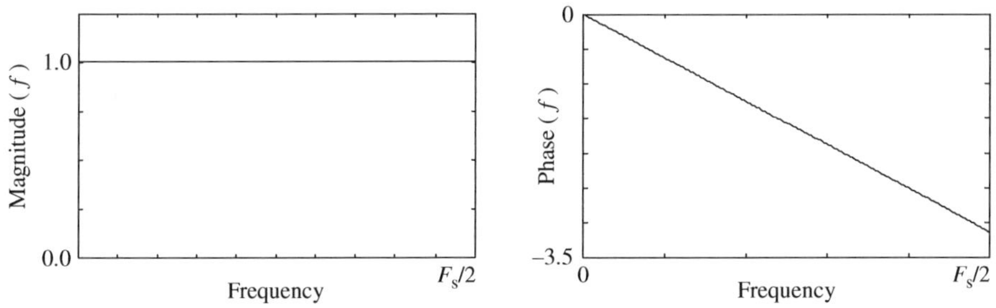

Figure 4.11 Magnitude frequency and phase response of a unit delay filter.

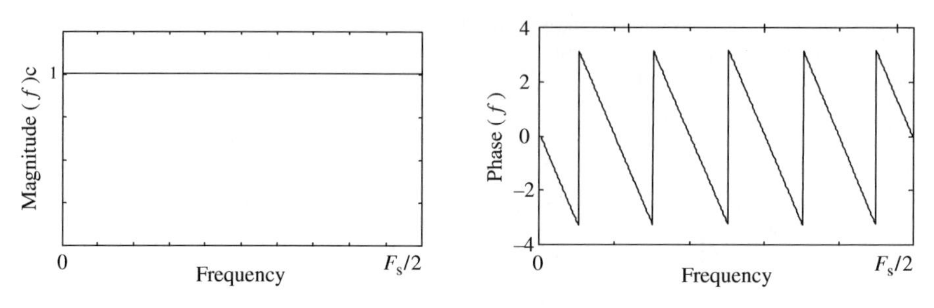

Figure 4.12 Magnitude and phase response of a delay filter with a delay of 10 units.

The filter magnitude response is

$$|H(f)| = |e^{-j2\pi K f}| = 1 \tag{4.48}$$

and the filter phase response is

$$\phi(f) = -2K\pi f \tag{4.49}$$

This filter has a unity magnitude response and a linear phase response. This is a pure delay filter as its only effect is to delay the input by K samples. In the phase plot of Figure 4.12, at the frequency of 0 Hz the phase $\phi(0) = 0$, at $f = 1/2K$ the phase is $\phi(1/2K) = -\pi$. The phase plot shows the periodic circular nature of the phase in that $e^{j(-\pi-\phi)} = e^{j(2\pi-\pi-\phi)} = e^{j(\pi-\phi)}$, so that there is a jump to π whenever the phase becomes smaller (more negative) than $-\pi$, as shown in Figure 4.12.

Example 4.4 Consider a filter with the following impulse response.

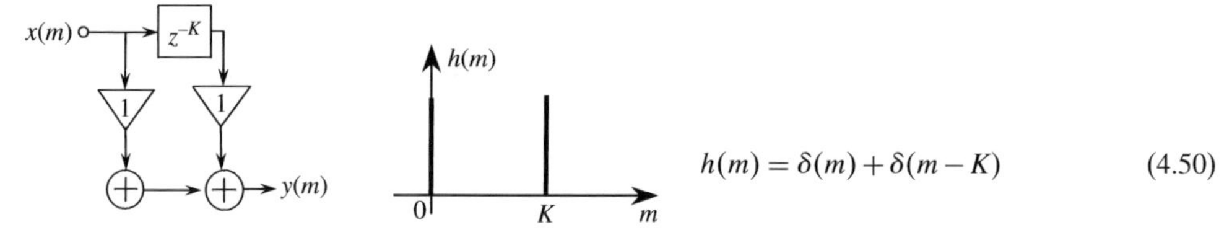

$$h(m) = \delta(m) + \delta(m-K) \tag{4.50}$$

The frequency response of the filter is obtained as the Fourier transform of the impulse response as

$$\begin{aligned} H(f) &= \sum_{m=0}^{M} h(m)e^{-j2\pi m f} = 1 + e^{-j2K\pi f} \\ &= e^{-jK\pi f}(e^{jK\pi f} + e^{-jK\pi f}) = e^{-jK\pi f} 2\cos(K\pi f) \end{aligned} \tag{4.51}$$

This filter's magnitude frequency response is comb-shaped and given by

$$|H(f)| = \left|e^{-jK\pi f} 2\cos(K\pi f)\right| = |2\cos(K\pi f)| \tag{4.52}$$

and the filter phase response is

$$\phi(f) = \begin{cases} -K\pi f & \text{if } \cos(K\pi f) \geq 0 \\ -K\pi f + \pi & \text{if } \cos(K\pi f) < 0 \end{cases} \tag{4.53}$$

Figure 4.13 shows the filter magnitude and the phase responses for a value of $K = 4$.

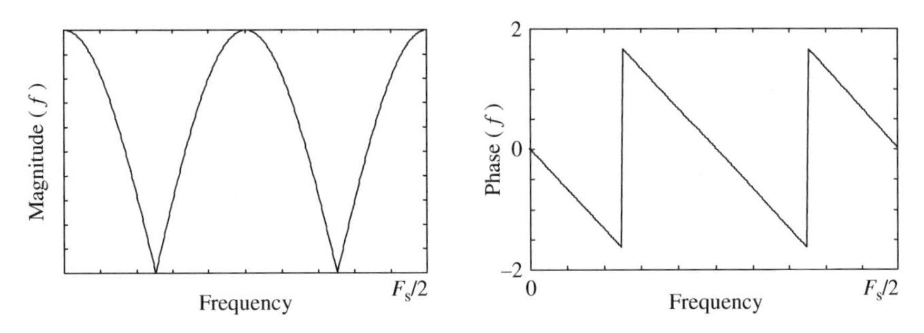

Figure 4.13 Magnitude frequency and phase response of a comb filter.

Example 4.5 Consider a filter with the following impulse response.

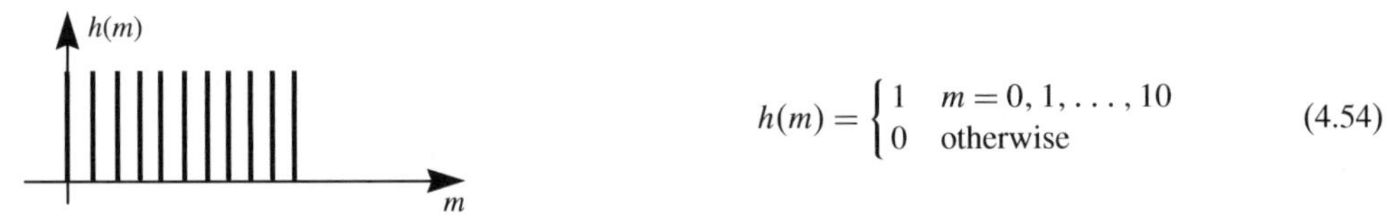

$$h(m) = \begin{cases} 1 & m = 0, 1, \ldots, 10 \\ 0 & \text{otherwise} \end{cases} \tag{4.54}$$

The filter length is 11 samples long. The frequency response of this filter is obtained from the Fourier transform of the impulse response and using the convergence formula for a geometric series, is

$$H(f) = \sum_{m=0}^{10} e^{-j2\pi f m T_s} = \frac{1 - e^{-j22\pi f}}{1 - e^{-j2\pi f}} = \frac{e^{-j11\pi f}}{e^{-j\pi f}} \times \frac{e^{j11\pi f} - e^{-j11\pi f}}{e^{j\pi f} - e^{-j\pi f}} = e^{-j10\pi f} \frac{\sin(11\pi f)}{\sin(\pi f)} \tag{4.55}$$

and the filter phase response is

$$\phi(f) = \begin{cases} -10\pi f & \text{if } H_a(f) \geq 0 \\ -10\pi f + \pi & \text{if } H_a(f) < 0 \end{cases} \tag{4.56}$$

where $H_a(f) = \sin(11\pi f)/\sin(\pi f)$. Note that at $f = 0$, $H(f) = \sin(Af)/\sin(Bf) = 0/0$ is undefined, and is obtained as $H(f) = A\cos(Af)/B\cos(Bf) = A/B$. Notice two trends in the phase response plot of Figure 4.14. First, at the points when $H_a(f)$ changes sign (a change of sign is equivalent to

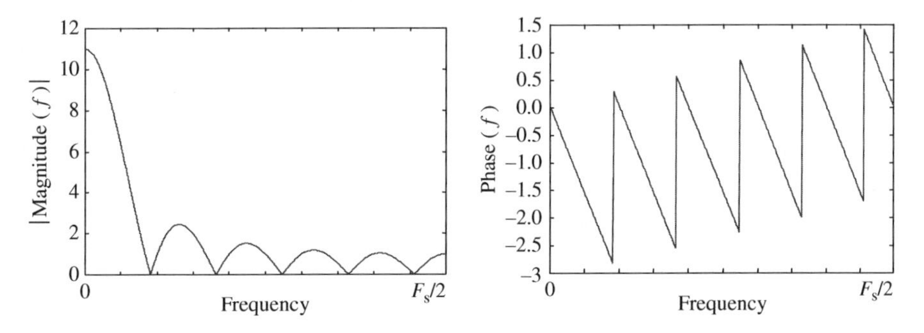

Figure 4.14 Magnitude and phase response of filter with a rectangular impulse response.

multiplying $H_a(f)$ by -1 or $e^{j\pi}$) there is a sudden jump of π in phase. Second, depending on the filter coefficients, $H_a(f)$ may change sign before $e^{-j10\pi f}$ attains a phase of $-\pi, -2\pi\ldots$ causing a gradual positive shift in the phase as shown in the ramp trend of Figure 4.14.

4.6.2 Design of FIR Filters by Windowing

A simple way to design an FIR filter is through the inverse Fourier transform of the desired frequency response. Consider an FIR filter described by the difference equation

$$y(m) = \sum_{k=0}^{M} b_k x(m-k) \tag{4.57}$$

where b_k are the filter coefficients, M is the filter order, and $x(m)$ and $y(m)$ are the filter's input and output signals respectively. Now, when an FIR filter equation (4.57) is excited with an impulse input then it is easy to see, as shown in Figure 4.2, that the impulse response of the FIR filter is identical to the coefficient sequence of the filter $\{b_k\}$ that is

$$h(k) = \begin{cases} b_k & 0 \le k \le M \\ 0 & \text{otherwise} \end{cases} \tag{4.58}$$

Hence the filtering equation (4.58) can be expressed as the convolution of the impulse response of the FIR filter $h(m)$ with the input signal $x(m)$ as

$$y(m) = \sum_{k=0}^{M} h(k) x(m-k) \tag{4.59}$$

This observation forms the basis for the method of FIR filter design by windowing. In the window design method we begin with the desired frequency response $H_d(f)$ and obtain the corresponding impulse response $h_d(m)$ which, for an FIR filter, is the filter coefficients.

The frequency response $H_d(f)$ and the impulse response $h_d(m)$ of a linear filter are related by the Fourier transform pair as

$$H_d(f) = \sum_{m=0}^{M} h_d(m) e^{-j2m\pi f} \tag{4.60}$$

$$h_d(m) = \int_{-1/2}^{1/2} H_d(f) e^{j2m\pi f} df \tag{4.61}$$

Thus given the desired filter frequency response $H_d(f)$ we can determine the impulse response $h_d(m)$ by evaluating the above Fourier integral. However, there are two problems here. First, the filter has infinite duration impulse response. Second, the filter is non-causal, as it will have non-zero values for coefficients with negative time index. Note that a negative-indexed coefficient, e.g. $h(-k)$, requires a future sample value $x(m+k)$; this makes the filter non-causal for real-time operations. These problems are solved by first multiplying $h_d(m)$ by a truncation window of length $(M+1)$ samples, and then shifting the truncated impulse response by $M/2$ samples in the positive time direction.

The FIR filter window design technique involves the following steps:

(1) Start with the desired frequency response $H_d(f)$.
(2) Choose a filter order M.
(3) Obtain $h_d(m)$ the inverse Fourier transform of $H_d(f)$, from Equation (4.61).
(4) Window $h_d(m)$ to obtain $M+1$ FIR filter coefficients centred at $m=0$.
(5) Shift the windowed impulse response by $M/2$ samples for causality.

4.6.3 Influence of the Choice of Window in FIR Filter Design

After windowing the impulse response of the FIR filter can be described as

$$h(m) = w(m)h_d(m) \tag{4.62}$$

Since multiplication in time is equivalent to convolution in the frequency domain, the frequency response $H(f)$ of the truncated FIR filter $h(k)$ is given by

$$H(f) = \int_{-1/2}^{1/2} H_d(\varphi)W(f-\varphi)d\varphi \tag{4.63}$$

where $W(f)$, the frequency response of a rectangular window $w(m)$, is obtained from a Fourier transform. Using the convergence formula for a geometric series, $W(f)$ can be expressed as

$$W(f) = \sum_{m=0}^{M} w(m)e^{-j2m\pi f} = \frac{1-e^{-j2(M+1)\pi f}}{1-e^{-j2\pi f}} = e^{-jM\pi f}\frac{\sin((M+1)\pi f)}{\sin(\pi f)} \tag{4.64}$$

Note that at $f=0$, $W(f) = \sin(Af)/\sin(Bf) = 0/0$ is undefined, and is obtained as $W(f) = A\cos(Af)/B\cos(Bf) = A/B$. The sinc-shaped magnitude spectrum of the rectangular window is illustrated in Figure 4.14(a) for $M = 32, 128$. The bandwidth of the main lobe is $2F_s/M$, where F_s is the sampling frequency.

The convolution of the rectangular window $W(f)$ with the filter $H_d(f)$ has the two main effects of (1) introducing ripples in the frequency spectrum of the filter and (2) extending the transition regions of the filter. As the filter order M increases the main lobe width and the amplitude of the side lobes decrease.

The side lobes in the frequency domain are a manifestation of the discontinuities in time domain at the edges of a rectangular window and can be alleviated by the use of windows that do not contain sharp discontinuities and roll down rather gently to zero, such as the raised cosine windows of Hamming, Hanning, and Blackman windows described in Section 2.4.3 and reproduced here for convenience. The general form of the raised cosine window equation is given by

$$w(m) = \begin{cases} \alpha - (1-\alpha)\cos\frac{2\pi m}{M+1} & 0 \le m \le M \\ 0 & \text{otherwise} \end{cases} \tag{4.65}$$

For $\alpha = 0.5$ we have the Hanning window

$$w_{\text{Hanning}}(m) = 0.5 - 0.5\cos\frac{2\pi m}{M+1} \quad 0 \le m \le M \tag{4.66}$$

For $\alpha = 0.54$ we have the Hamming window

$$w_{\text{Hamming}}(m) = 0.54 - 0.46\cos\frac{2\pi m}{M+1} \quad 0 \le m \le M \tag{4.67}$$

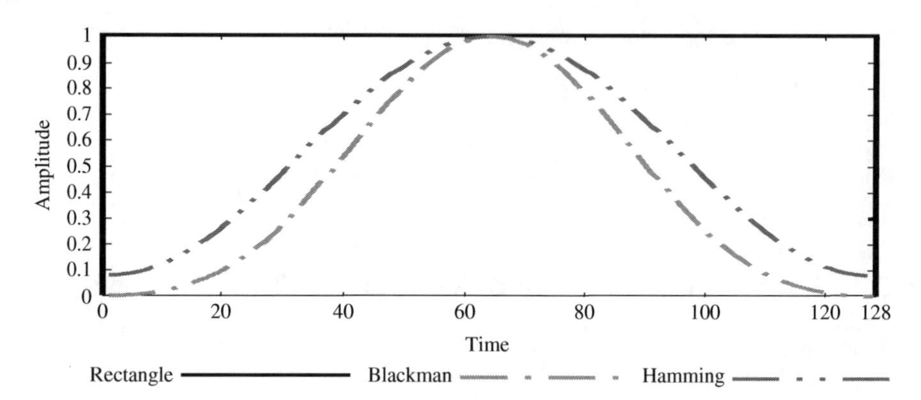

Figure 4.15 Plot of (a) rectangular window, (b) Hamming window, (c) Blackman window for a window 128 samples.

The Blackman window is given by

$$w_{\text{Blackman}}(m) = 0.42 - 0.5\cos\frac{2\pi m}{M+1} - 0.08\cos\frac{4\pi m}{M+1} \quad 0 \le m \le M \tag{4.68}$$

Figure 4.15 shows the rectangular, the Hamming and the Blackman windows. Figure 4.16 shows the frequency spectrum of these widows for window lengths of 32 samples and 128 samples. As shown, the Blackman window has considerably smaller side-lobes but at the expense of a higher bandwidth of the main-lobe.

4.6.4 Design of a Low Pass Linear-Phase FIR Filter Using Windows

Example 4.6 Consider the design of a low-pass linear-phase digital FIR filter operating at a sampling rate of F_s Hz and with a cut-off frequency of F_c Hz. The frequency response of the filter is given by

$H(f)$, 1, $-F_c$, F_c, f

$$H_d(f) = \begin{cases} 1.0 & |f| < F_c \\ 0 & \text{otherwise} \end{cases} \tag{4.69}$$

The impulse response of this filter is obtained via the inverse Fourier integral as

$$h_d(m) = \int_{-F_c/F_s}^{F_c/F_s} 1.0\, e^{j2\pi fm} df = \left.\frac{e^{j2\pi fm}}{j2\pi m}\right|_{-F_c/F_s}^{F_c/F_s} = \frac{\sin(2\pi m F_c/F_s)}{\pi m} \tag{4.70}$$

where F_c/F_s is the normalised cut-off frequency. Clearly $h_d(m)$ is of infinite duration and non-causal as it is non-zero for $m < 0$. Note that in the FIR filter equation (4.12) a negative index value of $h(-m)$ such as $h(-10)$ would require a future sample value $x(m+10)$ hence it is non-causal.

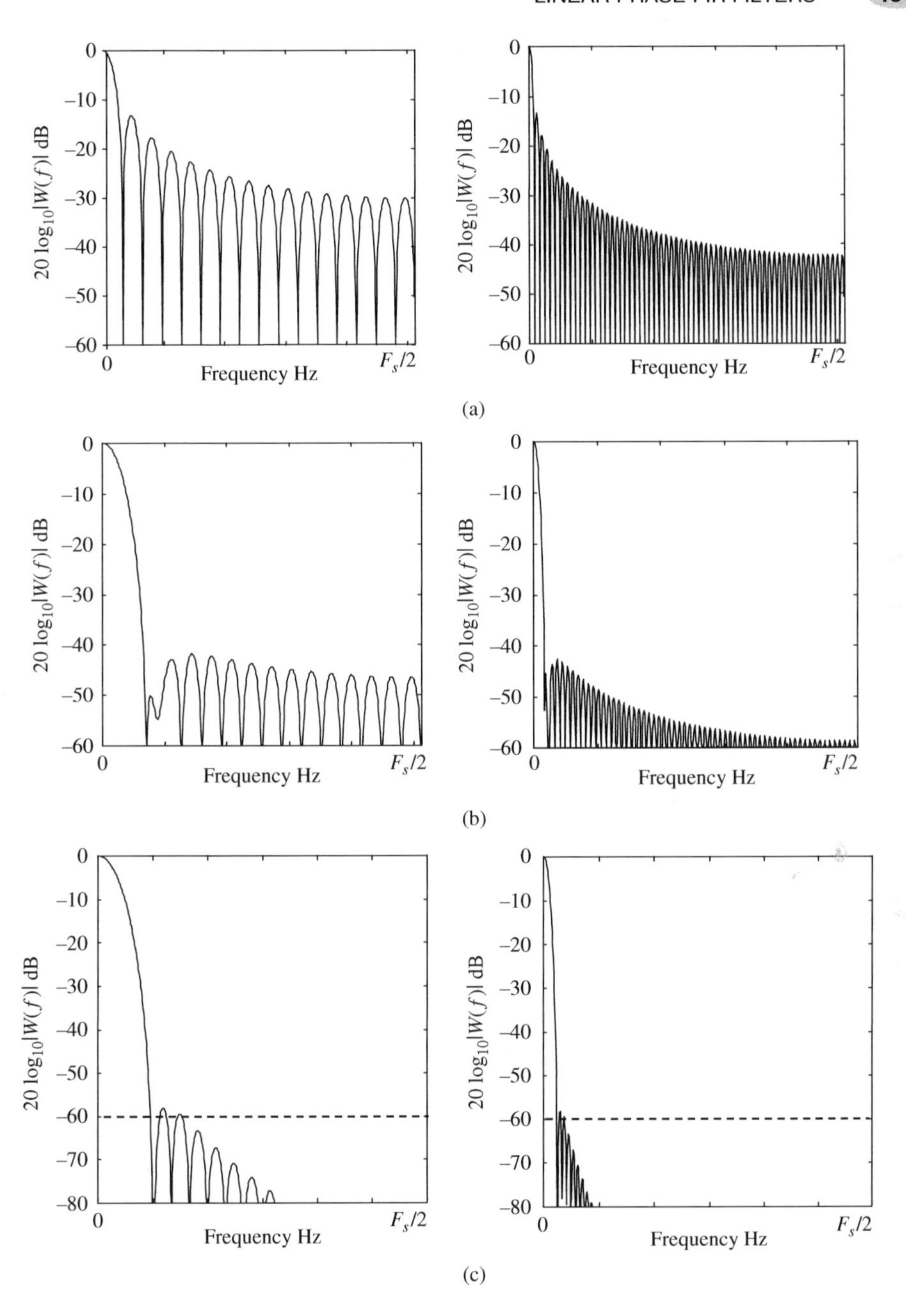

Figure 4.16 The spectra of (a) rectangular window, (b) Hamming window, (c) Blackman window for a window length of 32 samples (left) and 128 samples (right).

To obtain an FIR filter of order M we multiply $h_d(m)$ by a rectangular window sequence of length $M+1$ samples centred at $m=0$. To introduce causality ($h(m)=0$ for $m<0$) shift $h(m)$ by $M/2$ samples

$$h(m) = \frac{\sin(2\pi(m-M/2)F_c/F_s)}{\pi(m-M/2)} \quad 0 \le m \le M \tag{4.71}$$

The impulse response and frequency response of a low pass filter for the filter orders of $M=30$ and $M=100$ are shown in Figure 4.17. Observe the relatively large oscillations or the ripples in the pass-band and the stop-band. These ripples increase in frequency as M the filter order is increased. These oscillations are a direct result of the convolution with the side-lobes of the rectangular window.

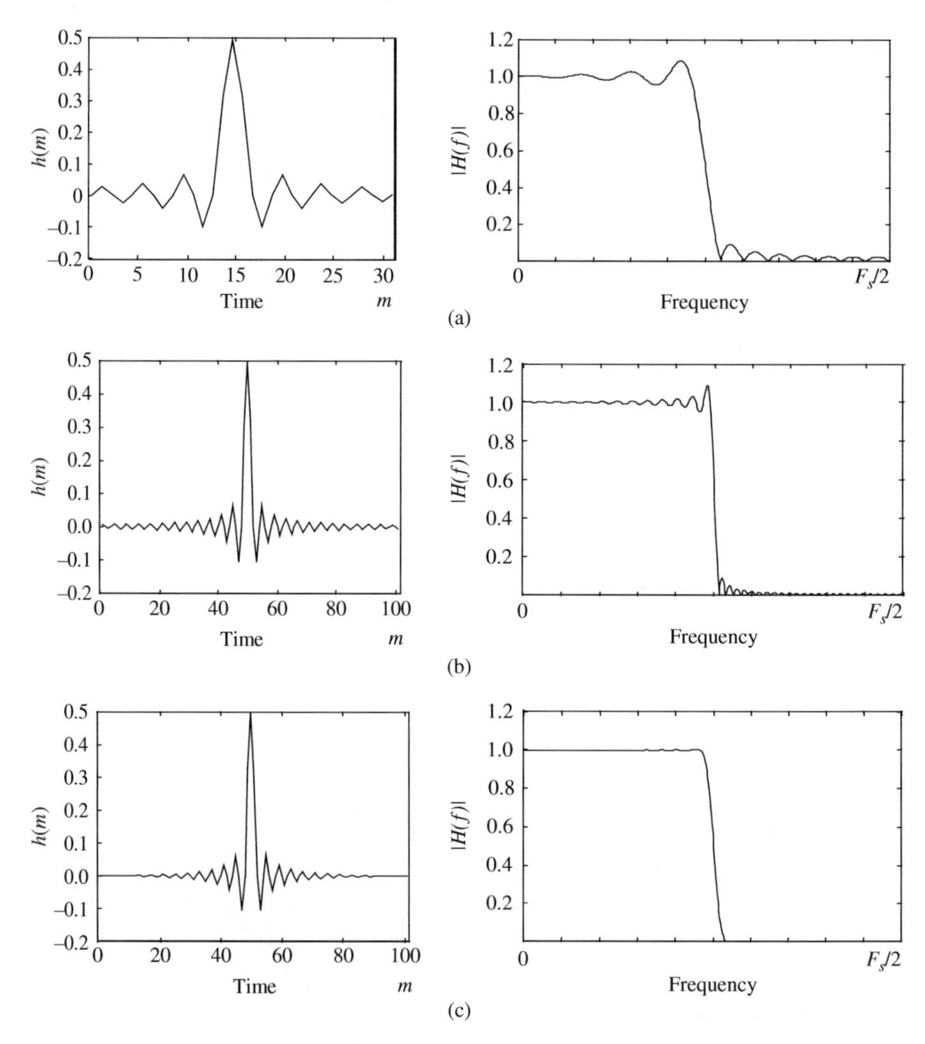

Figure 4.17 The impulse response and the magnitude frequency response of an FIR filter for the following cases: (a) filter order 30 with a rectangular window, (b) filter order 100 with a rectangular window, (c) filter order 100 with a Hamming window.

Matlab function FIR_Low-pass( )
Animation experiment on the design and response of low-pass filter using equation h(m) = sin(2*pi*Fc*(m-M/2-1))/(pi*(m-M/2 − 1)). The effect of different windows on the frequency response of the pass band, stop band and transition band are explored. M = filter order, Fc = cut-off frequency. The effect of varying filter order M is animated.

To reduce the ripples in both the pass-band and the stop-band we can use a window function that decays to zero gradually instead of the abrupt discontinuous transition that happens at the edges of a rectangular window. Figure 4.17(c) illustrates how the use of a Hamming window reduces the ripples in the pass-band and the stop-band. The 'price' paid is an increase in the bandwidth of the transition band.

4.6.5 Design of High-Pass FIR Filters

Example 4.7 Consider the design of a high-pass linear-phase digital FIR filter operating at a sampling rate of F_s Hz and with a cut-off frequency of F_c Hz. The frequency response of the filter is given by

$$H_d(f) = \begin{cases} 0 & |f| < F_c \\ 1.0 & \text{otherwise} \end{cases} \tag{4.72}$$

The impulse response of this filter is obtained via the inverse Fourier integral as

$$\begin{aligned} h_d(m) &= \int_{-0.5}^{-F_c/F_s} 1.0 e^{j2m\pi f} df + \int_{F_c/F_s}^{0.5} 1.0 e^{j2m\pi f} df = \left. \frac{e^{j2m\pi f}}{j2m\pi} \right|_{-0.5}^{-F_c/F_s} + \left. \frac{e^{j2m\pi f}}{j2m\pi} \right|_{F_c/F_s}^{0.5} \\ &= \frac{\sin(\pi m)}{m\pi} - \frac{\sin(2\pi m F_c/F_s)}{m\pi} \end{aligned} \tag{4.73}$$

where F_c/F_s is the normalised cut-off frequency. Note from Equation (4.74) that the impulse response of a high-pass filter is the impulse response of an all-pass filter minus the impulse response of a low-pass filter. The impulse response $h_d(m)$ is non-causal (it is non-zero for $m < 0$) and infinite in duration. To obtain an FIR filter of order M we multiply $h_d(m)$ by a rectangular window sequence of length $M + 1$ samples. To introduce causality ($h(m) = 0$ for $m < 0$) shift truncated $h(m)$ by $M/2$ samples

$$h(m) = \underbrace{\frac{\sin(\pi(m - M/2))}{(m - M/2)\pi}}_{\text{All-passfilter}} - \underbrace{\frac{\sin(2\pi(m - M/2)F_c/F_s)}{(m - M/2)\pi}}_{\text{Low-passfilter}} \quad 0 \le m \le M \tag{4.74}$$

Note that a high-pass FIR filter is also equivalent to the configuration shown in Figure 4.18.

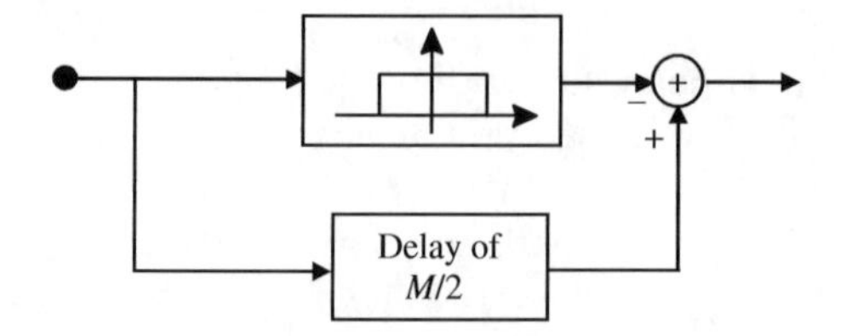

Figure 4.18 Design of a high-pass FIR filter from a low-pass filter.

Matlab function FIR_High-pass( )
Animation demonstration of the design and response of high-pass filter using

h(m) = sin(pi*(m-M/2))/(pi*(m-M/2-1))- sin(2*pi*Fc*(m-M/2-1))/(pi*(m-M/2-1))

The effect of different windows on the frequency response of the pass-band, stop-band and transition-band are explored. M = filter order, Fc = cut-off frequency. The effect of varying filter order M is animated.

4.6.6 Design of Band-Pass FIR Filters

Example 4.8 Consider the design of a band-pass linear-phase digital FIR filter operating at a sampling rate of F_s Hz and with lower and higher cut-off frequencies of F_L and F_H Hz. The frequency response of the filter is given by

$$H_d(f) = \begin{cases} 1 & F_L < |f| < F_H \\ 0 & \text{otherwise} \end{cases} \tag{4.75}$$

The impulse response of this filter is obtained via the inverse Fourier integral as

$$\begin{aligned} h_d(m) &= \int_{-F_H/F_S}^{-F_L/F_S} 1.0\, e^{j2m\pi f}\, df + \int_{F_L/F_S}^{F_H/F_S} 1.0\, e^{j2m\pi f}\, df = \left.\frac{e^{j2m\pi f}}{j2m\pi}\right|_{-F_H/F_S}^{-F_L/F_S} + \left.\frac{e^{j2m\pi f}}{j2m\pi}\right|_{F_L/F_S}^{F_H/F_S} \\ &= \frac{\sin(2\pi m F_H/F_S)}{m\pi} - \frac{\sin(2\pi m F_L/F_S)}{m\pi} \end{aligned} \tag{4.76}$$

where F_L/F_S and F_H/F_S are normalised low and high cut-off frequencies. The impulse response $h_d(m)$ is non-causal (it is non-zero for $m < 0$) and infinite in duration. To obtain an FIR filter of order M multiply $h_d(m)$ by a rectangular window sequence of length $M+1$ samples. To introduce causality ($h(m) = 0$ for $m < 0$) shift truncated $h(m)$ by $M/2$ samples

$$h(m) = \underbrace{\frac{\sin(2\pi(m-M/2)F_H/F_s)}{(m-M/2)\pi}}_{\text{Low-passfilter}} - \underbrace{\frac{\sin(2\pi(m-M/2)F_L/F_s)}{(m-M/2)\pi}}_{\text{Low-passfilter}} \qquad 0 \le m \le M \tag{4.77}$$

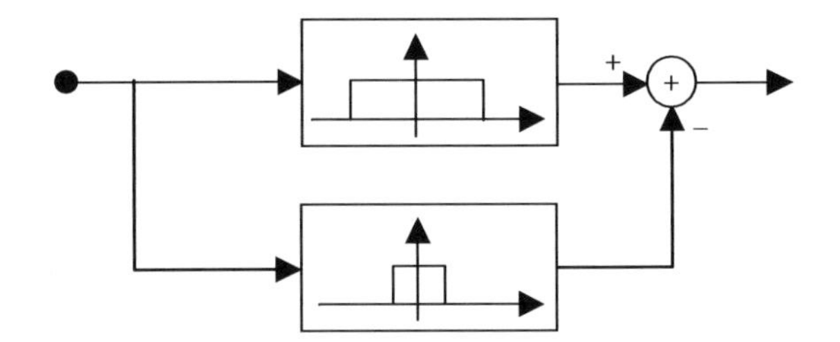

Figure 4.19 Design of a band-pass FIR filter from low-pass FIR filters.

Note from Equation (4.77) that the impulse response of a band pass FIR filter is the difference between the impulse responses of two low-pass FIR filters. A band-pass FIR filter is equivalent to the configuration shown in Figure 4.19.

Matlab function FIR_Bandpass( )
Animation demonstration of the design and response of band-pass filter using

h(m) = sin(2*pi*(m-M/2)*Fh/Fs)/(pi*(m-M/2-1))
- sin(2*pi*(m-M/2-1)*Fl/Fs)/(pi*(m-M/2-1))

The effect of different windows on the frequency response of the pass-band, stop-bands and transition-bands are explored. M = filter order, Fh = higher cut-off frequency, Fl = lower cut-off frequency, Fh = sampling Frequency. The effect of varying filter order M is animated.

4.6.7 Design of Band-Stop FIR Filters

Example 4.9 Consider the design of a band-stop linear-phase digital FIR filter operating at a sampling rate of F_s Hz and with a lower and higher cut-off frequencies of F_L and F_H Hz. The frequency response of the filter is given by

$H(f)$, 1, $-F_H$, $-F_L$, F_L, F_H, f

$$H_d(f) = \begin{cases} 0 & F_L < |f| < F_H \\ 1 & \text{otherwise} \end{cases} \tag{4.78}$$

The impulse response of this filter is obtained via the inverse Fourier integral as

$$\begin{aligned} h_d(m) &= \int_{-0.5}^{-F_H/F_s} 1.0\, e^{j2m\pi f} df + \int_{-F_L/F_s}^{F_L/F_s} 1.0\, e^{j2m\pi f} df + \int_{F_H/F_s}^{0.5} 1.0 e^{j2m\pi f} df \\ &= \frac{\sin \pi m}{m\pi} - \frac{\sin 2\pi m F_H/F_s}{\pi m} + \frac{\sin 2\pi m F_L/F_s}{\pi m} \end{aligned} \tag{4.79}$$

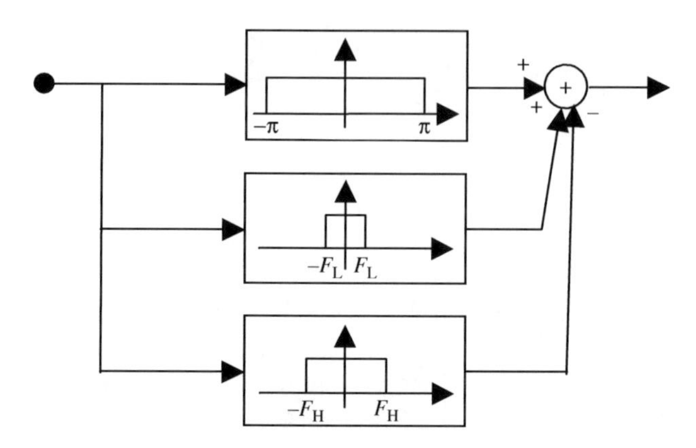

Figure 4.20 Design of a band-stop FIR filter from low-pass FIR filters.

where F_L/F_S and F_H/F_S are normalised low and high cut-off frequencies. The impulse response $h_d(m)$ is non-causal (it is non-zero for $m < 0$) and infinite in duration. To obtain an FIR filter of order M we multiply $h_d(m)$ by a rectangular window sequence of length $M+1$ samples. To introduce causality ($h(m) = 0$ for $m < 0$) shift $h(m)$ by $M/2$ samples

$$h(m) = \underbrace{\frac{\sin \pi(m - M/2)}{(m - M/2)\pi}}_{\text{All-pass}} - \underbrace{\left(\frac{\sin 2\pi(m - M/2)F_H/F_s}{(m - M/2)\pi} - \frac{\sin 2\pi(m - M/2)F_L/F_s}{(m - M/2)\pi} \right)}_{\text{Band-pass}} \tag{4.80}$$

where $0 \le m \le M$. From Equation (4.80) note that the impulse response of a band-stop FIR filter is the difference between the impulse responses of an all-pass filter and a band-pass FIR filter. Note that a band-stop FIR filter is also equivalent to the configuration shown in Figure 4.20.

Matlab function FIR_Bandstop( )
Animation demonstration of the design and response of band-stop filter using equation

h(m) = sin(pi*(m-M/2))/(pi*(m-M/2-1))-(sin(2*pi*(m-M/2)*Fh/Fs)/(pi*(m-M/2-1))-
sin(2*pi*(m-M/2-1)*Fl/Fs)/(pi*(m-M/2-1)))

The effect of different windows on the frequency response of the pass-band, stop-band and transition-band are explored. M = filter order, Fc = cut-off frequency. The effect of varying filter order M is animated.

4.7 Design of Digital FIR Filter-banks

Filter-banks have many applications in signal processing and communication such as in sub-band signal coding, music coders (MPEG and ATRAC), multi-rate communication systems, frequency multiplexing, noise reduction systems and audio graphic equalisers.

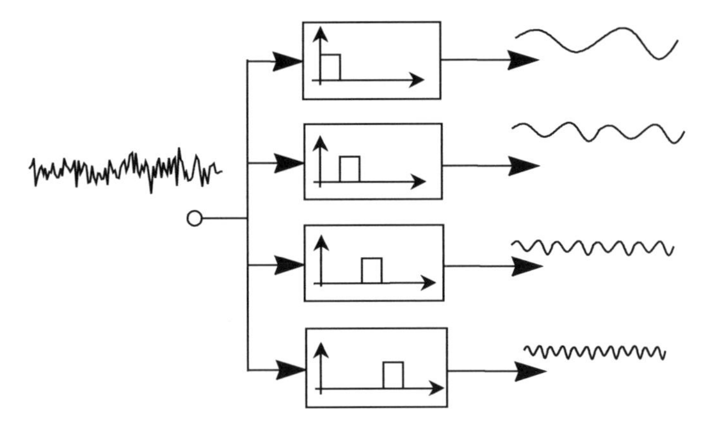

Figure 4.21 A block diagram illustration of a four-band filter-bank.

In its conventional form a filter-bank is a parallel configuration of band-pass filters as shown in Figure 4.21. The sub-band filters may be uniformly spaced in the frequency domain or they may be spaced non-uniformly according to the energy contents of the various bands and/or the human perception of the signals in critical bands of hearing. For example in speech processing it is common to use a perceptually based non-uniform spacing of the sub-band filters such that increasingly higher frequency bands have increasingly larger bandwidths.

The band-pass filters that form a filter-bank can be FIR filters, or IIR filters such as band-pass Butterworth filters. The following example demonstrates a simple procedure for the design of a four-band filter-bank using FIR filters.

Example 4.10 *Design of an FIR Filter-bank*
Design a bank of digital FIR filters for a telephony speech application to split a total bandwidth of 4 kHz into four equal-bandwidth sub-bands.

Matlab function filterbank( )
This function designs a bank of N equal bandwidth FIR filters using a low-pass filter form. The low-pass filter is modulated to act as band-pass filters as described in this section.

Solution: Assume the sampling frequency is 8 kHz. For uniformly spaced band-pass filters the filter cut-off points for the four band-pass filters are 0 and 1 kHz, 1 and 2 kHz, 2 and 3 kHz, and 3 and 4 kHz respectively.

First, we design a low-pass filter with a bandwidth of 1 kHz. At a sampling rate of $F_s = 8$ kHz, a frequency of 1 kHz corresponds to a normalised angular frequency of $\omega_N = 2\pi \times f/F_s = \pi/4$ radians or a normalised frequency of $f_N = f/F_s = 1/8 = 0.124$. Using the window design technique the FIR impulse response is obtained from the inverse Fourier transform as

$$h_d(m) = \int_{-0.125}^{0.125} 1.0 e^{j2m\pi f} df \tag{4.81}$$

Using the solution described in Section 4.6.4 we obtain the windowed FIR filter response as

$$h_1(m) = w(m) \times 0.25\,\mathrm{sinc}(0.25\pi(m - M/2)) \qquad 0 \leq m \leq M \tag{4.82}$$

To design the band-pass filters we can use the amplitude modulation (AM) method to translate a low-pass filter to a band-pass filter. For a band-pass width of 1 kHz the low-pass filter should have cut-off frequencies of ±500 Hz (note that from −500 to +500 Hz we have a bandwidth of 1 kHz). Thus the required low-pass FIR filter equation is similar to Equation (4.82), but with half the bandwidth, and its impulse response is given by

$$h(m) = w(m) \times 0.125\,\mathrm{sinc}\,(0.125\pi(m - M/2)) \qquad 0 \leq m \leq M \tag{4.83}$$

To translate this low-pass filter to the specified band-pass filters we need AM sinusoidal carriers with frequencies 1.5 kHz, 2.5 kHz and 3.5 kHz. The angular frequencies of the sinusoidal carriers are

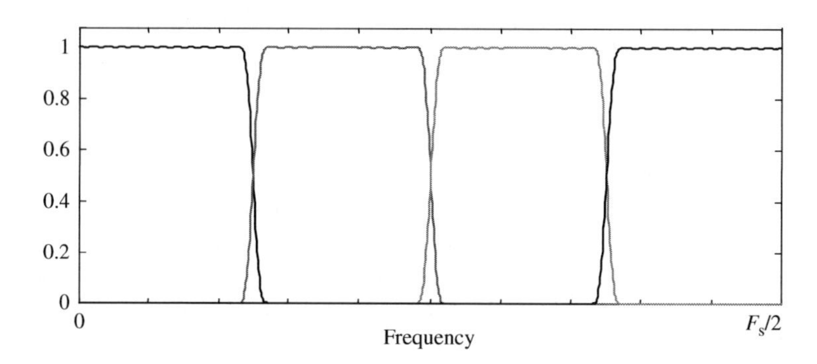

Figure 4.22 The frequency spectrum of a four-band FIR filter.

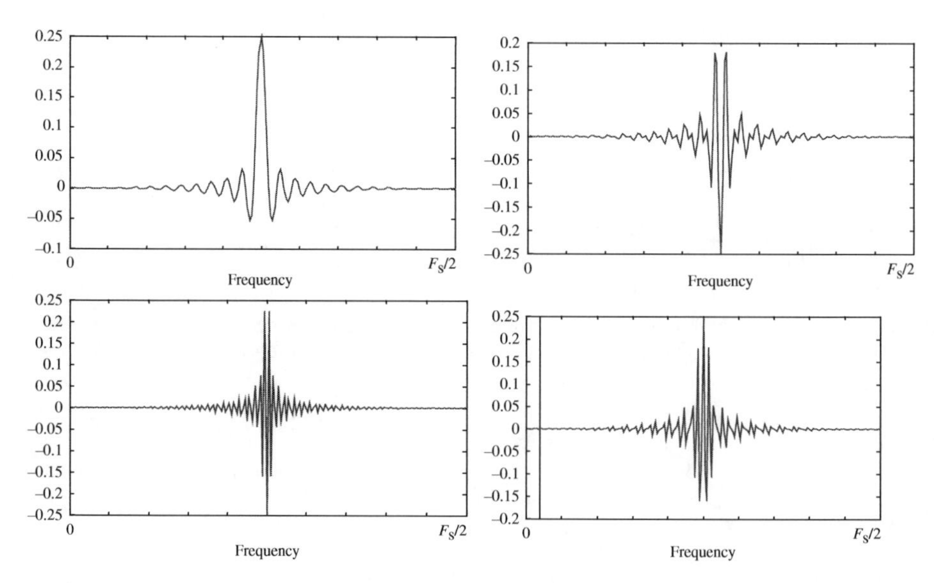

Figure 4.23 The impulse response of the individual band-pass filters of a 4-band filter-bank, plotted clockwise from the top left corner for bands 1 to 4.

obtained by multiplying them by $2\pi/8$ to give $3\pi/8, 5\pi/8$ and $7\pi/8$. The modulated band-pass filter equations are given by

$$h_2(m) = 2 \times w(m) \times 0.125\,\mathrm{sinc}\,(0.125\pi(m - M/2))\sin(3\pi m/8)\ \ 0 \le m \le M \tag{4.84}$$

$$h_3(m) = 2 \times w(m) \times 0.125\,\mathrm{sinc}\,(0.125\pi(m - M/2))\sin(5\pi m/8)\ \ 0 \le m \le M \tag{4.85}$$

$$h_4(m) = 2 \times w(m)0.125\,\mathrm{sinc}\,(0.125\pi(m - M/2))\sin(7\pi m/8)\ \ 0 \le m \le M \tag{4.86}$$

The process of amplitude modulation halves the amplitude of each sideband, hence we have the multiplying factor of 2 in Equations (4.84)–(4.86). Figures 4.22 and 4.23 illustrates the impulse and frequency responses of the individual filters of the four-band filter bank defined by Equations (4.84)–(4.86).

4.8 Quadrature Mirror Sub-band Filters

Quadrature mirror filters are 'perfect reconstruction' filters for sub-band processing. They are used to divide a signal into two equal-width sub-bands with a low-pass and (its mirror) high-pass filter. The signals are then down-sampled. The same low-pass and high-pass filters followed by down-sampling may be applied a number of times to progressively divide the input signal into a desired number of sub-bands as shown in Figure 4.24. During merging, the signals are up-sampled and filtered with

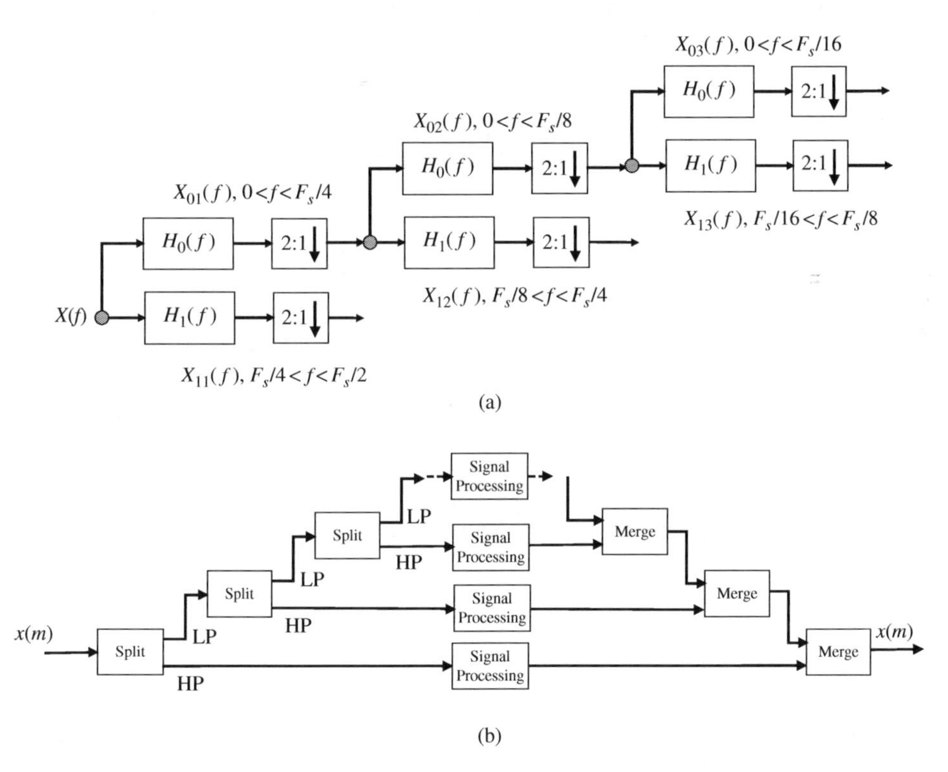

Figure 4.24 Illustration of (a) a tree-structure filterbank that progressively splits the input signal into two equal-width bands using a low-pass filter $H_0(f)$ and a high-pass filter $H_1(f)$ at each stage. (b) The outputs of the filterbank are processed and then merged to reconstitute the processed signal.

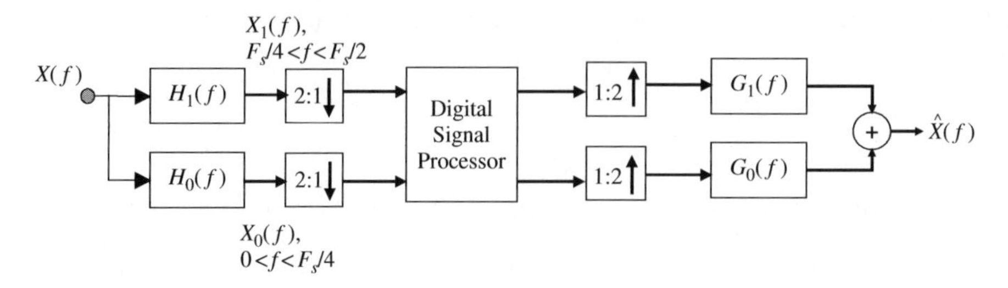

Figure 4.25 A QMF filter splitting a signal into two equal-bandwidth sub-bands. For reconstruction the individual sub-band signals are upsampled and filtered by antialiasing filters.

perfect reconstruction without introducing aliasing noise. QMFs are used for applications such as sub-band music and speech coding or wavelet analysis.

4.8.1 The Basic Two-Band QMF Unit

A two-band quadrature mirror filter unit, shown in Figure 4.25, consists of a low-pass filter $H_0(z)$, a high-pass filter $H_1(z)$, down-samplers, a DSP processor (for such functions as signal coding, noise

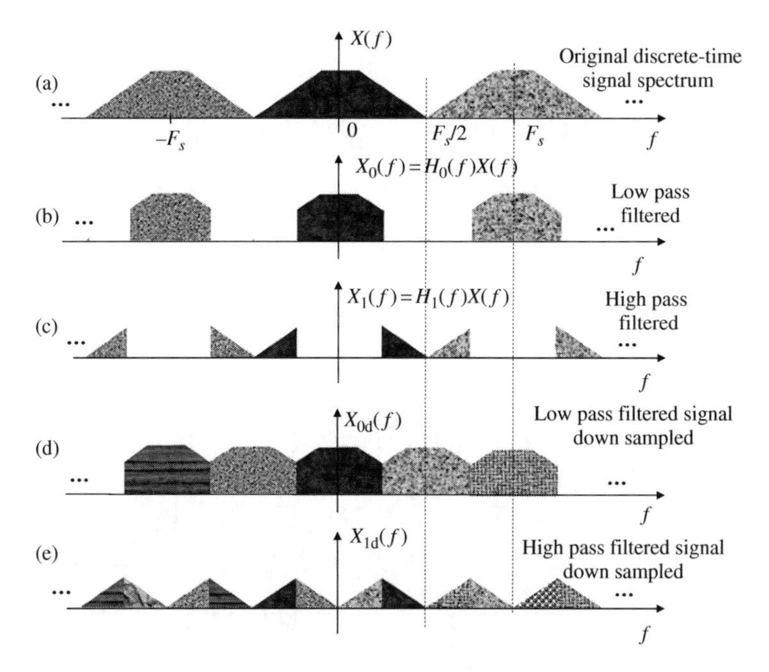

Figure 4.26 Illustration of dividing a signal into two sub-bands: (a) the original full-band signal, (b) the lower half-band, (c) the upper half-band, (d) the lower half-band after down-sampling, and (e) the upper half-band after down-sampling. Note that as a result of down-sampling a signal by a factor of 2 and the consequent aliasing $X(F_s/2)$ to $X(F_s)$ is aliased with $X(f)$. Hence after down-sampling the higher frequencies of the upper band signal (c) appear at the lower frequencies in (e) and vice versa.

reduction etc.), up-samplers and anti-aliasing filters $G_0(z)$ and $G_1(z)$. The low-pass filter $H_0(z)$ and the high-pass filter $H_1(z)$, split the input signal into low frequency and high frequency bands. Both sub-band signals are then down-sampled by a factor of 2. Figure 4.26 shows an illustration of the effect of a two-band QMF filter on the spectrum of the signals. Note that after down-sampling, the upper band has folded frequencies, i.e. the low frequencies appear at the high frequencies and vice versa as illustrated in Figure 4.26(d) and (e).

MATLAB function QMF_Unit ( )
Demonstrates the impulse response, the frequency response and the pole–zero diagram of a low-pass filter and its mirrored high-pass filter.

4.8.2 Theory and Design of QMF Filter

Once a suitable low-pass filter $H_0(z)$, with impulse response $h_0(m)$, is designed, the high-pass filter can be derived by time-reversing the filter and negating the sign of the odd-indexed coefficients as

$$h_1(m) = (-1)^m h_0(N-1-m) \quad \text{for } 0 \le m \le N-1 \tag{4.87}$$

In the z-plane the change of sign of odd-indexed coefficients of an FIR filter switches the position of the zeros of the filter to the opposite half plane (mirrored about the vertical jω-axis) and turns a low-pass filter into a high-pass filter and vice versa, whereas the time-reversal of the impulse response mirrors the position of the zeros about the unit circle from a radius of r to $1/r$, Figure 4.27.

In z-transform domain time Equation (4.87) is equivalent to

$$H_1(z) = H_0(-z^{-1}) \tag{4.88}$$

For prefect reconstruction we require that the low-pass and high-pass filters are power complementary, that is their responses add to one

$$|H_0(z)|^2 + |H_1(z)|^2 = |H_0(z)|^2 + |H_0(-z^{-1})|^2 = 1 \tag{4.89}$$

From Equation (4.88) we have

$$H_1\left(z = e^{j\omega}\right) = H_0\left(-e^{-j\omega}\right) = H_0\left(e^{-j(\pi-\omega)}\right) \tag{4.90}$$

A classical method for designing a QMF filter is to find an $H(\omega)$ that minimises the following criterion

$$J(H(\omega)) = \underbrace{\mu \int_{\omega_c}^{\pi} |H(\omega)|^2 d\omega}_{\text{Stopband energy}} + \underbrace{(1-\mu) \int_{0}^{\pi} \left[|H(\omega)|^2 + |H(\pi-\omega)|^2 - 1\right] d\omega}_{\text{Power-complementaryconstraint}} \tag{4.91}$$

Note that as shown in Figures 4.24 and 4.25, after filtering the signals are down-sampled. The z-transform of a signal down-sampled by a factor of 2 (with the odd-indexed samples removed) can be expressed in terms of the z-transform of the original signal as

$$X_d(z) = \frac{1}{2}\left[X(z^{1/2}) + X(-z^{1/2})\right] \tag{4.92}$$

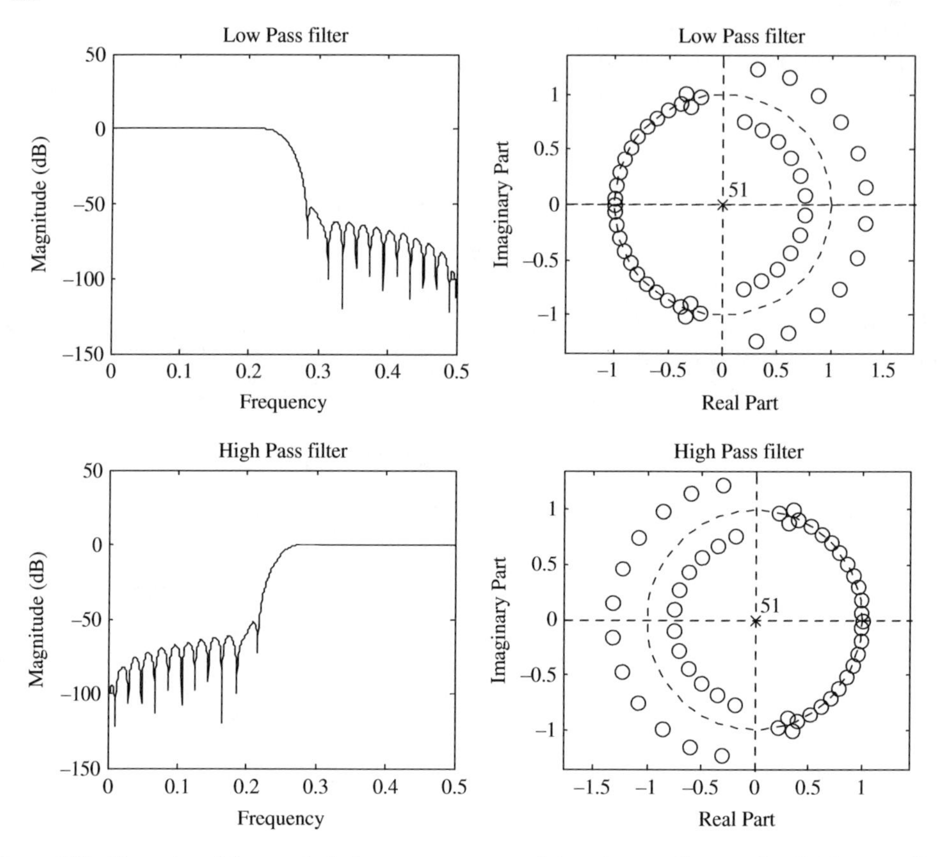

Figure 4.27 Illustration of the magnitude frequency response and the pole–zero diagram of an FIR low-pass filter and its mirror high-pass filter.

For example, for a sequence $[x(0), x(1), x(2), x(3), x(4), x(5), \ldots]$ we have

$$X(z) = x(0) + x(1)z^{-1} + x(2)z^{-2} + x(3)z^{-3} + x(4)z^{-4} + x(5)z^{-5} + \ldots$$

$$X(z^{1/2}) = x(0) + x(1)z^{-1/2} + x(2)z^{-1} + x(3)z^{-3/2} + x(4)z^{-2} + x(5)z^{-5/2} + \ldots$$

$$X(-z^{1/2}) = x(0) - x(1)z^{-1/2} + x(2)z^{-1} - x(3)z^{-3/2} + x(4)z^{-2} - x(5)z^{-5/2} + \ldots$$

hence $X(z^{1/2}) + X(-z^{1/2}) = x(0) + x(2)z^{-1} + x(4)z^{-2} + \ldots = X_d(z)$.

Note that Equation (4.92) is true because

$$(z^{1/2})^k + (-z^{1/2})^k = \begin{cases} 2z^{k/2} & \text{for } k \text{ even} \\ 0 & \text{otherwise} \end{cases} \tag{4.93}$$

For signal reconstruction the down-sampled signal need to be up-sampled. As explained in Chapter 5, the up-sampling, or interpolation, involves zero-insertion (i.e. inserting zeros between samples) followed by low-pass filtering.

Consider a signal $X(z)$ that has been low-pass filtered to $H_0(z)X(z)$ and then down-sampled by a factor of 2. In the interpolation stage, the process of zero-insertion (by a factor of 2) of the down-sampled signal is equivalent to the zeroing of the odd-indexed coefficients of $H_0(z)X(z)$ before down-sampling and this can be expressed in z-transform as

$$X_{\text{Zeros-Inserted}}(z) = \frac{1}{2}\left[H_0(z)X(z) + \underbrace{H_0(-z)X(-z)}_{\text{Aliasing term}}\right] \tag{4.94}$$

Note that in general for any signal $y(m)$ we have $Y(z) + Y(-z) = 2(\,y(0) + 0 + y(2)z^{-2} + 0 + y(4)z^{-4} + \ldots\,)$ and that in Equation (4.94) the even-indexed coefficients of $H_0(z)X(z)$ are those samples that would have been retained after down-sampling.

If we substitute $z = e^{j\omega}$ in Equation (4.94) we have

$$X_{\text{Zero Inserted}}(\omega) = \frac{1}{2}\left[H_0(\omega)X(\omega) + \underbrace{H_0(\omega+\pi)X(\omega+\pi)}_{\text{Aliasing terms}}\right] \tag{4.95}$$

Note from Equations (4.94) and (4.95) that the z-transform terms with $'-z'$ arguments such as $X(-(z = e^{j\omega})) \rightarrow X(e^{j(\pi+\omega)}) \rightarrow X(\pi+\omega)$ correspond to out-of-band aliasing terms. Hence, after filtering, down-sampling and the subsequent up-sampling process, the z-transforms of the low-pass and high-pass sub-band signals, denoted here as $X_0(z)$ and $X_1(z)$ respectively, can be expressed as

$$X_0(z) = \frac{1}{2}G_0(z)\left[H_0(z)X(z) + H_0(-z)X(-z)\right] \tag{4.96}$$

$$X_1(z) = \frac{1}{2}G_1(z)\left[H_1(z)X(z) + H_1(-z)X(-z)\right] \tag{4.97}$$

Hence, the z-transform of the reconstructed signal is given by

$$\begin{aligned} X(z) &= X_0(z) + X_1(z) \\ &= \frac{1}{2}\left[G_0(z)H_0(z) + G_1(z)H_1(z)\right]X(z) + \underbrace{\frac{1}{2}\left[G_0(z)H_0(-z) + G_1(z)H_1(-z)\right]X(-z)}_{\text{Aliasing Term}} \end{aligned} \tag{4.98}$$

For perfect reconstruction we require that the aliasing term is zero, hence

$$G_0(z)H_0(-z) + G_1(z)H_1(-z) = 0 \tag{4.99}$$

and

$$\frac{1}{2}\left[G_0(z)H_0(z) + G_1(z)H_1(z)\right] = 1 \tag{4.100}$$

For Equations (4.99) and (4.100) to be true we need to have

$$G_0(z) = 2H_1(-z) = 2H_0(z^{-1}) \text{ and } G_1(z) = -2H_0(-z) \tag{4.101}$$

Hence to summarise, the equations for one stage of a QMF are

$$|H_0(\omega)|^2 + |H_0(\pi - \omega)|^2 = 1 \tag{4.102.a}$$

$$h_1(m) = (-1)^m h_0(N-1-m) \tag{4.102.b}$$

$$g_0(m) = 2h_0(N-1-m) \tag{4.102.c}$$

$$g_1(m) = -2(-1)^m h_0(m) \tag{4.102.d}$$

Note that the spectrum of the down-sampled signals, in Figure 4.25, can be expressed as

$$X_{0d}(f) = \sum_{k=-\infty}^{\infty} X(f + kF_s/2) H_0(f + kF_s/2) \tag{4.103}$$

$$X_{1d}(f) = \sum_{k=-\infty}^{\infty} X(f + kF_s/2) H_1(f + kF_s/2) \tag{4.104}$$

Figure 4.26 illustrates the spectrum of a full-band signal and also the spectra of the low-pass and high-pass sub-bands before and after down-sampling by a factor of 2 to 1. Note that whereas the spectrum of the full-band signal has a bandwidth of F_s and a repetition period of F_s, the spectra of the sub-band signals after down-sampling by a factor of 2 have a bandwidth of $F_s/2$ and a repetition period of $F_s/2$.

To reconstruct the signal, both sub-band signals are up-sampled by a factor of 2, and filtered with anti-aliasing filters $G_0(f)$ and $G_1(f)$. Thus, the reconstructed output can be written as

$$\begin{aligned} \hat{X}(f) = & \frac{1}{2}[H_0(f)G_0(f) + H_1(f)G_1(f)]X(f) + \\ & \underbrace{\frac{1}{2}[H_0(f + F_s/2)G_0(f) + H_1(f + F_s/2)G_1(f)]X(f + F_s/2)}_{\text{aliasing}} \end{aligned} \tag{4.105}$$

where for notational convenience the sampling frequency F_s (e.g. 44,100 Hz for music) is usually normalised to 1. Note that as a result of down-sampling by a factor of 2 and then later up-sampling by a factor of 2, each frequency f will have an aliasing term $X(f + F_s/2)$. For cancellation of the aliasing terms, we require the following relationships between the frequency response of the sub-band filters $H_0(f)$ and $H_1(f)$ and the anti-aliasing filters $G_0(f)$ and $G_1(f)$:

$$G_0(f) = 2H_1(f + F_s/2) \tag{4.106}$$

$$G_1(f) = -2H_0(f + F_s/2) \tag{4.107}$$

We also require

$$\frac{1}{2}[G_0(f)H_0(f) + G_1(f)H_1(f)] = 1 \tag{4.108}$$

so that perfect reconstruction is achieved.

The aliasing cancellation may not work perfectly in the presence of quantisation noise. To get good reconstruction without relying on the aliasing cancellation, the QMF filters need to have a steep pass-to-stop-band transition.

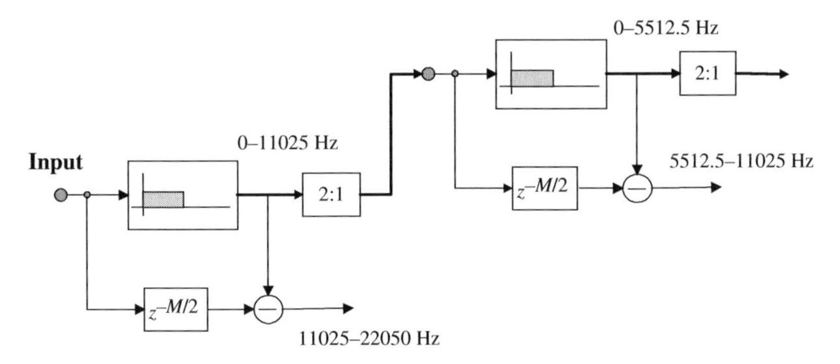

Figure 4.28 The configuration of a simple filter bank to split a total bandwidth of 22.05 kHz into three bands of 0–4.5125 kHz, 4.5125–11.025 kHz and 11.025–22.05 kHz.

Example 4.11 *Design of Tree-structured FIR Filter-bank*
Using the digital FIR low-pass filter and a binary-tree structure, design a bank of filters, operating at a sampling rate of 44,100, to divide a total bandwidth of 22.05 kHz into three bands of 0–5.5125 kHz, 5.5125–11.025 kHz and 11.025–22.05 kHz.

Solution: We start by designing a low-pass filter with a cut-off frequency of 11.025 kHz, which at a sampling rate of 44.1 kHz translates to a normalised cut-off frequency of

$$f_c = 11.025/44.1 = 0.25$$

Using the window design technique the FIR filter impulse response is obtained from the inverse Fourier transform as

$$h_d(m) = \int_{-fc}^{fc} 1.0 e^{j2m\pi f} df \tag{4.109}$$

We obtain the FIR low-pass filter response as

$$h_1(m) = w(m) \times 0.5 \,\mathrm{sinc}\,(0.5\pi(m - M/2)) \quad 0 \le m \le M \tag{4.110}$$

Figure 4.28 shows the configuration of a simple tree-structured filter bank constructed from a combination of a low-pass filter unit, delays and down-samplers. Note that after down-sampling, the low-pass band is further subdivided into two bands by the repeated application of the low-pass filter.

4.9 Design of Infinite Impulse Response (IIR) Filters by Pole–zero Placements

Filter design by pole–zero placements is a useful method for simple applications such as the design of a notch filter and pre-emphasis and de-emphasis filters. Filter design by pole–zero placements is also illustrative of the effects of zeros or poles on shaping the spectrum of the filter response. As described in Section 3.5 the effect of a pair of complex zeros is to introduce a trough in the frequency domain and the effect of a pair of complex poles is to introduce a resonance. In the following examples we describe simple filter design techniques by pole–zero placements.

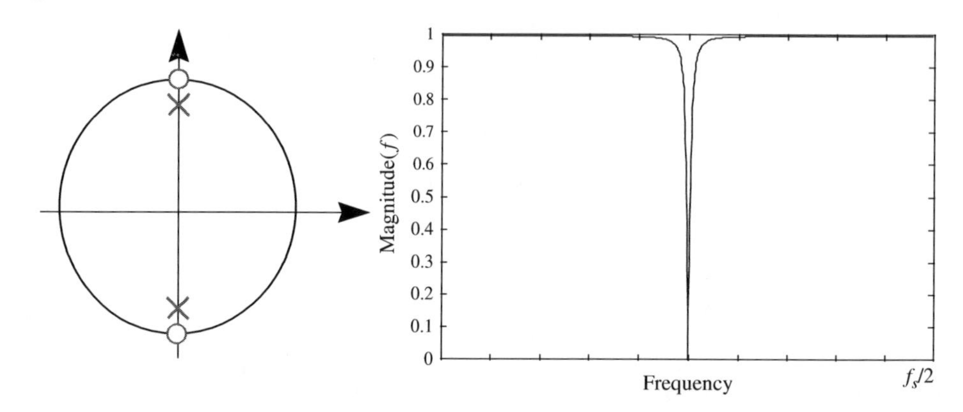

Figure 4.29 The pole–zero diagram and the magnitude spectrum of a second order notch filter.

Example 4.12 *Design of an IIR Notch Filter*
Design a notch filter, operating at a sampling rate of 20 kHz, with a zero magnitude response at 5 kHz, a 3 dB stop-band bandwidth of 100 Hz and unity response at frequencies of 0 kHz and 10 kHz.

Solution: The normalised angular frequency can be obtained from $\omega = 2\pi f/F_s$, with $F_s = 20$ kHz. The frequencies of 0, 5 and 10 kHz correspond to 0, $\pi/2$ and π radians respectively.

The specification given is $H(0) = H(\pi) = 1$ and $H(\pi/2) = 0$. The 3 dB point frequencies are 4950 Hz and 5050 Hz corresponding to the angular frequencies of

$$\phi_{\text{low}}^{3dB} = \frac{2\pi}{20000}4950 = \frac{99\pi}{200} \quad \text{and} \quad \phi_{\text{upper}}^{3dB} = \frac{2\pi}{20000}5050 = \frac{101\pi}{200} \tag{4.111}$$

To obtain a zero magnitude response at an angular frequency of $\pi/2$, corresponding to a frequency of 5 kHz at a sampling rate of 20 kHz, we need a pair of unit-radius complex zeros at $e^{\pm j\pi/2}$. The corresponding z-transfer function is given by

$$H(z) = (1 - e^{j\pi/2}z^{-1})(1 - e^{-j\pi/2}z^{-1}) = 1 + z^{-2} \tag{4.112}$$

To control the bandwidth of the notch filter we place a pair of poles on the same frequency as the notch frequency ($\pi/2$) as shown in Figure 4.29. The resulting transfer function is

$$H(z) = g\frac{1 + z^{-2}}{1 + r^2 z^{-2}} \tag{4.113}$$

The radius of the pole controls the bandwidth of the notch filter, the nearer the position of the pole to the unit circle the smaller the bandwidth, resulting in a sharper notch filter. Figure 4.29 shows the angular position and the frequency response of a complex pair of poles and zeros at an angle of $\pi/2$.

The frequency response of the notch filter is obtained by substituting $z = e^{j\omega}$ as

$$H(e^{j\omega}) = g\frac{1 + e^{-j2\omega}}{1 + r^2 e^{-j2\omega}} \tag{4.114}$$

Substituting the magnitude response specifications $H(\omega = 0) = 1$ or $H(\omega = \pi) = 1$ yields

$$H(0) = g\frac{1 + e^0}{1 + r^2 e^0} = 1 \rightarrow g = 0.5(1 + r^2) \tag{4.115}$$

The 3 dB point, the frequency at which the magnitude response falls by $1/\sqrt{2} = 0.7071$, is at $\omega = 101\pi/200$ radians, hence

$$H(101\pi/200) = g\frac{1+e^{-j101\pi/100}}{1+r^2e^{-j101\pi/100}} = \frac{1}{\sqrt{2}} \rightarrow g^2\left|\frac{1+e^{-j101\pi/100}}{1+r^2e^{-j101\pi/100}}\right|^2 = \frac{1}{2} \tag{4.116}$$

$$\frac{9.8688\times 10^{-4}g^2}{(1-0.9995r^2)^2+9.8596\times 10^{-4}r^4} = \frac{1}{2} \tag{4.117}$$

Substituting for g from Equation (4.115) into Equation (4.117) we have

$$\frac{9.8688\times 10^{-4}\times 0.25(1+r^2)^2}{(1-0.9995r^2)^2+9.8596\times 10^{-4}r^4} = \frac{1}{2} \tag{4.118}$$

The solution of Equation (4.118) yields $r = 0.9891$. Substituting r in Equation (4.115) yields $g = 0.9892$.

Example 4.13 Design of a Comb Filter
A comb filter is used to filter out the harmonics of a periodic interference. A comb filter would have a pair of zeros positioned on the fundamental frequency of the periodic signal and other zeros uniformly spaced at harmonic frequencies as illustrated in Figure 4.30.

Write a Matlab program for a comb filter operating at a sampling rate of F_s kHz to remove the fundamental and the harmonics of a periodic interference with a fundamental frequency of F_0 Hz.

Solution: The angular position of the fundamental frequency is $\omega = 2\pi F_0/F_s$. We place pairs of complex zeros at $z = e^{\pm jk\omega_0}$, $k = 1, \ldots, \text{fix}(\pi/\omega_0)$. We also place pairs of complex poles inside the unit circle at the same angular frequencies as those of the zeros at $z = re^{\pm jk\omega_0}$, $k = 1, \ldots, \text{fix}(\pi/\omega_0)$. Note fix($x$) rounds the elements of x to the nearest integers towards zero.

Matlab CombFilter( )
Demonstration of the frequency response of a comb filter designed by pole–zero placement. The graphs show the pole–zero plot and the frequency response as a function of decreasing fundamental frequency, increasing number of notches and increasing radius of poles.

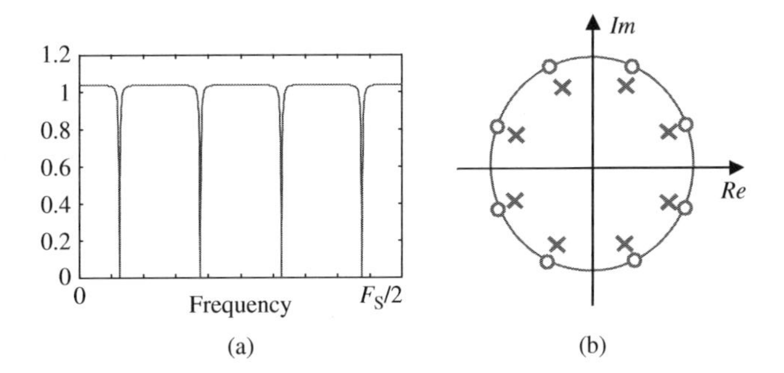

Figure 4.30 The pole–zero diagram (a), and the magnitude spectrum (b), of an 8th order notch filter.

4.10 Issues in the Design and Implementation of a Digital Filter

There are essentially two steps in the filter design process: (a) determination of the filter coefficients a_k and b_k that produce a desired frequency or impulse response and (b) implementation of the digital filter given a set of coefficients $\{a_k, b_k\}$. Essentially a digital filter may be viewed as a computational system that performs the computation of the output signal $y(m)$ from the input signal $x(m)$. However, the difference equation and the computational procedure required to implement a filter can be arranged in different (but essentially equivalent) forms (e.g. direct form or cascade form). These different forms require different configuration and interconnections of memory elements, multipliers and adders. We refer to each distinct configuration as a realisation or equivalently as a structure for realising the digital filter of the equation. The following factors are used for comparing different filter structures.

Computational complexity This is the number of arithmetic operations (multiplications, and additions) required to compute an output value $y(m)$ for the system. In the past, multiplications and additions were the only factors that were used to measure the computational complexity of a digital filter. Recently many advanced digital signal processors have been developed which can be software programmed to perform the type of computations indicated by the filter's difference equation. When measuring the computational complexity of a filter program other factors such as the number of fetch operations of various data from memory per output sample become important. The computational complexity gives a measure of the speed of the algorithm in terms of numbers of addition, multiplication and fetch operations. The time required to compute an output sample must be less than the sampling interval of the input signals otherwise the system would be too slow to cope with the input rate.

Memory requirement refers to the number of memory locations required to store the filter coefficients, past inputs, past outputs and any intermediate values.

Finite-word-length effects In any implementation of a digital system, either in hardware or on a digital computing system, the filter coefficients must necessarily be represented with finite precision. The results of the computations that are performed at each stage must be rounded off or truncated to fit within the limited precision constraints of the system. The accumulated effect of the finite precision computations is referred to as finite-word-length effects.

4.11 Summary

This chapter presented methods for analysis, design and operation of digital filters. Filters, in their various forms, are a basic component of all digital signal processing and telecommunication systems. This chapter began with an introduction to digital filters and introduced the general form of the difference equation for description of a linear time-invariant filter. The various methods of description of a filter in time and frequency domains were considered and specifically the concepts of convolution, impulse response, frequency response, and z-transfer function were introduced. The different filter forms and structures covered included FIR and IIR forms and direct and cascade structures. The design of low-pass filters, band-pass filters, band-stop filters, filter banks and quadrature mirror filters were explained. Practical issues in filter design such as the effect of filter length and windows were explored.

Further Reading

Antoniou A. (1993) Digital Filters: Analysis, Design, and Applications McGraw-Hill, New York.
Mitra S.K. (1998) Digital Signal Processing: A Computer-Based Approach McGraw-Hill, New York.
Oppenheim A.V. and Schafer R.W. (1999) Discrete-Time Signal Processing Prentice-Hall, Upper Saddle River, NJ.

Exercises

4.1 (a) State the main effects of (i) a complex conjugate pair of poles, and (ii) a complex conjugate pair of zeros, on the frequency response of a filter.
Briefly list the signal processing steps required in the window design technique, to obtain the coefficients of a causal finite impulse response (FIR) discrete-time filter given the desired frequency response of the filter $H(f)$.
Using the inverse Fourier transform method, design two digital filters, a low-pass filter and a high-pass filter, to divide the input signal into two equal-bandwidth signals.

4.2 State three different applications of filters.
Explain why the impulse response function can completely describe the characteristics of a linear time-invariant filter.
Write the relationship between the impulse response and the frequency response of a linear time-invariant filter.
A second-order digital filter is given by

$$y(m) = 1.386y(m-1) - 0.9604y(m-2) + gx(m)$$

(i) Obtain the z-transfer function of this filter and write the z-transfer function equation in polar form in terms of the radius and angular frequencies of its zeros.

(ii) Sketch the pole–zero diagram and the frequency response of the filter.

4.3 Explain the relationships between the Laplace transform, the z-transform and the Fourier transform.
Show how z-transform can be derived from Laplace transform and how discrete Fourier transform (DFT) can be derived from z-transform of a signal.
(c) Figure 4.31 shows the discrete-time input signal and the impulse response of a linear time-invariant filter. Using the principles of linearity and superposition, obtain the output of the filter in response to the input discrete-time input signal shown in the figure.

4.4 (a) State the reasons why it is preferable to design IIR filters as a cascade of second-order units.
(b) The difference equation relating the input and output of the infinite duration impulse response (IIR) filter, shown in Figure 4.32, is

$$y(m) = ay(m-4) + gx(n) - gx(m-4)$$

Taking the z-transform of the difference equation, find the transfer function of the filter.
Describe the transfer function in the polar form, and for a value of $a = 0.6561$ obtain the pole and zeros of the filter and sketch its pole–zero diagram.

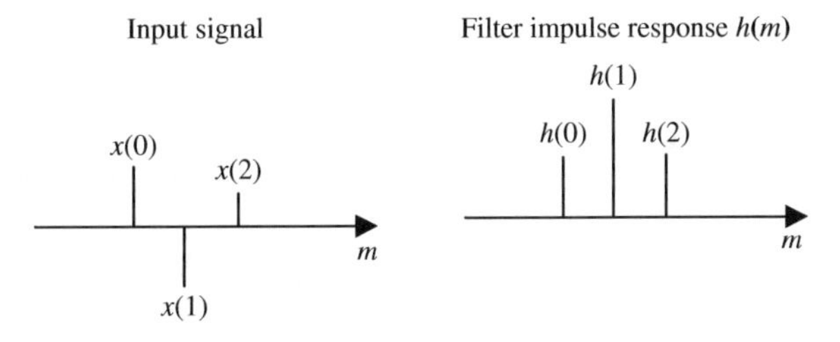

Figure 4.31 The discrete-time input signal and the impulse response of a linear time-invariant filter.

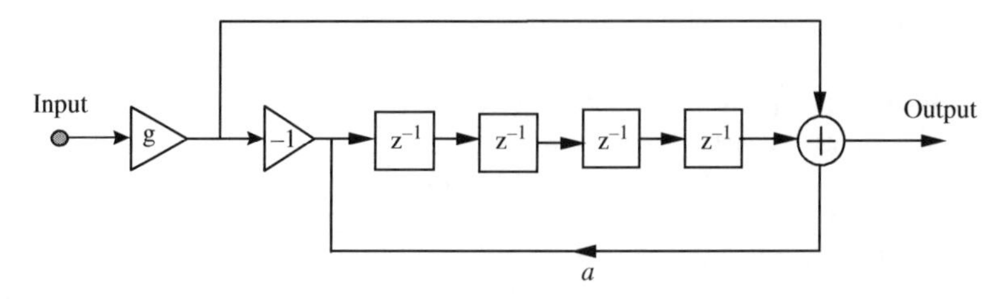

Figure 4.32 Illustration of the IIR filter described in exercise 4.4.

Sketch the frequency response of the filter, and suggest an application for this filter. Discuss the effect of varying the value of a on the frequency response of the filter.
Calculate the value of the gain variable g such that the gain for the band-pass regions is 1.0.

4.5 A first-order digital pre-emphasis filter is given by

$$y(m) = x(m) - ax(m-1)$$

Obtain the z-transfer function of this filter.
Assuming a value of $a = 0.98$ draw its pole–zero and frequency response diagrams.
State the range of values of the parameter a where this filter acts as a pre-emphasis filter.
Give an example of an application of a pre-emphasis filter.
Explain what is a de-emphasis filter.

4.6 The equation describing the general form of the z-transfer function of a second-order system in polar form is given by

$$H(z) = \frac{1 - 2r_z \cos(\varphi_z)z^{-1} + r_z^2 z^{-2}}{1 - 2r_p \cos(\varphi_p)z^{-1} + r_p^2 z^{-2}}$$

Using the transfer function equation, obtain the values of the coefficients r_z, φ_z and φ_p of a second-order notch filter for removing a sinusoidal interference with a frequency of 100 Hz. Assume a sampling frequency of 10 kHz.
Sketch the pole–zero diagram and the frequency response of the notch filter.
Explain the effect of varying r_p on the frequency response of the notch filter.
Calculate the values of filter coefficients for which the second-order system acts as a digital oscillator operating at 10 kHz and with an oscillation frequency of 1 kHz.

4.7 (a) Determine and sketch the impulse response $h(n)$ of the low-pass filter described by the equation

$$H(f) = \begin{cases} 1 & |f| < f_c \\ 0 & |f| \geq f_c \end{cases}$$

Figure 4.33 shows the block diagram of a high-pass digital filter with frequency response $H_{HP}(f)$, incorporating a low-pass filter $H_{LP}(f)$.
(i) Obtain the frequency response and the impulse response of the filter $H_{HP}(f)$.
Assuming that the low-pass filter $H_{LP}(f)$ has N taps, for what value of the delay variable D would the high-pass filter $H_{HP}(f)$ perform correctly?

4.8 Design a system for converting the sampling rate of a digital signal $x(m)$, originally sampled at a rate of 44 kHz, to a sampling rate of 16 kHz. Draw a block diagram of this system and describe the operation of each subsystem.

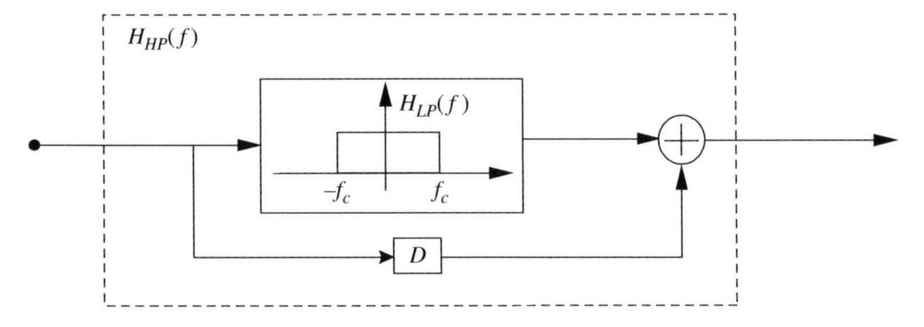

Figure 4.33 The block diagram of a high-pass digital filter, exercise 4.7.

4.9 Determine the z-transform, and hence the Fourier transform of the following function

$$h(m) = \delta(m-4) + \delta(m+4)$$

Suggest a use for a filter with $h(m)$ as its impulse response.

$$H(z) = z^{-4} + z^{+4}$$

What modification of the equation for $h(m)$ is necessary before it can be used as the impulse response of a causal filter?

4.10 (a) Describe in detail the window design techniques for the design of a finite impulse response (FIR) low-pass filter with the frequency response specification as shown in Figure 4.34.

(b) Obtain and sketch the impulse response of this filter.

(c) Explain the main effects of varying the filter order on its frequency response.

(d) (d) Using the window design technique and the filter impulse response from part (a) above, design a bank of digital filters for a telephony speech application to split a total bandwidth of 4 kHz into four equal-width sub-bands.

Write the cut-off frequencies and the equations for the response of the filter in each band. Your equations should include the numerical parameters needed to achieve the required cut-off frequencies.

Choose a value for the number of taps of the filter in each band and explain how the number of filter coefficients affects the filter response.

Suggest an application for this filter bank.

4.11 Determine the values of a_1 and a_2 for which the second-order system shown in Figure 4.35 and operating at a sampling rate of 40 kHz becomes a sinusoidal oscillator with an oscillation frequency of 5 kHz. Plot a pole–zero diagram for this oscillator.

4.12 State the factors that affect the choice of a filter.

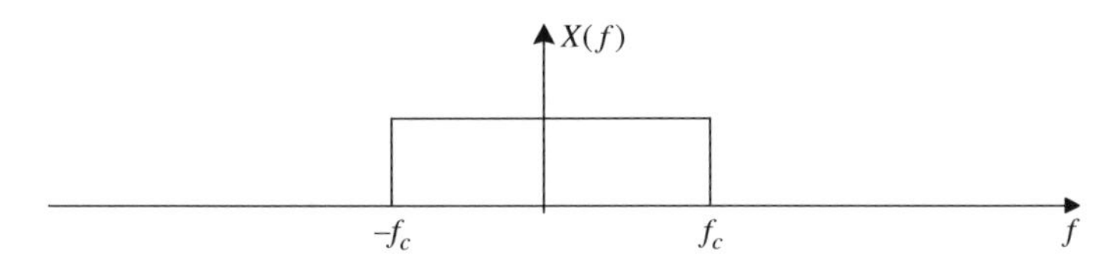

Figure 4.34 The frequency response specification of a low pass filter, exercise 4.10.

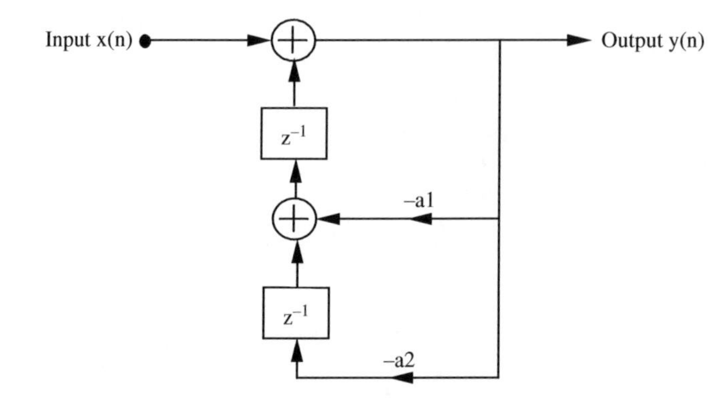

Figure 4.35 The block diagram for a second order IIR filter oscillator, exercise 4.11.

The z-transfer function of a finite impulse response filter $H(z)$ is given by

$$H(z) = \left(1 - 0.5z^{-1} + 0.25z^{-2}\right)\left(1 - 2z^{-1} + 4z^{-2}\right)$$

Obtain the equation for $H(\omega)$ the frequency response of this filter.
Prove that $H(\omega)$ is a linear phase filter.
Obtain the zeros of this filter and plot the pole–zero diagram for this filter.

4.13 Consider a second-order infinite impulse response (IIR) system operating at a sampling rate of 40 KHz and with the z-transfer function given as

$$H(z) = G\frac{1 + 2r_1 \cos\theta_1 z^{-1} + r_2^2 z^{-2}}{1 + 2r_2 \cos\theta_2 z^{-1} + r_2^2 z^{-2}}$$

Assume that this system has a complex conjugate pair of poles and zeros.
Rewrite the transfer function in terms of the radii and angular frequencies of the poles and zeros of the filter. Find the poles and the zeros for which the transfer function has a notch frequency response at a frequency of 10 kHz, and unit response at 0 Hz and 20 kHz. Obtain the frequency spectrum and write the difference equation of the filter. Plot its pole–zero diagram, and sketch its spectrum.

4.14 Consider a second-order all-zero infinite impulse response (IIR) system operating at 20 KHz and with the z-transfer function given as

$$H(z) = \frac{G}{1 + b_1 z^{-1} + b_2 z^{-2}}$$

Assume that this system has a complex conjugate pair of poles. Rewrite the transfer function in terms of pole radius and pole resonance frequency.
Find the values of the coefficients b_1, b_2 and the gain G for which this filter has a resonance frequency of 5 kHz, a gain G of unity at this frequency, and a bandwidth of 100 Hz.
Plot its pole–zero diagram, and write its frequency response equation. Write the difference equation of this system.

4.15 With the aid of a block diagram, sketch and explain the relationship between Fourier series, the Fourier transform of continuous-time signals, the Fourier transform of discrete-time signals and the discrete Fourier transform (DFT).

Explain the advantages and disadvantages of the use of windowing in DFT analysis. Name two popular windows.

The input–output relationship of a discrete Fourier transform (DFT) is given by

$$X(k) = \sum_{m=0}^{N-1} x(m)e^{-j\frac{2\pi}{N}mk} \quad k = 0, \ldots, N-1$$

Assuming a sampling rate of $F_s = 20\,\text{kHz}$, and a DFT length of $N = 512$, what is the frequency resolution and time resolution?

Describe a method for improving the 'apparent' resolution of a DFT of N samples of a signal by a factor of four.

Using the DFT equation, given above, derive the DFT transform $X_z(k)$ of a zero-inserted signal in terms of the DFT of the original signal $X(k)$. Assume zero-insertion by a factor of L (i.e. $L-1$ zeros inserted between every two samples). Briefly explain what happens to the frequency domain spectrum of the signal after it is zero-inserted.

5 Sampling and Quantisation

Sampling (i.e. the process of discretization of time or space) and quantisation (i.e. the process of discretization of the values of the samples of a signal) are the first two steps in all digital signal processing and digital communication systems which have analogue inputs. Most signals such as speech, image and electromagnetic waves, are not naturally in a digital format but need to be digitised (i.e. sampled and quantised) for subsequent processing and storage in a digital system such as in a computer or in a mobile DSP chip or a digital music player.

A signal needs to be sampled at a rate of more than twice the highest frequency content of the signal; otherwise the sampling process will result in loss of information and distortion. Hence, prior to sampling, the input signal needs to be filtered by an anti-aliasing filter to remove the unwanted signal frequencies above a preset value of less than half the sampling frequency. Each sample value is subsequently quantised to the nearest of 2^n quantisation levels and coded with an n-bit word.

The digitisation process should be performed such that the original continuous signal can be recovered from its digital version with no loss of information, and with as high a fidelity as is required in an application.

In this chapter the processes of sampling and quantisation, and the subsequent recovery of continuous-time signals is presented. The process of re-sampling of a discrete-time signal to a new sampling rate is also introduced.

5.1 Introduction

A digital signal is a sequence of discrete real-valued or complex-valued numbers, representing the fluctuations of an information-bearing quantity with time, space or some other variable.

Digital signal processing involves the processing of signals by a computer or by a purpose-built signal processing microchip. The signal is represented in the computer's memory in a binary format in terms of a sequence of n-bit words. Hence, to digitally process signals that are not already in a digital format, the signals need to be converted into a digital format that can be stored and processed in a computing device.

Multimedia Signal Processing: Theory and Applications in Speech, Music and Communications Saeed V. Vaseghi

The most elementary unit of a discrete-time (or discrete-space) signal is the unit-sample signal δm defined as

$$\delta(m) = \begin{cases} 1 & m = 0 \\ 0 & m \neq 0 \end{cases} \tag{5.1}$$

where m is the discrete-time index.

A discrete-time signal $x(m)$ can be expressed as the sum of a number of amplitude-scaled and time-shifted unit samples as

$$x(m) = \sum_{k=-\infty}^{\infty} x(k)\delta(m-k) \tag{5.2}$$

Figure 5.1 illustrates a discrete-time signal. Many signals such as speech, music, image, video, radar, sonar and bio-signals and medical signals are analogue, in that in their original form they appear to vary continuously with time (and/or space) and are sensed by analogue sensors such as microphones, optical devices and antennas. Other signals such as stock market prices are inherently discrete-time and/or discrete amplitude signals. Continuous signals are termed analogue signals because their fluctuations with time are analogous to the variations of the signal source.

For digital processing of continuous signals, the signals are first sampled and then each sample is converted into an n-bit binary digit. The sampling and digitisation process should be performed such that the original continuous signal can be recovered from its digital version with no loss of information, and with as high a fidelity as is required in an application.

Analogue-to-digital conversion, that is the conversion of an analogue signal into a sequence of n-bit words, consists of the two basic steps of sampling and quantisation:

(a) *Sampling*. The first step is to sample a signal to produce a discrete-time and/or discrete-space signal. The sampling process, when performed with sufficiently high frequency (greater than twice the highest frequency), can capture the fastest fluctuations of the signal, and can be a loss-less operation in that the original analogue signal can be recovered through interpolation of the sampled sequence.

(b) *Quantisation*. The second step is quantisation of each sample value into an n-bit word. Quantisation involves some irrevocable errors and possible loss of information. However, in practice the quantisation error (aka quantisation noise) can be made negligible by using an appropriately high number of bits as in a digital audio hi-fi.

Figure 5.2 illustrates a block diagram configuration of a digital signal processor with an analogue input. The anti-aliasing low-pass filter (LPF) removes the out-of-band signal frequencies above a pre-selected cut-off frequency which should be set to less than half the intended sampling frequency. The sample-and-hold (S/H) unit periodically samples the signal to convert the continuous-time signal into a discrete-time, continuous-amplitude, signal.

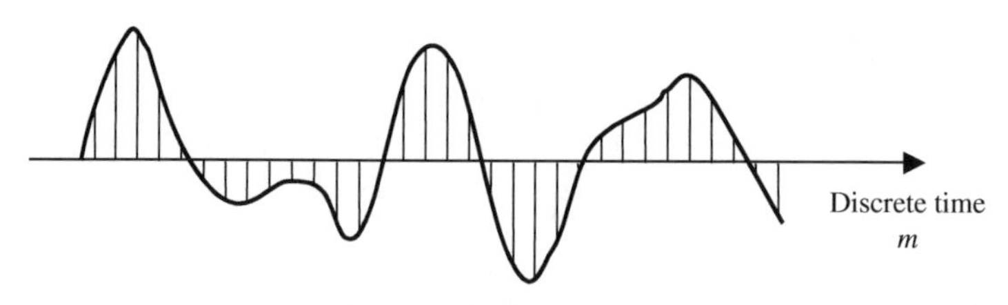

Figure 5.1 A discrete-time signal and its continuous-time envelope of variation.

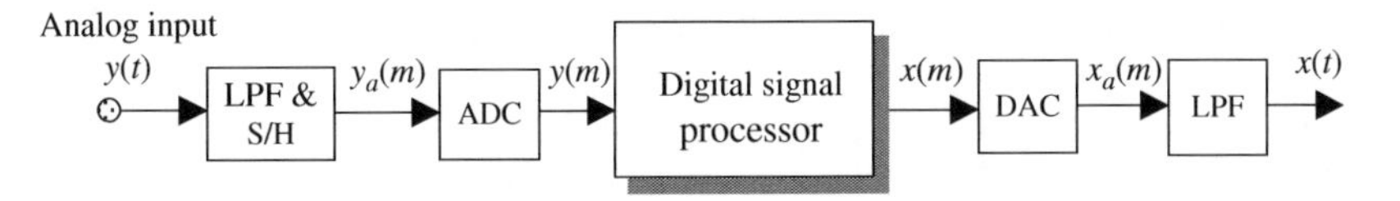

Figure 5.2 Configuration of a digital signal processing system.

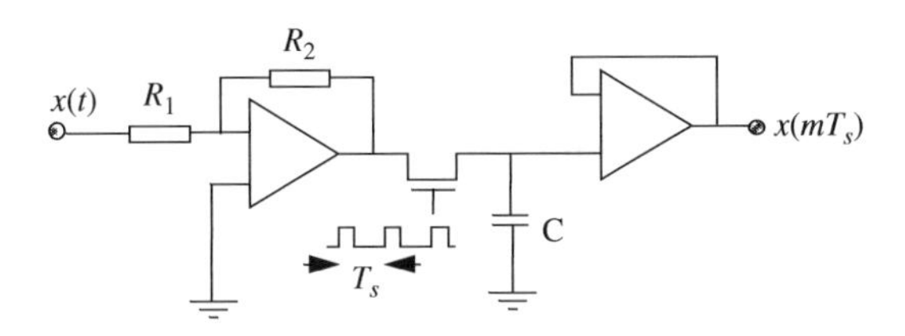

Figure 5.3 A simplified sample-and-hold circuit diagram; when the switch closes the capacitor charges or discharges to the input level.

The analogue-to-digital converter (ADC) follows the S/H unit and maps each continuous-amplitude sample into an n-bit word. After the signal is processed, the digital output of the processor can be converted back into an analogue signal using a digital-to-analogue converter (DAC) and a low-pass filter as illustrated in Figure 5.2. Figure 5.3 shows a sample-and-hold circuit diagram where a transistor switch is turned 'on' and 'off' thereby allowing the capacitor to charge up/down to the level of the input signal during the 'on' periods and then holding the sample value during the 'off' period.

Advantages of Digital Format

The advantages of the digital format are as follows:

(a) *Digital devices such as mobile phones are pervasive.*
(b) *Transmission bandwidth and storage space savings.* Digital compression techniques, such as MP3, can be used to compress a digital signal. When combined with error-control coding and efficient digital modulation methods the required overall bandwidth is less than that of say an FM-modulated analogue signal of similar noise robustness and quality. There is a similar reduction in storage requirement.
(c) *Power savings.* Power saving depends on the compression rate and the modulation method. In general digital systems can achieve power efficiency compared with analogue systems.
(d) *Noise robustness.* Digital waveforms are inherently robust to noise and additional robustness can be provided through error-control coding methods.
(e) *Security.* Digital systems can be encrypted for security, and in particular the code division multiple access (CDMA) method, employed for sharing of time/bandwidth resources in mobile phone networks, is inherently secure.
(f) *Recovery and restoration.* Digital signals are more amenable to recovery of lost segments.
(g) *Noise reduction.* Digital noise reduction methods can be used to substantially reduce noise and interference and hence improve the perceived quality and intelligibility of a signal.
(h) *Editing and mixing* of audio/video and other signals in digital format is relatively easy.
(i) *Internet and multimedia systems.* Digital communication, pattern recognition, Internet and multimedia communication would not have been possible without the digital format.

Digital Signals Stored and Transmitted in Analogue Format

Digital signals are actually stored and transmitted in analogue format. For example, a binary-state transistor stores a one or a zero as a quantity of electronic charge, in bipolar baseband signalling a '1' or a '0' is signalled with a pulse of $\pm V$ volts and in digital radio-frequency signalling binary digits are converted to modulated sinusoidal carriers for transmission over the airwaves. Also the digital data on a CD track consists of a sequence of bumps of micro to nanometre size arranged as a single, continuous, long spiral track of data.

The Effect of Digitisation on Signal Bandwidth

In its simplest form each binary bit ('1' or '0') in a bit-stream representation of a signal can be viewed as pulse of duration T seconds, resulting in a bit rate of $r_b = 1/T$ bps. In Chapter 2 on Fourier transform, it is shown that the bandwidth of such a pulse sequence is about $2/T = 2r_b$ Hz.

For example, the digitisation of stereo music at a rate of 44.1 kHz and with each sample quantised to 16 bits generates a bit rate r_b of (2 channels × 44,100 samples/second × 16 bits per sample) 1411.2 kbps. This would require a bandwidth of $2r_b = 2822.4$ kHz. However, using advanced compression and modulation methods the number of bits per second and the required bandwidth can be greatly reduced by a factor of more than 10.

5.2 Sampling a Continuous-Time Signal

This section describes the sampling process, the aliasing distortion and the Nyquist sampling theorem.

5.2.1 Sampling Results in a Periodic Spectrum

Figure 5.4 illustrates the process of sampling of a continuous-time signal in time and its effect on the frequency spectrum of the signal. In time-domain a sampled signal can be modelled as the product of a continuous-time signal $x(t)$ multiplied by a periodic impulse train sampler $p(t)$ as

$$\begin{aligned} x_{\text{sampled}}(t) &= x(t)p(t) \\ &= \sum_{m=-\infty}^{\infty} x(t)\delta(t - mT_s) \end{aligned} \tag{5.3}$$

where T_s is the sampling interval (the sampling frequency is $F_s = 1/T_s$), $\delta(t)$ is the discrete-time delta (unit-sample) function and the sampling train function $p(t)$ is defined as

$$p(t) = \sum_{m=-\infty}^{\infty} \delta(t - mT_s) \tag{5.4}$$

As shown in Example 2.7, the spectrum, $P(f)$, of a periodic train of sampling impulses in time $p(t)$, is a periodic train of impulses in frequency given by

$$P(f) = \sum_{k=-\infty}^{\infty} \delta(f - kF_s) \tag{5.5}$$

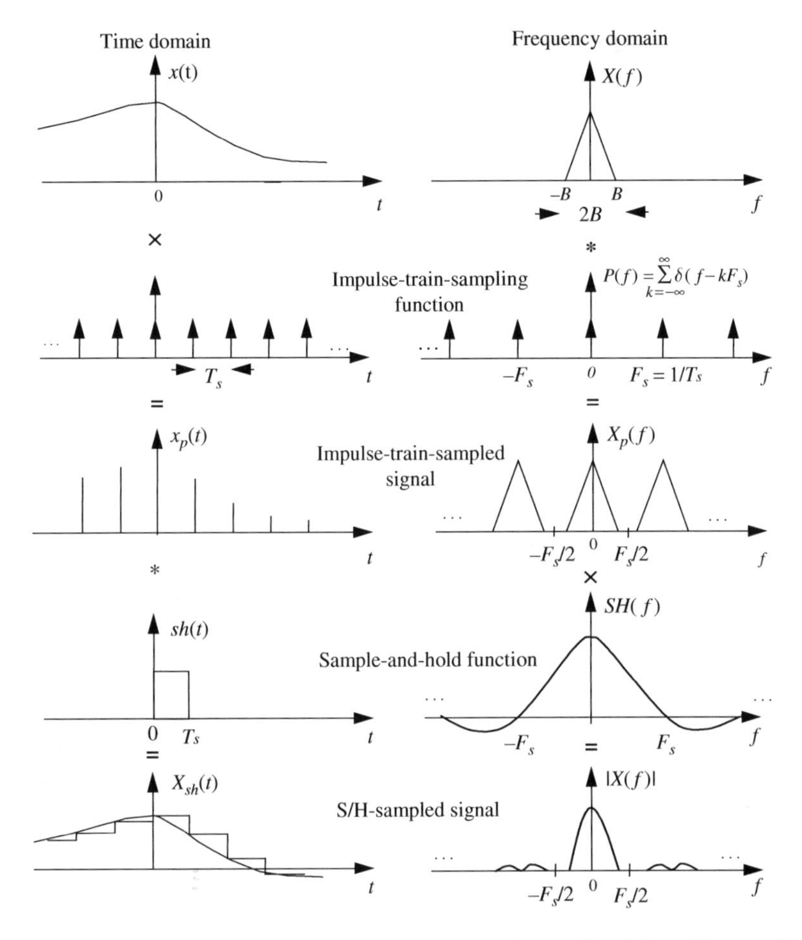

Figure 5.4 A sample-and-hold signal is modelled as an impulse-train sampling followed by convolution with a rectangular pulse.

where $F_s = 1/T_s$ is the sampling frequency.

Since multiplication of two time-domain signals is equivalent to the convolution of their frequency spectra we have

$$X_{\text{sampled}}(f) = FT\,[x(t).p(t)] = X(f)^{*}P(f) = \sum_{k=-\infty}^{\infty} X(f - kF_s) \tag{5.6}$$

where the operator $FT\,[.]$ denotes the Fourier transform.

Note from Equation (5.6) that the convolution of a signal spectrum $X(f)$ with each impulse $\delta(f - kF_s)$, shifts $X(f)$ and centres it on kF_s. Hence, Equation (5.6) shows that the sampling of a signal $x(t)$ results in a periodic repetition of its spectrum $X(f)$ with the 'images' of the baseband spectrum $X(f)$ centred on frequencies $\pm F_s, \pm 2F_s, \ldots$ as shown in Figure 5.4.

Note in Figure 5.4 that a sample-and-hold process produces a sampled signal which is in the shape of an amplitude-modulated staircase function. Also note that the sample-and-hold staircase function can itself be modelled as the output of a filter, with a rectangular impulse response, excited by an idealised sampling impulse train as shown in Figure 5.4.

5.2.2 Aliasing Distortion

The process of sampling results in a periodic spectra. When the sampling frequency F_s is higher than twice the maximum frequency content of the signal B Hz (i.e. $F_s > 2B$), then the repetitions ('images') of the signal spectra are separated as shown in Figure 5.4. In this case, the analogue signal can be recovered by passing the sampled signal through an analogue low-pass filter with a cut-off frequency of just above B Hz. If the sampling frequency is less than $2B$ (i.e. $F_s < 2B$), then the adjacent repetitions of the spectrum overlap and in this case the original spectrum cannot be recovered. The distortion, due to an insufficiently high sampling rate, is irrevocable and is known as *aliasing*. Note in Figure 5.5 that the aliasing distortion results in the high frequency components of the signal folding and appearing at the lower frequencies, hence the name aliasing. Figure 5.6 shows the sum of two sine waves sampled at above and below the Nyquist sampling rate. Note that below the Nyquist rate a frequency of F_0 may appear at $-kF_s + F_0$ or $kF_s - F_0$, where k is an integer, as shown in Figure 5.6.

5.2.3 Nyquist Sampling Theorem

The above observation on aliasing distortion is the basis of the Nyquist sampling theorem, which states: *a band-limited continuous-time signal, with a highest frequency content (bandwidth) of B Hz, can be recovered from its samples provided that the sampling frequency F_s is greater than 2B samples per second so that there is no aliasing*. Note that the sampling frequency F_s needs to be greater than $2B$ to avoid aliasing distortion and to allow space for a low-pass filter to recover the original (baseband) continuous signal from its sampled version.

In practice sampling is achieved using an electronic switch that allows a capacitor to charge or discharge to the level of the input voltage once every T_s seconds as illustrated in Figure 5.3. The sample-and-hold signal can be modelled as the output of a filter with a rectangular impulse response, and with the impulse-train-sampled signal as the input as illustrated in Figure 5.4.

5.2.4 Recovery (Interpolation) of Continuous-Time Signal from Discrete-Time Sampled Signal

The continuous-time signal is recovered from the discrete-time signal by passing the latter through a continuous-time (analogue) low-pass filter with a cut-off frequency of less than $F_s/2$. Viewed

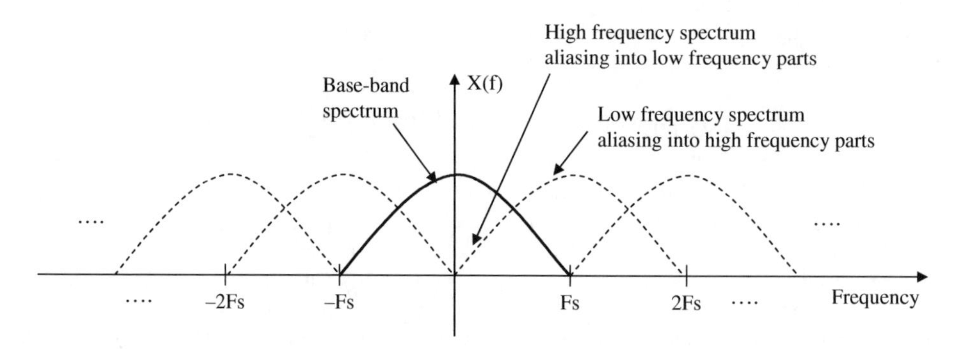

Figure 5.5 Aliasing distortion results from the overlap of spectral images (dashed curves) with the baseband spectrum. Note high frequency aliases itself as low frequency and vice versa. In this example the signal is sampled at half the required rate.

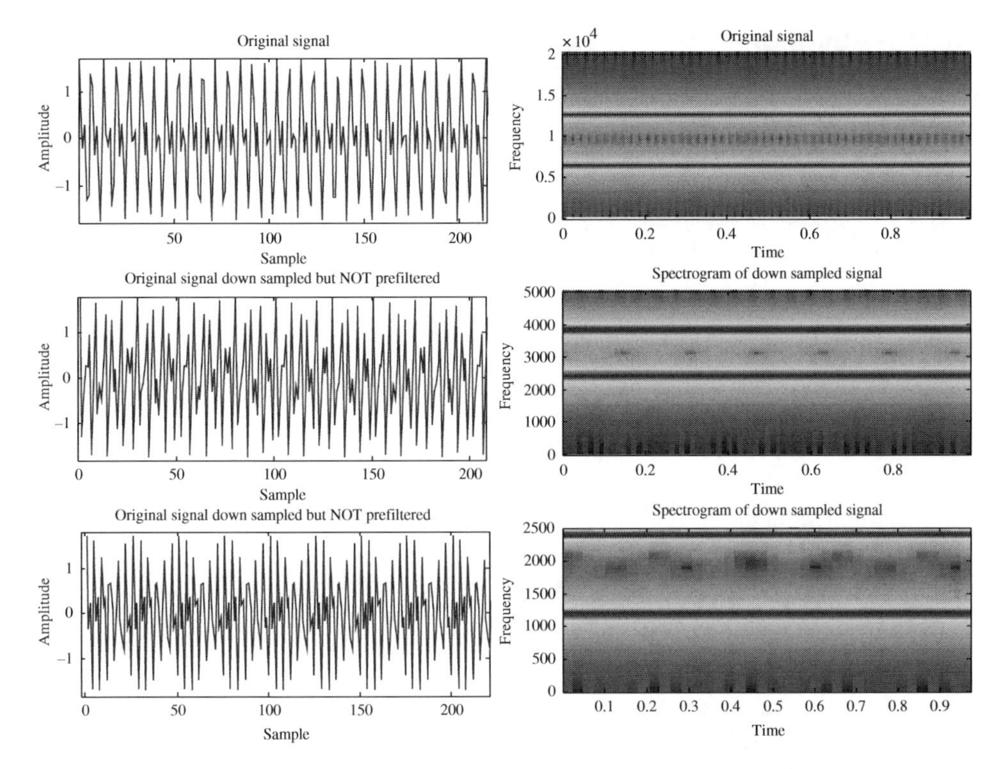

Figure 5.6 Illustration of aliasing. Top panel: the sum of two sinewaves, the assumed frequencies of the sinewaves are 6200 Hz and 12,400 Hz, the sampling frequency is 40,000 Hz. Middle panel: the sine waves down-sampled by a factor of 4 to a sampling frequency 10,000 Hz; note the aliased frequencies appear at $10,000 - 6200 = 3800$ Hz and $-10,000 + 12,400 = 2400$ Hz. Bottom panel: the sine waves down-sampled by a factor of 8 to a sampling frequency of 5000 Hz; note the aliased frequencies appear at $-5000 + 6200 = 1200$ Hz and at $-2 \times 5000 + 12,400 = 2400$ Hz.

from a frequency domain perspective, the low-pass filter removes the spectral images introduced as a result of sampling. Viewed from a time-domain perspective the low-pass filter acts as an ideal interpolator.

MATLAB function sinewave_aliasing_demo( )

Demonstrates spectrograms showing aliasing of the sum of two sine waves when down-sampled below the Nyquist rate. Note that the aliased frequencies appear as Fs_new ± F0, for example for Fs_new = 10,000 and sine wave frequencies of F0 = 6200 and 12,400 the new frequencies will appear at 3800 and 2400.

MATLAB function speech_aliasing_demo( )

Demonstration of the effect of aliasing on speech. A speech signal originally sampled at 16 kHz is down-sampled to 2 kHz sampling rate, first with an anti-aliasing filter and then without it. The spectrograms of the original speech and the down-sampled versions are displayed.

5.3 Quantisation

Quantisation is the process of converting each continuous-valued sample of a signal into a discrete value that can be assigned a unique digital codeword. For digital signal processing, discrete-time continuous-amplitude samples, from the sample-and-hold, are quantised and mapped into n-bit binary code words before being stored and processing.

Figure 5.7 illustrates an example of the quantisation of a signal into four discrete quantisation levels with each quantisation level represented by a 2-bit codeword. For quantisation to n-bit codewords, the amplitude range of the signal is divided into 2^n quantisation levels. Each continuous-amplitude sample is quantised to the nearest quantisation level and then mapped to the n-bit binary code assigned to that level.

Quantisation is a many-to-one mapping; this means that all the infinite number of values that fall within the continuum of the infinite values of a quantisation band are mapped to one single value at the centre of the band. The mapping is hence an irreversible process in that we cannot recover the exact value of the quantised sample. The mapping between an analogue sample $x_a(m)$ and its quantised value $x(m)$ can be expressed as

$$x(m) = Q[x_a(m)] \tag{5.7}$$

where $Q[\cdot]$ is the quantising function.

The performance of a quantiser is measured by signal-to-quantisation noise ratio (SQNR). The quantisation noise is defined as the difference between the analogue value of a sample and its quantised value as

$$e(m) = x(m) - x_a(m) \tag{5.8}$$

Now consider an n-bit quantiser with an amplitude range of $\pm V$ volts. The quantisation step size is $\Delta = 2V/2^n$. Assuming that the quantisation noise is a zero-mean random process with a uniform probability distribution (i.e. a probability of $1/\Delta$ and with an amplitude range of $\pm\Delta/2$) we can express the noise power as

$$\begin{aligned} E\left[e^2(m)\right] &= \int_{-\Delta/2}^{\Delta/2} p(e(m))\, e^2(m) de(m) = \frac{1}{\Delta} \int_{-\Delta/2}^{\Delta/2} e^2(m) de(m) \\ &= \frac{\Delta^2}{12} = \frac{V^2 2^{-2n}}{3} \end{aligned} \tag{5.9}$$

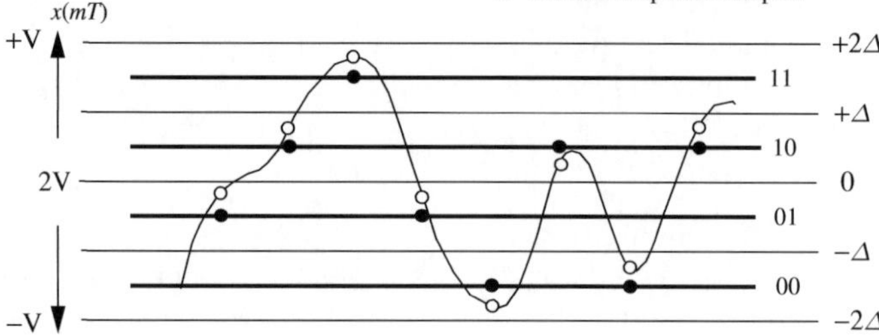

Figure 5.7 Illustration of offset-binary scalar quantisation.

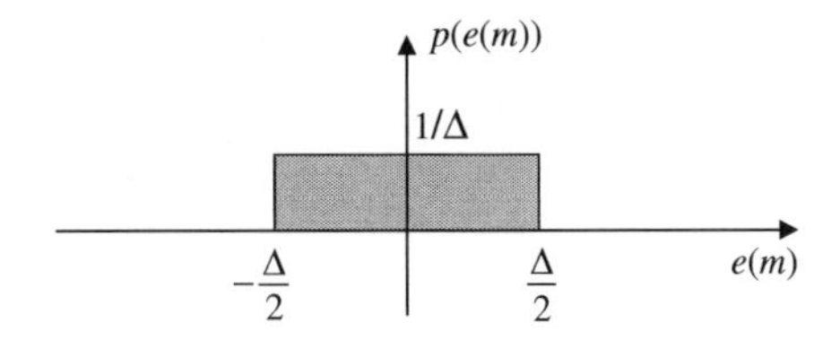

Figure 5.8 Illustration of the uniform probability distribution of the quantization noise.

where $p(e(m)) = 1/\Delta$, shown in Figure 5.8, is the uniform probability density function of the noise and $\Delta = 2V2^{-n}$. Using Equation (5.9) the SQNR is given by

$$
\begin{aligned}
SQNR(n) &= 10\log_{10}\left(\frac{\mathcal{E}[x^2(m)]}{\mathcal{E}[e^2(m)]}\right) = 10\log_{10}\left(\frac{P_{\text{signal}}}{V^2 2^{-2n}/3}\right) \\
&= 10\log_{10}3 - 10\log_{10}\left(\frac{V^2}{P_{\text{signal}}}\right) + 10\log_{10}2^{2n} \\
&= 4.77 - \alpha + 6n
\end{aligned}
\tag{5.10}
$$

where P_{signal} is the mean signal power, and α is the ratio in decibels of the peak signal power V^2 to the mean signal power P_{signal}, which for a sine wave α is 3. Therefore, from Equation (5.10) every additional bit in an analogue-to-digital converter results in a 6 dB improvement in signal-to-quantisation noise ratio.

5.3.1 Non-Linear Quantisation, Companding

A uniform quantiser is only optimal, in the sense of achieving the minimum mean squared quantization error, when the input signal is uniformly distributed within the full range of the quantiser, so that the uniform probability distribution of the signal sample values and the uniform distribution of the quantiser levels are matched and hence different quantisation levels are used with equal probability.

When a signal has a non-uniform probability distribution then a non-uniform quantisation scheme matched to the probability distribution of the signal is more appropriate. This can also be achieved through a transformation of the input signal to change the distribution of the input signal towards a uniform distribution prior to application of a uniform quantiser.

For speech signals, non-uniform qunatisation is achieved through a logarithmic compression of speech, a process known as companding, Figure 5.9. Companding (derived from compressing-expanding) refers to the process of first compressing an analogue signal at the transmitter, and then expanding this signal back to its original size at the receiver. During the companding process, continuous-amplitude input samples are compressed logarithmically and then quantised and coded using a uniform quantiser. The assumption is that speech has an exponential distribution and that the logarithm of speech has a more uniform distribution.

Figure 5.10 shows the effect of logarithmic compression on the distribution of a Gaussian signal. Note from Figure 5.10(b) that the distribution of the Gaussian signal is more spread after logarithmic compression. Figure 5.11 shows three sets of plots of speech and their respective histogram for speech quantised with 16 bits, 8 bits uniform quantisation and logarithmic compression followed by 8 bits quantisation respectively. Note that the histogram of the logarithm of absolute value of speech has a relatively uniform distribution suitable for uniform quantisation.

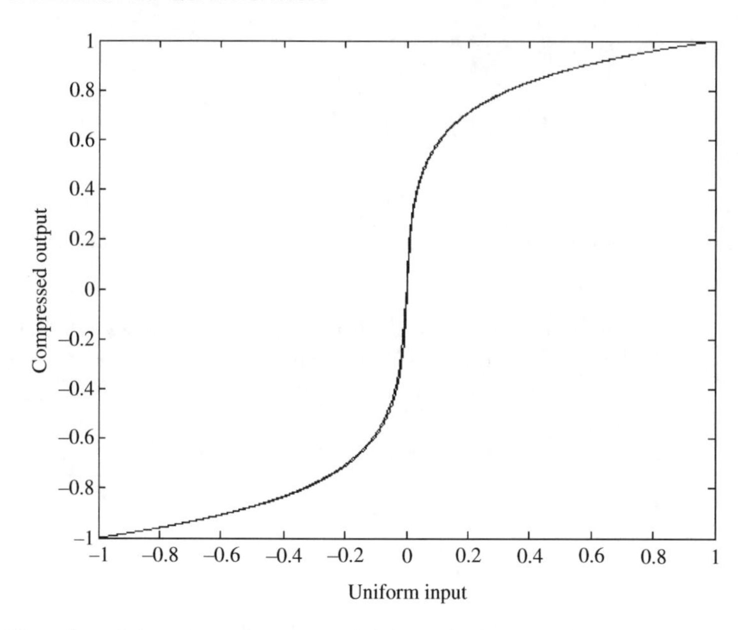

Figure 5.9 Illustration of the compression curves of A-law and u-law quantisers. Note that the curves almost coincide.

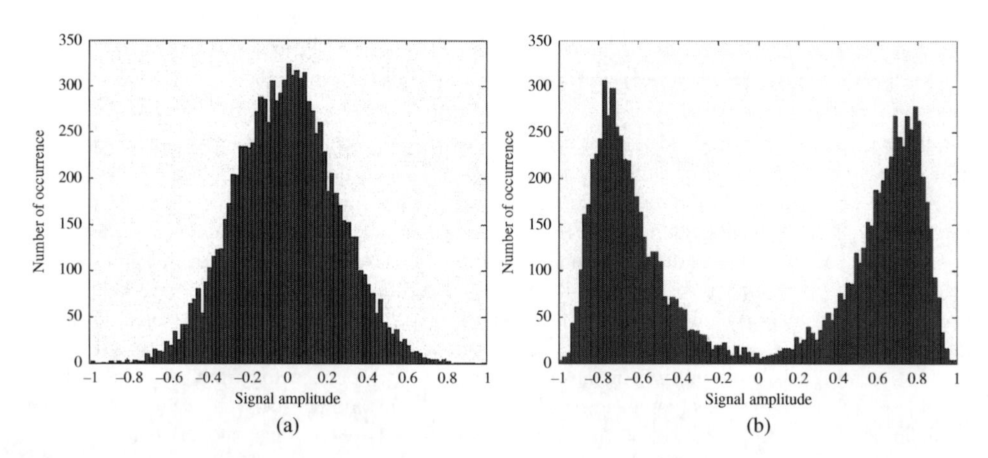

Figure 5.10 (a) The histogram of a Gaussian input signal to a u-law logarithmic function, (b) the histogram of the output of the u-law function.

The International Telecommunication Union (ITU) standards for companding are called u-law (in the USA) and A-law (in Europe). The u-law compression of a normalised sample value x is given by the relation

$$F(x) = \text{sign}(x)\frac{\ln(1+\mu|x|)}{\ln(1+\mu)} \tag{5.11}$$

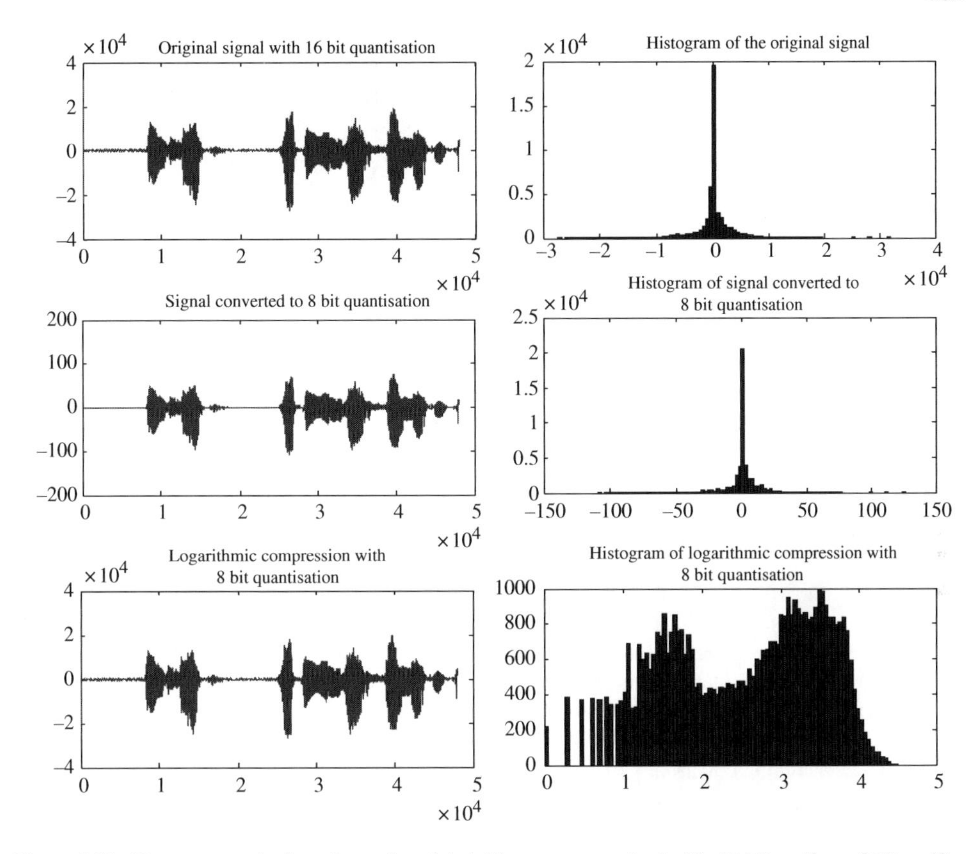

Figure 5.11 From top panel, plots of speech and their histograms quantised with: 16 bits uniform, 8 bits uniform and 8 bits logarithmic respectively. Note logarithmic compression of speech renders a uniform distribution.

where for the parameter μ a value of 255 is typically used. The A-law compression of a sample x is given by the relation

$$F(x) = \begin{cases} \dfrac{A|x|}{1+\ln A} & \text{if } 0 \le |x| < \dfrac{1}{A} \\ \dfrac{1+A|x|}{1+\ln A} & \text{if } \dfrac{1}{A} \le |x| \le 1 \end{cases} \tag{5.12}$$

A value of 78.6 is used for A. A-law and u-law methods are implemented using 8-bit codewords per sample (256 quantisation levels). At a speech sampling rate of 8 kHz this results in a bit rate of 64 kbps. An implementation of the coding methods may divide a dynamic range into a total of 16 segments: 8 positive and 8 negative segments. The segment range increase logarithmically; each segment is twice the range of the preceding one. Each segment is coded with 4 bits and a further 4-bit uniform quantisation is used within each segment. At 8 bits per sample, A-law and u-law quantisation methods can achieve the equivalent quality of 13-bit uniform quantisation.

MATLAB function QuantisationDemo( )

Quantises a speech signal originally quantised with 16 bits with two schemes: (1) uniform quantisation down to 8 bits, (2) logarithmic compression followed by uniform quantisation with 8 bits. The three versions of speech (16 bits, 8 bits uniform, 8 bits logarithmic) are played back and displayed.

5.4 Sampling Rate Conversion: Interpolation and Decimation

In some communication signal processing applications it is required to change the sampling rate of a discrete-time signal to a new sampling rate, for example from one audio sampling format to another with a different rate. Sampling rate conversion is achieved by a combination of interpolation and decimation as explained in this section.

5.4.1 Interpolation (Up-sampling) by an Integer Factor of *L*

Interpolation is the estimation of the values of a function or a sequence in-between the known values. Interpolation does not add any new information to a signal; it only changes the sampling rate by interpolating in-between the available information.

Applications of digital interpolators include sampling rate conversion in multi-rate communication systems and up-sampling for improved smoothness in image/video signals and graphical representation.

An interpolated signal would have a higher sampling rate and would seem smoother than the original signal. A higher sampling rate usually means a higher bandwidth, however in the case of an interpolated signal the frequency band between the original maximum frequency of the signal and half the new sampling rate would contain no energy and no information.

For example, consider a telephone quality audio signal with a bandwidth of 4 kHz originally sampled at 8 kHz and later interpolated to a new sampling rate of say 20 kHz. The new signal has a bandwidth equal to half the new sampling rate of 20 kHz (i.e. a bandwidth of 10 kHz), however the interpolated signal will not have any energy or information in-between 4 kHz and 10 kHz. This interpolated signal played back at the new sampling rate of 20 KHz would not sound different from the original signal.

Consider a band-limited discrete-time signal $x(m)$ with a base-band spectrum $X(f)$ as shown in Figure 5.12. The sampling rate can be increased by a factor of L through the interpolation of $L-1$ samples between every two known samples of $x(m)$. In the following it is shown that digital interpolation by a factor of L can be achieved through a two-stage process of:

(a) insertion of $L-1$ zeros in-between every two samples and
(b) low-pass filtering of the zero-inserted signal by a filter with a cut-off frequency of $F_s/2L$, where F_s is the sampling rate.

Consider the zero-inserted signal $x_z(m)$ obtained by inserting $L-1$ zeros between every two samples of $x(m)$. Therefore we have $x_z(0)=x(0)$, $x_z(1)=0, \ldots, x_z(L-1)=0$, $x_z(L)=x(1)$, $x_z(L+1)=0, \ldots,$ $x_z(2L-1)=0$, $x_z(2L)=x(2)$, $x_z(2L+1)=0, \ldots, x_z(3L-1)=0$, $x_z(3L)=x(3)$, $x_z(3L+1)=0, \ldots$ and so on. In general we have

$$x_z(m) = \begin{cases} x\left(\dfrac{m}{L}\right), & m = 0, \pm L, \pm 2L, \ldots \\ 0, & \text{otherwise} \end{cases} \tag{5.13}$$

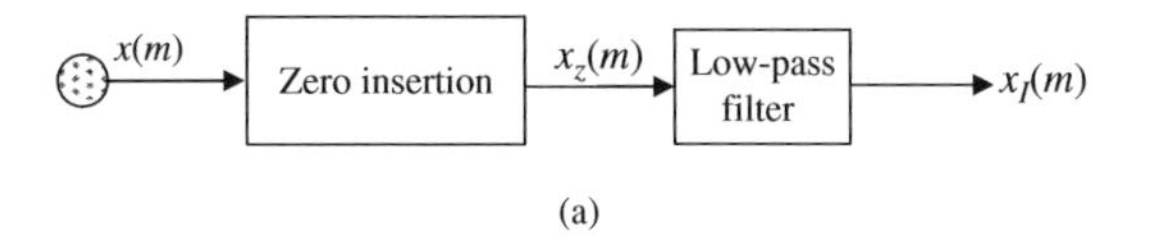

(a)

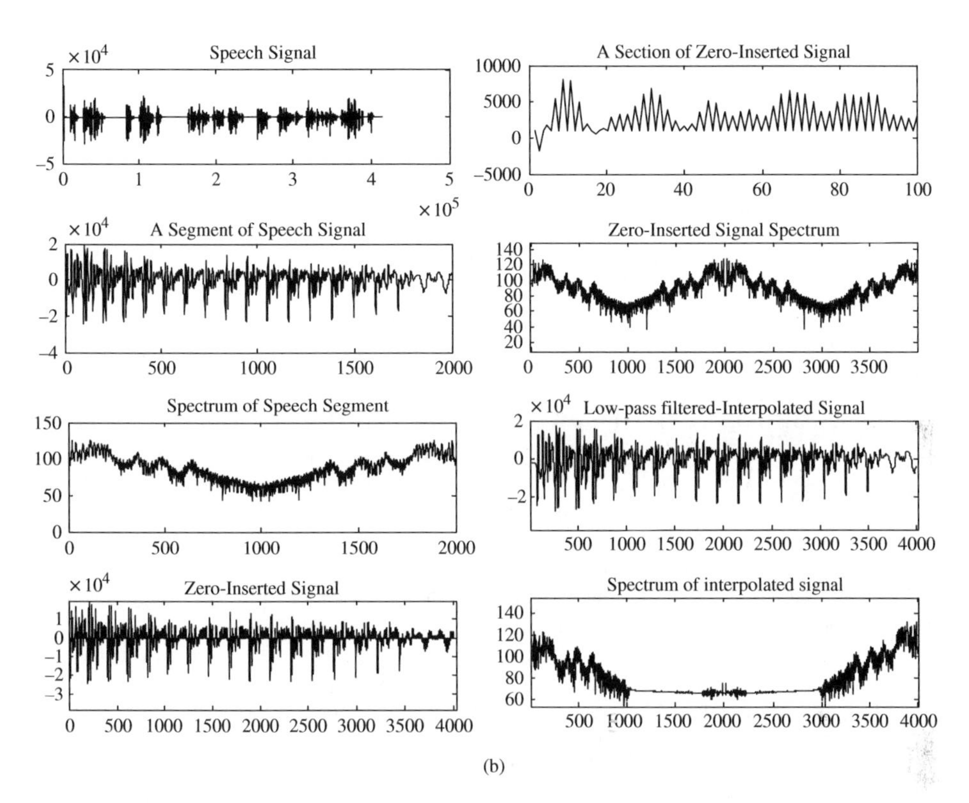

(b)

Figure 5.12 (a) Block diagram of interpolation method, (b) left-hand side: speech signal, a segment of speech, its spectrum and zero-inserted signal (zero-inserted by a factor of 2); right-hand side: zoom into a section of zero-inserted signal, the spectrum of zero-inserted signal, interpolated signal and its spectrum. Note that x-axis is sample number in time or frequency and y-axis is amplitude or magnitude.

The spectrum of the zero-inserted signal is related to the spectrum of the original discrete-time signal by

$$\begin{aligned} X_z(f) &= \sum_{m=-\infty}^{\infty} x_z(m)e^{-j2\pi fm} \\ &= \sum_{m=-\infty}^{\infty} x(m)e^{-j2\pi fmL} \\ &= X(L.f) \end{aligned} \tag{5.14}$$

where the second line of Equation (5.14) was obtained using Equation (5.13). Equation (5.14) states that the spectrum of the zero-inserted signal $X_z(f)$ is a frequency-scaled version of the spectrum of the original signal $X(f)$. Note that zero-insertion of a signal effectively stretches the time duration of the signal and it follows from the inverse relationship between time and frequency that the signal's bandwidth shrinks.

Figure 5.12 shows an example of a signal interpolated by a factor of five times the original sampling rate. The base-band spectrum of the zero-inserted signal is composed of L (in this example five) repetitions of the based band spectrum of the original signal. The interpolation of the zero-inserted signal is therefore equivalent to filtering out the repetitions of $X(f)$ in the base band of $X_z(f)$, as illustrated in Figure 5.12. Note that in order to maintain the real-time duration of the signal the play-back sampling rate of the interpolated signal $x_z(m)$ needs to be increased by a factor of L.

MATLAB function InterpolationDemo( )
This function demonstrates the process of interpolation or up-sampling of a signal by an integer factor of I. The interpolation process by a factor of I involves insertion of I – 1 zeros between every two samples followed by low-pass filtering the spectrum by a low-pass filter with a cut-off frequency of Fs/2I where Fs is the sampling rate.

5.4.2 Decimation (Down-sampling) by an Integer Factor of *L*

Down-sampling a signal by a factor of L results in a loss of bandwidth by the same factor of L, for example, down-sampling by a factor of 2 results in losing the upper half of the total bandwidth and down-sampling by a factor of 4 results in losing three-quarters of the spectrum. Consider a band-limited discrete-time signal $x(m)$ with a base-band spectrum $X(f)$. The sampling rate can be decreased by a factor of L through discarding $L-1$ samples for every L samples of $x(m)$. In the following it is shown that digital decimation (down-sampling) by a factor of L can be achieved through a two-stage process of:

(a) Low-pass filtering of the signal by a filter with a cut-off frequency of $F_s/2L$, where F_s is the input sampling rate. This is the anti-aliasing process for removing the part of the spectrum that would be lost as a result of down-sampling.
(b) Discarding $L-1$ samples for every L samples.

Mathematical Model of Decimation

Consider a re-sampled signal $x_r(m)$ expressed as the product of the original sampled signal $x(m)$ and the re-sampling pulse train $p(m)$ as

$$x_r(m) = x(m).p(m) = \sum_{k=-\infty}^{\infty} x(k)\delta(m-kL) = \begin{cases} x(m), & m = 0, \pm L, \pm 2L, \ldots \\ 0, & \text{otherwise} \end{cases} \tag{5.15}$$

Note that the re-sampling train of discrete-time impulses $p(m)$ periodically retains one out of every L samples and sets the remaining $L-1$ samples to zero. The re-sampling impulse train $p(m)$ and its Fourier series are given by

$$p(m) = \sum_{k=-\infty}^{\infty} \delta(m-kL) \quad \Leftrightarrow \quad P(f) = \sum_{k=-\infty}^{\infty} \delta\left(f - k\frac{F_s}{L}\right) \tag{5.16}$$

The re-sampling process is illustrated in Figure 5.13. Note that the period LTs of the re-sampling impulse train in time and its fundamental frequency F_s/L are inversely related. From the Fourier

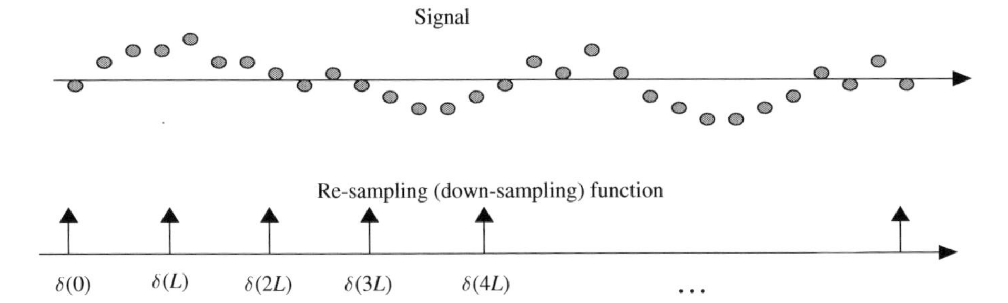

Figure 5.13 Illustration of a model for the process of down-sampling a signal. The sampling train of impulses periodically retains one sample and discards $L-1$.

properties (see Chapter 2), the spectrum of a re-sampled signal is the convolution of the spectra of the original signal $x(m)$ and the sampling train $p(m)$ and can be expressed as

$$X_r(f) = \sum_{k=-\infty}^{\infty} X\left(f + k\frac{F_s}{L}\right) \tag{5.17}$$

Note that the original discrete-time signal (before re-sampling) has its base-band frequency spectrum $X(f)$ and the periodic images of the base-band that result from sampling, $X(f+kF_s)$, centred at multiples of F_s (and separated by $F_s/2$ as shown in Figure 5.4), whereas the re-sampled signal's base-band spectrum and images $X(f+kF_s/L)$ are centred at multiples of F_s/L (and separated by $F_s/2L$).

The down-sampled signal can be obtained from the re-sampled signal as

$$x_d(m) = x_r(mL) \tag{5.18}$$

> **MATLAB function DecimationDemo( )**
> Demonstrates the process of decimation or down-sampling of a signal by an integer factor of D. The decimation process by a factor of D involves low-pass filtering by a low-pass filter with a cut-off frequency of Fs/2D where Fs is the sampling rate, followed by discarding every other D − 1 samples.

Note that whereas the re-sampled signal had $L-1$ zeros between every two samples, the down-sampled signal discards the zeroed samples. The spectrum of the decimated (down-sampled) signal can be related to the spectrum of the re-sampled discrete-time signal, Figure 5.14, as

$$\begin{aligned} X_D(f) &= \sum_{m=-\infty}^{\infty} x_d(m)e^{-j2\pi fm} \\ &= \sum_{m=-\infty}^{\infty} x_r(mL)e^{-j2\pi fm} \\ &= \sum_{m'=-\infty}^{\infty} x_r(m')e^{-j2\pi \frac{f}{L}m'} = X_r(f/L) \end{aligned} \tag{5.19}$$

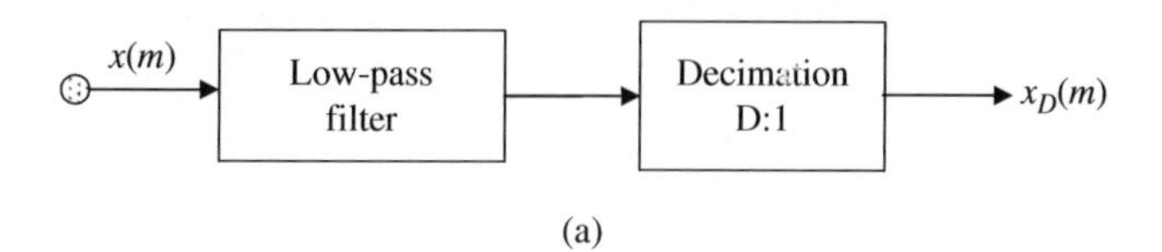

(a)

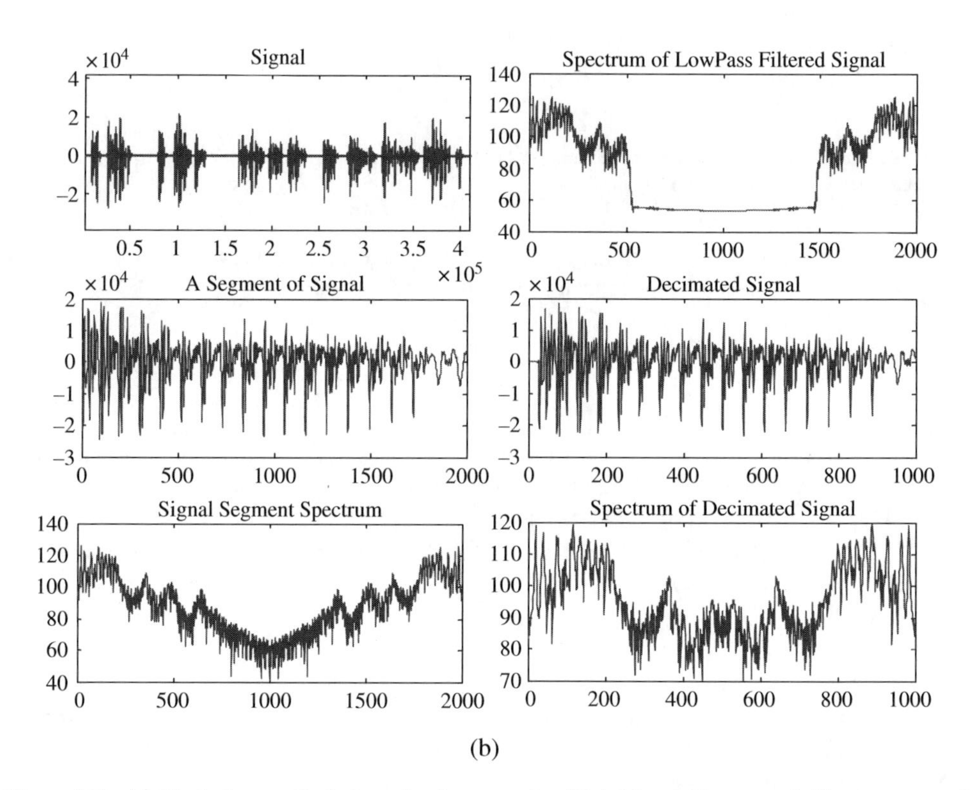

(b)

Figure 5.14 (a) Block diagram illustration of a down-sampler, (b) left-hand side: a speech file, a segment of the speech file and its spectrum; right-hand side the spectrum of low-pass filtered speech prior to decimation by a factor of 2, the decimated signal and its spectrum. Note that x-axis is sample number in time or frequency and y-axis is amplitude or magnitude.

where $m' = mL$. Note that in developing the third line of Equation (5.19) from the second line, we used the fact that in between every two signal samples of $x_r(m)$ there are $L-1$ zeros. Equation (5.19) shows that the spectrum of the decimated signal is $1/L$ of that of the original signal.

5.4.3 Sampling Rate Conversion by a Non-integer Factor

Figure 5.15 illustrates a system for conversion of the sampling rate of a signal by a non-integer value. To change a sampling rate by a non-integer factor of $V = L/D$ (where L and D are integers), the signal is first interpolated by a factor of L, and then decimated by a factor of D. Before decimation the signal must be filtered by a low-pass filter which acts as an interpolator and also as an anti-aliasing filter for the decimation stage. The cut-off frequency of the low-pass filter should be set to $\min(F_s/2L, F_s/2D)$ where F_s is the input sampling rate.

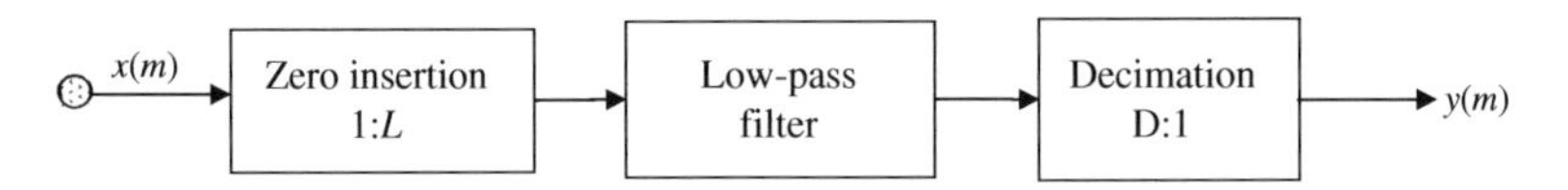

Figure 5.15 Block diagram illustration of a system for conversion of the sampling rate of a signal by a non-integer value.

5.5 Summary

This chapter considered the basic and essential processes of sampling and quantisation of a signal and the process of sampling rate conversion. It was shown that sampling can be an invertible process, in that the original signal can be recovered from its samples, provided that the sampling rate is more than twice the highest frequency content of the signal. Two widely used forms of quantisation, namely uniform and logarithmic quantisation, were introduced and it was explained that although quantisation always incurs some distortion known as quantisation noise, it is important that the information is not lost and that the quantisation noise is kept at an imperceptible level. Sampling rate conversion, namely up-sampling, down-sampling and the combination of the two, was also studied.

Further Reading

Nyquist H. (1928) Certain Topics in Telegraph Transmission Theory. AIEE Trans, 47, pp. 617–644, (reprinted as classic paper in: Proc. IEEE, 90, 2).

Shannon, C.E. (1949) Communications in the Presence of Noise. Proc. IRE 37, pp. 10–21.

Exercises

5.1 State the minimum number of samples per cycle required to convert a continuous-time sine wave to a discrete-time sine wave.

Hence state the Nyquist–Shannon sampling theorem for sampling a continuous-time signal with a maximum frequency content of B Hz.

5.2 Consider a sine wave with a frequency of F_0 Hz. What is the outcome of sampling this sine wave at a rate of (i) 2.1 F_0, (ii) F_0, (iii) 0.5 F_0?

5.3 Calculate the bit rate per second and the bandwidth required to transmit a digital stereo music signal sampled at a rate of 44,100 samples per second and with each sample represented with a 16-bit word.

5.4 Calculate the signal-to-quantisation noise ratio at the output of a 16-bit PCM quantiser.

Explain why 16-bit PCM quantisation is sufficient to obtain high fidelity music.

5.5 State the qualitative effects on the perceived quality of an audio sound of (i) interpolation by a factor of I and (ii) decimation by a factor of D.

5.6 With the aid of a diagram, describe the outline of a system composed of an interpolator and a decimator for re-sampling a hi-fi digital audio signal $x(m)$ originally sampled at a rate of 44.1 kHz to a new sampling rate of 16 kHz.

Part II

Model-Based Signal Processing

6 Information Theory and Probability Models

Information theory and probability models provide the mathematical foundation for the analysis, modelling and design of telecommunication and signal processing systems.

What constitutes information, news, data or knowledge may be somewhat of a philosophical question. However, information theory is concerned with quantifiable variables, hence, information may be defined as knowledge or data about the states or values of a random process, such as the number of states of the process, the likelihood of each state, the probability of the observable outputs in each state, the conditional probabilities of the state sequences (i.e. how the random process moves across its various states) and the process history (i.e. its past, current and likely future states).

For example, the history of fluctuations of random variables, such as the various states of weather/climate, the demand on a cellular mobile phone system at a given time of day and place or the fluctuations of stock market prices, may be used to obtain a finite-state model of these random variables.

Information theory allows prediction and estimation of the values or states of a random process, from related observations that may be incomplete and/or noisy. This is facilitated through utilisation of the probability models of the information-bearing process and noise and the history of the dependencies of the state sequence of the random process.

Probability models form the foundation of information theory. Information is quantified in units of 'bits' in terms of a logarithmic function of probability. Probability models are used in communications and signal processing systems to characterise and predict random signals in diverse areas of applications such as speech/image recognition, audio/video coding, bio-engineering, weather forecasting, financial data modelling, noise reduction, communication networks and prediction of the call demand on a service facility such as a mobile phone network.

This chapter introduces the concept of random processes and probability models and explores the relationship between probability and information. The concept of entropy is introduced as a measure for quantification of information and its application to Huffman coding is presented. Finally, several different forms of probability models and their applications in communication signal processing are considered.

Multimedia Signal Processing: Theory and Applications in Speech, Music and Communications Saeed V. Vaseghi

6.1 Introduction: Probability and Information Models

All areas of information processing and decision making deal with signals that are random, may carry multiple layers of information (e.g. speech signals convey words, meaning, gender, emotion, state of health, accent etc.) and are often noisy and perhaps incomplete.

Figure 6.1 illustrates a simplified bottom-up illustration of the information processing hierarchy from the signal generation level to information decoding and decision making. At all levels of information flow there is some randomness or uncertainty and the observation may contain mixed signals and/or hidden parameters and noise.

Probability models form the foundation of information theory and decision making. As shown in Figure 6.2, many applications of information theory such as data compression, pattern recognition, decision making, search engines and artificial intelligence are based on the use of probability models of the signals.

As explained later in this chapter the information content of a source is measured and quantified in units of 'bits' in terms of a logarithmic function of probability known as entropy. It would be practically impossible to design and develop large-scale efficient reliable advanced communication systems without the use of probability models and information theory.

Information theory deals with information-bearing signals that are random such as text, speech, image, noise and stock market time series. Indeed, a signal cannot convey information without being random in the sense that a predictable signal has no information and conversely the future values of an information-bearing signal are not entirely predictable from the past values.

The modelling, quantification, estimation and ranking of information in communication systems and search engines requires appropriate mathematical tools for modelling the randomness and uncertainty in information-bearing signals and the main mathematical tools for modelling randomness/uncertainty in a signal are those offered by statistics and probability theory.

This chapter begins with a study of random processes and probability models. Probability models are used in communications and signal processing systems to characterise and predict random signals in diverse areas of applications such as speech/image recognition, audio/video coding, bio-medical engineering, weather forecasting, financial data modelling, noise reduction, communication networks and prediction of the call demand on a service facility such as a mobile phone network.

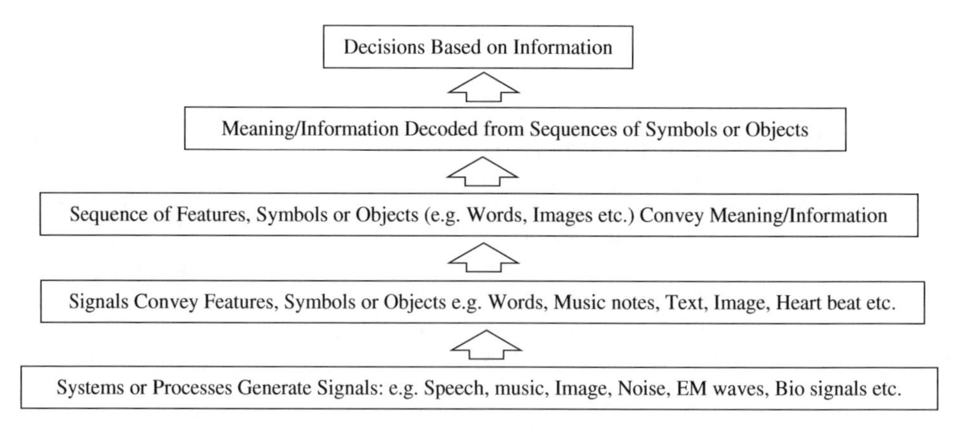

Figure 6.1 A bottom-up illustration of the information processing hierarchy from the signal generation to information extraction and decision making. At all levels of signal/information processing there is some randomness or uncertainty or noise that may be modelled with probability functions.

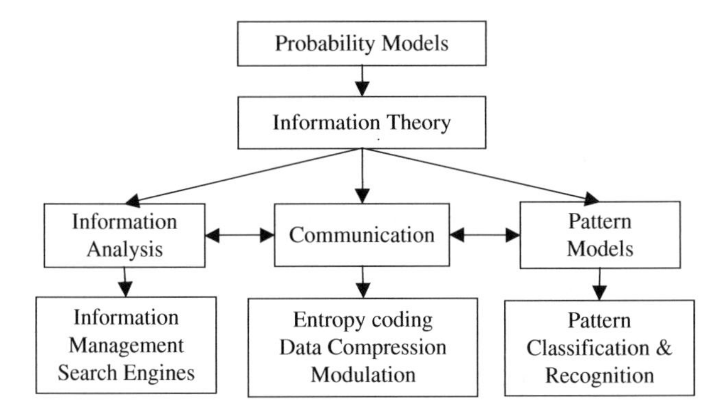

Figure 6.2 A simplified tree-structure illustration of probability models leading to information theory and its applications in information analysis, communication and pattern recognition.

The concepts of randomness, information and entropy, are introduced and their close relationships explored. A random process can be completely described in terms of a probability model, but may also be partially characterised with relatively simple statistics, such as the mean, the correlation and the power spectrum. We study stationary, non-stationary and finite-state processes. We consider some widely used classes of random processes, and study the effect of filtering or transformation of a signal on its probability distribution. Finally, several applications of probability models in communication signal processing are considered.

6.2 Random Processes

This section introduces the concepts of random and stochastic processes and describes a method for generation of random numbers.

6.2.1 Information-bearing Random Signals vs Deterministic Signals

Signals, in terms of one of their most fundamental characteristics, i.e. their ability to convey information, can be classified into two broad categories:

(a) *Deterministic* signals such as sine waves that on their own convey no information but can act as carriers of information when modulated by a random information-bearing signal.
(b) *Random* signals such as speech and image that contain information.

In each class, a signal may be continuous or discrete in time, may have continuous-valued or discrete-valued amplitudes and may be one-dimensional or multi-dimensional.

A deterministic signal has a predetermined trajectory in time and/or space. The exact fluctuations of a deterministic signal can be described in terms of a function of time, and its exact value at any time is predictable from the functional description and the past history of the signal. For example, a sine wave $x(t)$ can be modelled, and accurately predicted either by a second-order linear predictive model or by the more familiar equation $x(t) = A\sin(2\pi ft + \phi)$. Note that a deterministic signal carries no information other than its own shape and characteristic parameters.

Random signals have unpredictable fluctuations; hence it is not possible to formulate an equation that can predict the *exact* future value of a random signal. Most signals of interest such as speech, music and noise are at least in part random. The concept of randomness is closely associated with the concepts of information, bandwidth and noise.

For a signal to have a capacity to convey information, it must have a degree of randomness: a predictable signal conveys no information.

Therefore the random part of a signal is either the information content of the signal, or noise, or a mixture of information and noise. In telecommunication it is wasteful of resources, such as time, power and bandwidth, to retransmit the predictable part of a signal. Hence signals are randomised (de-correlated) before transmission.

Although a random signal is not predictable, it often exhibits a set of well-defined statistical values such as maximum, minimum, mean, median, variance, correlation, power spectrum and higher order statistics. A random process is described in terms of its statistics, and most completely in terms of a probability model from which its statistics can be calculated.

Example 6.1 *A deterministic signal model*
Figure 6.3(a) shows a model of a deterministic discrete-time signal. The model generates an output signal $x(m)$ from the P past output samples as

$$x(m) = h_1\left(x(m-1), x(m-2), \ldots, x(m-P)\right) + \delta(m) \tag{6.1}$$

where the function h_1 may be a linear or a non-linear model and $\delta(m)$ is a delta function that acts as an initial 'kick-start' impulse input. Note that there is no sustained input. A functional description of the model h_1 together with the P initial sample values are all that is required to generate, or predict the future values of the signal $x(m)$.

For example, for a discrete-time sinusoidal signal generator (i.e. a *digital oscillator*) Equation (6.1) becomes

$$x(m) = a_1 x(m-1) + a_2 x(m-2) + \delta(m) \tag{6.2}$$

where the parameter $a_1 = 2\cos(2\pi F_0/F_s)$ determines the oscillation frequency F_0 of the sinusoid at a sampling frequency of F_s and $a_2 = 1$.

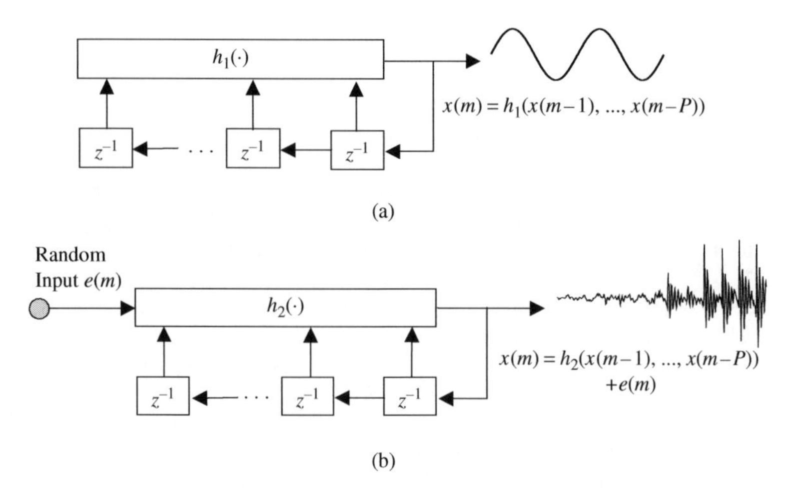

Figure 6.3 (a) A deterministic signal model, (b) a random signal model.

Example 6.2 *A random signal model*
Figure 6.3(b) illustrates a model for a random signal given by

$$x(m) = h_2 (x(m-1), x(m-2), \ldots, x(m-P)) + e(m) \tag{6.3}$$

where the random input $e(m)$ models the part of the signal $x(m)$ that is unpredictable, and the function h_2 models the part of the signal that is correlated with and predictable from the past samples. For example, a narrowband, second-order autoregressive process can be modelled as

$$x(m) = a_1 x(m-1) + a_2 x(m-2) + e(m) \tag{6.4}$$

where the choice of the model parameters a_1 and a_2 will determine the centre frequency and the bandwidth of the process.

6.2.2 Pseudo-Random Number Generators (PRNG)

Random numbers are generated by a feedback system such as $x(m) = f(x(m-1), x(m-2), \ldots)$ as shown in Figure 6.4. The random number generation starts with an initial value, $x(-1)$, known as the 'seed'.

A 'random' number generator implemented on a deterministic computer system, with finite memory, will exhibit periodic properties and hence it will not be purely random. This is because a computer is a finite-state memory system and will, given sufficient time, eventually revisit previous internal states, after which it will repeat the state sequence. For this reason computer random number generators are known as pseudo-random number generators (PRNG).

The outputs of pseudo-random number generators only approximate some of the statistical properties of random numbers. However, in practice the undesirable periodicity can be 'avoided' if the period is extremely large so that no repetitions would be practically observed. Note that the length of the maximum period typically doubles with each additional bit, and hence it is not difficult to build PRNGs with periods so long that the repetitions will not observed.

A PRNG can be started from an arbitrary starting point, using a 'random' seed state; it will then always produce the same sequence when initialised with that state.

Most pseudo-random generator algorithms produce sequences which are uniformly distributed. Common classes of PRNG algorithms are linear feedback shift registers (Figure 6.4(b)), linear congruential generators and lagged Fibonacci generators. Recent developments of PRNG algorithms include Blum Blum Shub, Fortuna, and the Mersenne twister. A description of these methods is outside the scope of this book.

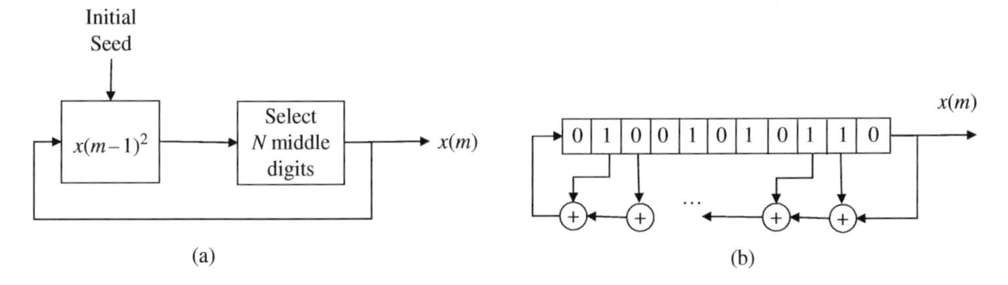

Figure 6.4 Illustration of two different PRNG methods: (a) a simple middle digits of power of 2 method and (b) a linear feedback shift register method.

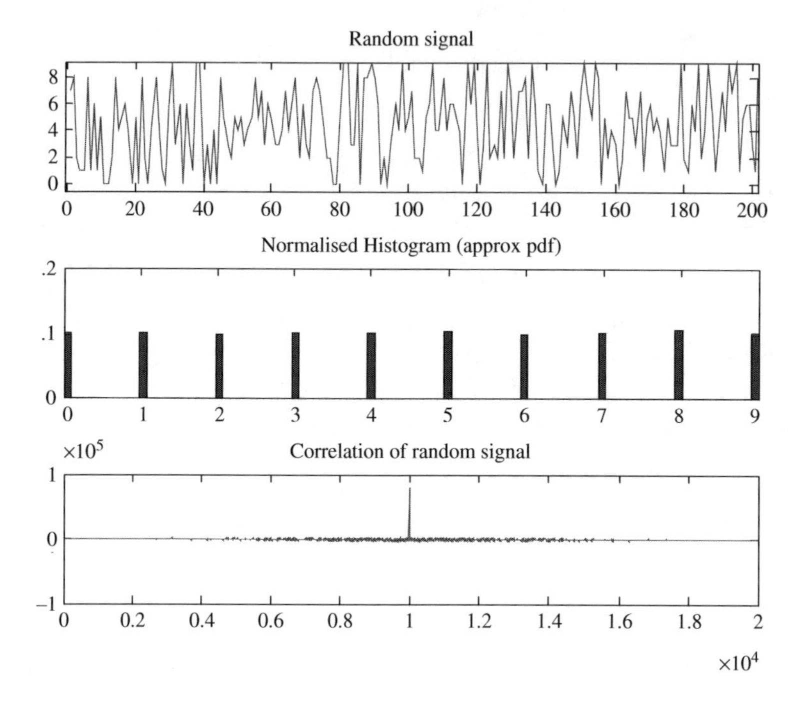

Figure 6.5 Illustration of a uniformly distributed discrete random process with 10 values/states 0, 1, 2, …, 9. Top: a segment of the process, middle: the normalised histogram of the process, bottom: the autocorrelation of the process. The histogram and autocorrelation were obtained from 10^5 samples.

However, a simple Matlab program is provided for illustrating the basic principles of PRNG. The program squares an N-digit number to yield a $2N$ digit output, the N middle digits of the output are fed back to the square function. At each cycle the program outputs M digits where $M \leq N$. Figure 6.5 shows the output of the PRNG together with its histogram and autocorrelation function.

Matlab function Pseudo-Random Number Generator function PRNG( )
Demonstration of a simple random number generator. In this method of 'middle digits of the Pth power' of an input, the middle N digits of the output are fed back to input and the middle M digits are taken as the output. The random signal, its histogram and its correlation function are displayed.

Note that the pseudo-numbers repeat themselves; this can be seen in experiment with small and large values of N.

6.2.3 Stochastic and Random Processes

A random process is any process or function that generates random signals. The term 'stochastic process' is broadly used to describe a random process that generates *sequential* random signals such as a sequence of speech samples, a video sequence, a sequence of noise samples, a sequence of stock market data fluctuations or a DNA sequence.

In signal processing terminology, a random or stochastic process is also a probability model of a class of random signals, e.g. Gaussian process, Markov process, Poisson process, binomial process, multinomial process etc.

In this chapter, we are mainly concerned with discrete-time random processes that may occur naturally or may be obtained by sampling a continuous-time band-limited random process. The term 'discrete-time stochastic process' refers to a class of discrete-time random signals, $X(m)$, characterised by a probabilistic model. Each realisation of a discrete-time stochastic process $X(m)$ may be indexed in time and space as $x(m, s)$, where m is the discrete time index, and s is an integer variable that designates a space index to each realisation of the process.

6.2.4 The Space of Variations of a Random Process

The collection of all realisations of a random process is known as the space, or the ensemble, of the process. For an illustration, consider a random noise process over a communication network as shown in Figure 6.6. The noise on each telephone line fluctuates randomly with time, and may be denoted as $n(m, s)$, where m is the discrete-time index and s denotes the line index. The collection of noise on different lines forms the space of the noise process denoted by $N(m) = \{n(m, s)\}$, where $n(m, s)$ denotes a realisation of the noise process $N(m)$ on line s.

The 'true' statistics of a random process are obtained from the averages taken over the space of many different realisations of the process. However, in many practical cases, only one or a finite number of realisations of a process are available. In Sections 6.6.9 and 6.6.10, we consider ergodic random processes in which time-averaged statistics, from a single realisation of a process, may be used instead of the ensemble-averaged statistics.

Notation: In this chapter $X(m)$, with upper case X, denotes a random process, the signal $x(m, s)$ is a particular realisation of the process $X(m)$, the signal $x(m)$ is any realisation of $X(m)$, and the collection of all realisations of $X(m)$, denoted by $\{x(m, s)\}$, forms the ensemble or the space of the process $X(m)$.

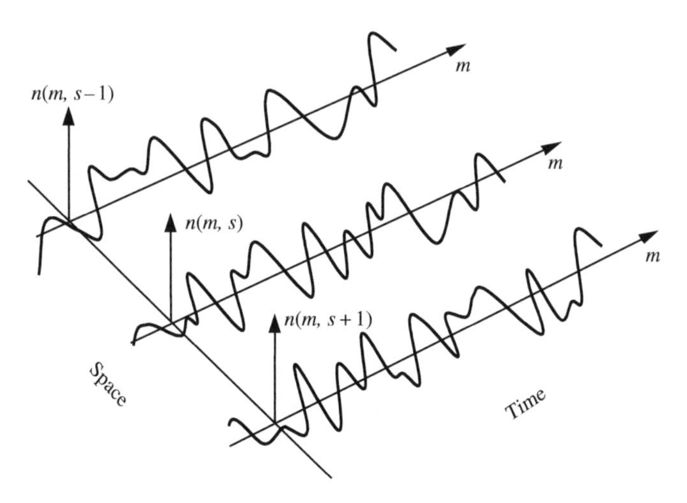

Figure 6.6 Illustration of three different realisations in the space of a random noise process $N(m)$.

6.3 Probability Models of Random Signals

Probability models, initially devised to calculate the odds for the different outcomes in a chance game, provide a complete mathematical description of the distribution of the likelihood of the different outcomes of a random process. In its simplest form a probability model provides a numerical value, between 0 and 1, for the likelihood of a discrete-valued random variable assuming a particular state or value. The probability of an outcome of a variable should reflect the fraction of times that the outcome is observed to occur.

6.3.1 Probability as a Numerical Mapping of Belief

It is useful to note that often people quantify their intuitive belief/feeling in the probability of the outcome of a process or a game in terms of a number between zero and one or in terms of its equivalent percentage. A probability of 'zero' expresses the impossibility of the occurrence of an event, i.e. it never happens, whereas a probability of 'one' means that the event is certain to happen, i.e. always happens. Hence a person's belief (perhaps formed by practical experience, intuitive feeling or deductive reasoning) is mapped into a number between one and zero.

6.3.2 The Choice of One and Zero as the Limits of Probability

The choice of zero for the probability of occurrence of an infinitely improbable event, is necessary if the laws of probability are to hold; for example the joint probability of an impossible event *and* a probable event, P(impossible, possible), should be the same as the probability of an impossible event; this requirement can only be satisfied with the use of zero to represent the probability of an impossible event. The choice of one for the probability of an event that happens with certainty is arbitrary but it is a convenient and established choice.

6.3.3 Discrete, Continuous and Finite-State Probability Models

Probability models enable the estimation of the likely values of a process from noisy or incomplete observations. As illustrated in Figure 6.7, probability models can describe random processes that

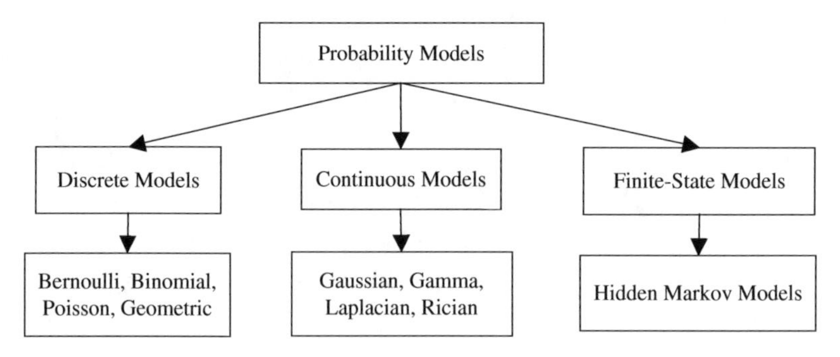

Figure 6.7 A categorisation of different classes and forms of probability models together with some examples in each class.

are discrete-valued, continuous-valued or finite-state continuous-valued. Figure 6.7 lists the most commonly used forms of probability models. Probability models are often expressed as functions of the statistical parameters of the random process; most commonly they are in the form of exponential functions of the mean value and covariance of the process.

6.3.4 Random Variables and Random Processes

At this point it is useful to define the difference between a random variable and a random process. A random variable is a variable that assumes random values such as the outcomes of a chance game or the values of a speech sample or an image pixel or the outcome of a sport match. A random process, such as a Markov process, generates random variables usually as functions of time and space. Also a time or space series, such as a sequence of speech or an image is often called a random process.

Consider a random process that generates a time-sequence of numbers $x(m)$. Let $\{x(m, s)\}$ denote a collection of different time-sequences generated by the same process where m denotes time and s is the sequence index for example as illustrated in Figure 6.6. For a given time instant m, the sample realisation of a random process $\{x(m, s)\}$ is a random variable that takes on various values across the space s of the process. The main difference between a random variable and a random process is that the latter generates random time/space series. Therefore the probability models used for random variables are also applied to random processes. We continue this section with the definitions of the probability functions for a random variable.

6.3.5 Probability and Random Variables

Probability models the behaviour of random variables. Classical examples of random variables are the random outcomes in a chance process, or gambling game, such as the outcomes of throwing a fair coin, Figure 6.8(a), or a pair of fair dice, Figure 6.8(b), or dealing cards in a game.

The Space and Subspaces of a Random Variable

The space of a random variable is the collection of all the values, or outcomes, that the variable can assume. For example the space of the outcomes of a coin is the set {H, T} and the space of the outcome of tossing a die is {1, 2, 3, 4, 5, 6}, Figure 6.8. The space of a random variable can be partitioned, according to some criteria, into a number of subspaces. A subspace is a collection of values with a common attribute, such as the set of outcomes bigger or smaller than a threshold or a cluster of closely spaced samples, or the collection of samples with their values within a given band of values.

Each subspace is called an event, and the probability of an event A, $P(A)$, is the ratio of the number of observed outcomes from the space of A, N_A, divided by the total number of observations:

$$P(A) = \frac{N_A}{\sum\limits_{\text{All events}\, i} N_i} \tag{6.5}$$

From Equation (6.5), it is evident that the sum of the probabilities of all likely events in an experiment is one.

$$\sum_{\text{All events } A} P(A) = 1 \tag{6.6}$$

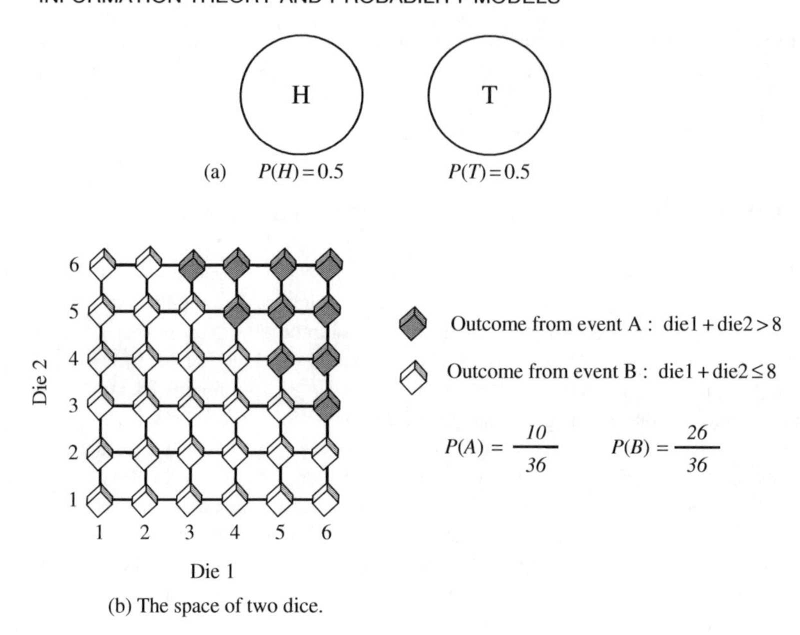

Figure 6.8 (a) The probability of two outcomes (Head or Tail) of tossing a coin; P(H) = P(T) = 0.5, (b) a two-dimensional representation of the outcomes of two dice, and the subspaces associated with the events corresponding to the sum of the dice being greater than 8, or less than or equal to 8; $P(A) + P(B) = 1$.

Example 6.3 The space of two discrete numbers obtained as outcomes of throwing a pair of dice is shown in Figure 6.8(b). This space can be partitioned in different ways; for example, the two subspaces A and B shown in Figure 6.8 are associated with the pair of numbers that in the case of subspace A add up to a value greater than 8, and in the case of subspace B add up to a value less than or equal to 8. In this example, assuming that the dice are not loaded, all numbers are equally likely and the probability of each event is proportional to the total number of outcomes in the space of the event as shown in the figure.

6.3.6 Probability Mass Function – Discrete Random Variables

For a discrete random variable X that can assume values from a finite set of N numbers or symbols $\{x_1, x_2, \ldots, x_N\}$, each outcome x_i may be considered an event and assigned a probability of occurrence. For example if the variable is the outcome of tossing a coin then the outcomes are Head (H) and Tail (T), hence $X = \{H, T\}$ and $P(X = H) = P(X = T) = 0.5$.

The probability that a discrete-valued random variable X takes on a value of x_i, $P(X = x_i)$, is called the *probability mass function* (*pmf*). For two such random variables X and Y, the probability of an outcome in which X takes on a value of x_i and Y takes on a value of y_j, $P(X = x_i, Y = y_j)$, is called the *joint probability mass function*.

The joint pmf can be described in terms of the conditional and the marginal probability mass functions as

$$\begin{aligned} P_{X,Y}(x_i, y_j) &= P_{Y|X}(y_j|x_i)P_X(x_i) \\ &= P_{X|Y}(x_i|y_j)P_Y(y_j) \end{aligned} \tag{6.7}$$

where $P_{Y|X}(y_j|x_i)$ is the *conditional* probability of the random variable Y taking on a value of y_j conditioned on the variable X having taken a value of x_i, and the so-called *marginal* pmf of X is obtained as

$$\begin{aligned} P_X(x_i) &= \sum_{j=1}^{M} P_{X,Y}(x_i, y_j) \\ &= \sum_{j=1}^{M} P_{X|Y}(x_i|y_j) P_Y(y_j) \end{aligned} \tag{6.8}$$

where M is the number of values, or outcomes, in the space of the discrete random variable Y.

6.3.7 Bayes' Rule

Assume we wish to find the probability that a random variable X takes a value of x_i given that a related variable Y has taken a value of y_j. From Equations (6.7) and (6.8), we have *Bayes' rule*, for the conditional probability mass function, given by

$$\begin{aligned} P_{X|Y}(x_i|y_j) &= \frac{1}{P_Y(y_j)} P_{Y|X}(y_j|x_i) P_X(x_i) \\ &= \frac{P_{Y|X}(y_j|x_i) P_X(x_i)}{\sum_{i=1}^{M} P_{Y|X}(y_j|x_i) P_X(x_i)} \end{aligned} \tag{6.9}$$

Bayes' rule forms the foundation of probabilistic estimation and classification. Bayesian inference is introduced in Chapter 7.

Example 6.4 *Probability of the sum of two random variables*
Figure 6.9(a) shows the pmf of a die. Now, let the variables (x, y) represent the outcomes of throwing a pair of dice. The probability that the sum of the outcomes of throwing two dice is equal to A, is given by

$$P(x + y = A) = \sum_{i=1}^{6} P(x = i) P(y = A - i) \tag{6.10}$$

The pmf of the sum of two dice is plotted in Figure 6.9(b). Note from Equation (6.10) that the probability of the sum of two random variables is the convolution sum of the probability functions of the individual variables. Note that from the central limit theorem, the sum of many independent random variables will have a normal distribution.

Matlab function: demonstration of central limit theorem Function Uniform_to_Gaussian( )
Shows that the sum of N random signals with uniform probability density tends to a Gaussian signal as N increases.

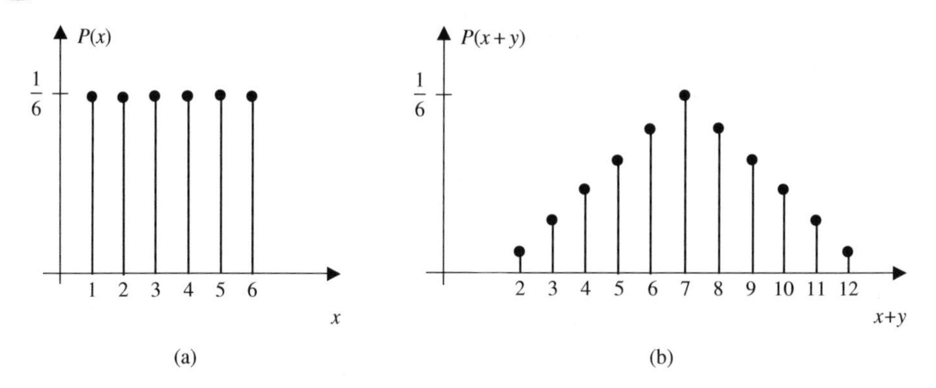

Figure 6.9 The probability mass function (pmf) of (a) a die, and (b) the sum of a pair of dice.

6.3.8 Probability Density Function – Continuous Random Variables

A continuous-valued random variable can assume an infinite number of values, even within an extremely small range of values, and hence the probability that it takes on any given value is infinitely small and vanishes to zero.

For a continuous-valued random variable X the cumulative distribution function (cdf) is defined as the probability that the outcome is less than x as:

$$F_X(x) = \text{Prob}\,(X \le x) \tag{6.11}$$

where Prob(·) denotes probability. The probability that a random variable X takes on a value within a range of values Δ centred on x can be expressed as

$$\begin{aligned}\frac{1}{\Delta} Prob(x-\Delta/2 \le X \le x+\Delta/2) &= \frac{1}{\Delta}[Prob(X \le x+\Delta/2) - Prob(X \le x-\Delta/2)] \\ &= \frac{1}{\Delta}[F_X(x+\Delta/2) - F_X(x-\Delta/2)]\end{aligned} \tag{6.12}$$

Note that both sides of Equation (6.12) are divided by Δ. As the interval Δ tends to zero we obtain the probability density function (pdf) as

$$f_X(x) = \lim_{\Delta \to 0} \frac{1}{\Delta}[F_X(x+\Delta/2) - F_X(x-\Delta/2)] = \frac{\partial F_X(x)}{\partial x} \tag{6.13}$$

Since $F_X(x)$ increases with x, the pdf of x is a non-negative-valued function, i.e. $f_X(x) \ge 0$. The integral of the pdf of a random variable X in the range $\pm\infty$ is unity:

$$\int_{-\infty}^{\infty} f_X(x)dx = 1 \tag{6.14}$$

The conditional and marginal probability functions and the Bayes' rule, of Equations (6.7)–(6.9), also apply to probability density functions of continuous-valued variables. Figure 6.10 shows the cumulative density function (cdf) and probability density function of a Gaussian variable.

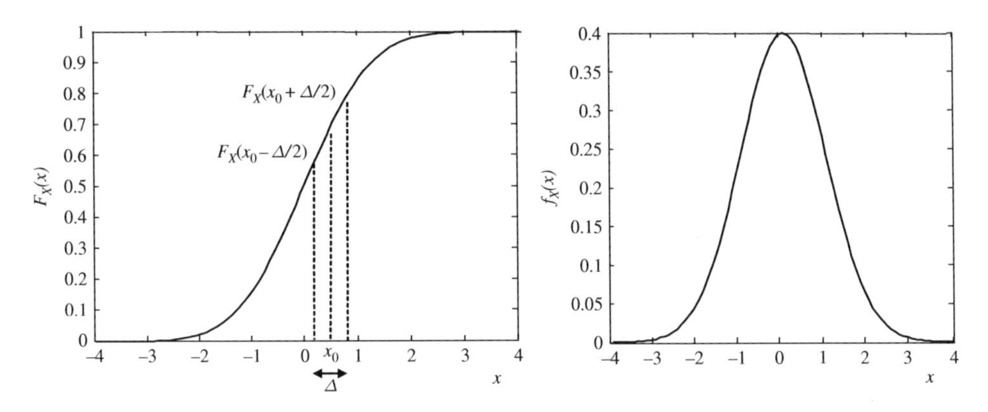

Figure 6.10 The cumulative density function (cdf) $F_X(x)$ and probability density function $f_X(x)$ of a Gaussian variable.

6.3.9 Probability Density Functions of Continuous Random Processes

The probability models obtained for random variables can be applied to random processes such as time series, speech, images etc. For a continuous-valued random process $X(m)$, the simplest probabilistic model is the univariate pdf $f_{X(m)}(x)$, which is the pdf of a sample from the random process $X(m)$ taking on a value of x. A bivariate pdf $f_{X(m)X(m+n)}(x_1, x_2)$ describes the pdf of two samples of the process X at time instants m and $m+n$ taking on the values x_1 and x_2 respectively.

In general, an M-variate pdf $f_{X(m_1)X(m_2)\cdots X(m_M)}(x_1, x_2, \ldots, x_M)$ describes the pdf of a vector of M samples of a random process taking on specific values at specific time instants. For an M-variate pdf, we can write

$$\int_{-\infty}^{\infty} f_{X(m_1)\cdots X(m_M)}(x_1, \ldots, x_M)dx_M = f_{X(m_1)\cdots X(m_{M-1})}(x_1, \ldots, x_{M-1}) \tag{6.15}$$

and the sum of the pdfs of all realisations of a random process is unity, i.e.

$$\int_{-\infty}^{\infty} \cdots \int_{-\infty}^{\infty} f_{X(m_1)\cdots X(m_M)}(x_1, \ldots, x_M)dx_1 \ldots dx_M = 1 \tag{6.16}$$

The probability of the value of a random process at a specified time instant may be conditioned on the value of the process at some other time instant, and expressed as a conditional probability density function as

$$f_{X(m)|X(n)}(x_m | x_n) = \frac{f_{X(n)|X(m)}(x_n | x_m) f_{X(m)}(x_m)}{f_{X(n)}(x_n)} \tag{6.17}$$

Equation (6.17) is Bayes' rule. If the outcome of a random process at any time is independent of its outcomes at other time instants, then the random process is uncorrelated. For an uncorrelated process a multivariate pdf can be written in terms of the products of univariate pdfs as

$$f_{[X(m_1)\cdots X(m_M)|X(n_1)\cdots X(n_N)]}(x_{m_1}, \ldots, x_{m_M} | x_{n_1}, \ldots, x_{n_N}) = \prod_{i=1}^{M} f_{X(m_i)}(x_{m_i}) \tag{6.18}$$

Discrete-valued random processes can only assume values from a finite set of allowable numbers $[x_1, x_2, \ldots, x_n]$. An example is the output of a binary digital communication system that generates a sequence of 1s and 0s. Discrete-time, discrete-valued, stochastic processes are characterised by multivariate probability mass functions (pmf) denoted as

$$P_{[x(m_1)\cdots x(m_M)]}\left(x(m_1) = x_i, \ldots, x(m_M) = x_k\right) \tag{6.19}$$

The probability that a discrete random process $X(m)$ takes on a value of x_m at a time instant m can be conditioned on the process taking on a value x_n at some other time instant n, and expressed in the form of a conditional pmf as

$$P_{X(m)|X(n)}\left(x_m \left| x_n\right.\right) = \frac{P_{X(n)|X(m)}\left(x_n \left| x_m\right.\right) P_{X(m)}(x_m)}{P_{X(n)}(x_n)} \tag{6.20}$$

and for a statistically independent process we have

$$P_{[X(m_1)\cdots X(m_M)|X(n_1)\cdots X(n_N)]}\left(x_{m_1}, \ldots, x_{m_M} \left| x_{n_1}, \ldots, x_{n_N}\right.\right) = \prod_{i=1}^{M} P_{X(m_i)}(X(m_i) = x_{m_i}) \tag{6.21}$$

6.3.10 Histograms – Models of Probability

A histogram is a bar graph that shows the number of times (or the normalised fraction of times) that a random variable takes values in each class or in each interval (bin) of values.

Given a set of observations of a random variable, the range of values of the variable between the minimum value and the maximum value are divided into N equal-width bins and the number of times that the variable falls within each bin is calculated.

A histogram is an estimate of the probability distribution of a variable derived from a set of observations. The Matlab routine hist (x, N) displays the histogram of the variable x in N uniform intervals. Figure 6.11 shows the histogram and the probability model of a Gaussian signal. Figure 6.12 shows the scatter plot of a two-dimensional Gaussian process superimposed on an ellipse which represents the standard-deviation contour.

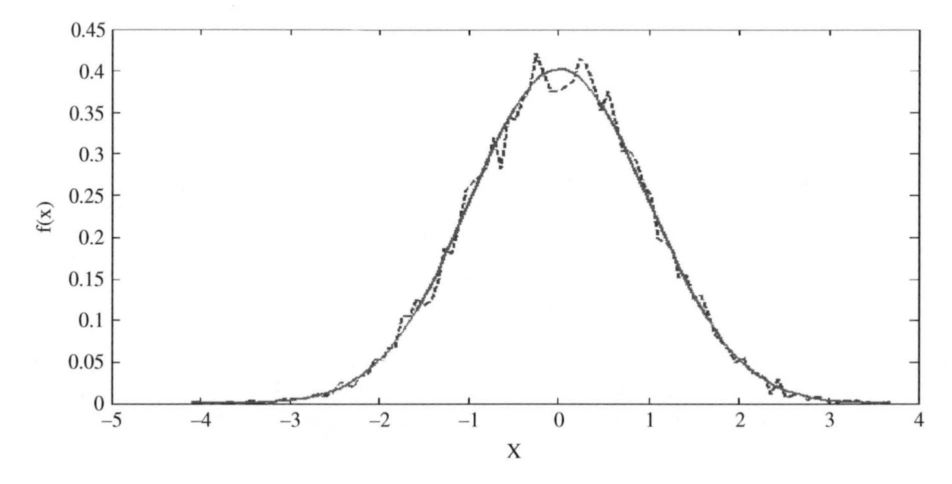

Figure 6.11 Histogram (dashed line) and probability model of a Gaussian signal.

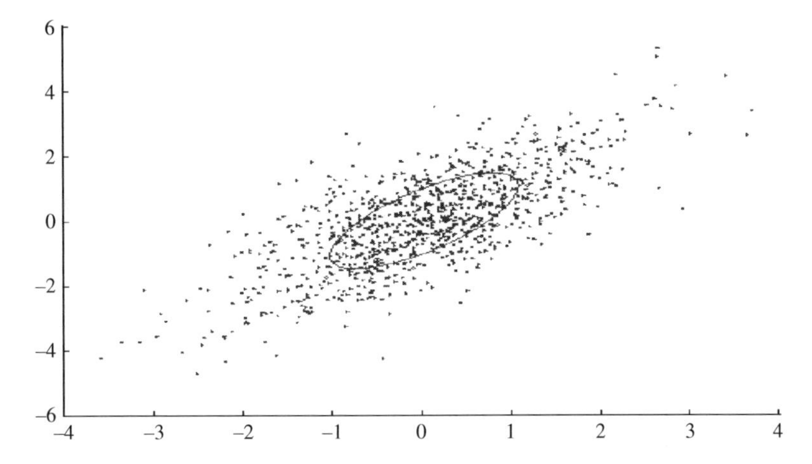

Figure 6.12 The scatter plot of a two-dimensional Gaussian distribution.

Matlab function TwoDimensionalGaussianPlot( )
Shows the scatter diagram and the elliptical contour of the covariance matrix of a two-dimensional Gaussian process.

Matlab function [xx_cord,yy_cord]=drawGauss(xx)
Assumes that the 2-D feature vectors 'xx' are distributed as a 2-D Gaussian distribution, and returns x- and y-coordinates of the ellipse which represents the standard-deviation contour. This ellipse is plotted by 'axis equal; plot(xx_cord,yy_cord)'.

6.4 Information Models

As explained, information is knowledge or data about the state sequence of a random process, such as how many states the random process has, how often each state is observed, the outputs of each state, how the process moves across its various states and what is its past, current or likely future states.

The information conveyed by a random process is associated with its state sequence. Examples are the states of weather, the states of someone's health, the states of market share price indices, communication symbols, DNA states or protein sequences.

Information is usually discrete (or quantised) and can be represented in a binary format in terms of M states of a variable. The states of an information-bearing variable may be arranged in a binary tree structure as shown later in Example 6.10.

In this section it is shown that information is measured in terms of units of bits. One bit of information is equivalent to the information conveyed by two equal-probability states. Note that the observation from which information is obtained may be continuous-valued or discrete-valued.

The concepts of information, randomness and probability models are closely related. For a signal to convey information it must satisfy two conditions:

(a) Possess two or more states or values.
(b) Move between the states in a random manner.

For example, the outcome of tossing a coin is an unpredictable binary state (Head/Tail) event, a digital communication system with N-bit codewords has 2^N states and the outcome of a weather forecast can be one or more of the following states: {sun, cloud, cold, warm, hot, rain, snow, storm etc.}.

Since random processes are modelled with probability functions, it is natural that information is also modelled by a function of probability. The negative logarithm of probability of a variable is used as a measure of the information content of the variable. Note that since probability is less than one, the negative log of probability has a positive value. Using this measure an event with a probability of one has zero information whereas the more unlikely an event the more the information it carries when the event is observed.

The *expected (average) information content* of a state x_i of a random variable is quantified as

$$I(x_i) = -P_X(x_i)\log P_X(x_i) \text{ bits} \tag{6.22}$$

where the base of logarithm is 2. For a binary source the information conveyed by the two states $[x_1, x_2]$ can be described as

$$\begin{aligned} H(X) &= I(x_1) + I(x_2) \\ &= -P(x_1)\log P(x_1) - P(x_2)\log P(x_2) \end{aligned} \tag{6.23}$$

Alternatively $H(X)$ in Equation (6.23) can be written as

$$H(X) = -P(x_i)\log P(x_i) - (1 - P(x_i))\log(1 - P(x_i)) \quad i = 1 \text{ or } 2 \tag{6.24}$$

Note from Figure 6.13 that the expected information content of a variable has a value of zero for an event whose probability is zero, i.e. an impossible event, and its value is also zero for an event that happens with probability of one.

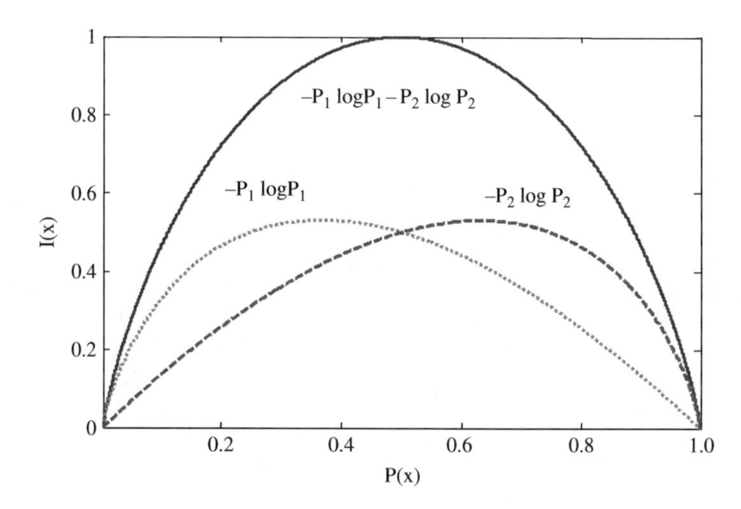

Figure 6.13 Illustration of $I(x_i)$ vs $P(x_i)$; for a binary source, the maximum information content is one bit, when the two states have equal probability of 0.5. $P(x_1) = P_1$ and $P(x_2) = P_2$.

6.4.1 Entropy: A Measure of Information and Uncertainty

Entropy provides a measure of the quantity of the information content of a random variable in terms of the minimum number of bits per symbol required to encode the variable. Entropy is an indicator of the amount of randomness or uncertainty of a discrete random process.

Entropy can be used to calculate the theoretical minimum capacity or bandwidth required for the storage or transmission of an information source such as text, image, music, etc.

In his pioneering work, *A Mathematical Theory of Communication*, Claude Elwood Shannon derived the entropy information measure, H, as a function that satisfies the following conditions:

(1) Entropy H should be a continuous function of probability P_i.
(2) For $P_i = 1/M$, H should be a monotonically increasing function of M.
(3) If the communication symbols are broken into two (or more) subsets, the entropy of the original set should be equal to the probability-weighted sum of the entropy of the subsets.

Consider a random variable X with M states $[x_1, x_2, \ldots, x_M]$ and state probabilities $[p_1, p_2, \ldots, p_M]$ where $PX(x_i) = p_i$, the entropy of X is defined as

$$H(X) = -\sum_{i=1}^{M} P(x_i) \log P(x_i) \text{ bits} \tag{6.25}$$

where the base of the logarithm is 2. The base 2 of the logarithm reflects the binary nature of information. Information is discrete and can be represented by a set of binary symbols.

$\log_2$ has several useful properties. $\log_2$ of 1 is zero, which is a useful mapping as an event with probability of one has zero information. Furthermore, with the use of logarithm the addition of a binary symbol (with two states) to M existing binary symbols doubles the number of possible outcomes from 2^M to 2^{M+1} but increases the logarithm (to the base 2) of the number of outcomes by one. With the entropy function two equal probability states corresponds to one bit of information, four equal probability states corresponds to two bits and so on. Note that the choice of base 2 gives one bit for a two-state equi-probable random variable.

Entropy is measured in units of bits. Entropy is bounded as

$$0 \le H(X) \le \log_2 M \tag{6.26}$$

where $H(X) = 0$ if one symbol x_i has a probability of one and all other symbols have probabilities of zero, and M denotes the number of symbols in the set X. *The entropy of a set attains a maximum value of* $\log_2 M$ *bits for a uniformly distributed M-valued variable with each outcome having an equal probability of* $1/M$. Figure 6.14 compares the entropy of two discrete-valued random variables; one is derived from a uniform process, the other is obtained from quantisation of a Gaussian process.

Entropy gives the minimum number of bits per symbol required for binary coding of different values of a random variable X. This theoretical minimum is usually approached by encoding N samples of the process simultaneously with K bits where the number of bits per sample $K/N \ge H(X)$. As N becomes large, for an efficient code the number of bits per sample K/N approaches the entropy $H(X)$ of X (see Huffman coding).

Shannon's source coding theorem states: N independent identically distributed (IID) random variables each with entropy H can be compressed into more than NH bits with negligible loss of quality as $N \to \infty$.

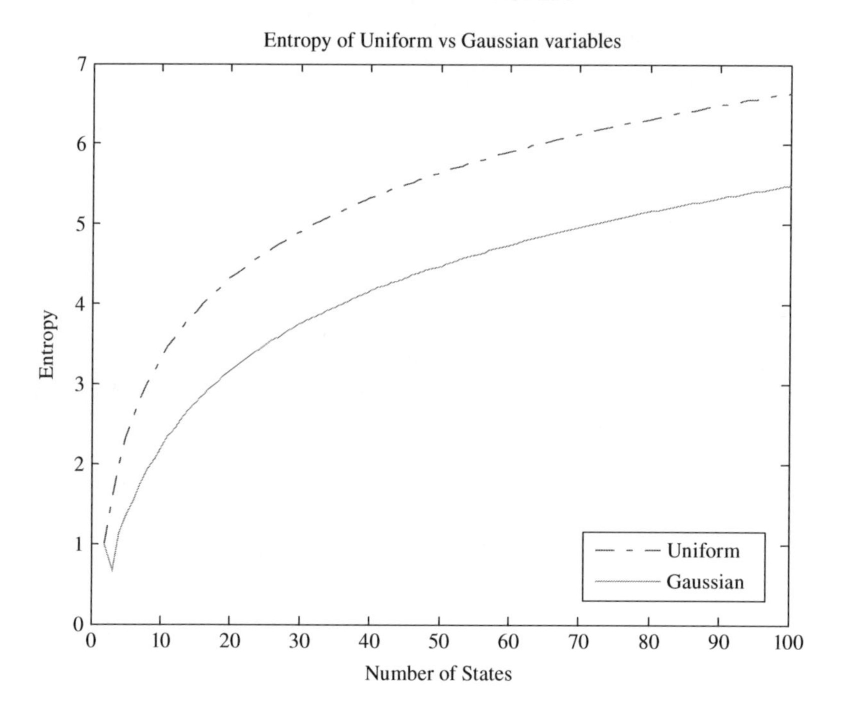

Figure 6.14 Entropy of two discrete random variables as a function of the number of states. One random variable has a uniform probability and the other has a Gaussian-shaped probability function.

Example 6.5 *Entropy of the English alphabet*
Calculate the entropy of the set letters in the English alphabet [A, B, C, D, …, Z], assuming that all letters are equally likely. Hence, calculate the theoretical minimum number of bits required to code a text file of 2000 words with an average of 5 letters per word.

> **Matlab function: demonstration of entropy of uniform and non-uniform random discrete processes**
> **function EntropyDemo( )**
> Demonstrates the calculation of the entropy of an N-valued discrete random process and compares the entropies of two discrete N-valued variables derived from the N-bin histograms of a uniform random process and a Gaussian random process respectively. The entropy functions are plotted with the increasing number of states N.

Solution: For the English alphabet the number of symbols $N = 26$, and assuming that all symbols are equally likely the probability of each symbol becomes $p_i = 1/26$. Using Equation (6.25) we have

$$H(X) = -\sum_{i=1}^{26} \frac{1}{26} \log_2 \frac{1}{26} = 4.7 \text{ bits} \tag{6.27}$$

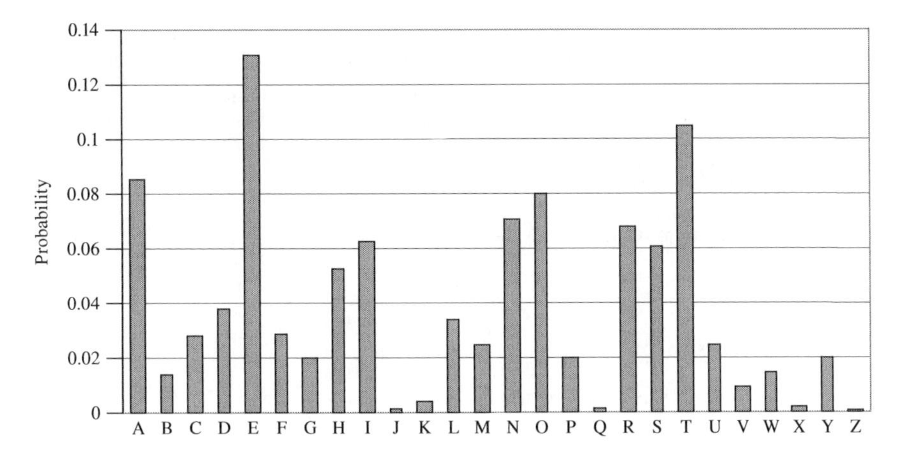

Figure 6.15 The probability (histogram) of the letters of the English alphabet, A to Z.

The total number of bits for encoding 2000 words $= 4.7 \times 2000 \times 5 = 47$ kbits. Note that different letter type cases (upper case, lower case etc.), font types (bold, italic etc.) and symbols (!, ?, etc) are not taken into account and also note that the actual distribution of the letters is non-uniform resulting in an entropy of less than 4.7 bits/symbol.

Example 6.6 *Entropy of the English alphabet using estimates of probabilities of letters of the alphabet*
Use the set of probabilities of the alphabet shown in Figure 6.15: $\mathrm{P(A)} = 0.0856, \mathrm{P(B)} = 0.0139, \mathrm{P(C)} = 0.0279, \mathrm{P(D)} = 0.0378, \mathrm{P(E)} = 0.1304, \mathrm{P(F)} = 0.0289, \mathrm{P(G)} = 0.0199, \mathrm{P(H)} = 0.0528, \mathrm{P(I)} = 0.0627, \mathrm{P(J)} = 0.0013, \mathrm{P(K)} = 0.0042, \mathrm{P(L)} = 0.0339, \mathrm{P(M)} = 0.0249, \mathrm{P(N)} = 0.0707, \mathrm{P(O)} = 0.0797, \mathrm{P(P)} = 0.0199, \mathrm{P(Q)} = 0.0012, \mathrm{P(R)} = 0.0677, \mathrm{P(S)} = 0.0607, \mathrm{P(T)} = 0.1045, \mathrm{P(U)} = 0.0249, \mathrm{P(V)} = 0.0092, \mathrm{P(W)} = 0.0149, \mathrm{P(X)} = 0.0017, \mathrm{P(Y)} = 0.0199, \mathrm{P(Z)} = 0.0008$.
The entropy of this set is given by

$$H(X) = -\sum_{i=1}^{26} P_i \log_2 P_i = 4.13 \text{ bits/symbol} \tag{6.28}$$

Example 6.7 *Entropy of the English phonemes*
Spoken English is constructed from about 40 basic acoustic symbols, known as phonemes (or phonetic units); these are used to construct words, sentences etc. For example the word '*signal*' is transcribed in phonemic form as '*s iy g n aa l*'. Assuming that all phonemes are equi-probable, and the average speaking rate is 120 words per minute, and the average word has 4 phonemes, calculate the minimum number of bits per second required to encode speech at the average speaking rate.

Solution: For speech $N = 40$, assume $P_i = 1/40$. The entropy of phonemes is given by

$$H(X) = -\sum_{i=1}^{40} \frac{1}{40} \log_2 \frac{1}{40} = 5.3 \text{ bits/symbol} \tag{6.29}$$

Number of bits/sec = (120/60 words per second) × (4 phonemes per word) × (5.3 bits per phoneme) = 46.4 bps

Note that the actual distribution of phonemes is non-uniform resulting in an entropy of less than 5.3 bits. Furthermore, the above calculation ignores the information (and hence the entropy) in speech due to contextual variations of phonemes, speaker identity, accent, pitch intonation and emotion signals.

6.4.2 Mutual Information

Consider two dependent random variables X and Y, the conditional entropy of X given Y is defined as

$$H(X|Y) = -\sum_{i=1}^{M_x}\sum_{j=1}^{M_y} P(x_i, y_j) \log P(x_i|y_j) \tag{6.30}$$

$H(X|Y)$ is equal to $H(X)$ if Y is independent of X and is equal to zero if Y has the same information as X. The information that the variable Y contains about the variable X is given by

$$I(X; Y) = H(X) - H(X|Y) \tag{6.31}$$

Substituting Equation (6.30) in Equation (6.31) and also using the relation

$$H(X) = -\sum_{i=1}^{M_x} P(x_i) \log P(x_i) = -\sum_{i=1}^{M_x}\sum_{j=1}^{M_y} P(x_i, y_j) \log P(x_i) \tag{6.32}$$

yields

$$I(X; Y) = \sum_{i=1}^{M_x}\sum_{j=1}^{M_y} P(x_i, y_j) \log \frac{P(x_i, y_j)}{P(x_i)P(y_j)} \tag{6.33}$$

Note from Equation (6.33) that $I(X; Y) = I(Y; X)$, that is the information that Y has about X is the same as the information that X has about Y, hence $I(X; Y)$ is called *mutual information*. As shown next, mutual information has a minimum of zero, $I(X; Y) = 0$, for independent variables and a maximum of $I(X; Y) = H(X) = H(Y)$ when X and Y have identical information.

Example 6.8 *Upper and lower bounds on mutual information*
Obtain the bounds on the mutual information of two random variables X and Y.

Solution: The upper bound is given when X and Y contain identical information, in this case substituting $P(x_i, y_j) = P(x_i)$ and $P(y_j) = P(x_i)$ in Equation (6.33) and assuming that each x_i has a mutual relation with only one y_j we have

$$I(X; Y) = \sum_{i=1}^{M_x} P(x_i) \log \frac{P(x_i)}{P(x_i)P(x_i)} = H(X) \tag{6.34}$$

The lower bound is given by the case when X and Y are independent, $P(x_i, y_j) = P(x_i)P(y_j)$, i.e. have no mutual information, hence

$$I(X; Y) = \sum_{i=1}^{M_x}\sum_{j=1}^{M_y} P(x_i)P(y_j) \log \frac{P(x_i)P(y_j)}{P(x_i)P(y_j)} = 0 \tag{6.35}$$

Example 6.9 Show that the *mutual entropy* of two independent variables X and Y are additive.

Solution: Assume X and Y are M-valued and N-valued variables respectively. The entropy of two random variables is given by

$$H(X,Y) = \sum_{i=1}^{M}\sum_{j=1}^{N} P_X(x_i, y_j) \log \frac{1}{P_{XY}(x_i, y_j)} \tag{6.36}$$

Substituting $P(x_i, y_j) = P(y_i)P(y_j)$ in Equation (6.34) yields

$$\begin{aligned} H(X,Y) &= \sum_{i=1}^{M}\sum_{j=1}^{N} P_X(x_i, y_j) \log \frac{1}{P_{XY}(x_i, y_j)} \\ &= -\sum_{i=1}^{M}\sum_{j=1}^{N} P_X(x_i)P_Y(y_j) \log P_X(x_i) - \sum_{i=1}^{M}\sum_{j=1}^{N} P_X(x_i)P_Y(y_j) \log P_Y(y_j) \\ &= -\sum_{i=1}^{M} P_X(x_i) \log P_X(x_i) - \sum_{j=1}^{N} P_Y(y_j) \log P_Y(y_j) \\ &= H(X) + H(Y) \end{aligned} \tag{6.37}$$

where we have used the following relations

$$\sum_{j=1}^{M} P_Y(y_j)P_X(x_i) \log P_X(x_i) = P_X(x_i) \log P_X(x_i) \sum_{j=1}^{M} P_Y(y_j) = P_X(x_i) \log P_X(x_i) \tag{6.38}$$

and for two independent variables

$$\log[1/P_{XY}(x_i, y_j)] = -\log P_X(x_i) - \log P_Y(y_j) \tag{6.39}$$

6.4.3 Entropy Coding – Variable Length Codes

As explained above, entropy gives the minimum number of bits required to encode an information source. This theoretical minimum may be approached by encoding N samples of a signal simultaneously with K bits where $K/N \geq H(X)$. As N becomes large, for an efficient coder K/N approaches the entropy $H(X)$ of X. The efficiency of a coding scheme in terms of its entropy is defined as $H(X)/(K/N)$. When $K/N = H(X)$ then the entropy coding efficiency of the code is $H(X)/(K/N) = 1$ or 100%.

For an information source X the average number of bits per symbol, aka the average code length CL, can be expressed as

$$CL(X) = \sum_{i=1}^{M} P(x_i)L(x_i) \text{ bits} \tag{6.40}$$

where $L(x_i)$ is the length of the binary codeword used to encode symbol x_i and $P(x_i)$ is the probability of x_i. A comparison of Equation (6.40) with the entropy equation (6.25) shows that for an optimal code $L(x_i)$ is $-\log_2 P(x_i)$. The aim of the design of a minimum length code is that the *average* code length should approach the entropy.

The simplest method to encode an M-valued variable is to use a fixed-length code that assigns N binary digits to each of the M values with $N = \text{Nint}(\log_2 M)$, where Nint denotes the nearest integer round up function. When the source symbols are *not equally probable*, a more efficient method is *entropy encoding*.

Entropy coding is a variable-length coding method which assigns codewords of variable lengths to communication alphabet symbols $[x_i]$ such that the more probable symbols, which occur more frequently, are assigned shorter codewords and the less probable symbols, which happen less frequently, are assigned longer code words. An example of such a code is the Morse code which dates back to the nineteenth century. Entropy coding can be applied to the coding of music, speech, image, text and other forms of communication symbols. If the entropy coding is ideal, the bit rate at the output of a uniform M-level quantiser can be reduced by an amount of $\log_2 M - H(X)$ compared with fixed length coding.

6.4.4 Huffman Coding

A simple and efficient form of entropy coding is the *Huffman code* which creates a set of prefix-free codes (no code is part of the beginning of another code) for a given discrete random source. The ease with which Huffman codes can be created and used makes this code a popular tool for data compression. Huffman devised his code while he was a student at MIT.

In one form of Huffman tree code, illustrated in Figure 6.16, the symbols are arranged in the decreasing order of probabilities in a column. The two symbols with the lowest probabilities are

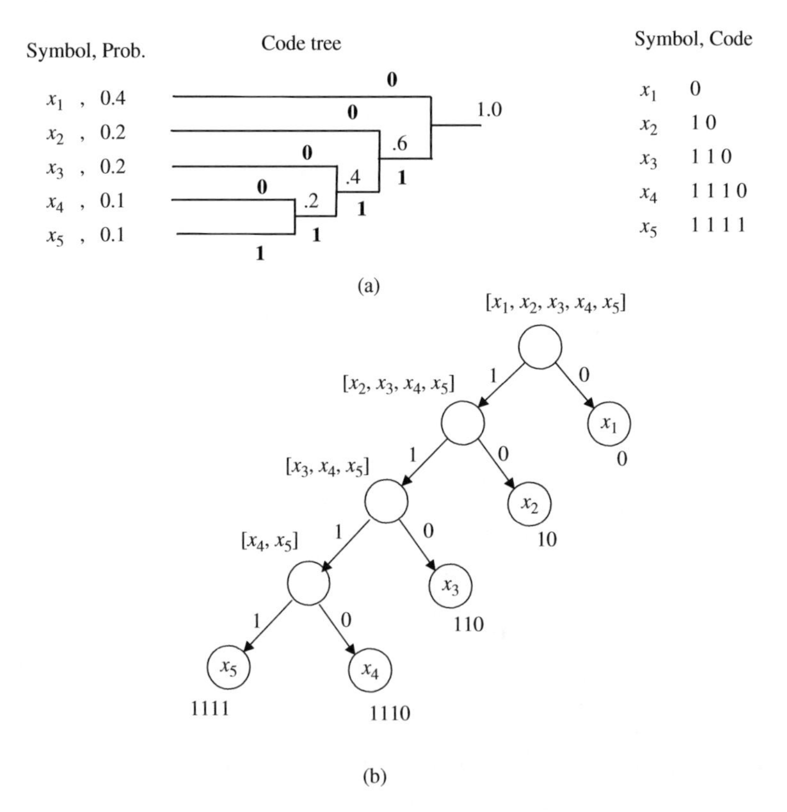

Figure 6.16 (a) Illustration of Huffman coding tree; the source entropy is 2.1219 bits/sample and Huffman code gives 2.2 bits/sample. (b) Alternative illustration of Huffman tree.

combined by drawing a straight line from each and connecting them. This combination is combined with the next symbol and the procedure is repeated to cover all symbols. Binary codewords are assigned by moving from the root of the tree at the right-hand side to left in the tree and assigning a 1 to the lower branch and a 0 to the upper branch where each pair of symbols have been combined.

Example 6.10 Given five symbols $x_1, x_2, \ldots, x_5$ with probabilities of $P(x_1) = 0.4$, $P(x_2) = P(x_3) = 0.2$ and $P(x_4) = P(x_5) = 0.1$, design a binary variable length code for this source.

Solution: The entropy of X is $H(X) = 2.122$ bits/symbol. Figure 6.16 illustrates the design of a Huffman code for this source. For this tree we have:
average codeword length $= 1 \times 0.4 + 2 \times 0.2 + 4 \times 0.1 + 4 \times 0.1 + 3 \times 0.2 = 2.2$ bits/symbol
The average codeword length of 2.2 bits/symbol is close to the entropy of 2.1219 bits/symbol. We can get closer to the minimum average codeword length by encoding pairs of symbols or blocks of more than two symbols at a time (with added complexity). The Huffman code has a useful *prefix condition* property whereby no codeword is a prefix or an initial part of another codeword. Thus codewords can be readily concatenated (in a comma-free fashion) and be uniquely (unambiguously) decoded.

Figure 6.17 illustrates a Huffman code tree created by a series of successive binary divisions of the symbols into two sets with as near set probabilities as possible. At each node the set splitting process is repeated until the leaf node containing a single symbol is reached. Binary bits of 0 and 1 are assigned to tree branches as shown. Each end-node with a single symbol represents a leaf node and is assigned a binary code which is read from the top (root) node to the leaf node.

Example 6.11 Given a communication system with four symbols x_1, x_2, x_3 and x_4 and with probabilities of $P(x_1) = 0.4$, $P(x_2) = 0.3$, $P(x_3) = 0.2$ and $P(x_4) = 0.1$, design a variable-length coder to encode two symbols at a time.

Solution: First we note that the entropy of the four symbols is obtained from Equation (6.25) as 1.8464 bits/symbol. Using a Huffman code tree, as illustrated in Figure 6.18(a), the codes are $x_1 \to 0$, $x_2 \to 10$, $x_3 \to 110$ and $x_4 \to 111$. The average length of this code is 1.9 bits/symbol.

Assuming the symbols are independent the probability of 16 pairs of symbols can be written as

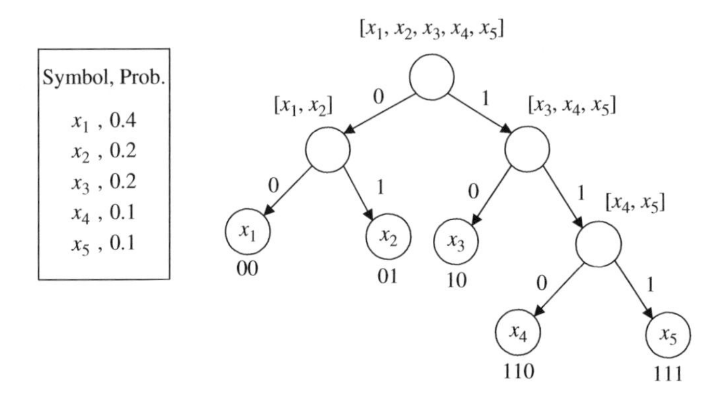

Figure 6.17 A binary Huffman coding tree. From the top node at each stage the set of symbols are divided into two sets with as near set probability (max entropy) as possible. Each end-node with a single symbol represents a leaf and is assigned a binary code which is read from the top node to the leaf node.

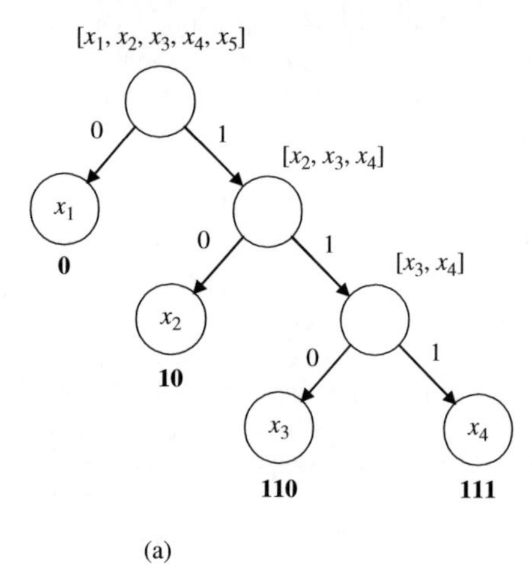

(a)

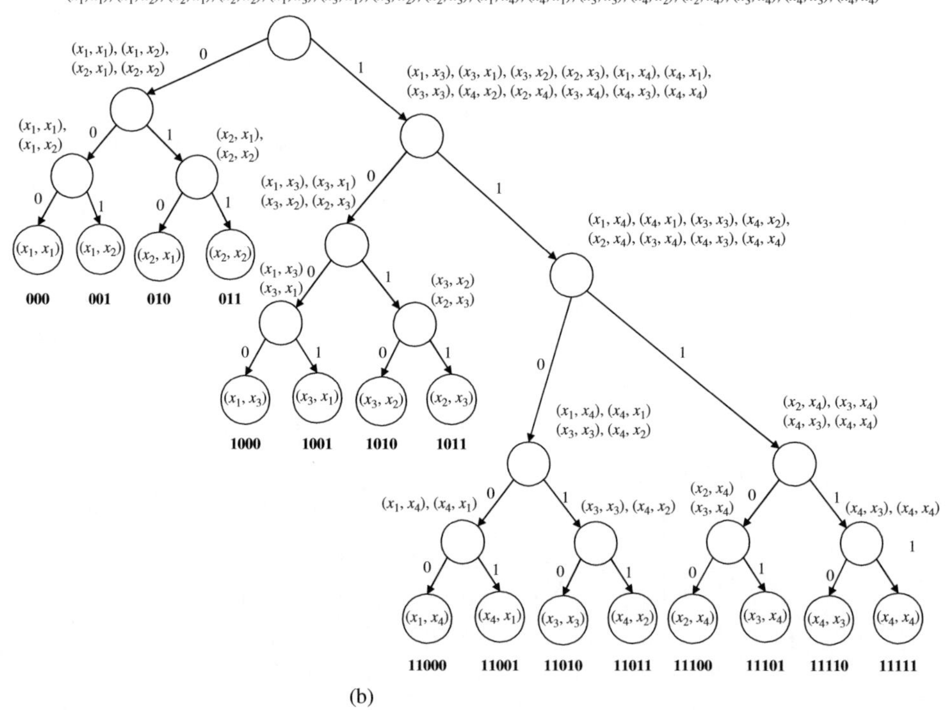

(b)

Figure 6.18 A Huffman coding binary tree for Example 6.11: (a) individual symbols are coded, (b) pairs of symbols are coded.

$$P(x_1, x_1) = 0.16, P(x_1, x_2) = 0.12, P(x_1, x_3) = 0.08, P(x_1, x_4) = 0.04$$

$$P(x_2, x_1) = 0.12, P(x_2, x_2) = 0.09, P(x_2, x_3) = 0.06, P(x_2, x_4) = 0.03$$

$$P(x_3, x_1) = 0.08, P(x_3, x_2) = 0.06, P(x_3, x_3) = 0.04, P(x_3, x_4) = 0.02$$

$$P(x_4, x_1) = 0.04, P(x_4, x_2) = 0.03, P(x_4, x_3) = 0.02, P(x_4, x_4) = 0.01$$

The entropy of pairs of symbols is 3.6929 which is exactly twice the entropy of the individual symbols. The 16 pairs of symbols and their probabilities can be used in a Huffman tree code as illustrated in Figure 6.18(b). The average code length for Huffman coding of pairs of symbols is 1.87 compared with the average code length of 1.9 for Huffman coding of individual symbols.

6.5 Stationary and Non-Stationary Random Processes

Although the amplitude of a signal fluctuates with time m, the parameters of the model or the process that generates the signal may be time-invariant (stationary) or time-varying (non-stationary). Examples of non-stationary processes are speech, Figure 6.19, and music whose loudness and spectral composition change continuously, image and video.

A process is stationary if the parameters of the probability model of the process are time-invariant; otherwise it is non-stationary. The stationary property implies that all the statistical parameters, such as the mean, the variance, the power spectral composition and the higher-order moments of the process, are constant. In practice, there are various degrees of stationarity: it may be that one set of the statistics of a process is stationary whereas another set is time-varying. For example, a random process may have a time-invariant mean, but a time-varying power.

Example 6.12 Consider the *time-averaged* values of the mean and the power of (a) a stationary signal $A \sin \omega t$ and (b) a transient exponential signal $Ae^{-\alpha t}$. The mean and power of the sinusoid, integrated over one period, are

$$\text{Mean}\,(A \sin \omega t) = \frac{1}{T} \int_T A \sin \omega t \; dt = 0, \;\; \text{constant} \tag{6.41}$$

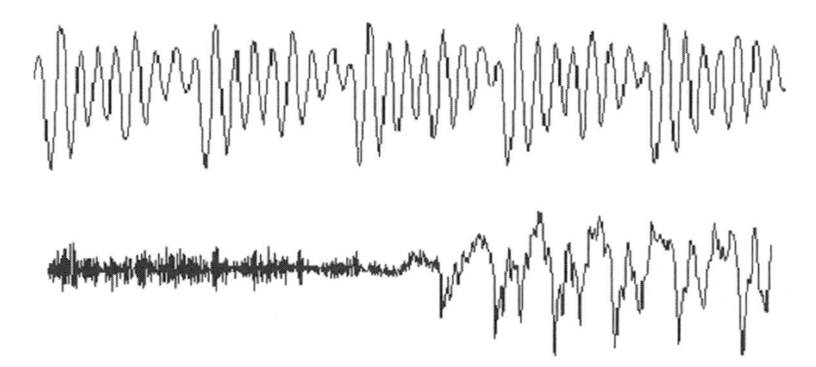

Figure 6.19 Examples of quasi-stationary voiced speech (above) and non-stationary speech composed of unvoiced and voiced speech segments.

$$\text{Power}\,(A\sin\omega t) = \frac{1}{T}\int_T A^2 \sin^2 \omega t\; dt = \frac{A^2}{2}, \text{ constant} \tag{6.42}$$

where T is the period of the sine wave.

The mean and the power of the transient signal are given by

$$\text{Mean}\,(Ae^{-\alpha t}) = \frac{1}{T}\int_t^{t+T} Ae^{-\alpha\tau}\, d\tau = \frac{A}{\alpha T}(1-e^{-\alpha T})e^{-\alpha t}, \text{ time-varying} \tag{6.43}$$

$$\text{Power}\,(Ae^{-\alpha t}) = \frac{1}{T}\int_t^{t+T} A^2 e^{-2\alpha\tau}\, d\tau = \frac{A^2}{2\alpha T}\left(1-e^{-2\alpha T}\right)e^{-2\alpha t}, \text{ time-varying} \tag{6.44}$$

In Equations (6.43) and (6.44), the signal mean and power are exponentially decaying functions of the time variable t.

Example 6.13 *A binary-state non-stationary random process*

Consider a non-stationary signal $y(m)$ generated by a binary-state random process, Figure 6.20, described by the equation

$$y(m) = \bar{s}(m)x_0(m) + s(m)x_1(m) \tag{6.45}$$

where $s(m)$ is a binary-valued state-indicator variable and $\bar{s}(m)$ is the binary complement of $s(m)$. From Equation (6.45), we have

$$y(m) = \begin{cases} x_0(m) & \text{if } s(m) = 0 \\ x_1(m) & \text{if } s(m) = 1 \end{cases} \tag{6.46}$$

Let μ_{x_0} and P_{x_0} denote the mean and the power of the signal $x_0(m)$, and μ_{x_1} and P_{x_1} the mean and the power of $x_1(m)$ respectively. The expectation of $y(m)$, given the state $s(m)$, is obtained as

$$\begin{aligned} \mathcal{E}[y(m)\,|s(m)] &= \bar{s}(m)\mathcal{E}[x_0(m)] + s(m)\mathcal{E}[x_1(m)] \\ &= \bar{s}(m)\mu_{x_0} + s(m)\mu_{x_1} \end{aligned} \tag{6.47}$$

In Equation (6.47), the mean of $y(m)$ is expressed as a function of the state of the process at time m. The power of $y(m)$ is given by

$$\begin{aligned} \mathcal{E}\left[y^2(m)\,|s(m)\right] &= \bar{s}(m)\mathcal{E}\left[x_0^2(m)\right] + s(m)\mathcal{E}\left[x_1^2(m)\right] \\ &= \bar{s}(m)P_{x_0} + s(m)P_{x_1} \end{aligned} \tag{6.48}$$

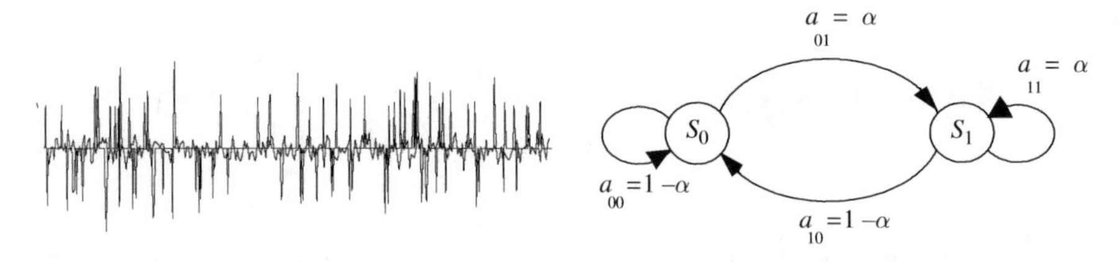

Figure 6.20 Illustration of a binary state process: an impulsive noise state observed in background noise.

Although most signals are non-stationary, the concept of a stationary process plays an important role in the development of signal processing methods. Furthermore, most non-stationary signals such as speech can be considered as approximately stationary for a short period of time. In signal processing theory, two classes of stationary processes are defined: (a) strict-sense stationary processes and (b) wide-sense stationary processes, which is a less strict form of stationarity, in that it only requires that the first-order and second-order statistics of the process should be time-invariant.

6.5.1 Strict-Sense Stationary Processes

A random process $X(m)$ is strict-sense stationary if all its distributions and statistics are time-invariant. Strict-sense stationarity implies that the nth order distribution is time-invariant (or translation-invariant) for all $n = 1, 2, 3, \ldots$:

$$\begin{aligned} &\text{Prob}\left[x(m_1) \le x_1, x(m_2) \le x_2, \ldots, x(m_n) \le x_n)\right] \\ &\quad = \text{Prob}\left[x(m_1+\tau) \le x_1, x(m_2+\tau) \le x_2, \ldots, x(m_n+\tau) \le x_n)\right] \end{aligned} \tag{6.49}$$

where m_i is the discrete-time index and τ is any arbitrary shift along the time axis. Equation (6.49) implies that the signal has the same probability function at all times.

From Equation (6.49) the statistics of a strict-sense stationary process are time invariant. In general for any of the moments of the process we have

$$\mathcal{E}[x^{k_1}(m_1), x^{k_2}(m_1+\tau_1), \ldots, x^{k_L}(m_1+\tau_L)] = \mathcal{E}[x^{k_1}(m_2), x^{k_2}(m_2+\tau_2), \ldots, x^{k_L}_{i_2+\tau_L}(m_2+\tau_L)] \tag{6.50}$$

where $k_1, \ldots, k_L$ are arbitrary powers. For a strict-sense stationary process, all the moments of a signal are time-invariant. The first-order moment, i.e. the mean, and the second-order moments, i.e. the correlation and power spectrum, of a stationary process are given by

$$\mathcal{E}[x(m)] = \mu_x \tag{6.51}$$

$$\mathcal{E}[x(m)x(m+k)] = r_{xx}(k) \tag{6.52}$$

and

$$\mathcal{E}[|X(f,m)|^2] = \mathcal{E}[|X(f)|^2] = P_{XX}(f) \tag{6.53}$$

where μ_x, $r_{xx}(k)$ and $P_{XX}(f)$ are the mean value, the autocorrelation and the power spectrum of the signal $x(m)$ respectively, and $X(f,m)$ denotes the frequency–time spectrum of $x(m)$.

6.5.2 Wide-Sense Stationary Processes

The strict-sense stationarity condition requires that all statistics of the process should be time-invariant. A less restrictive form of a stationary process is called wide-sense stationarity. A process is said to be wide-sense stationary if the mean and the autocorrelation functions (first- and second-order statistics) of the process are time invariant:

$$\mathcal{E}[x(m)] = \mu_x \tag{6.54}$$

$$\mathcal{E}[x(m)x(m+k)] = r_{xx}(k) \tag{6.55}$$

From the definitions of strict-sense and wide-sense stationary processes, it is clear that a strict-sense stationary process is also wide-sense stationary, whereas the reverse is not necessarily true.

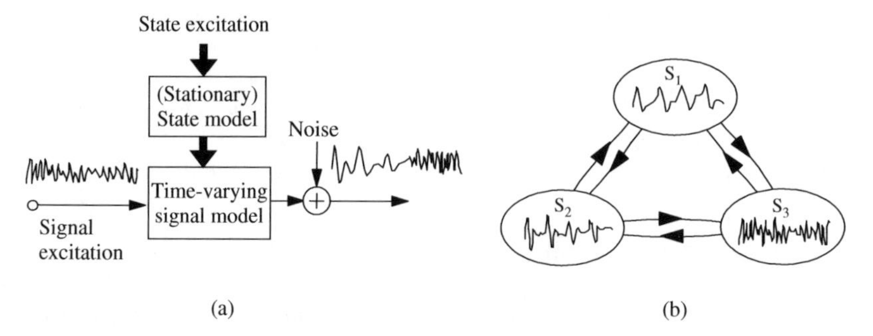

Figure 6.21 Two models for non-stationary processes: (a) a stationary process drives the parameters of a continuously time-varying model; (b) a finite-state model with each state having a different set of statistics.

6.5.3 Non-Stationary Processes

A random process is a non-stationary process if its statistics vary with time. Most stochastic processes such as video and audio signals, financial data, meteorological data, biomedical signals are non-stationary as they are generated by systems whose contents, environments and parameters vary or evolve over time. For example, speech is a non-stationary process generated by a time-varying articulatory system. The loudness and the frequency composition of speech change over time.

Time-varying processes may be modelled by some combination of stationary random models as illustrated in Figure 6.21. In Figure 6.21(a) a non-stationary process is modelled as the output of a time-varying system whose parameters are controlled by a stationary process. In Figure 6.21(b) a time-varying process is modelled by a Markov chain of time-invariant states, with each state having a different set of statistics or probability distributions. Finite-state statistical models for time-varying processes are discussed in detail in Chapter 11.

6.6 Statistics (Expected Values) of a Random Process

The expected values of a random process, also known as its statistics or moments, are the mean, variance, correlation, power spectrum and the higher-order moments of the process.

Expected values play an indispensable role in signal processing. Furthermore, the probability models of a random process are usually expressed as functions of the expected values. For example, a Gaussian pdf is defined as an exponential function centred about the mean and with its volume and orientation determined by the covariance of the process, and a Poisson pdf is defined in terms of the mean of the process.

In signal processing applications, we often use our prior experience or the available data or perhaps our intuitive feeling to select a suitable statistical model for a process, e.g. a Gaussian or Poisson pdf. To complete the model we need to specify the parameters of the model which are usually the expected values of the process such as its mean and covariance. Furthermore for many algorithms, such as noise reduction filters or linear prediction, what we essentially need is an estimate of the mean or the correlation function of the process.

The expected value of a function $h(X)$ of random process X, $h(X(m_1), X(m_2), \ldots, X(m_M))$, is defined as

$$\mathcal{E}[h(X(m_1), \ldots, X(m_M))] = \int_{-\infty}^{\infty} \cdots \int_{-\infty}^{\infty} h(x_1, \ldots, x_M) f_{X(m_1)\cdots X(m_M)}(x_1, \ldots, x_M) dx_1 \ldots dx_M \quad (6.56)$$

The most important, and widely used, expected values are the first-order moment, namely the mean value, and the second-order moments, namely the correlation, the covariance, and the power spectrum.

6.6.1 The Mean (or Average) Value

The mean value of a process is its first-order moment. The mean value of a process plays an important part in signal processing and parameter estimation from noisy observations. For example, the maximum likelihood linear estimate of a Gaussian signal observed in additive Gaussian noise is a weighted interpolation between the mean value and the observed value of the noisy signal. The mean value of a vector process $[X(m_1), \ldots, X(m_M)]$ is its average value across the space of the process defined as

$$\mathcal{E}[X(m_1), \ldots, X(m_M)] = \int_{-\infty}^{\infty} \cdots \int_{-\infty}^{\infty} (x_1, \ldots, x_M) f_{X(m_1),\ldots,X(m_M)}(x_1, \ldots, x_M) dx_1 \ldots dx_M \tag{6.57}$$

For a segment of N samples of a signal $x(m)$, an estimate of the mean value of the segment is obtained as

$$\hat{\mu}_x = \frac{1}{N} \sum_{m=0}^{N-1} x(m) \tag{6.58}$$

Note that the estimate of the mean $\hat{\mu}_x$ in Equation (6.58), from a finite number of N samples, is itself a random variable with its own mean value, variance and probability distribution.

Figure 6.22 shows the histograms of the mean and the variance of a random process obtained from 10,000 segments for a uniform process. Each segment is 1000 samples long. Note that from the central limit theorem the estimates of the mean and variance have Gaussian distributions.

6.6.2 Correlation, Similarity and Dependency

The correlation is a second-order moment or statistic. The correlation of two signals is a measure of the similarity or the dependency of their fluctuations in time or space. Figure 6.23 shows the scatter diagram of two random variables for different values of their correlation.

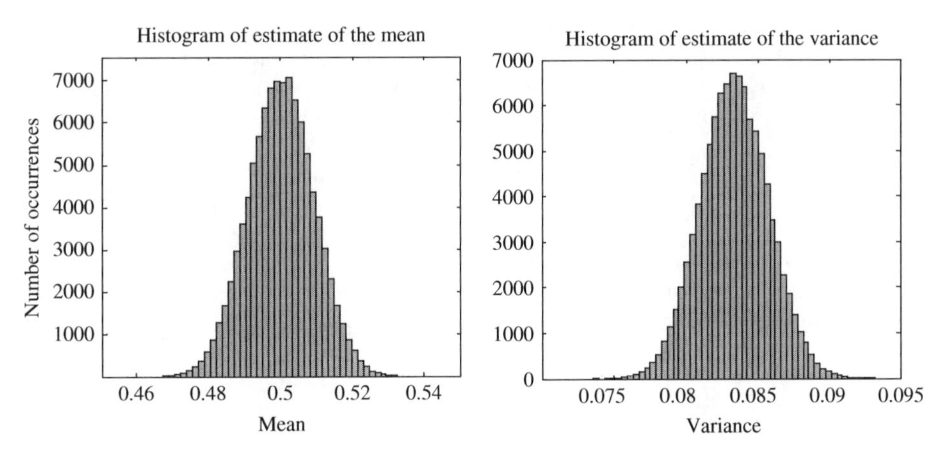

Figure 6.22 The histogram of the estimates of the mean and variance of a uniform random process.

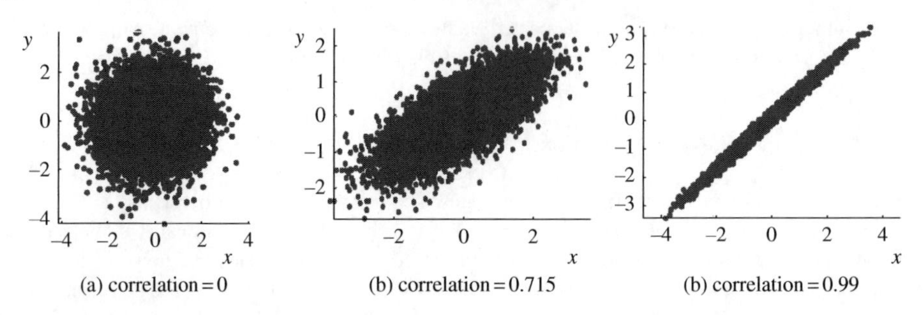

Figure 6.23 The scatter plot of two random signals x and y for different values of their cross-correlation. Note that when two variables are uncorrelated their scatter diagram is a circle. As the correlation approaches one the scatter diagram approaches a line.

The values or the courses of action of two correlated signals can be at least partially predicted from each other. Two independent signals have zero correlation. However, two dependent non-Gaussian signals may have zero correlation but non-zero higher order moments as shown in Chapter 12 on independent component analysis.

The correlation function, and its Fourier transform, the power spectral density, are extensively used in modelling and identification of patterns and structures in a signal process. Correlators play a central role in signal processing and telecommunication systems, including detectors, digital decoders, predictive coders, digital equalisers, delay estimators, classifiers and signal restoration systems. In Chapter 12 the use of correlations in principal component analysis of speech and image is described.

The correlation of a signal with a delayed version of itself is known as the *autocorrelation*. The autocorrelation function of a random process $X(m)$, denoted by $r_{xx}(m_1, m_2)$, is defined as

$$\begin{aligned} r_{xx}(m_1, m_2) &= \mathcal{E}[x(m_1)x(m_2)] \\ &= \int_{-\infty}^{\infty}\int_{-\infty}^{\infty} x(m_1)x(m_2)f_{X(m_1),X(m_1)}(x(m_1), x(m_2))\, dx(m_1)dx(m_2) \end{aligned} \tag{6.59}$$

The autocorrelation function $r_{xx}(m_1, m_2)$ is a measure of the self-similarity, dependency, or the mutual relation, of the outcomes of the process Xat time instants m_1 and m_2. If the outcome of a random process at time m_1 bears no relation to that at time m_2 then $X(m_1)$ and $X(m_2)$ are said to be independent or uncorrelated and $r_{xx}(m_1, m_2) = 0$. White noise is an example of such an uncorrelated signal.

For a wide-sense stationary process, the autocorrelation function is time-invariant and depends on the time difference, or time lag, $m = m_1 - m_2$:

$$r_{xx}(m_1 + \tau, m_2 + \tau) = r_{xx}(m_1, m_2) = r_{xx}(m_1 - m_2) = r_{xx}(k) \tag{6.60}$$

where $k = m_1 - m_2$ is the *autocorrelation lag*.

The autocorrelation function of a real-valued wide-sense stationary process is a symmetric function with the following properties:

$$r_{xx}(-k) = r_{xx}(k) \tag{6.61}$$

$$r_{xx}(k) \le r_{xx}(0) \tag{6.62}$$

Equation (6.61) says that $x(m)$ has the same relation with $x(m+k)$ as with $x(m-k)$.

For a segment of N samples of signal $x(m)$, the autocorrelation function is obtained as

$$r_{xx}(k) = \frac{1}{N} \sum_{m=0}^{N-1-k} x(m)x(m+k) \tag{6.63}$$

Note that for a zero-mean signal, $r_{xx}(0)$ is the signal power. Autocorrelation of a signal can be obtained as the inverse Fourier transform of the magnitude spectrum as

$$r_{xx}(k) = \frac{1}{N} \sum_{l=0}^{N-1} |X(l)|^2 e^{j2\pi kl/N} \tag{6.64}$$

Example 6.14 *Autocorrelation of a periodic signal: estimation of period*;
Autocorrelation can be used to calculate the repetition period T of a periodic signal such as the heartbeat pulses shown in Figure 6.24(a). Figures 6.24(b) and (c) show the estimate of the periods and the autocorrelation function of the signal in Figure 6.24(a) respectively. Note that the largest peak

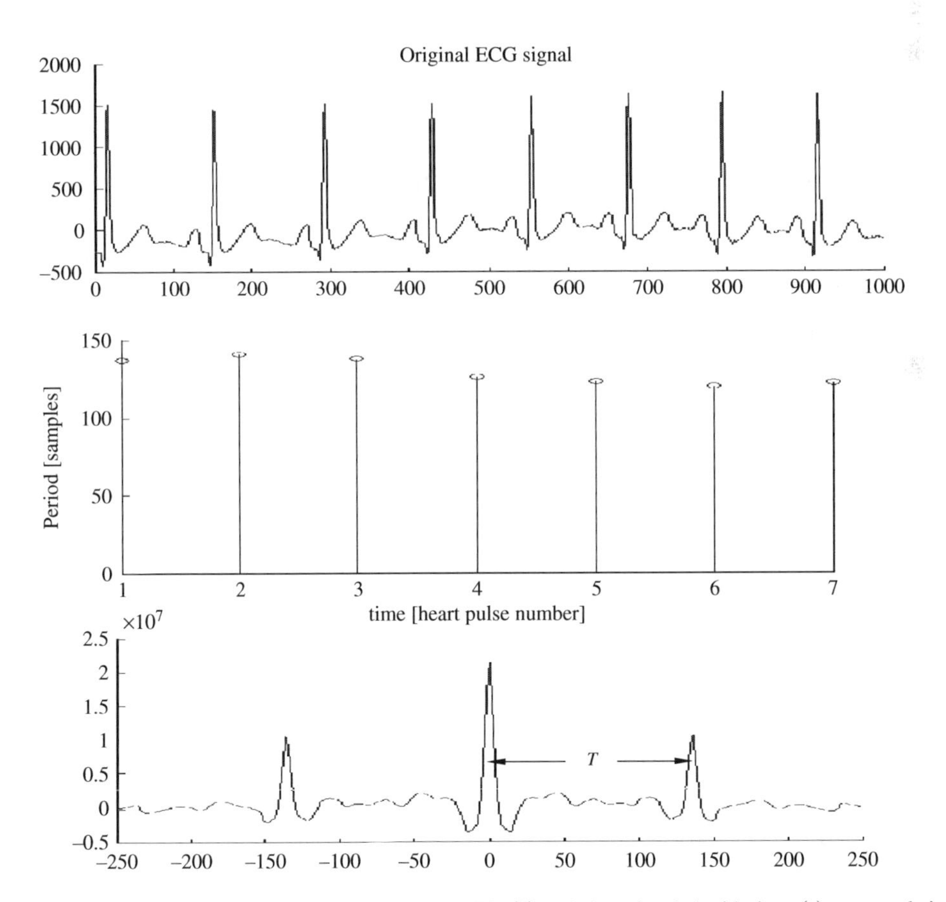

Figure 6.24 (a) Heartbeat signal, electrocardiograph ECG; (b) variation of period with time; (c) autocorrelation function of ECG.

of the autocorrelation function occurs at a lag of zero at $r_{xx}(0)$ and the second largest peak occurs at a lag of T at $r_{xx}(T)$. Hence the difference of the time indices of the first and second peaks of the autocorrelation function provides an estimate of the period of a signal.

Example 6.15 *Autocorrelation of the output of a linear time-invariant (LTI) system*
Let $x(m)$, $y(m)$ and $h(m)$ denote the input, the output and the impulse response of a LTI system respectively. The input–output relation is given by

$$y(m) = \sum_i h(i)x(m-i) \tag{6.65}$$

The autocorrelation function of the output signal $y(m)$ can be related to the autocorrelation of the input signal $x(m)$ by

$$\begin{aligned} r_{yy}(k) &= \mathcal{E}[y(m)y(m+k)] \\ &= \sum_i \sum_j h(i)h(j)\mathcal{E}[x(m-i)x(m+k-j)] \\ &= \sum_i \sum_j h(i)h(j)r_{xx}(k+i-j) \end{aligned} \tag{6.66}$$

When the input $x(m)$ is an uncorrelated zero-mean random signal with a unit variance, its autocorrelation is given by

$$r_{xx}(l) = \begin{cases} 1 & l = 0 \\ 0 & l \neq 0 \end{cases} \tag{6.67}$$

Then $r_{xx}(k+i-j) = 1$ only when $j = k+i$ and Equation (6.66) becomes

$$r_{yy}(k) = \sum_i h_i h_{k+i} \tag{6.68}$$

6.6.3 Autocovariance

The autocovariance function $c_{xx}(m_1, m_2)$ of a random process $X(m)$ is a measure of the scatter, or the dispersion, of the process about the mean value, and is defined as

$$\begin{aligned} c_{xx}(m_1, m_2) &= \mathcal{E}[(x(m_1) - \mu_x(m_1))\,(x(m_2) - \mu_x(m_2))] \\ &= r_{xx}(m_1, m_2) - \mu_x(m_1)\mu_x(m_2) \end{aligned} \tag{6.69}$$

where $\mu_x(m)$ is the mean of $X(m)$. Note that for a zero-mean process the autocorrelation and the autocovariance functions are identical. Note also that $c_{xx}(m_1, m_1)$ is the variance of the process. For a stationary process the autocovariance function of Equation (6.69) becomes

$$c_{xx}(m_1, m_2) = c_{xx}(m_1 - m_2) = r_{xx}(m_1 - m_2) - \mu_x^2 \tag{6.70}$$

6.6.4 Power Spectral Density

The power spectral density (PSD) function, also called the power spectrum, of a process gives the spectrum of the distribution of power at different frequencies of vibrations along the frequency axis. It can be shown that the power spectrum of a wide-sense stationary process $X(m)$ is the Fourier transform of the autocorrelation function

$$\begin{aligned} P_{XX}(f) &= E[X(f)X^*(f)] \\ &= \sum_{k=-\infty}^{\infty} r_{xx}(k)e^{-j2\pi fk} \end{aligned} \tag{6.71}$$

where $r_{xx}(k)$ and $P_{XX}(f)$ are the autocorrelation and power spectrum of $x(m)$ respectively, and f is the frequency variable. For a real-valued stationary process, the autocorrelation is a symmetric function, and the power spectrum may be written as

$$P_{XX}(f) = r_{xx}(0) + \sum_{k=1}^{\infty} 2r_{xx}(k)\cos(2\pi fk) \tag{6.72}$$

The power spectral density is a real-valued non-negative function, expressed in units of watts per hertz. From Equation (6.71), the autocorrelation sequence of a random process may be obtained as the inverse Fourier transform of the power spectrum as

$$r_{xx}(k) = \int_{-1/2}^{1/2} P_{XX}(f)e^{j2\pi fk}df \tag{6.73}$$

Note that the autocorrelation and the power spectrum represent the second-order statistics of a process in time and frequency domains respectively.

Example 6.16 *Power spectrum and autocorrelation of white noise* (Figure 6.25)
A noise process with uncorrelated independent samples is called a white noise process. The autocorrelation of a stationary white noise $n(m)$ is defined as

$$r_{nn}(k) = \mathcal{E}[n(m)n(m+k)] = \begin{cases} \text{Noise power} & k=0 \\ 0 & k \neq 0 \end{cases} \tag{6.74}$$

Equation (6.74) is a mathematical statement of the definition of an uncorrelated white noise process. The equivalent description in the frequency domain is derived by taking the Fourier transform of $r_{nn}(k)$:

$$P_{NN}(f) = \sum_{k=-\infty}^{\infty} r_{nn}(k)e^{-j2\pi fk} = r_{nn}(0) = \text{Noise power} \tag{6.75}$$

From Equation (6.75), the power spectrum of a stationary white noise process is spread equally across all time instances and across all frequency bins. White noise is one of the most difficult types of noise to remove, because it does not have a localised structure either in the time domain or in the frequency domain.

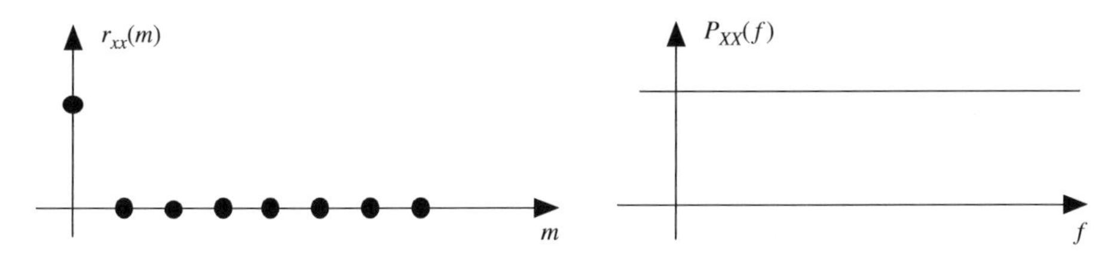

Figure 6.25 Autocorrelation and power spectrum of white noise.

Example 6.17 *Power spectrum and autocorrelation of a discrete-time impulse*
The autocorrelation of a discrete-time impulse with amplitude A, $A\delta(m)$, is defined as

$$r_{\delta\delta}(k) = \mathcal{E}[A^2\delta(m)\delta(m+k)] = \begin{cases} A^2 & k=0 \\ 0 & k\neq 0 \end{cases} \tag{6.76}$$

The power spectrum of the impulse is obtained by taking the Fourier transform of $r_{\delta\delta}(k)$ as

$$P_{\Delta\Delta}(f) = \sum_{k=-\infty}^{\infty} r_{\delta\delta}(k)e^{-j2\pi fk} = A^2 \tag{6.77}$$

Example 6.18 *Autocorrelation and power spectrum of impulsive noise*
Impulsive noise is a random, binary-state ('on/off') sequence of impulses of random amplitudes and random times of occurrence. A random impulsive noise sequence $n_i(m)$ can be modelled as an amplitude-modulated random binary sequence as

$$n_i(m) = n(m)b(m) \tag{6.78}$$

where $b(m)$ is a binary-state random sequence that indicates the presence or the absence of an impulse, and $n(m)$ is a random noise process. Assuming that impulsive noise is an uncorrelated process, the autocorrelation of impulsive noise can be defined as a binary-state process as

$$r_{nn}(k,m) = \mathcal{E}[n_i(m)n_i(m+k)] = \sigma_n^2\delta(k)b(m) \tag{6.79}$$

where σ_n^2 is the noise variance. Note that in Equation (6.79), the autocorrelation is expressed as a binary-state function that depends on the on/off state of impulsive noise at time m. When $b(m) = 1$, impulsive noise is present and its autocorrelation is equal to $\sigma_n^2\delta(k)$; when $b(m) = 0$, impulsive noise is absent and its autocorrelation is equal to 0. The power spectrum of an impulsive noise sequence is obtained by taking the Fourier transform of the autocorrelation function as

$$P_{NN}(f,m) = \sigma_n^2 b(m) \tag{6.80}$$

6.6.5 Joint Statistical Averages of Two Random Processes

In many signal processing problems, for example in the processing of the outputs of an array of radar sensors, or in smart antenna arrays for mobile phones, we have two or more random processes which may or may not be independent. Joint statistics and joint distributions are used to describe the statistical relationship between two or more random processes.

For two discrete-time random processes $x(m)$ and $y(m)$, the joint pdf is denoted by

$$f_{X(m_1),\ldots,X(m_M),Y(n_1),\ldots,Y(n_N)}(x_1,\ldots,x_M,y_1,\ldots,y_N) \tag{6.81}$$

The joint probability gives the likelihood of two or more variables assuming certain states or values. When two random processes $X(m)$ and $Y(m)$ are uncorrelated, the joint pdf can be expressed as product of the pdfs of each process as

$$\begin{aligned} &f_{X(m_1),\ldots,X(m_M),Y(n_1),\ldots,Y(n_N)}(x_1,\ldots,x_M,y_1,\ldots,y_N) \\ &= f_{X(m_1),\ldots,X(m_M)}(x_1,\ldots,x_M)f_{Y(n_1),\ldots,Y(n_N)}(y_1,\ldots,y_N) \end{aligned} \tag{6.82}$$

6.6.6 Cross-Correlation and Cross-Covariance

The cross-correlation of two random processes $x(m)$ and $y(m)$ is defined as

$$\begin{aligned} r_{xy}(m_1, m_2) &= \mathcal{E}[x(m_1)y(m_2)] \\ &= \int_{-\infty}^{\infty}\int_{-\infty}^{\infty} x(m_1)y(m_2)f_{X(m_1)Y(m_2)}(x(m_1), y(m_2))\,dx(m_1)dy(m_2) \end{aligned} \tag{6.83}$$

For wide-sense stationary processes, the cross-correlation function $r_{xy}(m_1, m_2)$ depends only on the time difference $m = m_1 - m_2$:

$$r_{xy}(m_1 + \tau, m_2 + \tau) = r_{xy}(m_1, m_2) = r_{xy}(m_1 - m_2) = r_{xy}(k) \tag{6.84}$$

where $k = m_1 - m_2$ is the *cross-correlation lag*.
The cross-covariance function is defined as

$$\begin{aligned} c_{xy}(m_1, m_2) &= \mathcal{E}\left[(x(m_1) - \mu_x(m_1))\left(y(m_2) - \mu_y(m_2)\right)\right] \\ &= r_{xy}(m_1, m_2) - \mu_x(m_1)\mu_y(m_2) \end{aligned} \tag{6.85}$$

Note that for zero-mean processes, the cross-correlation and the cross-covariance functions are identical. For a wide-sense stationary process the cross-covariance function of Equation (6.85) becomes

$$c_{xy}(m_1, m_2) = c_{xy}(m_1 - m_2) = r_{xy}(m_1 - m_2) - \mu_x\mu_y \tag{6.86}$$

Example 6.19 *Time-delay estimation*
Consider two signals $y_1(m)$ and $y_2(m)$, each composed of an information bearing signal $x(m)$ and an additive noise, given by

$$y_1(m) = x(m) + n_1(m) \tag{6.87}$$

$$y_2(m) = Ax(m - D) + n_2(m) \tag{6.88}$$

where A is an amplitude factor and D is a time-delay variable. The cross-correlation of the signals $y_1(m)$ and $y_2(m)$ yields

$$\begin{aligned} r_{y_1y_2}(k) &= \mathcal{E}[y_1(m)y_2(m+k)] \\ &= \mathcal{E}\{[x(m) + n_1(m)][Ax(m - D + k) + n_2(m + k)]\} \\ &= Ar_{xx}(k - D) + r_{xn_2}(k) + Ar_{xn_1}(k - D) + r_{n_1n_2}(k) \end{aligned} \tag{6.89}$$

Assuming that the signal and noise are uncorrelated, we have $r_{y_1y_2}(k) = Ar_{xx}(k - D)$. As shown in Figure 6.26, the cross-correlation function has its maximum at the lag D.

6.6.7 Cross-Power Spectral Density and Coherence

The cross-power spectral density of two random processes $X(m)$ and $Y(m)$ is defined as the Fourier transform of their cross-correlation function:

$$\begin{aligned} P_{XY}(f) &= \mathcal{E}[X(f)Y^*(f)] \\ &= \sum_{k=-\infty}^{\infty} r_{xy}(k)e^{-j2\pi fk} \end{aligned} \tag{6.90}$$

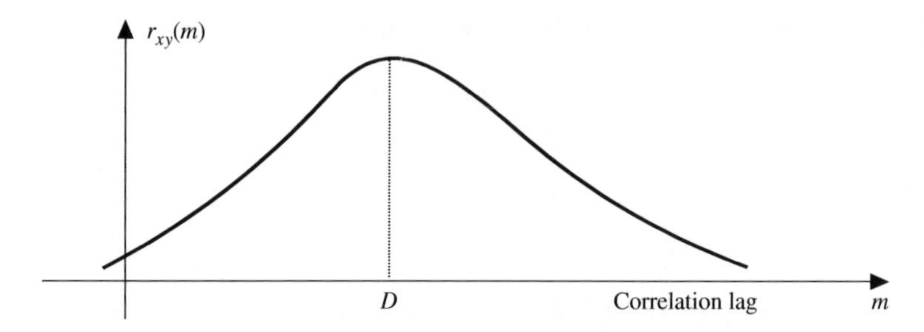

Figure 6.26 The peak of the cross-correlation of two delayed signals can be used to estimate the time delay D.

Cross-power spectral density of two processes is a measure of the similarity, or coherence, of their power spectra. The coherence, or spectral coherence, of two random processes is a normalised form of the cross-power spectral density, defined as

$$C_{XY}(f) = \frac{P_{XY}(f)}{\sqrt{P_{XX}(f)P_{YY}(f)}} \tag{6.91}$$

The coherence function is used in applications such as time-delay estimation and signal-to-noise ratio measurements.

6.6.8 Ergodic Processes and Time-Averaged Statistics

Ergodic theory has its origins in the work of Boltzmann in statistical mechanics problems where time- and space-distribution averages are equal.

In many signal processing problems, there is only a single realisation of a random process from which its statistical parameters, such as the mean, the correlation and the power spectrum, can be estimated. In these cases, time-averaged statistics, obtained from averages along the time dimension of a single realisation of the process, are used instead of the ensemble averages obtained across the space of different realisations of the process. This section considers ergodic random processes for which time-averages can be used instead of ensemble averages.

A stationary stochastic process is said to be *ergodic* if it exhibits the same statistical characteristics along the time dimension of a single realisation as across the space (or ensemble) of different realisations of the process.

Over a very long time, a single realisation of an ergodic process takes on all the values, the characteristics and the configurations exhibited across the entire space of the process. For an ergodic process $\{x(m, s)\}$, we have

$$\underset{\text{along time } m}{\text{statistical averages}}[x(m, s)] = \underset{\text{across space } s}{\text{statistical averages}}[x(m, s)] \tag{6.92}$$

where the statistical averages[$\cdot$] function refers to any statistical operation such as the mean, the variance, the power spectrum, etc.

6.6.9 Mean-Ergodic Processes

The time-averaged estimate of the mean of a signal $x(m)$ obtained from N samples is given by

$$\hat{\mu}_X = \frac{1}{N}\sum_{m=0}^{N-1} x(m) \tag{6.93}$$

A stationary process is mean-ergodic if the time-averaged value of an infinitely long realisation of the process is the same as the ensemble-mean taken across the space of the process. Therefore, for a mean-ergodic process, we have

$$\lim_{N\to\infty} \mathcal{E}[\hat{\mu}_X] = \mu_X \tag{6.94}$$

$$\lim_{N\to\infty} \text{var}\,[\hat{\mu}_X] = 0 \tag{6.95}$$

where μ_X is the ensemble average of the process. The time-averaged estimate of the mean of a signal, obtained from a random realisation of the process, is itself a random variable, with it is own mean, variance and probability density function. If the number of observation samples N is relatively large then, from the central limit theorem, the probability density function of the estimate $\hat{\mu}_X$ is Gaussian. The expectation of $\hat{\mu}_X$ is given by

$$\mathcal{E}[\hat{\mu}_x] = E\left[\frac{1}{N}\sum_{m=0}^{N-1} x(m)\right] = \frac{1}{N}\sum_{m=0}^{N-1}\mathcal{E}[x(m)] = \frac{1}{N}\sum_{m=0}^{N-1}\mu_x = \mu_x \tag{6.96}$$

From Equation (6.96), the time-averaged estimate of the mean is unbiased. The variance of $\hat{\mu}_X$ is given by

$$\begin{aligned}\text{Var}[\hat{\mu}_x] &= \mathcal{E}[\hat{\mu}_x^2] - \mathcal{E}^2[\hat{\mu}_x] \\ &= \mathcal{E}[\hat{\mu}_x^2] - \mu_x^2\end{aligned} \tag{6.97}$$

Now the term $\mathcal{E}[\hat{\mu}_x^2]$ in Equation (6.97) may be expressed as

$$\begin{aligned}\mathcal{E}[\hat{\mu}_x^2] &= \mathcal{E}\left[\left(\frac{1}{N}\sum_{m_1=0}^{N-1} x(m_1)\right)\left(\frac{1}{N}\sum_{m_2=0}^{N-1} x(m_2)\right)\right] \\ &= \frac{1}{N}\sum_{k=-(N-1)}^{N-1}\left(1-\frac{|k|}{N}\right) r_{xx}(k)\end{aligned} \tag{6.98}$$

Substitution of Equation (6.98) in Equation (6.97) yields

$$\begin{aligned}\text{Var}\,[\hat{\mu}_x^2] &= \frac{1}{N}\sum_{k=-(N-1)}^{N-1}\left(1-\frac{|k|}{N}\right) r_{xx}(k) - \mu_x^2 \\ &= \frac{1}{N}\sum_{k=-(N-1)}^{N-1}\left(1-\frac{|k|}{N}\right) c_{xx}(k)\end{aligned} \tag{6.99}$$

The condition for a process to be mean-ergodic in the mean square error sense is

$$\lim_{N\to\infty}\frac{1}{N}\sum_{k=-(N-1)}^{N-1}\left(1-\frac{|k|}{N}\right) c_{xx}(k) = 0 \tag{6.100}$$

6.6.10 Correlation-Ergodic Processes

The time-averaged estimate of the autocorrelation of a random process, estimated from a segment of N samples, is given by

$$\hat{r}_{xx}(k) = \frac{1}{N} \sum_{k=0}^{N-1} x(m)x(m+k) \tag{6.101}$$

The estimate of autocorrelation $\hat{r}_{xx}(k)$ is itself a random variable with its own mean, variance and probability distribution. A process is correlation-ergodic, in the mean square error sense, if

$$\lim_{N\to\infty} \mathcal{E}[\hat{r}_{xx}(k)] = r_{xx}(k) \tag{6.102}$$

$$\lim_{N\to\infty} \mathrm{Var}\,[\hat{r}_{xx}(k)] = 0 \tag{6.103}$$

where $r_{xx}(k)$ is the ensemble-averaged autocorrelation. Taking the expectation of $\hat{r}_{xx}(k)$ shows that it is an unbiased estimate, since

$$\mathcal{E}[\hat{r}_{xx}(k)] = \mathcal{E}\left[\frac{1}{N}\sum_{m=0}^{N-1} x(m)x(m+k)\right] = \frac{1}{N}\sum_{m=0}^{N-1} \mathcal{E}[x(m)x(m+k)] = r_{xx}(k) \tag{6.104}$$

The variance of $\hat{r}_{xx}(k)$ is given by

$$\mathrm{Var}\,[\hat{r}_{xx}(k)] = \mathcal{E}[\hat{r}^2_{xx}(k)] - r^2_{xx}(k) \tag{6.105}$$

The term $\mathcal{E}[\hat{r}^2_{xx}(m)]$ in Equation (6.105) may be expressed as

$$\begin{aligned}
\mathcal{E}[\hat{r}^2_{xx}(m)] &= \frac{1}{N^2}\sum_{k=0}^{N-1}\sum_{j=0}^{N-1} \mathcal{E}[x(k)x(k+m)x(j)x(j+m)] \\
&= \frac{1}{N^2}\sum_{k=0}^{N-1}\sum_{j=0}^{N-1} \mathcal{E}[z(k,m)z(j,m)] \\
&= \frac{1}{N}\sum_{k=-N+1}^{N-1}\left(1-\frac{|k|}{N}\right) r_{zz}(k,m)
\end{aligned} \tag{6.106}$$

where $z(i,m) = x(i)x(i+m)$. The condition for correlation ergodicity, in the mean square error sense, is given by

$$\lim_{N\to\infty}\left[\frac{1}{N}\sum_{k=-N+1}^{N-1}\left(1-\frac{|k|}{N}\right) r_{zz}(k,m) - r^2_{xx}(m)\right] = 0 \tag{6.107}$$

6.7 Some Useful Practical Classes of Random Processes

In this section, we consider some important classes of random processes that are extensively used in communication signal processing for such applications as modelling traffic, decoding, channel equalisation, modelling of noise and fading and pattern recognition.

6.7.1 Gaussian (Normal) Process

The Gaussian process, also called the normal process, is the most widely applied of all probability models. Some advantages of Gaussian probability models are the following:

(a) Gaussian pdfs can model the distribution of many processes including some important classes of signals and noise in communication systems.
(b) Non-Gaussian processes can be approximated by a weighted combination (i.e. a mixture) of a number of Gaussian pdfs of appropriate means and variances.
(c) Optimal estimation methods based on Gaussian models often result in linear and mathematically tractable solutions.

The sum of many independent random processes has a Gaussian distribution. This is known as the *central limit theorem.*

A scalar-valued Gaussian random variable is described by the following probability density function:

$$f_X(x) = \frac{1}{\sqrt{2\pi}\sigma_x} \exp\left(-\frac{(x-\mu_x)^2}{2\sigma_x^2}\right) \tag{6.108}$$

where μ_x and σ_x^2 are the mean and the variance of the random variable x. *Note that the argument of the exponential of a Gaussian function, $(x-\mu_x)^2/2\sigma_x^2$, is a variance-normalised distance.*

The Gaussian process of Equation (6.108) is also denoted by $N(x, \mu_x, \sigma_x^2)$. The maximum of a Gaussian pdf occurs at the mean μ_x, and is given by

$$f_X(\mu_x) = \frac{1}{\sqrt{2\pi}\sigma_x} \tag{6.109}$$

From Equation (6.108), the Gaussian pdf of x decreases exponentially with the distance of the variable x from the mean value μ_x. The cumulative distribution function (cdf) $F(x)$ is given by

$$F_X(x) = \frac{1}{\sqrt{2\pi}\sigma_x} \int_{-\infty}^{x} \exp\left(-\frac{(\chi-\mu_x)^2}{2\sigma_x^2}\right) d\chi \tag{6.110}$$

Figure 6.27 shows the bell-shaped pdf and the cdf of a Gaussian model. The most probable values of a Gaussian process happen around the mean value and the probability of a value decreases exponentially with the increasing distance from the mean value. The total area under the pdf curve is one. Note that the area under the pdf curve one standard deviation on each side of the mean value $(\mu \pm \sigma)$ is 0.682, the area two standard deviations on each side of the mean value $(\mu \pm 2\sigma)$ is 0.955 and the area three standard deviations on each side of the mean value $(\mu \pm 3\sigma)$ is 0.997.

6.7.2 Multivariate Gaussian Process

Multivariate probability densities model vector-valued processes. Consider a P-variate Gaussian vector process $\boldsymbol{x} = [x(m_0), x(m_1), \ldots, x(m_{P-1})]^{\mathrm{T}}$ with mean vector μ_x, and covariance matrix Σ_{xx}. The multivariate Gaussian pdf of $\boldsymbol{x}$ is given by

$$f_X(\boldsymbol{x}) = \frac{1}{(2\pi)^{P/2} |\Sigma_{xx}|^{1/2}} \exp\left(-\frac{1}{2}(\boldsymbol{x}-\boldsymbol{\mu}_x)^{\mathrm{T}} \Sigma_{xx}^{-1} (\boldsymbol{x}-\boldsymbol{\mu}_x)\right) \tag{6.111}$$

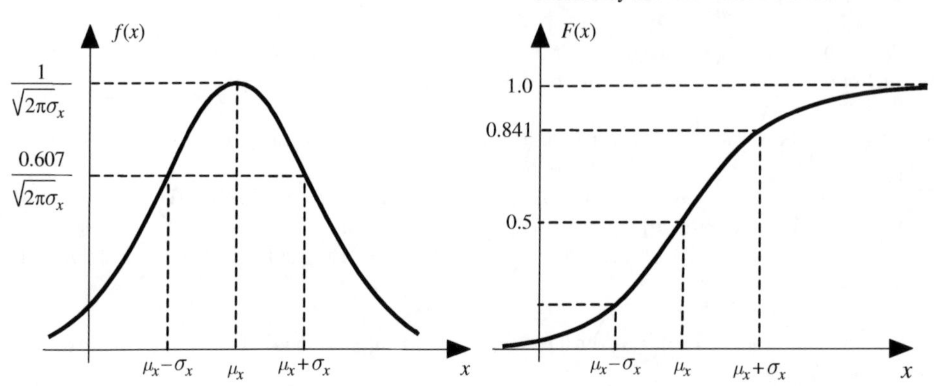

Figure 6.27 Gaussian probability density and cumulative density functions.

where the mean vector μ_x is defined as

$$\mu_x = \begin{pmatrix} \mathcal{E}[x(m_0)] \\ \mathcal{E}[x(m_2)] \\ \vdots \\ \mathcal{E}[x(m_{P-1})] \end{pmatrix} \tag{6.112}$$

and the covariance matrix Σ_{xx} is given by

$$\Sigma_{xx} = \begin{pmatrix} c_{xx}(m_0, m_0) & c_{xx}(m_0, m_1) & \cdots & c_{xx}(m_0, m_{P-1}) \\ c_{xx}(m_1, m_0) & c_{xx}(m_1, m_1) & \cdots & c_{xx}(m_1, m_{P-1}) \\ \vdots & \vdots & \ddots & \vdots \\ c_{xx}(m_{P-1}, m_0) & c_{xx}(m_{P-1}, m_1) & \cdots & c_{xx}(m_{P-1}, m_{P-1}) \end{pmatrix} \tag{6.113}$$

where $c(m_i, m_j)$ is the covariance of elements m_i and m_j of the vector process. The Gaussian process of Equation (6.111) is also denoted by $N(\boldsymbol{x}, \mu_x, \Sigma_{xx})$. If the elements of a vector process are uncorrelated then the covariance matrix is a diagonal matrix with zeros in the off-diagonal elements. In this case the multivariate pdf may be described as the product of the pdfs of the individual elements of the vector:

$$f_X\left(\boldsymbol{x} = [x(m_0), \ldots, x(m_{P-1})]^{\mathrm{T}}\right) = \prod_{i=0}^{P-1} \frac{1}{\sqrt{2\pi}\sigma_{xi}} \exp\left\{-\frac{[x(m_i) - \mu_{xi}]^2}{2\sigma_{xi}^2}\right\} \tag{6.114}$$

Example 6.20 *Conditional multivariate Gaussian probability density function*
This is useful in situations when we wish to estimate the expectation of vector $\boldsymbol{x}(m)$ given a related vector $\boldsymbol{y}(m)$.

Consider two vector realisations $\boldsymbol{x}(m)$ and $\boldsymbol{y}(m)$ from two vector-valued correlated stationary Gaussian processes $N(\boldsymbol{x}, \mu_x, \Sigma_{xx})$ and $N(\boldsymbol{y}, \mu_y, \Sigma_{yy})$. The joint probability density function of $\boldsymbol{x}(m)$ and $\boldsymbol{y}(m)$ is a multivariate Gaussian density function $N([\boldsymbol{x}(m), \boldsymbol{y}(m)], \mu_{(x,y)}, \Sigma_{(x,y)})$, with mean vector and covariance matrix given by

$$\mu_{(x,y)} = \begin{bmatrix} \mu_x \\ \mu_y \end{bmatrix} \tag{6.115}$$

$$\Sigma_{(x,y)} = \begin{bmatrix} \Sigma_{xx} & \Sigma_{xy} \\ \Sigma_{yx} & \Sigma_{yy} \end{bmatrix} \tag{6.116}$$

The conditional density of $\boldsymbol{x}(m)$ given $\boldsymbol{y}(m)$ is given from Bayes' rule as

$$f_{X|Y}\left(\boldsymbol{x}(m)\,|\boldsymbol{y}(m)\,\right) = \frac{f_{X,Y}\left(\boldsymbol{x}(m), \boldsymbol{y}(m)\right)}{f_Y\left(\boldsymbol{y}(m)\right)} \tag{6.117}$$

It can be shown that the conditional density is also a multivariate Gaussian density function with its mean vector and covariance matrix given by

$$\begin{aligned} \mu_{(x|y)} &= \mathcal{E}[\boldsymbol{x}(m)\,|\boldsymbol{y}(m)\,] \\ &= \mu_x + \Sigma_{xy}\Sigma_{yy}^{-1}(\boldsymbol{y} - \mu_y) \end{aligned} \tag{6.118}$$

$$\Sigma_{(x|y)} = \Sigma_{xx} - \Sigma_{xy}\Sigma_{yy}^{-1}\Sigma_{yx} \tag{6.119}$$

6.7.3 Gaussian Mixture Process

The probability density functions of many random processes, such as speech, are non-Gaussian. A non-Gaussian pdf may be approximated by a weighted sum (i.e. a mixture) of a number of Gaussian densities of appropriate mean vectors and covariance matrices. A mixture Gaussian density with M components is defined as

$$f_X(\boldsymbol{x}) = \sum_{i=1}^{M} P_i \mathcal{N}_i(\boldsymbol{x}, \mu_{x_i}, \Sigma_{xx_i}) \tag{6.120}$$

where $\mathcal{N}_i(\boldsymbol{x}, \mu_{x_i}, \Sigma_{xx_i})$ is a multivariate Gaussian density with mean vector μ_{x_i} and covariance matrix Σ_{xx_i}, and P_i are the mixing coefficients. The parameter P_i is the prior probability of the ith component of the mixture, given by

$$P_i = \frac{N_i}{\sum_{j=1}^{M} N_j} \tag{6.121}$$

where N_i is the number of observations of the process associated with the mixture i. Figure 6.28 shows a non-Gaussian pdf modelled as a mixture of five Gaussian pdfs. Algorithms developed for Gaussian processes can be extended to mixture Gaussian densities.

6.7.4 Binary-State Gaussian Process

A simple example of a binary-state process is the observations at the output of a communication system with the input signal consisting of a binary sequence ('0' and '1') process.
Consider a random process $x(m)$ with two states s_0 and s_1 such that in the state s_0 the process has a Gaussian pdf with mean $\mu_{x,0}$ and variance $\sigma_{x,0}^2$, and in the state s_1 the process is also Gaussian with mean $\mu_{x,1}$ and variance $\sigma_{x,0}^2$. The state-dependent pdf of $x(m)$ can be expressed as

$$f_{X|S}\left(x(m)\,|s_i\,\right) = \frac{1}{\sqrt{2\pi}\sigma_{x,i}} \exp\left(-\frac{1}{2\sigma_{x,i}^2}\left(x(m) - \mu_{x,i}\right)^2\right), \quad i = 0, 1 \tag{6.122}$$

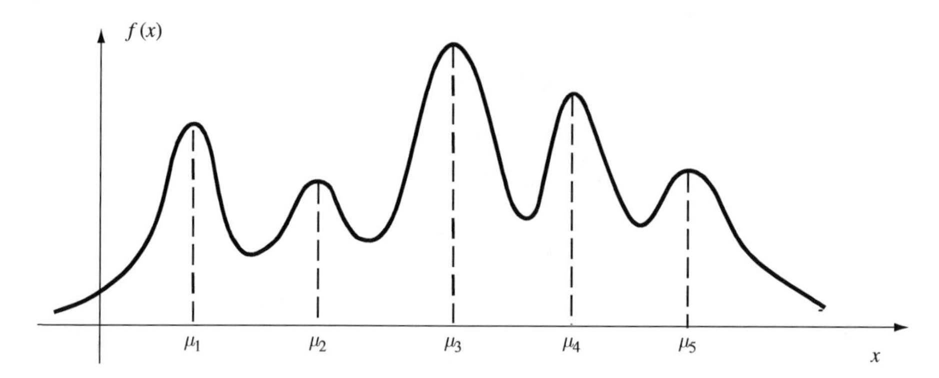

Figure 6.28 A Gaussian mixture model (GMM) pdf.

The joint probability distribution of the binary-valued state s_i and the continuous-valued signal $x(m)$ can be expressed as

$$\begin{aligned} f_{X,S}(x(m), s_i) &= f_{X|S}(x(m)|s_i) P_S(s_i) \\ &= \frac{1}{\sqrt{2\pi}\sigma_{x,i}} \exp\left(-\frac{1}{2\sigma_{x,i}^2}\left(x(m)-\mu_{x,i}\right)^2\right) \times P_S(s_i) \end{aligned} \tag{6.123}$$

where $P_S(s_i)$ is the state probability. For a multi-state process we have the following probabilistic relations between the joint and marginal probabilities:

$$\sum_S f_{X,S}(x(m), s_i) = f_X(x(m)) \tag{6.124}$$

$$\int_X f_{X,S}(x(m), s_i)\, dx = P_S(s_i) \tag{6.125}$$

and

$$\sum_S \int_X f_{X,S}(x(m), s_i)\, dx = 1 \tag{6.126}$$

Note that in a multi-state model, the statistical parameters of the process *switch* between a number of different states, whereas in a single-state mixture pdf, a *weighted* combination of a number of pdfs models the process. In Chapter 11 on hidden Markov models we consider multi-state models with a mixture Gaussian pdf per state.

6.7.5 Poisson Process – Counting Process

The Poisson process is a continuous-time integer-valued counting process, used for modelling the probability of the number of occurrences of a random discrete event in various time intervals.

An important area of application of the Poisson process is in the queuing theory for the analysis and modelling of the distribution of demand on a service facility such as a telephone network, a computer network, a financial service, a transport network, a petrol station, etc. Other applications of the Poisson distribution include the counting of the number of particles emitted in particle physics, the number of times that a component may fail in a system, and the modelling of radar clutter, shot noise and impulsive noise.

Consider an event-counting process $X(t)$, in which the probability of occurrence of the event is governed by a rate function $\lambda t)$, such that the probability that an event occurs in a small time interval Δt is

$$\text{Prob}\,(1 \text{ occurrence in the interval } (t, t+\Delta t)) = \lambda(t)\Delta t \tag{6.127}$$

Assuming that in the small interval Δt, no more than one occurrence of the event is possible, the probability of no occurrence of the event in a time interval of Δt is given by

$$\text{Prob}\,(0 \text{ occurrence in the interval } (t, t+\Delta t)) = 1 - \lambda(t)\Delta t \tag{6.128}$$

When the parameter $\lambda(t)$ is independent of time, $\lambda(t) = \lambda$, the process is called a homogeneous Poisson process. Now, for a homogeneous Poisson process, consider the probability of k occurrences of an event in a time interval of $t+\Delta t$, denoted by $P(k, (0, t+\Delta t))$:

$$\begin{aligned} P\,(k, (0, t+\Delta t)) &= P\,(k, (0, t))\,P\,(0, (t, t+\Delta t)) + P\,(k-1, (0, t))\,P\,(1, (t, t+\Delta t)) \\ &= P\,(k, (0, t))\,(1 - \lambda\Delta t) + P\,(k-1, (0, t))\,\lambda\Delta t \end{aligned} \tag{6.129}$$

Rearranging Equation (6.129), and letting Δt tend to zero, we obtain the linear differential equation

$$\lim_{\Delta t \to 0} \frac{P(k, (0, t+\Delta t)) - P(k, (0, t))}{\Delta t} = \frac{dP(k, t)}{dt} = -\lambda P(k, t) + \lambda P(k-1, t) \tag{6.130}$$

where $P(k, t) = P(k, (0, t))$. The solution of this differential equation is given by

$$P(k, t) = \lambda e^{-\lambda t} \int_0^t P(k-1, \tau) e^{\lambda \tau} d\tau \tag{6.131}$$

Equation (6.131) can be solved recursively: starting with $P(0, t) = e^{-\lambda t}$ and $P(1, t) = \lambda t e^{-\lambda t}$, we obtain the Poisson density

$$P(k, t) = \frac{(\lambda t)^k}{k!} e^{-\lambda t} \tag{6.132}$$

Figure 6.29 illustrates Poisson pdf. From Equation (6.132), it is easy to show that for a homogenous Poisson process, the probability of k occurrences of an event in a time interval (t_1, t_2) is given by

$$P[k, (t_1, t_2)] = \frac{[\lambda(t_2 - t_1)]^k}{k!} e^{-\lambda(t_2 - t_1)} \tag{6.133}$$

Equation (6.133) states the probability of k events in a time interval of $t_2 - t_1$, in terms of the rate at which the process happens.

A Poisson counting process $X(t)$ is incremented by one every time the event occurs. From Equation (6.132), the mean and variance of a Poisson counting process $X(t)$ are

$$\mathcal{E}[X(t)] = \lambda t \tag{6.134}$$

$$r_{XX}(t_1, t_2) = \mathcal{E}[X(t_1)X(t_2)] = \lambda^2 t_1 t_2 + \lambda \min(t_1, t_2) \tag{6.135}$$

$$\text{Var}[X(t)] = \mathcal{E}\left[X^2(t)\right] - E^2\left[X(t)\right] = \lambda t \tag{6.136}$$

Note that the variance of a Poisson process is equal to its mean value.

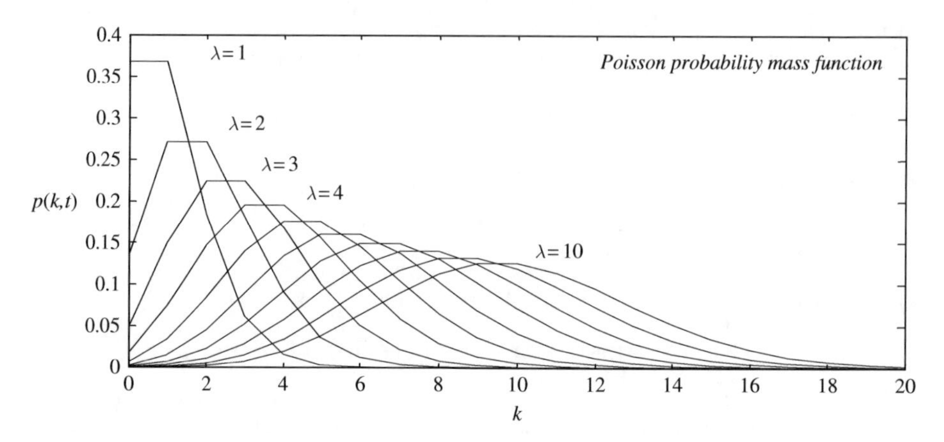

Figure 6.29 Illustration of Poisson process with increasing rate λ from 1 to 10. At each curve the maximum happens at the expected number of events i.e. rate $\times$ time interval. Note the probability exists for discrete values of k only.

6.7.6 Shot Noise

Shot noise results from randomness in directional flow of particles, for example in the flow of electrons from the cathode to the anode of a cathode ray tube, the flow of photons in a laser beam, the flow and recombination of electrons and holes in semiconductors, and the flow of photoelectrons emitted in photodiodes. Shot noise has the form of a random pulse sequence that may be modelled as the response of a linear filter excited by a Poisson-distributed binary impulse sequence (Figure 6.30).

Consider a Poisson-distributed binary-valued impulse process $x(t)$. Divide the time axis into uniform short intervals of Δt such that only one occurrence of an impulse is possible within each time interval. Let $x(m\Delta t)$ be '1' if an impulse is present in the interval $m\Delta t$ to $(m+1)\Delta t$, and '0' otherwise. For $x(m\Delta t)$, we obtain the mean and correlation functions as

$$\mathcal{E}[x(m\Delta t)] = 1 \times P(x(m\Delta t) = 1) + 0 \times P(x(m\Delta t) = 0) = \lambda\Delta t \tag{6.137}$$

and

$$\mathcal{E}[x(m\Delta t)x(n\Delta t)] = \begin{cases} 1 \times P(x(m\Delta t) = 1) = \lambda\Delta t & m = n \\ 1 \times P(x(m\Delta t) = 1)) \times P(x(n\Delta t) = 1) = (\lambda\Delta t)^2 & m \neq n \end{cases} \tag{6.138}$$

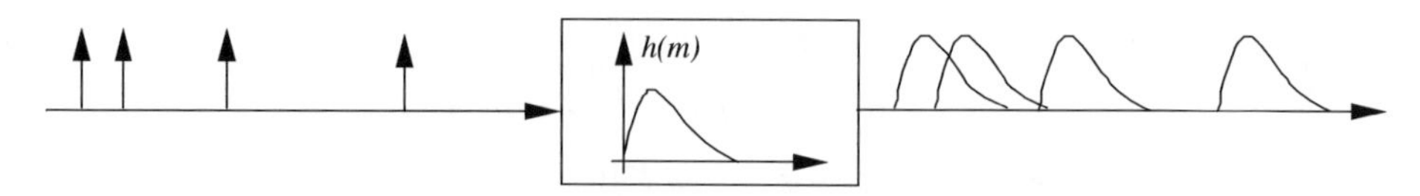

Figure 6.30 Shot noise is modelled as the output of a filter excited with a process.

A shot noise process $y(m)$ is modelled as the output of a linear system with an impulse response $h(t)$, excited by a Poisson-distributed binary impulse input $x(t)$:

$$\begin{aligned} y(t) &= \int_{-\infty}^{\infty} x(\tau)h(t-\tau)d\tau \\ &= \sum_{k=-\infty}^{\infty} x(m\Delta t)h(t-m\Delta t) \end{aligned} \tag{6.139}$$

where the binary signal $x(m\Delta t)$ can assume a value of 0 or 1. In Equation (6.139) it is assumed that the impulses happen at the beginning of each interval. This assumption becomes more valid as Δt becomes smaller. The expectation of $y(t)$ is obtained as

$$\begin{aligned} \mathcal{E}[y(t)] &= \sum_{k=-\infty}^{\infty} \mathcal{E}[x(m\Delta t)]\, h(t-m\Delta t) \\ &= \sum_{k=-\infty}^{\infty} \lambda\Delta t h(t-m\Delta t) \end{aligned} \tag{6.140}$$

and

$$\begin{aligned} r_{yy}(t_1,t_2) &= \mathcal{E}[y(t_1)y(t_2)] \\ &= \sum_{m=-\infty}^{\infty}\sum_{n=-\infty}^{\infty} (\mathcal{E}[x(m\Delta t)x(n\Delta t)]\, t)h(t_2-m\Delta t) \end{aligned} \tag{6.141}$$

Using Equation (6.138) in (6.141) the autocorrelation of $y(t)$ can be obtained as

$$\begin{aligned} r_{yy}(t_1,t_2) = & \sum_{m=-\infty}^{\infty} (\lambda\Delta t)h(t_1-m\Delta t)h(t_2-m\Delta t) \\ & + \sum_{m=-\infty}^{\infty}\sum_{\substack{n=-\infty \\ n\neq m}}^{\infty} (\lambda\Delta t)^2 h(t_1-m\Delta t)h(t_2-n\Delta t) \end{aligned} \tag{6.142}$$

6.7.7 Poisson–Gaussian Model for Clutters and Impulsive Noise

An impulsive noise process consists of a sequence of short-duration pulses of random amplitude and random time of occurrence whose shape and duration depends on the characteristics of the channel through which the impulse propagates. A Poisson process can be used to model the random time of occurrence of impulsive noise, and a Gaussian process can model the random amplitude of the impulses. Finally, the finite duration character of real impulsive noise may be modelled by the impulse response of a linear filter. The Poisson–Gaussian impulsive noise model is given by

$$x(m) = \sum_{k=-\infty}^{\infty} A_k h(m-\tau_k) \tag{6.143}$$

where $h(m)$ is the response of a linear filter that models the shape of impulsive noise, A_k is a zero-mean Gaussian process of variance σ_x^2 and τ_k denotes the instances of occurrences of impulses modelled by a Poisson process. The output of a filter excited by a Poisson-distributed sequence of Gaussian amplitude impulses can also be used to model clutters in radar. Clutters are due to reflection of radar pulses from a multitude of background surfaces and objects other than the intended radar target.

6.7.8 Markov Processes

Markov processes are used to model the trajectory of a random process and to describe the dependency of the outcome of a process at any given time on the past outcomes of the process. Applications of Markov models include modelling the trajectory of a process in signal estimation and pattern recognition for speech, image and biomedical signal processing.

A first-order discrete-time Markov process is defined as one in which the state or the value of the process at time m depends only on its state or value at time $m-1$ and is independent of the states or values of the process before time $m-1$. In probabilistic terms, a first-order Markov process can be defined as

$$\begin{aligned} f_X\left(x(m) = x_m \,|x(m-1) = x_{m-1}, \ldots, x(m-N) = x_{m-N}\right) \\ = f_X\left(x(m) = x_m \,|x(m-1) = x_{m-1}\right) \end{aligned} \tag{6.144}$$

The marginal density of a Markov process at time m can be obtained by integrating the conditional density over all values of $x(m-1)$:

$$f_X(x(m) = x_m) = \int_{-\infty}^{\infty} f_X\left(x(m) = x_m \,|x(m-1) = x_{m-1})\right) f_X\left(x(m-1) = x_{m-1}\right) dx_{m-1} \tag{6.145}$$

A process in which the present state of the system depends on the past n states may be described in terms of n first-order Markov processes and is known as an nth order Markov process. The term 'Markov process' usually refers to a first-order process.

Example 6.21 A simple example of a Markov process is a first-order auto-regressive process (Figure 6.31) defined as

$$x(m) = ax(m-1) + e(m) \tag{6.146}$$

In Equation (6.146), $x(m)$ depends on the previous value $x(m-1)$ and the input $e(m)$. The conditional pdf of $x(m)$ given the previous sample value can be expressed as

$$\begin{aligned} f_X\left(x(m)\,|x(m-1), \ldots, x(m-N)\right) = f_X\left(x(m)\,|x(m-1)\right) \\ = f_E\left(e(m) = x(m) - ax(m-1)\right) \end{aligned} \tag{6.147}$$

where $f_E(e(m))$ is the pdf of the input signal. Assuming that input $e(m)$ is a zero-mean Gaussian process with variance σ_e^2, we have

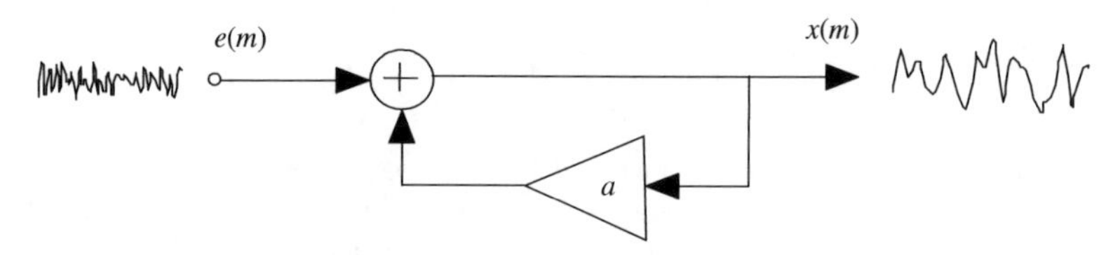

Figure 6.31 A first-order autoregressive (Markov) process; in this model the value of the process at time m, $x(m)$ depends only on $x(m-1)$ and a radon input.

$$\begin{aligned} f_X(x(m)\,|x(m-1),\ldots,x(m-N)) &= f_X(x(m)\,|x(m-1)) \\ &= f_E(x(m)-ax(m-1)) \\ &= \frac{1}{\sqrt{2\pi}\sigma_e}\exp\left[-\frac{1}{2\sigma_e^2}(x(m)-ax(m-1))^2\right] \end{aligned} \tag{6.148}$$

When the input to a Markov model is a Gaussian process the output is known as a Gauss–Markov process.

6.7.9 Markov Chain Processes

A discrete-time Markov process $x(m)$ with N allowable states may be modelled by a Markov chain of N states (Figure 6.32). Each state can be associated with one of the N values that $x(m)$ may assume. In a Markov chain, the Markovian property is modelled by a set of state transition probabilities defined as

$$a_{ij}(m-1,m) = \text{Prob}\,(x(m)=j\,|x(m-1)=i) \tag{6.149}$$

where $a_{ij}(m-1,m)$ is the probability that at time $m-1$ the process is in the state i and then at time m it moves to state j. In Equation (6.149), the transition probability is expressed in a general time-dependent form. The marginal probability that a Markov process is in state j at time m, $P_j(m)$, can be expressed as

$$P_j(m) = \sum_{i=1}^{N} P_i(m-1)a_{ij}(m-1,m) \tag{6.150}$$

A Markov chain is defined by the following set of parameters:
number of states, N state probability vector

$$\boldsymbol{p}^{\mathrm{T}}(m) = [p_1(m), p_2(m), \ldots, p_N(m)]$$

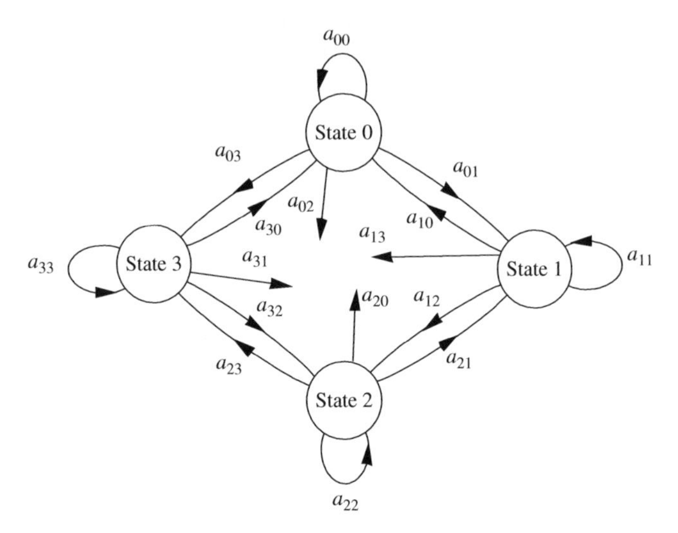

Figure 6.32 A Markov chain model of a four-state discrete-time Markov process.

and the state transition matrix.

$$\boldsymbol{A}(m-1,m)=\begin{bmatrix} a_{11}(m-1,m) & a_{12}(m-1,m) & \dots & a_{1N}(m-1,m) \\ a_{21}(m-1,m) & a_{22}(m-1,m) & \dots & a_{2N}(m-1,m) \\ \vdots & \vdots & \ddots & \vdots \\ a_{N1}(m-1,m) & a_{N2}(m-1,m) & \dots & a_{NN}(m-1,m) \end{bmatrix}$$

Homogenous and Inhomogeneous Markov Chains

A Markov chain with time-invariant state transition probabilities is known as a homogenous Markov chain. For a homogenous Markov process, the probability of a transition from state i to state j of the process is independent of the time of the transition m, as expressed in the equation

$$\text{Prob}\,(x(m)=j\,|x(m-1)=i)=a_{ij}(m-1,m)=a_{ij} \tag{6.151}$$

Inhomogeneous Markov chains have time-dependent transition probabilities. In most applications of Markov chains, homogenous models are used because they usually provide an adequate model of the signal process, and because homogenous Markov models are easier to train and use. Markov models are considered in Chapter 11.

6.7.10 Gamma Probability Distribution

The Gamma pdf is defined as

$$\text{gamma}\,(x,a,b)=\begin{cases} \dfrac{1}{b^a\Gamma(a)}x^{a-1}e^{-x/b} & \text{for } x\geq 0 \\ 0 & \text{otherwise} \end{cases} \tag{6.152}$$

where a and b are both greater than zero and the Gamma function $\Gamma(a)$ is defined as

$$\Gamma(a)=\int_0^\infty x^{a-1}e^{-x}\,dx \tag{6.153}$$

Note that $\Gamma(1)=1$. The Gamma pdf is sometimes used in modelling speech and image signals. Figure 6.33 illustrates the Gamma pdf.

Matlab function GammaProb( )
Draws the probability density of a Gamma process for various values of the parameter 'a'.

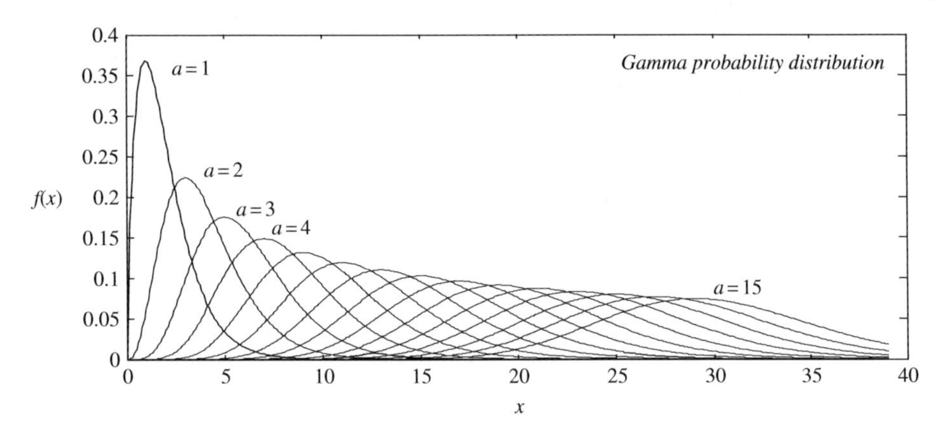

Figure 6.33 Illustration of Gamma pdf with increasing value of a, $b = 1$.

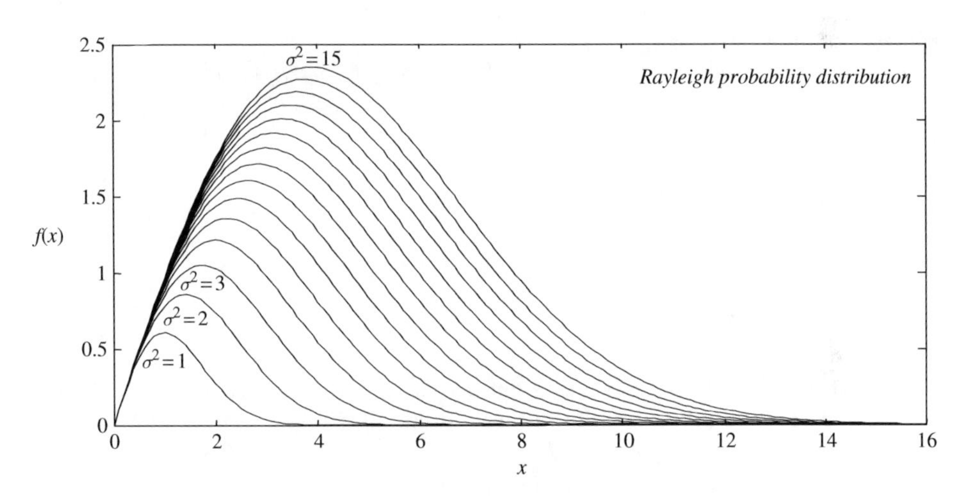

Figure 6.34 Illustration of Rayleigh process with increasing variance σ^2 from 1 to 15.

6.7.11 Rayleigh Probability Distribution

The Rayleigh pdf, show in Figure 6.34, is defined as

$$p(x) = \begin{cases} \frac{x}{\sigma^2} \exp\left(-\frac{x^2}{2\sigma^2}\right) & x \geq 0 \\ 0 & x < 0 \end{cases} \tag{6.154}$$

The Rayleigh pdf is often employed to describe the amplitude spectrum of signals. In mobile communication, channel fading is usually modelled with a Rayleigh distribution.

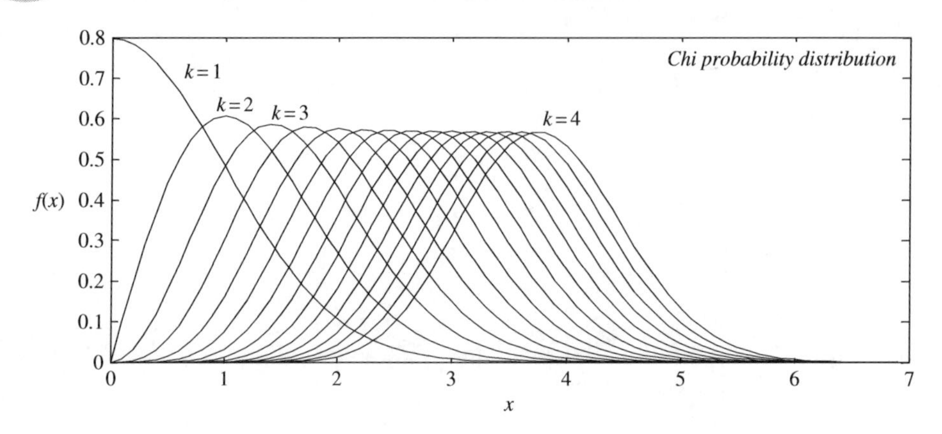

Figure 6.35 Illustration of a Chi process $p_k(x)$ with increasing value of variable k.

Matlab function Rayleigh( )
Draws the probability density of a Rayleigh process for various values of the parameter 'a'.

6.7.12 Chi Distribution

A Chi pdf, Figure 6.35, is defined as

$$p_k(x) = \frac{2^{1-k/2} x^{k-1}}{\Gamma(k/2)} \exp\left(-\frac{x^2}{2}\right) \quad (6.155)$$

As shown in Figure 6.35 the parameter k controls the shape of the distribution. For $k = 2$ the Chi distribution is the same as the Rayleigh distribution (Equation (6.154)) with variance $\sigma^2 = 1$.

Matlab function Chi( )
Draws the probability density of a Gamma process for various values of the parameter 'a'.

6.7.13 Laplacian Probability Distribution

A Laplacian pdf, Figure 6.36, is defined as

$$p(x) = \frac{1}{2\sigma} \exp\left(-\frac{|x|}{\sigma}\right) \quad (6.156)$$

where σ is the standard deviation. Speech signal samples in time domain have a distribution that can be approximated by a Laplacian pdf. The Laplacian pdf is also used for modelling image signals.

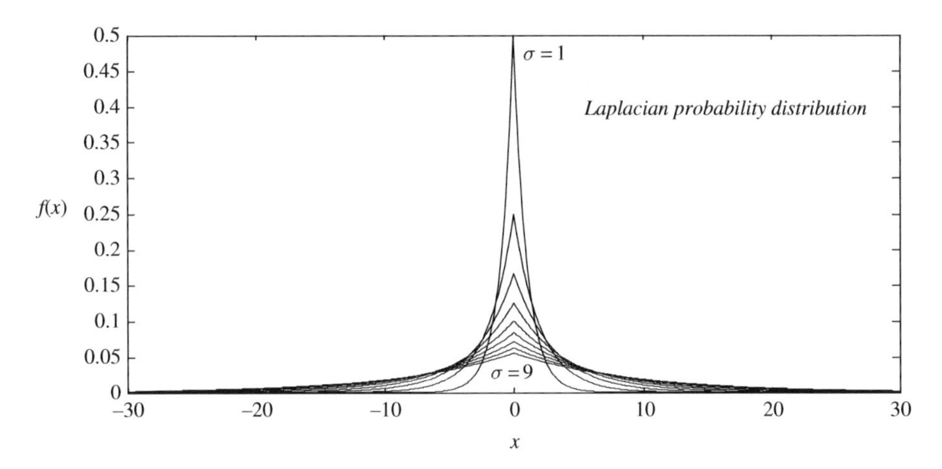

Figure 6.36 Illustration of Laplacian pdfs with standard deviation from 1 to 9.

Matllab function Laplacian( )
Draws the probability density of a Laplacian process for various values of the parameter 'a'.

6.8 Transformation of a Random Process

In this section we consider the effect of filtering or transformation of a random process on its probability density function. Figure 6.37 shows a generalised mapping operator $h(\cdot)$ that transforms a random input process X into an output process Y. The input and output signals $x(m)$ and $y(m)$ are realisations of the random processes X and Y respectively. If $x(m)$ and $y(m)$ are both discrete-valued such that $x(m) \in \{x_1, \ldots, x_N\}$ and $y(m) \in \{y_1, \ldots, y_M\}$ then we have

$$P_Y\left(y(m) = y_j\right) = \sum_{x_i \to y_j} P_X\left(x(m) = x_i\right) \tag{6.157}$$

where the summation is taken over all values of $x(m)$ that map to $y(m) = y_j$. An example of discrete-valued mapping is quantisation from a larger codebook to a smaller codebook. Consider the transformation of a discrete-time continuous-valued process. The probability that the output process

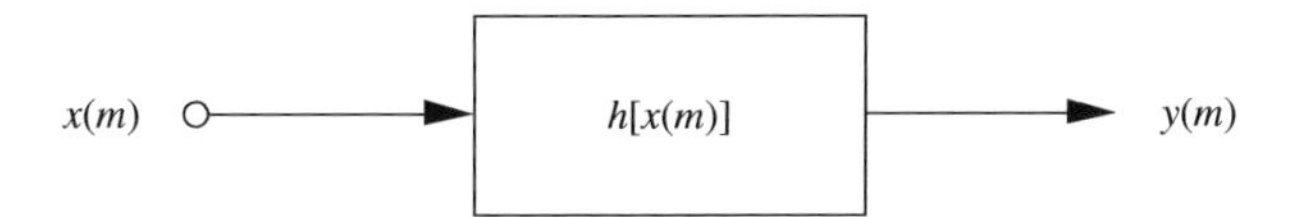

Figure 6.37 Transformation of a random process $x(m)$ to an output process $y(m)$.

Y has a value in the range $y(m) < Y < y(m) + \Delta y$ is

$$\text{Prob}\,[y(m) < Y < y(m) + \Delta y] = \int_{x(m)|y(m)<Y<y(m)+\Delta y} f_X(x(m))dx(m) \tag{6.158}$$

where the integration is taken over all the values of $x(m)$ that yield an output in the range $y(m)$ to $y(m) + \Delta y$.

6.8.1 Monotonic Transformation of Random Processes

Now for a monotonic one-to-one transformation $y(m) = h[x(m)]$ (e.g. as in Figure 6.38) Equation (6.158) becomes

$$\text{Prob}\,(y(m) < Y < y(m) + \Delta y) = \text{Prob}\,(x(m) < X < x(m) + \Delta x) \tag{6.159}$$

or, in terms of the cumulative distribution functions

$$F_Y\,(y(m) + \Delta y) - F_Y\,(y(m)) = F_X\,(x(m) + \Delta x) - F_X\,(x(m)) \tag{6.160}$$

Multiplication of the left-hand side of Equation (6.160) by $\Delta y/\Delta y$ and the right-hand side by $\Delta x/\Delta x$ and re-arrangement of the terms yields

$$\frac{F_Y\,(y(m) + \Delta y) - F_Y\,(y(m))}{\Delta y} = \frac{\Delta x}{\Delta y}\,\frac{F_X\,(x(m) + \Delta x) - F_X\,(x(m))}{\Delta x} \tag{6.161}$$

Now as the intervals Δx and Δy tend to zero, Equation (6.161) becomes

$$f_Y\,(y(m)) = \left|\frac{\partial x(m)}{\partial y(m)}\right| f_X\,(x(m)) \tag{6.162}$$

where $f_Y(y(m))$ is the probability density function. In Equation (6.162), substitution of $x(m) = h^{-1}(y(m))$ yields

$$f_Y\,(y(m)) = \left|\frac{\partial h^{-1}\,(y(m))}{\partial y(m)}\right| f_X\left(h^{-1}\,(y(m))\right) \tag{6.163}$$

Equation (6.163) gives the pdf of the output signal in terms of the pdf of the input signal and the transformation.

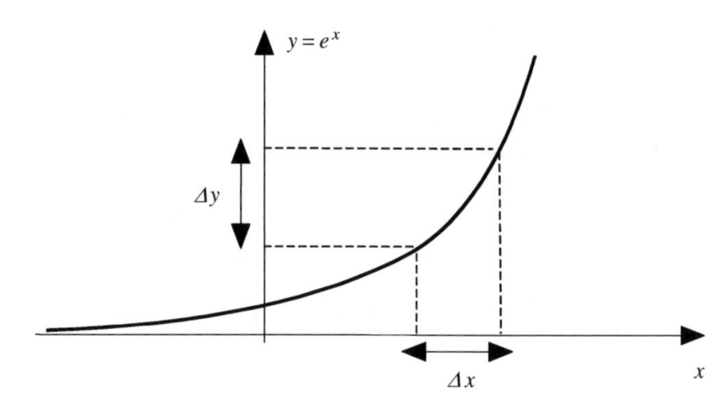

Figure 6.38 An example of a monotonic one-to-one mapping.

Example 6.22 *Probability density of a scaled variable*
Consider the simple scaled function

$$y(m) = \alpha x(m) \tag{6.164}$$

From Equation (6.163), noting that $h^{-1}(y(m)) = \frac{1}{\alpha} y(m)$, we have

$$f_Y(y(m)) = \left|\frac{1}{\alpha}\right| f_X(y(m)/\alpha) \tag{6.165}$$

Example 6.23 *Probability density of frequency spectrum: Cartesian to polar transformation*
Consider the kth spectral component of the discrete Fourier transform of a discrete-time Gaussian distributed variable

$$X(k) = X_R(k) + X_I(k) = r(k)e^{j\varphi(k)} \tag{6.166}$$

The Fourier transform is a linear operation, hence if the input is Gaussian it follows that the real part of the spectrum $X_R(f)$ and the imaginary part of the spectrum $X_I(f)$ are also Gaussian with a pdf that can be described as

$$\begin{aligned} f_{X_R,X_I}(X_R, X_I) &= f(X_R) f(X_I) \\ &= \frac{1}{\sqrt{2\pi}\sigma} \exp\left(\frac{X_R^2}{2\sigma^2}\right) \frac{1}{\sqrt{2\pi}\sigma} \exp\left(\frac{X_I^2}{2\sigma^2}\right) \\ &= \frac{1}{2\pi\sigma^2} \exp\left(\frac{X_R^2 + X_I^2}{2\sigma^2}\right) \end{aligned} \tag{6.167}$$

Now the relation between transforming differential areas in Cartesian and polar co-ordinates is obtained as

$$dX_R dX_I = r dr d\varphi \tag{6.168}$$

Hence from Equation (6.162) we have

$$f_{R,\Phi}(r, \varphi) = \frac{dX_R dX_I}{dr d\varphi} f_{X_R,X_I}(X_R, X_I) = r f_{X_R,X_I}(X_R, X_I) \tag{6.169}$$

From Equations (6.167) and (6.169) we have

$$f_{R,\Phi}(r, \varphi) = \frac{r}{2\pi\sigma^2} \exp\left(\frac{r^2}{2\sigma^2}\right) \tag{6.170}$$

The probability of the phase φ can be obtained from

$$f_\Phi(\varphi) = \int_0^\infty f_{R,\Phi}(r, \varphi)\, dr = \frac{1}{2\pi} \int_0^\infty \frac{r}{\sigma^2} \exp\left(\frac{r^2}{2\sigma^2}\right) dr = \frac{1}{2\pi} \tag{6.171}$$

Hence φ has a uniform distribution pdf with a probability of $1/2\pi$ and the pdf of r is Rayleigh distributed as

$$f_R(r) = \begin{cases} \frac{r}{\sigma^2} \exp\left(-\frac{r^2}{2\sigma^2}\right) & r \geq 0 \\ 0 & r < 0 \end{cases} \tag{6.172}$$

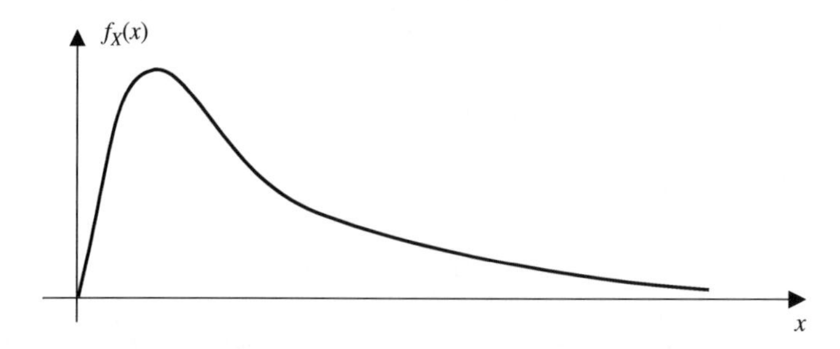

Figure 6.39 A log-normal probability density function.

Example 6.24 *Transformation of a Gaussian process to a log-normal process*
Log-normal pdfs are used for modelling positive-valued processes such as power spectra. If a random variable $x(m)$ has a Gaussian pdf then the non-negative valued variable $y(m) = \exp(x(m))$ has a log-normal distribution (Figure 6.39) obtained from Equation (6.162) as

$$f_Y(y) = \frac{1}{\sqrt{2\pi}\sigma_x y(m)} \exp\left\{-\frac{(\ln y(m) - \mu_x)^2}{2\sigma_x^2}\right\} \tag{6.173}$$

Conversely, if the input y to a logarithmic function has a log-normal distribution then the output $x = \ln y$ is Gaussian. The mapping functions for translating the mean and variance of a log-normal distribution to a normal distribution can be derived as

$$\mu_x = \ln(\mu_y) - \frac{1}{2}\ln\left(1 + \sigma_y^2/\mu_y^2\right) \tag{6.174}$$

$$\sigma_x^2 = \ln\left(1 + \sigma_y^2/\mu_y^2\right) \tag{6.175}$$

(μ_x, σ_x^2), and (μ_y, σ_y^2) are the mean and variance of x and y respectively. The inverse mapping relations for the translation of means and variances of normal to log-normal variables are

$$\mu_y = \exp(\mu_x + \sigma_x^2/2) \tag{6.176}$$

$$\sigma_y^2 = \mu_x^2[\exp(\sigma_y^2) - 1] \tag{6.177}$$

6.8.2 Many-to-One Mapping of Random Signals

Now consider the case when the transform $h(\cdot)$ is a non-monotonic function such as that shown in Figure 6.40. Assuming that the equation $y(m) = h(x(m))$ has K roots, there are K different values of $x(m)$ that map to the same $y(m)$. The probability that a realisation of the output process Y has a value in the range $y(m)$ to $y(m) + \Delta y$ is given by

$$\text{Prob}\,(y(m) < Y < y(m) + \Delta y) = \sum_{k=1}^{K} \text{Prob}\,(x_k(m) < X < x_k(m) + \Delta x_k) \tag{6.178}$$

where x_k is the kth root of $y(m) = h(x(m))$. Similar to the development in Section 6.7.1, Equation (6.178) can be written as

$$\frac{F_Y\,(y(m) + \Delta y) - F_Y\,(y(m))}{\Delta y}\Delta y = \sum_{k=1}^{K} \frac{F_X\,(x_k(m) + \Delta x_k) - F_X\,(x_k(m))}{\Delta x_k}\Delta x_k \tag{6.179}$$

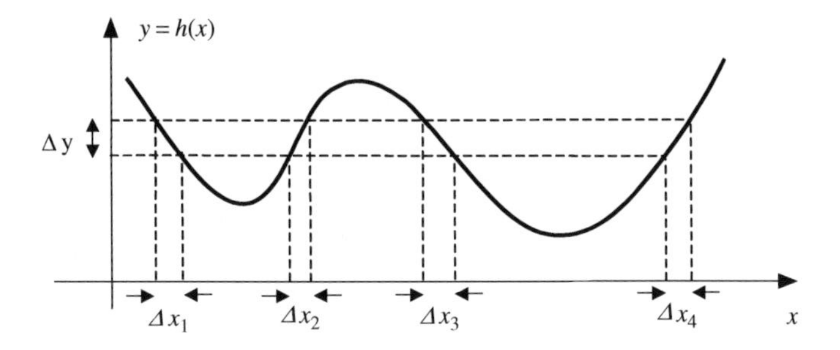

Figure 6.40 Illustration of a many-to-one transformation.

Equation (6.179) can be rearranged as

$$\frac{F_Y(y(m)+\Delta y)-F_Y(y(m))}{\Delta y}=\sum_{k=1}^{K}\frac{\Delta x_k}{\Delta y}\frac{F_X(x_k(m)+\Delta x_k)-F_X(x_k(m))}{\Delta x_k} \tag{6.180}$$

Now as the intervals Δx and Δy tend to zero Equation (6.180) becomes

$$\begin{aligned} f_Y(y(m)) &= \sum_{k=1}^{K}\left|\frac{\partial x_k(m)}{\partial y(m)}\right| f_X(x_k(m)) \\ &= \sum_{k=1}^{K}\frac{1}{|h'(x_k(m))|} f_X(x_k(m)) \end{aligned} \tag{6.181}$$

where $h'(x_k(m)) = \partial h(x_k(m))/\partial x_k(m)$. Note that Equation (6.179) is a generalised form of Equation (6.162); for a monotonic function, $K = 1$ and Equation (6.179) becomes the same as Equation (6.162). Equation (6.181) can be expressed as

$$f_Y(y(m))=\sum_{k=1}^{K}|J(x_k(m))|^{-1} f_X(x_k(m)) \tag{6.182}$$

where $J(x_k(m)) = h'(x_k(m))$ is called the *Jacobian* of the transformation.

For a multivariate transformation of a vector-valued process such as

$$\boldsymbol{y}(m)=\boldsymbol{H}(\boldsymbol{x}(m)) \tag{6.183}$$

the pdf of the output $\boldsymbol{y}(m)$ is given by

$$f_Y(\boldsymbol{y}(m))=\sum_{k=1}^{K}|\boldsymbol{J}(\boldsymbol{x}_k(m))|^{-1} f_X(\boldsymbol{x}_k(m)) \tag{6.184}$$

where $|\boldsymbol{J}(\boldsymbol{x})|$, the Jacobian of the transformation $\boldsymbol{H}(\cdot)$, is the determinant of a matrix of derivatives

$$|\boldsymbol{J}(\boldsymbol{x})| = \begin{vmatrix} \frac{\partial y_1}{\partial x_1} & \frac{\partial y_1}{\partial x_2} & \cdots & \frac{\partial y_1}{\partial x_P} \\ \vdots & \vdots & \ddots & \vdots \\ \frac{\partial y_P}{\partial x_1} & \frac{\partial y_P}{\partial x_2} & \cdots & \frac{\partial y_P}{\partial x_P} \end{vmatrix} \tag{6.185}$$

For a monotonic linear vector transformation such as

$$y = Hx \tag{6.186}$$

the pdf of y becomes

$$f_Y(y) = |J(x)|^{-1} f_X(H^{-1}y) = |H|^{-1} f_X(H^{-1}y) \tag{6.187}$$

where $|J(x)| = H$ is the Jacobian of the transformation and $|H| = |\det(H)|$ where $\det(\cdot)$ is the abbreviation of determinant of a matrix.

Example 6.25 The input–output relation of a $P \times P$ linear transformation matrix H is given by

$$y = H\,x \tag{6.188}$$

The Jacobian of the linear transformation H is $|H|$. Assume that the input x is a zero-mean Gaussian P-variate process with a covariance matrix of Σ_{xx} and a probability density function given by

$$f_X(x) = \frac{1}{(2\pi)^{P/2}|\Sigma_{xx}|^{1/2}} \exp\left[-\frac{1}{2}x^{\mathrm{T}}\Sigma_{xx}^{-1}x\right] \tag{6.189}$$

From Equations (6.162) and (6.188)–(6.189), the pdf of the output y is given by

$$\begin{aligned} f_Y(y) &= \frac{1}{(2\pi)^{P/2}|\Sigma_{xx}|^{1/2}} \exp\left(-\frac{1}{2}y^{\mathrm{T}}H^{-1^{\mathrm{T}}}\Sigma_{xx}^{-1}H^{-1}y\right)\|H\|^{-1} \\ &= \frac{1}{(2\pi)^{P/2}|\Sigma_{xx}|^{1/2}\|H\|} \exp\left(-\frac{1}{2}y^{\mathrm{T}}\Sigma_{yy}^{-1}y\right) \end{aligned} \tag{6.190}$$

where $\Sigma_{yy} = H\Sigma_{xx}H^{\mathrm{T}}$. Note that a linear transformation of a Gaussian process yields another scaled Gaussian process.

6.9 Search Engines: Citation Ranking

A search engine is a system designed to retrieve documents and files stored on computer systems such as on the World Wide Web, or a corporate computer network, or in a personal computer. The search engine allows the user to specify a set of keywords for searching a *list of items* (indexes) for contents in the database that match the keywords. This list of items is often sorted with respect to some measure of relevance of the results. Search engines use regularly updated indexes to operate efficiently.

Internet search engines sort and index the text information in many billions of web pages on the World Wide Web. A good set of search keywords will help to focus the search on the documents and websites that contain the input keywords. However, the problem remains that often the contents of many websites are not of the required quality and furthermore there are misleading websites containing popular keywords aimed to attract visitors to increase the hit rates and the advertising revenues of the sites.

For efficient information management the websites and their information content need to be rank ordered using an objective quality measure. A well-established objective measure of quality of published information on any medium is citation ranking, which for a long time has been used as a ranking method in academic research. A map of hyperlinks and pointers to websites on the World Wide Web allows rapid calculation of a web page's rank in terms of citation. The page rank measure based on citation is a good way to organise the order of presentation of the results of an Internet search.

Citation Ranking in Web Page Rank Calculation

Search engines usually find many web pages that contain the search keywords. The problem is how to present the web links containing the search keywords in a rank-ordered form, such that the rank of a page represents a measure of the quality of information on the page. The relevance of a web page containing the search text string, can be determined from the following analysis of the web page:

(a) Page title containing the search words is an indicator of the relevance of the topic of the page, but not of its quality.
(b) Number of times the search words are mentioned in the web page is also an indicator.
(c) Number of citations of the web page from other web pages is an objective indicator of the quality as perceived by web users.
(d) Each citation to a web page can be weighted by its importance which itself is a weighted citation of the citation.

The simplest way to rank a web page is to count the total number of citation links pointing to that page and then divide this by the total number of citations links on the web. This method would rank a web page using a simple probability measure defined as the frequency of citation links. However, as with the tradition of academic research, a citation itself needs to be weighted by the quality of the citation source i.e. by the citation ranking of the source itself. A weighted citation gives some approximation of a page's importance or quality, where each source of citation is weighted by its own citation ranking.

Let $PR(A)$ define the page rank for a web page A. Assume that page A has pages $T_1, \ldots, T_n$ pointing to it. Page rank of A can be defined as

$$PR(A) = (1-d) + d(PR(T_1)/C(T_1) + \ldots + PR(T_n)/C(T_n))$$

where $C(T)$ is defined as the number of links going out of page T. The parameter d is a damping factor which can be set between 0 and 1; usually d is set to 0.85. Note that the page ranks form a probability distribution over web pages, so the sum of all web pages' page ranks will be one. Page rank $PR(A)$ can be calculated using a simple iterative algorithm, and corresponds to the principal eigenvector of the normalised link matrix of the web.

6.10 Summary

The theories of stochastic processes and statistical modelling are central to the development of signal processing methods and communication systems. We began this chapter with the definitions of some important concepts such as probability, information, deterministic signals, random signals and random processes. Probabilistic models and statistical measures, originally developed for random variables, were introduced and extended to model random signals.

Probability models were used to quantify information in terms of the entropy of a random process. The use of entropy in modelling information and in variable length coding were explained and several examples were presented.

We considered the concepts of stationary, ergodic-stationary and non-stationary processes. The concept of a stationary process is central to the theory of linear time-invariant systems, and furthermore even non-stationary processes can be modelled by a chain of stationary sub-processes as described in Chapter 11 on hidden Markov models. For signal processing applications, a number of useful pdfs, including the Gaussian, the mixture Gaussian, the Markov and the Poisson process, were considered. These pdf models are extensively employed in the remainder of this book. Signal processing normally involves the filtering or transformation of an input signal to an output signal. We derived general expressions for the pdf of the output of a system in terms of the pdf of the input. We also considered some applications of stochastic processes for modelling random noise such as white noise, clutters, shot noise and impulsive noise.

Further Reading

Anderson O.D. (1976) Time Series Analysis and Forecasting. The Box–Jenkins Approach. Butterworth, London.
Ayre A.J. (1972) Probability and Evidence. Columbia University Press, New York.
Bartlett M.S. (1960) Stochastic Processes. Cambridge University Press.
Box G.E.P and Jenkins G.M. (1976) Time Series Analysis: Forecasting and Control. Holden-Day, San Francisco.
Breiphol A.M. (1970) Probabilistic System Analysis. Wiley, New York.
Carter G. (1987) Coherence and Time Delay Estimation. Proc. IEEE, 75, 2, pp. 236–255.
Clark A.B. and Disney R.L. (1985) Probability and Random Processes, 2nd edn. Wiley, New York.
Cooper G.R. and McGillem C.D. (1986) Probabilistic Methods of Signal and System Analysis Holt. Rinehart and Winston, New York.
Davenport W.B. and Root W.L. (1958) Introduction to Random Signals and Noise. McGraw-Hill, New York.
Davenport W.B. and Wilbur B. (1970) Probability and Random Processes: An Introduction for Applied Scientists and Engineers. McGraw-Hill, New York.
Einstein A. (1956) Investigation on the Theory of Brownian Motion. Dover, New York.
Gardener W.A. (1986) Introduction to Random Processes: With Application to Signals and Systems. Macmillan, New York.
Gauss C.F. (1963) Theory of Motion of Heavenly Bodies. Dover, New York.
Helstrom C.W. (1991) Probability and Stochastic Processes for Engineers. Macmillan, New York.
Isaacson D. and Masden R. (1976) Markov Chains Theory and Applications. Wiley, New York.
Jeffrey H. (1961) Scientific Inference, 3rd edn. Cambridge University Press.
Jeffrey H. (1973) Theory of Probability, 3rd edn. Clarendon Press, Oxford.
Kay S.M. (1993) Fundamentals of Statistical Signal Processing. Estimation Theory. Prentice-Hall, Englewood Cliffs, NJ.
Kendall M. and Stuart A. (1977) The Advanced Theory of Statistics. Macmillan.
Kolmogorov A.N. (1956) Foundations of the Theory of Probability. Chelsea Publishing Company, New York.
Leon-Garcia A. (1994) Probability and Random Processes for Electrical. Engineering. Addison Wesley, Reading, MA.
Markov A.A. (1913) An Example of Statistical Investigation in the Text of *Eugen Onyegin* Illustrating Coupling of Tests in Chains. Proc. Acad. Sci. St Petersburg VI Ser., 7, pp. 153–166.
Meyer P.L. (1970) Introductory Probability and Statistical Applications. Addison-Wesley, Reading, MA.
Papopulis A. (1977) Signal Analysis, McGraw-Hill, New York.
Papopulis A. (1984) Probability, Random Variables and Stochastic Processes. McGraw-Hill, New York.
Parzen E. (1962) Stochastic Processes. Holden-Day, San Francisco.
Peebles P.Z. (1987) Probability, Random Variables and Random Signal Principles. McGraw-Hill, New York.
Rao C.R. (1973) Linear Statistical Inference and Its Applications. Wiley, New York.
Rozanov Y.A. (1969) Probability Theory: A Concise Course. Dover, New York.
Shanmugan K.S. and Breipohl A.M. (1988) Random Signals: Detection, Estimation and Data Analysis. Wiley, New York.
Thomas J.B. (1988) An Introduction to Applied Probability and Random Processes. Huntington, Krieger Publishing, New York.
Wozencraft J.M. and Jacobs I.M. (1965) Principles of Communication Engineering. Wiley, New York.

7 Bayesian Inference

Bayesian inference provides the general framework for formulation of statistical estimation and classification problems. Inference is the process of deriving conclusions from evidence. In statistical estimation, the value of a signal or a parameter is inferred from observations that may be incomplete and noisy. The Bayesian philosophy, for the prediction or estimation of a random process from a related signal, is based on combining the *evidence*, contained in the signal and its likelihood, with the *prior* knowledge of the probability distribution of the process and a cost of error function.

It can be argued that Bayesian inference resembles the human reasoning process since people often weight the evidence of an observation with their prior beliefs or dispositions. Indeed, the form of the prior functions used in a Bayesian estimation can be an objective description of the distribution of the variables or it can be a subjective description based on a person's belief, experience or intuitive feeling.

Bayesian inference is based on minimisation of the Bayes' risk function, which includes a probability model of the unknown parameters conditional on the given observation and a cost-of-error function. Bayesian methodology includes the classical estimators such as maximum a posteriori (MAP), maximum-likelihood (ML), minimum mean square error (MMSE) and minimum mean absolute value of error (MAVE) as its special cases, Figure 7.1. The hidden Markov model, widely used in pattern recognition, is an example of a Bayesian model.

This chapter begins with an introduction to the basic concepts of the estimation theory, and considers the statistical measures that are used to quantify the performance of an estimator. We study Bayesian estimation methods and consider the effect of using a prior model on the mean and the variance of an estimate. The estimate–maximise (EM) method for the estimation of a set of unknown parameters from an incomplete observation is studied, and applied to estimation of the mixture Gaussian model of the space of a continuous random variable. This chapter concludes with Bayesian classification of discrete or finite-state signals, and the K-means clustering method.

7.1 Bayesian Estimation Theory: Basic Definitions

Estimation theory is concerned with the determination of the *best* estimate of an unknown parameter vector from a related observation signal, or the recovery of a clean signal degraded by noise and distortion. For example, given a noisy sine wave, we may be interested in estimating its basic parameters

Multimedia Signal Processing: Theory and Applications in Speech, Music and Communications Saeed V. Vaseghi

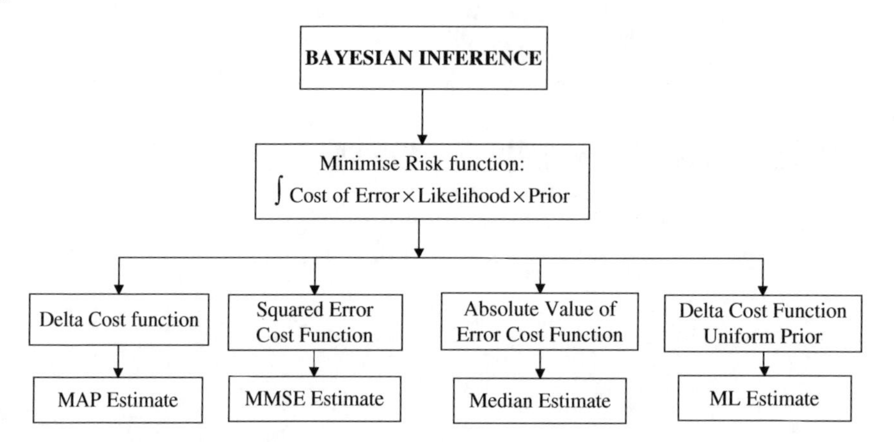

Figure 7.1 Bayesian inference involves a cost function, a prior function and a likelihood function. As illustrated, other probabilistic estimation methods can be considered as special cases of Bayesian estimation.

(i.e. amplitude, frequency or phase) as in radar signal processing applications, or we may wish to recover the signal itself.

Note that the *best* estimate is the one that minimises a risk function or a cost of error function and that often involves the use of probability of the unknown vector given the known observation as explained in this chapter.

An estimator takes as input a set of noisy or incomplete observations, and, using a dynamic model (e.g. a linear predictive model) and/or a probabilistic model (e.g. Gaussian model) of the process, estimates the unknown parameters. The estimation accuracy depends on the available information, the accuracy of the models and the efficiency of the estimator. In this chapter, the Bayesian estimation of stationary parameters is studied. The modelling and estimation of non-stationary finite-state processes is covered in Chapter 11 on hidden Markov models (HMMs).

Bayesian theory is a general inference framework for the derivation of statistical estimation methods as illustrated in Figure 7.1. The various forms of statistical estimators can be derived as special cases of Bayesian estimation.

7.1.1 Bayes' Rule

Consider the estimation of the value of a random parameter vector $\boldsymbol{\theta}$, given a related observation vector $\boldsymbol{y}$. From .i.Bayes' rule (introduce in Chapter 6) the posterior probability density function (pdf) of the parameter vector $\boldsymbol{\theta}$ given $\boldsymbol{y}$, $f_{\Theta|Y}(\boldsymbol{\theta}|\boldsymbol{y})$, can be expressed as

$$f_{\Theta|Y}(\boldsymbol{\theta}|\boldsymbol{y}) = \frac{1}{f_Y(\boldsymbol{y})} \underbrace{f_{Y|\Theta}(\boldsymbol{y}|\boldsymbol{\theta})}_{\text{Likelihood}} \underbrace{f_{\Theta}(\boldsymbol{\theta})}_{\text{Prior}} \tag{7.1}$$

where for a given observation, $f_Y(\boldsymbol{y})$ is a constant and has only a normalising effect. Thus there are two variable terms in Equation (7.1): one term $f_{Y|\Theta}(\boldsymbol{y}|\boldsymbol{\theta})$ is the *likelihood* that the observation signal $\boldsymbol{y}$ was generated by the parameter vector $\boldsymbol{\theta}$ and the second term is the *prior* probability of the parameter vector having a value of $\boldsymbol{\theta}$. Hence

Posterior Probability $\propto$ Likelihood $\times$ Prior

The relative influence of the likelihood pdf $f_{Y|\Theta}(y|\boldsymbol{\theta})$ and the prior pdf $f_{\Theta}(\boldsymbol{\theta})$ on the posterior pdf $f_{\Theta|Y}(\boldsymbol{\theta}|y)$ depends on the shape of these functions, i.e. on how relatively peaked each pdf is. In general the more peaked a probability density function, the more it will influence the outcome of the estimation process. Conversely, a uniform prior pdf will have no influence. The roles of the posterior, prior and likelihood functions are explained in more detail in Section 7.1.5.

Elements of Bayesian Inference

The Bayesian inference can be derived from the minimisation of a risk function defined as

$$\text{Risk}(\hat{\boldsymbol{\theta}}) = \int_{\boldsymbol{\theta}}\int_{Y} \text{Cost}(\hat{\boldsymbol{\theta}}, \boldsymbol{\theta}) \underbrace{f_{Y|\Theta}(y|\boldsymbol{\theta})}_{\text{Likelihood}} \underbrace{f_{\Theta}(\boldsymbol{\theta})}_{\text{Prior}} \, dy \, d\boldsymbol{\theta} \tag{7.2}$$

The elements of Bayesian inference are as follows:

Bayesian Risk: The risk in making an estimate $\hat{\boldsymbol{\theta}}$ of the unknown value of the parameter, is related to the probability of error and the cost of error. Since the true value of the parameters $\boldsymbol{\theta}$ is unknown, the cost needs to be averaged over all possible values of $\boldsymbol{\theta}$ as shown in Equation (7.2).

Cost function: The cost of error determines the type of Bayesian solution. A popular cost function is the minimum mean squared error cost. The cost function can be chosen to give very high costs to unacceptable errors with catastrophic consequences.

Likelihood: The likelihood, $f_{Y|\Theta}(y|\boldsymbol{\theta})$, is the conditional probability of the observation $\boldsymbol{y}$ given the parameter $\boldsymbol{\theta}$. In other words, it gives the likelihood that the observation signal $\boldsymbol{y}$ is due to parameter $\boldsymbol{\theta}$. The likelihood does not take into account the probability of $\boldsymbol{\theta}$.

Prior probability: The prior gives the probability of $\boldsymbol{\theta}$. The prior acts as a 'moderating' influence on the likelihood function. It will be shown later that the extent to which the likelihood and the prior functions influence the estimate depends on such factors as the signal-to-noise ratio, the length of the observation and the shape of the prior.

Posterior probability: The posterior probability, $f_{Y|\Theta}(\boldsymbol{\theta}|y) = f_{Y|\Theta}(y|\boldsymbol{\theta})f_{\Theta}(\boldsymbol{\theta})/f_Y(y)$, is proportional to the product of the likelihood and the prior as expressed by Bayes' rule in Equation (7.1).

The remainder of this chapter is concerned with different forms of Bayesian estimation and its applications. First, in this section, some basic concepts of estimation theory are introduced.

7.1.2 Dynamic and Probability Models in Estimation

Optimal estimation algorithms, such as Kalman filters, utilise both *dynamic* and *probabilistic* models of the observation signals.

A dynamic predictive model captures the correlation structure of a signal, and models the dependence of the present and future values of the signal on its past trajectory and the input stimulus. Examples of estimation methods employing dynamic models are sinusoidal models and linear prediction models.

A statistical probability model characterises the *space* of different realisations of a random signal in terms of its statistics, such as the mean and the covariance, and most completely in terms of a probability model. Conditional probability models, in addition to modelling the random fluctuations of a signal, can also model the dependence of the signal on its past values or on the values of some other related parameters or process.

Dynamic and probability models can be combined; for example a finite-state model may be constructed from a combination of hidden Markov models (HMMs, introduced in Chapter 11) of the

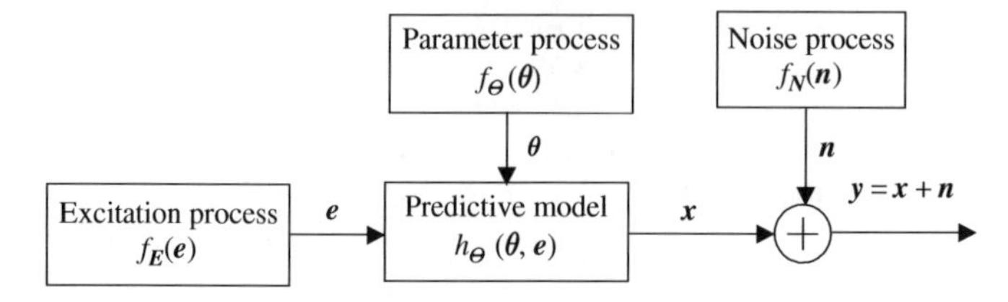

Figure 7.2 A random process $\boldsymbol{y}$ is described in terms of a predictive model $h(\cdot)$, and statistical models $f_E(\cdot), f_\Theta(\cdot)$ and $f_N(\cdot)$.

probability distribution of the signal and sinusoidal or linear prediction models of the dynamics of the signal. A further example is a Kalman filter, Chapter 9, that utilises state-space equations for the dynamics of the signal and noise and assumes that the random processes have Gaussian distributions.

As an illustration consider the estimation of a P-dimensional parameter vector $\boldsymbol{\theta} = [\theta_0, \theta_1, \ldots, \theta_{P-1}]$ from a noisy observation vector $\boldsymbol{y} = [y(0), y(1), \ldots, y(N-1)]$ modelled as

$$\boldsymbol{y} = \boldsymbol{x} + \boldsymbol{n} = h(\boldsymbol{\theta}, \boldsymbol{e}) + \boldsymbol{n} \tag{7.3}$$

where, as illustrated in Figure 7.2, it is assumed that the clean signal $\boldsymbol{x}$ is the output of a predictive model $h(\cdot)$ with a random input $\boldsymbol{e}$ and parameter vector $\boldsymbol{\theta}$ and $\boldsymbol{n}$ is an additive random noise process. In Figure 7.2, the distributions of the random input $\boldsymbol{e}$, the parameter vector $\boldsymbol{\theta}$ and the random noise $\boldsymbol{n}$ are modelled by probability density functions $f_E(\boldsymbol{e}), f_\Theta(\boldsymbol{\theta})$ and $f_N(\boldsymbol{n})$ respectively. The pdf model most often used is the Gaussian model.

Predictive and statistical models of a process *guide* the estimator towards the set of values of the unknown parameters that are most consistent with both the prior distribution of the model parameters and the observation. In general, the more modelling information employed in an estimation process, the better the results, provided that the models are an accurate characterisation of the observation and the parameter process. The drawback is that if the models are not accurate then more harm than good may result from using them.

7.1.3 Parameter Space and Signal Space

Consider a random process with a parameter vector $\boldsymbol{\theta}$. For example, each instance of $\boldsymbol{\theta}$ could be the parameter vector for a dynamic model, such as a harmonic model, of a speech sound or a musical note. The parameter space of a process $\boldsymbol{\Theta}$ is the collection of all the values that the parameter vector $\boldsymbol{\theta}$ can assume.

The parameters of a random process determine the 'characteristics' (i.e. the mean, the variance, the power spectrum, etc.) of the signals generated by the process. As the process parameters change, so do the characteristics of the signals generated by the process. Each value of the parameter vector $\boldsymbol{\theta}$ of a process has an associated signal space $\boldsymbol{Y}$; this is the collection of all the signal realisations of the process with the parameter value $\boldsymbol{\theta}$.

For example, consider a three-dimensional vector-valued Gaussian process with parameter vector $\boldsymbol{\theta} = [\boldsymbol{\mu}, \boldsymbol{\Sigma}]$, where $\boldsymbol{\mu}$ is the mean vector and $\boldsymbol{\Sigma}$ is the covariance matrix of the Gaussian process. Figure 7.3 illustrates three mean vectors in a three-dimensional parameter space. Also shown is the signal space associated with each parameter. As shown, the signal space of each parameter vector of a Gaussian process contains an infinite number of points, centred on the mean vector $\boldsymbol{\mu}$, and with a spatial volume and orientation that is determined by the covariance matrix $\boldsymbol{\Sigma}$. For simplicity, the variances are not shown in the parameter space, although they are evident in the shape of the Gaussian signal clusters in the signal space.

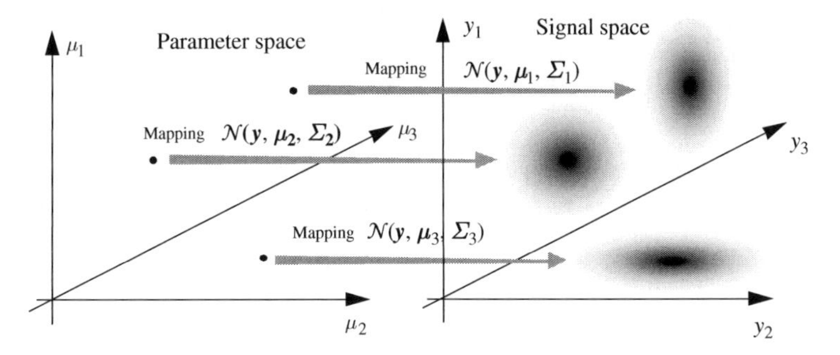

Figure 7.3 Illustration of three points in the parameter space of a Gaussian process and the associated signal spaces; for simplicity the variances are not shown in parameter space.

Parameter Estimation and Signal Restoration

Parameter estimation and signal restoration are closely related problems. The main difference is due to the rapid fluctuations of most signals in comparison with the relatively slow variations of most parameters. For example, speech sounds fluctuate at speeds of up to 20 kHz, whereas the underlying vocal tract and pitch parameters vary at a relatively lower rate of less than 100 Hz. This observation implies that normally more averaging can be done in parameter estimation than in signal restoration.

As a simple example, consider a signal observed in a zero-mean random noise process. Assume we wish to estimate (a) the average of the clean signal and (b) the clean signal itself. As the observation length increases, the estimate of the signal mean approaches the mean value of the clean signal, whereas the estimate of the clean signal samples depends on the correlation structure of the signal and the signal-to-noise ratio as well as on the estimation method used.

As a further example, consider the interpolation of a sequence of lost samples of a signal given N recorded samples, as illustrated in Figure 7.4. Assume that an autoregressive (AR) process is used to model the signal as

$$y = X\theta + e + n \tag{7.4}$$

where y is the observation signal, X is the signal matrix, θ is the AR parameter vector, e is the random input of the AR model and n is the random noise. Using Equation (7.3), the signal restoration process involves the estimation of both the model parameter vector θ and the random input e for the lost samples. Assuming the parameter vector θ is time-invariant, the estimate of θ can be averaged over the

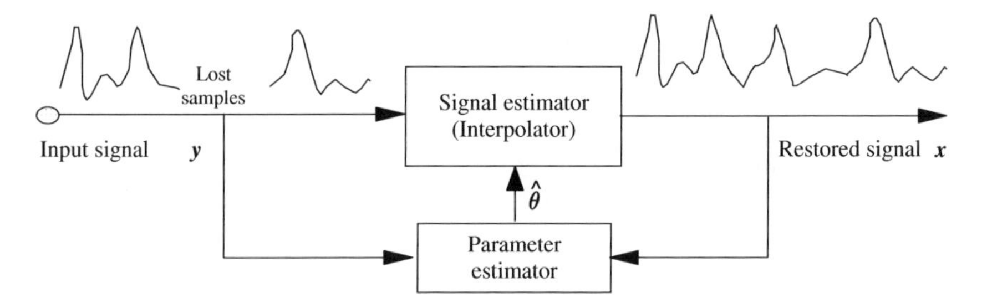

Figure 7.4 Illustration of signal restoration using a parametric model of the signal.

entire N observation samples, and as N becomes infinitely large, a consistent estimate should approach the true parameter value. The difficulty in signal interpolation is that the underlying excitation $\boldsymbol{e}$ of the signal $\boldsymbol{x}$ is purely random and, unlike $\boldsymbol{\theta}$, it cannot be estimated through an averaging operation.

7.1.4 Performance Measures and Desirable Properties of Estimators

In estimation of a parameter vector $\boldsymbol{\theta}$ from N observation samples $\boldsymbol{y}$, a set of performance measures is used to quantify and compare the characteristics of different estimators. In general an estimate of a parameter vector is a function of the observation vector $\boldsymbol{y}$, the length of the observation N and the process model $\mathcal{M}$. This dependence may be expressed as

$$\hat{\boldsymbol{\theta}} = f(\boldsymbol{y}, N, \mathcal{M}) \tag{7.5}$$

Different parameter estimators produce different results depending on the estimation method and utilisation of the observation and the influence of the prior information. Due to randomness of the observations, even the same estimator would produce different results with different observations from the same process. Therefore an estimate is itself a random variable, it has a mean and a variance, and it may be described by a probability density function. However, for most cases, it is sufficient to characterise an estimator in terms of the mean and the variance of the estimation error. The most commonly used performance measures for an estimator are the following:

(a) *Expected value* of estimate: $\mathcal{E}[\hat{\boldsymbol{\theta}}]$
(b) *Bias* of estimate: $\mathcal{E}[\hat{\boldsymbol{\theta}} - \boldsymbol{\theta}] = \mathcal{E}[\hat{\boldsymbol{\theta}}] - \boldsymbol{\theta}$
(c) *Covariance* of estimate: $\text{Cov}[\hat{\boldsymbol{\theta}}] = \mathcal{E}[(\hat{\boldsymbol{\theta}} - \mathcal{E}[\hat{\boldsymbol{\theta}}])(\hat{\boldsymbol{\theta}} - \mathcal{E}[\hat{\boldsymbol{\theta}}])^{\mathrm{T}}]$

Optimal estimators aim for zero bias and minimum estimation error covariance. The desirable properties of an estimator can be listed as follows:

(a) *Unbiased estimator*: an estimator of $\boldsymbol{\theta}$ is unbiased if the expectation of the estimate is equal to the true parameter value:

$$\mathcal{E}[\hat{\boldsymbol{\theta}}] = \boldsymbol{\theta} \tag{7.6}$$

An estimator is *asymptotically unbiased* if for increasing length of observations N we have

$$\lim_{N\to\infty} \mathcal{E}[\hat{\boldsymbol{\theta}}] = \boldsymbol{\theta} \tag{7.7}$$

(b) *Efficient estimator*: an unbiased estimator of $\boldsymbol{\theta}$ is an efficient estimator if it has the smallest covariance matrix compared with all other unbiased estimates of $\boldsymbol{\theta}$:

$$\text{Cov}[\hat{\boldsymbol{\theta}}_{\text{Efficient}}] \le \text{Cov}[\hat{\boldsymbol{\theta}}] \tag{7.8}$$

where $\hat{\boldsymbol{\theta}}$ is any other estimate of $\boldsymbol{\theta}$.

(c) *Consistent estimator*: an estimator is consistent if the estimate improves with the increasing length of the observation N, such that the estimate $\hat{\theta}$ converges probabilistically to the true value θ as N becomes infinitely large:

$$\lim_{N\to\infty} P[|\hat{\boldsymbol{\theta}} - \boldsymbol{\theta}| > \varepsilon] = 0 \tag{7.9}$$

where ε is arbitrary small.

Example 7.1 *Estimation of the mean and variance of a signal*
Consider the bias in the time-averaged estimates of the mean μ_y and the variance of σ_y^2 of N observation samples $[y(0), \ldots, y(N-1)]$, of an ergodic random process, given as

$$\hat{\mu}_y = \frac{1}{N}\sum_{m=0}^{N-1} y(m) \tag{7.10}$$

$$\hat{\sigma}_y^2 = \frac{1}{N}\sum_{m=0}^{N-1} \left(y(m) - \hat{\mu}_y\right)^2 \tag{7.11}$$

It is easy to show that $\hat{\mu}_y$ is an unbiased estimate, since

$$\mathcal{E}\left[\hat{\mu}_y\right] = \frac{1}{N}\sum_{m=0}^{N-1} \mathcal{E}[y(m)] = \mu_y \tag{7.12}$$

The expectation of the estimate of the variance can be expressed as

$$\begin{aligned}\mathcal{E}\left[\hat{\sigma}_y^2\right] &= \mathcal{E}\left[\frac{1}{N}\sum_{m=0}^{N-1}\left(y(m) - \frac{1}{N}\sum_{k=0}^{N-1} y(k)\right)^2\right] \\ &= \sigma_y^2 - \frac{2}{N}\sigma_y^2 + \frac{1}{N}\sigma_y^2 \\ &= \sigma_y^2 - \frac{1}{N}\sigma_y^2\end{aligned} \tag{7.13}$$

From Equation (7.13), the bias in the estimate of the variance is inversely proportional to the signal length N, and vanishes as N tends to infinity; hence the estimate is asymptotically unbiased. In general, the bias and the variance of an estimate decrease with increasing number of observation samples N and with improved modelling. Figure 7.5 illustrates the general dependence of the distribution and the bias and the variance of an asymptotically unbiased estimator on the number of observation samples N.

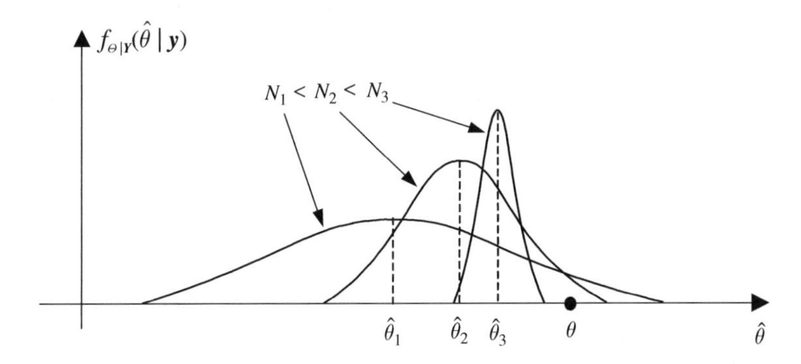

Figure 7.5 Illustration of the decrease in the bias and variance of an asymptotically unbiased estimate of the parameter θ with increasing length of observation.

7.1.5 Prior and Posterior Spaces and Distributions

The Bayesian inference method weights the likelihood that a value of parameter $\boldsymbol{\theta}$ underlies an observation $\boldsymbol{y}$, $f(\boldsymbol{y}|\boldsymbol{\theta})$, with the prior probability of the value of the parameters $f(\boldsymbol{\theta})$. We may say that there is some similarity between Bayesian inference and human cognitive inference in that we often weight evidence with prior experience or disposition.

The *prior space* of a signal or a parameter vector is the collection of all possible values that the signal or the parameter vector can assume. Within the prior space, all the values of a parameter vector may have the same probability, in which case the prior space would have a uniform probability distribution, or the probability of the parameter vectors may have a non-uniform distribution. If the prior distribution is non-uniform (e.g. Gaussian, Gamma etc.) then the non-uniform prior can be used to weight the inference drawn from the observation; this would give more weight to the values that have a higher prior probability of occurrence.

The ***evidence*** of the value of a signal $\boldsymbol{x}$, or a parameter vector $\boldsymbol{\theta}$, is contained in the observation signal $\boldsymbol{y}$ which is used in a likelihood or cost function from which the signal or the parameter vector is estimated. For example a noisy speech signal $\boldsymbol{y}$ may be used to obtain an estimate of the clean speech $\boldsymbol{x}$ and/or the parameter vector $\boldsymbol{\theta}$ of a linear prediction of model of speech.

The *posterior signal* or *parameter space* is the subspace of all the likely values of a signal $\boldsymbol{x}$, or a parameter vector $\boldsymbol{\theta}$, that are consistent with *both* the *prior* information on signal $\boldsymbol{x}$ (or parameter $\boldsymbol{\theta}$) and the *evidence* contained in the observation $\boldsymbol{y}$. The significance of posterior probability is that the likelihood $f_{Y|\Theta}(y|\boldsymbol{\theta})$ of each value of a parameter $\boldsymbol{\theta}$ is weighted with the prior probability of the value $f_{\Theta}(\boldsymbol{\theta})$.

For example, the likelihood that a variable, such as the state of tomorrow's weather $\boldsymbol{\theta}$, takes a particular value given some meteorological observations $\boldsymbol{y}$, (i.e. probability of $\boldsymbol{\theta}$ given $\boldsymbol{y}$, $f_{\Theta|Y}(\boldsymbol{\theta}|y)$), can be obtained using the likelihood that meteorological observation $\boldsymbol{y}$ can be seen in the weather state $\boldsymbol{\theta}$ (i.e. $\boldsymbol{y}$ given $\boldsymbol{\theta}$, $f_{Y|\Theta}(y|\boldsymbol{\theta})$), weighted with the prior likelihood of the weather state (i.e. $f_{\Theta}(\boldsymbol{\theta})$) (irrespective of the observation) which itself could be obtained from previous years' weather data and would be also conditional on the time of the year.

Consider a random process with a parameter space $\boldsymbol{\Theta}$, observation space $\boldsymbol{Y}$ and a joint pdf $f_{Y,\Theta}(y, \boldsymbol{\theta})$. From.i.Bayes' rule the posterior pdf of the parameter vector $\boldsymbol{\theta}$, given an observation vector $\boldsymbol{y}$, $f_{\Theta|Y}(\boldsymbol{\theta}|y)$, can be expressed as

$$f_{\Theta|Y}(\boldsymbol{\theta}|y) = \frac{f_{Y|\Theta}(y|\boldsymbol{\theta})\, f_{\Theta}(\boldsymbol{\theta})}{f_Y(y)} \tag{7.14}$$

where, for a given observation vector $\boldsymbol{y}$, the pdf $f_Y(y)$ is a constant and has only a normalising effect. From Equation (7.14), the posterior pdf is proportional to the weighted likelihood; that is the product of the likelihood $f_{Y|\Theta}(y|\boldsymbol{\theta})$ that the observation $\boldsymbol{y}$ was generated by the parameter vector $f_{\Theta}(\boldsymbol{\theta})$, and the prior pdf $f_{\theta}(\boldsymbol{\theta})$. The prior pdf gives the unconditional parameter distribution *averaged* over the entire observation space as

$$f_{\Theta}(\boldsymbol{\theta}) = \int_{Y} f_{Y,\Theta}(y, \boldsymbol{\theta})dy \tag{7.15}$$

For most applications, it is relatively convenient to obtain the likelihood function $f_{Y|\Theta}(y|\boldsymbol{\theta})$. The *prior* pdf *influences* the inference drawn from the likelihood function by weighting it with $f_{\Theta}(\boldsymbol{\theta})$. The influence of the prior is particularly important for short-length and/or noisy observations, where the confidence in the estimate is limited by the lack of a sufficiently long observation and by the noise. The influence of the prior on the bias and the variance of an estimate are considered in Section 7.7.1.

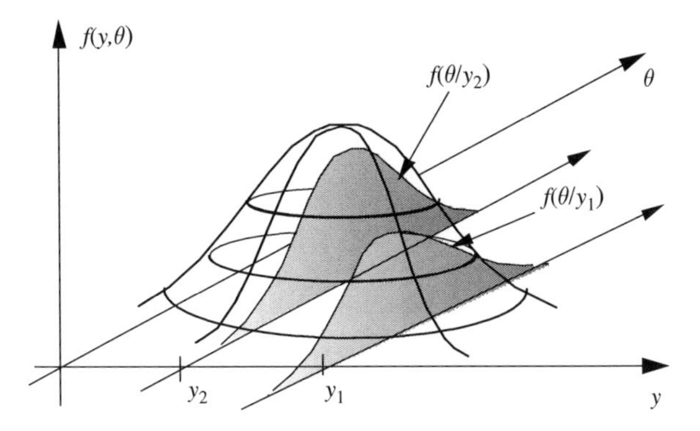

Figure 7.6 Illustration of joint distribution of a signal y and a parameter θ and the posterior distribution of θ given y.

A prior knowledge of the signal distribution can be used to confine the estimate to the prior signal space. The observation then *guides* the estimator to focus on the posterior space: that is the subspace consistent with both the prior belief and the evidence contained in the observation. Figure 7.6 illustrates the joint pdf of a scalar signal $y(m)$ and a scalar parameter θ. As shown, an observation y cuts a posterior pdf $f_{\Theta|Y}(\boldsymbol{\theta}|y)$ through the joint distribution.

Example 7.2 A noisy signal vector of length N samples is modelled as

$$\boldsymbol{y}(m) = \boldsymbol{x}(m) + \boldsymbol{n}(m) \tag{7.16}$$

Assume that the signal $\boldsymbol{x}(m)$ is Gaussian with mean vector $\boldsymbol{\mu}_x$ and covariance matrix $\boldsymbol{\Sigma}_{xx}$, and that the noise $\boldsymbol{n}(m)$ is also Gaussian with mean vector $\boldsymbol{\mu}_n$ and covariance matrix $\boldsymbol{\Sigma}_{nn}$. The signal and noise pdfs model the prior spaces of the signal and the noise respectively. Given an observation vector $\boldsymbol{y}(m)$, the underlying signal $\boldsymbol{x}(m)$ would have a likelihood distribution with a mean vector of $\boldsymbol{y}(m) - \boldsymbol{\mu}_n$ and a covariance matrix $\boldsymbol{\Sigma}_{nn}$ as shown in Figure 7.7. The likelihood function is given by

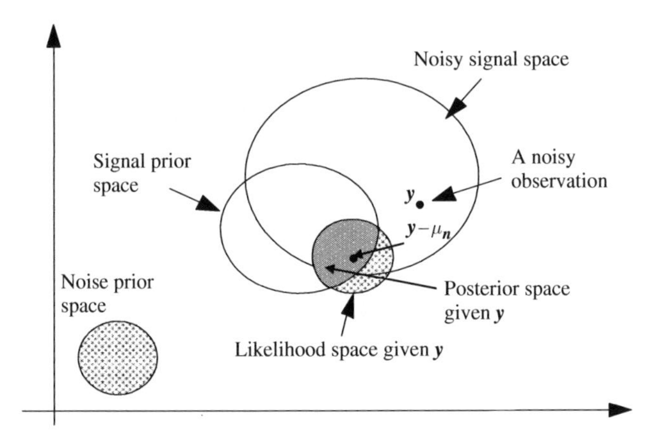

Figure 7.7 Sketch of two-dimensional signal and noise prior spaces and the likelihood and posterior spaces of a noisy observation **y**.

$$f_{Y|X}(\boldsymbol{y}(m)|\boldsymbol{x}(m)) = f_N(\boldsymbol{y}(m)-\boldsymbol{x}(m))$$
$$= \frac{1}{(2\pi)^{N/2}|\boldsymbol{\Sigma}_{nn}|^{1/2}} \exp\left\{-\frac{1}{2}[(\boldsymbol{y}(m)-\boldsymbol{\mu}_n)-\boldsymbol{x}(m)]^{\mathrm{T}}\boldsymbol{\Sigma}_{nn}^{-1}[(\boldsymbol{y}(m)-\boldsymbol{\mu}_n)-\boldsymbol{x}(m)]\right\} \tag{7.17}$$

where the terms in the exponential function have been rearranged to emphasize the illustration of the likelihood space in Figure 7.7. Hence the posterior pdf can be expressed as

$$\underbrace{f_{X|Y}(\boldsymbol{x}(m)|\boldsymbol{y}(m))}_{\text{Posterior}} = \frac{1}{f_Y(\boldsymbol{y}(m))}\underbrace{f_{Y|X}(\boldsymbol{y}(m)|\boldsymbol{x}(m))}_{\text{Likelihood}}\underbrace{f_X(\boldsymbol{x}(m))}_{\text{Prior}}$$
$$= \frac{1}{f_Y(\boldsymbol{y}(m))}\frac{1}{(2\pi)^N|\boldsymbol{\Sigma}_{nn}|^{1/2}|\boldsymbol{\Sigma}_{xx}|^{1/2}}$$
$$\times\exp\left(-\frac{1}{2}\left\{\underbrace{[(\boldsymbol{y}(m)-\boldsymbol{\mu}_n)-\boldsymbol{x}(m)]^{\mathrm{T}}\boldsymbol{\Sigma}_{nn}^{-1}[(\boldsymbol{y}(m)-\boldsymbol{\mu}_n)-\boldsymbol{x}(m)]}_{\text{Likelihood}}+\underbrace{(\boldsymbol{x}(m)-\boldsymbol{\mu}_x)^{\mathrm{T}}\boldsymbol{\Sigma}_{xx}^{-1}(\boldsymbol{x}(m)-\boldsymbol{\mu}_x)}_{\text{Prior}}\right\}\right) \tag{7.18}$$

For a two-dimensional signal and noise process, the prior spaces of the signal, the noise, and the noisy signal are illustrated in Figure 7.7. Note in this figure that the mean (i.e. centre) and the variance (related to the size of the ellipse) of the noisy signal space are the sum of the means and variances of the signal and noise spaces respectively. The likelihood and the posterior spaces for a noisy observation vector $\boldsymbol{y}$ are also illustrated in Figure 7.7. The likelihood space, i.e. the space of the likely values of the signal, is centred at $\boldsymbol{y}-\boldsymbol{\mu}_n$ and has a size proportional to the variance of the noise $\boldsymbol{\Sigma}$. The clean signal is then somewhere within the posterior subspace determined by the intersection of the likelihood and the prior space of the signal.

7.2 Bayesian Estimation

The Bayesian estimation of a parameter vector $\boldsymbol{\theta}$, from a related observation vector $\boldsymbol{y}$, is based on minimisation of a Bayesian risk function defined as an average cost-of-error function:

$$\begin{aligned}\mathcal{R}(\hat{\boldsymbol{\theta}}) &= \mathcal{E}[C(\hat{\boldsymbol{\theta}},\boldsymbol{\theta})]\\ &= \int_{\boldsymbol{\theta}}\int_{Y} C(\hat{\boldsymbol{\theta}},\boldsymbol{\theta})f_{Y,\Theta}(\boldsymbol{y},\boldsymbol{\theta})d\boldsymbol{y}d\boldsymbol{\theta}\\ &= \int_{\boldsymbol{\theta}}\int_{Y} C(\hat{\boldsymbol{\theta}},\boldsymbol{\theta})f_{Y|\Theta}(\boldsymbol{y}|\boldsymbol{\theta})f_{\Theta}(\boldsymbol{\theta})d\boldsymbol{y}d\boldsymbol{\theta}\end{aligned} \tag{7.19}$$

where the cost-of-error function $C(\hat{\boldsymbol{\theta}},\boldsymbol{\theta})$ allows the appropriate weighting of the various outcomes to achieve desirable objective or subjective properties of the estimator. The cost function can be chosen to associate higher costs with those outcomes that are undesirable or disastrous, such as a high cost for a false diagnosis in a medical test. In Equation (7.19) the Bayes' risk function is averaged over the space of all values of the parameter $\boldsymbol{\theta}$ and the observation $\boldsymbol{y}$.

For a given $\boldsymbol{y}$, $f_Y(\boldsymbol{y})$ is a constant and has no effect on the risk-minimisation process. Hence substituting $f_{Y,\Theta}(\boldsymbol{y},\boldsymbol{\theta}) = f_{\Theta|Y}(\boldsymbol{\theta}|\boldsymbol{y})f_Y(\boldsymbol{y})$ in the (middle) line of Equation (7.19) a conditional risk function is obtained as

$$\mathcal{R}(\hat{\boldsymbol{\theta}}|\boldsymbol{y}) = \int_{\boldsymbol{\theta}} C(\hat{\boldsymbol{\theta}},\boldsymbol{\theta})f_{\Theta|Y}(\boldsymbol{\theta}|\boldsymbol{y})d\boldsymbol{\theta} \tag{7.20}$$

The Bayesian estimate $\hat{\boldsymbol{\theta}}$ is obtained as the minimum-risk parameter vector given by

$$\hat{\boldsymbol{\theta}}_{\text{Bayesian}} = \arg\min_{\hat{\boldsymbol{\theta}}} \mathcal{R}(\hat{\boldsymbol{\theta}}|\boldsymbol{y}) = \arg\min_{\hat{\boldsymbol{\theta}}} \left[\int_{\boldsymbol{\theta}} C(\hat{\boldsymbol{\theta}}, \boldsymbol{\theta}) f_{\boldsymbol{\Theta}|Y}(\boldsymbol{\theta}|\boldsymbol{y}) d\boldsymbol{\theta} \right] \tag{7.21}$$

Using Bayes' rule, Equation (7.21) can be written as

$$\hat{\boldsymbol{\theta}}_{\text{Bayesian}} = \arg\min_{\hat{\boldsymbol{\theta}}} \left[\int_{\boldsymbol{\theta}} C(\hat{\boldsymbol{\theta}}, \boldsymbol{\theta}) f_{Y|\boldsymbol{\Theta}}(\boldsymbol{y}|\boldsymbol{\theta}) f_{\boldsymbol{\Theta}}(\boldsymbol{\theta}) d\boldsymbol{\theta} \right] \tag{7.22}$$

Assuming that the risk function is differentiable, and has a well-defined minimum, the Bayesian estimate can be obtained as

$$\hat{\boldsymbol{\theta}}_{\text{Bayesian}} = \arg \operatorname*{zero}_{\hat{\boldsymbol{\theta}}} \frac{\partial \mathcal{R}(\hat{\boldsymbol{\theta}}|\boldsymbol{y})}{\partial \hat{\boldsymbol{\theta}}} = \arg \operatorname*{zero}_{\hat{\boldsymbol{\theta}}} \left[\frac{\partial}{\partial \hat{\boldsymbol{\theta}}} \int_{\boldsymbol{\theta}} C(\hat{\boldsymbol{\theta}}, \boldsymbol{\theta}) f_{Y|\boldsymbol{\Theta}}(\boldsymbol{y}|\boldsymbol{\theta}) f_{\boldsymbol{\Theta}}(\boldsymbol{\theta}) d\boldsymbol{\theta} \right) \tag{7.23}$$

7.2.1 Maximum A Posteriori Estimation

The maximum a posteriori (MAP) estimate $\hat{\boldsymbol{\theta}}_{\text{MAP}}$ is obtained as the parameter vector that maximises the posterior pdf $f_{\boldsymbol{\Theta}}|\boldsymbol{Y}(\boldsymbol{\theta}|\boldsymbol{y})$. The MAP estimate corresponds to a Bayesian estimate with a so-called uniform cost function (in fact, as shown in Figure 7.8 the cost function is notch-shaped) defined as

$$C(\hat{\boldsymbol{\theta}}, \boldsymbol{\theta}) = 1 - \delta(\hat{\boldsymbol{\theta}}, \boldsymbol{\theta}) \tag{7.24}$$

where $\delta(\hat{\boldsymbol{\theta}}, \boldsymbol{\theta})$ is the Kronecker delta function. Substitution of the cost function in the Bayesian risk equation yields

$$\begin{aligned} \mathcal{R}_{\text{MAP}}(\hat{\boldsymbol{\theta}}|\boldsymbol{y}) &= \int_{\boldsymbol{\theta}} [1 - \delta(\hat{\boldsymbol{\theta}} - \boldsymbol{\theta})] f_{\boldsymbol{\Theta}|Y}(\boldsymbol{\theta}|\boldsymbol{y}) d\boldsymbol{\theta} \\ &= 1 - f_{\boldsymbol{\Theta}|Y}(\hat{\boldsymbol{\theta}}|\boldsymbol{y}) \end{aligned} \tag{7.25}$$

From Equation (7.25), the minimum Bayesian risk estimate corresponds to the parameter value where the posterior function attains a maximum. Hence the MAP estimate of the parameter vector $\boldsymbol{\theta}$ is obtained from a minimisation of the risk equation (7.25) or equivalently maximisation of the posterior function

$$\begin{aligned} \hat{\boldsymbol{\theta}}_{\text{MAP}} &= \arg \max_{\boldsymbol{\theta}} f_{\boldsymbol{\Theta}|Y}(\boldsymbol{\theta}|\boldsymbol{y}) \\ &= \arg \max_{\boldsymbol{\theta}} [f_{Y|\boldsymbol{\theta}}(\boldsymbol{y}|\boldsymbol{\theta}) f_{\boldsymbol{\Theta}}(\boldsymbol{\theta})] \end{aligned} \tag{7.26}$$

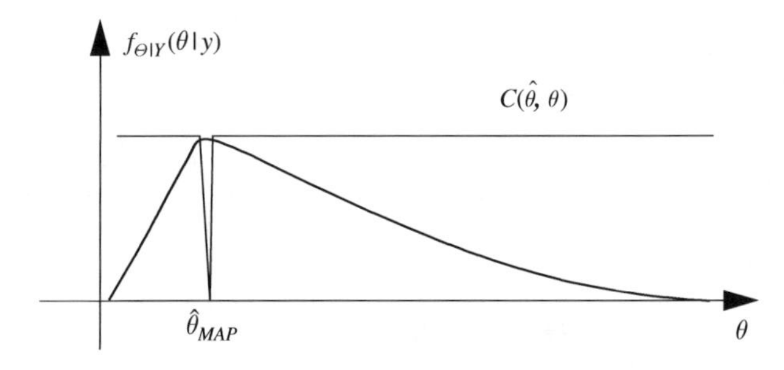

Figure 7.8 Illustration of the Bayesian cost function for the MAP estimate.

7.2.2 Maximum-Likelihood (ML) Estimation

The maximum-likelihood (ML) estimate $\hat{\boldsymbol{\theta}}_{\mathrm{ML}}$ is obtained as the parameter vector that maximises the likelihood function $f_{Y|\Theta}(y|\boldsymbol{\theta})$. The ML estimator corresponds to a Bayesian estimator with a notch-shaped cost function and a uniform parameter prior pdf:

$$\begin{aligned}\mathcal{R}_{\mathrm{ML}}(\hat{\boldsymbol{\theta}}|y) &= \int_{\boldsymbol{\theta}} \underbrace{[1-\delta(\hat{\boldsymbol{\theta}}-\boldsymbol{\theta})]}_{\text{cost function}} \underbrace{f_{Y|\Theta}(y|\boldsymbol{\theta})}_{\text{Likelihood}} \underbrace{f_{\Theta}(\boldsymbol{\theta})}_{\text{prior}} \, d\boldsymbol{\theta} \\ &= \text{const.}[1 - f_{Y|\Theta}(y|\hat{\boldsymbol{\theta}})]\end{aligned} \tag{7.27}$$

where the prior function $f_{\Theta}(\boldsymbol{\theta}) = \text{const.}$

From a Bayesian perspective the main difference between the ML and MAP estimators is that the ML estimator assumes that the prior pdf of $\boldsymbol{\theta}$ is uniform. Note that a uniform prior, in addition to modelling genuinely uniform pdfs, is also used when the parameter prior pdf is unknown, or when the parameter is an unknown constant.

From Equation (7.27), it is evident that minimisation of the risk function is achieved by maximisation of the likelihood function

$$\hat{\boldsymbol{\theta}}_{\mathrm{ML}} = \arg\max_{\boldsymbol{\theta}} f_{Y|\Theta}(y|\boldsymbol{\theta}) \tag{7.28}$$

In practice it is convenient to maximise the log-likelihood function instead of the likelihood:

$$\boldsymbol{\theta}_{\mathrm{ML}} = \arg\max_{\boldsymbol{\theta}} \log f_{Y|\boldsymbol{\theta}}(y|\boldsymbol{\theta}) \tag{7.29}$$

The log-likelihood is usually chosen in practice because of the following properties:

(a) the logarithm is a monotonic function, and hence the log-likelihood has the same turning points as the likelihood function;
(b) the joint log-likelihood of a set of independent variables is the sum of the log-likelihood of individual variables; and
(c) unlike the likelihood function, the log-likelihood has a dynamic range that does not cause computational underflow.

Note that dynamic range is the range of minimum and maximum values. The multiplication of extremely small numbers can yield a result that is too small for computer numerical representation resulting in an underflow. Conversely the multiplication of very large numbers can yield a number that is too big for computer representation resulting in an overflow.

Example 7.3 *ML estimation of the mean and variance of a Gaussian process*
Consider the problem of maximum likelihood estimation of the mean vector $\boldsymbol{\mu}_y$ and the covariance matrix $\boldsymbol{\Sigma}_{yy}$ of a P-dimensional Gaussian vector process $\boldsymbol{y}(m)$ from N observation vectors $[\boldsymbol{y}(0), \boldsymbol{y}(1), \ldots, \boldsymbol{y}(N-1)]$. Assuming the observation vectors are uncorrelated, the pdf of the observation sequence is given by

$$f_Y(\boldsymbol{y}(0), \cdots, \boldsymbol{y}(N-1)) = \prod_{m=0}^{N-1} \frac{1}{(2\pi)^{P/2} \left|\boldsymbol{\Sigma}_{yy}\right|^{1/2}} \exp\left\{-\frac{1}{2}\left[\boldsymbol{y}(m) - \boldsymbol{\mu}_y\right]^{\mathrm{T}} \boldsymbol{\Sigma}_{yy}^{-1} \left[\boldsymbol{y}(m) - \boldsymbol{\mu}_y\right]\right\} \tag{7.30}$$

and the log-likelihood equation is given by

$$\ln f_Y(\boldsymbol{y}(0), \ldots, \boldsymbol{y}(N-1)) = \sum_{m=0}^{N-1} \left\{-\frac{P}{2}\ln(2\pi) - \frac{1}{2}\ln\left|\boldsymbol{\Sigma}_{yy}\right| - \frac{1}{2}\left[\boldsymbol{y}(m) - \boldsymbol{\mu}_y\right]^{T} \boldsymbol{\Sigma}_{yy}^{-1} \left[\boldsymbol{y}(m) - \boldsymbol{\mu}_y\right]\right\} \tag{7.31}$$

Taking the derivative of the log-likelihood equation with respect to the mean vector $\boldsymbol{\mu}_y$ yields

$$\frac{\partial \ln f_Y(\boldsymbol{y}(0), \ldots, \boldsymbol{y}(N-1))}{\partial \boldsymbol{\mu}_y} = \sum_{m=0}^{N-1} \left[-2\boldsymbol{\Sigma}_{yy}^{-1}\boldsymbol{y}(m) + 2\boldsymbol{\Sigma}_{yy}^{-1}\boldsymbol{\mu}_y\right] = 0 \tag{7.32}$$

From Equation (7.32), we have

$$\hat{\boldsymbol{\mu}}_y = \frac{1}{N}\sum_{m=0}^{N-1} \boldsymbol{y}(m) \tag{7.33}$$

To obtain the ML estimate of the covariance matrix we take the derivative of the log-likelihood equation with respect to Σ_{yy}^{-1}:

$$\frac{\partial \ln f_Y(\boldsymbol{y}(0), \cdots, \boldsymbol{y}(N-1))}{\partial \boldsymbol{\Sigma}_{yy}^{-1}} = \sum_{m=0}^{N-1} \left\{\frac{1}{2}\boldsymbol{\Sigma}_{yy} - \frac{1}{2}\left[\boldsymbol{y}(m) - \boldsymbol{\mu}_y\right]\left[\boldsymbol{y}(m) - \boldsymbol{\mu}_y\right]^{\mathrm{T}}\right\} = 0 \tag{7.34}$$

From Equation (7.34), we have an estimate of the covariance matrix as

$$\hat{\boldsymbol{\Sigma}}_{yy} = \frac{1}{N}\sum_{m=0}^{N-1} \left[\boldsymbol{y}(m) - \hat{\boldsymbol{\mu}}_y\right]\left[\boldsymbol{y}(m) - \hat{\boldsymbol{\mu}}_y\right]^{\mathrm{T}} \tag{7.35}$$

Example 7.4 *ML and MAP estimation of a Gaussian random parameter*
Consider the estimation of a P-dimensional random parameter vector $\boldsymbol{\theta}$ from an N-dimensional observation vector $\boldsymbol{y}$. Assume that the relation between the signal vector $\boldsymbol{y}$ and the parameter vector $\boldsymbol{\theta}$ is described by a linear model as

$$\boldsymbol{y} = \boldsymbol{G}\boldsymbol{\theta} + \boldsymbol{e} \tag{7.36}$$

where $\boldsymbol{e}$ is a random excitation input signal and $\boldsymbol{G}$ is a data matrix. For example for an autoregressive process, $\boldsymbol{G}$ would be composed of past values of $\boldsymbol{y}$. The pdf of the parameter vector $\boldsymbol{\theta}$ given an observation vector $\boldsymbol{y}$ can be described, using Bayes' rule, as

$$f_{\Theta|Y}(\boldsymbol{\theta}|\boldsymbol{y}) = \frac{1}{f_Y(\boldsymbol{y})} f_{Y|\Theta}(\boldsymbol{y}|\boldsymbol{\theta}) f_{\Theta}(\boldsymbol{\theta}) \tag{7.37}$$

Assuming that the matrix $\boldsymbol{G}$ in Equation (7.36) is known, the likelihood of the signal $\boldsymbol{y}$ given the parameter vector $\boldsymbol{\theta}$ is the pdf of the random vector $\boldsymbol{e}$:

$$f_{Y|\Theta}(\boldsymbol{y}|\boldsymbol{\theta}) = f_E(\boldsymbol{e} = \boldsymbol{y} - \boldsymbol{G}\boldsymbol{\theta}) \tag{7.38}$$

Now assume the input $\boldsymbol{e}$ is a zero-mean, Gaussian-distributed, random process with a diagonal covariance matrix, and the parameter vector $\boldsymbol{\theta}$ is also a Gaussian process with mean of $\boldsymbol{\mu}_\theta$ and covariance matrix $\boldsymbol{\Sigma}_{\theta\theta}$. Therefore, the likelihood function equation (7.38) can be written as

$$f_{Y|\Theta}(\boldsymbol{y}|\boldsymbol{\theta}) = f_E(\boldsymbol{e}) = \frac{1}{(2\pi\sigma_e^2)^{N/2}} \exp\left[-\frac{1}{2\sigma_e^2}(\boldsymbol{y} - \boldsymbol{G}\boldsymbol{\theta})^{\mathrm{T}}(\boldsymbol{y} - \boldsymbol{G}\boldsymbol{\theta})\right] \tag{7.39}$$

and

$$f_\Theta(\boldsymbol{\theta}) = \frac{1}{(2\pi)^{P/2}\left|\boldsymbol{\Sigma}_{\theta\theta}\right|^{1/2}} \exp\left[-\frac{1}{2}(\boldsymbol{\theta} - \boldsymbol{\mu}_\theta)^T \boldsymbol{\Sigma}_{\theta\theta}^{-1}(\boldsymbol{\theta} - \boldsymbol{\mu}_\theta)\right] \tag{7.40}$$

The ML estimate, obtained from maximisation of the log-likelihood function $\ln\left[f_{Y|\Theta}(\boldsymbol{y}|\boldsymbol{\theta})\right]$ with respect to $\boldsymbol{\theta}$, is given by

$$\hat{\boldsymbol{\theta}}_{\mathrm{ML}}(\boldsymbol{y}) = \left(\boldsymbol{G}^{\mathrm{T}}\boldsymbol{G}\right)^{-1}\boldsymbol{G}^{\mathrm{T}}\boldsymbol{y} \tag{7.41}$$

To obtain the MAP estimate we first form the posterior probability distribution by substituting Equations (7.39) and (7.40) in Equation (7.37):

$$\begin{aligned} f_{\Theta|Y}(\boldsymbol{\theta}|\boldsymbol{y}) = &\frac{1}{f_Y(\boldsymbol{y})} \frac{1}{(2\pi\sigma_e^2)^{N/2}} \frac{1}{(2\pi)^{P/2}\left|\boldsymbol{\Sigma}_{\theta\theta}\right|^{1/2}} \\ &\times \exp\left(-\frac{1}{2\sigma_e^2}(\boldsymbol{y} - \boldsymbol{G}\boldsymbol{\theta})^{\mathrm{T}}(\boldsymbol{y} - \boldsymbol{G}\boldsymbol{\theta}) - \frac{1}{2}(\boldsymbol{\theta} - \mu_\theta)^{\mathrm{T}}\boldsymbol{\Sigma}_{\theta\theta}^{-1}(\boldsymbol{\theta} - \mu_\theta)\right) \end{aligned} \tag{7.42}$$

The MAP parameter estimate, obtained by differentiating the log-likelihood function $\ln\left[f_{\Theta|Y}(\boldsymbol{\theta}|\boldsymbol{y})\right]$ and setting the derivative to zero, is given by

$$\hat{\boldsymbol{\theta}}_{\mathrm{MAP}}(\boldsymbol{y}) = \left(\boldsymbol{G}^T\boldsymbol{G} + \sigma_e^2\boldsymbol{\Sigma}_{\theta\theta}^{-1}\right)^{-1}\left(\boldsymbol{G}^T\boldsymbol{y} + \sigma_e^2\boldsymbol{\Sigma}_{\theta\theta}^{-1}\mu_\theta\right) \tag{7.43}$$

Note that as the covariance of the Gaussian-distributed parameter increases, or equivalently as $\boldsymbol{\Sigma}_{\theta\theta}^{-1} \to 0$, the Gaussian prior tends to a uniform prior and the MAP solution equation (7.43) tends to the ML solution given by Equation (7.41). Conversely as the pdf of the parameter vector $\boldsymbol{\theta}$ becomes more peaked, i.e. as $\boldsymbol{\Sigma}_{\theta\theta} \to 0$, the estimate tends towards $\boldsymbol{\mu}_\theta$ the mean of the prior pdf.

7.2.3 Minimum Mean Square Error Estimation

The Bayesian minimum mean square error (MMSE) estimate is obtained as the parameter vector that minimises a mean square error cost function (Figure 7.9) defined as

$$\begin{aligned} \mathcal{R}_{MSE}(\hat{\boldsymbol{\theta}}|\boldsymbol{y}) &= \mathcal{E}\left[(\hat{\boldsymbol{\theta}} - \boldsymbol{\theta})^2 \,|\boldsymbol{y}\right] \\ &= \int_{\boldsymbol{\theta}} (\hat{\boldsymbol{\theta}} - \boldsymbol{\theta})^2 f_{\theta|Y}(\boldsymbol{\theta}|\boldsymbol{y})\,d\boldsymbol{\theta} \end{aligned} \tag{7.44}$$

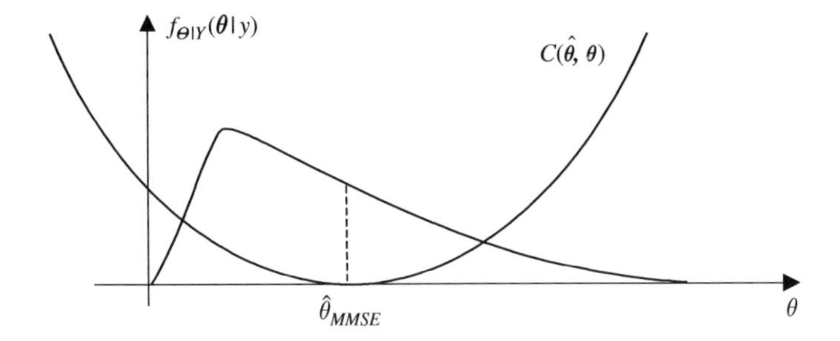

Figure 7.9 Illustration of the mean square error cost function and estimate.

In the following, it is shown that *the Bayesian MMSE estimate is the conditional mean of the posterior pdf.* Assuming that the mean square error risk function is differentiable and has a well-defined minimum, the MMSE solution can be obtained by setting the gradient of the mean square error risk function to zero:

$$\frac{\partial \mathcal{R}_{MSE}(\hat{\boldsymbol{\theta}}|\mathbf{y})}{\partial \hat{\boldsymbol{\theta}}} = 2\hat{\boldsymbol{\theta}} \int_{\boldsymbol{\theta}} f_{\Theta|Y}(\boldsymbol{\theta}|\mathbf{y})d\boldsymbol{\theta} - 2\int_{\boldsymbol{\theta}} \boldsymbol{\theta} f_{\Theta|Y}(\boldsymbol{\theta}|\mathbf{y})d\boldsymbol{\theta} \tag{7.45}$$

Since the first integral on the right-hand side of Equation (7.45) is equal to 1, we have

$$\frac{\partial R_{MSE}(\hat{\boldsymbol{\theta}}|\mathbf{y})}{\partial \hat{\boldsymbol{\theta}}} = 2\hat{\boldsymbol{\theta}} - 2\int_{\boldsymbol{\theta}} \boldsymbol{\theta} f_{\Theta|Y}(\boldsymbol{\theta}|\mathbf{y})d\boldsymbol{\theta} \tag{7.46}$$

The MMSE solution is obtained by setting Equation (7.46) to zero:

$$\hat{\boldsymbol{\theta}}_{MMSE}(\mathbf{y}) = \int_{\boldsymbol{\theta}} \boldsymbol{\theta} f_{\Theta|Y}(\boldsymbol{\theta}|\mathbf{y})d\boldsymbol{\theta} \tag{7.47}$$

For cases where we do not have the pdf models of the parameter process $\boldsymbol{\theta}$ and the signal $\mathbf{y}$, the minimum mean square error (known as the least square error, LSE) estimate is obtained through minimisation of a mean square error function $\boldsymbol{E}[e^2(\boldsymbol{\theta}|\mathbf{y})]$:

$$\hat{\boldsymbol{\theta}}_{\mathrm{LSE}} = \arg\min_{\boldsymbol{\theta}} \mathcal{E}[e^2(\boldsymbol{\theta}|\mathbf{y})] \tag{7.48}$$

where $e(\boldsymbol{\theta}|\mathbf{y})$ is an error signal. Least squared error estimation is considered in some depth in Chapter 8 on Wiener filters, in Chapter 9 on adaptive filters and in Chapter 10 on linear prediction. For a process with a Gaussian likelihood and a uniform parameter prior the MMSE estimate is the same as the LSE estimate. The LSE estimation of Equation (7.48) does not use any prior knowledge of the distribution of the signals and the parameters. This can be considered as a strength of the LSE method in situations where the prior pdfs are unknown, but it can also be considered as a weakness in cases where fairly accurate models of the pdfs are available but not utilised. In the following example it is shown that the least squared error solution of Equation (7.48) is equivalent to ML solution equation (7.41) when the signal has a Gaussian distribution.

Example 7.5 Consider the LSE estimation of the parameter vector $\boldsymbol{\theta}$ of a linear signal model from an observation signal vector $\boldsymbol{y}$ given as

$$\boldsymbol{y} = \boldsymbol{G}\boldsymbol{\theta} + \boldsymbol{e} \tag{7.49}$$

where $\boldsymbol{G}$ is a matrix. The LSE estimate is obtained as the parameter vector at which the gradient of the mean squared error, $\boldsymbol{e}^{\mathrm{T}}\boldsymbol{e}$, with respect to $\boldsymbol{\theta}$ is zero:

$$\frac{\partial \boldsymbol{e}^{\mathrm{T}}\boldsymbol{e}}{\partial \boldsymbol{\theta}} = \frac{\partial}{\partial \theta}(\boldsymbol{y}^{\mathrm{T}}\boldsymbol{y} - 2\boldsymbol{\theta}^{\mathrm{T}}\boldsymbol{G}^{\mathrm{T}}\boldsymbol{y} + \boldsymbol{\theta}^{\mathrm{T}}\boldsymbol{G}^{\mathrm{T}}\boldsymbol{G}\boldsymbol{\theta})\Bigg|_{\boldsymbol{\theta}_{\mathrm{LSE}}} = 0 \tag{7.50}$$

From Equation (7.50) the LSE parameter estimate is given by

$$\boldsymbol{\theta}_{\mathrm{LSE}} = [\boldsymbol{G}^{\mathrm{T}}\boldsymbol{G}]^{-1}\boldsymbol{G}^{\mathrm{T}}\boldsymbol{y} \tag{7.51}$$

Note that the MMSE estimate is given by equation (7.46). Also note that for a Gaussian likelihood function, the LSE solution is the same as the ML solution of Equation (7.41).

7.2.4 Minimum Mean Absolute Value of Error Estimation

The minimum mean absolute value of error (MAVE) estimate (Figure 7.10) is obtained through minimisation of a Bayesian risk function defined as

$$\mathcal{R}_{\mathrm{MAVE}}(\hat{\boldsymbol{\theta}}|\boldsymbol{y}) = \mathcal{E}[(|\hat{\boldsymbol{\theta}} - \boldsymbol{\theta}|)|\boldsymbol{y}] = \int_{\boldsymbol{\theta}} |\hat{\boldsymbol{\theta}} - \boldsymbol{\theta}| f_{\boldsymbol{\theta}|Y}(\boldsymbol{\theta}|\boldsymbol{y})\,d\boldsymbol{\theta} \tag{7.52}$$

In the following it is shown that the minimum mean absolute value estimate is the median of the parameter process. Equation (7.52) can be expressed as

$$\mathcal{R}_{\mathrm{MAVE}}(\hat{\boldsymbol{\theta}}|\boldsymbol{y}) = \int_{-\infty}^{\hat{\boldsymbol{\theta}}} [\hat{\boldsymbol{\theta}} - \boldsymbol{\theta}] f_{\Theta|Y}(\boldsymbol{\theta}|\boldsymbol{y})\,d\boldsymbol{\theta} + \int_{\hat{\boldsymbol{\theta}}}^{\infty} [\boldsymbol{\theta} - \hat{\boldsymbol{\theta}}] f_{\Theta|Y}(\boldsymbol{\theta}|\boldsymbol{y})\,d\theta \tag{7.53}$$

Taking the derivative of the risk function with respect to $\hat{\boldsymbol{\theta}}$ yields

$$\frac{\partial \mathcal{R}_{\mathrm{MAVE}}(\hat{\boldsymbol{\theta}}|\boldsymbol{y})}{\partial \hat{\boldsymbol{\theta}}} = \int_{-\infty}^{\hat{\boldsymbol{\theta}}} f_{\Theta|Y}(\boldsymbol{\theta}|\boldsymbol{y})\,d\boldsymbol{\theta} - \int_{\hat{\boldsymbol{\theta}}}^{\infty} f_{\Theta|Y}(\boldsymbol{\theta}|\boldsymbol{y})\,d\boldsymbol{\theta} \tag{7.54}$$

The minimum absolute value of error is obtained by setting Equation (7.54) to zero:

$$\int_{-\infty}^{\hat{\boldsymbol{\theta}}_{\mathrm{MAVE}}} f_{\Theta|Y}(\boldsymbol{\theta}|\boldsymbol{y})\,d\boldsymbol{\theta} = \int_{\hat{\boldsymbol{\theta}}_{\mathrm{MAVE}}}^{\infty} f_{\Theta|Y}(\boldsymbol{\theta}|\boldsymbol{y})\,d\boldsymbol{\theta} \tag{7.55}$$

From Equation (7.55) we note the MAVE estimate is the median of the posterior density.

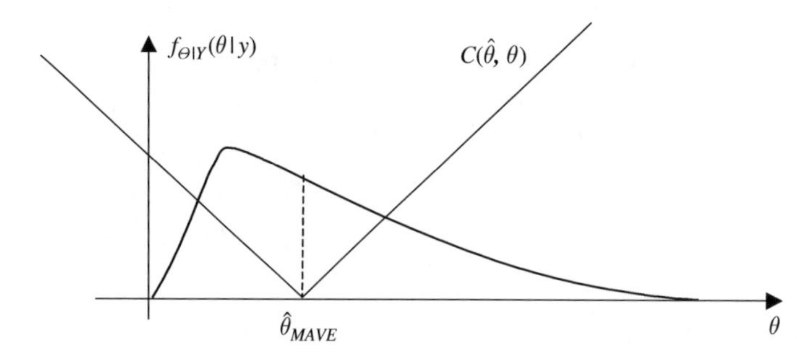

Figure 7.10 Illustration of mean absolute value of error cost function. Note that the MAVE estimate coincides with the conditional median of the posterior function.

7.2.5 Equivalence of the MAP, ML, MMSE and MAVE Estimates for Gaussian Processes with Uniform Distributed Parameters

Example 7.5 shows that for a Gaussian-distributed process the LSE estimate and the ML estimate are identical. Furthermore, Equation (7.43), for the MAP estimate of a Gaussian-distributed parameter, shows that as the parameter variance increases, or equivalently as the parameter prior pdf tends to a uniform distribution, the MAP estimate tends to the ML and LSE estimates. In general, for any symmetric distribution, centred round the maximum, the mode, the mean and the median are identical. Hence, for a process with a symmetric pdf, if the prior distribution of the parameter is uniform then the MAP, the ML, the MMSE and the MAVE parameter estimates are identical. Figure 7.11 illustrates a symmetric pdf, an asymmetric pdf, and the relative positions of various estimates.

7.2.6 Influence of the Prior on Estimation Bias and Variance

The use of a prior pdf introduces a bias in the estimate towards the range of parameter values with a relatively high values of prior pdf, and reduces the variance of the estimate. To illustrate the effects of the prior pdf on the bias and the variance of an estimate, we consider the following examples in which the bias and the variance of the ML and the MAP estimates of the mean value of a process are compared.

Example 7.6 *Estimation of bias and variance of an ML estimator*
Consider the ML estimation of a random scalar parameter θ, observed in a zero-mean additive white Gaussian noise (AWGN) $n(m)$, and expressed as

$$y(m) = \theta + n(m), \quad m = 0, \ldots, N-1 \tag{7.56}$$

It is assumed that, for each realisation of the parameter θ, N observation samples are available. Note that, since the noise is assumed to be a zero-mean process, this problem is equivalent to estimation of the mean of the process $y(m)$. The likelihood of an observation vector $\boldsymbol{y} = [y(0), y(1), \ldots, y(N-1)]$ and a parameter value of θ is given by

$$f_{Y|\Theta}(\boldsymbol{y}|\theta) = \prod_{m=0}^{N-1} f_N\left(y(m) - \theta\right) = \frac{1}{(2\pi\sigma_n^2)^{N/2}} \exp\left\{ -\frac{1}{2\sigma_n^2} \sum_{m=0}^{N-1} [y(m) - \theta]^2 \right\} \tag{7.57}$$

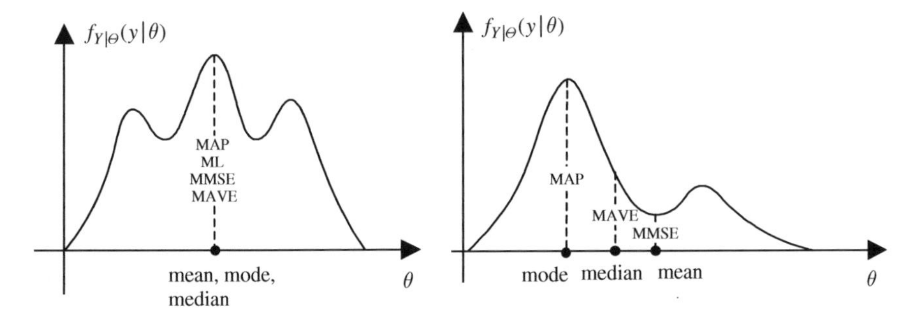

Figure 7.11 Illustration of a symmetric and an asymmetric pdf and their respective mode, mean and median and the relations to MAP, MAVE and MMSE estimates.

From Equation (7.57) the log-likelihood function is given by

$$\ln f_{Y|\Theta}(\mathbf{y}|\theta) = -\frac{N}{2}\ln(2\pi\sigma_n^2) - \frac{1}{2\sigma_n^2}\sum_{m=0}^{N-1}[y(m)-\theta]^2 \tag{7.58}$$

The ML estimate of θ, obtained by setting the derivative of $\ln f_{Y|\Theta}(\mathbf{y}|\boldsymbol{\theta})$ to zero, is given by

$$\hat{\theta}_{\mathrm{ML}} = \frac{1}{N}\sum_{m=0}^{N-1} y(m) = \bar{y} \tag{7.59}$$

where $\bar{y}$ denotes the time average of $y(m)$. From Equation (7.59), we note that the ML solution is an unbiased estimate

$$\mathcal{E}[\hat{\theta}_{\mathrm{ML}}] = \mathcal{E}\left(\frac{1}{N}\sum_{m=0}^{N-1} y(m)\right) = \mathcal{E}\left(\frac{1}{N}\sum_{m=0}^{N-1}[\theta + n(m)]\right) = \theta \tag{7.60}$$

and the variance of the ML estimate is given by

$$\mathrm{Var}[\hat{\theta}_{\mathrm{ML}}] = \mathcal{E}[(\hat{\theta}_{\mathrm{ML}}-\theta)^2] = \mathcal{E}\left[\left(\frac{1}{N}\sum_{m=0}^{N-1} y(m) - \theta\right)^2\right] = \frac{\sigma_n^2}{N} \tag{7.61}$$

Note that the variance of the ML estimate decreases with increasing length of observation.

Example 7.7 *Estimation of a uniformly-distributed scalar parameter observed in AWGN*
Consider the effects of using a uniform parameter prior on the mean and the variance of the estimate in Example 7.6. Assume that the prior for the parameter θ is given by

$$f_\Theta(\theta) = \begin{cases} 1/(\theta_{\max}-\theta_{\min}) & \theta_{\min} \le \theta \le \theta_{\max} \\ 0 & \text{otherwise} \end{cases} \tag{7.62}$$

as illustrated in Figure 7.12. From Bayes' rule, the posterior pdf of uniformly distributed parameter θ is given by

$$\begin{aligned} f_{\Theta|Y}(\theta|\mathbf{y}) &= \frac{1}{f_Y(\mathbf{y})} f_{Y|\Theta}(\mathbf{y}|\theta) f_\Theta(\theta) \\ &= \begin{cases} \frac{1}{f_Y(\mathbf{y})}\frac{1}{(2\pi\sigma_n^2)^{N/2}} \exp\left(-\frac{1}{2\sigma_n^2}\sum_{m=0}^{N-1}(y(m)-\theta)^2\right) & \theta_{\min} \le \theta \le \theta_{\max} \\ 0 & \text{otherwise} \end{cases} \end{aligned} \tag{7.63}$$

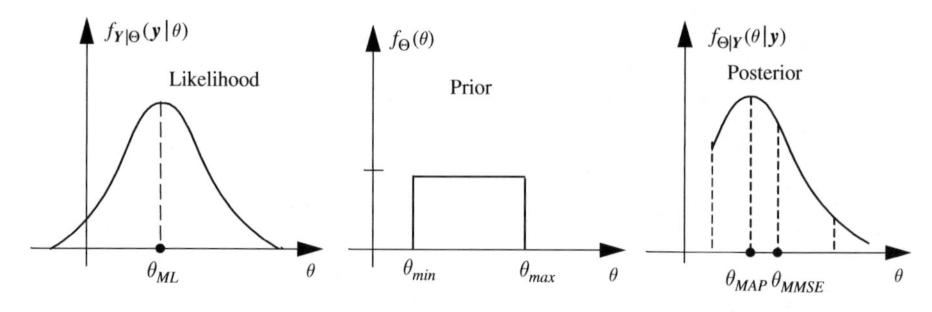

Figure 7.12 Illustration of the effects of a uniform prior on the estimate of a parameter observed in AWGN, where it is assumed that $\theta_{\min} \le \theta_{\mathrm{ML}} \le \theta_{\max}$.

The MAP estimate is obtained by maximising the posterior pdf:

$$\hat{\theta}_{\text{MAP}}(\mathbf{y}) = \begin{cases} \theta_{\min} & \text{if } \hat{\theta}_{\text{ML}}(\mathbf{y}) < \theta_{\min} \\ \hat{\theta}_{\text{ML}}(y) & \text{if } \theta_{\min} \le \hat{\theta}_{\text{ML}}(\mathbf{y}) \le \theta_{\max} \\ \theta_{\max} & \text{if } \hat{\theta}_{\text{ML}}(\mathbf{y}) > \theta_{\max} \end{cases} \tag{7.64}$$

Note that the MAP estimate is constrained to the range θ_{min} to θ_{max}. This constraint is desirable and moderates the estimates that, due to say low signal-to-noise ratio, fall outside the range of possible values of θ. It is easy to see that the variance of an estimate constrained to a range of $\theta_{\min}$ to $\theta_{\max}$ is less than the variance of the ML estimate in which there is no constraint on the range of the parameter estimate:

$$\text{Var}[\hat{\theta}_{\text{MAP}}] = \int_{\theta_{\min}}^{\theta_{\max}} (\hat{\theta}_{\text{MAP}} - \theta)^2 f_{Y|\Theta}(\mathbf{y}|\theta) d\theta \quad \le \quad \text{Var}[\hat{\theta}_{\text{ML}}] = \int_{-\infty}^{\infty} (\hat{\theta}_{\text{ML}} - \theta)^2 f_{Y|\Theta}(\mathbf{y}|\theta) d\theta \tag{7.65}$$

Example 7.8 *Estimation of a Gaussian-distributed scalar parameter observed in AWGN*
In this example, Figure 7.13, we consider the effect of a Gaussian prior on the mean and the variance of the MAP estimate. Assume that the parameter θ is Gaussian-distributed with a mean μ_θ and a variance σ_θ^2 as

$$f_\Theta(\theta) = \frac{1}{(2\pi\sigma_\theta^2)^{1/2}} \exp\left[-\frac{(\theta - \mu_\theta)^2}{2\sigma_\theta^2}\right] \tag{7.66}$$

From Bayes' rule the posterior pdf is given as the product of the likelihood and the prior pdfs as

$$\begin{aligned} f_{\Theta|Y}(\theta|\mathbf{y}) &= \frac{1}{f_Y(\mathbf{y})} f_{Y|\Theta}(\mathbf{y}|\theta) f_\Theta(\theta) \\ &= \frac{1}{f_Y(\mathbf{y})} \frac{1}{(2\pi\sigma_n^2)^{N/2}(2\pi\sigma_\theta^2)^{1/2}} \exp\left(-\frac{1}{2\sigma_n^2}\sum_{m=0}^{N-1}(y(m)-\theta)^2 - \frac{1}{2\sigma_\theta^2}(\theta-\mu_\theta)^2\right) \end{aligned} \tag{7.67}$$

The maximum posterior solution is obtained by setting the derivative of the log-posterior function, $\ln f_{\Theta|Y}(\theta|\mathbf{y})$, with respect to θ to zero:

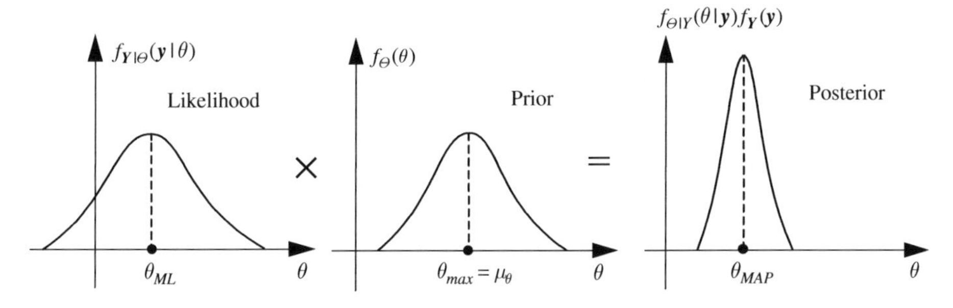

Figure 7.13 Illustration of the posterior pdf as product of the likelihood and the prior.

$$\hat{\theta}_{\text{MAP}}(\mathbf{y}) = \frac{\sigma_\theta^2}{\sigma_\theta^2 + \sigma_n^2/N}\bar{y} + \frac{\sigma_n^2/N}{\sigma_\theta^2 + \sigma_n^2/N}\mu_\theta \tag{7.68}$$

where $\bar{y} = \sum_{m=0}^{N-1} y(m)/N$.

Note that the MAP estimate is an interpolation between the ML estimate $\bar{y}$ and the mean of the prior pdf μ_θ, as shown in Figure 7.13. Also note that in Equation (7.68) the interpolation weights are dependent on signal-to-noise ratio and the length of the observation. As the variance (i.e. power) of noise decreases relative to the variance of the parameter and/or as the number of observations increases, the influence of the prior decreases; conversely, as the variance (i.e. power) of noise increases and/or as the number of observations decreases, the influence of the prior increases.

The expectation of the MAP estimate is obtained by noting that the only random variable on the right-hand side of Equation (7.68) is the term $\bar{y}$, and that $E[\bar{y}] = \theta$

$$\mathcal{E}[\hat{\theta}_{MAP}(\mathbf{y})] = \frac{\sigma_\theta^2}{\sigma_\theta^2 + \sigma_n^2/N}\theta + \frac{\sigma_n^2/N}{\sigma_\theta^2 + \sigma_n^2/N}\mu_\theta \tag{7.69}$$

and the variance of the MAP estimate is given as

$$\text{Var}[\hat{\theta}_{MAP}(\mathbf{y})] = \frac{\sigma_\theta^2}{\sigma_\theta^2 + \sigma_n^2/N} \times \text{Var}[\bar{y}] = \frac{\sigma_n^2/N}{1 + \sigma_n^2/N\sigma_\theta^2} \tag{7.70}$$

Substitution of Equation (7.61) in Equation (7.70) yields

$$\text{Var}[\hat{\theta}_{MAP}(\mathbf{y})] = \frac{\text{Var}[\hat{\theta}_{\text{ML}}(\mathbf{y})]}{1 + \text{Var}[\hat{\theta}_{\text{ML}}(\mathbf{y})]/\sigma_\theta^2} \tag{7.71}$$

Note that as σ_θ^2, the variance of the parameter θ, increases the influence of the prior decreases, and the variance of the MAP estimate tends towards the variance of the ML estimate.

7.2.7 Relative Importance of the Prior and the Observation

A fundamental issue in the Bayesian inference method is the relative influence of the observation signal and the prior pdf on the outcome. The importance of the observation depends on the confidence in the observation, and the confidence in turn depends on the length of the observation and on the signal-to-noise ratio (SNR). In general, as the number of observation samples and the SNR increase, the variance of the estimate and the influence of the prior decrease. From Equation (7.68) for the estimation of a Gaussian distributed parameter observed in AWGN, as the length of the observation N increases, the importance of the prior decreases, and the MAP estimate tends to the ML estimate:

$$\lim_{N\to\infty} \hat{\theta}_{\text{MAP}}(\mathbf{y}) = \lim_{N\to\infty}\left(\frac{\sigma_\theta^2}{\sigma_\theta^2 + \sigma_n^2/N}\bar{y} + \frac{\sigma_n^2/N}{\sigma_\theta^2 + \sigma_n^2/N}\mu_\theta\right) = \bar{y} = \hat{\theta}_{\text{ML}} \tag{7.72}$$

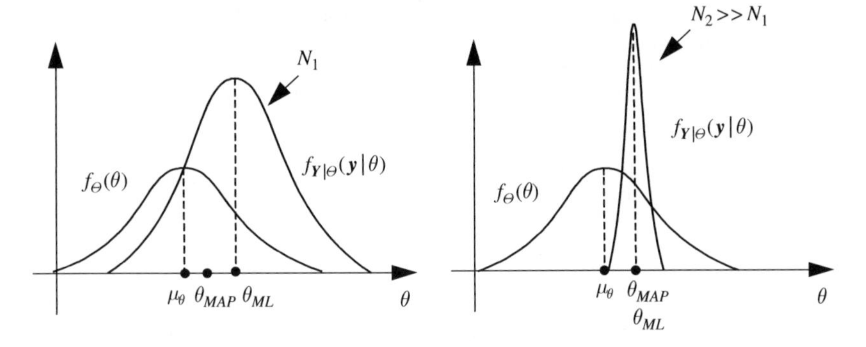

Figure 7.14 Illustration of the effect of the increasing length of observation on the variance an estimator.

As illustrated in Figure 7.14, as the length of the observation N tends to infinity then both the MAP and the ML estimates of the parameter should tend to its true value θ.

Example 7.9 *MAP estimation of a scalar Gaussian signal in additive noise*
Consider the estimation of a scalar-valued Gaussian signal $x(m)$, observed in an additive Gaussian white noise $n(m)$, and modelled as

$$y(m) = x(m) + n(m) \tag{7.73}$$

The posterior pdf of the signal $x(m)$ is given by

$$\begin{aligned} f_{X|Y}(x(m)|y(m)) &= \frac{1}{f_Y(y(m))} f_{Y|X}(y(m)|x(m)) f_X(x(m)) \\ &= \frac{1}{f_Y(y(m))} \underbrace{f_N(y(m)-x(m))}_{\text{Likelihood}} \underbrace{f_X(x(m))}_{\text{Prior}} \end{aligned} \tag{7.74}$$

where $f_X(x(m)) = \mathcal{N}(x(m), \mu_x, \sigma_x^2)$ and $f_N(n(m)) = \mathcal{N}(n(m), \mu_n, \sigma_n^2)$ are the Gaussian pdfs of the signal and noise respectively. Substitution of the signal and noise pdfs in Equation (7.74) yields

$$\begin{aligned} f_{X|Y}(x(m)|y(m)) = &\frac{1}{f_Y(y(m))} \frac{1}{\sqrt{2\pi}\sigma_n} \exp\left(-\frac{[y(m)-x(m)-\mu_n]^2}{2\sigma_n^2}\right) \\ &\times \frac{1}{\sqrt{2\pi}\sigma_x} \exp\left(-\frac{[x(m)-\mu_x]^2}{2\sigma_x^2}\right) \end{aligned} \tag{7.75}$$

This equation can be rewritten as

$$\begin{aligned} f_{X|Y}(x(m)|y(m)) = &\frac{1}{f_Y(y(m))} \frac{1}{2\pi\sigma_n\sigma_x} \times \\ &\exp\left\{-\frac{\sigma_x^2[y(m)-x(m)-\mu_n]^2 + \sigma_n^2[x(m)-\mu_x]^2}{2\sigma_x^2\sigma_n^2}\right\} \end{aligned} \tag{7.76}$$

To obtain the MAP estimate we set the derivative of the log-likelihood function $\ln f_{X|Y}(x(m)|y(m))$ with respect to $x(m)$ to zero as

$$\frac{\partial[\ln f_{X|Y}(x(m)|y(m))]}{\partial\hat{x}(m)} = -\frac{-2\sigma_x^2[y(m)-x(m)-\mu_n]+2\sigma_n^2[x(m)-\mu_x]}{2\sigma_x^2\sigma_n^2} = 0 \tag{7.77}$$

From Equation (7.77) the MAP signal estimate is given by

$$\hat{x}(m) = \frac{\sigma_x^2}{\sigma_x^2+\sigma_n^2}[y(m)-\mu_n] + \frac{\sigma_n^2}{\sigma_x^2+\sigma_n^2}\mu_x \tag{7.78}$$

Note that the estimate $\hat{x}(m)$ is a weighted linear interpolation between the unconditional mean of $x(m)$, μ_x, and the observed value $(y(m)-\mu_n)$. At a very poor SNR i.e. when $\sigma_x^2 << \sigma_n^2$ we have $\hat{x}(m) \approx \mu_x$; and, on the other hand, for a noise-free signal $\sigma_n^2 = 0$ and $\mu_n = 0$ and we have $\hat{x}(m) = y(m)$.

Example 7.10 *MAP estimate of a Gaussian–AR process observed in AWGN*
Consider a vector $\boldsymbol{x}$ of N samples from an autoregressive (AR) process observed in an additive Gaussian noise, and modelled as

$$\boldsymbol{y} = \boldsymbol{x} + \boldsymbol{n} \tag{7.79}$$

From Chapter 10, a vector $\boldsymbol{x}$ from an AR process may be expressed as

$$\boldsymbol{e} = \boldsymbol{A}\boldsymbol{x} \tag{7.80}$$

where $\boldsymbol{A}$ is a matrix of the AR model coefficients, and the vector $\boldsymbol{e}$ is the input signal of the AR model. Assuming that the signal $\boldsymbol{x}$ is Gaussian, and that the P initial samples $\boldsymbol{x}_0$ are known, the pdf of the signal $\boldsymbol{x}$ is given by

$$f_X(\boldsymbol{x}|\boldsymbol{x}_0) = f_E(\boldsymbol{e}) = \frac{1}{(2\pi\sigma_e^2)^{N/2}}\exp\left(-\frac{1}{2\sigma_e^2}\boldsymbol{x}^T\boldsymbol{A}^T\boldsymbol{A}\boldsymbol{x}\right) \tag{7.81}$$

where it is assumed that the input signal $\boldsymbol{e}$ of the AR model is a zero-mean uncorrelated process with variance σ_e^2. The pdf of a zero-mean Gaussian noise vector $\boldsymbol{n}$, with covariance matrix Σ_{nn}, is given by

$$f_N(\boldsymbol{n}) = \frac{1}{(2\pi)^{N/2}\left|\Sigma_{nn}\right|^{1/2}}\exp\left(-\frac{1}{2}\boldsymbol{n}^T\Sigma_{nn}^{-1}\boldsymbol{n}\right) \tag{7.82}$$

From Bayes' rule, the pdf of the signal given the noisy observation is

$$f_{X|Y}(\boldsymbol{x}|\boldsymbol{y}) = \frac{f_{Y|X}(\boldsymbol{y}|\boldsymbol{x})f_X(\boldsymbol{x})}{f_Y(\boldsymbol{y})} = \frac{1}{f_Y(\boldsymbol{y})}f_N(\boldsymbol{y}-\boldsymbol{x})f_X(\boldsymbol{x}) \tag{7.83}$$

Substitution of the pdfs of the signal and noise in Equation (7.83) yields

$$f_{X|Y}(\boldsymbol{x}|\boldsymbol{y}) = \frac{1}{f_Y(\boldsymbol{y})(2\pi)^N\sigma_e^{N/2}\left|\Sigma_{nn}\right|^{1/2}}\exp\left\{-\frac{1}{2}\left[(\boldsymbol{y}-\boldsymbol{x})^T\Sigma_{nn}^{-1}(\boldsymbol{y}-\boldsymbol{x})+\frac{\boldsymbol{x}^T\boldsymbol{A}^T\boldsymbol{A}\boldsymbol{x}}{\sigma_e^2}\right]\right\} \tag{7.84}$$

The MAP estimate corresponds to the minimum of the argument of the exponential function in Equation (7.84). Assuming that the argument of the exponential function is differentiable, and has a well-defined minimum, we can obtain the MAP estimate from

$$\hat{\boldsymbol{x}}_{\text{MAP}}(\boldsymbol{y}) = \arg\underset{\boldsymbol{x}}{\text{zero}}\left\{\frac{\partial}{\partial\boldsymbol{x}}\left((\boldsymbol{y}-\boldsymbol{x})^T\Sigma_{nn}^{-1}(\boldsymbol{y}-\boldsymbol{x})+\frac{\boldsymbol{x}^T\boldsymbol{A}^T\boldsymbol{A}\boldsymbol{x}}{\sigma_e^2}\right]\right\} \tag{7.85}$$

The MAP estimate is

$$\hat{x}_{\text{MAP}}(y) = \left(\mathbf{I} + \frac{1}{\sigma_e^2}\Sigma_{nn}A^{\text{T}}A\right)^{-1} y \tag{7.86}$$

where **I** is the identity matrix.

7.3 Expectation Maximisation Method

The EM algorithm is used for estimation of a parameter vector θ given an incomplete observation y. The EM algorithm (EM stands for expectation maximisation or estimate–maximise) is an iterative 'hill climbing' method for maximisation of the *expectation* of a likelihood function $f_{Y;\Theta}(y|\theta)$. It has applications in blind de-convolution, clustering, training of hidden Markov models, model-based signal interpolation, spectral estimation from noisy observations and signal restoration.

The EM is a framework for solving problems where it is difficult to obtain a direct ML estimate either because the data is *incomplete* (Figure 7.15), e.g. when there are missing samples or missing labels, or because the problem is difficult. Hence instead of a direct ML estimate, the expectation of the ML estimate is maximised. For example, in clustering applications usually the raw data do not have a cluster label attached to them and hence an iterative EM process is employed consisting of (a) labelling of data (expectation) and (b) calculation of means and variances of clusters.

To define the notions of incomplete and complete data, consider a signal x from a random process X with an unknown parameter vector θ and a joint pdf $f_{X;\Theta}(x, \theta)$. The signal x is the so-called *complete data* and the ML estimate of the parameter vector θ may be obtained from $f_{X|\Theta}(x|\theta)$. Now assume that the signal x goes through a many-to-one non-invertible transformation (e.g. when a number of samples of the vector x are lost) and is observed as y. The observation y is the so-called *incomplete data.*

Maximisation of the likelihood of the incomplete data, $f_{Y;\Theta}(y|\theta)$ with respect to the parameter vector θ is often a difficult task, whereas maximisation of the likelihood of the complete data $f_{X;\Theta}(x|\theta)$ is relatively easy. Since the complete data is unavailable, the parameter estimate is obtained through maximisation of the *conditional expectation* of the log-likelihood of the complete data defined as

$$\mathcal{E}\left[\ln f_{X;\Theta}(x|\theta)\,|y\right] = \int_X f_{X|Y;\Theta}(x\,|y;\theta)\ln f_{X;\Theta}(x|\theta)dx \tag{7.87}$$

In Equation (7.87), the computation of the term $f_{X|Y;\Theta}(x|y, \theta)$ requires an estimate of the unknown parameter vector θ. For this reason, the expectation of the likelihood function is maximised iteratively starting with an initial estimate of θ, and updating the estimate as described in the following. Note that the r.h.s. of Equation (7.87) is similar to an entropy function.

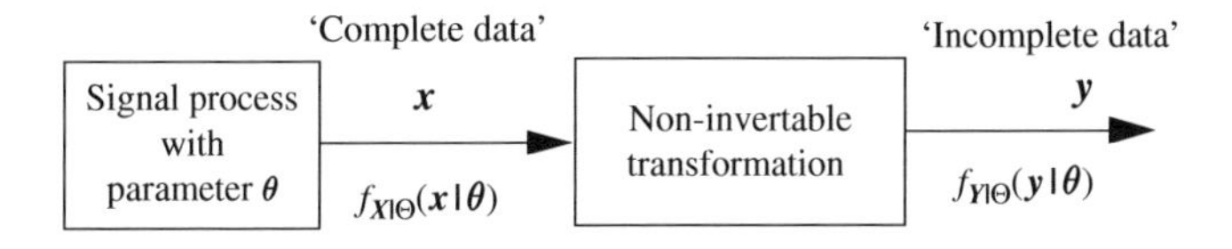

Figure 7.15 Illustration of transformation of complete data to incomplete data.

EM Algorithm

Step 1: *Initialisation* – Select an initial parameter estimate $\boldsymbol{\theta}_0$, and
For $i = 0, 1, \ldots$ until convergence:
Step 2: *Expectation* – Compute

$$U(\boldsymbol{\theta}, \hat{\boldsymbol{\theta}}_i) = E[\ln f_{X;\Theta}(\boldsymbol{x}; \boldsymbol{\theta})|\boldsymbol{y}; \hat{\boldsymbol{\theta}}_i] = \int_X f_{X|Y;\Theta}(\boldsymbol{x}|\boldsymbol{y}; \hat{\boldsymbol{\theta}}_i) \ln f_{X;\Theta}(\boldsymbol{x}; \boldsymbol{\theta}) d\boldsymbol{x} \tag{7.88}$$

Step 3: *Maximisation* – Select

$$\hat{\boldsymbol{\theta}}_{i+1} = \arg\max_{\boldsymbol{\theta}} U(\boldsymbol{\theta}, \hat{\boldsymbol{\theta}}_i) \tag{7.89}$$

Step 4: *Convergence test* – If not converged then go to Step 2.

7.3.1 Convergence of the EM Algorithm

In this section, it is shown that the EM algorithm converges to a maximum of the likelihood of the incomplete data $f_{Y;\Theta}(\boldsymbol{y}|\boldsymbol{\theta})$. The likelihood of the complete data can be written as

$$f_{X,Y;\Theta}(\boldsymbol{x}, \boldsymbol{y}|\boldsymbol{\theta}) = f_{X|Y;\Theta}(\boldsymbol{x}|\boldsymbol{y}, \boldsymbol{\theta}) f_{Y;\Theta}(\boldsymbol{y}|\boldsymbol{\theta}) \tag{7.90}$$

where $f_{X,Y;\Theta}(\boldsymbol{x}, \boldsymbol{y}|\boldsymbol{\theta})$ is the likelihood of $\boldsymbol{x}$ and $\boldsymbol{y}$ with $\boldsymbol{\theta}$ as a parameter. From Equation (7.90), the log-likelihood of the incomplete data is obtained as

$$\ln f_{Y|\Theta}(\boldsymbol{y}|\boldsymbol{\theta}) = \ln f_{X,Y|\Theta}(\boldsymbol{x}, \boldsymbol{y}|\boldsymbol{\theta}) - \ln f_{X|Y,\Theta}(\boldsymbol{x}|\boldsymbol{y}, \boldsymbol{\theta}) \tag{7.91}$$

Using an estimate of $\boldsymbol{\theta}$, $\hat{\boldsymbol{\theta}}_i$, and taking the expectation of Equation (7.91) – this is done by multiplying both sides of Equation (7.91) by $f_{X|Y;\Theta}(\boldsymbol{x}|\boldsymbol{y}, \hat{\boldsymbol{\theta}}_i)$ and integrating w.r.t. $\boldsymbol{x}$ – we have

$$\ln f_{Y;\Theta}(\boldsymbol{y}, \boldsymbol{\theta}) = U(\boldsymbol{\theta}, \hat{\boldsymbol{\theta}}_i) - V(\boldsymbol{\theta}, \hat{\boldsymbol{\theta}}_i) \tag{7.92}$$

Since expectation is taken w.r.t. $\boldsymbol{x}$ $\mathcal{E}[\ln f_{Y|\Theta}(\boldsymbol{y}|\boldsymbol{\theta})] = \ln f_{Y|\Theta}(\boldsymbol{y}|\boldsymbol{\theta})$. The function $U(\boldsymbol{\theta}, \hat{\boldsymbol{\theta}}_i)$ is the conditional expectation of $\ln f_{X,Y;\Theta}(\boldsymbol{x}, \boldsymbol{y}|\boldsymbol{\theta})$, conditioned on $\boldsymbol{y}$ and $\hat{\boldsymbol{\theta}}_i$:

$$U(\boldsymbol{\theta}; \hat{\boldsymbol{\theta}}_\iota) = \mathcal{E}[\ln f_{X,Y|\Theta}(\boldsymbol{x}, \boldsymbol{y}|\boldsymbol{\theta})|\boldsymbol{y}, \hat{\boldsymbol{\theta}}_i] = \int_X f_{X|Y,\Theta}(\boldsymbol{x}|\boldsymbol{y}, \hat{\boldsymbol{\theta}}_i) \ln f_{X;\Theta}(\boldsymbol{x}|\boldsymbol{\theta}) d\boldsymbol{x} \tag{7.93}$$

Note that $f_{X,Y|\Theta}(\boldsymbol{x},\boldsymbol{y}|\boldsymbol{\theta}) = f_{X,Y|\Theta}(\boldsymbol{x}|\boldsymbol{\theta})$ as the complete information $\boldsymbol{x}$ includes the incomplete information $\boldsymbol{y}$. The function $V(\boldsymbol{\theta},\hat{\boldsymbol{\theta}})$ is the conditional expectation of $\ln f_{X|Y;\Theta}(\boldsymbol{x}|\boldsymbol{y},\boldsymbol{\theta})$ conditioned on $\boldsymbol{y}$ and $\hat{\boldsymbol{\theta}}_i$:

$$\begin{aligned} V(\boldsymbol{\theta},\hat{\boldsymbol{\theta}}_i) &= \mathcal{E}\left[\ln f_{X|Y,\Theta}(\boldsymbol{x}|\boldsymbol{y},\boldsymbol{\theta}) \,\Big|\, \boldsymbol{y},\hat{\boldsymbol{\theta}}_i\right] \\ &= \int_X f_{X|Y,\Theta}(\boldsymbol{x}|\boldsymbol{y},\hat{\boldsymbol{\theta}}_i)\ln f_{X|Y,\Theta}(\boldsymbol{x}|\boldsymbol{y},\boldsymbol{\theta})d\boldsymbol{x} \end{aligned} \tag{7.94}$$

Now, from Equation (7.92), the log-likelihood of the incomplete data $\boldsymbol{y}$ with parameter estimate $\hat{\boldsymbol{\theta}}_i$ at iteration i is

$$\ln f_{Y;\Theta}(\boldsymbol{y},\hat{\boldsymbol{\theta}}_i) = U(\hat{\boldsymbol{\theta}}_i,\hat{\boldsymbol{\theta}}_i) - V(\hat{\boldsymbol{\theta}}_i,\hat{\boldsymbol{\theta}}_i) \tag{7.95}$$

It can be shown (see Dempster *et al.*, 1977) that the function V satisfies the inequality

$$V(\hat{\boldsymbol{\theta}}_{i+1},\hat{\boldsymbol{\theta}}_i) \le V(\hat{\boldsymbol{\theta}}_i,\hat{\boldsymbol{\theta}}_i) \tag{7.96}$$

and in the maximisation step of EM we choose $\hat{\theta}_{i+1}$ such that

$$U(\hat{\boldsymbol{\theta}}_{i+1},\hat{\boldsymbol{\theta}}_i) \ge U(\hat{\boldsymbol{\theta}}_i,\hat{\boldsymbol{\theta}}_i) \tag{7.97}$$

From Equation (7.95) and the inequalities (7.96) and (7.97), it follows that

$$\ln f_{Y|\Theta}(\boldsymbol{y}|\hat{\boldsymbol{\theta}}_{i+1}) \ge \ln f_{Y|\Theta}(\boldsymbol{y}|\hat{\boldsymbol{\theta}}_i) \tag{7.98}$$

Therefore at every iteration of the EM algorithm, the conditional likelihood of the estimate increases until the estimate converges to a local maximum of the log-likelihood function $\ln f_{Y|\Theta}(\boldsymbol{y}|\boldsymbol{\theta})$.

The EM algorithm is applied to the solution of a number of problems in this book. In Section 7.5, the estimation of the parameters of a mixture Gaussian model for the signal space of a recorded process is formulated in an EM framework. In Chapter 11, the EM is used for estimation of the parameters of a hidden Markov model.

7.4 Cramer–Rao Bound on the Minimum Estimator Variance

An important measure of the performance of an estimator is the variance of the estimate with different values of the observation signal $\boldsymbol{y}$ and the parameter vector $\boldsymbol{\theta}$. The minimum estimation variance depends on the distributions of the parameter vector $\boldsymbol{\theta}$ and on the observation signal $\boldsymbol{y}$. In this section, we first consider the lower bound on the variance of the estimates of a constant parameter, and then extend the results to random parameters.

The Cramer–Rao lower bound on the variance of the estimate of the ith coefficient θ_i of a parameter vector $\boldsymbol{\theta}$ is given as

$$\operatorname{Var}[\hat{\theta}_i(\boldsymbol{y})] \ge \frac{\left(1+\dfrac{\partial\theta_{\text{Bias}}}{\partial\theta_i}\right)^2}{\mathcal{E}\left[\left(\dfrac{\partial \ln f_{Y|\Theta}(\boldsymbol{y}|\boldsymbol{\theta})}{\partial\theta_i}\right)^2\right]} \tag{7.99}$$

An estimator that achieves the lower bound on the variance is called the minimum variance, or the most efficient, estimator.

Proof The bias in the estimate $\hat{\theta}_i(\boldsymbol{y})$ of the ith coefficient of the parameter vector $\boldsymbol{\theta}$, averaged over the observation space $\boldsymbol{Y}$ is defined as

$$E[\hat{\theta}_i(\boldsymbol{y}) - \theta_i] = \int_{-\infty}^{\infty} [\hat{\theta}_i(\boldsymbol{y}) - \theta_i] f_{Y|\Theta}(\boldsymbol{y}|\boldsymbol{\theta}) d\boldsymbol{y} = \theta_{\text{Bias}} \tag{7.100}$$

Differentiation of Equation (7.100) with respect to θ_i yields

$$\int_{-\infty}^{\infty} \left\{ [\hat{\theta}_i(\boldsymbol{y}) - \theta_i] \frac{\partial f_{Y|\Theta}(\boldsymbol{y}|\boldsymbol{\theta})}{\partial \theta_i} - f_{Y|\Theta}(\boldsymbol{y}|\boldsymbol{\theta}) \right\} d\boldsymbol{y} = \frac{\partial \theta_{\text{Bias}}}{\partial \theta_i} \tag{7.101}$$

For a probability density function we have

$$\int_{-\infty}^{\infty} f_{Y|\Theta}(\boldsymbol{y}|\boldsymbol{\theta}) d\boldsymbol{y} = 1 \tag{7.102}$$

Therefore Equation (7.101) can be written as

$$\int_{-\infty}^{\infty} [\hat{\theta}_i(\boldsymbol{y}) - \theta_i] \frac{\partial f_{Y|\Theta}(\boldsymbol{y}|\boldsymbol{\theta})}{\partial \theta_i} d\boldsymbol{y} = 1 + \frac{\partial \theta_{\text{Bias}}}{\partial \theta_i} \tag{7.103}$$

Now, since the derivative of the integral of pdf equation (7.102) is zero, taking the derivative of Equation (7.102) and multiplying the result by θ_{Bias} yields

$$\theta_{\text{Bias}} \int_{-\infty}^{\infty} \frac{\partial f_{Y|\Theta}(\boldsymbol{y}|\boldsymbol{\theta})}{\partial \theta_i} d\boldsymbol{y} = 0 \tag{7.104}$$

Substituting $\partial f_{Y|\Theta}(\boldsymbol{y}|\boldsymbol{\theta})/\partial \theta_i = f_{Y|\Theta}(\boldsymbol{y}|\boldsymbol{\theta}) \partial \ln f_{Y|\Theta}(\boldsymbol{y}|\boldsymbol{\theta})/\partial \theta_i$ into Equation (7.103), and using Equation (7.104), we obtain

$$\int_{-\infty}^{\infty} [\hat{\theta}_i(\boldsymbol{y}) - \theta_{\text{Bias}} - \theta_i] \frac{\partial \ln f_{Y|\Theta}(\boldsymbol{y}|\boldsymbol{\theta})}{\partial \theta_i} f_{Y|\Theta}(\boldsymbol{y}|\boldsymbol{\theta}) d\boldsymbol{y} = 1 + \frac{\partial \theta_{\text{Bias}}}{\partial \theta_i} \tag{7.105}$$

Now squaring both sides of Equation (7.105), we obtain

$$\left(\int_{-\infty}^{\infty} [\hat{\theta}_i(\boldsymbol{y}) - \theta_{\text{Bias}} - \theta_i] \frac{\partial \ln f_{Y|\Theta}(\boldsymbol{y}|\boldsymbol{\theta})}{\partial \theta_i} f_{Y|\Theta}(\boldsymbol{y}|\boldsymbol{\theta}) d\boldsymbol{y} \right)^2 = \left(1 + \frac{\partial \theta_{\text{Bias}}}{\partial \theta_i} \right)^2 \tag{7.106}$$

For the left-hand side of Equation (7.106) application of the Schwartz inequality

$$\left(\int_{-\infty}^{\infty} f(y) g(y) dy \right)^2 \leq \int_{-\infty}^{\infty} (f(y))^2 \, dy \times \int_{-\infty}^{\infty} (g(y))^2 \, dy \tag{7.107}$$

yields

$$\left\{ \int_{-\infty}^{\infty} \left([\hat{\theta}_i(\mathbf{y}) - \theta_{\text{Bias}} - \theta_i] f_{Y|\Theta}^{1/2}(\mathbf{y}|\boldsymbol{\theta}) \right) \left(\frac{\partial \ln f_{Y|\Theta}(\mathbf{y}|\boldsymbol{\theta})}{\partial \theta_i} f_{Y|\Theta}^{1/2}(\mathbf{y}|\boldsymbol{\theta}) \right) d\mathbf{y} \right\}^2$$
$$\leq \left\{ \int_{-\infty}^{\infty} \left([\hat{\theta}_i(\mathbf{y}) - \theta_{\text{Bias}} - \theta_i]^2 f_{Y|\Theta}(\mathbf{y}|\boldsymbol{\theta}) \right) d\mathbf{y} \right\} \left\{ \int_{-\infty}^{\infty} \left(\frac{\partial \ln f_{Y|\Theta}(\mathbf{y}|\boldsymbol{\theta})}{\partial \theta_i} \right)^2 f_{Y|\Theta}(\mathbf{y}|\boldsymbol{\theta}) d\mathbf{y} \right\} \tag{7.108}$$

From Equations (7.106) and (7.108), we have

$$\text{Var}[\hat{\theta}_i(\mathbf{y})] \geq \frac{\left(1 + \dfrac{\partial \theta_{\text{Bias}}}{\partial \theta_i} \right)^2}{\mathcal{E}\left[\left(\dfrac{\partial \ln f_{Y|\Theta}(\mathbf{y}|\boldsymbol{\theta})}{\partial \theta_i} \right)^2 \right]} \tag{7.109}$$

The Cramer–Rao inequality (7.99) results directly from the inequality (7.109). □

7.4.1 Cramer–Rao Bound for Random Parameters

For random parameters the Cramer–Rao bound may be obtained using the same procedure as above, with the difference that in Equation (7.99) instead of the likelihood $f_{Y|\Theta}(\mathbf{y}|\boldsymbol{\theta})$ we use the joint pdf $f_{Y|\Theta}(\mathbf{y}|\boldsymbol{\theta})$, and we also use the logarithmic relation

$$\frac{\partial \ln f_{Y,\Theta}(\mathbf{y}, \boldsymbol{\theta})}{\partial \theta_i} = \frac{\partial \ln f_{Y|\Theta}(\mathbf{y}|\boldsymbol{\theta})}{\partial \theta_i} + \frac{\partial \ln f_{\Theta}(\boldsymbol{\theta})}{\partial \theta_i} \tag{7.110}$$

The Cramer–Rao bound for random parameters is obtained as

$$\text{Var}[\hat{\theta}_i(\mathbf{y})] \geq \frac{\left(1 + \dfrac{\partial \theta_{\text{Bias}}}{\partial \theta_i} \right)^2}{\mathcal{E}\left[\left(\dfrac{\partial \ln f_{Y|\Theta}(\mathbf{y}|\boldsymbol{\theta})}{\partial \theta_i} \right)^2 + \left(\dfrac{\partial \ln f_{\Theta}(\boldsymbol{\theta})}{\partial \theta_i} \right)^2 \right]} \tag{7.111}$$

where the second term in the denominator of Equation (7.111) describes the effect of the prior pdf of $\boldsymbol{\theta}$. As expected the use of the prior $f_{\Theta}(\boldsymbol{\theta})$ can result in a decrease in the variance of the estimate. An alternative form of the minimum bound on estimation variance can be obtained by using the likelihood relation

$$\mathcal{E}\left[\left(\frac{\partial \ln f_{Y,\Theta}(\mathbf{y}, \boldsymbol{\theta})}{\partial \theta_i} \right)^2 \right] = -\mathcal{E}\left[\frac{\partial^2 \ln f_{Y,\Theta}(\mathbf{y}, \boldsymbol{\theta})}{\partial \theta_i^2} \right] \tag{7.112}$$

as

$$\text{Var}[\hat{\theta}_i(\mathbf{y})] \geq -\frac{\left(1 + \dfrac{\partial \theta_{\text{Bias}}}{\partial \theta_i} \right)^2}{\mathcal{E}\left[\dfrac{\partial^2 \ln f_{Y|\Theta}(\mathbf{y}|\boldsymbol{\theta})}{\partial \theta_i^2} + \dfrac{\partial^2 \ln f_{\Theta}(\boldsymbol{\theta})}{\partial \theta_i^2} \right]} \tag{7.113}$$

7.4.2 Cramer–Rao Bound for a Vector Parameter

For real-valued P-dimensional vector parameters, the Cramer–Rao bound for the covariance matrix of an unbiased estimator of $\boldsymbol{\theta}$ is given by

$$\mathrm{Cov}[\hat{\boldsymbol{\theta}}] \geq \mathbf{J}^{-1}(\boldsymbol{\theta}) \tag{7.114}$$

where $\mathbf{J}$ is the $P \times P$ Fisher information matrix, with its elements given by

$$[\mathbf{J}(\boldsymbol{\theta})]_{ij} = -\mathcal{E}\left[\frac{\partial^2 \ln f_{Y,\Theta}(\boldsymbol{y}, \boldsymbol{\theta})}{\partial\theta_i \partial\theta_j}\right] \tag{7.115}$$

The lower bound on the variance of the ith element of the vector θ is given by

$$\mathrm{Var}(\hat{\theta}_i) \geq \left[\mathbf{J}^{-1}(\boldsymbol{\theta})\right]_{ii} = \frac{1}{\mathcal{E}\left[\dfrac{\partial^2 \ln f_{Y,\Theta}(\boldsymbol{y}, \boldsymbol{\theta})}{\partial\theta_i^2}\right]} \tag{7.116}$$

where $\left[\mathbf{J}^{-1}(\boldsymbol{\theta})\right]_{ii}$ is the ith diagonal element of the inverse of the Fisher matrix.

7.5 Design of Gaussian Mixture Models (GMM)

A practical method for modelling the probability density function of an arbitrary signal space is to fit (or 'tile') the space with a mixture of a number of Gaussian probability density functions. The Gaussian functions hence act as elementary pdfs from which other pdfs can be constructed.

Figure 7.16 illustrates the cluster modelling of a two-dimensional signal space with a number of circular and elliptically-shaped Gaussian processes. Note that the Gaussian densities can be overlapping, with the result that in an area of overlap, a data point can be associated with different components of the Gaussian mixture.

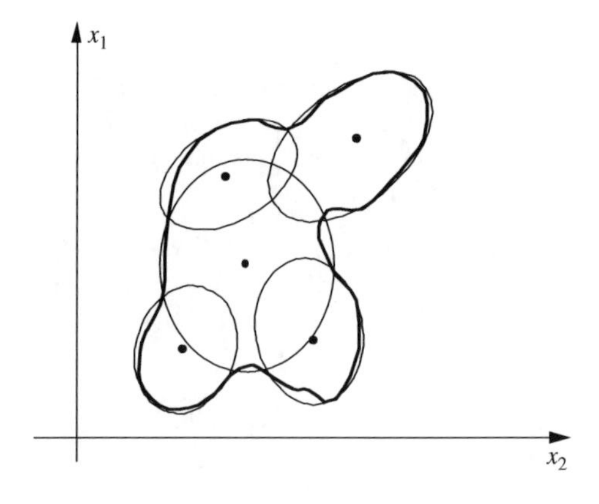

Figure 7.16 Illustration of probabilistic modelling of a two-dimensional signal space with a mixture of five bivariate Gaussian densities.

A main advantage of the use of a mixture Gaussian model is that it results in mathematically tractable signal processing solutions. A mixture Gaussian pdf model for a process $\boldsymbol{X}$ is defined as

$$f_X(\boldsymbol{x}) = \sum_{k=1}^{K} P_k \mathcal{N}_k(\boldsymbol{x}; \boldsymbol{\mu}_k, \boldsymbol{\Sigma}_k) \tag{7.117}$$

where $\mathcal{N}_k(\boldsymbol{\mu}_k, \boldsymbol{\Sigma}_k)$ denotes the kth component of the mixture Gaussian pdf, with mean vector μ_k and covariance matrix $\boldsymbol{\Sigma}_k$. The parameter P_k is the prior probability of the kth mixture, and it can be interpreted as the expected fraction of the number of vectors from the process $\boldsymbol{X}$ associated with the kth mixture.

In general, there are an infinite number of different K-mixture Gaussian densities that can be used to 'tile up' a signal space. Hence the modelling of a signal space with a K-mixture pdf space can be regarded as a many-to-one mapping, and the expectation–maximisation (EM) method can be applied for the estimation of the parameters of the Gaussian pdf models.

7.5.1 EM Estimation of Gaussian Mixture Model

The EM algorithm, discussed in Section 7.3, is an iterative maximum-likelihood (ML) estimation method, and can be employed to calculate the parameters of a K-mixture Gaussian pdf model for a given data set. To apply the EM method we first need to define the so-called complete and incomplete data sets. As usual the observation vectors $[\boldsymbol{y}(m),\ m = 0, \ldots, N-1]$ form the incomplete data. The complete data may be viewed as the observation vectors with a *label* k attached to each vector $\boldsymbol{y}(m)$ to indicate the component of the mixture Gaussian model that generated the vector. Note that if each signal vector $\boldsymbol{y}(m)$ had a mixture component label attached, then the computation of the mean vector and the covariance matrix of each component of the mixture would be a relatively simple exercise. Therefore the complete and incomplete data can be defined as follows:

The *incomplete* data $\quad \boldsymbol{y}(m), m = 0, \ldots, N-1$
The *complete* data $\quad \boldsymbol{x}(m) = [\boldsymbol{y}(m), k] = \boldsymbol{y}_k(m), m = 0, \ldots, N-1, k \in (1, \ldots, K)$

The probability of the complete data is the probability that an observation vector $\boldsymbol{y}(m)$ has a label k associating it with the kth component of the mixture density. The main step in application of the EM method is to define the expectation of the complete data, given the observations and a current estimate of the parameter vector, as

$$\begin{aligned} U(\boldsymbol{\Theta}, \hat{\Theta}_i) &= \mathcal{E}[\ln f_{Y,K;\boldsymbol{\Theta}}(\boldsymbol{y}(m), k; \boldsymbol{\Theta}) | \boldsymbol{y}(m); \hat{\boldsymbol{\Theta}}_i] \\ &= \sum_{m=0}^{N-1} \sum_{k=1}^{K} \frac{f_{Y,K|\boldsymbol{\Theta}}(\boldsymbol{y}(m), k | \hat{\boldsymbol{\Theta}}_i)}{f_{Y|\boldsymbol{\Theta}}(\boldsymbol{y}(m) | \hat{\boldsymbol{\Theta}}_i)} \ln f_{Y,K;\boldsymbol{\Theta}}(\boldsymbol{y}(m), k; \boldsymbol{\Theta}) \end{aligned} \tag{7.118}$$

where $\boldsymbol{\Theta} = \{\boldsymbol{\theta}_k = [P_k, \boldsymbol{\mu}_k, \boldsymbol{\Sigma}_k], k = 1, \ldots, K\}$, are the parameters of the Gaussian mixture as in Equation (7.117). Now the joint pdf of $\boldsymbol{y}(m)$ and the kth Gaussian component of the mixture density can be written as

$$\begin{aligned} f_{Y,K|\boldsymbol{\Theta}}\left(\boldsymbol{y}(m), k \middle| \hat{\boldsymbol{\theta}}_i\right) &= P_{k_i} f_k\left(\boldsymbol{y}(m) \middle| \hat{\boldsymbol{\theta}}_{k_i}\right) \\ &= P_{k_i} \mathcal{N}_k\left(\boldsymbol{y}(m); \hat{\boldsymbol{\mu}}_{k_i}, \hat{\boldsymbol{\Sigma}}_{k_i}\right) \end{aligned} \tag{7.119}$$

where $\mathcal{N}_k\left(\boldsymbol{y}(m); \hat{\boldsymbol{\mu}}_k, \hat{\boldsymbol{\Sigma}}_k\right)$ is a Gaussian density with mean vector $\boldsymbol{\mu}_k$ and covariance matrix $\boldsymbol{\Sigma}_k$:

$$N_k\left(\boldsymbol{y}(m); \boldsymbol{\mu}_k, \boldsymbol{\Sigma}_k\right) = \frac{1}{(2\pi)^{P/2}|\boldsymbol{\Sigma}_k|^{1/2}} \exp\left(-\frac{1}{2}(\boldsymbol{y}(m)-\boldsymbol{\mu}_k)^T \boldsymbol{\Sigma}_k^{-1}(\boldsymbol{y}(m)-\boldsymbol{\mu}_k)\right) \tag{7.120}$$

The pdf of $\boldsymbol{y}(m)$ as a mixture of K Gaussian densities is given by

$$\begin{aligned} f_{Y|\boldsymbol{\theta}}\left(\boldsymbol{y}(m)\middle|\hat{\boldsymbol{\theta}}_i\right) &= \mathcal{N}\left(\boldsymbol{y}(m)\middle|\hat{\boldsymbol{\theta}}_i\right) \\ &= \sum_{k=1}^{K} \hat{P}_{k_i} \mathcal{N}_k\left(\boldsymbol{y}(m); \hat{\boldsymbol{\mu}}_{k_i}, \hat{\boldsymbol{\Sigma}}_{k_i}\right) \end{aligned} \tag{7.121}$$

Substitution of the Gaussian densities of Equation (7.119) and Equation (7.121) in Equation (7.118) yields

$$\begin{aligned} U\left[(\boldsymbol{\mu}; \boldsymbol{\Sigma}, \boldsymbol{P}), (\hat{\boldsymbol{\mu}}_i, \hat{\boldsymbol{\Sigma}}_i, \hat{\boldsymbol{P}}_i)\right] &= \sum_{m=0}^{N-1}\sum_{k=1}^{K} \frac{\hat{P}_{k_i}\mathcal{N}_k\left(\boldsymbol{y}(m); \hat{\boldsymbol{\mu}}_{k_i}, \hat{\boldsymbol{\Sigma}}_{k_i}\right)}{\mathcal{N}(\boldsymbol{y}(m)|\hat{\boldsymbol{\Theta}}_i)} \ln\left[P_k \mathcal{N}_k\left(\boldsymbol{y}(m); \boldsymbol{\mu}_k, \boldsymbol{\Sigma}_k\right)\right] \\ &= \sum_{m=0}^{N-1}\sum_{k=1}^{K}\left(\frac{\hat{P}_{k_i}\mathcal{N}_k\left(\boldsymbol{y}(m); \hat{\boldsymbol{\mu}}_{k_i}, \hat{\boldsymbol{\Sigma}}_{k_i}\right)}{\mathcal{N}(\boldsymbol{y}(m)|\hat{\boldsymbol{\Theta}}_i)} \ln P_{k_i}\right. \\ &\quad \left. + \frac{\hat{P}_{k_i}\mathcal{N}_k\left(\boldsymbol{y}(m); \hat{\boldsymbol{\mu}}_{k_i}, \hat{\boldsymbol{\Sigma}}_{k_i}\right)}{\mathcal{N}(\boldsymbol{y}(m)|\hat{\boldsymbol{\Theta}}_i)} \ln \mathcal{N}\left(\boldsymbol{y}(m); \boldsymbol{\mu}_k, \boldsymbol{\Sigma}_k\right)\right) \end{aligned} \tag{7.122}$$

Equation (7.122) is maximised with respect to the parameter P_k using the constrained optimisation method. This involves subtracting the constant term $\Sigma P_k = 1$ from the right-hand side of Equation (7.122) and then setting the derivative of this equation with respect to P_k to zero; this yields

$$\hat{P}_{k_{i+1}} = \arg\max_{P_k} U\left[(\boldsymbol{\mu}; \boldsymbol{\Sigma}, \boldsymbol{P}), (\hat{\boldsymbol{\mu}}_i, \hat{\boldsymbol{\Sigma}}_i, \hat{\boldsymbol{P}}_i)\right] = \frac{1}{N}\sum_{m=0}^{N-1} \frac{\hat{P}_{k_i}\mathcal{N}_k\left(\boldsymbol{y}(m); \hat{\boldsymbol{\mu}}_{k_i}, \hat{\boldsymbol{\Sigma}}_{k_i}\right)}{\mathcal{N}(\boldsymbol{y}(m)|\hat{\boldsymbol{\Theta}}_i)} \tag{7.123}$$

The parameters $\boldsymbol{\mu}_k$ and $\boldsymbol{\Sigma}_k$ that maximise the function U are obtained by setting the derivative of the function with respect to these parameters to zero:

$$\begin{aligned} \hat{\boldsymbol{\mu}}_{k_{i+1}} &= \arg\max_{\hat{\boldsymbol{\mu}}_k} U\left[(\boldsymbol{\mu}; \boldsymbol{\Sigma}, \boldsymbol{P}), (\hat{\boldsymbol{\mu}}_i, \hat{\boldsymbol{\Sigma}}_i, \hat{\boldsymbol{P}}_i)\right] \\ &= \frac{\displaystyle\sum_{m=0}^{N-1} \frac{\hat{P}_{k_i}\mathcal{N}_k\left(\boldsymbol{y}(m); \hat{\boldsymbol{\mu}}_{k_i}, \hat{\boldsymbol{\Sigma}}_{k_i}\right)}{\mathcal{N}(\boldsymbol{y}(m)|\hat{\boldsymbol{\Theta}}_i)}\boldsymbol{y}(m)}{\displaystyle\sum_{m=0}^{N-1} \frac{\hat{P}_{k_i}\mathcal{N}_k\left(\boldsymbol{y}(m); \hat{\boldsymbol{\mu}}_{k_i}, \hat{\boldsymbol{\Sigma}}_{k_i}\right)}{\mathcal{N}(\boldsymbol{y}(m)|\hat{\boldsymbol{\Theta}}_i)}} \end{aligned} \tag{7.124}$$

and

$$\hat{\Sigma}_{k_{i+1}} = \arg\max_{\hat{\Sigma}_k} U\left[(\boldsymbol{\mu}; \boldsymbol{\Sigma}, \boldsymbol{P}), (\hat{\boldsymbol{\mu}}_i, \hat{\boldsymbol{\Sigma}}_i, \hat{\boldsymbol{P}}_i)\right]$$

$$= \frac{\sum_{m=0}^{N-1} \frac{\hat{P}_{k_i} \mathcal{N}_k\left(\boldsymbol{y}(m); \hat{\boldsymbol{\mu}}_{k_i}, \hat{\boldsymbol{\Sigma}}_{k_i}\right)}{\mathcal{N}(\boldsymbol{y}(m)|\hat{\boldsymbol{\Theta}}_i)} \left(\boldsymbol{y}(m) - \hat{\boldsymbol{\mu}}_{k_i}\right)\left(\boldsymbol{y}(m) - \hat{\boldsymbol{\mu}}_{k_i}\right)^{\mathrm{T}}}{\sum_{m=0}^{N-1} \frac{\hat{P}_{k_i} \mathcal{N}_k\left(\boldsymbol{y}(m); \hat{\boldsymbol{\mu}}_{k_i}, \hat{\boldsymbol{\Sigma}}_{k_i}\right)}{\mathcal{N}(\boldsymbol{y}(m)|\hat{\boldsymbol{\Theta}}_i)}} \tag{7.125}$$

Equations (7.123)–(7.125) are the estimates of the parameters of a mixture Gaussian pdf model. These equations can be used in further iterations of the EM method until the parameter estimates converge.

7.6 Bayesian Classification

Classification is the process of *labelling* of an observation sequence $\{\boldsymbol{y}(m)\}$ with one of M classes of signals $\{C_k; k = 1, \ldots, M\}$ that could have generated the observation. Classifiers are present in all modern digital communication systems and in applications such as the decoding of discrete-valued symbols in digital communication receivers, speech compression, video compression, speech recognition, image recognition, character recognition, signal/noise classification and detectors.

For example, in an M-symbol digital communication system, the channel output signal is classified as one of the M signalling symbols; in speech recognition systems that are based on acoustic models of phonemes, each speech segment is labelled with one of the phonemes; and in speech or video compression, a segment of speech samples or a block of image pixels are quantised and labelled with one of a number of prototype signal vectorsin a codebook. In the design of a classifier, the aim is to reduce the classification error given the constraints on the signal-to-noise ratio, available training data, bandwidth and the computational resources.

Classification errors are due to overlap of the distributions of different classes of signals. This is illustrated in Figure 7.17 for a binary classification problem with two Gaussian distributed signal classes C_1 and C_2. In the shaded region, where the signal distributions overlap, a sample x could belong to either of the two classes. The shaded area gives a measure of the classification error. The obvious solution suggested by Figure 7.17 for reducing the classification error is to reduce the overlap of the distributions. The overlap can be reduced in two ways: (a) by increasing the distance between the mean values of different classes, and (b) by reducing the variance of each class. In

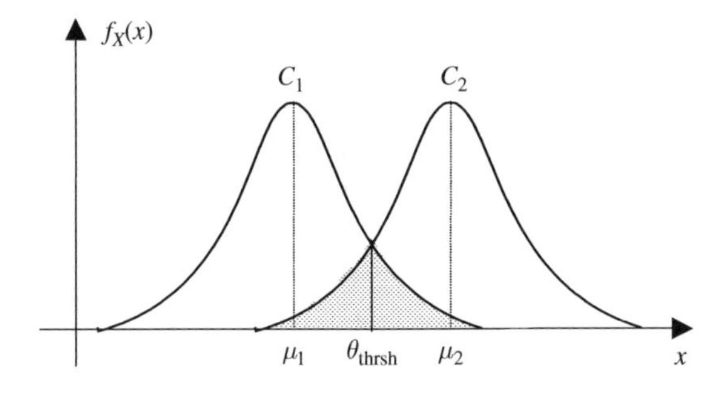

Figure 7.17 Illustration of the overlap of the distribution of two classes of signals.

telecommunication systems the overlap between the signal classes is reduced using a combination of several methods including increasing the signal-to-noise ratio, increasing the distance between signal patterns by adding redundant error control coding bits, and signal shaping and post-filtering operations. In pattern recognition, where it is not possible to control the signal generation process (as in speech and image recognition), the choice of the pattern features and models affects the classification error. The design of an efficient classification for pattern recognition depends on a number of factors, which can be listed as follows:

(1) Extraction and transformation of a set of discriminative features from the signal that can aid the classification process. The features need to adequately characterise each class and emphasise the differences between various classes.
(2) Statistical modelling of the observation features for each class. For Bayesian classification, a posterior probability model for each class should be obtained.
(3) Labelling of an unlabelled signal with one of the N classes.

7.6.1 Binary Classification

The simplest form of classification is the labelling of an observation with one of two classes of signals. Figures 7.18(a) and 7.18(b) illustrate two examples of a simple binary classification problem in a two-dimensional signal space. In each case, the observation is the result of a random mapping (e.g. signal plus noise) from the binary source to the continuous observation space. In Figure 7.18(a), the binary sources and the observation space associated with each source are well separated, and it is possible to make an error-free classification of each observation. In Figure 7.18(b) there is less distance

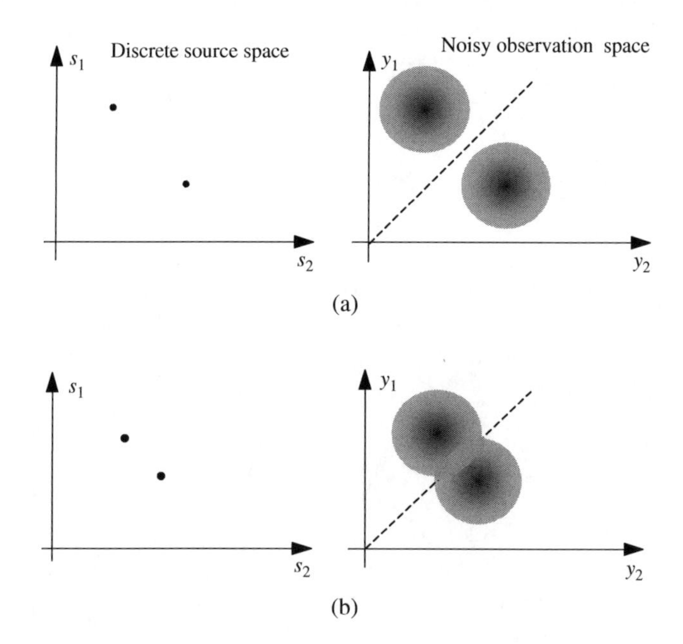

Figure 7.18 Illustration of binary classification: (a) the source and observation spaces are well separated, (b) the observation spaces overlap.

between the mean of the sources, and the observation signals have a greater spread. These result in some overlap of the signal spaces and classification error can occur.

In binary classification, a signal $\boldsymbol{x}$ is labelled with the class that scores the higher a posterior probability:

$$P_{C|X}\left(C_1\left|\boldsymbol{x}\right.\right)\underset{C_2}{\overset{C_1}{\gtrless}}P_{C|X}\left(C_2\left|\boldsymbol{x}\right.\right) \tag{7.126}$$

Note the above notation means that a signal $\boldsymbol{x}$ is classified as C_1 if $P_{C|X}\left(C_1\left|\boldsymbol{x}\right.\right) > P_{C|X}\left(C_2\left|\boldsymbol{x}\right.\right)$ otherwise it is classified as C_2. Using Bayes' rule Equation (7.126) can be rewritten as

$$P_C\left(C_1\right) f_X \left|C\right. \left(\boldsymbol{x}\left|C_1\right.\right)\underset{C_2}{\overset{C_1}{\gtrless}}P_C\left(C_2\right) f_{X|C}\left(\boldsymbol{x}\left|C_2\right.\right) \tag{7.127}$$

Letting $P_C(C_1)=P_1$ and $P_C(C_2)=P_2$, Equation (7.127) is often written in terms of a *likelihood ratio test* as

$$\frac{f_{X|C}\left(\boldsymbol{x}\left|C_1\right.\right)}{f_{X|C}\left(\boldsymbol{x}\left|C_2\right.\right)}\underset{C_2}{\overset{C_1}{\gtrless}}\frac{P_2}{P_1} \tag{7.128}$$

Taking the likelihood ratio yields the discriminant function

$$h(\boldsymbol{x}) = \ln f_{X|C}\left(\boldsymbol{x}\left|C_1\right.\right) - \ln f_{X|C}\left(\boldsymbol{x}\left|C_2\right.\right)\underset{C_2}{\overset{C_1}{\gtrless}}\ln\frac{P_2}{P_1} \tag{7.129}$$

Now assume that the signal in each class has a Gaussian distribution with a probability distribution function given by

$$f_{X|C}\left(\boldsymbol{x}\left|c_i\right.\right) = \frac{1}{\sqrt{2\pi}\left|\boldsymbol{\Sigma}_i\right|}\exp\left[-\frac{1}{2}(\boldsymbol{x}-\boldsymbol{\mu}_i)^{\mathrm{T}}\boldsymbol{\Sigma}_i^{-1}(\boldsymbol{x}-\boldsymbol{\mu}_i)\right],\quad i=1,2 \tag{7.130}$$

From Equations (7.129) and (7.130), the discriminant function $h(\boldsymbol{x})$ becomes

$$h\left(\boldsymbol{x}\right) = -\frac{1}{2}(\boldsymbol{x}-\boldsymbol{\mu}_1)^{\mathrm{T}}\boldsymbol{\Sigma}_1^{-1}(\boldsymbol{x}-\boldsymbol{\mu}_1)+\frac{1}{2}(\boldsymbol{x}-\boldsymbol{\mu}_2)^{\mathrm{T}}\boldsymbol{\Sigma}_2^{-1}(\boldsymbol{x}-\boldsymbol{\mu}_2)+\ln\frac{\left|\boldsymbol{\Sigma}_2\right|}{\left|\boldsymbol{\Sigma}_1\right|}\underset{C_2}{\overset{C_1}{\gtrless}}\ln\frac{P_2}{P_1} \tag{7.131}$$

Example 7.11 For two Gaussian-distributed classes of scalar-valued signals with distributions given by $\mathcal{N}(x(m), \mu_1, \sigma^2)$ and $\mathcal{N}(x(m), \mu_2, \sigma^2)$, and equal class probability $P_1 = P_2 = 0.5$, the discrimination function of Equation (7.131) becomes

$$h(x(m)) = \frac{\mu_2-\mu_1}{\sigma^2}x(m)+\frac{1}{2}\frac{\mu_2^2-\mu_1^2}{\sigma^2}\underset{C_2}{\overset{C_1}{\gtrless}}0 \tag{7.132}$$

Hence the rule for signal classification becomes

$$x(m)\underset{C_2}{\overset{C_1}{\gtrless}}\frac{\mu_1+\mu_2}{2} \tag{7.133}$$

The signal is labelled with class C_1 if $x(m) < (\mu_1+\mu_2)/2$ and as class C_2 otherwise.

7.6.2 Classification Error

Classification errors are due to the overlap of the distributions of different classes of signals. This is illustrated in Figure 7.17 for the binary classification of a scalar-valued signal and in Figure 7.18 for the binary classification of a two-dimensional signal. In each figure the overlapped area gives a measure of the classification error. The obvious solution for reducing the classification error is to reduce the overlap of the distributions. This may be achieved by increasing the distance between the mean values of various classes or by reducing the variance of each class. In the binary classification of a scalar-valued variable x, the probability of classification error is given by

$$P(\text{Error}\,|x) = P(C_1)\,P(x > \text{Thrsh}|x \in C_1) + P(C_2)\,P(x < \text{Thrsh}|x \in C_2) \tag{7.134}$$

For two Gaussian-distributed classes of scalar-valued signals with pdfs $\mathcal{N}(x(m), \mu_1, \sigma_1^2)$ and $\mathcal{N}(x(m), \mu_2, \sigma_2^2)$, Equation (7.134) becomes

$$\begin{aligned} P(\text{Error}\,|x) =& P(C_1) \int_{\text{Thrsh}}^{\infty} \frac{1}{\sqrt{2\pi}\sigma_1} \exp\left(-\frac{(x-\mu_1)^2}{2\sigma_1^2}\right) dx \\ &+ P(C_2) \int_{-\infty}^{\text{Thrsh}} \frac{1}{\sqrt{2\pi}\sigma_2} \left(-\frac{(x-\mu_2)^2}{2\sigma_2^2}\right) dx \end{aligned} \tag{7.135}$$

where the parameter Thrsh is the classification threshold.

7.6.3 Bayesian Classification of Discrete-Valued Parameters

Let the set $\boldsymbol{\Theta} = \{\boldsymbol{\theta}_i, i = 1, \ldots, M\}$ denote the values that a discrete P-dimensional parameter vector $\boldsymbol{\theta}$ can assume. In general, the observation space $\boldsymbol{Y}$ associated with a discrete parameter space $\boldsymbol{\Theta}$ may be a discrete-valued space or a continuous-valued space, an example of the latter is a discrete-valued parameter observed in continuous-valued noise.

Assuming the observation space is continuous, the pdf of the parameter vector $\boldsymbol{\theta}_i$, given observation $\boldsymbol{y}$, may be expressed, using Bayes' rule, as

$$P_{\boldsymbol{\Theta}|\boldsymbol{Y}}(\boldsymbol{\theta}_i|\boldsymbol{y}) = \frac{f_{\boldsymbol{Y}|\boldsymbol{\Theta}}(\boldsymbol{y}|\boldsymbol{\theta}_i) P_{\boldsymbol{\Theta}}(\boldsymbol{\theta}_i)}{f_{\boldsymbol{Y}}(\boldsymbol{y})} \tag{7.136}$$

For the case when the observation space $\boldsymbol{Y}$ is discrete-valued, the probability density functions are replaced by the appropriate probability mass functions. The Bayesian risk in selecting the parameter vector $\boldsymbol{\theta}_i$ given the observation $\boldsymbol{y}$ is defined as

$$\mathcal{R}(\boldsymbol{\theta}_i|\boldsymbol{y}) = \sum_{\mathrm{j}=1}^{M} C(\boldsymbol{\theta}_i|\boldsymbol{\theta}_\mathrm{j}) P_{\boldsymbol{\Theta}|\boldsymbol{Y}}(\boldsymbol{\theta}_\mathrm{j}|\boldsymbol{y}) \tag{7.137}$$

where $C(\boldsymbol{\theta}_i|\boldsymbol{\theta}_\mathrm{j})$ is the cost of selecting the parameter $\boldsymbol{\theta}_i$ when the true parameter is $\boldsymbol{\theta}_\mathrm{j}$. The Bayesian classification equation (7.137) can be employed to obtain the maximum a posteriori, the maximum likelihood or the minimum mean square error classifiers as described next.

7.6.4 Maximum A Posteriori Classification

MAP classification corresponds to Bayesian classification with a uniform cost function defined as

$$C(\boldsymbol{\theta}_i|\boldsymbol{\theta}_j) = 1 - \delta(\boldsymbol{\theta}_i, \boldsymbol{\theta}_j) \tag{7.138}$$

where $\delta(\cdot)$ is the delta function. Substitution of this cost function in the Bayesian risk function yields

$$\begin{aligned}\mathcal{R}_{\mathrm{MAP}}(\boldsymbol{\theta}_i|\boldsymbol{y}) &= \sum_{j=1}^{M}[1 - \delta(\boldsymbol{\theta}_i, \boldsymbol{\theta}_j)]P_{\Theta|y}(\boldsymbol{\theta}_j|\boldsymbol{y}) \\ &= 1 - P_{\Theta|y}(\boldsymbol{\theta}_i|\boldsymbol{y})\end{aligned} \tag{7.139}$$

Note that the MAP risk in selecting $\boldsymbol{\theta}_i$ is the classification error probability; that is the sum of the probabilities of all other candidates. From Equation (7.139) minimisation of the MAP risk function is achieved by maximisation of the posterior pmf:

$$\begin{aligned}\hat{\boldsymbol{\theta}}_{\mathrm{MAP}}(\boldsymbol{y}) &= \arg\max_{\boldsymbol{\theta}_i} P_{\Theta|Y}(\boldsymbol{\theta}_i|\boldsymbol{y}) \\ &= \arg\max_{\boldsymbol{\theta}_i} P_{\Theta}(\boldsymbol{\theta}_i) f_{Y|\Theta}(\boldsymbol{y}|\boldsymbol{\theta}_i)\end{aligned} \tag{7.140}$$

7.6.5 Maximum-Likelihood Classification

The ML classification corresponds to Bayesian classification when the parameter $\boldsymbol{\theta}$ has a uniform prior pmf and the cost function is also uniform:

$$\begin{aligned}\mathcal{R}_{\mathrm{ML}}(\boldsymbol{\theta}_i|\boldsymbol{y}) &= \sum_{j=1}^{M}[1 - \delta(\boldsymbol{\theta}_i, \boldsymbol{\theta}_j)]\frac{1}{f_Y(\boldsymbol{y})} f_{Y|\Theta}(\boldsymbol{y}|\boldsymbol{\theta}_j)\, P_{\Theta}(\boldsymbol{\theta}_j) \\ &= 1 - \frac{1}{f_Y(\boldsymbol{y})} f_{Y|\theta}(\boldsymbol{y}|\boldsymbol{\theta}_i)\, P_{\Theta}\end{aligned} \tag{7.141}$$

where P_{Θ} is the uniform pmf of $\boldsymbol{\theta}$. Minimisation of the ML risk function (7.141) is equivalent to maximisation of the likelihood $f_{Y|\Theta}(\boldsymbol{y}|\boldsymbol{\theta}_i)$:

$$\hat{\boldsymbol{\theta}}_{\mathrm{ML}}(\boldsymbol{y}) = \arg\max_{\boldsymbol{\theta}_i} f_{Y|\Theta}(\boldsymbol{y}|\boldsymbol{\theta}_i) \tag{7.142}$$

7.6.6 Minimum Mean Square Error Classification

The Bayesian minimum mean square error classification results from minimisation of the risk function

$$\mathcal{R}_{MMSE}(\boldsymbol{\theta}_i|\boldsymbol{y}) = \sum_{j=1}^{M} \left|\boldsymbol{\theta}_i - \boldsymbol{\theta}_j\right|^2 P_{\Theta|Y}(\boldsymbol{\theta}_j|\boldsymbol{y}) \tag{7.143}$$

For the case when $P_{\Theta|Y}(\boldsymbol{\theta}_j|\boldsymbol{y})$ is not available, the MMSE classifier is given by

$$\hat{\boldsymbol{\theta}}_{MMSE}(\boldsymbol{y}) = \arg\min_{\boldsymbol{\theta}_i} |\boldsymbol{\theta}_i - \boldsymbol{\theta}(\boldsymbol{y})|^2 \tag{7.144}$$

where $\boldsymbol{\theta}(\boldsymbol{y})$ is an estimate based on the observation $\boldsymbol{y}$.

7.6.7 Bayesian Classification of Finite State Processes

In this section, the classification problem is formulated within the framework of a finite state random process. A finite state process is composed of a probabilistic chain of a number of different random processes. Finite state processes are used for modelling non-stationary signals such as speech, image, background acoustic noise, and impulsive noise as discussed in Chapter 11.

Consider a process with a set of M states denoted as $S = \{s_1, s_2, \ldots, s_M\}$, where each state has some distinct statistical property. In its simplest form, a state is just a single vector, and the finite state process is equivalent to a discrete-valued random process with M outcomes. In this case the Bayesian state estimation is identical to the Bayesian classification of a signal into one of M discrete-valued vectors. More generally, a state generates continuous-valued, or discrete-valued vectors from a pdf, or a pmf, associated with the state. Figure 7.19 illustrates an M-state process, where the output of the ith state is expressed as

$$\mathbf{x}(m) = h_i(\boldsymbol{\theta}_i, \boldsymbol{e}(m)), \quad i = 1, \ldots, M \tag{7.145}$$

where in each state the signal $\boldsymbol{x}(m)$ is modelled as the output of a state-dependent function $h_i(\cdot)$ with parameter θ_i, input $\boldsymbol{e}(m)$ and an input pdf $f_{Ei}(\boldsymbol{e}(m))$. Each state may be a model of a segment of speech or image. The prior probability of each state is given by

$$P_S(s_i) = \mathcal{E}[N(s_i)] \Big/ \mathcal{E}\left[\sum_{j=1}^{M} N(s_j)\right] \tag{7.146}$$

where $E[N(s_i)]$ is the expected number of observation from state s_i. The pdf of the output of a finite state process is a weighted combination of the pdf of each state and is given by

$$f_X(\boldsymbol{x}(m)) = \sum_{i=1}^{M} P_S(s_i) f_{X|S}(\boldsymbol{x}|s_i) \tag{7.147}$$

In Figure 7.19, the noisy observation $\boldsymbol{y}(m)$ is the sum of the process output $\boldsymbol{x}(m)$ and an additive noise $\boldsymbol{n}(m)$. From Bayes' rule, the posterior probability of the state s_i given the observation $\boldsymbol{y}(m)$ can be expressed as

$$P_{S|Y}(s_i|\boldsymbol{y}(m)) = \frac{f_{Y|S}(\boldsymbol{y}(m)|s_i) P_S(s_i)}{\sum_{j=1}^{M} f_{Y|S}(\boldsymbol{y}(m)|s_j) P_S(s_j)} \tag{7.148}$$

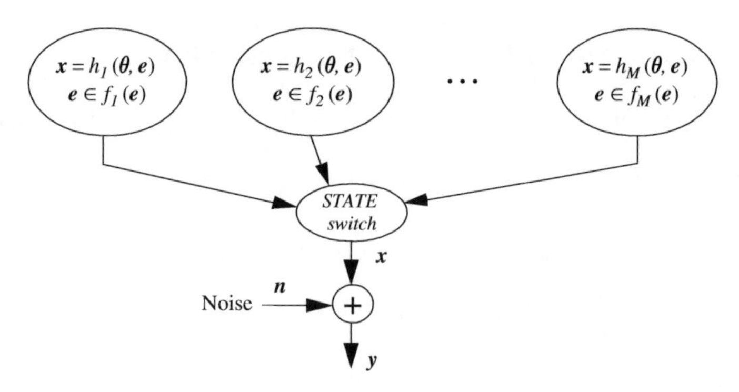

Figure 7.19 Illustration of a random process generated by a finite state system.

In MAP classification, the state with the maximum posterior probability is selected as

$$S_{\mathrm{MAP}}(\mathbf{y}(m)) = \arg\max_{s_i} P_{S|\mathbf{Y}}(s_i|\mathbf{y}(m)) \tag{7.149}$$

The Bayesian state classifier assigns a misclassification cost function $C(s_i|s_j)$ to the action of selecting the state s_i when the true state is s_j. The risk function for the Bayesian classification is given by

$$\mathcal{R}(s_i|\mathbf{y}(m)) = \sum_{j=1}^{M} C(s_i|s_j) P_{S|\mathbf{Y}}(s_j|\mathbf{y}(m)) \tag{7.150}$$

7.6.8 Bayesian Estimation of the Most Likely State Sequence

Consider the estimation of the most likely state sequence $\boldsymbol{s} = [s_{i_0}, s_{i_1}, \ldots, s_{i_{T-1}}]$ of a finite state process, given a sequence of T observation vectors $\boldsymbol{Y} = [\mathbf{y}_0, \mathbf{y}_1, \ldots, \mathbf{y}_{T-1}]$. A state sequence $\boldsymbol{s}$, of length T, is itself a random integer-valued vector process with N^T possible values. From Bayes' rule, the posterior pmf of a state sequence $\boldsymbol{s}$, given an observation sequence $\boldsymbol{Y}$, can be expressed as

$$P_{S|Y}(s_{i_0}, \ldots, s_{i_{T-1}}|\mathbf{y}_0, \ldots, \mathbf{y}_{T-1}) = \frac{f_{Y|S}(\mathbf{y}_0, \ldots, \mathbf{y}_{T-1}|s_{i_0}, \ldots, s_{i_{T-1}}) P_S(s_{i_0}, \ldots, s_{i_{T-1}})}{f_Y(\mathbf{y}_0, \ldots, \mathbf{y}_{T-1})} \tag{7.151}$$

where $P_S(\boldsymbol{s})$ is the pmf of the state sequence $\boldsymbol{s}$, and for a given observation sequence, the denominator $f_Y(\mathbf{y}_0, \ldots, \mathbf{y}_{T-1})$ is a constant. The Bayesian risk in selecting a state sequence $\boldsymbol{s}_i$ is expressed as

$$\mathcal{R}(\boldsymbol{s}_i|\boldsymbol{Y}) = \sum_{j=1}^{N^T} C(\boldsymbol{s}_i|\boldsymbol{s}_j) P_{S|Y}(\boldsymbol{s}_j|\boldsymbol{Y}) \tag{7.152}$$

For a statistically independent process, the state of the process at any time is independent of the previous states, and hence the conditional probability of a state sequence can be written as

$$P_{S|Y}(s_{i_0}, \ldots, s_{i_{T-1}}|\mathbf{y}_0, \ldots, \mathbf{y}_{T-1}) = \prod_{k=0}^{T-1} f_{Y|S}(\mathbf{y}_k|s_{i_k}) P_S(s_{i_k}) \tag{7.153}$$

where s_{ik} denotes state s_i at time instant k. A particular case of a finite state process is the Markov chain, Figure 7.20, where the state transition is governed by a Markovian process such that the probability of the state i at time m depends on the state of the process at time $m-1$. The conditional pmf of a Markov state sequence can be expressed as

$$P_{S|Y}(s_{i_0}, \ldots, s_{i_{T-1}}|\mathbf{y}_0, \ldots, \mathbf{y}_{T-1}) = \prod_{k=0}^{T-1} a_{i_{k-1}i_k} f_{S|Y}(s_{i_k}|\mathbf{y}_k) \tag{7.154}$$

where $a_{i_{k-1}i_k}$ is the probability that the process moves from state $s_{i_{k-1}}$ to state s_{i_k}. Finite state random processes and computationally efficient methods of state sequence estimation are described in detail in Chapter 11.

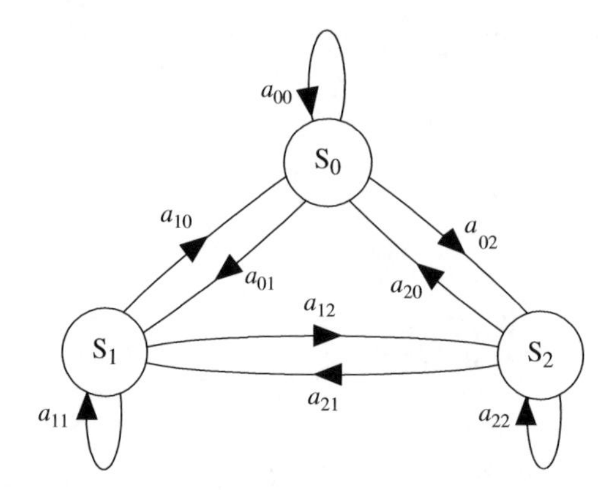

Figure 7.20 A three-state Markov process.

7.7 Modelling the Space of a Random Process

In this section, we consider the training of statistical models for a database of P-dimensional vectors of a random process. The vectors in the database can be visualised as forming a number of clusters in a P-dimensional space. The statistical modelling method consists of two steps: (a) the partitioning of the database into a number of regions, or clusters, and (b) the estimation of the parameters of a statistical model for each cluster. A simple method for modelling the space of a random signal is to use a set of prototype vectors that represent the centroids of the signal space. This method effectively quantises the space of a random process into a relatively small number of typical vectors, and is known as *vector quantisation* (VQ). In the following, we first consider a VQ model of a random process, and then extend this model to a pdf model, based on a mixture of Gaussian densities.

7.7.1 Vector Quantisation of a Random Process

Vector quantisation is used in signal compression and pattern recognition, such as in the coding or recognition of speech, music or image signals.

In vector quantisation, the space of the training data, from a random vector process $\boldsymbol{X}$, is partitioned into K clusters or regions $[\boldsymbol{X}_1, \boldsymbol{X}_2, \ldots, \boldsymbol{X}_K]$ and each cluster $\boldsymbol{X}_i$ is represented by a cluster centroid $\boldsymbol{c}_i$. The set of centroid vectors $[\boldsymbol{c}_1, \boldsymbol{c}_2, \ldots, \boldsymbol{c}_K]$ forms a VQ codebook model for the process $\boldsymbol{X}$.

The VQ codebook can then be used to classify an unlabelled vector $\boldsymbol{x}$ with the nearest centroid. The codebook is searched to find the centroid vector with the minimum distance from $\boldsymbol{x}$, then $\boldsymbol{x}$ is labelled with the index of the minimum distance centroid as

$$\text{Label}(\boldsymbol{x}) = \arg\min_i d(\boldsymbol{x}, \boldsymbol{c}_i) \tag{7.155}$$

where $d(\boldsymbol{x}, \boldsymbol{c}_i)$ is a measure of distance between the vectors $\boldsymbol{x}$ and $\boldsymbol{c}_i$. The most commonly used distance measure is the mean squared distance.

7.7.2 Vector Quantisation using Gaussian Models of Clusters

In vector quantisation, instead of using only the centre or mean value of each cluster, a Gaussian pdf model of each cluster comprising the mean of the cluster, its covariance matrix and its probability may be used. In this way, the space of the training data, from a random vector process $\boldsymbol{X}$, is partitioned into K clusters or regions $[\boldsymbol{X}_1, \boldsymbol{X}_2, \ldots, \boldsymbol{X}_K]$ and each cluster $\boldsymbol{X}_i$ is represented by a cluster centroid $\boldsymbol{c}_i$, the cluster covariance matrix Σ_i and the cluster probability p_i as $[N(\boldsymbol{c}_i, \boldsymbol{\Sigma}_i), p_i]$. The set of VQ pdfs $\{[N(\boldsymbol{c}_1, \boldsymbol{\Sigma}_1), p_1], [N(\boldsymbol{c}_2, \boldsymbol{\Sigma}_2), p_2], \ldots, [N(\boldsymbol{c}_K, \boldsymbol{\Sigma}_K), p_K]\}$ forms a VQ codebook model for the process $\boldsymbol{X}$. The VQ codebook can then be used to classify an unlabelled vector $\boldsymbol{x}$ with the nearest pdf. The codebook is searched to find the VQ pdf with the maximum probability of membership for $\boldsymbol{x}$, then $\boldsymbol{x}$ is labelled with the index of the pdf as

$$\text{Label}(\boldsymbol{x}) = \arg\max_i p_i \mathcal{N}(\mathbf{x}, \boldsymbol{c}_i, \boldsymbol{\Sigma}_i) \tag{7.156}$$

where the weighted Gaussian pdf distance $p_i \mathcal{M}(\boldsymbol{x}, \boldsymbol{c}_i, \boldsymbol{\Sigma}_i)$, is a measure of membership of the input vector $\boldsymbol{x}$ and the VQ class i.

7.7.3 Design of a Vector Quantiser: *K*-Means Clustering

The K-means algorithm, illustrated in Figures 7.21 and 7.22, is an iterative method for the design of a VQ codebook. Each iteration consists of two basic steps: (a) partition the training signal space into K regions or clusters and (b) compute the centroid of each region. The steps in the K-means method are as follows:

Step 1: *Initialisation* Use a suitable method to choose a set of K initial centroids $[\boldsymbol{c}_i]$ for $m = 1, 2, \ldots$

Step 2: *Classification* Classify the training vectors $\{\boldsymbol{x}\}$ into K clusters $\{[\boldsymbol{x}_1], [\boldsymbol{x}_2], \ldots, [\boldsymbol{x}_K]\}$ using the so-called nearest-neighbour rule, Equation (7.154).

Step 3: *Centroid computation* Use the vectors $[\boldsymbol{x}_i]$ associated with the ith cluster to compute an updated cluster centroid $\boldsymbol{c}_i$, and calculate the cluster distortion defined as

$$D_i(m) = \frac{1}{N_i} \sum_{j=1}^{N_i} d(\boldsymbol{x}_i(j), \boldsymbol{c}_i(m)) \tag{7.157}$$

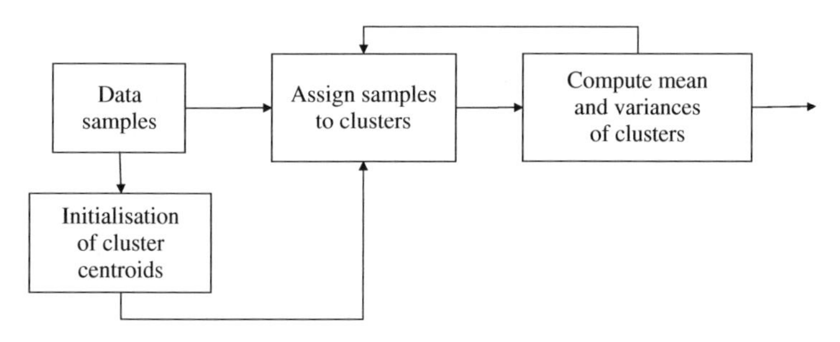

Figure 7.21 Illustration of the K-means algorithm.

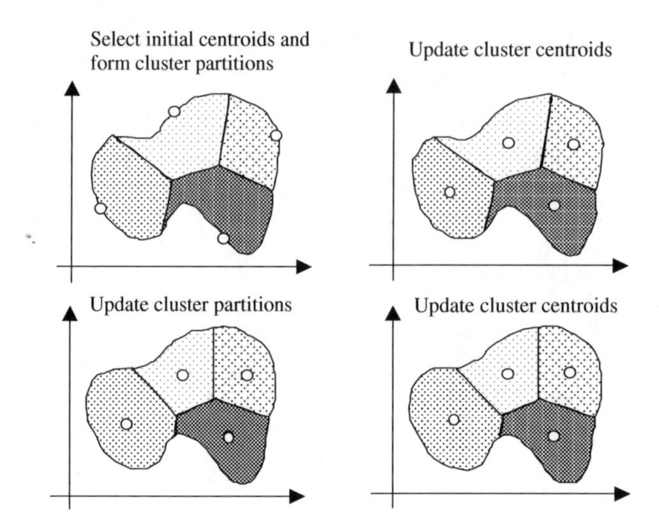

Figure 7.22 Illustration of the working of the K-means clustering method.

where it is assumed that a set of N_i vectors $[\boldsymbol{x}_i(j),\ j=0,\ldots,N_i]$ are associated with cluster i. The total distortion is given by

$$D(m)=\sum_{i=1}^{K}D_i(m) \tag{7.158}$$

Step 4: *Convergence test*:
if
$D(m-1)-D(m)\geq$ Threshold stop,
else
goto Step 2.

A vector quantiser models the regions, or the clusters, of the signal space with a set of cluster centroids. A more complete description of the signal space can be achieved by modelling each cluster with a Gaussian density as described in Chapter 11.

MATLAB function K_means_2DGaussian( )

Demonstrates the k-means clustering method for a sequence of vector-valued signals. The program generates a 2-D Gaussian mixture signal composed of a mix of M 2-D Gaussian random signals with mean vectors $[m_1, m_2, \ldots, m_M]$ and covariance matrices $[s_1, s_2, \ldots, s_M]$ and probabilities $[p_1, p_2, \ldots, p_M]$. The program then uses the k-means method to estimate the mean vector and covariance matrices of the modes of the mixture; a scatter diagram shows the signal samples and the estimated mean and covariance values.

7.8 Summary

This chapter began with an introduction to the basic concepts in estimation theory, such as the signal space and the parameter space, the prior and posterior spaces, and the statistical measures that are used to quantify the performance of an estimator. The Bayesian inference method, with its ability to include as much information as is available, provides a general framework for statistical signal processing problems. The minimum mean square error, the maximum-likelihood, the maximum a posteriori, and the minimum absolute value of error methods were derived from the Bayesian formulation. Further examples of the applications of Bayesian type models in this book include the hidden Markov models for non-stationary processes studied in Chapter 11, and blind equalisation of distorted signals studied in Chapter 17.

We considered a number of examples of the estimation of a signal observed in noise, and derived expressions for the effects of using prior pdfs on the mean and the variance of the estimates. The choice of the prior pdf is an important consideration in Bayesian estimation. Many processes, for example speech or the response of a telecommunication channel, are not uniformly distributed in space, but are constrained to a particular region of signal or parameter space. The use of a prior pdf can guide the estimator to focus on the posterior space, that is the subspace consistent with both the likelihood and the prior pdfs. The choice of the prior, depending on how well it fits the process, can have a significant influence on the solutions.

The iterative estimate–maximise method, studied in Section 7.3, provides a practical framework for solving many statistical signal processing problems, such as the modelling of a signal space with a mixture Gaussian densities, and the training of hidden Markov models in Chapter 11. In Section 7.4 the Cramer–Rao lower bound on the variance of an estimator was derived, and it was shown that the use of a prior pdf can reduce the minimum estimator variance.

Finally we considered the problem of vector quantisation, the popular K-means clustering methods and the modelling of a data space with a mixture Gaussian process. We used the EM method to derive a solution for the parameters of the mixture Gaussian model.

Further Reading

Abramson N. (1963) Information Theory and Coding. McGraw Hill, New York.
Andergerg M.R. (1973) Cluster Analysis for Applications. Academic Press, New York.
Baum L.E., Petrie T., Soules G. and Weiss N. (1970) A Maximisation Technique Occurring in the Statistical Analysis of Probabilistic Functions of Markov Chains. Ann. Math. Stat. 41, pp. 164–171.
Bayes T. (1763) An Essay Towards Solving a Problem in the Doctrine of Changes, Phil. Trans. Royal Society of London, 53, pp. 370–418, (reprinted in 1958 in Biometrika, 45, pp. 293–315).
Bezdek J.C. (1981) Pattern Recognition with Fuzzy Objective Function Algorithms. Plenum Press, New York.
Chou P., Lookabaugh T. and Gray R. (1989) Entropy-Constrained Vector Quantisation. IEEE Trans. Acoustics, Speech and Signal Processing, ASSP-37, pp. 31–42.
Cramer H. (1974) Mathematical Methods of Statistics. Princeton University Press.
Dempster A.P., Laird N.M. and Rubin D.B. (1977) Maximum Likelihood from Incomplete Data via the EM Algorithm. J. R. Stat. Soc. Ser. B, 39, pp. 1–38.
Deutsch R. (1965) Estimation Theory. Prentice-Hall, Englewood Cliffs, NJ.
Duda R.O. and Hart R.E. (1973) Pattern Classification. Wiley, New York.
Feder M. and Weinstein E. (1988) Parameter Estimation of Superimposed Signals using the EM algorithm. IEEE Trans. Acoustics, Speech and Signal Processing, ASSP-36(4), pp. 477, 489.
Fisher R.A. (1922) On the Mathematical Foundations of the Theoretical Statistics. Phil Trans. Royal. Soc. London, 222, pp. 309–368.
Gersho A. (1982) On the Structure of Vector Quantisers. IEEE Trans. Information Theory, IT-28, pp. 157–166.

Gray R.M. (1984) Vector Quantisation. IEEE ASSP Magazine, p. 4-29.

Gray R.M. and Karnin E.D (1982), Multiple Local Optima in Vector Quantisers. IEEE Trans. Information Theory, IT-28, pp. 256–261.

Jeffrey H. (1961) Scientific Inference, 3rd edn. Cambridge University Press.

Larson H.J. and Bruno O.S. (1979) Probabilistic Models in Engineering Sciences. I and II. Wiley, New York.

Linde Y., Buzo A. and Gray R.M. (1980) An Algorithm for Vector Quantiser Design. IEEE Trans. Comm. COM-28, pp. 84–95.

Makhoul J., Roucos S., and Gish H. (1985) Vector Quantisation in Speech Coding. Proc. IEEE, 73, pp. 1551–1588.

Mohanty N. (1986) Random Signals, Estimation and Identification. Van Nostrand, New York.

Rao C.R. (1945) Information and Accuracy Attainable in the Estimation of Statistical Parameters. Bull Calcutta Math. Soc., 37, pp. 81–91.

Render R.A. and Walker H.F. (1984) Mixture Densities, Maximum Likelihood and the EM algorithm. SIAM Review, 26, pp. 195–239.

Scharf L.L. (1991) Statistical Signal Processing: Detection, Estimation, and Time Series Analysis. Addison Wesley, Reading, MA.

8 Least Square Error, Wiener–Kolmogorov Filters

Least squared error filter theory was initially formulated independently by Andrei Kolmogorov (1941) and Norbert Wiener (1949). While Kolmogorov's method was based on time-domain analysis, Wiener's method was based on frequency domain analysis.

Least squared error filter theory forms the foundation of data-dependent adaptive linear filters. Least squared error filters play a central role in a wide range of applications such as linear prediction, echo cancellation, signal restoration, channel equalisation, radar, and system identification.

The coefficients of a least squared error filter are calculated to minimise the average squared distance between the filter output and a desired or target signal. The solution for the Wiener filter coefficients requires estimates of the autocorrelation of the input and the cross-correlation of the input and the desired signal.

In its basic form, the least squared error filter theory assumes that the signals are stationary processes. However, if the filter coefficients are periodically recalculated and updated for every block of N signal samples then the filter adapts itself to the average characteristics of the signals within each block and becomes block-adaptive. A block-adaptive (or frame-adaptive) filter can be used for signals such as speech and image that may be considered as almost stationary over a relatively small block of samples.

In this chapter, we study the least square error filter theory, and consider alternative methods of formulation of the filtering problem. We consider the application of least square error filters in channel equalisation, time-delay estimation and additive noise reduction. A case study of the frequency response of the least square error filter, for additive noise reduction, provides useful insight into the operation of the filter. We also deal with some implementation issues of filters.

8.1 Least Square Error Estimation: Wiener–Kolmogorov Filter

Norbert Wiener, and independently Andrei N. Kolmogorov, formulated the continuous-time least mean square error estimation problem of smoothing, interpolation and prediction of signals. Wiener's work

Multimedia Signal Processing: Theory and Applications in Speech, Music and Communications Saeed V. Vaseghi

is described in his classic work on interpolation, extrapolation and smoothing of time series (Wiener 1949). The extension of the Wiener filter theory from continuous time to discrete time is simple, and of more practical use for implementation on digital signal processors.

The typical scenarios in which Wiener filters are used is in the context of estimation or prediction of a signal observed in noise and system identification/estimation (such as channel estimation) given the inputs and the outputs of a system.

The Wiener filter can be used for signal enhancement to remove the effect of linear distortions such as the de-blurring of distorted or unfocused images or equalisation of the distortion of a telecommunication channel, or noise reduction. Wiener filter can also be used to predict the trajectory of a projectile; a problem during the second world war on which Norbert Wiener worked. Predicting the fluctuations of a signal from its past values has a wide range of applications from speech and video coding to economic data analysis. The Wiener filter formulation is the basis of least squared error applications such as linear prediction and adaptive filters.

A Wiener filter can be an infinite-duration impulse response (IIR) or a finite-duration impulse response (FIR) filter. In this chapter, we consider FIR Wiener filters, since they are relatively simple to compute, inherently stable and more practical. The main drawback of FIR filters compared with IIR filters is that they may need a large number of coefficients to approximate a desired response.

8.1.1 Derivation of Wiener Filter Equation

Figure 8.1 illustrates a Wiener filter represented by the filter's coefficient vector $\boldsymbol{w}$. The filter takes as the input a signal $y(m)$, usually a distorted version of a desired signal $x(m)$, and produces an output signal $\hat{x}(m)$, where $\hat{x}(m)$ is the least mean square error estimate of the desired or target signal $x(m)$. The filter input–output relation is given by

$$\begin{aligned}\hat{x}(m) &= \sum_{k=0}^{P-1} w_k y(m-k) \\ &= \boldsymbol{w}^{\mathrm{T}}\boldsymbol{y}\end{aligned} \tag{8.1}$$

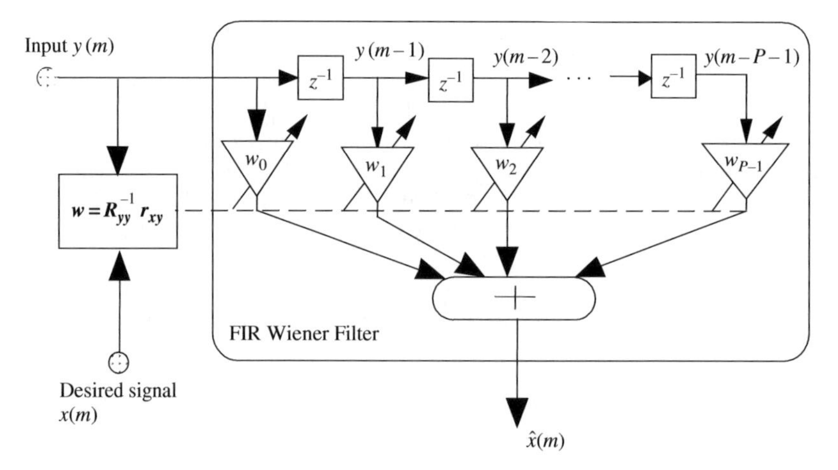

Figure 8.1 Illustration of a Wiener filter. The output signal $\hat{x}(m)$ is an estimate of the desired signal $x(m)$. It is obtained as the product of the input vector $[y(m-1)\ldots y(m-P-1)]$ and the coefficients vector $[w_0\ldots w_{P-1}]$.

where m is the discrete-time index, vector $\boldsymbol{y}^{\mathrm{T}} = [y(m), y(m-1), \ldots, y(m-P-1)]$ is the filter input signal, $\hat{x}(m)$ is the filter output and the parameter vector $\boldsymbol{w}^{\mathrm{T}} = [w_0, w_1, \ldots, w_{P-1}]$ is the Wiener filter coefficient vector. In Equation (8.1), the filtering operation is expressed in two alternative and equivalent representations of a convolutional sum and an inner vector product.

The Wiener filter error signal, $e(m)$ is defined as the difference between the desired (or target) signal $x(m)$ and the filter output $\hat{x}(m)$:

$$\begin{aligned} e(m) &= x(m) - \hat{x}(m) \\ &= x(m) - \boldsymbol{w}^{\mathrm{T}}\boldsymbol{y} \end{aligned} \tag{8.2}$$

where, as expressed in Equation (8.1), $\hat{x}(m)$ is the convolution of the input signal vector $\boldsymbol{y}$ and Wiener filter $\boldsymbol{w}$. In Equation (8.2), for a given input signal $y(m)$ and a desired signal $x(m)$, the filter error $e(m)$ depends on the filter coefficient vector $\boldsymbol{w}$. The Wiener filter is the best filter in the sense of minimising the mean squared error signal.

To explore the relation between the filter coefficient vector $\boldsymbol{w}$ and the error signal $e(m)$ we write Equation (8.2) N times in a matrix for a segment of N samples of the signals $[x(0), x(1), \ldots, x(N-1)]$ and signals $[y(0), y(1), \ldots, y(N-1)]$ as

$$\begin{pmatrix} e(0) \\ e(1) \\ e(2) \\ \vdots \\ e(N-1) \end{pmatrix} = \begin{pmatrix} x(0) \\ x(1) \\ x(2) \\ \vdots \\ x(N-1) \end{pmatrix} - \begin{pmatrix} y(0) & y(-1) & y(-2) & \cdots & y(1-P) \\ y(1) & y(0) & y(-1) & \cdots & y(2-P) \\ y(2) & y(1) & y(0) & \cdots & y(3-P) \\ \vdots & \vdots & \vdots & \ddots & \vdots \\ y(N-1) & y(N-2) & y(N-3) & \cdots & y(N-P) \end{pmatrix} \begin{pmatrix} w_0 \\ w_1 \\ w_2 \\ \vdots \\ w_{P-1} \end{pmatrix} \tag{8.3}$$

In a compact vector notation this matrix equation may be written as

$$\boldsymbol{e} = \boldsymbol{x} - \boldsymbol{Y}\boldsymbol{w} \tag{8.4}$$

where $\boldsymbol{e}$ is the error vector, $\boldsymbol{x}$ is the desired signal vector, $\boldsymbol{Y}$ is the input signal matrix and $\boldsymbol{Y}\boldsymbol{w} = \hat{\boldsymbol{x}}$ is the Wiener filter output signal vector. It is assumed that the P initial input signal samples $[y(-1), \ldots, y(-P-1)]$ are either known or set to zero.

At this point we explore the dependency of the solution of Equation (8.3) on the number of available samples N, which is also the number of linear equations in (8.3). In Equation (8.3), if the number of given signal samples is equal to the number of unknown filter coefficients $N = P$, then we have a *square* matrix equation, with as many equations as there are unknowns, and theoretically there is a unique filter solution $\boldsymbol{w}$, with a zero estimate on error $\boldsymbol{e} = \boldsymbol{0}$, such that $\hat{\boldsymbol{x}} = \boldsymbol{Y}\boldsymbol{w} = \boldsymbol{x}$.

If $N < P$ then the number of signal samples N, and hence the number of linear equations, is insufficient to obtain a unique solution for the filter coefficients; in this case there are an infinite number of solutions with zero estimation error, and the matrix equation is said to be *under-determined.*

In practice, there are two issues: (i) the target signal $x(m)$ is not available and (ii) the number of signal samples is larger than the filter length. When $N > P$ the matrix equation is said to be *over-determined* and has a unique solution, usually with a non-zero error. When $N > P$, the filter coefficients are calculated to minimise an average error cost function, such as the mean square error $E[e^2(m)]$, or the average absolute value of error $E[|e(m)|]$, where $E[.]$ is the expectation (averaging) operator. The choice of the error function affects the optimality and the computational complexity of the solution.

In Wiener theory, the objective criterion is the least mean square error (LSE) between the filter output and the desired signal. The least square error criterion is optimal for Gaussian distributed signals. As shown in the following, for FIR filters the LSE criterion leads to a linear and closed-form solutions. The Wiener filter coefficients are obtained by minimising an average squared error function $\mathcal{E}[e^2(m)]$ with respect to the filter coefficient vector $\boldsymbol{w}$, where $\mathcal{E}$ is expectation or average. From Equation (8.2), the mean square estimation error is given by

$$\begin{aligned}\mathcal{E}[e^2(m)] &= \mathcal{E}[(x(m) - \boldsymbol{w}^{\mathrm{T}}\boldsymbol{y})^2] \\ &= \mathcal{E}[x^2(m)] - 2\boldsymbol{w}^{\mathrm{T}}\mathcal{E}[\boldsymbol{y}x(m)] + \boldsymbol{w}^{\mathrm{T}}\mathcal{E}[\boldsymbol{y}\boldsymbol{y}^{\mathrm{T}}]\boldsymbol{w} \\ &= r_{xx}(0) - 2\boldsymbol{w}^{\mathrm{T}}\boldsymbol{r}_{yx} + \boldsymbol{w}^{\mathrm{T}}\boldsymbol{R}_{yy}\boldsymbol{w}\end{aligned} \tag{8.5}$$

where $\boldsymbol{R}_{yy} = E[\boldsymbol{y}(m)\boldsymbol{y}^{\mathrm{T}}(m)]$ is the autocorrelation matrix of the input signal and $\boldsymbol{r}_{xy} = E[x(m)\boldsymbol{y}(m)]$ is the cross-correlation vector of the input and the desired signals. An expanded form of Equation (8.5) can be obtained as

$$\mathcal{E}[e^2(m)] = r_{xx}(0) - 2\sum_{k=0}^{P-1} w_k r_{yx}(k) + \sum_{k=0}^{P-1} w_k \sum_{j=0}^{P-1} w_j r_{yy}(k-j) \tag{8.6}$$

where $r_{yy}(k)$ and $r_{yx}(k)$ are the elements of the autocorrelation matrix $\boldsymbol{R}_{yy}$ and the cross-correlation vector $\boldsymbol{r}_{xy}$ respectively.

From Equation (8.5), the mean square error for an FIR filter is a quadratic function of the filter coefficient vector $\boldsymbol{w}$ and has a single minimum point. For example, for a filter with two coefficients (w_0, w_1), the mean square error function is a bowl-shaped surface, with a single minimum point, as illustrated in Figure 8.2. The least mean square error point corresponds to the minimum error power. At this operating point the mean square error surface has zero gradient. From Equation (8.5), the gradient of the mean square error function with respect to the filter coefficient vector is given by

$$\begin{aligned}\frac{\partial}{\partial \boldsymbol{w}}\mathcal{E}[e^2(m)] &= -2\mathcal{E}[x(m)\boldsymbol{y}(m)] + 2\boldsymbol{w}^{\mathrm{T}}\mathcal{E}[\boldsymbol{y}(m)\boldsymbol{y}^{\mathrm{T}}(m)] \\ &= -2\boldsymbol{r}_{yx} + 2\boldsymbol{w}^{\mathrm{T}}\boldsymbol{R}_{yy}\end{aligned} \tag{8.7}$$

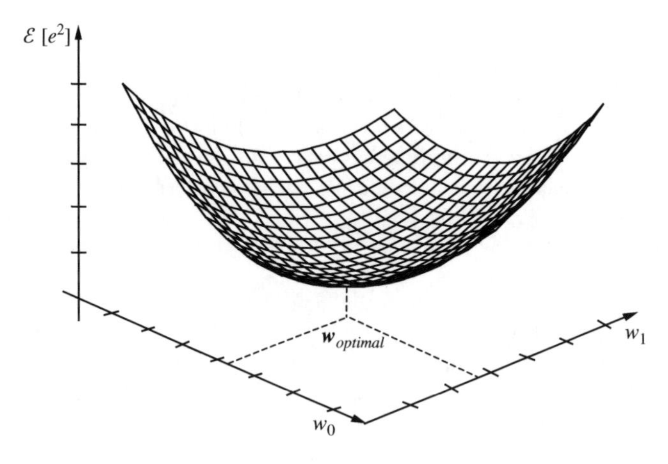

Figure 8.2 Mean square error surface for a two-tap FIR filter.

where the gradient vector is defined as

$$\frac{\partial}{\partial \boldsymbol{w}} = \left[\frac{\partial}{\partial w_0}, \frac{\partial}{\partial w_1}, \frac{\partial}{\partial w_2}, \ldots, \frac{\partial}{\partial w_{P-1}}\right]^{\mathrm{T}} \tag{8.8}$$

The minimum mean square error Wiener filter is obtained by setting Equation (8.7) to zero:

$$\boldsymbol{R}_{yy}\boldsymbol{w} = \boldsymbol{r}_{yx} \tag{8.9}$$

or, equivalently,

$$\boldsymbol{w} = \boldsymbol{R}_{yy}^{-1}\boldsymbol{r}_{yx} \tag{8.10}$$

In an expanded form, the Wiener filter solution Equation (8.10) can be written as

$$\begin{pmatrix} w_0 \\ w_1 \\ w_2 \\ \vdots \\ w_{P-1} \end{pmatrix} = \begin{pmatrix} r_{yy}(0) & r_{yy}(1) & r_{yy}(2) & \cdots & r_{yy}(P-1) \\ r_{yy}(1) & r_{yy}(0) & r_{yy}(1) & \cdots & r_{yy}(P-2) \\ r_{yy}(2) & r_{yy}(1) & r_{yy}(0) & \cdots & r_{yy}(P-3) \\ \vdots & \vdots & \vdots & \ddots & \vdots \\ r_{yy}(P-1) & r_{yy}(P-2) & r_{yy}(P-3) & \cdots & r_{yy}(0) \end{pmatrix}^{-1} \begin{pmatrix} r_{yx}(0) \\ r_{yx}(1) \\ r_{yx}(2) \\ \vdots \\ r_{yx}(P-1) \end{pmatrix} \tag{8.11}$$

8.1.2 Calculation of Autocorrelation of Input and Cross-Correlation of Input and Desired Signals

From Equation (8.11), the calculation of the Wiener filter coefficients requires the autocorrelation matrix of the input signal and the cross-correlation vector of the input and the desired signals.

In statistical signal processing theory, the correlation values of a random process are obtained as the averages taken across the ensemble of different realisations of the process as described in Chapter 3. However in many practical situations there are only one or two finite-duration realisations of the signals $x(m)$ and $y(m)$. Furthermore most signals are non-stationary and need to be segmented in quasi-stationary short segments. In such cases, assuming the signals are correlation-ergodic, we can use time averages instead of ensemble averages.

For a signal record of length N samples, the time-averaged correlation values are computed as

$$r_{yy}(k) = \frac{1}{N}\sum_{m=0}^{N-1} y(m)x(m+k) \tag{8.12}$$

Note from Equation (8.11) that the autocorrelation matrix $\boldsymbol{R}_{yy}$ has a highly regular Toeplitz structure. A Toeplitz matrix has identical elements along the left–right diagonals of the matrix. Furthermore, the correlation matrix is also symmetric about the main diagonal elements. There are a number of efficient methods for solving the linear matrix Equation (8.11), including the Cholesky decomposition, the singular value decomposition and the QR decomposition (Section 8.2.1) methods.

The cross-correlation values are estimated as

$$r_{yx}(k) = \frac{1}{N}\sum_{m=0}^{N-1} y(m)x(m+k) \tag{8.13}$$

The calculation of the autocorrelation of the input signal $r_{yy}(k)$ using Equation (8.12) is straightforward. The main challenge is in the calculation of the cross-correlation $r_{yx}(k)$ since the signal $x(m)$ is not available. However, this problem is resolved by either using pre-calculated values or taking advantage of the specific form of the problem as in Section 8.6.

8.2 Block-Data Formulation of the Wiener Filter

In this section we consider an alternative formulation of a Wiener filter for a segment of N samples of the input signal $[y(0), y(1), \ldots, y(N-1)]$ and the desired signal $[x(0), x(1), \ldots, x(N-1)]$. The set of N linear equations describing the input/output relationship of Wiener filter can be written in matrix form as

$$\begin{pmatrix} \hat{x}(0) \\ \hat{x}(1) \\ \hat{x}(2) \\ \vdots \\ \hat{x}(N-2) \\ \hat{x}(N-1) \end{pmatrix} = \begin{pmatrix} y(0) & y(-1) & y(-2) & \cdots & y(2-P) & y(1-P) \\ y(1) & y(0) & y(-1) & \cdots & y(3-P) & y(2-P) \\ y(2) & y(1) & y(0) & \cdots & y(4-P) & y(3-P) \\ \vdots & \vdots & \vdots & \ddots & \vdots & \vdots \\ y(N-2) & y(N-3) & y(N-4) & \cdots & y(N-P) & y(N-1-P) \\ y(N-1) & y(N-2) & y(N-3) & \cdots & y(N+1-P) & y(N-P) \end{pmatrix} \begin{pmatrix} w_0 \\ w_1 \\ w_2 \\ \vdots \\ w_{P-2} \\ w_{P-1} \end{pmatrix} \tag{8.14}$$

Equation (8.14) can be rewritten in compact matrix notation as

$$\hat{\boldsymbol{x}} = \boldsymbol{Y}\boldsymbol{w} \tag{8.15}$$

The Wiener filter error is the difference between the desired signal and the filter output defined as

$$\begin{aligned} \boldsymbol{e} &= \boldsymbol{x} - \hat{\boldsymbol{x}} \\ &= \boldsymbol{x} - \boldsymbol{Y}\boldsymbol{w} \end{aligned} \tag{8.16}$$

The energy of the error vector, that is the sum of the squared elements of the error vector $\boldsymbol{e}$, is given by the inner vector product as

$$\begin{aligned} \boldsymbol{e}^{\mathrm{T}}\boldsymbol{e} &= (\boldsymbol{x} - \boldsymbol{Y}\boldsymbol{w})^{\mathrm{T}}(\boldsymbol{x} - \boldsymbol{Y}\boldsymbol{w}) \\ &= \boldsymbol{x}^{\mathrm{T}}\boldsymbol{x} - \boldsymbol{x}^{\mathrm{T}}\boldsymbol{Y}\boldsymbol{w} - \boldsymbol{w}^{\mathrm{T}}\boldsymbol{Y}^{\mathrm{T}}\boldsymbol{x} + \boldsymbol{w}^{\mathrm{T}}\boldsymbol{Y}^{\mathrm{T}}\boldsymbol{Y}\boldsymbol{w} \end{aligned} \tag{8.17}$$

The gradient of the squared error function with respect to the Wiener filter coefficients is obtained by differentiating Equation (8.17) w.r.t $\boldsymbol{w}$ as

$$\frac{\partial \boldsymbol{e}^{\mathrm{T}}\boldsymbol{e}}{\partial \boldsymbol{w}} = -2\boldsymbol{x}^{\mathrm{T}}\boldsymbol{Y} + 2\boldsymbol{w}^{\mathrm{T}}\boldsymbol{Y}^{\mathrm{T}}\boldsymbol{Y} \tag{8.18}$$

The Wiener filter coefficients are obtained by setting the gradient of the squared error function of Equation (8.18) to zero; this yields

$$\left(\boldsymbol{Y}^{\mathrm{T}}\boldsymbol{Y}\right)\boldsymbol{w} = \boldsymbol{Y}^{\mathrm{T}}\boldsymbol{x} \tag{8.19}$$

or

$$\boldsymbol{w} = \left(\boldsymbol{Y}^{\mathrm{T}}\boldsymbol{Y}\right)^{-1}\boldsymbol{Y}^{\mathrm{T}}\boldsymbol{x} \tag{8.20}$$

Note that the matrix Y^TY is a time-averaged estimate of the autocorrelation matrix of the filter input signal R_{yy}, and that the vector Y^Tx is a time-averaged estimate of r_{xy} the cross-correlation vector of the input and the desired signals. Since the least square error method described in this section requires a block of N samples of the input and the desired signals, it is also referred to as the block least square (BLS) error estimation method. The block estimation method is appropriate for processing of signals that can be considered as time-invariant over the duration of the block.

Theoretically, the Wiener filter is obtained from minimisation of the squared error across the ensemble of different realisations of a process as described in the previous section. For a correlation-ergodic process, as the signal length Napproaches infinity the block-data Wiener filter of Equation (8.20) approaches the Wiener filter of Equation (8.10):

$$\lim_{N\to\infty}\left[\boldsymbol{w}=\left(\boldsymbol{Y}^{\mathrm{T}}\boldsymbol{Y}\right)^{-1}\boldsymbol{Y}^{\mathrm{T}}\boldsymbol{x}\right]=\boldsymbol{R}_{yy}^{-1}\boldsymbol{r}_{xy} \tag{8.21}$$

Matlab Function Time Domain Wiener Filter
function TD_Wiener_Demo( )
A demonstration of time domain implementation of a Wiener filter for recovery of a signal in noise. The demonstration shows: noisy signal and its spectrum, the superposition of actual signal and its estimate (in time and frequency) and the impulse response and frequency response of the Wiener filter.

8.2.1 QR Decomposition of the Least Square Error Equation

An efficient and robust method for solving the least square error Equation (8.20) is the QR decomposition (QRD) method. In this method, the $N \times P$ signal matrix $\boldsymbol{Y}$ (shown in Equation (8.14)) is decomposed into the product of an $N \times N$ orthonormal matrix $\boldsymbol{Q}$ and a $P \times P$ upper-triangular matrix $\boldsymbol{R}$ as

$$\boldsymbol{Q}\boldsymbol{Y}=\begin{pmatrix}\boldsymbol{R}\\ \boldsymbol{0}\end{pmatrix} \tag{8.22}$$

where $\boldsymbol{0}$ is the $(N-P)\times P$ null matrix, $\boldsymbol{Q}$ is an orthonormal matrix

$$\boldsymbol{Q}^{\mathrm{T}}\boldsymbol{Q}=\boldsymbol{Q}\boldsymbol{Q}^{\mathrm{T}}=\boldsymbol{I} \tag{8.23}$$

and the upper-triangular matrix $\boldsymbol{R}$ is of the form

$$\boldsymbol{R}=\begin{pmatrix} r_{00} & r_{01} & r_{02} & r_{03} & \cdots & r_{0P-1}\\ 0 & r_{11} & r_{12} & r_{13} & \cdots & r_{1P-1}\\ 0 & 0 & r_{22} & r_{23} & \cdots & r_{2P-1}\\ 0 & 0 & 0 & r_{33} & \cdots & r_{3P-1}\\ \vdots & \vdots & \vdots & \vdots & \ddots & \vdots\\ 0 & 0 & 0 & 0 & \cdots & r_{P-1P-1}\end{pmatrix} \tag{8.24}$$

From Equations (8.22) and (8.23) we have

$$\boldsymbol{Y} = \boldsymbol{Q}^T \begin{pmatrix} \boldsymbol{R} \\ \boldsymbol{0} \end{pmatrix} \tag{8.25}$$

Substitution of Equation (8.25) in Equation (8.19) yields

$$\begin{pmatrix} \boldsymbol{R} \\ \boldsymbol{0} \end{pmatrix}^{\mathrm{T}} \boldsymbol{Q}\boldsymbol{Q}^{\mathrm{T}} \begin{pmatrix} \boldsymbol{R} \\ \boldsymbol{0} \end{pmatrix} \boldsymbol{w} = \begin{pmatrix} \boldsymbol{R} \\ \boldsymbol{0} \end{pmatrix}^{\mathrm{T}} \boldsymbol{Q}\boldsymbol{x} \tag{8.26}$$

From Equation (8.26) we have

$$\begin{pmatrix} \boldsymbol{R} \\ \boldsymbol{0} \end{pmatrix} \boldsymbol{w} = \boldsymbol{Q}\boldsymbol{x} \tag{8.27}$$

From Equation (8.27) we have

$$\boldsymbol{R}\boldsymbol{w} = \boldsymbol{x}_Q \tag{8.28}$$

where the vector $\boldsymbol{x}_Q$ on the right-hand side of Equation (8.28) is composed of the first P elements of the product $\boldsymbol{Q}\boldsymbol{x}$. Since the matrix $\boldsymbol{R}$ is upper-triangular, the coefficients of the least square error filter can be obtained easily through a process of back substitution from Equation (8.28), starting with the coefficient $w_{P-1} = x_Q(P-1)/r_{P-1P-1}$.

The main computational steps in the QR decomposition are the determination of the orthonormal matrix $\boldsymbol{Q}$ and of the upper triangular matrix $\boldsymbol{R}$. The decomposition of a matrix into $\boldsymbol{QR}$ matrices can be achieved using a number of methods, including the Gram–Schmidt orthogonalisation method, the Householder method and the Givens rotation method.

8.3 Interpretation of Wiener Filter as Projection in Vector Space

In this section, we consider an alternative formulation of Wiener filters where the least square error estimate is visualised as the perpendicular minimum distance *projection* of the desired signal vector onto the vector space of the input signal. A vector space is the collection of an infinite number of vectors that can be obtained from linear combinations of a number of independent vectors.

In order to develop a vector space interpretation of the least square error estimation problem, we rewrite the matrix Equation (8.11) and express the filter output signal vector $\hat{\boldsymbol{x}}$ as a linear weighted combination of the column vectors of the input signal matrix as

$$\begin{pmatrix} \hat{x}(0) \\ \hat{x}(1) \\ \hat{x}(2) \\ \vdots \\ \hat{x}(N-2) \\ \hat{x}(N-1) \end{pmatrix} = w_0 \begin{pmatrix} y(0) \\ y(1) \\ y(2) \\ \vdots \\ y(N-2) \\ y(N-1) \end{pmatrix} + w_1 \begin{pmatrix} y(-1) \\ y(0) \\ y(1) \\ \vdots \\ y(N-3) \\ y(N-2) \end{pmatrix} + \cdots + w_{P-1} \begin{pmatrix} y(1-P) \\ y(2-P) \\ y(3-P) \\ \vdots \\ y(N-1-P) \\ y(N-P) \end{pmatrix} \tag{8.29}$$

In compact notation, Equation (8.29) may be written as

$$\hat{\boldsymbol{x}} = w_0 \boldsymbol{y}_0 + w_1 \boldsymbol{y}_1 + \cdots + w_{P-1} \boldsymbol{y}_{P-1} \tag{8.30}$$

In Equation (8.30) the Wiener filter output $\hat{\boldsymbol{x}}$ is expressed as a linear combination of P basis vectors $[\boldsymbol{y}_0, \boldsymbol{y}_1, \ldots, \boldsymbol{y}_{P-1}]$, and hence it can be said that the estimate $\hat{\boldsymbol{x}}$ is in the vector subspace formed by the input signal vectors $[\boldsymbol{y}_0, \boldsymbol{y}_1, \ldots, \boldsymbol{y}_{P-1}]$.

In general, the N-dimensional input signal vectors $[\boldsymbol{y}_0, \boldsymbol{y}_1, \ldots, \boldsymbol{y}_{P-1}]$ in Equation (8.30) define the *basis* vectors for a subspace in an N-dimensional signal space. If the number of basis vectors P is equal to the vector dimension N, then the subspace encompasses the entire N-dimensional signal space and includes the desired signal vector $\boldsymbol{x}$. In this case, the signal estimate $\hat{\boldsymbol{x}} = \boldsymbol{x}$ and the estimation error is zero. However, in practice, $N > P$, and the signal space defined by the P input signal vectors of Equation (8.30) is only a subspace of the N-dimensional signal space. In this case, the estimation error is zero only if the desired signal $\boldsymbol{x}$ happens to be in the subspace of the input signal, otherwise the best estimate of $\boldsymbol{x}$ is the perpendicular projection of the vector $\boldsymbol{x}$ onto the vector space of the input signal $[\boldsymbol{y}_0, \boldsymbol{y}_1, \ldots, \boldsymbol{y}_{P-1}]$, as explained in the following example.

Example 8.1 Figure 8.3 illustrates a vector space interpretation of a simple least square error estimation problem, where $\boldsymbol{y}^{\mathrm{T}} = [y(m), y(m-1), y(m-2), y(m-3)]$ is the input observation signal, $\boldsymbol{x}^{\mathrm{T}} = [x(m), x(m-1), x(m-2)]$ is the desired signal and $\boldsymbol{w}^{\mathrm{T}} = [w_0, w_1]$ is the filter coefficient vector. As in Equation (8.26), the filter output can be written as

$$\begin{pmatrix} \hat{x}(m) \\ \hat{x}(m-1) \\ \hat{x}(m-2) \end{pmatrix} = w_0 \begin{pmatrix} y(m) \\ y(m-1) \\ y(m-2) \end{pmatrix} + w_1 \begin{pmatrix} y(m-1) \\ y(m-2) \\ y(m-3) \end{pmatrix} \tag{8.31}$$

In Equation (8.29), the filter input signal vectors $\boldsymbol{y}_1^{\mathrm{T}} = [y(m), y(m-1), y(m-2)]$ and $\boldsymbol{y}_2^{\mathrm{T}} = [y(m-1), y(m-2), y(m-3)]$ are 3-dimensional vectors. The subspace defined by the linear combinations of the two input vectors $[\boldsymbol{y}_1, \boldsymbol{y}_2]$ is a 2-dimensional plane in a 3-dimensional space. The filter output is a

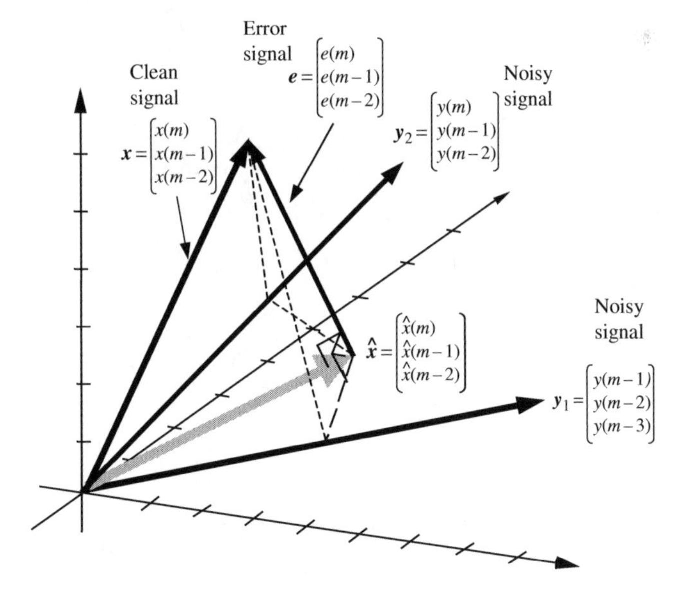

Figure 8.3 The least square error projection of a desired signal vector $\boldsymbol{x}$ onto a plane containing the input signal vectors $\boldsymbol{y}_1$ and $\boldsymbol{y}_2$ is the perpendicular projection of $\boldsymbol{x}$ shown as the shaded vector.

linear combination of $\boldsymbol{y}_1$ and $\boldsymbol{y}_2$, and hence it is confined to the plane containing these two vectors. The least square error estimate of $\boldsymbol{x}$ is the orthogonal projection of $\boldsymbol{x}$ on the plane of $[\boldsymbol{y}_1, \boldsymbol{y}_2]$ as shown by the shaded vector $\hat{\boldsymbol{x}}$. If the desired vector happens to be in the plane defined by the vectors $\boldsymbol{y}_1$ and $\boldsymbol{y}_2$ then the estimation error will be zero, otherwise the estimation error will be the perpendicular distance of $\boldsymbol{x}$ from the plane containing $\boldsymbol{y}_1$ and $\boldsymbol{y}_2$.

8.4 Analysis of the Least Mean Square Error Signal

The optimality criterion in the formulation of the Wiener filter is the least mean square distance between the filter output and the desired signal. In this section, the variance of the filter error signal is analysed. Substituting the Wiener equation $\boldsymbol{R}_{yy}\boldsymbol{w} = \boldsymbol{r}_{yx}$ in the mean squared error Equation (8.5) gives the least mean square error

$$\begin{aligned} \mathcal{E}[e^2(m)] &= r_{xx}(0) - \boldsymbol{w}^{\mathrm{T}}\boldsymbol{r}_{yx} \\ &= r_{xx}(0) - \boldsymbol{w}^{\mathrm{T}}\boldsymbol{R}_{yy}\boldsymbol{w} \end{aligned} \tag{8.32}$$

Now, for zero-mean signals, it is easy to show that the term $\boldsymbol{w}^{\mathrm{T}}\boldsymbol{R}_{yy}\boldsymbol{w}$ in Equation (8.32) is the variance of the Wiener filter output $\hat{x}(m)$:

$$\sigma_{\hat{x}}^2 = \mathcal{E}[\hat{x}^2(m)] = \boldsymbol{w}^{\mathrm{T}}\boldsymbol{R}_{yy}\boldsymbol{w} \tag{8.33}$$

Therefore Equation (8.32) may be written as

$$\sigma_e^2 = \sigma_x^2 - \sigma_{\hat{x}}^2 \tag{8.34}$$

where $\sigma_x^2 = \mathcal{E}[x^2(m)] = r_{xx}(0)$, $\sigma_{\hat{x}}^2 = \mathcal{E}[\hat{x}^2(m)]$, $\sigma_e^2 = \mathcal{E}[e^2(m)]$ are the variances of the desired signal, the filter output i.e. the estimate of the desired signal and the error signal respectively.

In general, the filter input $y(m)$ is composed of a signal component $x_c(m)$ and a random noise $n(m)$:

$$y(m) = x_c(m) + n(m) \tag{8.35}$$

where the signal $x_c(m)$ is that part of the observation $y(m)$ that is correlated with the desired signal $x(m)$; it is this part of the input signal that may be transformable through a Wiener filter to the desired signal. Using Equation (8.35) the Wiener filter error may be decomposed into two distinct components:

$$\begin{aligned} e(m) &= x(m) - \sum_{k=0}^{P} w_k y(m-k) \\ &= \left(x(m) - \sum_{k=0}^{P} w_k x_c(m-k) \right) - \sum_{k=0}^{P} w_k n(m-k) \end{aligned} \tag{8.36}$$

or

$$e(m) = e_x(m) + e_n(m) \tag{8.37}$$

where $e_x(m)$ is the difference between the desired signal $x(m)$ and the output of the filter in response to the input signal component $x_c(m)$, i.e.

$$e_x(m) = x(m) - \sum_{k=0}^{P-1} w_k x_c(m-k) \tag{8.38}$$

and $e_n(m)$ is the error in the filter output due to the presence of noise $n(m)$ in the input signal:

$$e_n(m) = -\sum_{k=0}^{P-1} w_k n(m-k) \tag{8.39}$$

The variance of filter error can be rewritten as

$$\sigma_e^2 = \sigma_{e_x}^2 + \sigma_{e_n}^2 \tag{8.40}$$

Note that in Equation (8.37), $e_x(m)$ is that part of the signal that cannot be recovered by the Wiener filter, and represents part of the distortion in the filter output signal, and $e_n(m)$ is that part of the noise that cannot be blocked by the Wiener filter. Ideally, $e_x(m) = 0$ and $e_n(m) = 0$, but this ideal situation is possible only if the following conditions are satisfied:

(a) The spectra of the signal and the noise are separable by a linear filter. The issue of signal and noise separability is addressed in Section 8.6.2.
(b) The signal component of the input, that is $x_c(m)$, is *linearly* transformable to $x(m)$.
(c) The filter length P is sufficiently large.

8.5 Formulation of Wiener Filters in the Frequency Domain

In the frequency domain, the Wiener filter output $\hat{X}(f)$ is the product of the input signal $Y(f)$ and the filter frequency response $W(f)$:

$$\hat{X}(f) = W(f)Y(f) \tag{8.41}$$

The estimation error signal $E(f)$ is defined as the difference between the desired signal $X(f)$ and the filter output $\hat{X}(f)$ as

$$\begin{aligned} E(f) &= X(f) - \hat{X}(f) \\ &= X(f) - W(f)Y(f) \end{aligned} \tag{8.42}$$

and the mean square error at a frequency f is given by

$$\mathcal{E}\left[|E(f)|^2\right] = \mathcal{E}\left[(X(f) - W(f)Y(f))^* (X(f) - W(f)Y(f))\right] \tag{8.43}$$

where $E[\cdot]$ is the expectation function, and the symbol * denotes the complex conjugate. Note from Parseval's theorem the mean square error in time and frequency domains are related by

$$\sum_{m=0}^{N-1} e^2(m) = \int_{-1/2}^{1/2} |E(f)|^2 \, df \tag{8.44}$$

To obtain the least mean square error filter we set the complex derivative of Equation (8.40) with respect to filter $W(f)$ to zero

$$\frac{\partial \mathcal{E}[|E(f)|^2]}{\partial W(f)} = 2W(f)P_{YY}(f) - 2P_{XY}(f) = 0 \tag{8.45}$$

where $P_{YY}(f) = E[Y(f)Y^*(f)]$ and $P_{XY}(f) = E[X(f)Y^*(f)]$ are the power spectrum of $Y(f)$, and the cross-power spectrum of $Y(f)$ and $X(f)$ respectively. From Equation (8.45), the least mean square error Wiener filter in the frequency domain is given as

$$W(f) = \frac{P_{XY}(f)}{P_{YY}(f)} \tag{8.46}$$

Alternatively, the frequency Wiener filter Equation (8.46) can be obtained from the Fourier transform of the time-domain Wiener Equation (8.9):

$$\sum_{m}\sum_{k=0}^{P-1} w_k r_{yy}(m-k)e^{-\mathrm{j}\omega m} = \sum_{m} r_{yx}(m)e^{-\mathrm{j}\omega m} \tag{8.47}$$

From the Wiener–Khinchine relation, correlation and power-spectral functions are Fourier transform pairs. Using this relation, and the Fourier transform property that convolution in time is equivalent to multiplication in frequency, Equation (8.47) can be transformed into frequency as

$$W(f)P_{YY}(f) = P_{XY}(f) \tag{8.48}$$

Re-arrangement of Equation (8.48) gives Equation (8.46).

Matlab Function Frequency Domain Wiener Filter
function FD_Wiener_Demo( )
A demonstration of frequency domain implementation of a Wiener filter for recovery of a signal in noise. The demonstration shows: noisy signal and its spectrum, the superposition of actual signal and its estimate (in time and frequency) and the impulse response and frequency response of the Wiener filter.

8.6 Some Applications of Wiener Filters

In this section, we consider some applications of the Wiener filter in reducing broadband additive noise, in time-alignment of signals in multi-channel or multi-sensor systems, and in communication channel equalisation.

8.6.1 Wiener Filter for Additive Noise Reduction

Consider a signal $x(m)$ observed in a broadband additive noise $n(m)$, and modelled as

$$y(m) = x(m) + n(m) \tag{8.49}$$

Assuming that the signal and the noise are uncorrelated i.e. $r_{xn}(m) = 0$, it follows that the autocorrelation matrix of the noisy signal is the sum of the autocorrelation matrix of the signal $x(m)$ and the noise $n(m)$:

$$\boldsymbol{R}_{yy} = \boldsymbol{R}_{xx} + \boldsymbol{R}_{nn} \tag{8.50}$$

and we can also write

$$\boldsymbol{r}_{xy} = \boldsymbol{r}_{xx} \tag{8.51}$$

where $\boldsymbol{R}_{yy}$, $\boldsymbol{R}_{xx}$ and $\boldsymbol{R}_{nn}$ are the autocorrelation matrices of the noisy signal, the noise-free signal and the noise respectively, and $\boldsymbol{r}_{xy}$ is the cross-correlation vector of the noisy signal and the noise-free signal. Substitution of Equations (8.50) and (8.51) in the Wiener filter, Equation (8.10), yields

$$\boldsymbol{w} = (\boldsymbol{R}_{xx} + \boldsymbol{R}_{nn})^{-1} \boldsymbol{r}_{xx} \tag{8.52}$$

Equation (8.52) is the optimal linear filter for the removal of additive noise. In the following, a study of the frequency response of the Wiener filter provides useful insight into the operation of the Wiener filter. In the frequency domain, the noisy signal $Y(f)$ is given by

$$Y(f) = X(f) + N(f) \tag{8.53}$$

where $X(f)$ and $N(f)$ are the signal and noise spectra. For a signal observed in additive random noise, the frequency Wiener filter is obtained as

$$W(f) = \frac{P_{XX}(f)}{P_{XX}(f) + P_{NN}(f)} \tag{8.54}$$

where $P_{XX}(f)$ and $P_{NN}(f)$ are the signal and noise power spectra. Dividing the numerator and the denominator of Equation (8.54) by the noise power spectra $P_{NN}(f)$ and substituting the variable $SNR(f) = P_{XX}(f)/P_{NN}(f)$ yields

$$W(f) = \frac{SNR(f)}{SNR(f) + 1} \tag{8.55}$$

where SNR is a signal-to-noise ratio measure. Note that in Equation (8.55) the variable $SNR(f)$ is expressed in terms of the power-spectral ratio, and not in the more usual terms of log power ratio or dB. Therefore $SNR(f) = 0$ corresponds to zero signal content or $10\log_{10}(0) = -\infty$ dB and $SNR(f) = 1$ corresponds to equal signal and noise power $P_{XX}(f) = P_{NN}(f)$ or $10\log_{10}(1) = 0$ dB and $SNR(f) = 0$ corresponds to $10\log_{10}(0.5) = -3$ dB. Figure 8.4 shows the variation of the Wiener filter response $W(f)$, with the signal-to-noise ratio $SNR(f)$.

From Equation (8.51), the following interpretation of the Wiener filter frequency response $W(f)$ in terms of the signal-to-noise ratio can be deduced. For additive noise, the Wiener filter frequency response is a real positive number in the range $0 \le W(f) \le 1$. Now consider the two limiting cases of (a) a noise-free signal $SNR(f) = \infty$ and (b) an extremely noisy signal $SNR(f) = 0$. At very high SNR, $W(f) \approx 1$, and the filter applies little or no attenuation to the noise-free frequency component. At the other extreme, when $SNR(f) = 0$, $W(f) = 0$. Therefore, *for additive noise, the Wiener filter attenuates each frequency component in proportion to an estimate of the signal-to-noise ratio.*

An alternative illustration of the variations of the Wiener filter frequency response with $SNR(f)$ is shown in Figure 8.5. It illustrates the similarity between the Wiener filter frequency response and the signal spectrum for the case of an additive white noise disturbance. Note that at a spectral peak of the signal spectrum, where the $SNR(f)$ is relatively high, the Wiener filter frequency response is also high, and the filter applies little attenuation. At a signal trough, the signal-to-noise ratio is low, and so is the Wiener filter response. Hence, for additive white noise, the Wiener filter response broadly follows the signal spectrum.

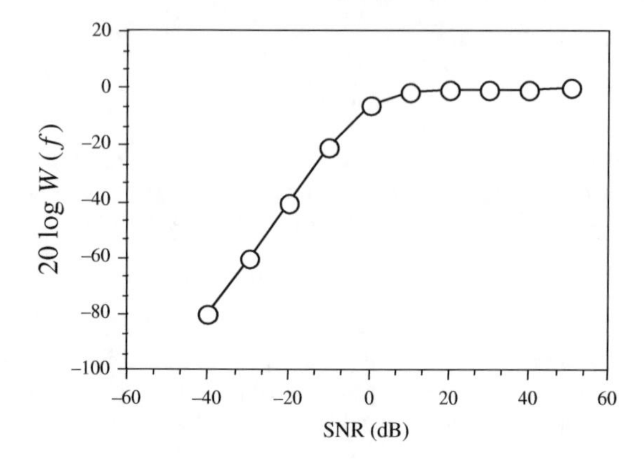

Figure 8.4 Variation of the gain of Wiener filter frequency response with SNR

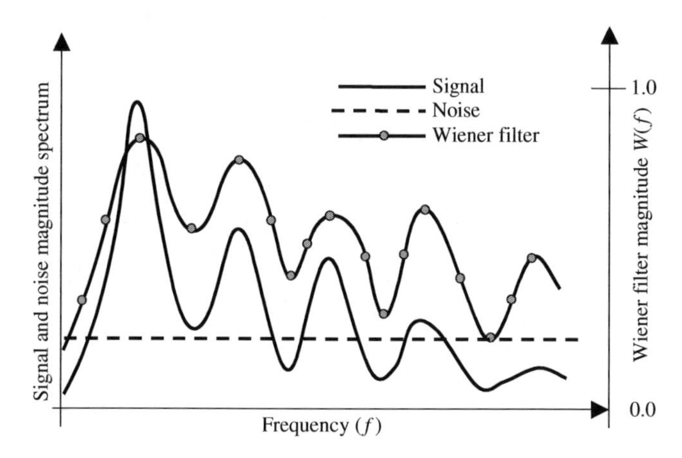

Figure 8.5 Illustration of the variation of Wiener frequency response with signal spectrum for additive white noise. The Wiener filter response broadly follows the signal spectrum.

8.6.2 Wiener Filter and Separability of Signal and Noise

In single-input noise reduction applications, where only one sensor is available (for example for speech enhancement on a mobile phone), the signal and noise cannot be perfectly separated unless their spectra are non-overlapping. A stochastic signal is completely recoverable from noise if and only if the spectra of the signal and the noise do not overlap.

An example of a noisy signal with separable signal and noise spectra is shown in Figure 8.6(a). In this case, the signal and the noise occupy different parts of the frequency spectrum, and can be separated with a low-pass or a high-pass filter. Figure 8.6(b) illustrates a more common example of a signal and noise process with overlapping spectra. For this case, it is not possible to completely separate the signal from the noise. However, the effects of the noise can be reduced by using a Wiener filter that attenuates each noisy signal frequency in proportion to an estimate of the signal-to-noise ratio as described by Equation (8.55).

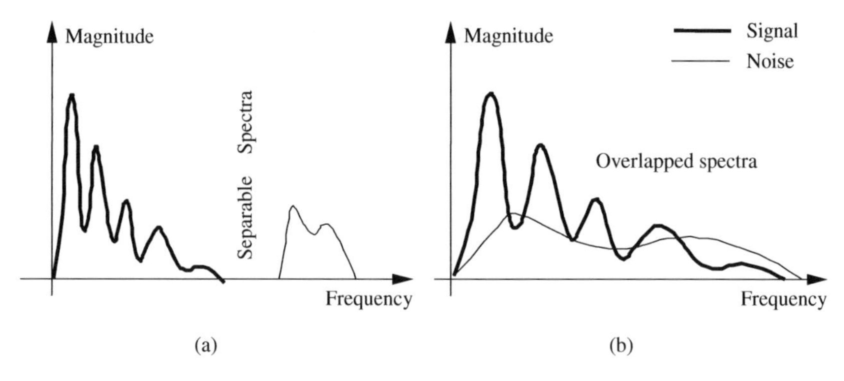

Figure 8.6 Illustration of separability: (a) the signal and noise spectra do not overlap, the signal can be recovered by a low-pass filter, (b) the signal and noise spectra overlap, the noise can be reduced but not completely removed.

8.6.3 The Square-Root Wiener Filter

In the frequency domain, the Wiener filter output $\hat{X}(f)$ is the product of the input frequency $Y(f)$ and the filter response $W(f)$ as expressed in Equation (8.41). Taking the expectation of the squared magnitude of both sides of Equation (8.41) yields the power spectrum of the filtered signal as

$$\begin{aligned}\mathcal{E}[|\hat{X}(f)|^2] &= |W(f)|^2\, \mathcal{E}[|Y(f)|^2] \\ &= |W(f)|^2\, P_{YY}(f)\end{aligned} \tag{8.56}$$

Substitution of $W(f)$ from Equation (8.46) in Equation (8.56) yields

$$\mathcal{E}[|\hat{X}(f)|^2] = \frac{P_{XY}^2(f)}{P_{YY}(f)} \tag{8.57}$$

Now, for a signal observed in an uncorrelated additive noise we have

$$P_{YY}(f) = P_{XX}(f) + P_{NN}(f) \tag{8.58}$$

and

$$P_{XY}(f) = P_{XX}(f) \tag{8.59}$$

Substitution of Equations (8.58) and (8.59) in Equation (8.57) yields

$$\mathcal{E}[|\hat{X}(f)|^2] = \frac{P_{XX}^2(f)}{P_{XX}(f) + P_{NN}(f)} \tag{8.60}$$

Now, in Equation (8.41) if instead of the Wiener filter, the square root of the Wiener filter magnitude frequency response is used, the result is

$$\hat{X}(f) = |W(f)|^{1/2}\, Y(f) \tag{8.61}$$

and the power spectrum of the signal, filtered by the square-root Wiener filter, is given by

$$\mathcal{E}[|\hat{X}(f)|^2] = \left(|W(f)|^{1/2}\right)^2 \mathcal{E}[|Y(f)|^2] = \frac{P_{XY}(f)}{P_{YY}(f)} P_{YY}(f) = P_{XY}(f) \tag{8.62}$$

Now, for uncorrelated signal and noise Equation (8.62) becomes

$$\mathcal{E}[|\hat{X}(f)|^2] = P_{XX}(f) \tag{8.63}$$

Thus, for additive noise the power spectrum of the output of the square-root Wiener filter is the same as the power spectrum of the desired signal.

8.6.4 Wiener Channel Equaliser

The distortions in a communication channel may be modelled by a combination of a linear filter and an additive random noise source as shown in Figure 8.7. The input/output signals of a linear time invariant channel can be modelled as

$$y(m) = \sum_{k=0}^{P-1} h_k x(m-k) + n(m) \tag{8.64}$$

where $x(m)$ and $y(m)$ are the transmitted and received signals, $[h_k]$ is the impulse response of a linear filter model of the channel, and $n(m)$ models the channel noise. In the frequency domain Equation((8.64) becomes

$$Y(f) = X(f)H(f) + N(f) \tag{8.65}$$

where $X(f)$, $Y(f)$, $H(f)$ and $N(f)$ are the signal, noisy signal, channel and noise spectra respectively. To remove the channel distortions, the receiver is followed by an equaliser. The input to the equaliser is the distorted signal from the channel output, and the desired signal is the clean signal at the channel input. Using Equation (8.46) it is easy to show that the frequency domain Wiener equaliser is given by

$$W(f) = \frac{P_{XX}(f)H^*(f)}{P_{XX}(f)\,|H(f)|^2 + P_{NN}(f)} \tag{8.66}$$

where it is assumed that the signal and the channel noise are uncorrelated. In the absence of channel noise, $P_{NN}(f) = 0$, and the Wiener filter is simply the inverse of the channel distortion model $W(f) = H^{-1}(f)$. The equalisation problem is treated in detail in Chapter 17.

8.6.5 Time-Alignment of Signals in Multichannel/Multi-sensor Systems

In multi-channel/multi-sensor signal processing there are a array of noisy and distorted versions of a signal $x(m)$, and the objective is to use all the observations in estimating $x(m)$, as illustrated in

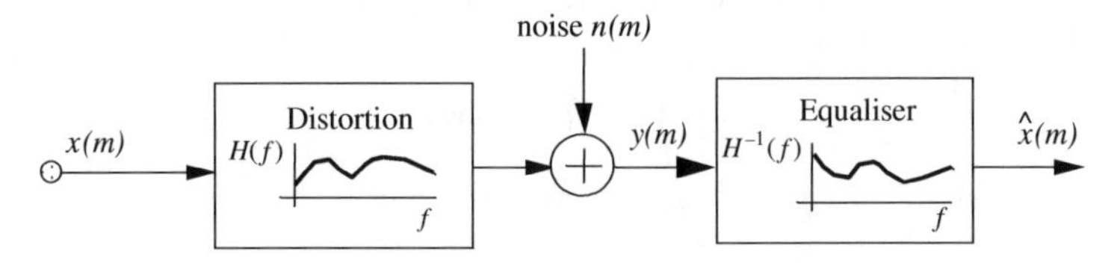

Figure 8.7 Illustration of a channel model followed by an equaliser.

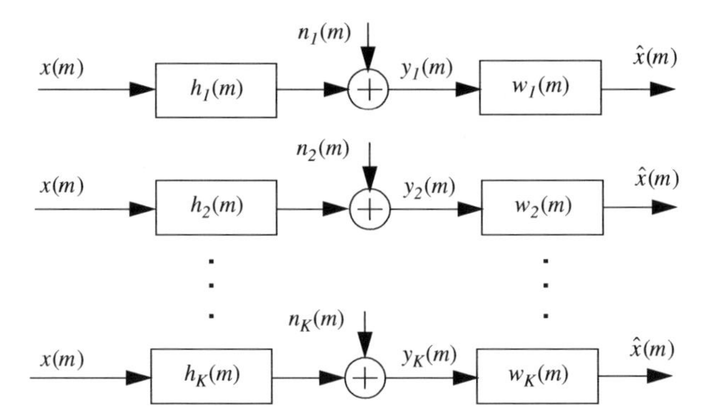

Figure 8.8 Illustration of a multi-channel system where Wiener filters are used to time-align the signals from different channels.

Figure 8.8, where the phase and frequency characteristics of each channel is modelled by a linear filter $h(m)$.

As a simple example, consider the problem of time-alignment of two noisy records of a signal given as

$$y_1(m) = x(m) + n_1(m) \tag{8.67}$$

$$y_2(m) = Ax(m-D) + n_2(m) \tag{8.68}$$

where $y_1(m)$ and $y_2(m)$ are the noisy observations from channels 1 and 2, $n_1(m)$ and $n_2(m)$ are uncorrelated noise in each channel, D is the relative time delay of arrival of the two signals, and A is an amplitude scaling factor. Now assume that $y_1(m)$ is used as the input to a Wiener filter and that, in the absence of the signal $x(m)$, $y_2(m)$ is used as the 'desired' signal. The error signal is given by

$$\begin{aligned} e(m) &= y_2(m) - \sum_{k=0}^{P-1} w_k y_1(m) \\ &= \left(Ax(m-D) - \sum_{k=0}^{P-1} w_k x(m) \right) + \left(\sum_{k=0}^{P-1} w_k n_1(m) \right) + n_2(m) \end{aligned} \tag{8.69}$$

The Wiener filter strives to minimise the terms shown inside the square brackets in Equation (8.69). Using the Wiener filter Equation (8.10), we have

$$\begin{aligned} \boldsymbol{w} &= \boldsymbol{R}_{y_1 y_1}^{-1} \boldsymbol{r}_{y_1 y_2} \\ &= \left(\boldsymbol{R}_{xx} + \boldsymbol{R}_{n_1 n_1} \right)^{-1} A \boldsymbol{r}_{xx}(D) \end{aligned} \tag{8.70}$$

where $\boldsymbol{r}_{xx}(D) = E[x(m-D)\boldsymbol{x}(m)]$. The frequency-domain equivalent of Equation (8.70) can be derived as

$$W(f) = \frac{P_{XX}(f)}{P_{XX}(f) + P_{N_1 N_1}(f)} A e^{-j\omega D} \tag{8.71}$$

Note that in the absence of noise, the Wiener filter becomes a pure phase (or a pure delay) filter, $W(f) = Ae^{-j\omega D}$, with a flat magnitude response.

8.7 Implementation of Wiener Filters

The implementation of a Wiener filter for additive noise reduction, using Equation (8.52) or (8.54), requires the autocorrelation functions, or equivalently the power spectra, of the signal and noise. In speech recognition the power spectra, or autocorrelation functions of signal and noise can be obtained from speech and noise models see Chapter 11 and Chapter 15.

When statistical models of speech and noise are not available, the noise power spectrum can be obtained from the signal-inactive, noise-only, periods. The assumption is that the noise is quasi-stationary, and that its power spectra remain relatively stationary between the update periods. This is a reasonable assumption for many noisy environments such as the noise inside a car emanating from the engine and wind, aircraft noise, office noise from computer machines, etc.

The main practical problem in the implementation of a Wiener filter is that the desired signal is often observed in noise, and that the autocorrelation or power spectra of the desired signal are not readily available. Figure 8.9 illustrates the block-diagram configuration of a system for implementation of a Wiener filter for additive noise reduction. The implementation of this filter requires estimates of the spectral signal-to-noise ratio $SNR(f)$.

The estimate of spectral signal-to-noise ratio is obtained from the estimates of the power spectra of the signal and noise. The noise estimate is obtained from speech-inactive periods. An estimate of the clean signal power spectra may be obtained by subtracting an estimate of the noise spectra from that of the noisy signal.

A filter bank implementation of the Wiener filter is shown in Figure 8.10, where the incoming signal is divided into N sub-bands. A first-order integrator, placed at the output of each band-pass filter, gives an estimate of the power spectra of the noisy signal. The power spectrum of the original signal is obtained by subtracting an estimate of the noise power spectrum from the noisy signal.

In a Bayesian implementation of the Wiener filter, prior models of speech and noise, such as hidden Markov models, are used to obtain the power spectra of speech and noise required for calculation of the filter coefficients.

8.7.1 Choice of Wiener Filter Order

The choice of Wiener filter order affects:

(a) the ability of the filter to model and remove distortions and reduce the noise
(b) the computational complexity of the filter
(c) the numerical stability of the Wiener filter solution; a large filter may produce an ill-conditioned large-dimensional correlation matrix in Equation (8.10).

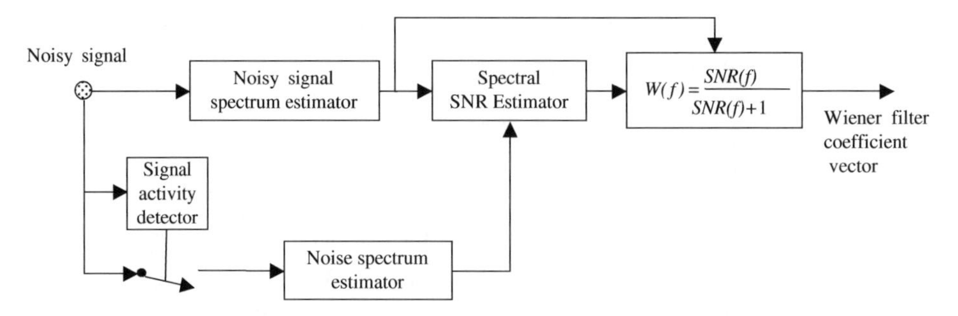

Figure 8.9 Configuration of a system for estimation of frequency Wiener filter.

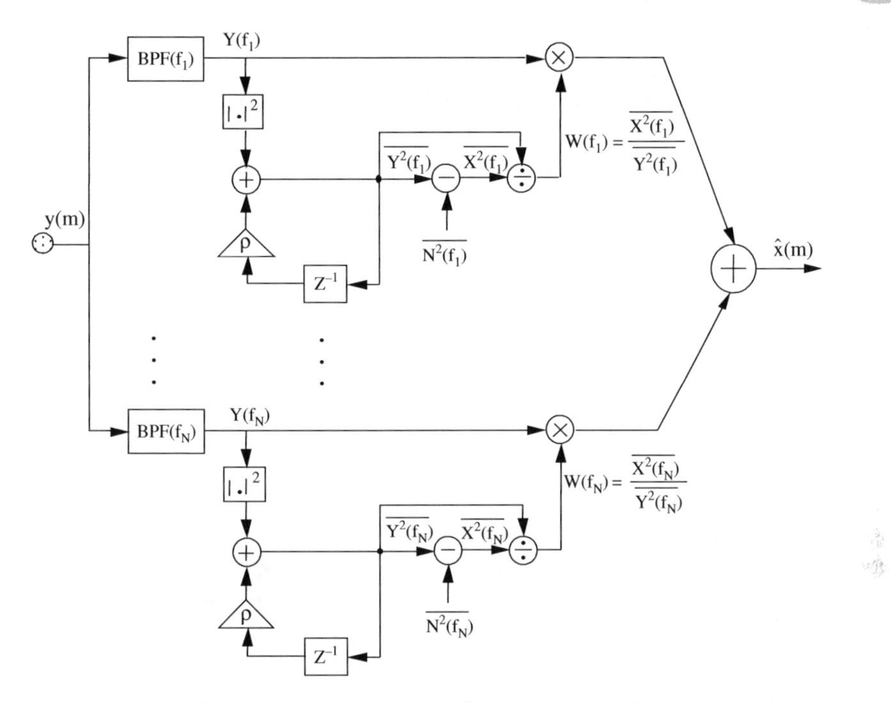

Figure 8.10 A filter-bank implementation of a Wiener filter for additive noise reduction.

The choice of the filter length also depends on the application and the method of implementation of the Wiener filter. For example, in a filter-bank implementation of the Wiener filter for additive noise reduction, Figure 8.10, the number of filter coefficients is equal to the number of filter banks, and typically the number of filter banks is between 16 to 64. On the other hand for many applications, a direct implementation of the time-domain Wiener filter requires a larger filter length, say between 64 and 256 taps.

A reduction in the required length of a time-domain Wiener filter can be achieved by dividing the time-domain signal into N sub-band signals. Each sub-band signal can then be down-sampled by a factor of N. The down-sampling results in a reduction, by a factor of N, in the required length of each sub-band Wiener filter. In Chapter 16, a sub-band echo canceller is described.

Improvements to Wiener Filters

The performance of Wiener filter can be limited by the following factors:

(a) The signal-to-noise ratio: generally the Wiener filter performance deteriorates with decreasing SNR.
(b) The signal non-stationarity: the Wiener filter theory assumes that the signal processes are stationary and any deviations from the assumption of stationarity will affect the ability of the filter to estimate and track the correlation or power spectrum functions needed for computation of the filter coefficients.
(c) The Wiener filter is a linear filter and the presence of significant non-linear distortion in the input will affect the filter performance.

The performance of Wiener filter can be improved by the use of a spectral time-tracking and smoothing process employed to track and smooth the variations of the spectral components of the signals over time. For example in noisy speech processing, the evolution over time of the significant spectral components of the signal and noise may be tracked in order to remove the fluctuations and errors in estimation of the correlation or spectral functions needed to compute Wiener filter coefficients.

8.8 Summary

A Wiener filter is formulated to transform an input signal to an output that is as close to a desired signal as possible. This chapter began with the derivation of the least square error Wiener filter. In Section 8.2, we derived the block-data least square error Wiener filter for applications where only finite-length realisations of the input and the desired signals are available. In such cases, the filter is obtained by minimising a time-averaged squared error function. In Section 8.3, we considered a vector space interpretation of the Wiener filters as the perpendicular projection of the desired signal onto the space of the input signal.

In Section 8.4, the least mean square error signal was analysed. The mean square error is zero only if the input signal is related to the desired signal through a linear and invertible filter. For most cases, owing to noise and/or non-linear distortions of the input signal, the minimum mean square error would be non-zero. In Section 8.5, we derived the Wiener filter in the frequency domain, and considered the issue of separability of signal and noise using a linear filter. Finally in Section 8.6, we considered some applications of Wiener filters in noise reduction, time-delay estimation and channel equalisation.

Further Reading

Akaike H. (1974) A New Look at Statistical Model Identification. IEEE Trans. Automatic Control, AC-19, pp. 716–723.

Alexander S.T. (1986) Adaptive Signal Processing Theory and Applications. Springer-Verlag, New York.

Anderson B.D. and Moor J.B. (1979) Linear Optimal Control. Prentice-Hall, Englewood Cliffs, NJ.

Dorny C.N. (1975) A Vector Space Approach to Models and Optimisation. Wiley, New York.

Durbin J. (1959) Efficient Estimation of Parameters in Moving Average Models. Biometrica, 46, pp. 306–318.

Giordano A.A. and Hsu F.M. (1985) Least Square Estimation with Applications to Digital Signal Processing. Wiley, New York.

Givens W. (1958) Computation of Plane Unitary Rotations Transforming a General Matrix to Triangular Form. SIAM J. Appl. Math. 6, pp. 26–50.

Golub G.H. and Reinsch (1970) Singular Value Decomposition and Least Squares Solutions. Numerical Mathematics, 14, pp. 403–420.

Golub G.H. and Van Loan C.F. (1980) An Analysis of the Total Least Squares Problem. SIAM Journal of Numerical Analysis, 17, pp. 883–893.

Golub G.H. and Van Loan C.F. (1983) Matrix Computations. Johns Hopkins University Press, Baltimore, MD.

Halmos P.R. (1974) Finite-Dimensional Vector Spaces. Springer-Verlag, New York.

Haykin S. (1991) Adaptive Filter Theory, 2nd Ed. Prentice-Hall, Englewood Cliffs, NJ.

Householder A.S. (1964) The Theory of Matrices in Numerical Analysis. Blaisdell, Waltham, MA.

Kailath T. (1974) A View of Three Decades of Linear Filtering Theory. IEEE Trans. Info. Theory, IT-20, pp. 146–181.

Kailath T. (1977) Linear Least Squares Estimation, Benchmark Papers in Electrical Engineering and Computer Science. Dowden, Hutchinson & Ross.

Kailath T. (1980) Linear Systems. Prentice-Hall, Englewood Cliffs, NJ.

Klema V.C. and Laub A.J. (1980) The Singular Value Decomposition: Its Computation and Some Applications. IEEE Trans. Automatic Control, AC-25, pp. 164–178.

Kolmogrov A.N. (1939) Sur l' Interpolation et Extrapolation des Suites Stationaires. Comptes Rendus de l'Academie des Sciences, 208, pp. 2043–2048.
Lawson C.L. and Hanson R.J. (1974) Solving Least Squares Problems. Prentice-Hall, Englewood Cliffs, NJ.
Orfanidis S.J. (1988) Optimum Signal Processing: An Introduction, 2nd edn. Macmillan, New York.
Scharf L.L. (1991) Statistical Signal Processing: Detection, Estimation, and Time Series Analysis. Addison Wesley, Reading, MA.
Strang G. (1976) Linear Algebra and Its Applications, 3rd edn. Harcourt Brace Jovanovich, San Diego, CA.
Whittle P.W. (1983) Prediction and Regulation by Linear Least-Squares Methods. University of Minnesota Press, Minneapolis, Minnesota.
Wiener N. (1949) Extrapolation, Interpolation and Smoothing of Stationary Time Series. MIT Press Cambridge, MA.
Wilkinson J.H. (1965) The Algebraic Eigenvalue Problem. Oxford University Press.
Wold H. (1954) The Analysis of Stationary Time Series, 2nd edn. Almquist and Wicksell, Uppsala.

9 Adaptive Filters: Kalman, RLS, LMS

Adaptive filters appear in many signal processing and communication systems for applications such as channel equalisation, echo cancellation, noise reduction, radar/sonar signal processing, beam-forming, space-time signal processing in mobile communication and low delay speech coding.

Adaptive filters work on the principle of minimising an objective error function, usually the mean squared difference (or error), between the filter output signal and a target (or desired) signal. Adaptive filters are used for estimation and identification of non-stationary signals, channels and systems or in applications where a sample-by-sample adaptation of a process and/or a low processing delay is required.

This chapter begins with a study of the theory of state-space Kalman filters. In Kalman filter theory a state equation models the dynamics of the signal generation process and an observation equation models the channel distortion and additive noise. Although Kalman filter theory assumes that the parameters of the signal, noise and channel are known and can be used for recursive signal estimation, in practice these parameters are unknown and Kalman filters have to adapt to the unknown parameters and statistics of signal, noise and channel.

We study recursive least square (RLS) error adaptive filters. The RLS filter is a sample-adaptive formulation of the Wiener filter, and for stationary signals should converge to the same solution as the Wiener filter. For least square error filtering, an alternative to using a Wiener-type closed-form recursive solution is an iterative gradient-based search for the optimal filter coefficients. The steepest-descent recursive search is a gradient-based method for searching the least square error performance curve for the minimum error filter coefficients. We study the steepest-descent method, and then consider the computationally inexpensive LMS gradient search method and its energy-normalised version.

9.1 Introduction

Adaptive filters are used in applications that involve a combination of three broad signal processing problems:

(a) *De-noising and channel equalisation*: for example the adaptive filtering of a non-stationary distorted or noisy signal, such as speech or image, to remove the effects of noise and channel distortions.
(b) *Trajectory estimation*: tracking and prediction of the trajectory of a non-stationary signal or parameter observed in noise. Kalman filters were used in the Apollo space programme for trajectory estimation.
(c) *System identification*: adaptive estimation of the parameters of an unknown time-varying system from a related observation, for example in the acoustic modelling of the impulse response of a room or a musical instrument.

Adaptive linear filters work on the principle that the desired signal or parameters can be extracted from the input through a filtering or estimation operation. The adaptation of the filter parameters is based on minimising an objective function; often the mean squared error between the filter output and a target (or desired) signal. The use of the least squared error (LSE) criterion is equivalent to the principal of orthogonality in which at any discrete-time m the estimator is expected to use all the available information such that the estimation error at time m is orthogonal to all the information available up to time m. In Chapter 12, on independent component analysis, adaptive filters with non-linear objective functions are considered.

An adaptive filter can be a combination of the following types of filters:

- single input or multi-input filters
- scalar-valued input or vector-valued input vectors
- linear or non-linear filters
- finite impulse response (FIR) or infinite impulse response (IIR) filters.

In this chapter we are mainly concerned with linear FIR filters which, because of their stability and relative ease of adaptation, are the most widely used type of adaptive filter. The adaptation algorithm can be based on a variant of one of the three most commonly used adaptive estimation methods, namely

- state-space Kalman filters
- recursive least squared (RLS) filters
- least mean squared (LMS) filters.

The different types of adaptation algorithms differ mainly in terms of the use of prior or estimated knowledge of system function and the covariance matrices of signal and noise and also in terms of the complexity of the solutions, see Table 9.1.

Table 9.1 Comparison of different adaptive/recursive filtering methods

Method	Operation/Adaptation	Input–output model	Statistics required	Gain vector
Kalman	Sample recursive	State-space models	Signal and noise covariance matrices, system and channel matrices	Kalman gain
LS/Wiener	Block adaptive	Filter	Correlation matrices	None
RLS	Recursive Wiener	Filter	Recursive calculation of correlation matrices	RLS gain
LMS	Gradient sample adaptive	Filter	None	LMS gain

9.2 State-Space Kalman Filters

Kalman filter, as illustrated in Figure 9.1, is a Bayesian recursive least square error method for estimation of a signal distorted in transmission through a channel and observed in noise. Kalman filters can be used with time-varying as well as time-invariant processes.

The filter is named after its inventor, Rudolf E. Kalman who formulated it in 1960. A similar filter was also formulated by Peter Swerling. Kalman filter was applied for trajectory estimation in the Apollo space programme; it has many applications, for example it is widely used in process control, in radio and other communication devices as phase-lock-loop systems and in signal denoising and system identifiation problems.

Kalman filter is a Bayesian filter in that it employs the prior probability distributions of the signal and noise processes which are assumed to be zero-mean Gaussian processes. The filter also assumes that the parameters of the models of signal and noise generation and channel distortion are known a priori.

Kalman filter formulation is based on a state-space approach in which a state equation models the dynamics of the signal generation process and an observation equation models the noisy and distorted observation signal. For a signal vector $\boldsymbol{x}(m)$ and noisy observation vector $\boldsymbol{y}(m)$, equations describing the state process model and the observation model are defined as

$$\boldsymbol{x}(m) = \boldsymbol{A}\,\boldsymbol{x}(m-1) + \boldsymbol{B}\,\boldsymbol{u}(m) + \boldsymbol{e}(m) \tag{9.1}$$

$$\boldsymbol{y}(m) = \boldsymbol{H}\,\boldsymbol{x}(m) + \boldsymbol{n}(m) \tag{9.2}$$

where

$\boldsymbol{x}(m)$ is the P-dimensional signal, or the state parameter vector at time m;
$\boldsymbol{A}$ is a $P \times P$ dimensional state transition matrix that relates the states of the process at times $m-1$ and m;
$\boldsymbol{B}$ is a $P \times P$ dimensional control matrix, used in process control modelling;
$\boldsymbol{u}(m)$ is the P-dimensional control input;
$\boldsymbol{e}(m)$ is the P-dimensional uncorrelated input excitation vector of the state equation, $\boldsymbol{e}(m)$ is a normal (Gaussian) process $p(\boldsymbol{e}(m)) \sim N(0, \boldsymbol{Q})$;
$\boldsymbol{n}(m)$ is an M-dimensional noise vector, also known as measurement noise, $\boldsymbol{n}(m)$ is a normal (Gaussian) process $p(\boldsymbol{n}(m)) \sim N(0, \boldsymbol{R})$;
$\boldsymbol{Q}$ is the $P \times P$ dimensional covariance matrix of $\boldsymbol{n}(m)$;
$\boldsymbol{y}(m)$ is the M-dimensional noisy and distorted observation vector;

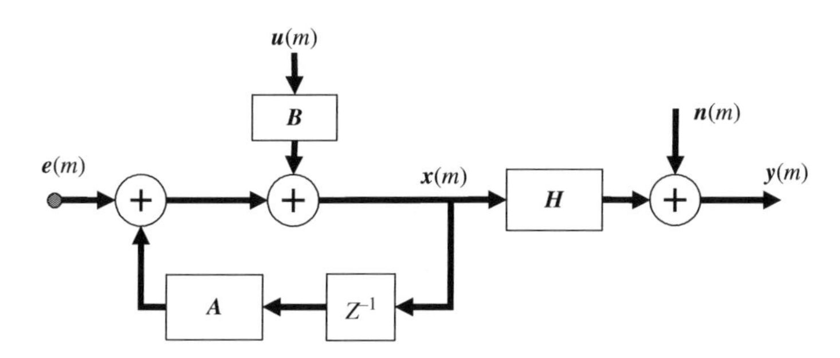

Figure 9.1 Illustration of signal and observation models in Kalman filter theory.

$\boldsymbol{H}$ is the $M \times P$ dimensional channel distortion matrix;
$\boldsymbol{R}$ is the $M \times M$ dimensional covariance matrix of $\boldsymbol{n}(m)$.

Note that the control matrix $\boldsymbol{B}$ and the control vector $\boldsymbol{u}$ are used in control applications where often an external input may be applied by a controller process to change, adjust or correct the trajectory of the vector process $\boldsymbol{x}$.

In communication signal processing applications, such as channel equalisation or speech enhancement, often there is no external control input vector $\boldsymbol{u}$ and the Kalman equations reduce to

$$\boldsymbol{x}(m) = \boldsymbol{A}\,\boldsymbol{x}(m-1) + \boldsymbol{e}(m) \tag{9.3}$$

$$\boldsymbol{y}(m) = \boldsymbol{H}\,\boldsymbol{x}(m) + \boldsymbol{n}(m) \tag{9.4}$$

Kalman Filter Algorithm

Input: observation vectors $\{\boldsymbol{y}(m)\}$
Output: state or signal vectors $\{\hat{\boldsymbol{x}}(m)\}$

Initial Conditions
Prediction error covariance matrix:

$$\boldsymbol{P}(0|-1) = \delta\mathbf{I} \tag{9.5}$$

Prediction:

$$\hat{\boldsymbol{x}}(0|-1) = 0 \tag{9.6}$$

For $m = 0, 1, \ldots$

Time-Update Process Prediction Equations
State prediction equation:

$$\hat{\boldsymbol{x}}(m|m-1) = \boldsymbol{A}\hat{\boldsymbol{x}}(m-1) \tag{9.7}$$

Covariance matrix of prediction error:

$$\boldsymbol{P}(m|m-1) = \boldsymbol{A}\boldsymbol{P}(m-1)\boldsymbol{A}^{\mathrm{T}} + \boldsymbol{Q} \tag{9.8}$$

Measurement-Update (Estimate) Equations
Kalman gain vector:

$$\boldsymbol{K}(m) = \boldsymbol{P}(m|m-1)\boldsymbol{H}^{\mathrm{T}}\left(\boldsymbol{H}\boldsymbol{P}(m|m-1)\boldsymbol{H}^{\mathrm{T}} + \boldsymbol{R}\right)^{-1} \tag{9.9}$$

State update:

$$\hat{\boldsymbol{x}}(m) = \hat{\boldsymbol{x}}(m|m-1) + \boldsymbol{K}(m)\left(\boldsymbol{y}(m) - \boldsymbol{H}\hat{\boldsymbol{x}}(m|m-1)\right) \tag{9.10}$$

Covariance matrix of estimation error:

$$\boldsymbol{P}(m) = [\boldsymbol{I} - \boldsymbol{KH}]\boldsymbol{P}(m|m-1) \tag{9.11}$$

9.2.1 Derivation of Kalman Filter Algorithm

Kalman filter can be derived as a recursive minimum mean square error estimator of a signal $\boldsymbol{x}(m)$ from a noisy observation $\boldsymbol{y}(m)$. The derivation of Kalman filter assumes that the state transition matrix $\boldsymbol{A}$, the channel distortion matrix $\boldsymbol{H}$, the covariance matrix $\boldsymbol{Q}$ of the input $\boldsymbol{e}(m)$ of the state equation and the covariance matrix $\boldsymbol{R}$ of the additive noise $\boldsymbol{n}(m)$ are given or estimated by some means.

The derivation of Kalman filter, described next, is based on the following methodology:

(1) Prediction (update) step: the signal state is predicted from the previous observations and a prediction error covariance matrix is obtained as expressed in Equations (9.7)–(9.8).
(2) Estimation (measurement) step: the results of prediction from step (1) and the innovation signal (note innovation is the difference between prediction and noise observation) are used to estimate the signal. At this stage the Kalman gain vector and estimation error covariance matrix are calculated as expressed in Equations (9.9)–(9.11).

In this chapter, we use the notation $\hat{\boldsymbol{y}}(m|m-i)$ to denote a prediction of $\boldsymbol{y}(m)$ based on the observation samples up to the time $m-i$. Assume that $\hat{\boldsymbol{x}}(m|m-1)$ is the least square error prediction of $\boldsymbol{x}(m)$ based on the observations $[\boldsymbol{y}(0), \ldots, \boldsymbol{y}(m-1)]$. Define a prediction equation as

$$\hat{\boldsymbol{x}}(m|m-1) = \boldsymbol{A}\hat{\boldsymbol{x}}(m-1) \tag{9.12}$$

An *innovation* signal composed of prediction error plus noise, may be defined as

$$\boldsymbol{v}(m) = \boldsymbol{y}(m) - \boldsymbol{H}\hat{\boldsymbol{x}}(m|m-1) \tag{9.13}$$

where $\hat{\boldsymbol{x}}(m|m-1)$ denotes the least square error prediction of the signal $\boldsymbol{x}(m)$. The innovation signal vector $\boldsymbol{v}(m)$ is the part of $\boldsymbol{y}(m)$ that is unpredictable from the past observations; it includes both the noise and the unpredictable information-bearing part of the signal $\boldsymbol{x}(m)$.

For an optimal linear least mean square error estimate, the innovation signal must be an uncorrelated process orthogonal to the past observation vectors; hence we have

$$\mathcal{E}\left[\boldsymbol{v}(m)\boldsymbol{y}^{\mathrm{T}}(m-k)\right] = 0 \quad k > 0 \tag{9.14}$$

and

$$\mathcal{E}\left[\boldsymbol{v}(m)\boldsymbol{v}^{\mathrm{T}}(k)\right] = 0, \quad m \neq k \tag{9.15}$$

The concept of innovations is central to the derivation of the Kalman filter. The least square error criterion is satisfied if the estimation error is *orthogonal* to the past samples.

In the following derivation of the Kalman filter, the orthogonality condition of Equation (9.14) is used as the starting point to derive an optimal linear filter whose innovations $\nu(m)$ are orthogonal to the past observations $\boldsymbol{y}(m)$.

Substituting the observation Equation (9.2) in Equation (9.13) yields

$$\begin{aligned}\boldsymbol{v}(m) &= \boldsymbol{H}\,\boldsymbol{x}(m) + \boldsymbol{n}(m) - \hat{\boldsymbol{x}}(m|m-1) \\ &= \boldsymbol{H}\tilde{\boldsymbol{x}}(m|m-1) + \boldsymbol{n}(m)\end{aligned} \tag{9.16}$$

where $\tilde{\boldsymbol{x}}(m|m-1)$ is the signal prediction error vector defined as

$$\tilde{\boldsymbol{x}}(m|m-1) = \boldsymbol{x}(m) - \hat{\boldsymbol{x}}(m|m-1) \tag{9.17}$$

From Equation (9.16) the covariance matrix of the innovation signal $\nu(m)$ is given by

$$\mathcal{E}\left[\boldsymbol{v}(m)\boldsymbol{v}^{\mathrm{T}}(m)\right] = \boldsymbol{H}\,\boldsymbol{P}(m|m-1)\boldsymbol{H}^{\mathrm{T}} + \boldsymbol{R} \tag{9.18}$$

where $\boldsymbol{P}(m|m-1)$ is the covariance matrix of the prediction error $\tilde{\boldsymbol{x}}(m|m-1)$. The estimation of $\boldsymbol{x}(m)$, based on the samples available up to the time m, can be expressed recursively as a linear combination of the prediction of $\boldsymbol{x}(m)$ based on the samples available up to the time $m-1$ and the innovation signal at time m as

$$\hat{\boldsymbol{x}}(m) = \hat{\boldsymbol{x}}(m|m-1) + \boldsymbol{K}(m)\boldsymbol{v}(m) \tag{9.19}$$

where the $P \times M$ matrix $\boldsymbol{K}(m)$ is the Kalman gain matrix. Now, from Equation (9.1), we have

$$\hat{\boldsymbol{x}}(m|m-1) = \boldsymbol{A}\hat{\boldsymbol{x}}(m-1) \tag{9.20}$$

Substitution of Equation (9.20) in (9.19) gives

$$\hat{\boldsymbol{x}}(m) = \boldsymbol{A}\hat{\boldsymbol{x}}(m-1) + \boldsymbol{K}(m)\boldsymbol{v}(m) \tag{9.21}$$

To obtain a recursive relation for the computation and update of the Kalman gain matrix, multiply both sides of Equation (9.19) by $\nu^{\mathrm{T}}(m)$ and take the expectation of the results to yield

$$\mathcal{E}\left[\hat{\boldsymbol{x}}(m)\,\boldsymbol{v}^{\mathrm{T}}(m)\right] = \mathcal{E}\left[\hat{\boldsymbol{x}}(m|m-1)\,\boldsymbol{v}^{\mathrm{T}}(m)\right] + \boldsymbol{K}(m)\mathcal{E}\left[\boldsymbol{v}(m)\boldsymbol{v}^{\mathrm{T}}(m)\right] \tag{9.22}$$

Owing to the required orthogonality of the innovation sequence $\nu(m)$ to the past samples, we have

$$\mathcal{E}\left[\hat{\boldsymbol{x}}(m|m-1)\,\boldsymbol{v}^{\mathrm{T}}(m)\right] = 0 \tag{9.23}$$

Hence, from Equations (9.2) and (9.23), the Kalman gain matrix is given by

$$\boldsymbol{K}(m) = \mathcal{E}\left[\hat{\boldsymbol{x}}(m)\,\boldsymbol{v}^{\mathrm{T}}(m)\right]\mathcal{E}\left[\boldsymbol{v}(m)\,\boldsymbol{v}^{\mathrm{T}}(m)\right]^{-1} \tag{9.24}$$

The first term on the right-hand side of Equation (9.24) can be expressed as

$$\begin{aligned}\mathcal{E}\left[\hat{\boldsymbol{x}}(m)\,\boldsymbol{v}^{\mathrm{T}}(m)\right] &= \mathcal{E}\left[(\boldsymbol{x}(m) - \tilde{\boldsymbol{x}}(m))\,\boldsymbol{v}^{\mathrm{T}}(m)\right] \\ &= \mathcal{E}\left[(\hat{\boldsymbol{x}}(m|m-1) + \tilde{\boldsymbol{x}}(m|m-1) - \tilde{\boldsymbol{x}}(m))\,\boldsymbol{v}^{\mathrm{T}}(m)\right] \\ &= \mathcal{E}\left[\tilde{\boldsymbol{x}}(m|m-1)\,\boldsymbol{v}^{\mathrm{T}}(m)\right] \\ &= \mathcal{E}\left[\tilde{\boldsymbol{x}}(m|m-1)\,(\boldsymbol{H}\,\boldsymbol{x}(m) + \boldsymbol{n}(m) - \boldsymbol{H}\hat{\boldsymbol{x}}(m|m-1))^{\mathrm{T}}\right] \\ &= \mathcal{E}\left[\tilde{\boldsymbol{x}}(m|m-1)\tilde{\boldsymbol{x}}^{\mathrm{T}}(m|m-1)\right]\boldsymbol{H}^{\mathrm{T}} \\ &= \boldsymbol{P}(m|m-1)\boldsymbol{H}^{\mathrm{T}}\end{aligned} \tag{9.25}$$

where $\tilde{\boldsymbol{x}}(m)$ is the estimation error vector. In developing the successive lines of Equation (9.25), we have used the following relations:

$$\boldsymbol{x}(m) = \hat{\boldsymbol{x}}(m|m-1) + \tilde{\boldsymbol{x}}(m|m-1) \tag{9.26}$$

$$\mathcal{E}\left[\tilde{\boldsymbol{x}}(m)\,\boldsymbol{v}^{\mathrm{T}}(m)\right] = 0 \tag{9.27}$$

$$\mathcal{E}\left[\hat{\boldsymbol{x}}(m|m-1)\,\boldsymbol{v}^{\mathrm{T}}(m)\right] = 0 \tag{9.28}$$

$$\mathcal{E}\left[\tilde{\boldsymbol{x}}(m|m-1)\,\boldsymbol{n}^{\mathrm{T}}(m)\right] = 0 \tag{9.29}$$

Substitution of Equations (9.18) and (9.25) in Equation (9.24) yields the following equation for the Kalman gain matrix:

$$\boldsymbol{K}(m) = \boldsymbol{P}(m|m-1)\boldsymbol{H}^{\mathrm{T}}\left[\boldsymbol{H}\boldsymbol{P}(m|m-1)\boldsymbol{H}^{\mathrm{T}} + \boldsymbol{R}\right]^{-1} \tag{9.30}$$

where $\boldsymbol{P}(m|m-1)$ is the covariance matrix of the signal prediction error $\tilde{\boldsymbol{x}}(m|m-1)$. Note that the Kalman gain vector can be interpreted as a function of signal-to-noise ratio of the innovation signal.

A recursive relation for calculation of $\boldsymbol{P}(m|m-1)$, the covariance matrix of prediction error $\tilde{\boldsymbol{x}}(m|m-1)$ is derived as follows:

$$\tilde{\boldsymbol{x}}(m|m-1) = \boldsymbol{x}(m) - \hat{\boldsymbol{x}}(m|m-1) \tag{9.31}$$

Substitution of Equation (9.1) and (9.20) in Equation (9.31) and rearrangement of the terms yields

$$\begin{aligned}\tilde{\boldsymbol{x}}(m|m-1) &= [\boldsymbol{A}\boldsymbol{x}(m-1) + \boldsymbol{e}(m)] - [\boldsymbol{A}\hat{\boldsymbol{x}}(m-1)] \\ &= \boldsymbol{A}\tilde{\boldsymbol{x}}(m-1) + \boldsymbol{e}(m)\end{aligned} \tag{9.32}$$

The covariance matrix of $\tilde{\boldsymbol{x}}(m|m-1)$ is obtained as

$$\mathcal{E}\left[\tilde{\boldsymbol{x}}(m|m-1)\,\tilde{\boldsymbol{x}}(m|m-1)^{\mathrm{T}}\right] = \boldsymbol{A}\mathcal{E}\left[\tilde{\boldsymbol{x}}(m-1)\,\tilde{\boldsymbol{x}}(m-1)^{\mathrm{T}}\right]\boldsymbol{A}^{\mathrm{T}} + \boldsymbol{Q} \tag{9.33}$$

or

$$\boldsymbol{P}(m|m-1) = \boldsymbol{A}\boldsymbol{P}(m-1)\boldsymbol{A}^{\mathrm{T}} + \boldsymbol{Q} \tag{9.34}$$

where $\boldsymbol{P}(m|m-1)$ and $\boldsymbol{P}(m)$ are the covariance matrices of the prediction error $\tilde{\boldsymbol{x}}(m|m-1)$ and estimation error $\tilde{\boldsymbol{x}}(m)$ respectively. A recursive relation for the covariance matrix of the signal estimation error vector $\tilde{\boldsymbol{x}}(m)$ can be derived as follows. Subtracting both sides of Equation (9.19) from $\boldsymbol{x}(m)$ we have

$$\tilde{\boldsymbol{x}}(m) = \tilde{\boldsymbol{x}}(m|m-1) - \boldsymbol{K}(m)\boldsymbol{v}(m) \tag{9.35}$$

From Equation (9.35) the covariance matrix of the estimation error vector can be expressed as

$$\begin{aligned}\mathcal{E}\left[\tilde{\boldsymbol{x}}(m)\,\tilde{\boldsymbol{x}}(m)^{T}\right] = \mathcal{E}\left[\tilde{\boldsymbol{x}}(m|m-1)\,\tilde{\boldsymbol{x}}(m|m-1)^{T}\right] + \boldsymbol{K}(m)\,\mathcal{E}\left[\boldsymbol{v}(m)\,\boldsymbol{v}(m)^{T}\right]\boldsymbol{K}(m)^{T} \\ - 2\mathcal{E}\left[\tilde{\boldsymbol{x}}(m|m-1)\,\boldsymbol{v}(m)^{T}\right]\boldsymbol{K}(m)^{T}\end{aligned} \tag{9.36}$$

From Equations (9.24) and (9.25) we have

$$\boldsymbol{K}(m)\mathcal{E}\left[\boldsymbol{v}(m)\boldsymbol{v}(m)^{\mathrm{T}}\right] = \boldsymbol{P}(m|m-1)\boldsymbol{H}^{T} \tag{9.37}$$

From Equation (9.25) we have

$$\mathcal{E}\left[\tilde{\boldsymbol{x}}(m|m-1)\,\boldsymbol{v}(m)^{\mathrm{T}}\right] = \boldsymbol{P}(m|m-1)\boldsymbol{H}^{\mathrm{T}} \tag{9.38}$$

Substitution of Equations (9.37) and (9.38) in (9.36) and rearranging yields

$$\boldsymbol{P}(m) = [\boldsymbol{I} - \boldsymbol{K}\boldsymbol{H}]\,\boldsymbol{P}(m|m-1) \tag{9.39}$$

Example 9.1 *Recursive estimation of a constant signal observed in noise*
Consider the estimation of a constant signal observed in a random noise. The state and observation equations for this problem are given by

$$x(m) = x(m-1) = x \tag{9.40}$$

$$y(m) = x + n(m) \tag{9.41}$$

Note that $a = 1$, the state excitation $e(m) = 0$, the variance of excitation $Q = 0$ and the variance of noise $R = \sigma_n^2$.

Using the Kalman algorithm, we have the following recursive solutions:

Initial Conditions

$$P(-1) = \delta \tag{9.42}$$

$$\hat{x}(0|-1) = 0 \tag{9.43}$$

For $m = 0, 1, \ldots$

Time-Update Equations
Signal prediction equation:

$$\hat{x}(m|m-1) = \hat{x}(m-1) \tag{9.44}$$

Covariance matrix of prediction error:

$$P(m|m-1) = P(m-1) \tag{9.45}$$

Measurement-Update Equations
Kalman gain vector:

$$K(m) = P(m|m-1)\left(P(m|m-1) + \sigma_n^2\right)^{-1} \tag{9.46}$$

Signal estimation equation:

$$\hat{x}(m) = \hat{x}(m|m-1) + K(m)\left(y(m) - \hat{x}(m|m-1)\right) \tag{9.47}$$

Covariance matrix of estimation error:

$$P(m) = [1 - K(m)]\,P(m|m-1) \tag{9.48}$$

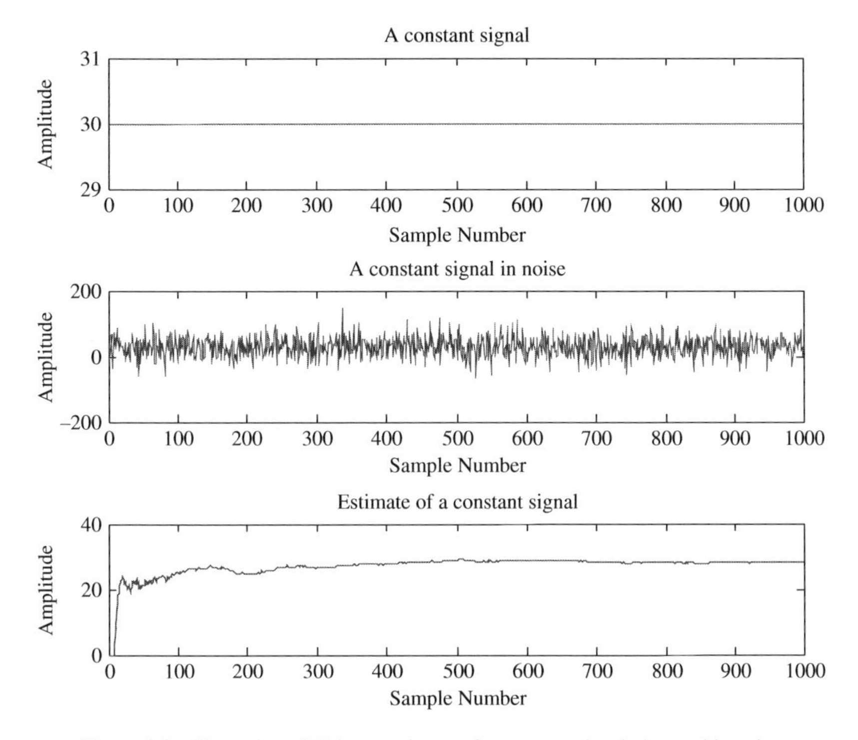

Figure 9.2 Illustration of Kalman estimate of a constant signal observed in noise.

Figure 9.2 shows the application of Kalman filter for estimation of a constant signal observed in noise.

Example 9.2 *Estimation of an autoregressive (AR) signal observed in noise*
Consider the Kalman filtering of a Pth order AR process $x(m)$ observed in an additive white Gaussian noise $n(m)$. Assume that the signal generation and the observation equations are given as

$$x(m) = \sum_{k=1}^{P} a_k x(m-k) + e(m) \tag{9.49}$$

$$y(m) = x(m) + n(m) \tag{9.50}$$

Matlab function Kalman_Noisy_Constant_Signal( )
Implements a Kalman filter for estimation of a constant (dc) signal observed in additive white Gaussian noise. The program displays the constant signal, the noise and the estimate of the constant signal.

Let $\sigma_e^2(m)$ and $\sigma_n^2(m)$ denote the variances of the excitation signal $e(m)$ and the noise $n(m)$ respectively. Equation (9.49) can be written in a vector form as

$$\underbrace{\begin{bmatrix} x(m) \\ x(m-1) \\ x(m-2) \\ \vdots \\ x(m-P+1) \end{bmatrix}}_{\substack{\boldsymbol{x}(m) \\ P\times 1}} = \underbrace{\begin{bmatrix} a_1 & a_2 & \cdots & a_{P-1} & a_P \\ 1 & 0 & \cdots & 0 & 0 \\ 0 & 1 & \cdots & 0 & 0 \\ \vdots & \vdots & \ddots & \vdots & \vdots \\ 0 & 0 & 0 & 1 & 0 \end{bmatrix}}_{\substack{A_x \\ P\times P}} \underbrace{\begin{bmatrix} x(m-1) \\ x(m-2) \\ x(m-3) \\ \vdots \\ x(m-P) \end{bmatrix}}_{\substack{\boldsymbol{x}(m-1) \\ P\times 1}} + \underbrace{\begin{bmatrix} e(m) \\ 0 \\ 0 \\ \vdots \\ 0 \end{bmatrix}}_{\substack{\boldsymbol{e}(m) \\ P\times 1}} \tag{9.51}$$

Using Equation (9.51) and assuming that $\boldsymbol{H}=\boldsymbol{I}$ in the Kalman filter equations yields the following Kalman filter algorithm:

Initial Conditions

$$\boldsymbol{P}(-1)=\delta\boldsymbol{I} \tag{9.52}$$

$$\hat{\boldsymbol{x}}(0|-1)=0 \tag{9.53}$$

For $m=0, 1, \ldots$

Time-Update Equations
Signal prediction equation:

$$\hat{\boldsymbol{x}}(m|m-1)=\boldsymbol{A}\hat{\boldsymbol{x}}(m) \tag{9.54}$$

The prediction error given by

$$\begin{aligned} \boldsymbol{e}(m|m-1) &= \boldsymbol{x}(m)-\boldsymbol{A}\hat{\boldsymbol{x}}(m-1) \\ &= \boldsymbol{x}(m)-\boldsymbol{A}\left(\boldsymbol{x}(m-1)-\tilde{\boldsymbol{x}}(m-1)\right) \\ &= \boldsymbol{e}(m)+\boldsymbol{A}\tilde{\boldsymbol{x}}(m-1) \end{aligned} \tag{9.55}$$

The covariance matrix of prediction error

$$\boldsymbol{P}(m|m-1)=\boldsymbol{A}\boldsymbol{P}(m-1)\boldsymbol{A}^T+\boldsymbol{Q} \tag{9.56}$$

where $\boldsymbol{Q}$ is a matrix with just one non-zero element at the top-left corner of the matrix.
Taking the covariance of Equation (9.55) results in Equation (9.56).

Measurement-Update Equations
Kalman gain:

$$\boldsymbol{K}(m)=\boldsymbol{P}(m|m-1)\left(\boldsymbol{P}(m|m-1)+\boldsymbol{R}\right)^{-1} \tag{9.57}$$

Signal estimation equation:

$$\hat{\boldsymbol{x}}(m)=\hat{\boldsymbol{x}}(m|m-1)+\boldsymbol{K}(m)\left(\boldsymbol{y}(m)-\hat{\boldsymbol{x}}(m|m-1)\right) \tag{9.58}$$

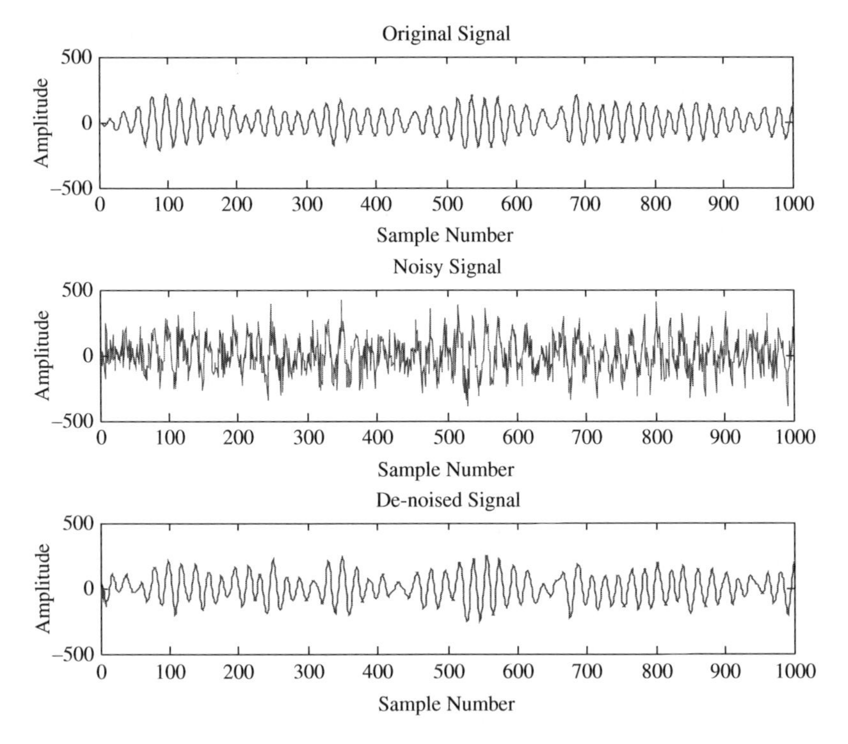

Figure 9.3 Illustration of Kalman estimation of an AR signal observed in additive white Gaussian noise.

Covariance matrix of estimation error:

$$\boldsymbol{P}(m) = [1 - \boldsymbol{K}(m)]\,\boldsymbol{P}(m|m-1) \tag{9.59}$$

Figure 9.3 shows an example of Kalman filtering of an AR signal observed in noise. In Chapter 15 the application of Kalman filter for restoration of speech signals in non-white noise is considered.

Matlab function Kalman_Autoregressive( )
Implements a Kalman filter for denoising of an autoregressive signal observed in additive white noise.

The autoregressive (AR) model is synthesised from user-specified frequencies and bandwidths of the poles of the model.

The function plots the frequency response of the all-pole model, the original AR signal, the noisy signal and the restored signal.

9.3 Sample Adaptive Filters

Adaptive filters, namely the RLS, the steepest descent and the LMS, are recursive formulations of the least square error Wiener filter. Sample-adaptive filters have a number of advantages over the block-adaptive filters of Chapter 8, including lower processing delay and better tracking of the trajectory of non-stationary signals. These are essential characteristics in applications such as echo cancellation, adaptive delay estimation, low-delay predictive coding, noise cancellation, radar, and channel equalisation in mobile telephony, where low delay and fast tracking of time-varying processes and time-varying environments are important objectives.

Figure 9.4 illustrates the configuration of a least square error adaptive filter. At each sampling time, an adaptation algorithm adjusts the P filter coefficients $\mathbf{w}(m) = [w_0(m), w_1(m), \ldots, w_{P-1}(m)]$ to minimise the difference between the filter output and a desired, or target, signal. An adaptive filter starts at some initial state, then the filter coefficients are periodically updated, usually on a sample-by-sample basis, to minimise the difference between the filter output and a desired or target signal. The adaptation formula has the general recursive form:

$$\text{Next parameter estimate} = \text{Previous parameter estimate} + \text{Update (error)}$$

where the update term is a function of the error signal. In adaptive filtering a number of decisions have to be made concerning the filter model and the adaptation algorithm:

(a) Filter type: This can be a finite impulse response (FIR) filter, or an infinite impulse response (IIR) filter. In this chapter we only consider FIR filters, since they have good stability and convergence properties and for these reasons are the type often used in practice.

(b) Filter order: Often the correct number of filter taps is unknown. The filter order is either set using *a priori* knowledge of the input and the desired signals, or it may be obtained by monitoring the changes in the error signal as a function of the increasing filter order.

(c) Adaptation algorithm: The two commonly used adaptation algorithms are the recursive least square (RLS) error and the least mean square error (LMS) methods. The factors that influence the choice of the adaptation algorithm are the computational complexity, the speed of convergence to optimal operating conditions, the minimum error at convergence, the numerical stability and the robustness of the algorithm to initial parameter states.

(d) Optimisation criteria: In this chapter two optimality criteria are used. One is based on the minimisation of the mean of squared error (used in LMS, RLS and Kalman filter) and the other is based on constrained minimisation of the norm of the incremental change in the filter coefficients which results in normalised LMS (NLMS). In Chapter 12, adaptive filters with non-linear objective functions are considered for independent component analysis.

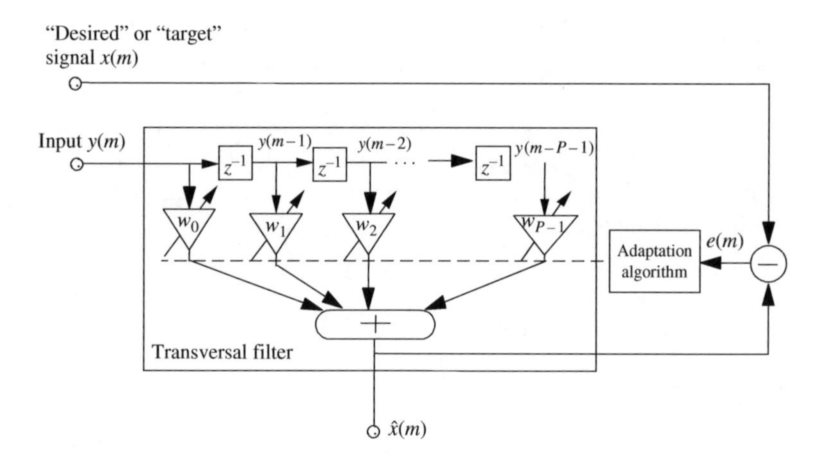

Figure 9.4 Illustration of the configuration of an adaptive filter.

9.4 Recursive Least Square (RLS) Adaptive Filters

The recursive least square error (RLS) filter is a sample-adaptive, time-update, version of the Wiener filter studied in Chapter 8. For stationary signals, the RLS filter converges to the same optimal filter coefficients as the Wiener filter. For non-stationary signals, the RLS filter tracks the time variations of the process. The RLS filter has a relatively fast rate of convergence to the optimal filter coefficients. This is useful in applications such as speech enhancement, channel equalisation, echo cancellation and radar where the filter should be able to track relatively fast changes in the signal process.

In the recursive least square algorithm, the adaptation starts with some initial filter state, and successive samples of the input signals are used to adapt the filter coefficients. Figure 9.4 illustrates the configuration of an adaptive filter where $y(m)$, $x(m)$ and $\boldsymbol{w}(m) = [w_0(m), w_1(m), \ldots, w_{P-1}(m)]$ denote the filter input, the desired (target) signal and the filter coefficient vector respectively. The filter output can be expressed as

$$\hat{x}(m) = \boldsymbol{w}^{\mathrm{T}}(m)\boldsymbol{y}(m) \tag{9.60}$$

where $\hat{x}(m)$ is an estimate of the desired signal $x(m)$. The filter error signal is defined as the difference between the filter output and the target signal as

$$\begin{aligned} e(m) &= x(m) - \hat{x}(m) \\ &= x(m) - \boldsymbol{w}^{\mathrm{T}}(m)\boldsymbol{y}(m) \end{aligned} \tag{9.61}$$

The adaptation process is based on the minimisation of the mean square error criterion defined as

$$\begin{aligned} \mathcal{E}[e^2(m)] &= \mathcal{E}\left\{\left[x(m) - \boldsymbol{w}^{\mathrm{T}}(m)\boldsymbol{y}(m)\right]^2\right\} \\ &= \mathcal{E}[x^2(m)] - 2\boldsymbol{w}^{\mathrm{T}}(m)\mathcal{E}[\boldsymbol{y}(m)x(m)] + \boldsymbol{w}^{\mathrm{T}}(m)\mathcal{E}[\boldsymbol{y}(m)\boldsymbol{y}^{\mathrm{T}}(m)]\boldsymbol{w}(m) \\ &= r_{xx}(0) - 2\boldsymbol{w}^{\mathrm{T}}(m)\boldsymbol{r}_{yx}(m) + \boldsymbol{w}^{\mathrm{T}}(m)\boldsymbol{R}_{yy}(m)\boldsymbol{w}(m) \end{aligned} \tag{9.62}$$

where $r_{xx}(0)$ is the autocorrelation at lag zero of the target signal $x(m)$, $\boldsymbol{R}_{yy}$ is the autocorrelation matrix of the input signal vector $\boldsymbol{y}(m)$ and $\boldsymbol{r}_{yx}$ is the cross-correlation vector of the input and the target signals.

The Wiener filter is obtained by minimising the mean square error with respect to the filter coefficients. For stationary signals, the result of this minimisation is given in Equation (8.10), as

$$\boldsymbol{w} = \boldsymbol{R}_{yy}^{-1}\,\boldsymbol{r}_{yx} \tag{9.63}$$

In the following, we formulate a recursive, time-update, adaptive formulation of Equation (9.61). From Section 8.2, for a block of N sample vectors, the correlation matrix can be written as

$$\boldsymbol{R}_{yy} = \boldsymbol{Y}^{\mathrm{T}}\boldsymbol{Y} = \sum_{m=0}^{N-1}\boldsymbol{y}(m)\boldsymbol{y}^{\mathrm{T}}(m) \tag{9.64}$$

where $\boldsymbol{y}(m) = [y(m), \ldots, y(m-P-1)]^{\mathrm{T}}$. Now, the sum of vector products in Equation (9.64) can be expressed in recursive fashion as

$$\boldsymbol{R}_{yy}(m) = \boldsymbol{R}_{yy}(m-1) + \boldsymbol{y}(m)\boldsymbol{y}^{\mathrm{T}}(m) \tag{9.65}$$

To introduce adaptability to the time variations of the signal statistics, the autocorrelation estimate in Equation (9.65) can be windowed by an exponentially decaying window:

$$\boldsymbol{R}_{yy}(m) = \lambda\boldsymbol{R}_{yy}(m-1) + \boldsymbol{y}(m)\boldsymbol{y}^{\mathrm{T}}(m) \tag{9.66}$$

where λ is the so-called adaptation, or forgetting factor, and is in the range $0 > \lambda > 1$. Similarly, the cross-correlation vector is given by

$$\boldsymbol{r}_{yx} = \sum_{m=0}^{N-1} \boldsymbol{y}(m)x(m) \tag{9.67}$$

The sum of products in Equation (9.65) can be calculated in recursive form as

$$\boldsymbol{r}_{yx}(m) = \boldsymbol{r}_{yx}(m-1) + \boldsymbol{y}(m)x(m) \tag{9.68}$$

Equation (9.68) can be made adaptive using a forgetting factor λ:

$$\boldsymbol{r}_{yx}(m) = \lambda \boldsymbol{r}_{yx}(m-1) + \boldsymbol{y}(m)x(m) \tag{9.69}$$

For a recursive solution of the least square error equation (9.63), we need to obtain a recursive time-update formula for the inverse matrix in the form

$$\boldsymbol{R}_{yy}^{-1}(m) = \boldsymbol{R}_{yy}^{-1}(m-1) + \text{Update}(m) \tag{9.70}$$

A recursive relation for the matrix inversion is obtained using the following lemma.

Matrix Inversion Lemma

Let $\boldsymbol{A}$ and $\boldsymbol{B}$ be two positive-definite $P \times P$ matrices related by

$$\boldsymbol{A} = \boldsymbol{B}^{-1} + \boldsymbol{C}\boldsymbol{D}^{-1}\boldsymbol{C}^{\mathrm{T}} \tag{9.71}$$

where $\boldsymbol{D}$ is a positive-definite $N \times N$ matrix and $\boldsymbol{C}$ is a $P \times N$ matrix. The matrix inversion lemma states that the inverse of the matrix $\boldsymbol{A}$ can be expressed as

$$\boldsymbol{A}^{-1} = \boldsymbol{B} - \boldsymbol{B}\boldsymbol{C}\left(\boldsymbol{D} + \boldsymbol{C}^{\mathrm{T}}\boldsymbol{B}\boldsymbol{C}\right)^{-1}\boldsymbol{C}^{\mathrm{T}}\boldsymbol{B} \tag{9.72}$$

This lemma can be proved by multiplying Equation (9.70) and Equation (9.71). The left- and right-hand sides of the results of multiplication are the identity matrix. The matrix inversion lemma can be used to obtain a recursive implementation for the inverse of the correlation matrix $\boldsymbol{R}_{yy}^{-1}(m)$. Let

$$\boldsymbol{R}_{yy}(m) = \boldsymbol{A} \tag{9.73}$$

$$\lambda^{-1}\boldsymbol{R}_{yy}^{-1}(m-1) = \boldsymbol{B} \tag{9.74}$$

$$\boldsymbol{y}(m) = \boldsymbol{C} \tag{9.75}$$

$$\boldsymbol{D} = \text{identity matrix} \tag{9.76}$$

Substituting Equations (9.73) to (9.76) in Equation (9.72), we obtain

$$\boldsymbol{R}_{yy}^{-1}(m) = \lambda^{-1}\boldsymbol{R}_{yy}^{-1}(m-1) - \frac{\lambda^{-2}\boldsymbol{R}_{yy}^{-1}(m-1)\boldsymbol{y}(m)\boldsymbol{y}^{\mathrm{T}}(m)\boldsymbol{R}_{yy}^{-1}(m-1)}{1 + \lambda^{-1}\boldsymbol{y}^{\mathrm{T}}(m)\boldsymbol{R}_{yy}^{-1}(m-1)\boldsymbol{y}(m)} \tag{9.77}$$

Now define the variables $\boldsymbol{\Phi}(m)$ and $\boldsymbol{k}(m)$ as

$$\boldsymbol{\Phi}_{yy}(m) = \boldsymbol{R}_{yy}^{-1}(m) \tag{9.78}$$

and

$$\boldsymbol{k}(m)=\frac{\lambda^{-1}\boldsymbol{R}_{yy}^{-1}(m-1)\boldsymbol{y}(m)}{1+\lambda^{-1}\boldsymbol{y}^{\mathrm{T}}(m)\boldsymbol{R}_{yy}^{-1}(m-1)\boldsymbol{y}(m)} \tag{9.79}$$

or

$$\boldsymbol{k}(m)=\frac{\lambda^{-1}\boldsymbol{\Phi}_{yy}(m-1)\boldsymbol{y}(m)}{1+\lambda^{-1}\boldsymbol{y}^{\mathrm{T}}(m)\boldsymbol{\Phi}_{yy}(m-1)\boldsymbol{y}(m)} \tag{9.80}$$

Using Equations (9.80) and (9.78), the recursive equation (9.77) for computing the inverse matrix can be written as

$$\boldsymbol{\Phi}_{yy}(m)=\lambda^{-1}\boldsymbol{\Phi}_{yy}(m-1)-\lambda^{-1}\boldsymbol{k}(m)\boldsymbol{y}^{\mathrm{T}}(m)\boldsymbol{\Phi}_{yy}(m-1) \tag{9.81}$$

From Equations (9.79) and (9.80), we have

$$\begin{aligned}\boldsymbol{k}(m)&=\left[\lambda^{-1}\boldsymbol{\Phi}_{yy}(m-1)-\lambda^{-1}\boldsymbol{k}(m)\boldsymbol{y}^{\mathrm{T}}(m)\boldsymbol{\Phi}_{yy}(m-1)\right]\boldsymbol{y}(m)\\&=\boldsymbol{\Phi}_{yy}(m)\boldsymbol{y}(m)\end{aligned} \tag{9.82}$$

Now Equations (9.81) and (9.82) are used in the following to derive the RLS adaptation algorithm.

Recursive Time-update of Filter Coefficients

The least square error filter coefficients are

$$\begin{aligned}\boldsymbol{w}(m)&=\boldsymbol{R}_{yy}^{-1}(m)\boldsymbol{r}_{yx}(m)\\&=\boldsymbol{\Phi}_{yy}(m)\boldsymbol{r}_{yx}(m)\end{aligned} \tag{9.83}$$

Substituting the recursive form of the correlation vector in Equation (9.84) yields

$$\begin{aligned}\boldsymbol{w}(m)&=\boldsymbol{\Phi}_{yy}(m)\left[\lambda\boldsymbol{r}_{yx}(m-1)+\boldsymbol{y}(m)x(m)\right]\\&=\lambda\boldsymbol{\Phi}_{yy}(m)\boldsymbol{r}_{yx}(m-1)+\boldsymbol{\Phi}_{yy}(m)\boldsymbol{y}(m)x(m)\end{aligned} \tag{9.84}$$

Now substitution of the recursive form of the matrix $\boldsymbol{\Phi}_{yy}(m)$ from Equation (9.81) and $\boldsymbol{k}(m)=\boldsymbol{\Phi}(m)\boldsymbol{y}(m)$ from Equation (9.82) in the right-hand side of Equation (9.84) yields

$$\boldsymbol{w}(m) = \left[\lambda^{-1}\boldsymbol{\Phi}_{yy}(m-1) - \lambda^{-1}\boldsymbol{k}(m)\boldsymbol{y}^{\mathrm{T}}(m)\boldsymbol{\Phi}_{yy}(m-1)\right]\lambda\boldsymbol{r}_{yx}(m-1) + \boldsymbol{k}(m)x(m) \tag{9.85}$$

or

$$\boldsymbol{w}(m) = \boldsymbol{\Phi}_{yy}(m-1)\boldsymbol{r}_{yx}(m-1) - \boldsymbol{k}(m)\boldsymbol{y}^{\mathrm{T}}(m)\boldsymbol{\Phi}_{yy}(m-1)\boldsymbol{r}_{yx}(m-1) + \boldsymbol{k}(m)x(m) \tag{9.86}$$

Substitution of $\boldsymbol{w}(m-1) = \boldsymbol{\Phi}(m-1)\boldsymbol{r}_{yx}(m-1)$ in Equation (9.86) yields

$$\boldsymbol{w}(m) = \boldsymbol{w}(m-1) + \boldsymbol{k}(m)\left[x(m) - \boldsymbol{y}^{\mathrm{T}}(m)\boldsymbol{w}(m-1)\right] \tag{9.87}$$

This equation can be rewritten in the form

$$\boldsymbol{w}(m) = \boldsymbol{w}(m-1) + \boldsymbol{k}(m)e(m) \tag{9.88}$$

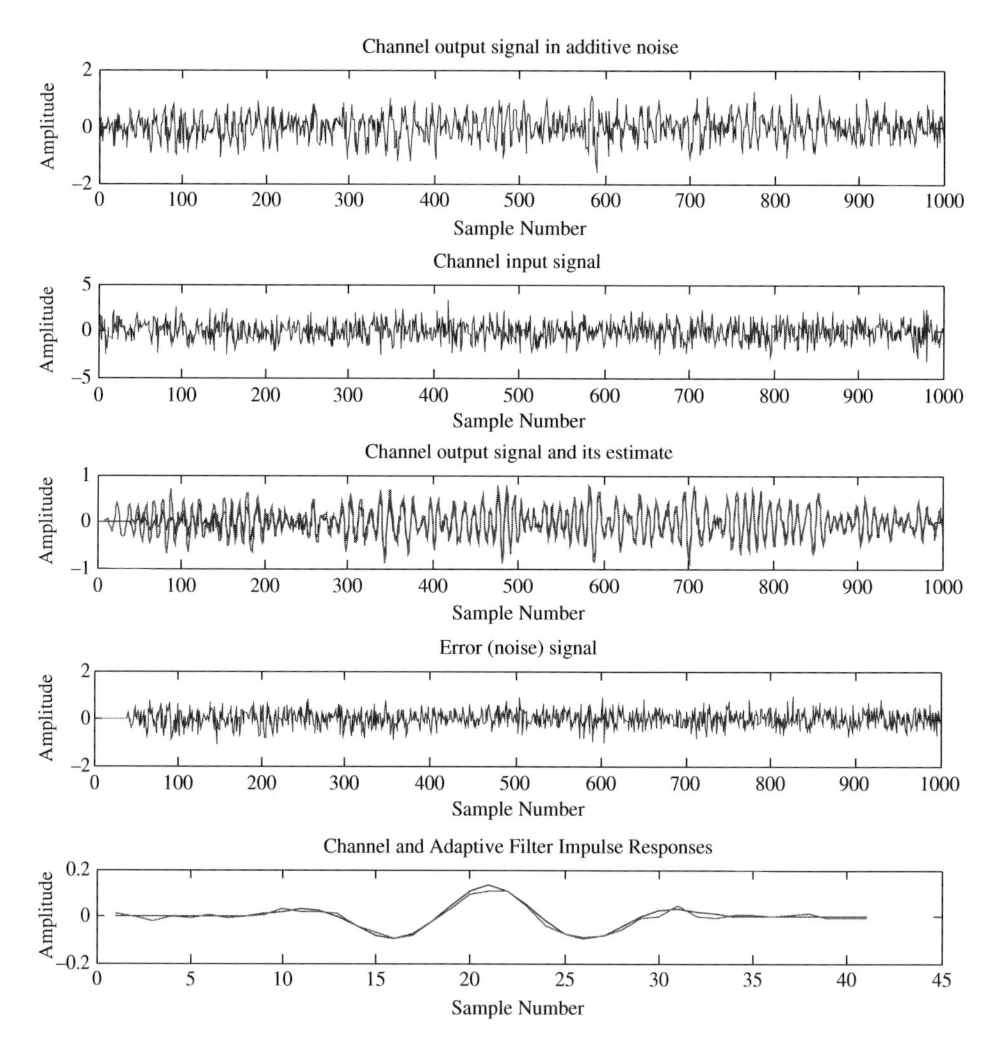

Figure 9.5 From top panel, reference signal (channel output observed in additive white Gaussian noise), input to channel, channel output and its estimate superimposed, error (and noise) signal, the channel response and its estimate.

Equation (9.88) is a recursive time-update implementation of the least square error Wiener filter.

Figure 9.5 shows the application of RLS filter to channel identification given the input signal and the noisy channel output signal.

RLS Adaptation Algorithm

Input signals: $y(m)$ and $x(m)$

Initial values:

$$\boldsymbol{\Phi}_{yy}(m) = \delta \mathbf{I}$$

$$\boldsymbol{w}(0) = \boldsymbol{w}_I$$

For $m = 1, 2, \ldots$

Filter gain vector update:

$$\boldsymbol{k}(m) = \frac{\lambda^{-1}\boldsymbol{\Phi}_{yy}(m-1)\boldsymbol{y}(m)}{1+\lambda^{-1}\boldsymbol{y}^{\mathrm{T}}(m)\boldsymbol{\Phi}_{yy}(m-1)\boldsymbol{y}(m)} \tag{9.89}$$

Error signal equation:

$$e(m) = x(m) - \boldsymbol{w}^{\mathrm{T}}(m-1)\boldsymbol{y}(m) \tag{9.90}$$

Filter coefficients adaptation:

$$\boldsymbol{w}(m) = \boldsymbol{w}(m-1) + \boldsymbol{k}(m)e(m) \tag{9.91}$$

Inverse correlation matrix update:

$$\boldsymbol{\Phi}_{yy}(m) = \lambda^{-1}\boldsymbol{\Phi}_{yy}(m-1) - \lambda^{-1}\boldsymbol{k}(m)\boldsymbol{y}^{T}(m)\boldsymbol{\Phi}_{yy}(m-1) \tag{9.92}$$

9.5 The Steepest-Descent Method

The surface of the mean square output error of an adaptive FIR filter, with respect to the filter coefficients, is a quadratic bowl-shaped curve, with a single global minimum that corresponds to the LSE filter coefficients.

Figure 9.6 illustrates the mean square error curve for a single coefficient filter. This figure also illustrates the steepest-descent search for the minimum mean square error coefficient. The search is based on taking a number of successive downward steps in the direction of the negative gradient of the error surface. Starting with a set of initial values, the filter coefficients are successively updated in the downward direction, until the minimum point, at which the gradient is zero, is reached. The steepest-descent adaptation method can be expressed as

$$\boldsymbol{w}(m+1) = \boldsymbol{w}(m) + \mu\left[-\frac{\partial \mathcal{E}[e^2(m)]}{\partial \boldsymbol{w}(m)}\right] \tag{9.93}$$

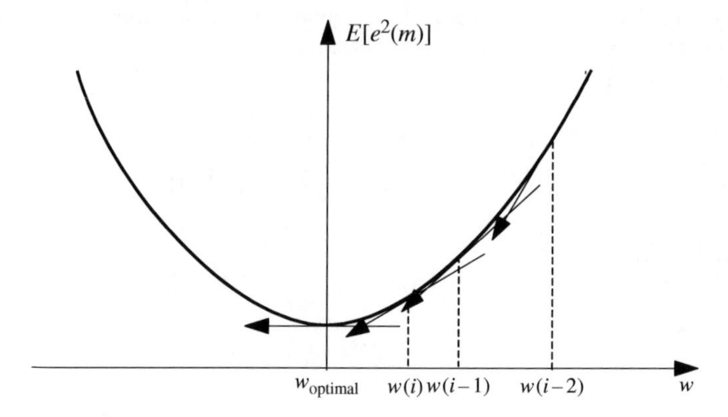

Figure 9.6 Illustration of gradient search of the mean square error surface for the minimum error point.

where μ is the adaptation step size.

From Equation (9.62), the gradient (derivative) of the mean square error function is given by

$$\frac{\partial \mathcal{E}[e^2(m)]}{\partial \boldsymbol{w}(m)} = -2\boldsymbol{r}_{yx} + 2\boldsymbol{R}_{yy}\boldsymbol{w}(m) \tag{9.94}$$

Substituting Equation (9.94) in Equation (9.93) yields

$$\boldsymbol{w}(m+1) = \boldsymbol{w}(m) + \mu\left[\boldsymbol{r}_{yx} - \boldsymbol{R}_{yy}\boldsymbol{w}(m)\right] \tag{9.95}$$

where the factor of 2 in Equation (9.94) has been absorbed in the adaptation step size μ. Let $\boldsymbol{w}_o$ denote the optimal LSE filter coefficient vector; we define a filter coefficients error vector $\tilde{\boldsymbol{w}}(m)$ as

$$\tilde{\boldsymbol{w}}(m) = \boldsymbol{w}(m) - \boldsymbol{w}_o \tag{9.96}$$

For a stationary process, the optimal LSE filter $\boldsymbol{w}_o$ is obtained from the Wiener filter, Equation (5.10), as

$$\boldsymbol{w}_o = \boldsymbol{R}_{yy}^{-1}\boldsymbol{r}_{yx} \tag{9.97}$$

Note from a comparison of Equations (9.94) and (9.96) that the recursive version does not need the computation of the inverse of the autocorrelation matrix.

Subtracting $\boldsymbol{w}_o$ from both sides of Equation (9.95), and then substituting $\boldsymbol{R}_{yy}\boldsymbol{w}_o$ for $\boldsymbol{r}_{yx}$, and using Equation (9.96) yields

$$\tilde{\boldsymbol{w}}(m+1) = \left[\mathbf{I} - \mu\boldsymbol{R}_{yy}\right]\tilde{\boldsymbol{w}}(m) \tag{9.98}$$

It is desirable that the filter error vector $\tilde{\boldsymbol{w}}(m)$ vanishes as rapidly as possible. The parameter μ, the adaptation step size, controls the stability and the rate of convergence of the adaptive filter. Too large a value for μ causes instability; too small a value gives a low convergence rate. The stability of the parameter estimation method depends on the choice of the adaptation parameter μ and the autocorrelation matrix.

From Equation (9.98), a recursive equation for the error in each individual filter coefficient can be obtained as follows. The correlation matrix can be expressed in terms of the matrices of eigenvectors and eigenvalues as

$$\boldsymbol{R}_{yy} = \boldsymbol{Q}\boldsymbol{\Lambda}\boldsymbol{Q}^{\mathrm{T}} \tag{9.99}$$

where $\boldsymbol{Q}$ is an orthonormal matrix of the eigenvectors of $\boldsymbol{R}_{yy}$, and Λ is a diagonal matrix with its diagonal elements corresponding to the eigenvalues of $\boldsymbol{R}_{yy}$. Substituting $\boldsymbol{R}_{yy}$ from Equation (9.99) in Equation (9.98) yields

$$\tilde{\boldsymbol{w}}(m+1) = \left[\mathbf{I} - \mu\boldsymbol{Q}\boldsymbol{\Lambda}\boldsymbol{Q}^{\mathrm{T}}\right]\tilde{\boldsymbol{w}}(m) \tag{9.100}$$

Multiplying both sides of Equation (9.99) by $\boldsymbol{Q}^{\mathrm{T}}$ and using the relation $\boldsymbol{Q}^{\mathrm{T}}\boldsymbol{Q} = \boldsymbol{Q}\boldsymbol{Q}^{\mathrm{T}} = \mathbf{I}$ yields

$$\boldsymbol{Q}^{\mathrm{T}}\tilde{\boldsymbol{w}}(m+1) = [\mathbf{I} - \mu\boldsymbol{\Lambda}]\boldsymbol{Q}^{\mathrm{T}}\tilde{\boldsymbol{w}}(m) \tag{9.101}$$

Let

$$\boldsymbol{\nu}(m) = \boldsymbol{Q}^{\mathrm{T}}\tilde{\boldsymbol{w}}(m) \tag{9.102}$$

Then

$$\boldsymbol{\nu}(m+1) = [\mathbf{I} - \mu\boldsymbol{\Lambda}]\,\boldsymbol{\nu}(m) \tag{9.103}$$

As $\boldsymbol{\Lambda}$ and $\mathbf{I}$ are both diagonal matrices, Equation (9.103) can be expressed in terms of the equations for the individual elements of the error vector $\nu(m)$ as

$$\nu_k(m+1) = [1 - \mu\lambda_k]\,\nu_k(m) \tag{9.104}$$

where λ_k is the kth eigenvalue of the autocorrelation matrix of the filter input $y(m)$. Figure 9.7 is a feedback network model of the time variations of the error vector. From Equation (9.103), the condition for the stability of the adaptation process and the decay of the coefficient error vector is

$$-1 < 1 - \mu\boldsymbol{\lambda}_k < 1 \tag{9.105}$$

Let $\lambda_{\max}$ denote the maximum eigenvalue of the autocorrelation matrix of $y(m)$, then from Equation (9.105) the limits on μ for stable adaptation are given by

$$0 < \mu < \frac{2}{\lambda_{\max}} \tag{9.106}$$

Convergence Rate

The convergence rate of the filter coefficients depends on the choice of the adaptation step size μ, where $0 < \mu < 2/\lambda_{\max}$. When the eigenvalues of the correlation matrix are unevenly spread, the filter coefficients converge at different speeds: the smaller the kth eigenvalue the slower the speed of convergence of the kth coefficients. The errors in filter coefficients with maximum and minimum eigenvalues, $\lambda_{\max}$ and $\lambda_{\min}$ converge according to the equations:

$$\nu_{\max}(m+1) = (1 - \mu\lambda_{\max})\,\nu_{\max}(m) \tag{9.107}$$

$$\nu_{\min}(m+1) = (1 - \mu\lambda_{\min})\,\nu_{\min}(m) \tag{9.108}$$

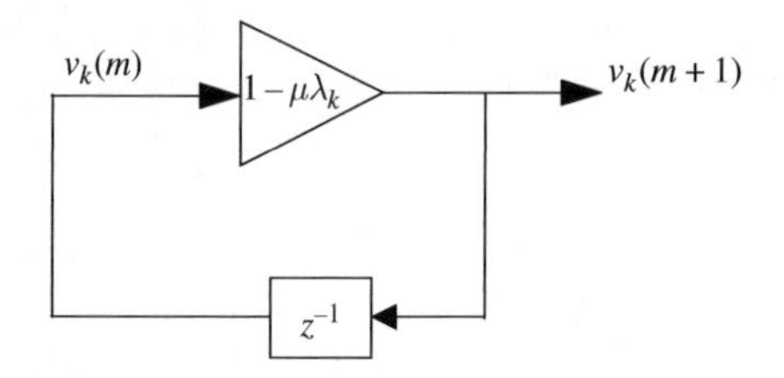

Figure 9.7 A feedback model of the variation of coefficient error with time.

The ratio of the maximum to the minimum eigenvalue of a correlation matrix is called the eigenvalue spread of the correlation matrix:

$$\text{eigenvalue spread} = \frac{\lambda_{\max}}{\lambda_{\min}} \tag{9.109}$$

Note that the differences in the speed of convergence of filter coefficients is proportional to the spread in eigenvalue of the autocorrelation matrix of the input signal. Also note that as the adaptation step size μ affects the convergence rate and as it must be less that $2/\lambda_{\max}$, if $\lambda_{\max}$ is reduced μ can be increased resulting in a more even and faster convergence rate. This can be achieved by pre-whitening the signal using an inverse linear prediction filter or by the application of the adaptive filters in sub-bands where the signal within each sub-band has a smaller eigenvalue spread.

Example 9.3 Assuming that the maximum eigenvalue of a signal is 2 and the minimum eigenvalue is 0.2, calculate:

(i) the eigenvalue spread
(ii) the bounds on adaptation step size
(iii) the decay factor of the error equations for the fastest and the slowest converging coefficients of the filter assuming that the adaptation step size is 0.4.

Solution:

(i) Eigenvalue spread $= \frac{\lambda_{\max}}{\lambda_{\min}} = 10$
(ii) The bounds on adaptation step size. $0 < \mu < \frac{2}{\lambda_{\max}} = 1$
(iii) The fastest decay factor $= (1-\mu\lambda_{\max}) = 1 - 0.4 \times 2 = 0.2$, and the slowest decay factor $= (1-\mu\lambda_{\min}) = 1 - 0.4 \times 0.2 = 0.92$.

Vector-Valued Adaptation Step Size

Instead of using a single scalar-valued adaptation step size μ we can use a vector-valued adaptation step size $\mu = [\mu_0, \mu_2, \ldots, \mu_{P-1}]$ with each filter coefficient, w_k, having its own adaptation step size μ_k. This is useful when the input signal has an eigenvalue spread of greater than one in which case, as shown in the preceding example, the use of a single-step adaptation size would cause an uneven rate of convergence of coefficients. With the use of a vector-valued step size the kth adaptation step size can be adjusted using the kth eigenvalue of the autocorrelation matrix of the input signal to the filter to ensure a more even rate of convergence of different filter coefficients.

9.6 LMS Filter

In its search for the least square error filter coefficients, the steepest-descent method employs the gradient of the *averaged* squared error. A computationally simpler version of the gradient search method is the least mean square (LMS) filter, in which the gradient of the *mean* square error is substituted with the gradient of the *instantaneous* squared error function. The LMS adaptation method is defined as

$$\mathbf{w}(m+1) = \mathbf{w}(m) + \mu \left(-\frac{\partial e^2(m)}{\partial \mathbf{w}(m)} \right) \tag{9.110}$$

where the error signal $e(m)$ is the difference between the adaptive filter output and the target (desired) signal $x(m)$, given by

$$e(m) = x(m) - \mathbf{w}^{\mathrm{T}}(m)\mathbf{y}(m) \tag{9.111}$$

The instantaneous gradient of the squared error can be re-expressed as

$$\begin{aligned} \frac{\partial e^2(m)}{\partial \mathbf{w}(m)} &= \frac{\partial}{\partial \mathbf{w}(m)}[x(m) - \mathbf{w}^{\mathrm{T}}(m)\mathbf{y}(m)]^2 \\ &= -2\mathbf{y}(m)[x(m) - \mathbf{w}^{\mathrm{T}}(m)\mathbf{y}(m)] \\ &= -2\mathbf{y}(m)e(m) \end{aligned} \tag{9.112}$$

Substituting Equation (9.112) into the recursion filter update equation (9.110) and absorbing the factor of 2 in the adaptation step size μ yields the LMS adaptation equation

$$\mathbf{w}(m+1) = \mathbf{w}(m) + \mu\,[\mathbf{y}(m)e(m)] \tag{9.113}$$

It can be seen that the filter update equation is very simple. The LMS filter is widely used in adaptive filter applications such as adaptive equalisation, echo cancellation, radar etc. The main advantage of the LMS algorithm is its simplicity both in terms of the memory requirement and the computational complexity which is $O(P)$, where P is the filter length.

9.6.1 Leaky LMS Algorithm

The stability and the adaptability of the recursive LMS adaptation equation (9.113) can be improved by introducing a so-called leakage factor α as

$$\mathbf{w}(m+1) = \alpha\mathbf{w}(m) + \mu\,[\mathbf{y}(m)e(m)] \tag{9.114}$$

Note that, as illustrated in Figure 9.8, the feedback equation for the time update of the filter coefficients is essentially a recursive (infinite impulse response) system with input $\mu\mathbf{y}(m)e(m)$ and its poles at α. When the parameter $\alpha < 1$, the effect is to introduce more stability and accelerate the filter adaptation to the changes in input signal characteristics.

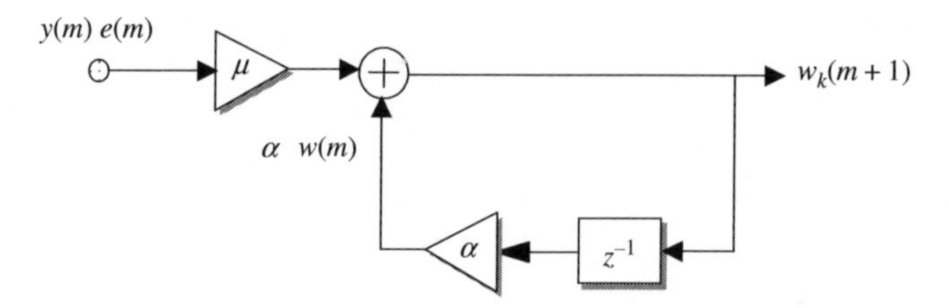

Figure 9.8 Illustration of leaky LMS adaptation of a filter coefficient.

9.6.2 Normalised LMS Algorithm

The normalised LMS (NLMS) adaptation equation is given by

$$\boldsymbol{w}(m+1) = \boldsymbol{w}(m) + \frac{\mu}{a + \sum_{k=0}^{P-1} y^2(m-k)} \boldsymbol{y}(m)e(m) \tag{9.115}$$

where $\sum_{k=0}^{P-1} y^2(m-k)$ is the input signal energy, μ controls the adaptation step size and a is a small constant employed to avoid the denominator of the update term becoming zero when the input signal $\boldsymbol{y}(m)$ is zero.

Derivation of the Normalised LMS Algorithm

In normalised LMS, instead of using the LMS criterion of minimising the difference between the filter output and the desired output, the criterion of minimising the Euclidean norm of incremental change $\delta\boldsymbol{w}(m+1)$ in the successive updates of the filter coefficient vector is used:

$$\|\delta\boldsymbol{w}(m+1)\| = \sum_{k=0}^{P-1} (w_k(m+1) - w_k(m))^2 \tag{9.116}$$

subject to the constraint that

$$\boldsymbol{w}^T(m+1)\boldsymbol{y}(m) = x(m) \tag{9.117}$$

The solution that satisfies the above criterion and the constraint can be obtained by the Lagrange multipliers method.

In the application of the Lagrange method to the above problem, we define an optimisation criterion $J(\cdot)$ as a combination of the squared error criterion expressed in Equation (9.116) and the constraint expressed in Equation (9.117) as

$$\begin{aligned} J(\boldsymbol{w}(m+1), \lambda) &= \|\delta\boldsymbol{w}(m+1)\| + \lambda[x(m) - \boldsymbol{w}^T(m+1)\boldsymbol{y}(m)] \\ &= \sum_{k=0}^{P-1} [w_k(m+1) - w_k(m)]^2 + \lambda[x(m) - \sum_{k=0}^{P-1} w_k(m+1)y(m-k)] \end{aligned} \tag{9.118}$$

where λ is a Lagrange multiplier. To obtain the minimum of the criterion $J(\boldsymbol{w}(m+1), \lambda)$ set the derivative of J with respect to each coefficient $w_i(m+1)$ to zero as

$$\frac{\partial J(\boldsymbol{w}(m+1), \lambda)}{\partial w_i(m+1)} = 2w_i(m+1) - 2w_i(m) - \lambda y(m-i) = 0 \quad i = 0, \ldots, P-1 \tag{9.119}$$

From Equation (9.119) we have

$$w_i(m+1) = w_i(m) + \frac{\lambda}{2} y(m-i) \tag{9.120}$$

From Equations (9.120) and (9.117) we have

$$\sum_{m=0}^{P-1} w_i(m) y(m-i) + \frac{\lambda}{2} \sum_{m=0}^{P-1} y(m-i)^2 = x(m) \tag{9.121}$$

Hence the Lagrange multiplier parameter λ is given by

$$\lambda = \frac{2[x(m) - \sum_{k=0}^{P-1} w_k(m) y(m-k)]}{\sum_{k=0}^{P-1} y^2(m-k)} = \frac{2e(m)}{\sum_{k=0}^{P-1} y^2(m-k)} \tag{9.122}$$

Substitution for λ from Equation (9.122) in Equation (9.120) yields the NLMS equation

$$w_i(m+1) = w_i(m) + \frac{e(m)}{\sum_{k=0}^{P-1} y^2(m-k)} y(m-i) \tag{9.123}$$

The NLMS equation (9.115) is obtained from (9.123) by introducing a variable μ to control the step size and a variable a to prevent the denominator of the update term equation (9.123) becoming zero when the input signal vector $\boldsymbol{y}$ has zero values.

Steady-State Error in LMS

The least mean square error (LSE), $E_{\min}$, is achieved when the filter coefficients approach the optimum value defined by the block least square error equation $\boldsymbol{w}_o = \boldsymbol{R}_{yy}^{-1} \boldsymbol{r}_{yx}$ derived in Chapter 8. The steepest-decent method employs the average gradient of the error surface for incremental updates of the filter coefficients towards the optimal value. Hence, when the filter coefficients reach the minimum point of the mean square error curve, the *averaged* gradient is zero and will remain zero so long as the error surface is stationary. In contrast, examination of the LMS equation shows that for applications in which the LSE is non-zero such as noise reduction, the incremental update term $\mu e(m)\boldsymbol{y}(m)$ would remain non-zero even when the optimal point is reached. Thus at the convergence, the LMS filter will randomly vary about the LSE point, with the result that the LSE for the LMS will be in excess of the LSE for Wiener or steepest-descent methods. Note that at, or near, convergence, a gradual decrease in μ would decrease the excess LSE at the expense of some loss of adaptability to changes in the signal characteristics.

function NLMS_ANC( )
Employs the normalised LMS (NLMS) adaptive filter method for adaptive noise cancellation and or time alignment. The function makes use of a noisy input signal and noisy target signal where both signals contain the clean signal.

Figure 9.9 illustrates the application of NLMS filter to adaptive noise cancellation where the reference and input both contain the desired signal with different levels of SNRs.

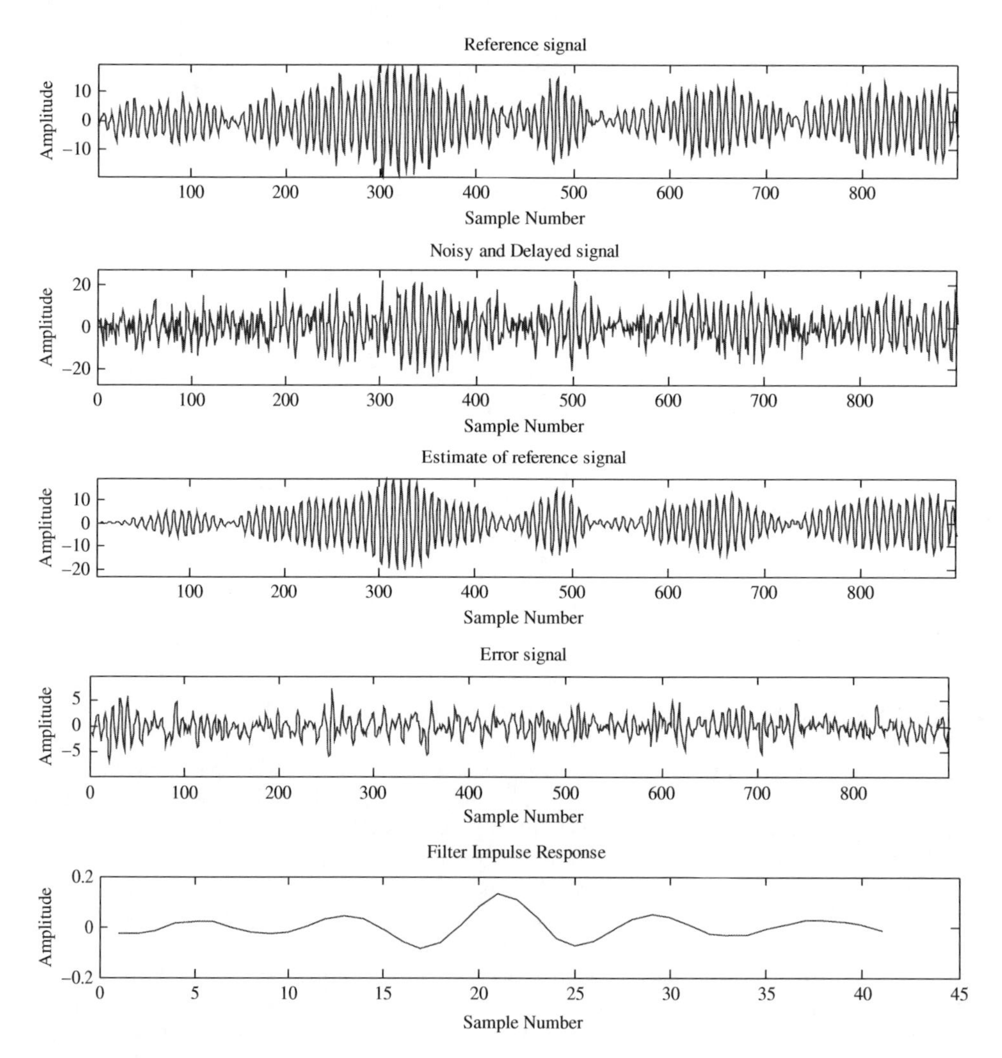

Figure 9.9 Illustration of the inputs and outputs of LMS adaptation. The filter input and the target/reference signal both contain the clean signal with different SNRs. The delay between target and reference is 20 samples as evident from the adaptive filter's impulse response.

9.7 Summary

This chapter began with an introduction to Kalman filter theory. The Kalman filter was derived using the orthogonality principle: for the optimal filter, the innovation sequence must be an uncorrelated process and orthogonal to the past observations. Note that the same principle can also be used to derive the Wiener filter coefficients. Although, like the Wiener filter, the derivation of the Kalman filter is based on the least squared error criterion, the Kalman filter differs from the Wiener filter in two respects. First, the Kalman filter can be applied to non-stationary processes, and second, the Kalman theory employs a model of the signal generation process in the form of the state equation. This is an important advantage in the sense that the Kalman filter can be used to explicitly model the dynamics of the signal process.

For many practical applications such as echo cancellation, channel equalisation, adaptive noise cancellation, time-delay estimation, etc., the RLS and LMS filters provide a suitable alternative to the Kalman filter. The RLS filter is a recursive implementation of the solution for the Wiener filter, and for stationary processes, it should converge to the same solution as the Wiener filter. The main advantage of the LMS filter is the relative simplicity of the algorithm. However, for signals with a large spectral dynamic range, or equivalently a large eigenvalue spread, the LMS has an uneven and slow rate of convergence. If, in addition to having a large eigenvalue spread a signal is also non-stationary with relatively high rate of change (e.g. speech and audio signals) then the LMS can be an unsuitable adaptation method, and the RLS method, with its better convergence rate and less sensitivity to the eigenvalue spread, becomes a more attractive alternative.

Further Reading

Alexander S.T. (1986) Adaptive Signal Processing: Theory and Applications. Springer-Verlag, New York.
Bellanger M.G. (1988) Adaptive Filters and Signal Analysis. Marcel-Dekker, New York.
Bershad N.J. (1986) Analysis of the Normalised LMS Algorithm with Gaussian Inputs. IEEE Trans. Acoustics Speech and Signal Processing, ASSP-34, pp. 793–809.
Bershad N.J. and Qu L.Z. (1989) On the Probability Density Function of the LMS Adaptive Filter Weights. IEEE Trans. Acoustics Speech and Signal Processing, ASSP–37, pp. 43–59.
Cioffi J.M. and Kailath T. (1984) Fast Recursive Least Squares transversal filters for adaptive filtering. IEEE Trans. Acoustics Speech and Signal Processing, ASSP-32, pp. 304–339.
Classen T.A. and Mecklanbrauker W.F. (1985) Adaptive Techniques for Signal Processing in Communications. IEEE Communications, 23, pp. 8–19.
Cowan C.F. and Grant P.M. (1985) Adaptive Filters. Prentice-Hall, Englewood Cliffs, NJ.
Eweda E. and Macchi O. (1985) Tracking Error Bounds of Adaptive Non-stationary Filtering. Automatica, 21, pp. 293–302.
Gabor D., Wilby W.P. and Woodcock R. (1960) A Universal Non-linear Filter, Predictor and Simulator which Optimises Itself by a Learning Process. IEE Proc. 108, pp. 422–438.
Gabriel W.F. (1976) Adaptive Arrays: An Introduction. Proc. IEEE, 64, pp. 239–272.
Haykin S. (2002) Adaptive Filter Theory. Prentice Hall, Englewood Cliffs, NJ.
Honig M.L. and Messerschmitt D.G. (1984) Adaptive filters: Structures, Algorithms and Applications. Kluwer Boston, Hingham, MA.
Kailath T. (1970) The Innovations Approach to Detection and Estimation Theory, Proc. IEEE, 58, pp. 680–965.
Kalman R.E. (1960) A New Approach to Linear Filtering and Prediction Problems. Trans. of the ASME, Series D, Journal of Basic Engineering, 82, pp. 34–45.
Kalman R.E. and Bucy R.S. (1961) New Results in Linear Filtering and Prediction Theory. Trans. ASME J. Basic Eng., 83, pp. 95–108.
Widrow B. (1990) 30 Years of Adaptive Neural Networks: Perceptron, Madaline, and Back Propagation. Proc. IEEE, Special Issue on Neural Networks I, 78.
Widrow B. and Sterns S.D. (1985) Adaptive Signal Processing. Prentice Hall, Englewood Cliffs, NJ.
Wilkinson J.H. (1965) The Algebraic Eigenvalue Problem, Oxford University Press, Oxford.
Zadeh L.A. and Desoer C.A. (1963) Linear System Theory: The State-Space Approach. McGraw-Hill, New York.

10 Linear Prediction Models

A linear prediction (LP) model predicts/forcasts the future values of a signal from a linear combination of its past values. A linear predictor model is an all-pole filter that models the resonances (poles) of the spectral envelope of a signal or a system.

LP models are used in diverse areas of applications, such as data forecasting, speech coding, video coding, speech recognition, model-based spectral analysis, model-based signal interpolation, signal restoration, noise reduction, impulse detection, and change detection. In the statistical literature, linear prediction models are often referred to as autoregressive (AR) processes.

In this chapter, we introduce the theory of linear prediction models and consider efficient methods for the computation of predictor coefficients. We study the forward, backward and lattice predictors, and consider various methods for the formulation and calculation of predictor coefficients, including the least square error and maximum a posteriori methods.

For modelling quasi-periodic signals, such as voiced-speech, an extended linear predictor that simultaneously utilises the short- and long-term correlation structures is introduced. We study sub-band linear predictors that are particularly useful for sub-band coding and processing of noisy signals. Finally, the application of linear prediction models in signal coding, enhancement, pattern recognition and watermarking are considered.

10.1 Linear Prediction Coding

As illustrated in Figures 10.1 and 10.2, a linear prediction model is an all-pole filter that forecasts the future values of a signal from a linear combination of its past values.

In terms of usefulness and the range of applications as a signal processing tool, the linear prediction (LP) model ranks alongside the Fourier transform. A major large-volume application of the LP model is in digital mobile phones where voice coders use linear prediction models for the efficient coding of speech. LP models are also used in inter-frame coding of music and image coding.

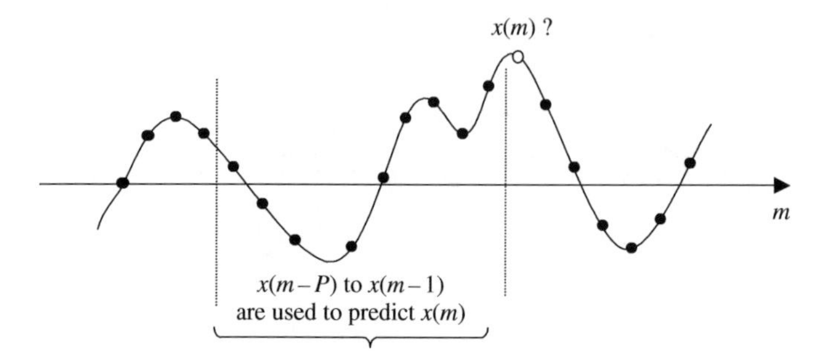

Figure 10.1 Illustration of prediction of a sample $x(m)$ from P past samples.

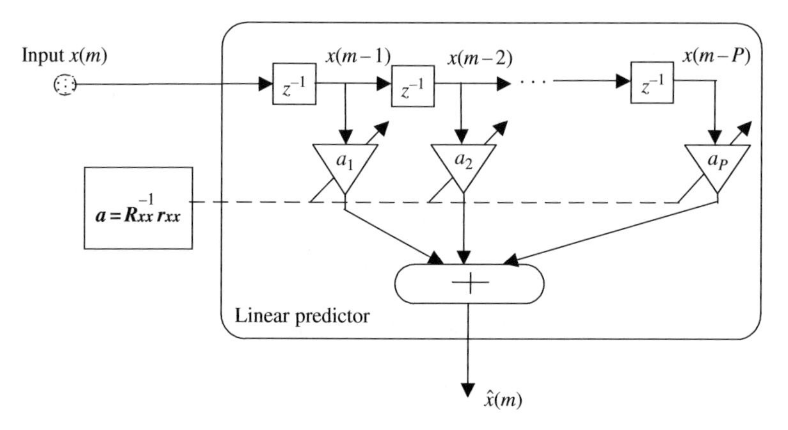

Figure 10.2 Illustration of a linear predictor: $x(m)$ is predicted as the product of the input vector $[x(m-1), \ldots, x(m-P)]$ and the predictor coefficients $[a_1, \ldots, a_P]$. The predictor coefficients are obtained from $\boldsymbol{a} = \boldsymbol{R}_{xx}^{-1}\boldsymbol{r}_{xx}$ as derived in this chapter.

There are two main motivations for the use of predictors in signal processing applications:

(a) To predict the trajectory of a signal. In the frequency domain the trajectory prediction is equivalent to the modelling of the spectrum of the signal.
(b) To remove the predictable part of a signal in order to avoid retransmitting 'redundant' parts of a signal that can be predicted at the receiver and thereby save storage, bandwidth, time and power.

10.1.1 Predictability, Information and Bandwidth

The accuracy by which a signal can be predicted from its past samples depends on the autocorrelation function, or equivalently the bandwidth and the power spectrum, of the signal. As illustrated in Figure 10.3, in the time domain a predictable signal has a smooth and correlated fluctuation, and in the frequency domain, the energy of a predictable signal is concentrated in extremely narrow band(s) of frequencies. In contrast, the energy of an unpredictable signal, such as a white noise, is spread over a wide band of frequencies.

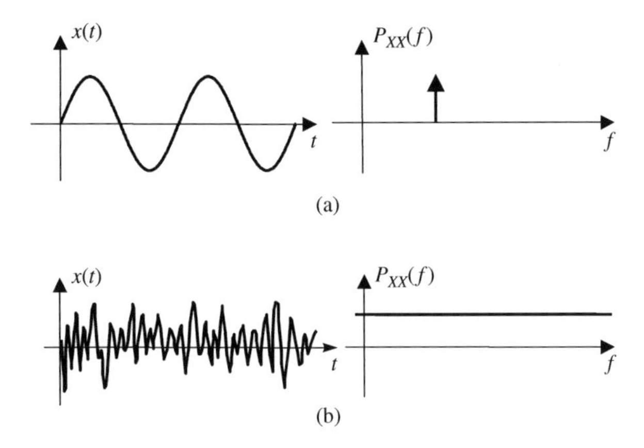

Figure 10.3 The concentration or spread of power in frequency indicates the predictable or random character of a signal: (a) a predictable signal; (b) a random signal.

For a signal to have a capacity to convey information it must have a degree of randomness. Most signals, such as speech, music and video signals, are partially predictable and partially random. These signals can be modelled as the output of a linear filter excited by an uncorrelated random input. The random input models the unpredictable part of the signal, whereas the filter models the predictable structure of the signal. The aim of linear prediction is to model the mechanism that introduces the correlation in a signal.

10.1.2 Applications of LP Model in Speech Processing

Linear prediction models are extensively used in speech processing applications such as in low-bit-rate speech coders, speech enhancement and speech recognition. Speech is generated by inhaling air and then exhaling it through the glottis and the vocal tract. The noise-like air, from the lung, is modulated and shaped by vibrations of the glottal cords and the resonance of the vocal tract. Figure 10.4 illustrates a source-filter model of speech. The source models the lung, and emits a random input excitation signal that is filtered by a pitch filter. The pitch filter models the vibrations of the glottal cords, and generates a sequence of quasi-periodic excitation pulses for voiced sounds as shown in Figure 10.4. The pitch filter model is also termed the 'long-term predictor' since it models the correlation of each sample with the samples a pitch period away. The main source of correlation and power amplification in speech is the vocal tract. The vocal tract is modelled by a linear predictor model, which is also termed the 'short-term predictor', as it models the correlation of each sample with the few (typically 8 to 20)

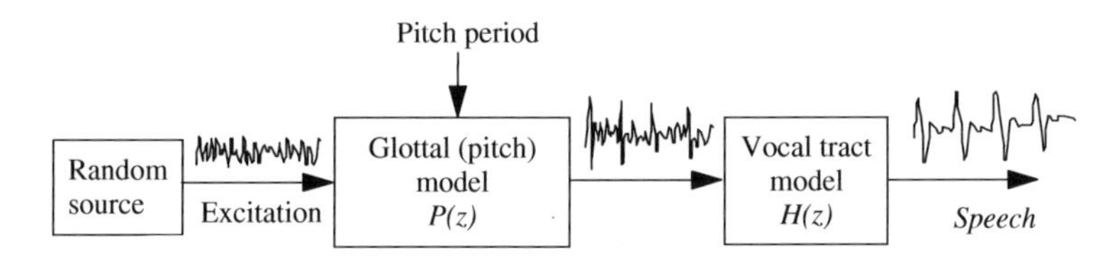

Figure 10.4 A source-filter model of speech production. The filter is usually modelled with a linear prediction model.

preceding samples. In this section, we study the short-term linear prediction model. In Section 10.3, the predictor model is extended to include long-term pitch period correlations.

10.1.3 Time-Domain Description of LP Models

A linear predictor model, Figures 10.1, 10.2 and 10.5, forecasts the amplitude of a signal at time m, $x(m)$, using a linearly weighted combination of P past samples $[x(m-1), x(m-2), \ldots, x(m-P)]$ as

$$\hat{x}(m) = \sum_{k=1}^{P} a_k x(m-k) \tag{10.1}$$

where the integer variable m is the discrete-time index, $\hat{x}(m)$ is the prediction of $x(m)$, and a_k are the predictor coefficients. A block diagram implementation of the predictor Equation (10.1) is illustrated in Figure 10.4.

The prediction error $e(m)$, defined as the difference between the actual sample value $x(m)$ and its predicted value $\hat{x}(m)$, is given by

$$\begin{aligned} e(m) &= x(m) - \hat{x}(m) \\ &= x(m) - \sum_{k=1}^{P} a_k x(m-k) \end{aligned} \tag{10.2}$$

For information-bearing signals, the prediction error $e(m)$ may be regarded as the information, or the innovation (i.e. 'new'), content of the sample $x(m)$. From Equation (10.2) a signal generated, or modelled, by a linear predictor can be described by the feedback equation

$$x(m) = \underbrace{\sum_{k=1}^{P} a_k x(m-k)}_{\substack{\text{Linearly predictable} \\ \text{part of } x(m)}} + \underbrace{e(m)}_{\substack{\text{Unpredictable} \\ \text{part of } x(m)}} \tag{10.3}$$

Figure 10.5 illustrates a linear predictor model of a signal $x(m)$. In this model, the random input excitation (i.e. the prediction error) is $e(m) = Gu(m)$, where $u(m)$ is a zero-mean, unit-variance random signal, and G, a gain term, is the square root of the variance (i.e. power) of $e(m)$:

$$G = \left(\mathcal{E}[e^2(m)]\right)^{1/2} \tag{10.4}$$

where $E[\cdot]$ is an averaging, or expectation, operator.

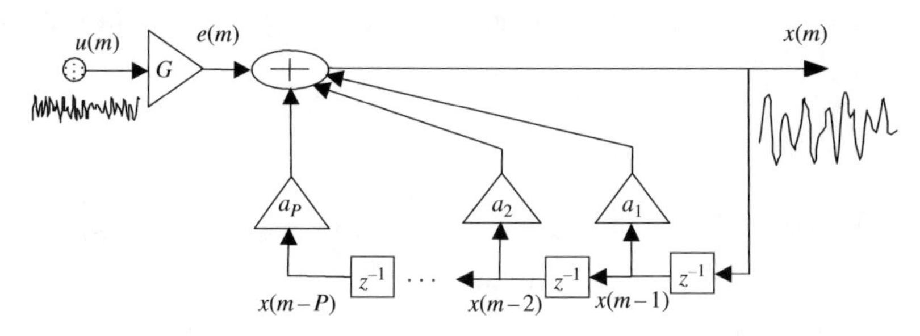

Figure 10.5 Illustration of a signal generated by a linear predictive model.

10.1.4 Frequency Response of LP Model and its Poles

The z-transform of LP equation (10.3) shows that the LP model is an all-pole digital filter with transfer function

$$H(z) = \frac{X(z)}{U(z)} = \underbrace{\frac{G}{1 - \sum_{k=1}^{P} a_k z^{-k}}}_{\text{Cartesian form}} = \underbrace{G \frac{1}{\prod_{k=1}^{N}(1 - r_k z^{-1})} \frac{1}{\prod_{k=1}^{M}(1 - 2r_k \cos\varphi_k z^{-1} + r_k^2 z^{-2})}}_{\text{Polar form}} \tag{10.5}$$

In Equation (10.5) it is assumed there are M complex pole pairs and N real poles with $P = N + 2M$ and r_k and ϕ_k are the radius and angle of the kth pole. The frequency response of an LP model is given by

$$H(f) = \frac{G}{1 - \sum_{k=1}^{P} a_k e^{-j2\pi kf}} = G \frac{1}{\prod_{k=1}^{N}(1 - r_k e^{-j2\pi f})} \frac{1}{\prod_{k=1}^{M}(1 - 2r_k \cos\varphi_k e^{-j2\pi f} + r_k^2 e^{-j4\pi f})} \tag{10.6}$$

Figure 10.6 illustrates the relation between the poles and the magnitude frequency response of an all-pole filter. The main features of the spectral resonance at a pole are the frequency, bandwidth and magnitude of the resonance. The roots of a complex pair of poles can be written in terms of the radius r_k and the angle φ_k of the pole as

$$z_k = r_k e^{\pm j\varphi_k} \tag{10.7}$$

The resonance frequency of a complex pair of poles is given by

$$F(\varphi_k) = \frac{F_s}{2\pi} \varphi_k \tag{10.8}$$

where F_s is the sampling frequency. The bandwidth of a pole is related to its radius r_k as

$$B_k = (-\log r_k)(F_s/\pi) \tag{10.9}$$

Note that the radius of a pole affect the bandwidth and damping of the resonance; these parameters are related to the amount of resistance to oscillations at the pole frequency. As the pole radius decreases the bandwidth and damping increase. When the radius of a pair of complex conjugate poles is equal to one, then the resistance to oscillations is zero, and the bandwidth of the resonance is zero, the damping is zero and the impulse response of the pole is a sine wave. The magnitude of the spectral resonance at the pole is given by $H(f = \varphi)$.

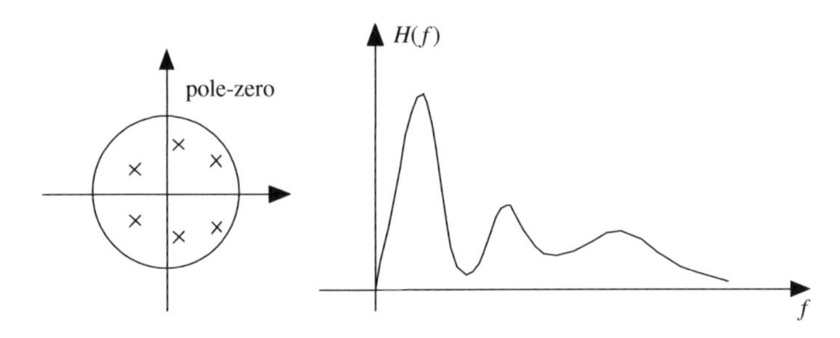

Figure 10.6 The pole–zero position and frequency response of a linear predictor.

10.1.5 Calculation of Linear Predictor Coefficients

Linear predictor coefficients are obtained by minimising the mean square prediction error as

$$
\begin{aligned}
\mathcal{E}[e^2(m)] &= \mathcal{E}\left[\left(x(m)-\sum_{k=1}^{P}a_k x(m-k)\right)^2\right] \\
&= \mathcal{E}[x^2(m)] - 2\sum_{k=1}^{P}a_k\mathcal{E}[x(m)x(m-k)] + \sum_{k=1}^{P}a_k\sum_{j=1}^{P}a_j\mathcal{E}\left[x(m-k)x(m-j)\right] \\
&= r_{xx}(0) - 2\sum_{k=1}^{P}a_k r_{xx}(k) + \sum_{k=1}^{P}a_k\sum_{j=1}^{P}a_j r_{xx}(k-j) \\
&= r_{xx}(0) - 2\boldsymbol{r}_{xx}^T\boldsymbol{a} + \boldsymbol{a}^T\boldsymbol{R}_{xx}\boldsymbol{a}
\end{aligned}
\tag{10.10}
$$

where $\boldsymbol{R}_{xx} = E[\boldsymbol{x}\boldsymbol{x}^{\mathrm{T}}]$ is the autocorrelation matrix of the input vector $\boldsymbol{x}^{\mathrm{T}} = [x(m-1), x(m-2), \ldots, x(m-P)]$, $\boldsymbol{r}_{xx} = E[x(m)\boldsymbol{x}]$ is the autocorrelation vector and $\boldsymbol{a}^{\mathrm{T}} = [a_1, a_2, \ldots, a_P]$ is the predictor coefficient vector. From Equation (10.10), the gradient of the mean square prediction error with respect to the predictor coefficient vector $\boldsymbol{a}$ is given by

$$\frac{\partial}{\partial \boldsymbol{a}}\mathcal{E}[e^2(m)] = -2\boldsymbol{r}_{xx}^{\mathrm{T}} + 2\boldsymbol{a}^{\mathrm{T}}\boldsymbol{R}_{xx} \tag{10.11}$$

where the gradient vector is defined as

$$\frac{\partial}{\partial \boldsymbol{a}} = \left(\frac{\partial}{\partial a_1}, \frac{\partial}{\partial a_2}, \ldots, \frac{\partial}{\partial a_{\mathrm{P}}}\right)^{\mathrm{T}} \tag{10.12}$$

The least mean square error solution, obtained by setting Equation (10.11) to zero, is given by

$$\boldsymbol{R}_{xx}\boldsymbol{a} = \boldsymbol{r}_{xx} \tag{10.13}$$

From Equation (10.13) the predictor coefficient vector is given by

$$\boldsymbol{a} = \boldsymbol{R}_{xx}^{-1}\boldsymbol{r}_{xx} \tag{10.14}$$

Equation (10.14) may also be written in an expanded form as

$$
\begin{pmatrix} a_1 \\ a_2 \\ a_3 \\ \vdots \\ a_P \end{pmatrix}
=
\begin{pmatrix}
r_{xx}(0) & r_{xx}(1) & r_{xx}(2) & \cdots & r_{xx}(P-1) \\
r_{xx}(1) & r_{xx}(0) & r_{xx}(1) & \cdots & r_{xx}(P-2) \\
r_{xx}(2) & r_{xx}(1) & r_{xx}(0) & \cdots & r_{xx}(P-3) \\
\vdots & \vdots & \vdots & \ddots & \vdots \\
r_{xx}(P-1) & r_{xx}(P-2) & r_{xx}(P-3) & \cdots & r_{xx}(0)
\end{pmatrix}^{-1}
\begin{pmatrix} r_{xx}(1) \\ r_{xx}(2) \\ r_{xx}(3) \\ \vdots \\ r_{xx}(P) \end{pmatrix}
\tag{10.15}
$$

An alternative formulation of the least square error problem is as follows. For a signal segment of N samples $[x(0), \ldots, x(N-1)]$, we can write a set of N linear prediction error equations as

$$
\begin{pmatrix} e(0) \\ e(1) \\ e(2) \\ \vdots \\ e(N-1) \end{pmatrix}
=
\begin{pmatrix} x(0) \\ x(1) \\ x(2) \\ \vdots \\ x(N-1) \end{pmatrix}
-
\begin{pmatrix}
x(-1) & x(-2) & x(-3) & \cdots & x(-P) \\
x(0) & x(-1) & x(-2) & \cdots & x(1-P) \\
x(1) & x(0) & x(-1) & \cdots & x(2-P) \\
\vdots & \vdots & \vdots & \ddots & \vdots \\
x(N-2) & x(N-3) & x(N-4) & \cdots & x(N-P-1)
\end{pmatrix}
\begin{pmatrix} a_1 \\ a_2 \\ a_3 \\ \vdots \\ a_P \end{pmatrix}
\tag{10.16}
$$

where $x^{\mathrm{T}} = [x(-1), \ldots, x(-P)]$ is the initial vector. In compact vector/matrix notation Equation (10.16) can be written as

$$e = x - Xa \tag{10.17}$$

Using Equation (10.17), the sum of squared prediction errors over a block of N samples can be expressed as

$$e^{\mathrm{T}}e = x^{\mathrm{T}}x - 2x^{\mathrm{T}}Xa + a^{\mathrm{T}}X^{\mathrm{T}}Xa \tag{10.18}$$

The least squared error predictor is obtained by setting the derivative of Equation (10.14) with respect to the parameter vector a to zero:

$$\frac{\partial e^{\mathrm{T}}e}{\partial a} = -2x^{\mathrm{T}}X + 2a^{\mathrm{T}}X^{\mathrm{T}}X = 0 \tag{10.19}$$

From Equation (10.19), the least square error predictor is given by

$$a = \left(X^{T}X\right)^{-1}\left(X^{T}x\right) \tag{10.20}$$

A comparison of Equations (10.15) and (10.20) shows that in Equation (10.20) the autocorrelation matrix and vector of Equation (10.15) are replaced by the time-averaged estimates as

$$\hat{r}_{xx}(m) = \frac{1}{N}\sum_{k=0}^{N-1} x(k)x(k-m) \tag{10.21}$$

Equations (10.15) or (10.20) may be solved efficiently by utilising the regular *Toeplitz* structure of the correlation matrix R_{xx}. In a Toeplitz matrix, all the elements on a left–right diagonal are equal. The correlation matrix is also cross-diagonal symmetric. Note that altogether there are only $P+1$ unique elements $[r_{xx}(0), r_{xx}(1), \ldots, r_{xx}(P)]$ in the correlation matrix and the cross-correlation vector. An efficient method for solution of Equation (10.15) or (10.20) is the Levinson–Durbin algorithm, introduced in Section 10.2.2.

10.1.6 Effect of Estimation of Correlation Function on LP Model Solution

Note that the term $\hat{r}_{xx}(m)$ in Equation (10.21) is only an *estimate* of the correlation function, obtained from a segment of N samples, and as such $\hat{r}_{xx}(m)$ is a random variable with its own mean, variance and probability distribution function. Indeed different segments of an even stationary signal will yield different values of $\hat{r}_{xx}(m)$. The goodness of an estimate depends on the number of samples N used in the estimation of the correlation function and on the signal-to-noise ratio.

10.1.7 The Inverse Filter: Spectral Whitening, De-correlation

The all-pole linear predictor model, in Figure 10.5, shapes the spectrum of the input signal by transforming an uncorrelated excitation signal $u(m)$ to a correlated output signal $x(m)$. In the frequency domain the input–output relation of the all-pole filter of Figure 10.5 is given by

$$X(f) = \frac{GU(f)}{A(f)} = \frac{E(f)}{1 - \sum\limits_{k=1}^{P} a_k e^{-\mathrm{j}2\pi fk}} \tag{10.22}$$

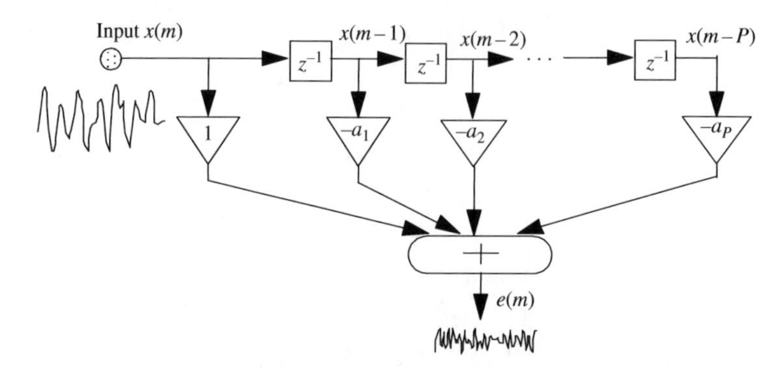

Figure 10.7 Illustration of the inverse (or whitening) filter.

where $X(f)$, $E(f)$ and $U(f)$ are the spectra of $x(m)$, $e(m)$ and $u(m)$ respectively, G is the input gain factor, and $A(f)$ is the frequency response of the inverse predictor. As the excitation signal $e(m)$ is assumed to have a flat spectrum, it follows that the shape of the signal spectrum $X(f)$ is due to the frequency response $1/A(f)$ of the all-pole predictor model. The inverse linear predictor, as the name implies, transforms a correlated signal $x(m)$ back to an uncorrelated flat-spectrum signal $e(m)$.

The inverse filter, Figure 10.7, also known as the prediction error filter, is an all-zero finite impulse response filter defined as

$$\begin{aligned} e(m) &= x(m) - \hat{x}(m) \\ &= x(m) - \sum_{k=1}^{P} a_k x(m-k) \\ &= (\boldsymbol{a}^{\text{inv}})^{\text{T}} \boldsymbol{x} \end{aligned} \tag{10.23}$$

where the inverse filter $(\boldsymbol{a}^{\text{inv}})^{\text{T}} = [1, -a_1, \ldots, -a_P] = [1, -\boldsymbol{a}]$, and $\boldsymbol{x}^{\text{T}} = [x(m), \ldots, x(m-P)]$. The z-transfer function of the inverse predictor model is given by

$$A(z) = 1 - \sum_{k=1}^{P} a_k z^{-k} \tag{10.24}$$

A linear predictor model is an all-pole filter, where the poles model the resonance of the signal spectrum. The inverse of an all-pole filter is an all-zero filter, with the zeros situated at the same positions in the pole–zero plot as the poles of the all-pole filter, as illustrated in Figure 10.8. Consequently, the zeros of the inverse filter introduce anti-resonance that cancels out the resonance of the poles of the predictor. The inverse filter has the effect of flattening the spectrum of the input signal, and is also known as a spectral whitening, or de-correlation, filter.

10.1.8 The Prediction Error Signal

In general, the prediction error signal is composed of three components:

(a) the input signal, also called the excitation signal;
(b) the errors due to the modelling inaccuracies;
(c) the noise.

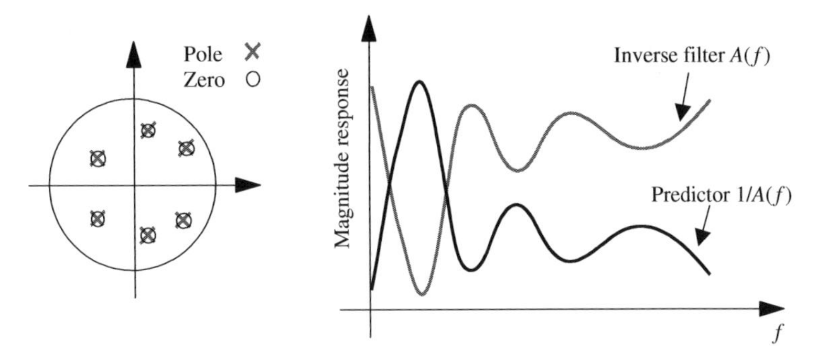

Figure 10.8 Illustration of the pole-zero diagram, and the frequency responses of an all-pole predictor and its all-zero inverse filter.

The mean square prediction error becomes zero only if the following three conditions are satisfied: (a) the signal is deterministic, (b) the signal is correctly modelled by a predictor of order P, and (c) the signal is noise-free. For example, a mixture of $P/2$ sine waves can be modelled by a predictor of order P, with zero prediction error. However, in practice, the prediction error is non-zero because information-bearing signals are random, often only approximately modelled by a linear system, and usually observed in noise. The least mean square prediction error, obtained from substitution of Equation (10.13) in Equation (10.10), is

$$E^{(P)} = \mathcal{E}[e^2(m)] = r_{xx}(0) - \sum_{k=1}^{P} a_k r_{xx}(k) \tag{10.25}$$

where $E^{(P)}$ denotes the prediction error for a predictor of order P. The prediction error decreases, initially rapidly and then slowly, with the increasing predictor order up to the correct model order. For the correct model order, the signal $e(m)$ is an uncorrelated zero-mean random process with an autocorrelation function defined as

$$\mathcal{E}\left[e(m)e(m-k)\right] = \begin{cases} \sigma_e^2 = G^2 & \text{if } m = k \\ 0 & \text{if } m \neq k \end{cases} \tag{10.26}$$

where σ_e^2 is the variance of $e(m)$.

Figure 10.9 shows an example of linear prediction analysis of a segment of speech.

Matlab function lpcdemo( )

Demonstrates linear prediction modelling and analysis of speech signals. The program analyses a segment of a speech signal and displays: the signal, its FFT-spectrum, LP-frequency response, pole-zero diagram, inverse filter output i.e. the estimated input/excitation signal to LP model, the autocorrelation function from which the pitch estimation function and the pitch may be obtained.

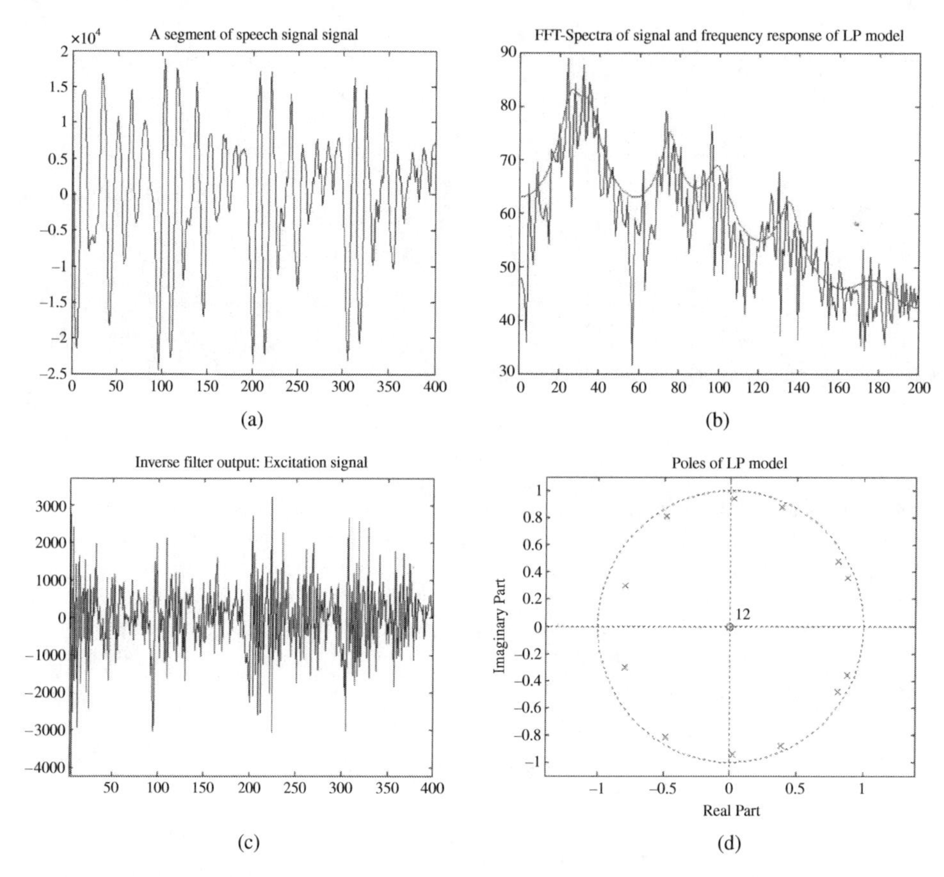

Figure 10.9 (a) A speech segment, (b) FFT-spectrum and LP frequency response of speech superimposed, (c) inverse filter output, (d) the poles of LP model, prediction order $P = 12$. Speech sampling rate = 16 kHz.

10.2 Forward, Backward and Lattice Predictors

The forward predictor model of Equation (10.1) predicts a sample $x(m)$ from a linear combination of P *past* samples $x(m-1), x(m-2), \ldots, x(m-P)$. Similarly, as shown in Figure 10.10, we can define a backward predictor, which predicts a sample $x(m-P)$ from P *future* samples $x(m-P+1), \ldots, x(m)$ as

$$\hat{x}(m-P) = \sum_{k=1}^{P} c_k x(m-k+1) \tag{10.27}$$

The backward prediction error is defined as the difference between the actual sample and its predicted value:

$$\begin{aligned} b(m) &= x(m-P) - \hat{x}(m-P) \\ &= x(m-P) - \sum_{k=1}^{P} c_k x(m-k+1) \end{aligned} \tag{10.28}$$

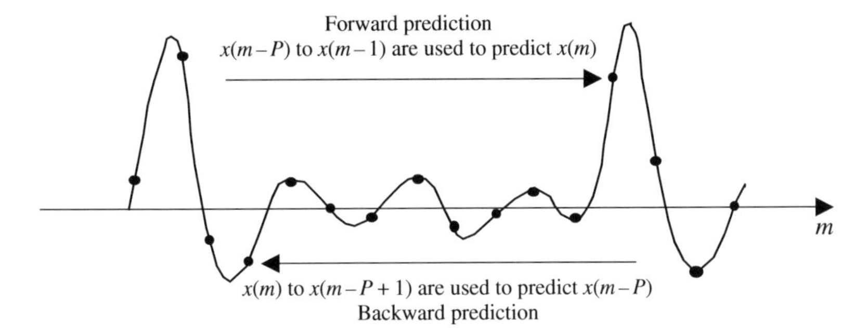

Figure 10.10 Illustration of forward and backward predictors.

From Equation (10.28), a signal generated by a backward predictor is given by

$$x(m-P)=\sum_{k=1}^{P} c_k x(m-k+1)+b(m) \tag{10.29}$$

The coefficients of the least square error backward predictor, obtained by a similar method to that of the forward predictor in Section 10.1.1, are given by

$$\begin{pmatrix} r_{xx}(0) & r_{xx}(1) & r_{xx}(2) & \cdots & r_{xx}(P-1) \\ r_{xx}(1) & r_{xx}(0) & r_{xx}(1) & \cdots & r_{xx}(P-2) \\ r_{xx}(2) & r_{xx}(1) & r_{xx}(0) & \cdots & r_{xx}(P-3) \\ \vdots & \vdots & \vdots & \ddots & \vdots \\ r_{xx}(P-1) & r_{xx}(P-2) & r_{xx}(P-3) & \cdots & r_{xx}(0) \end{pmatrix} \begin{pmatrix} c_1 \\ c_2 \\ c_3 \\ \vdots \\ c_P \end{pmatrix} = \begin{pmatrix} r_{xx}(P) \\ r_{xx}(P-1) \\ r_{xx}(P-2) \\ \vdots \\ r_{xx}(1) \end{pmatrix} \tag{10.30}$$

Note the main difference between Equations (10.30) and (10.15) is that the correlation vector on the right-hand side of the backward predictor Equation (10.30) is upside-down compared with the forward predictor Equation (10.15). Since the correlation matrix is Toeplitz and symmetric, Equation (10.15) for the forward predictor may be rearranged and rewritten in the form

$$\begin{pmatrix} r_{xx}(0) & r_{xx}(1) & r_{xx}(2) & \cdots & r_{xx}(P-1) \\ r_{xx}(1) & r_{xx}(0) & r_{xx}(1) & \cdots & r_{xx}(P-2) \\ r_{xx}(2) & r_{xx}(1) & r_{xx}(0) & \cdots & r_{xx}(P-3) \\ \vdots & \vdots & \vdots & \ddots & \vdots \\ r_{xx}(P-1) & r_{xx}(P-2) & r_{xx}(P-3) & \cdots & r_{xx}(0) \end{pmatrix} \begin{pmatrix} a_P \\ a_{P-1} \\ a_{P-2} \\ \vdots \\ a_1 \end{pmatrix} = \begin{pmatrix} r_{xx}(P) \\ r_{xx}(P-1) \\ r_{xx}(P-2) \\ \vdots \\ r_{xx}(1) \end{pmatrix} \tag{10.31}$$

Comparison of Equations (10.31) and (10.30) shows that the coefficients of the backward predictor are the time-reversed version of the forward predictor

$$\boldsymbol{c}=\begin{pmatrix} c_1 \\ c_2 \\ c_3 \\ \vdots \\ c_P \end{pmatrix} = \begin{pmatrix} a_P \\ a_{P-1} \\ a_{P-2} \\ \vdots \\ a_1 \end{pmatrix} = \boldsymbol{a}^{\mathrm{B}} \tag{10.32}$$

where the vector $\boldsymbol{a}^{\mathrm{B}}$ is the reversed version of the vector $\boldsymbol{a}$. The relation between the backward and forward predictors is employed in the Levinson–Durbin algorithm to derive an efficient method for calculation of the predictor coefficients as described in Section 10.2.2.

10.2.1 Augmented Equations for Forward and Backward Predictors

The inverse forward predictor coefficient vector is $[1, -a_1, \ldots, -a_P] = [1, -\boldsymbol{a}^{\mathrm{T}}]$. Equations (10.15) and (10.26) may be combined to yield a matrix equation for the inverse forward predictor coefficients:

$$\begin{pmatrix} r(0) & \boldsymbol{r}_{xx}^{\mathrm{T}} \\ \boldsymbol{r}_{xx} & \boldsymbol{R}_{xx} \end{pmatrix} \begin{pmatrix} 1 \\ -\boldsymbol{a} \end{pmatrix} = \begin{pmatrix} E^{(P)} \\ \boldsymbol{0} \end{pmatrix} \tag{10.33}$$

Equation (10.33) is called the augmented forward predictor equation. Similarly, for the inverse backward predictor, we can define an augmented backward predictor equation as

$$\begin{pmatrix} \boldsymbol{R}_{xx} & \boldsymbol{r}_{xx}^{\mathrm{B}} \\ \boldsymbol{r}_{xx}^{\mathrm{BT}} & r(0) \end{pmatrix} \begin{pmatrix} -\boldsymbol{a}^{\mathrm{B}} \\ 1 \end{pmatrix} = \begin{pmatrix} \boldsymbol{0} \\ E^{(P)} \end{pmatrix} \tag{10.34}$$

where $\boldsymbol{r}_{xx}^{\mathrm{T}} = [r_{xx}(1), \cdots, r_{xx}(P)]$ and $\boldsymbol{r}_{xx}^{\mathrm{BT}} = [r_{xx}(P), \cdots, r_{xx}(1)]$. Note that the superscript BT denotes backward and transposed. The augmented forward and backward matrix equations (10.33) and (10.34) are used to derive an order-update solution for the linear predictor coefficients as follows.

10.2.2 Levinson–Durbin Recursive Solution

The Levinson–Durbin algorithm was developed by N. Levinson in 1947 and modified by J. Durbin in 1959. The Levinson–Durbin algorithm is a recursive order-update method for calculation of linear predictor coefficients. A forward-prediction error filter of order i can be described in terms of the forward and backward prediction error filters of order $i-1$ as

$$\begin{pmatrix} 1 \\ -a_1^{(i)} \\ \vdots \\ -a_{i-1}^{(i)} \\ -a_i^{(i)} \end{pmatrix} = \begin{pmatrix} 1 \\ -a_1^{(i-1)} \\ \vdots \\ -a_{i-1}^{(i-1)} \\ 0 \end{pmatrix} + k_i \begin{pmatrix} 0 \\ -a_{i-1}^{(i-1)} \\ \vdots \\ -a_1^{(i-1)} \\ 1 \end{pmatrix} \tag{10.35}$$

or in more compact vector notation as

$$\begin{pmatrix} 1 \\ -\boldsymbol{a}^{(i)} \end{pmatrix} = \begin{pmatrix} 1 \\ -\boldsymbol{a}^{(i-1)} \\ 0 \end{pmatrix} + k_i \begin{pmatrix} 0 \\ -\boldsymbol{a}^{(i-1)\mathrm{B}} \\ 1 \end{pmatrix} \tag{10.36}$$

where k_i is the reflection coefficient. The proof of Equation (10.36) and the derivation of the value of the reflection coefficient for k_i, follows shortly. Similarly, a backward prediction error filter of order i is described in terms of the forward and backward prediction error filters of order $i-1$ as

$$\begin{pmatrix} -\boldsymbol{a}^{(i)\mathrm{B}} \\ 1 \end{pmatrix} = \begin{pmatrix} 0 \\ -\boldsymbol{a}^{(i-1)\mathrm{B}} \\ 1 \end{pmatrix} + k_i \begin{pmatrix} 1 \\ -\boldsymbol{a}^{(i-1)} \\ 0 \end{pmatrix} \tag{10.37}$$

To prove the order-update Equation (10.36) (or alternatively Equation (10.37)), we multiply both sides of the equation by the $(i+1)\times(i+1)$ augmented matrix $\boldsymbol{R}_{xx}^{(i+1)}$ and use the equality

$$\boldsymbol{R}_{xx}^{(i+1)} = \begin{pmatrix} \boldsymbol{R}_{xx}^{(i)} & \boldsymbol{r}_{xx}^{(i)\mathrm{B}} \\ \boldsymbol{r}_{xx}^{(i)\mathrm{BT}} & r_{xx}(0) \end{pmatrix} = \begin{pmatrix} r_{xx}(0) & \boldsymbol{r}_{xx}^{(i)\mathrm{T}} \\ \boldsymbol{r}_{xx}^{(i)} & \boldsymbol{R}_{xx}^{(i)} \end{pmatrix} \tag{10.38}$$

to obtain

$$\begin{pmatrix} \boldsymbol{R}_{xx}^{(i)} & \boldsymbol{r}_{xx}^{(i)\mathrm{B}} \\ \boldsymbol{r}_{xx}^{(i)\mathrm{BT}} & r_{xx}(0) \end{pmatrix} \begin{pmatrix} 1 \\ -\boldsymbol{a}^{(i)} \end{pmatrix} = \begin{pmatrix} \boldsymbol{R}_{xx}^{(i)} & \boldsymbol{r}_{xx}^{(i)\mathrm{B}} \\ \boldsymbol{r}_{xx}^{(i)\mathrm{BT}} & r_{xx}(0) \end{pmatrix} \begin{pmatrix} 1 \\ -\boldsymbol{a}^{(i-1)} \\ 0 \end{pmatrix} + k_i \begin{pmatrix} r_{xx}(0) & \boldsymbol{r}_{xx}^{(i)\mathrm{T}} \\ \boldsymbol{r}_{xx}^{(i)} & \boldsymbol{R}_{xx}^{(i)} \end{pmatrix} \begin{pmatrix} 0 \\ -\boldsymbol{a}^{(i-1)\mathrm{B}} \\ 1 \end{pmatrix} \tag{10.39}$$

where in Equation (10.38) and Equation (10.39) $\boldsymbol{r}_{xx}^{(i)\mathrm{T}} = [r_{xx}(1), \cdots, r_{xx}(i)]$, and $\boldsymbol{r}_{xx}^{(i)BT} = [r_{xx}(i), \cdots, r_{xx}(1)]$ is the reversed version of $\boldsymbol{r}_{xx}^{(i)\mathrm{T}}$. Matrix–vector multiplication of both sides of Equation (10.39) and the use of Equations (10.33) and (10.34) yields

$$\begin{pmatrix} E^{(i)} \\ \boldsymbol{0}^{(i)} \end{pmatrix} = \begin{pmatrix} E^{(i-1)} \\ \boldsymbol{0}^{(i-1)} \\ \Delta^{(i-1)} \end{pmatrix} + k_i \begin{pmatrix} \Delta^{(i-1)} \\ \boldsymbol{0}^{(i-1)} \\ E^{(i-1)} \end{pmatrix} \tag{10.40}$$

where

$$\begin{aligned} \Delta^{(i-1)} &= \begin{bmatrix} 1 & -\boldsymbol{a}^{(i-1)^{\mathrm{T}}} \end{bmatrix} \boldsymbol{r}_{xx}^{(i)B} \\ &= r_{xx}(i) - \sum_{k=1}^{i-1} a_k^{(i-1)} r_{xx}(i-k) \end{aligned} \tag{10.41}$$

If Equation (10.40) is true, it follows that Equation (10.36) must also be true. The conditions for Equation (10.40) to be true are

$$E^{(i)} = E^{(i-1)} + k_i \Delta^{(i-1)} \tag{10.42}$$

and

$$0 = \Delta^{(i-1)} + k_i E^{(i-1)} \tag{10.43}$$

From Equation (10.43),

$$k_i = -\frac{\Delta^{(i-1)}}{E^{(i-1)}} \tag{10.44}$$

Substitution of $\Delta^{(i-1)}$ from Equation (10.44) into Equation (10.42) yields

$$\begin{aligned} E^{(i)} &= E^{(i-1)}(1-k_i^2) \\ &= E^{(0)} \prod_{\mathrm{j}=1}^{i} (1-k_{\mathrm{j}}^2) \end{aligned} \tag{10.45}$$

Note that it can be shown that $\Delta^{(i)}$ is the cross-correlation of the forward and backward prediction errors:

$$\Delta^{(i-1)} = \mathcal{E}[b^{(i-1)}(m-1)e^{(i-1)}(m)] \tag{10.46}$$

The parameter $\Delta^{(i-1)}$ is known as the partial correlation.

Levinson–Durbin Algorithm

The Durbin algorithm starts with a predictor of order zero for which $E^{(0)} = r_{xx}(0)$. The algorithm then computes the coefficients of a predictor of order i, using the coefficients of a predictor of order $i-1$. In the process of solving for the coefficients of a predictor of order P, the solutions for the predictor coefficients of all orders less than P are also obtained:

$$E^{(0)} = r_{xx}(0) \tag{10.47}$$

For $i = 1, \ldots, P$

$$\Delta^{(i-1)} = r_{xx}(i) - \sum_{k=1}^{i-1} a_k^{(i-1)} r_{xx}(i-k) \tag{10.48}$$

$$k_i = -\frac{\Delta^{(i-1)}}{E^{(i-1)}} \tag{10.49}$$

$$a_i^{(i)} = -k_i \tag{10.50}$$

$$a_j^{(i)} = a_j^{(i-1)} + k_i a_{i-j}^{(i-1)} \quad 1 \le j \le i-1 \tag{10.51}$$

$$E^{(i)} = (1-k_i^2)E^{(i-1)} \tag{10.52}$$

10.2.3 Lattice Predictors

The lattice structure, shown in Figure 10.11, is a cascade connection of similar units, with each unit specified by a single parameter k_i, known as the *reflection* coefficient. A major attraction of a lattice structure is its modular form and the relative ease with which the model order can be extended. A further advantage is that, for a stable model, the magnitude of k_i is bounded by unity ($|k_i| < 1$), and therefore it is relatively easy to check a lattice structure for stability. The lattice structure is derived from the forward and backward prediction errors as follows. An order-update recursive equation can be obtained for the forward prediction error by multiplying both sides of Equation (10.32) by the input vector $[x(m), x(m-1), \ldots, x(m-i)]$:

$$e^{(i)}(m) = e^{(i-1)}(m) - k_i b^{(i-1)}(m-1) \tag{10.53}$$

Similarly, we can obtain an order-update recursive equation for the backward prediction error by multiplying both sides of Equation (10.37) by the input vector $[x(m-i), x(m-i+1), \ldots, x(m)]$ as

$$b^{(i)}(m) = b^{(i-1)}(m-1) - k_i e^{(i-1)}(m) \tag{10.54}$$

Equations (10.53) and (10.54) are interrelated and may be implemented by a lattice network as shown in Figure 10.11. Minimisation of the squared forward prediction error of Equation (10.53) over N samples yields

$$k_i = \frac{\sum\limits_{m=0}^{N-1} e^{(i-1)}(m) b^{(i-1)}(m-1)}{\sum\limits_{m=0}^{N-1} \left(e^{(i-1)}(m)\right)^2} \tag{10.55}$$

Note that a similar relation for k_i can be obtained through minimisation of the squared backward prediction error of Equation (10.54) over N samples. The reflection coefficients are also known as the normalised partial correlation (PARCOR) coefficients.

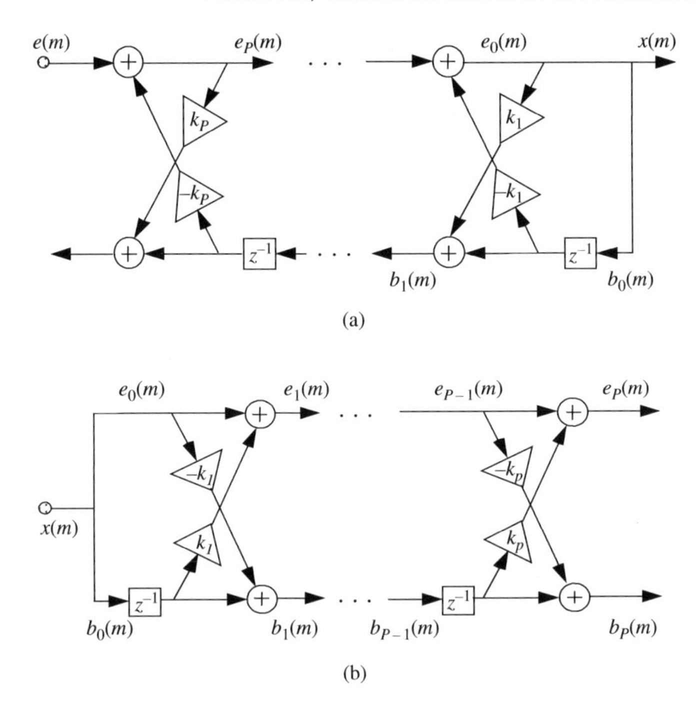

Figure 10.11 (a) Lattice implementation of linear predictor and (b) the inverse lattice linear predictor.

10.2.4 Alternative Formulations of Least Square Error Prediction

The methods described above for derivation of the predictor coefficients are based on minimisation of either the forward or the backward prediction error. In this section, we consider alternative methods based on the minimisation of the sum of the forward and backward prediction errors.

Burg's Method

Burg's method is based on minimisation of the sum of the forward and backward squared prediction errors. The squared error function is defined as

$$E_{fb}^{(i)} = \sum_{m=0}^{N-1} \left\{ \left[e^{(i)}(m)\right]^2 + \left[b^{(i)}(m)\right]^2 \right\} \tag{10.56}$$

Substitution of Equations (10.53) and (10.54) in Equation (10.56) yields

$$E_{fb}^{(i)} = \sum_{m=0}^{N-1} \left\{ \left[e^{(i-1)}(m) - k_i b^{(i-1)}(m-1)\right]^2 + \left[b^{(i-1)}(m-1) - k_i e^{(i-1)}(m)\right]^2 \right\} \tag{10.57}$$

Minimisation of $E_{fb}^{(i)}$ with respect to the reflection coefficients k_i yields

$$k_i = \frac{2\sum_{m=0}^{N-1} e^{(i-1)}(m)b^{(i-1)}(m-1)}{\sum_{m=0}^{N-1}\left\{\left[e^{(i-1)}(m)\right]^2 + [b^{(i-1)}(m-1)]^2\right\}} \tag{10.58}$$

Simultaneous Minimisation of the Backward and Forward Prediction Errors

From Equation (10.32) we have that the backward predictor coefficient vector is the reversed version of the forward predictor coefficient vector. Hence a predictor of order P can be obtained through simultaneous minimisation of the sum of the squared backward and forward prediction errors defined by

$$\begin{aligned} E_{fb}^{(P)} &= \sum_{m=0}^{N-1}\left\{\left[e^{(P)}(m)\right]^2 + \left[b^{(P)}(m)\right]^2\right\} \\ &= \sum_{m=0}^{N-1}\left\{\left[x(m) - \sum_{k=1}^{P} a_k x(m-k)\right]^2 + \left[x(m-P) - \sum_{k=1}^{P} a_k x(m-P+k)\right]^2\right\} \\ &= (\boldsymbol{x} - \boldsymbol{X}\boldsymbol{a})^{\mathrm{T}}(\boldsymbol{x} - \boldsymbol{X}\boldsymbol{a}) + \left(\boldsymbol{x}^{\mathrm{B}} - \boldsymbol{X}^{\mathrm{B}}\boldsymbol{a}\right)^{\mathrm{T}}\left(\boldsymbol{x}^{\mathrm{B}} - \boldsymbol{X}^{\mathrm{B}}\boldsymbol{a}\right) \end{aligned} \tag{10.59}$$

where $\boldsymbol{X}$ and $\boldsymbol{x}$ are the signal matrix and vector defined by Equations (10.16) and (10.17), and similarly $\boldsymbol{X}^{\mathrm{B}}$ and $\boldsymbol{x}^{\mathrm{B}}$ are the signal matrix and vector for the backward predictor. Using an approach similar to that used in derivation of Equation (10.20), the minimisation of the mean squared error function of Equation (10.59) yields

$$\boldsymbol{a} = \left(\boldsymbol{X}^{\mathrm{T}}\boldsymbol{X} + \boldsymbol{X}^{\mathrm{BT}}\boldsymbol{X}^{\mathrm{B}}\right)^{-1}\left(\boldsymbol{X}^{\mathrm{T}}\boldsymbol{x} + \boldsymbol{X}^{\mathrm{BT}}\boldsymbol{x}^{\mathrm{B}}\right) \tag{10.60}$$

where $\boldsymbol{X}^{BT}$ is the transpose of $\boldsymbol{X}^{B}$. Note that for an ergodic signal as the signal length N increases Equation (10.60) converges to the so-called normal Equation (10.14).

10.2.5 Predictor Model Order Selection

One procedure for the determination of the correct model order is to increment the model order, and monitor the differential change in the error power, until the change levels off. The incremental change in error power with the increasing model order from $i-1$ to i is defined as

$$\Delta E^{(i)} = E^{(i-1)} - E^{(i)} \tag{10.61}$$

Figure 10.12 illustrates the decrease in the normalised mean square prediction error with the increasing predictor length for a speech signal. The order P beyond which the decrease in the error power $\Delta E^{(P)}$ becomes less than a threshold is taken as the model order.

In linear prediction two coefficients are required for modelling each spectral peak of the signal spectrum. For example, the modelling of a signal with K dominant resonance in the spectrum needs $P = 2K$ coefficients. Hence a procedure for model selection is to examine the power spectrum of the signal process, and to set the model order to twice the number of significant spectral peaks in the spectrum.

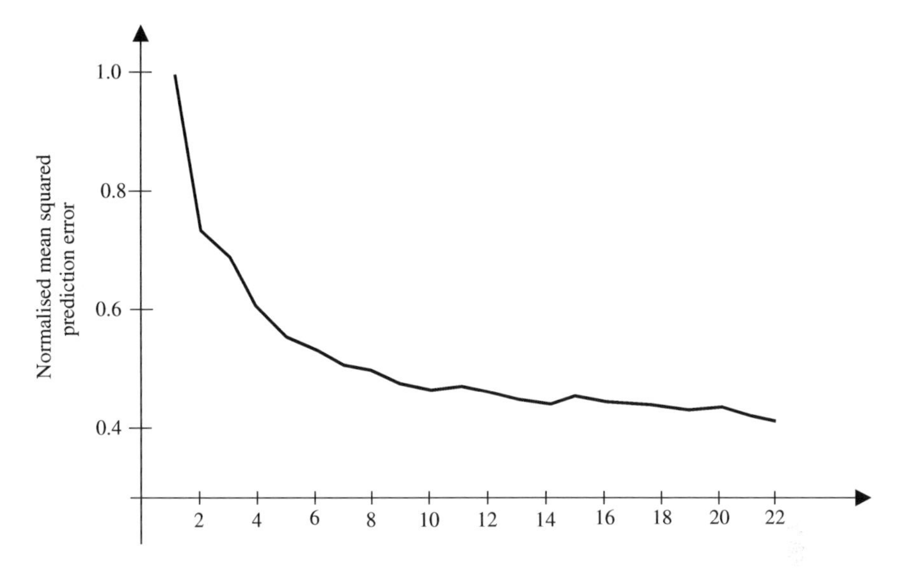

Figure 10.12 Illustration of the decrease in the normalised mean squared prediction error with the increasing predictor length for a speech signal.

When the model order is less than the correct order, the signal is under-modelled. In this case the prediction error is not well de-correlated and will be more than the optimal minimum. A further consequence of under-modelling is a decrease in the spectral resolution of the model: adjacent spectral peaks of the signal could be merged and appear as a single spectral peak when the model order is too small. When the model order is larger than the correct order, the signal is over-modelled. An over-modelled problem can result in an ill-conditioned matrix equation, unreliable numerical solutions and the appearance of spurious spectral peaks in the model.

10.3 Short-Term and Long-Term Predictors

For quasi-periodic signals, such as voiced speech, two types of correlation structures can be utilised for a more accurate prediction:

(a) The short-term correlation, which is the correlation of each sample with the P immediate past samples: $x(m-1), \ldots, x(m-P)$.
(b) The long-term correlation, which is the correlation of a sample $x(m)$ with say $2Q+1$ similar samples a pitch period T away: $x(m-T+Q), \ldots, x(m-T-Q)$.

Figure 10.13 is an illustration of the short-term relation of a sample with the P immediate past samples and its long-term relation with the samples a pitch period away. The short-term correlation of a signal may be modelled by the linear prediction Equation (10.3). The remaining correlation, in the prediction error signal $e(m)$, is called the long-term correlation. The long-term correlation in the prediction error signal may be modelled by a pitch predictor defined as

$$\hat{e}(m) = \sum_{k=-Q}^{Q} p_k e(m-T-k) \tag{10.62}$$

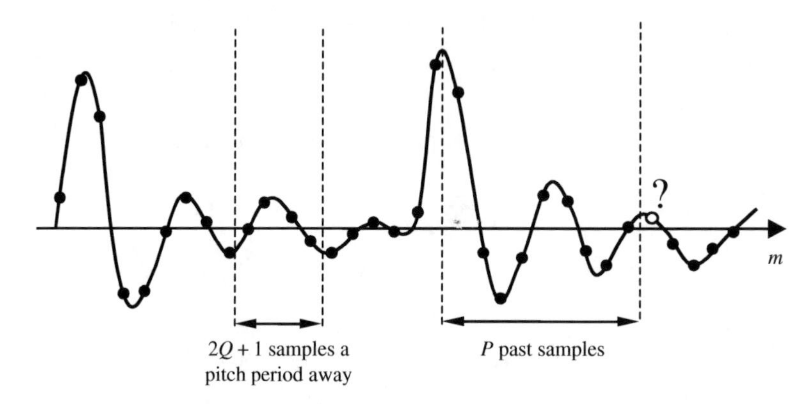

Figure 10.13 Illustration of the short-term relation of a sample with the P immediate past samples and the long-term relation with the samples a pitch period away.

where p_k are the coefficients of a long-term predictor of order $2Q+1$. The pitch period T can be obtained from the autocorrelation function of $x(m)$ or that of $e(m)$: it is the first non-zero time-lag where the autocorrelation function attains a maximum. Assuming that the long-term correlation is correctly modelled, the prediction error of the long-term filter is a completely random signal with a white spectrum, and is given by

$$\begin{aligned}\varepsilon(m) &= e(m) - \hat{e}(m) \\ &= e(m) - \sum_{k=-Q}^{Q} p_k e(m-T-k)\end{aligned} \tag{10.63}$$

Minimisation of $\mathrm{E}[e^2(m)]$ results in the following solution for the pitch predictor:

$$\begin{pmatrix} p_{-Q} \\ p_{-Q+1} \\ \vdots \\ p_{Q-1} \\ p_Q \end{pmatrix} = \begin{pmatrix} r_{xx}(0) & r_{xx}(1) & r_{xx}(2) & \cdots & r_{xx}(2Q) \\ r_{xx}(1) & r_{xx}(0) & r_{xx}(1) & \cdots & r_{xx}(2Q-1) \\ r_{xx}(2) & r_{xx}(1) & r_{xx}(0) & \cdots & r_{xx}(2Q-2) \\ \vdots & \vdots & \vdots & \ddots & \vdots \\ r_{xx}(2Q) & r_{xx}(2Q-1) & r_{xx}(2Q-2) & \cdots & r_{xx}(0) \end{pmatrix}^{-1} \begin{pmatrix} r_{xx}(T-Q) \\ r_{xx}(T-Q+1) \\ \vdots \\ r_{xx}(T+Q-1) \\ r_{xx}(T+Q) \end{pmatrix} \tag{10.64}$$

An alternative to the separate, cascade, modelling of the short- and long-term correlations is to combine the short- and long-term predictors into a single model described as

$$x(m) = \underbrace{\sum_{k=1}^{P} a_k x(m-k)}_{\text{short-term prediction}} + \underbrace{\sum_{k=-Q}^{Q} p_k x(m-k-T)}_{\text{long-term prediction}} + \varepsilon(m) \tag{10.65}$$

In Equation (10.65), each sample is expressed as a linear combination of P immediate past samples and $2Q+1$ samples a pitch period away.

Minimisation of $\mathrm{E}[e^2(m)]$ results in the following solution for the pitch predictor:

$$\begin{pmatrix} a_1 \\ a_2 \\ a_3 \\ \vdots \\ a_P \\ p_{-Q} \\ p_{-Q+1} \\ \vdots \\ p_{+Q} \end{pmatrix} = \begin{pmatrix} r(0) & r(1) & \dots & r(P-1) & r(T-Q-1) & r(T-Q) & \dots & r(T+Q-1) \\ r(1) & r(0) & \dots & r(P-2) & r(T-Q-2) & r(T-Q-1) & \dots & r(T+Q-2) \\ r(2) & r(1) & \dots & r(P-3) & r(T-Q-3) & r(T+Q-2) & \dots & r(T+Q-3) \\ \vdots & \vdots & \ddots & \vdots & \vdots & \vdots & \ddots & \vdots \\ r(P-1) & r(P-2) & \dots & r(0) & r(T-Q-P) & r(T-Q-P+1) & \dots & r(T+Q-P) \\ r(T-Q-1) & r(T-Q-2) & \dots & r(T-Q-P) & r(0) & r(1) & \dots & r(2Q) \\ r(T-Q) & r(T-Q-1) & \dots & r(T-Q-P+1) & r(1) & r(0) & \dots & r(2Q-1) \\ \vdots & \vdots & \ddots & \vdots & \vdots & \vdots & \ddots & \vdots \\ r(T+Q-1) & r(T+Q-2) & \cdots & r(T+Q-P) & r(2Q) & r(2Q-1) & \dots & r(0) \end{pmatrix}^{-1} \begin{pmatrix} r(1) \\ r(2) \\ r(3) \\ \vdots \\ r(P) \\ r(T-Q) \\ r(T-Q+1) \\ \vdots \\ r(T+Q) \end{pmatrix} \tag{10.66}$$

In Equation (10.66), for simplicity the subscript xx of $r_{xx}(k)$ has been omitted.

10.4 MAP Estimation of Predictor Coefficients

The posterior probability density function of a predictor coefficient vector $\boldsymbol{a}$, given a signal $\boldsymbol{x}$ and the initial samples $\boldsymbol{x}_\mathrm{I}$, can be expressed, using Bayes' rule, as

$$f_{A|X,X_\mathrm{I}}(\boldsymbol{a}|\boldsymbol{x},\boldsymbol{x}_\mathrm{I}) = \frac{f_{X|A,X_\mathrm{I}}(\boldsymbol{x}|\boldsymbol{a},\boldsymbol{x}_\mathrm{I})\, f_{A|X_\mathrm{I}}(\boldsymbol{a}|\boldsymbol{x}_\mathrm{I})}{f_{X|X_I}(\boldsymbol{x}|\boldsymbol{x}_\mathrm{I})} \tag{10.67}$$

In Equation (10.67), the pdfs are conditioned on P initial signal samples $\boldsymbol{x}_\mathrm{I} = [x(-P), x(-P+1), \dots, x(-1)]$. Note that for a given set of samples $[\boldsymbol{x}, \boldsymbol{x}_\mathrm{I}]$, $f_{X|X_\mathrm{I}}(\boldsymbol{x}|\boldsymbol{x}_\mathrm{I})$ is a constant, and it is reasonable to assume that $f_{A|X_I}(\boldsymbol{a}|\boldsymbol{x}_\mathrm{I}) = f_A(\boldsymbol{a})$.

10.4.1 Probability Density Function of Predictor Output

The pdf $f_{X|A,X_\mathrm{I}}(\boldsymbol{x}|\boldsymbol{a},\boldsymbol{x}_\mathrm{I})$ of the signal $\boldsymbol{x}$, given the predictor coefficient vector $\boldsymbol{a}$ and the initial samples $\boldsymbol{x}_\mathrm{I}$, is equal to the pdf of the input signal $\boldsymbol{e}$:

$$f_{X|A,X_\mathrm{I}}(\boldsymbol{x}|\boldsymbol{a},\boldsymbol{x}_\mathrm{I}) = f_E(\boldsymbol{x} - \boldsymbol{X}\boldsymbol{a}) \tag{10.68}$$

where the input signal vector is given by

$$\boldsymbol{e} = -\boldsymbol{X}\boldsymbol{a} \tag{10.69}$$

and $f_E(\boldsymbol{e})$ is the pdf of $\boldsymbol{e}$. Equation (10.64) can be expanded as

$$\begin{pmatrix} e(0) \\ e(1) \\ e(2) \\ \vdots \\ e(N-1) \end{pmatrix} = \begin{pmatrix} x(0) \\ x(1) \\ x(2) \\ \vdots \\ x(N-1) \end{pmatrix} - \begin{pmatrix} x(-1) & x(-2) & x(-3) & \cdots & x(-P) \\ x(0) & x(-1) & x(-2) & \cdots & x(1-P) \\ x(1) & x(0) & x(-1) & \cdots & x(2-P) \\ \vdots & \vdots & \vdots & \ddots & \vdots \\ x(N-2) & x(N-3) & x(N-4) & \cdots & x(N-P-1) \end{pmatrix} \begin{pmatrix} a_1 \\ a_2 \\ a_3 \\ \vdots \\ a_P \end{pmatrix} \tag{10.70}$$

Assuming that the input excitation signal $e(m)$ is a zero-mean, uncorrelated, Gaussian process with a variance of σ_e^2, the likelihood function in Equation (10.68) becomes

$$f_{X|A,X_I}(x|a,x_I) = f_E(x - Xa)$$
$$= \frac{1}{(2\pi\sigma_e^2)^{N/2}} \exp\left(\frac{1}{2\sigma_e^2}(x - Xa)^T (x - Xa)\right) \tag{10.71}$$

An alternative form of Equation (10.71) can be obtained by rewriting Equation (10.70) in the form:

$$\begin{pmatrix} e(0) \\ e(1) \\ e(2) \\ e(3) \\ \vdots \\ e(N-1) \end{pmatrix} = \begin{pmatrix} -a_P & \cdots & -a_2 & -a_1 & 1 & 0 & 0 & 0 & 0 & 0 \\ 0 & -a_P & \cdots & -a_2 & -a_1 & 1 & 0 & 0 & 0 & 0 \\ 0 & 0 & -a_P & \cdots & -a_2 & -a_1 & 1 & 0 & 0 & 0 \\ 0 & 0 & 0 & -a_P & \cdots & -a_2 & -a_1 & 1 & 0 & 0 \\ \vdots & \vdots & \vdots & \vdots & \vdots & \ddots & \vdots & \vdots & \vdots & \vdots \\ 0 & 0 & 0 & 0 & 0 & -a_P & \cdots & -a_2 & -a_1 & 1 \end{pmatrix} \begin{pmatrix} x(-P) \\ x(-P+1) \\ x(-P+2) \\ x(-P+3) \\ \vdots \\ x(N-1) \end{pmatrix} \tag{10.72}$$

In compact notation Equation (10.72) can be written as

$$e = Ax \tag{10.73}$$

Using Equation (10.73), and assuming that the excitation signal $e(m)$ is a zero mean, uncorrelated process with variance σ_e^2, the likelihood function of Equation (10.71) can be written as

$$f_{X|A,X_I}(x|a,x_I) = \frac{1}{(2\pi\sigma_e^2)^{N/2}} \exp\left(-\frac{1}{2\sigma_e^2} x^T A^T A x\right) \tag{10.74}$$

10.4.2 Using the Prior pdf of the Predictor Coefficients

The prior pdf of the predictor coefficient vector is assumed to have a Gaussian distribution with a mean vector μ_a and a covariance matrix Σ_{aa}:

$$f_A(a) = \frac{1}{(2\pi)^{P/2} |\Sigma_{aa}|^{1/2}} \exp\left[-\frac{1}{2}(a - \mu_a)^T \Sigma_{aa}^{-1} (a - \mu_a)\right] \tag{10.75}$$

Substituting Equations (10.72) and (10.76) in Equation (10.67), the posterior pdf of the predictor coefficient vector $f_{A|X,X_I}(a|x,x_I)$ can be expressed as

$$f_{A|X,X_I}(a|x,x_I) = \frac{1}{f_{X|X_I}(x|x_I)} \frac{1}{(2\pi)^{(N+P)/2} \sigma_e^N |\Sigma_{aa}|^{1/2}} \times \exp\left\{-\frac{1}{2}\left(\frac{1}{\sigma_e^2}(x - Xa)^T (x - Xa) + (a - \mu_a)^T \Sigma_{aa}^{-1} (a - \mu_a)\right]\right\} \tag{10.76}$$

The maximum a posteriori estimate is obtained by maximising the log-likelihood function:

$$\frac{\partial}{\partial a}\left[\ln f_{A|X,X_I}(a|x,x_I)\right] = \frac{\partial}{\partial a}\left[\frac{1}{\sigma_e^2}(x - Xa)^T (x - Xa) + (a - \mu_a)^T \Sigma_{aa}^{-1} (a - \mu_a)\right] = 0 \tag{10.77}$$

This yields

$$\hat{a}^{MAP} = \left(\Sigma_{aa} X^T X + \sigma_e^2 I\right)^{-1} \Sigma_{aa} X^T x + \sigma_e^2 \left(\Sigma_{aa} X^T X + \sigma_e^2 \mathbf{I}\right)^{-1} \mu_a \tag{10.78}$$

Note that as the Gaussian prior tends to a uniform prior, the determinant covariance matrix Σ_{aa} of the Gaussian prior increases, and the MAP solution tends to the least square error solution

$$\hat{\boldsymbol{a}}^{\mathrm{LS}} = \left(\boldsymbol{X}^{\mathrm{T}}\boldsymbol{X}\right)^{-1}\left(\boldsymbol{X}^{\mathrm{T}}\boldsymbol{x}\right) \tag{10.79}$$

Similarly as the observation length N increases the signal matrix term $\boldsymbol{X}^{\mathrm{T}}\boldsymbol{X}$ becomes more significant than Σ_{aa} and again the MAP solution tends to a least squared error solution.

10.5 Formant-Tracking LP Models

Formants are the resonance frequencies of speech. In the application of linear prediction to speech, the poles of linear prediction model the resonance at formants of speech. The z-transfer function of the linear prediction model of speech can be expressed in a cascade form as

$$X(z) = G(z)V(z)L(z) \tag{10.80}$$

where $G(z)$, $V(z)$ and $L(z)$ are the z-transforms of glottal pulse, vocal tract and lip radiation. The vocal tract can be modelled by formant-tracking LP models which may be expressed as

$$V(z, m) = G(m)\prod_{k=1}^{N}\frac{1}{1-2r_k\varphi_k(m)z^{-1}+r_k^2(m)z^{-2}} \tag{10.81}$$

where $\phi_k(m)$, $r_k(m)$ and $G(m)$ are the time-varying angular frequency and radii of the poles and the gain of a second-order section of the linear prediction model. The relationship between poles of linear prediction models and formants is not one-to-one. In fact depending on the model order the linear prediction model may associate more than one pole with a formant or conversely more than one formant with a pole. The poles of the linear prediction model are the formant candidates: the raw data from which the formants and their models are estimated.

The spectral resonance at formant can be characterised by a parameter vector comprising of the frequency F_k, bandwidth B_k and magnitude of the resonance M_k and their temporal slope of variation as

$$F_k = [F_k, B_k, M_k, \Delta F_k, \Delta B_k, \Delta M_k] \; k = 1, \ldots, M \tag{10.82}$$

where Δ denotes the slope of the trajectory of a feature vector over time, e.g. $\Delta F_k(t)$ for the kth formant at frame t is obtained as

$$\Delta F_k(t) = \frac{\sum_{m=1}^{N} m\left(F_k(t+m) - F_k(t-m)\right)}{\sum_{m=1}^{N} 2m^2} \qquad k = 1, \ldots, M \tag{10.83}$$

There are three issues in modelling formants: (1) modelling the distribution of formants using a probability model, (2) tracking the trajectory of each formant using a classifier, and (3) smoothing the trajectory of formants. Formant tracking methods are the subject of current research.

Using the formant tracks a formant-tracking LP model can be constructed. Formant tracking models provide a framework for modelling the inter-correlation of LP models across successive frames speech. Formant-tracking LP models can be used for speech synthesis and for speech enhancement through de-noising the parameters of LP model of speech.

Figure 10.14 illustrates the spectrogram of the frequency response of linear prediction model of a segment of speech with the estimates of the formant tracks superimposed.

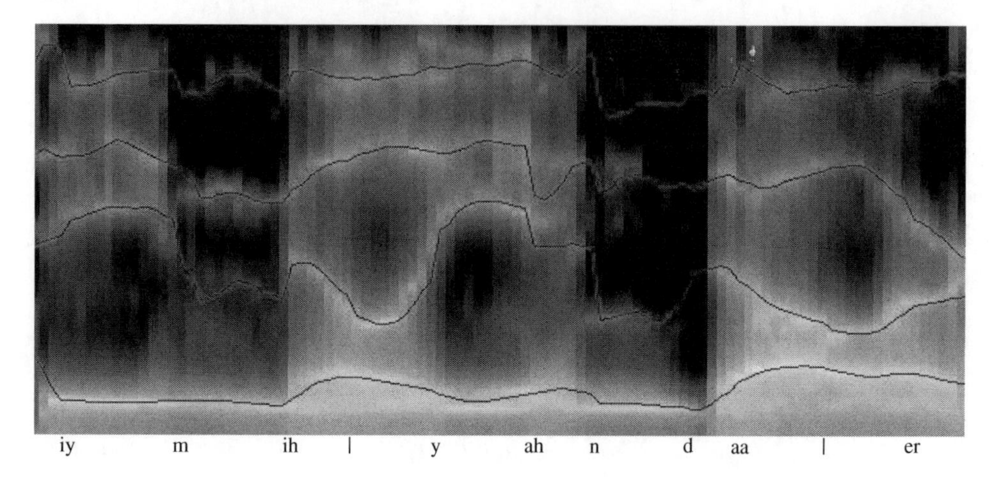

Figure 10.14 An example of formant tracks superimposed on an LP spectrogram. Lighter regions on the spectrogram correspond to high energy.

10.6 Sub-Band Linear Prediction Model

The poles of a linear predictor equation model the signal spectrum over its full bandwidth. The distribution of the poles of the LP model over the signal spectrum depends on the signal correlation and spectral structure. Generally, the poles redistribute themselves over the spectrum to minimise the mean square prediction error criterion. An alternative to a conventional LP model is to divide the signal into a number of sub-bands and to model the signal within each band with a lower-order linear prediction model as shown in Figure 10.15.

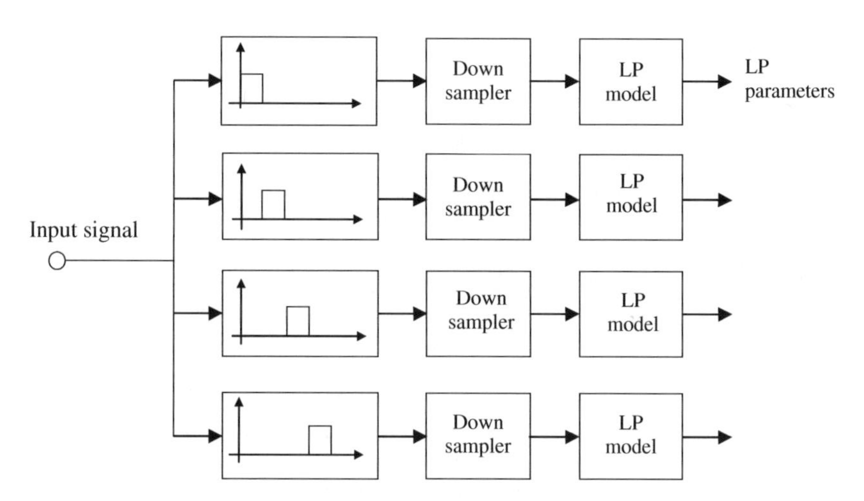

Figure 10.15 Configuration of a sub-band linear prediction (LP) model.

The advantages of using a sub-band LP model are:

(1) Sub-band linear prediction allows the designer to allocate different numbers of parameters to different bands.
(2) The solution of a full-band linear predictor equation (10.10) or (10.16) requires the inversion of a relatively large correlation matrix, whereas the solutions of the sub-band LP models require the inversion of a number of smaller correlation matrices with better numerical stability properties. For example, a predictor of order 18 requires the inversion of an 18×18 matrix, whereas three sub-band predictors of order 6 require the inversion of three 6×6 matrices.
(3) Sub-band linear prediction is useful for applications such as noise reduction where a sub-band approach can offer more flexibility and better performance.

In sub-band linear prediction, the signal $x(m)$ is passed through a bank of N band-pass filters, and is split into N sub-band signals $x_k(m)$, $k = 1, \ldots, N$. The kth sub-band signal is modelled using a low-order linear prediction model as

$$x_k(m) = \sum_{i=1}^{P_k} a_k(i) x_k(m-i) + g_k e_k(m) \tag{10.84}$$

where $[a_k, g_k]$ are the coefficients and the gain of the predictor model for the kth sub-band. The choice of the model order P_k depends on the width of the sub-band and on the signal correlation structure within each sub-band. The power spectrum of the input excitation of an ideal LP model for the kth sub-band signal can be expressed as

$$P_{\text{EE}}(f,k) = \begin{cases} 1 & f_{k,\text{start}} < f < f_{k,\text{end}} \\ 0 & \text{otherwise} \end{cases} \tag{10.85}$$

where $f_{k,\text{start}}, f_{k,\text{end}}$ are the start and end frequencies of the kth sub-band signal. The autocorrelation function of the excitation function in each sub-band is a sinc function given by

$$r_{ee}(m) = B_k \text{sinc}\left[m(B_k - f_{k0})/2\right] \tag{10.86}$$

where B_k and f_{k0} are the bandwidth and the centre frequency of the kth sub-band respectively. To ensure that sub-band LP parameters only model the signal within that sub-band (and do not model the bandpass filters) the sub-band signals are down-sampled as shown in Figure 10.15.

Note that a problem with sub-band linear prediction is that the pole frequencies may happen at or near the cut-off frequency of sub-bands. To avoid this problem formant trajectory estimation, described in the previous section, may be used to track the frequencies of the poles of the signal. The centre frequency of the sub-bands follows the formant tracks.

10.7 Signal Restoration Using Linear Prediction Models

Linear prediction models are extensively used in speech and audio signal restoration. For a noisy signal, linear prediction analysis models the combined spectra of the signal and the noise processes. For example, the frequency spectrum of a linear prediction model of speech, observed in additive white noise, would be flatter than the spectrum of the noise-free speech, owing to the influence of the flat spectrum of white noise. In this section we consider the estimation of the coefficients of a predictor model from noisy observations, and the use of linear prediction models in signal restoration. The noisy signal $y(m)$ is modelled as

$$y(m) = x(m) + n(m)$$
$$= \sum_{k=1}^{P} a_k x(m-k) + e(m) + n(m) \quad (10.87)$$

where the signal $x(m)$ is modelled by a linear prediction model with coefficients a_k and random input $e(m)$, and it is assumed that the noise $n(m)$ is additive. The least square error predictor model of the noisy signal $y(m)$ is given by

$$\boldsymbol{R}_{yy}\hat{\boldsymbol{a}} = \boldsymbol{r}_{yy} \quad (10.88)$$

where $\boldsymbol{R}_{yy}$ and $\boldsymbol{r}_{yy}$ are the autocorrelation matrix and vector of the noisy signal $y(m)$. For an additive noise model, Equation (10.88) can be written as

$$(\boldsymbol{R}_{xx} + \boldsymbol{R}_{nn})(\boldsymbol{a} + \tilde{\boldsymbol{a}}) = (\boldsymbol{r}_{xx} + \boldsymbol{r}_{nn}) \quad (10.89)$$

where $\tilde{\boldsymbol{a}}$ is the error in the predictor coefficient vector due to the noise. A simple method for removing the effects of noise is to subtract an estimate of the autocorrelation of the noise from that of the noisy signal. The drawback of this approach is that, owing to random variations of noise, correlation subtraction can cause numerical instability in Equation (10.88) and result in spurious solutions. In the following, we formulate the pdf of the noisy signal and describe an iterative signal-restoration/parameter-estimation procedure developed by Lee and Oppenheim.

From Bayes' rule, the MAP estimate of the predictor coefficient vector $\boldsymbol{a}$, given an observation signal vector $\boldsymbol{y} = [y(0), y(1), \ldots, y(N-1)]$, and the initial samples vector $\boldsymbol{x}_\text{I}$ is

$$f_{A|Y,X_I}(\boldsymbol{a}|\boldsymbol{y}, \boldsymbol{x}_\text{I}) = \frac{f_{Y|A,X_\text{I}}(\boldsymbol{y}|\boldsymbol{a}, \boldsymbol{x}_\text{I})\, f_{A,X_\text{I}}(\boldsymbol{a}, \boldsymbol{x}_\text{I})}{f_{Y,X_\text{I}}(\boldsymbol{y}, \boldsymbol{x}_\text{I})} \quad (10.90)$$

Consider the variance of the signal $\boldsymbol{y}$ in the argument of the term $f_{Y|A,X_I}(\boldsymbol{y}|\boldsymbol{a}, \boldsymbol{x}_I)$ in Equation (10.90). The innovation (i.e. prediction error) of $y(m)$ can be defined as

$$\varepsilon(m) = y(m) - \sum_{k=1}^{P} \hat{a}_k y(m-k)$$
$$= e(m) + \varepsilon(m) + n(m) - \sum_{k=1}^{P} \hat{a}_k n(m-k) \quad (10.91)$$

The variance of $y(m)$, given the previous P samples and the coefficient vector $\hat{\boldsymbol{a}}$, is the variance of the innovation signal given by

$$\text{Var}\,[y(m)\,|y(m-1), \ldots, y(m-P), \hat{\boldsymbol{a}}\,] = \sigma_e^2 + \sigma_\varepsilon^2 + \sigma_n^2 - \sigma_n^2 \sum_{k=1}^{P} \hat{a}_k^2 \quad (10.92)$$

where σ_e^2, σ_ε^2 and σ_n^2 are the variance of the excitation signal, the error in innovation due to noise and the noise respectively. From Equation (10.92), the variance of $[y(m)|y(m-1), \ldots, y(m-P), \hat{\boldsymbol{a}}]$ is a function of the coefficient vector $\hat{\boldsymbol{a}}$. Consequently, maximisation of $f_{Y|A,XI}(\boldsymbol{y}|\hat{\boldsymbol{a}}, \boldsymbol{x}_\text{I})$ with respect to the vector $\hat{\boldsymbol{a}}$ is a non-linear and non-trivial exercise.

Lim and Oppenheim proposed the following iterative process where an estimate $\hat{\boldsymbol{a}}$ of the predictor coefficient vector is used to make an estimate $\hat{\boldsymbol{x}}$ of the signal vector, and the signal estimate $\hat{\boldsymbol{x}}$ is then used to improve the estimate of the parameter vector $\hat{\boldsymbol{a}}$, and the process is iterated until convergence.

The posterior pdf of the noise-free signal $\boldsymbol{x}$ given the noisy signal $\boldsymbol{y}$ and an estimate of the parameter vector $\hat{\boldsymbol{a}}$ is given by

$$f_{X|A,Y}(\boldsymbol{x}|\hat{\boldsymbol{a}},\boldsymbol{y}) = \frac{f_{Y|A,X}(\boldsymbol{y}|\hat{\boldsymbol{a}},\boldsymbol{x})\, f_{X|A}(\boldsymbol{x}|\hat{\boldsymbol{a}})}{f_{Y|A}(\boldsymbol{y}|\hat{\boldsymbol{a}})} \tag{10.93}$$

Consider the likelihood term $f_{Y|A,X}(\boldsymbol{y}|\hat{\boldsymbol{a}},\boldsymbol{x})$. Since the noise is additive, we have

$$\begin{aligned} f_{Y|A,X}(\boldsymbol{y}|\hat{\boldsymbol{a}},\boldsymbol{x}) &= f_N(\boldsymbol{y}-\boldsymbol{x}) \\ &= \frac{1}{(2\pi\sigma_n^2)^{N/2}} \exp\left[-\frac{1}{2\sigma_n^2}(\boldsymbol{y}-\boldsymbol{x})^{\mathrm{T}}(\boldsymbol{y}-\boldsymbol{x})\right] \end{aligned} \tag{10.94}$$

Assuming that the input of the predictor model is a zero-mean Gaussian process with variance σ_e^2, the pdf of the signal $\boldsymbol{x}$ given an estimate of the predictor coefficient vector $\boldsymbol{a}$ is

$$\begin{aligned} f_{Y|A,X}(\boldsymbol{x}|\hat{\boldsymbol{a}}) &= \frac{1}{(2\pi\sigma_e^2)^{N/2}} \exp\left(-\frac{1}{2\sigma_e^2}\boldsymbol{e}^{\mathrm{T}}\boldsymbol{e}\right) \\ &= \frac{1}{(2\pi\sigma_e^2)^{N/2}} \exp\left(-\frac{1}{2\sigma_e^2}\boldsymbol{x}^{\mathrm{T}}\hat{\mathbf{A}}^{\mathrm{T}}\hat{\mathbf{A}}\boldsymbol{x}\right) \end{aligned} \tag{10.95}$$

where $\boldsymbol{e} = \hat{\mathbf{A}}\boldsymbol{x}$ as in Equation (10.73). Substitution of Equations (10.94) and (10.95) in Equation (10.93) yields

$$f_{X|A,Y}(\boldsymbol{x}|\hat{\boldsymbol{a}},\boldsymbol{y}) = \frac{1}{f_{Y|A}(\boldsymbol{y}|\hat{\boldsymbol{a}})} \frac{1}{(2\pi\sigma_n\sigma_e)^N} \exp\left[-\frac{1}{2\sigma_n^2}(\boldsymbol{y}-\boldsymbol{x})^{\mathrm{T}}(\boldsymbol{y}-\boldsymbol{x}) - \frac{1}{2\sigma_e^2}\boldsymbol{x}^{\mathrm{T}}\hat{\mathbf{A}}^{\mathrm{T}}\hat{\mathbf{A}}\boldsymbol{x}\right] \tag{10.96}$$

In Equation (10.92), for a given signal $\boldsymbol{y}$ and coefficient vector $\hat{\boldsymbol{a}}$, $f_{Y|A}(\boldsymbol{y}|\hat{\boldsymbol{a}})$ is a constant. From Equation (10.92), the ML signal estimate is obtained by maximising the log-likelihood function as

$$\frac{\partial}{\partial \boldsymbol{a}}\left(\ln f_{X|A,Y}(\boldsymbol{x}|\hat{\boldsymbol{a}},\boldsymbol{y})\right) = \frac{\partial}{\partial \boldsymbol{x}}\left(-\frac{1}{2\sigma_e^2}\boldsymbol{x}^{\mathrm{T}}\hat{\mathbf{A}}^{\mathrm{T}}\hat{\mathbf{A}}\boldsymbol{x} - \frac{1}{2\sigma_n^2}(\boldsymbol{y}-\boldsymbol{x})^{\mathrm{T}}(\boldsymbol{y}-\boldsymbol{x})\right) = \mathbf{0} \tag{10.97}$$

which gives

$$\hat{\boldsymbol{x}} = \sigma_e^2\left(\sigma_n^2\hat{\mathbf{A}}^{\mathrm{T}}\hat{\mathbf{A}} + \sigma_e^2\mathbf{I}\right)^{-1}\boldsymbol{y} \tag{10.98}$$

The signal estimate of Equation (10.98) can be used to obtain an updated estimate of the predictor parameter. Assuming that the signal is a zero-mean Gaussian process, the estimate of the predictor parameter vector $\boldsymbol{a}$ is given by

$$\hat{\boldsymbol{a}}(\hat{\boldsymbol{x}}) = \left(\hat{\boldsymbol{X}}^{\mathrm{T}}\hat{\boldsymbol{X}}\right)^{-1}\left(\hat{\boldsymbol{X}}^{\mathrm{T}}\hat{\boldsymbol{x}}\right) \tag{10.99}$$

Equations (10.98) and (10.99) form the basis for an iterative signal restoration/parameter estimation method.

10.7.1 Frequency-Domain Signal Restoration Using Prediction Models

The following algorithm is a frequency-domain implementation of the linear prediction model-based restoration of a signal observed in additive white noise.

Initialisation: Set the initial signal estimate to noisy signal $\hat{\boldsymbol{x}}_0 = \boldsymbol{y}$

For iterations $i = 0, 1, \ldots$

Step 1 Estimate the predictor parameter vector $\hat{\boldsymbol{a}}_i$:

$$\hat{\boldsymbol{a}}_i(\hat{\boldsymbol{x}}_i) = \left(\hat{\boldsymbol{X}}_i^{\mathrm{T}} \hat{\boldsymbol{X}}_i\right)^{-1} \left(\hat{\boldsymbol{X}}_i^{\mathrm{T}} \hat{\boldsymbol{x}}_i\right) \tag{10.100}$$

Step 2 Calculate an estimate of the model gain G using Parseval's theorem:

$$\frac{1}{N} \sum_{f=0}^{N-1} \frac{\hat{G}^2}{\left|1 - \sum_{k=1}^{P} \hat{a}_{k,i} e^{-\mathrm{j}2\pi fk/N}\right|^2} = \sum_{m=0}^{N-1} y^2(m) - N\hat{\sigma}_n^2 \tag{10.101}$$

where $\hat{a}_{k,i}$ are the coefficient estimates at iteration i, and $N\hat{\sigma}_n^2$ is the energy of white noise over N samples.

Step 3 Calculate an estimate of the power spectrum of speech model:

$$\hat{P}_{X_i X_i}(f) = \frac{\hat{G}^2}{\left|1 - \sum_{k=1}^{P} \hat{a}_{k,i} e^{-\mathrm{j}2\pi fk/N}\right|^2} \tag{10.102}$$

Step 4 Calculate the Wiener filter frequency response:

$$\hat{W}_i(f) = \frac{\hat{P}_{X_i X_i}(f)}{\hat{P}_{X_i X_i}(f) + \hat{P}_{N_i N_i}(f)} \tag{10.103}$$

where $\hat{P}_{N_i N_i}(f) = \hat{\sigma}_n^2$ is an estimate of the noise power spectrum.

Step 5 Filter the magnitude spectrum of the noisy speech as

$$\hat{X}_{i+1}(f) = \hat{W}_i(f) Y(f) \tag{10.104}$$

Restore the time domain signal $\hat{\boldsymbol{x}}_{i+1}$ by combining $\hat{X}_{i+1}(f)$ with the phase of noisy signal and the complex signal to time domain.

Step 6 Go to step 1 and repeat until convergence, or for a specified number of iterations.

Figure 10.16 illustrates a block diagram configuration of a Wiener filter using a linear prediction estimate of the signal spectrum. Figure 10.17 illustrates the result of an iterative restoration of the spectrum of a noisy speech signal.

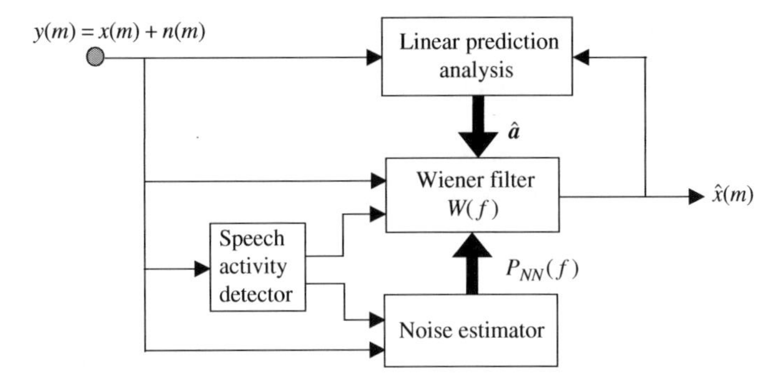

Figure 10.16 Iterative signal restoration based on linear prediction model of speech.

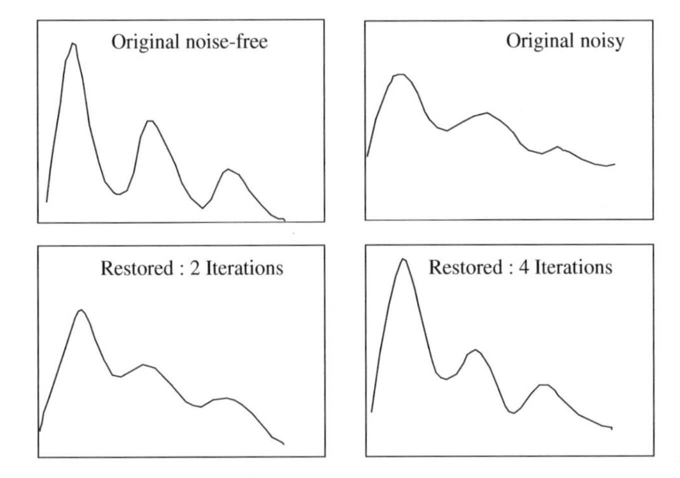

Figure 10.17 Illustration of restoration of a noisy signal with iterative linear prediction based method.

10.7.2 Implementation of Sub-Band Linear Prediction Wiener Filters

Assuming that the noise is additive, the noisy signal in each sub-band is modelled as

$$y_k(m) = x_k(m) + n_k(m) \tag{10.105}$$

The Wiener filter in the frequency domain can be expressed in terms of the power spectra, or in terms of LP model frequency responses, of the signal and noise process as

$$\begin{aligned} W_k(f) &= \frac{P_{X,k}(f)}{P_{Y,k}(f)} \\ &= \frac{g_{X,k}^2}{\left|A_{X,k}(f)\right|^2} \frac{\left|A_{Y,k}(f)\right|^2}{g_{Y,k}^2} \end{aligned} \tag{10.106}$$

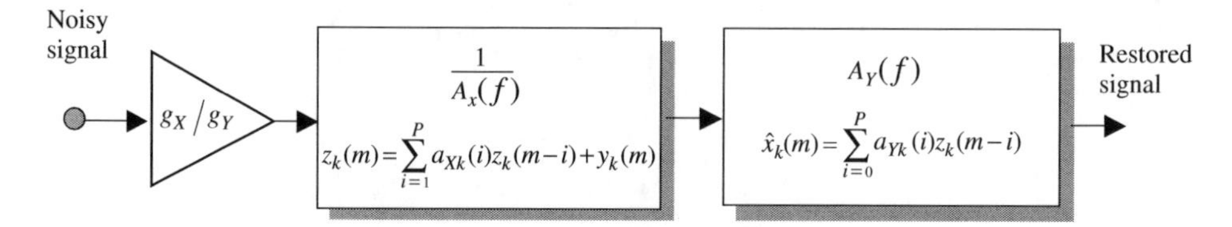

Figure 10.18 A cascade implementation of the LP squared-root Wiener filter.

where $P_{X,k}(f)$ and $P_{Y,k}(f)$ are the power spectra of the clean signal and the noisy signal for the kth sub-band respectively. From Equation (10.112) the square-root Wiener filter is given by

$$W_k^{1/2}(f) = \frac{g_{X,k}}{|A_{X,k}(f)|} \frac{|A_{Y,k}(f)|}{g_{Y,k}} \tag{10.107}$$

The linear prediction Wiener filter of Equation (10.107) can be implemented in the time domain with a cascade of a linear predictor of the clean signal, followed by an inverse predictor filter of the noisy signal as expressed by the following relations (see Figure 10.18):

$$z_k(m) = \sum_{i=1}^{P} a_{Xk}(i) z_k(m-i) + \frac{g_X}{g_Y} y_k(m) \tag{10.108}$$

$$\hat{x}_k(m) = \sum_{i=0}^{P} a_{Yk}(i) z_k(m-i) \tag{10.109}$$

where $\hat{x}_k(m)$ is the restored estimate of the clean speech signal $x_k(m)$, $z_k(m)$ is an intermediate signal, $a_{Yk}(i)$ are the coefficients of the linear predictor model of the noisy signal for the kth sub-band and $a_{Xk}(i)$ are an estimate of the coefficients of the linear prediction model of the kth sub-band of clean speech.

10.8 Summary

Linear prediction models are used in a wide range of signal processing applications from low-bit-rate speech/video coding to model-based spectral analysis. We began this chapter with an introduction to linear prediction theory, and considered different methods of formulation of the prediction problem and derivations of the predictor coefficients. The main attraction of the linear prediction method is the closed-form solution of the predictor coefficients, and the availability of a number of efficient and relatively robust methods for solving the prediction equation such as the Levinson–Durbin method.

In Section 10.2, we considered forward, backward and lattice predictors. Although the direct-form implementation of the linear predictor is the most convenient method, for many applications, such as transmission of the predictor coefficients in speech coding, it is advantageous to use the lattice form of the predictor. This is because the lattice form can be conveniently checked for stability, and furthermore a perturbation of the parameter of any section of the lattice structure has a limited and more localised effect. In Section 10.3, we considered a modified form of linear prediction that models the short-term and long-term correlations of the signal. This method can be used for the modelling of signals with a quasi-periodic structure such as voiced speech. In Section 10.4, we considered MAP estimation and the use of a prior pdf for derivation of the predictor coefficients. Section 10.5 introduced formant-tracking linear predictors which are used in Chapter 15 for speech enhancement. In Section 10.6, the sub-band linear prediction method was formulated. Finally in Section 10.7, a linear prediction model was applied to the restoration of a signal observed in additive noise.

Further Reading

Akaike H. (1970) Statistical Predictor Identification, Annals of the Institute of Statistical Mathematics. 22, pp. 203–217.

Akaike H. (1974) A New Look at Statistical Model Identification, IEEE Trans. on Automatic Control, AC-19, pp. 716–723, Dec.

Anderson O.D. (1976) Time Series Analysis and Forecasting, The Box-Jenkins Approach. Butterworth, London.

Ayre A.J. (1972) Probability and Evidence. Columbia University Press.

Box G.E.P and Jenkins G.M. (1976) Time Series Analysis: Forecasting and Control. Holden-Day, San Francisco, CA.

Burg J.P. (1975) Maximum Entropy Spectral Analysis. PhD thesis, Stanford University, Stanford, CA.

Cohen J. and Cohen, P. (1975) Applied Multiple Regression/Correlation Analysis for the Behavioural Sciences. Halsted, New York.

Draper N.R. and Smith, H. (1981). Applied Regression Analysis (2nd edn.). Wiley, New York.

Durbin J. (1959) Efficient Estimation of Parameters in Moving Average Models. Biometrica, 46, pp. 306–3110.

Durbin J. (1960) The Fitting of Time Series Models. Rev. Int. Stat. Inst., 28, pp. 233–244.

Fuller W.A (1976) Introduction to Statistical Time Series. Wiley, New York.

Hansen J.H. and Clements M.A. (1987) Iterative Speech Enhancement with Spectral Constrains. IEEE Proc. Int. Conf. on Acoustics, Speech and Signal Processing ICASSP-87, 1, pp. 189–192, Dallas, April.

Hansen J.H. and Clements M.A. (1988) Constrained Iterative Speech Enhancement with Application to Automatic Speech Recognition. IEEE Proc. Int. Conf. on Acoustics, Speech and Signal Processing, ICASSP-88, 1, pp. 561–564, New York, April.

Hocking R.R. (1996) The Analysis of Linear Models. Wiley.

Kobatake H., Inari J. and Kakuta S. (1978) Linear Prediction Coding of Speech Signals in a High Ambient Noise Environment. IEEE Proc. Int. Conf. On Acoustics, Speech And Signal Processing, Pp. 472–475, April.

Lim J.S. and Oppenheim A.V. (1978) All-Pole Modelling of Degraded Speech. IEEE Trans. Acoustics, Speech and Signal Processing, ASSP-26, 3, pp. 197–210, June.

Lim J.S. and Oppenheim A.V. (1979) Enhancement and Bandwidth Compression of Noisy Speech, Proc. IEEE, 67, pp. 1586–1604.

Makoul J. (1975) Linear Prediction: A Tutorial Review. Proceedings of the IEEE, 63, pp. 561–580.

Markel J.D. and Gray A.H. (1976) Linear Prediction of Speech. Springer Verlag, New York.

Rabiner L.R. and Schafer R.W. (1976) Digital Processing of Speech Signals. Prentice-Hall, Englewood Cliffs, NJ.

Stockham T.G., Cannon T.M. and Ingebretsen R.B (1975) Blind Deconvolutiopn Through Digital Signal Processing. IEEE Proc., 63, 4, pp. 678–692.

Tong H. (1975) Autoregressive Model Fitting with Noisy Data by Akaike's Information Criterion, IEEE Trans. Information Theory, IT-23, pp. 409–410.

11 Hidden Markov Models

Hidden Markov models (HMMs) are used for statistical modelling of non-stationary signal processes such as speech signals, image sequences and time-varying noise. A Markov process, developed by Andrei Markov, is a process whose state or value at any time t depends on its previous state or values at time $t-1$ and is independent of the history of the process before time $t-1$. An HMM is a double-layered process with a hidden Markov layer controlling the state of an observable layer.

An HMM models the time variations (and/or the space variations) of the statistics of a random process with a Markovian chain of state-dependent stationary sub-processes. An HMM is essentially a Bayesian finite state process, with a Markovian prior for modelling the transitions between the states, and a set of state probability density functions for modelling the random variations of the signal process within each state. This chapter begins with a brief introduction to continuous and finite state non-stationary models, before concentrating on the theory and applications of hidden Markov models. We study the various HMM structures, the Baum–Welch method for the maximum-likelihood training of the parameters of an HMM, and the use of HMMs and the Viterbi decoding algorithm for the classification and decoding of an unlabelled observation signal sequence. Finally, applications of HMMs for the enhancement of noisy signals are considered.

11.1 Statistical Models for Non-Stationary Processes

A non-stationary process can be defined as one whose statistical parameters, such as its moments, vary over time. Most 'naturally generated' signals, such as audio signals, video signals, biomedical signals and seismic signals, are non-stationary, in that the parameters of the systems that generate the signals, and the environments in which the signals propagate, change with time and/or space.

A non-stationary process can be modelled as a double-layered stochastic process, with a hidden process that controls the time variations of the statistics of an observable process, as illustrated in Figure 11.1. In general, non-stationary processes can be classified into one of two broad categories:

Multimedia Signal Processing: Theory and Applications in Speech, Music and Communications Saeed V. Vaseghi

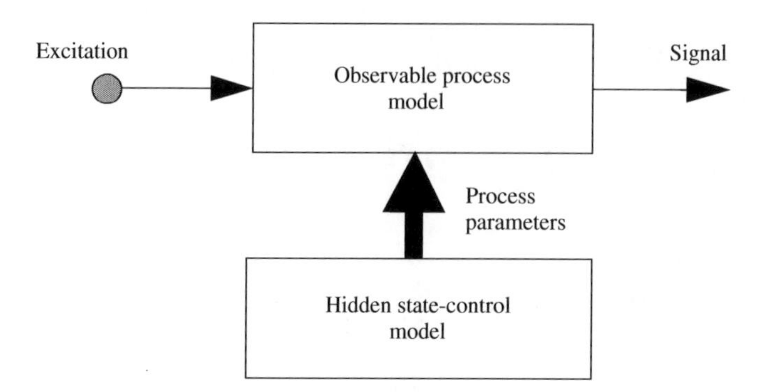

Figure 11.1 Illustration of a two-layered model of a non-stationary process.

(a) *continuously variable state* processes
(b) *finite state* processes.

A continuously variable state process is defined as one whose underlying statistics vary continuously with time. Examples of this class of random processes are most audio signals, such as speech, the power and spectral composition of which vary continuously with time. A finite state process is one whose statistical characteristics can *switch* between a finite number of stationary or non-stationary states. For example, impulsive noise is a binary-state process and across different phonetic segments speech is a finite state process. Note that a continuously variable process can be approximately expressed through 'quantisation' of its time-varying statistical variations in terms of a chain of finite state processes.

Figure 11.2(a) illustrates a non-stationary first-order autoregressive (AR) process. This process is modelled as the combination of a *hidden* stationary AR model of the signal parameters, and an observable time-varying AR model of the signal. The hidden model controls the time variations of the parameters of the non-stationary AR model. For this model, the observation signal equation and the hidden parameter state equation can be expressed as

$$x(m) = a(m)x(m-1) + e(m) \quad \text{Observation equation} \tag{11.1}$$

$$a(m) = \beta a(m-1) + \varepsilon(m) \quad \text{Hidden state equation} \tag{11.2}$$

where $a(m)$ is the time-varying coefficient of the observable AR process and β is the coefficient of the hidden state-control process.

A simple example of a finite-state non-stationary model is the binary-state autoregressive (AR) process illustrated in Figure 11.2(b), where at each time instant a random switch selects one of the two AR models for connection to the output terminal. For this model, the output signal $x(m)$ can be expressed as

$$x(m) = \bar{s}(m)x_0(m) + s(m)x_1(m) \tag{11.3}$$

where the binary switch $s(m)$ selects the state of the process at time m, and $\bar{s}(m)$ denotes the Boolean complement of $s(m)$.

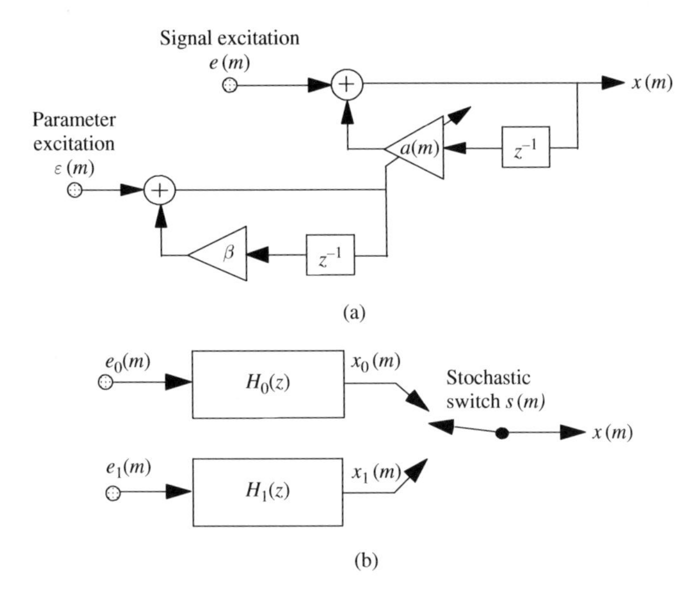

Figure 11.2 (a) A continuously variable state AR process. (b) A binary-state AR process.

11.2 Hidden Markov Models

Comparison of Markov and Hidden Markov Models

A Markov process is defined as stochastic process whose probability of being in a given state at time m depends on the previous state of the system at time $m-1$ and is independent of the states of the process before $m-1$.

Observable-State Markov Process

Consider a Markov process whose state sequence is observable from its output sequence. A simple example of a two-state Markov process is illustrated in Figure 11.3(a) which shows two containers (states): in state 1 the process always (with a probability of $P_B = 1$) outputs black balls and in state 2 the process always (with a probability of $P_W = 1$) outputs white balls. Now assume that at successive time intervals a random selection process selects one of the two containers to release a ball. The state selection process is probabilistic such that in state 1 the probability of staying in state 1 at time m is $p(s_{1,m}|s_{1,m-1}) = 0.8$ whereas the probability of moving to state 2 is $p(s_{2,m}|s_{1,m-1}) = 0.2$, where $s_{i,m}$ denotes state i at time m. In state 2 the probability of staying in state 2 is $p(s_{2,m}|s_{2,m-1}) = 0.6$ whereas the probability of moving from state 2 to state 1 is $p(s_{1,m}|s_{2,m-1}) = 0.4$. Note that the Markov process output sequence is the same as the state sequence.

Hidden-State Markov Process

A hidden Markov model (HMM) is a double-layered finite-state process, with a hidden Markovian process that controls the selection of the states of an observable process. As a simple illustration of

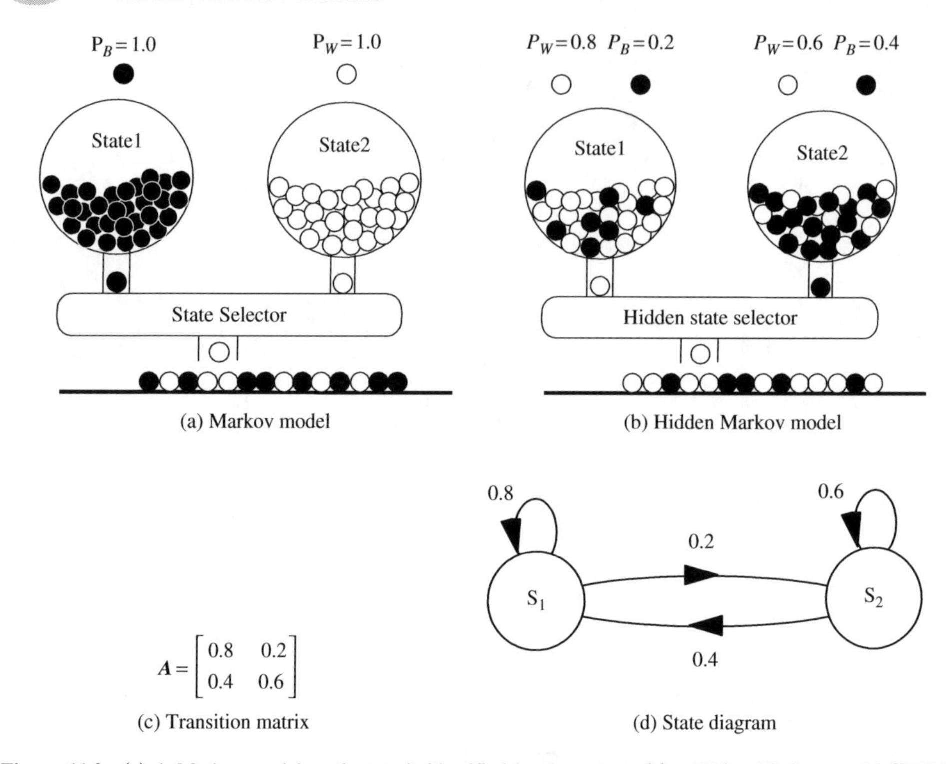

Figure 11.3 (a) A Markov model, each state is identified by the output; (b) a hidden Markov model (HMM), states are 'hidden' as both states can produce the same output with different probability; (c) the assumed transition matrix for (a) and (b); (d) a state diagram for Markov model and HMM.

a binary-state Markovian process, consider Figure 11.3(b), which shows two containers of different mixtures of black and white balls. The probability of the black and the white balls in each container, denoted as P_B and P_W respectively, are as shown in the paragraph above. Assume that at successive time intervals a hidden selection process selects one of the two containers to release a ball. The balls released are replaced so that the mixture density of the black and the white balls in each container remains unaffected. Each container can be considered as an underlying state of the output process. Now for an example assume that the hidden container-selection process is governed by the following rule: at any time, if the output from the currently selected container is a white ball then the same container is selected to output the next ball, otherwise the other container is selected. This is an example of a Markovian process because the next state of the process depends on the current state as shown in the binary state model of Figure 11.3(d). Note that in this example the observable outcome does not unambiguously indicate the underlying hidden state, because both states are capable of releasing black and white balls.

In general, a hidden Markov model has N states, with each state trained to model a distinct segment of a signal process. A hidden Markov model can be used to model a time-varying random process as a probabilistic Markovian chain of N stationary, or quasi-stationary, elementary sub-processes. The general form of a three-state HMM is shown in Figure 11.4. This structure is known as an *ergodic* HMM. In the context of an HMM, the term 'ergodic' implies that there are no structural constraints for connecting any one state to any other state.

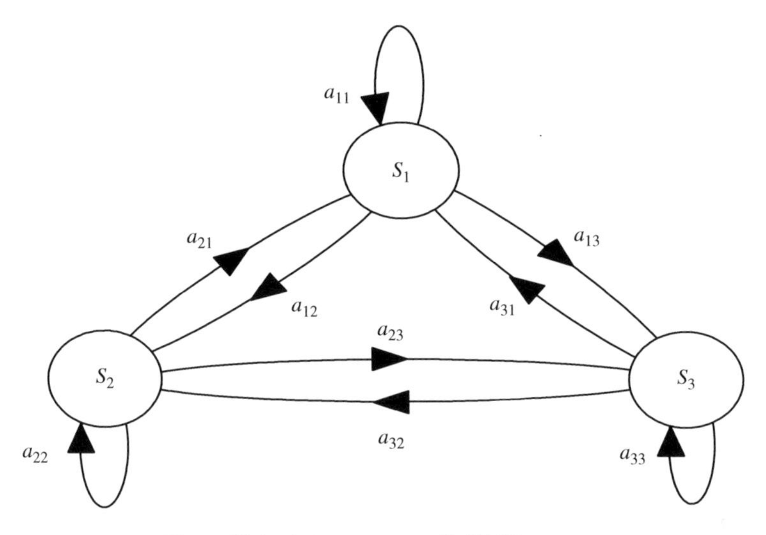

Figure 11.4 A three-state ergodic HMM structure.

A more constrained form of an HMM is the left–right model of Figure 11.5, so-called because the allowed state transitions are those from a left state to a right state and the self-loop transitions. The left–right constraint is useful for the characterisation of temporal or sequential structures of stochastic signals such as speech and musical signals, because time may be visualised as having a direction from left to right.

11.2.1 A Physical Interpretation: HMMs of Speech

For a physical interpretation of the use of HMMs in modelling a signal process, consider the illustration of Figure 11.5 which shows a left–right HMM of the spoken letter '*C*', phonetically transcribed as '*s-iy*', together with a plot of the speech signal waveform for '*C*'. In general, there are two main types of variation in speech and other stochastic signals: variations in the spectral composition, and variations in the time-scale or the articulation rate. In a hidden Markov model, these variations are modelled by the state observation and the state transition probabilities.

A useful way of interpreting and using HMMs is to consider each state of an HMM as a model of a segment of a stochastic process. For example, in Figure 11.5, state S_1 models the first segment of the spoken letter '*C*', state S_2 models the second segment, and so on. Each state must have a mechanism to accommodate the random variations in different realisations of the segments that it models. The state transition probabilities provide a mechanism for connection of various states, and for modelling the variations in the duration and time-scales of the signals in each state. For example if a segment of a speech utterance is elongated, owing, say, to slow articulation, then this can be accommodated by more self-loop transitions into the state that models the segment. Conversely, if a segment of a word is omitted, owing, say, to fast speaking, then the skip-next-state connection accommodates that situation. The state observation pdfs model the space of the probability distributions of the spectral composition of the signal segments associated with each state.

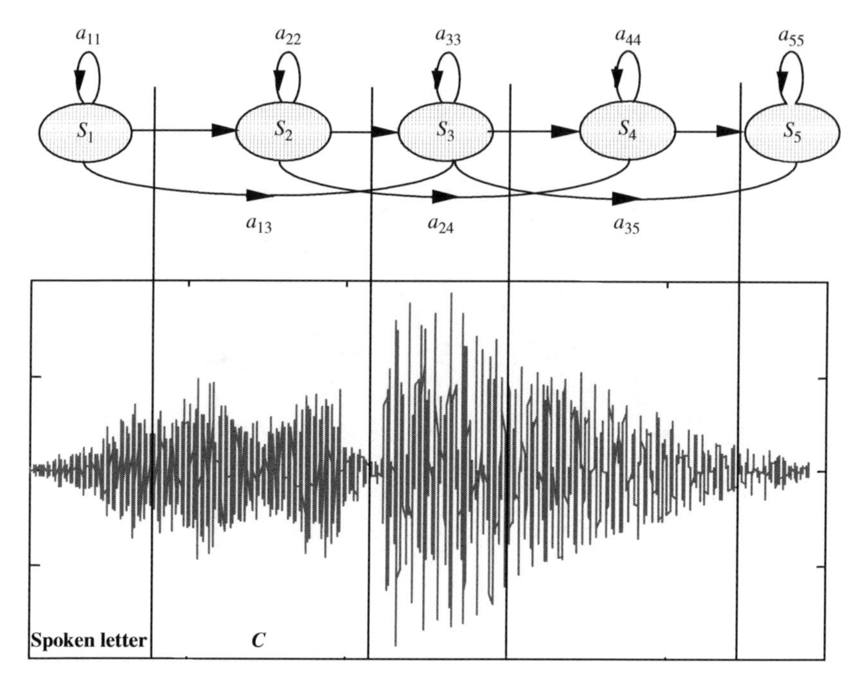

Figure 11.5 A five-state left–right HMM speech model.

11.2.2 Hidden Markov Model as a Bayesian Model

A hidden Markov model $\mathcal{M}$ is a Bayesian structure with a Markovian state transition probability and a state observation likelihood that can be either a discrete pmf or a continuous pdf.

The *posterior* probability of a state sequence $\boldsymbol{s}$ of a model $\mathcal{M}$, given a sequence of observation vectors $\boldsymbol{X} = [\boldsymbol{x}(0), \boldsymbol{x}(1), \ldots, \boldsymbol{x}(T-1)]$, can be expressed using Bayes' rule as the product of the *prior* probability of the state sequence $\boldsymbol{s}$ and the *likelihood* of the observation $\boldsymbol{X}$ as

$$P_{S|X,\mathcal{M}}(\boldsymbol{s}|\boldsymbol{X}, \mathcal{M}) = \frac{1}{f_X(\boldsymbol{X})} \underbrace{P_{S|\mathcal{M}}(\boldsymbol{s}|\mathcal{M})}_{\text{State prior}} \underbrace{f_{X|S,\mathcal{M}}(\boldsymbol{X}|\boldsymbol{s}, \mathcal{M})}_{\text{Observation likelihood}} \tag{11.4}$$

where the observation sequence $\boldsymbol{X}$ is modelled by a probability density function $P_{X|S,\mathcal{M}}(\boldsymbol{X}|\boldsymbol{s}, \mathcal{M})$.

The posterior probability that an observation signal sequence $\boldsymbol{X}$ was generated by the model $\mathcal{M}$ is summed over all likely state sequences, and may also be weighted by the model prior $P_{\mathcal{M}}(\mathcal{M})$:

$$P_{\mathcal{M}|X}(\mathcal{M}|\boldsymbol{X}) = \frac{1}{f_X(\boldsymbol{X})} \underbrace{P_{\mathcal{M}}(\mathcal{M})}_{\text{Model Prior}} \sum_{s} \underbrace{P_{S|\mathcal{M}}(\boldsymbol{s}|\mathcal{M})}_{\text{State Prior}} \underbrace{f_{X|S,\mathcal{M}}(\boldsymbol{X}|\boldsymbol{s}, \mathcal{M})}_{\text{Observation likelihood}} \tag{11.5}$$

The Markovian state transition prior can be used to model the time variations and the sequential dependence of most non-stationary processes. However, for many applications, such as speech recognition, the state observation likelihood has far more influence on the posterior probability than the state transition prior.

11.2.3 Parameters of a Hidden Markov Model

A hidden Markov model has the following parameters:

Number of states N. This is usually set to the total number of distinct, or elementary, stochastic events in a signal process. For example, in modelling a binary-state process such as impulsive noise, N is set to 2, and in phoneme-based speech modelling, N, the number of states for each phoneme, is set between 3 to 11.

State transition-probability matrix $\boldsymbol{A} = \{a_{ij}, i, j = 1, \ldots, N\}$. This provides a Markovian connection network between the states, and models the variations in the duration of the signals associated with each state. For a left–right HMM (see Figure 11.5), $a_{ij} = 0$ for $I > j$, and hence the transition matrix $\boldsymbol{A}$ is upper-triangular.

State observation vectors $\{\boldsymbol{\mu}_{i1}, \boldsymbol{\mu}_{i2}, \ldots, \boldsymbol{\mu}_{iM}, i = 1, \ldots, N\}$. For each state a set of M prototype vectors model the centroids of the signal space associated with that state.

State observation vector probability model. This can be either a discrete model composed of M prototype vectors and their associated probability $\boldsymbol{P} = \{P_{ij}(\cdot); i = 1, \ldots, N, j = 1, \ldots, M\}$, or it may be a continuous (usually Gaussian) pdf model $\boldsymbol{F} = \{f_{ij}(\cdot); i = 1, \ldots, N, j = 1, \ldots, M\}$.

Initial state probability vector $\pi = [\pi_1, \pi_2, \ldots, \pi_N]$.

11.2.4 State Observation Probability Models

Depending on whether a signal process is discrete-valued or continuous-valued, the state observation model for the process can be either a discrete-valued probability mass function (pmf), or a continuous-valued probability density function (pdf). The discrete models can also be used for modelling the space of a continuous-valued process quantised into a number of discrete points.

First, consider a discrete state observation density model. Assume that associated with the ith state of an HMM there are M discrete centroid vectors $[\boldsymbol{\mu}_{i1}, \ldots, \boldsymbol{\mu}_{iM}]$ with a pmf $[P_{i1}, \ldots, P_{iM}]$. These centroid vectors and their probabilities are normally obtained through clustering of a set of training signals associated with each state.

For the modelling of a continuous-valued process, the signal space associated with each state is partitioned into a number of clusters as in Figure 11.6. If the signals within each cluster are modelled by a uniform distribution then each cluster is described by the centroid vector and the cluster probability, and the state observation model consists of M cluster centroids and the associated pmf

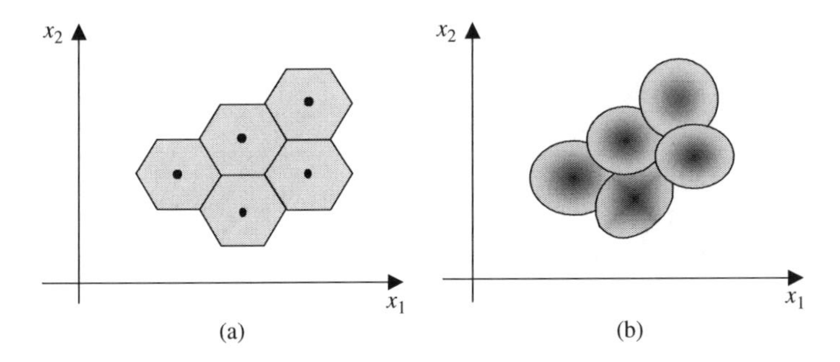

Figure 11.6 Modelling a random signal space using (a) a discrete-valued pmf and (b) a continuous-valued mixture Gaussian density.

$\{\mu_{ik}, P_{ik}; i = 1, \ldots, N, k = 1, \ldots, M\}$. In effect, this results in a discrete observation HMM for a continuous-valued process.

Figure 11.6(a) shows a partitioning and quantisation of a signal space into a number of centroids.

Now if each cluster of the state observation space is modelled by a continuous pdf, such as a Gaussian pdf, then a continuous density HMM results. The most widely used state observation pdf for an HMM is the Gaussian mixture density defined as

$$f_{X|S}(\boldsymbol{x}|s=i) = \sum_{k=1}^{M} P_{ik} \mathcal{N}(\boldsymbol{x}, \boldsymbol{\mu}_{ik}, \boldsymbol{\Sigma}_{ik}) \tag{11.6}$$

where $\mathcal{N}(\boldsymbol{x}, \boldsymbol{\mu}_{ik}, \boldsymbol{\Sigma}_{ik})$ is a Gaussian density with mean vector $\boldsymbol{\mu}_{ik}$ and covariance matrix $\boldsymbol{\Sigma}_{ik}$, and P_{ik} is a mixture weighting factor for the kth Gaussian pdf of the state i. Note that P_{ik} is the prior probability of the kth mode of the pdf mixture for the state i. Figure 11.6(b) shows the space of a Gaussian mixture model of an observation signal space. A five-mode Gaussian mixture pdf is shown in Figure 11.7.

11.2.5 State Transition Probabilities

The first-order Markovian property of an HMM entails that the transition probability to any state $s(t)$ at time t depends only on the state of the process at time $t-1$, $s(t-1)$, and is independent of the previous states of the HMM. This can be expressed as

$$\begin{aligned} &\text{Prob}(\, s(t) = j | s(t-1) = i, s(t-2) = k, \ldots, s(t-N) = l \,) \\ &= \text{Prob}(s(t) = j | s(t-1) = i) = a_{ij} \end{aligned} \tag{11.7}$$

where $s(t)$ denotes the state of the HMM at time t. The transition probabilities provide a probabilistic mechanism for connecting the states of an HMM, and for modelling the variations in the duration of the signals associated with each state. The probability of occupancy of a state i for d consecutive time units, $P_i(d)$, can be expressed in terms of the state self-loop transition probabilities a_{ii} as

$$P_i(d) = a_{ii}^{d-1}(1 - a_{ii}) \tag{11.8}$$

From Equation (11.8), using the geometric series conversion formula, the mean occupancy duration for each state of an HMM can be derived as

$$\text{Mean occupancy of state} i = \sum_{d=0}^{\infty} dP_i(d) = \frac{1}{1 - a_{ii}} \tag{11.9}$$

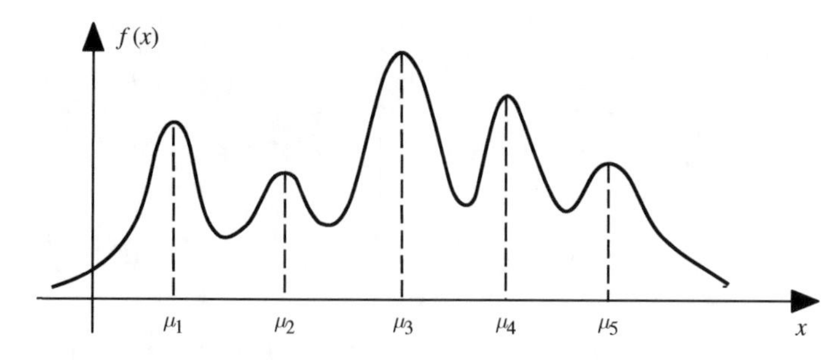

Figure 11.7 A mixture Gaussian probability density function.

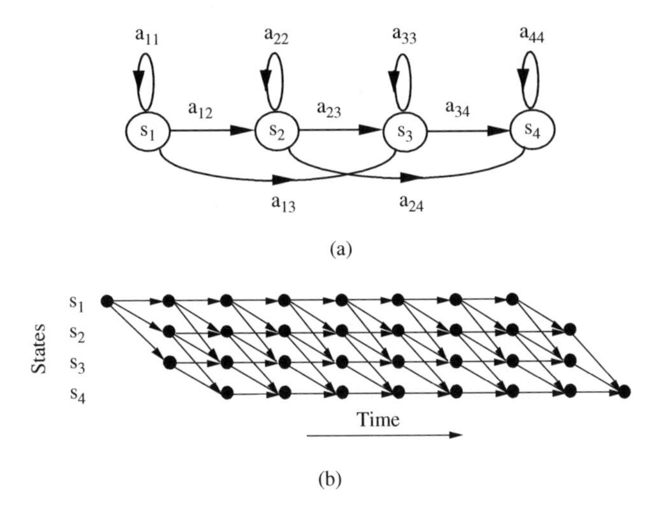

Figure 11.8 (a) A four-state left-right HMM, and (b) its state-time trellis diagram.

11.2.6 State–Time Trellis Diagram

A state–time trellis diagram shows the HMM states together with all the different paths that can be taken through various states as time unfolds. Figures 11.8(a) and 11.8(b) illustrate a four-state HMM and its state–time diagram. Since the number of states and the state parameters of an HMM are time-invariant, a state–time diagram is a repetitive and regular trellis structure. Note that in Figure 11.8 for a left–right HMM the state–time trellis has to diverge from the first state and converge into the last state. In general, there are many different state sequences that start from the initial state and end in the final state. Each state sequence has a prior probability that can be obtained by multiplication of the state transition probabilities of the sequence. For example, the probability of the state sequence $s = [S_1, S_1, S_2, S_2, S_3, S_3, S_4]$ is $P(s) = \pi_1 a_{11} a_{12} a_{22} a_{23} a_{33} a_{34}$. Since each state has a different set of prototype observation vectors, different state sequences model different observation sequences. In general, over T time units, an N-state HMM can reproduce N^T different realisations of the random process of length T.

11.3 Training Hidden Markov Models

The first step in training the parameters of an HMM is to collect a training database of a sufficiently large number of different examples of the random process to be modelled. Assume that the examples in a training database consist of L vector-valued sequences $[X] = [X_k; k = 0, \ldots, L-1]$, with each sequence $X_k = [x(t); t = 0, \ldots, T_k - 1]$ having a variable number of T_k vectors. The objective is to train the parameters of an HMM to model the statistics of the signals in the training data set. In a probabilistic sense, the fitness of a model is measured by the posterior probability $P_{M|X}(\mathcal{M}|X)$ of the model $\mathcal{M}$ given the training data X. The training process aims to maximise the posterior probability of the model $\mathcal{M}$ and the training data $[X]$, expressed using Bayes' rule as

$$P_{\mathcal{M}|X}(\mathcal{M}|X) = \frac{1}{f_X(X)} f_{X|\mathcal{M}}(X|\mathcal{M}) P_{\mathcal{M}}(\mathcal{M}) \tag{11.10}$$

where the denominator $f_X(X)$ on the right-hand side of Equation (11.10) has only a normalising effect and $P_{\mathcal{M}}(\mathcal{M})$ is the prior probability of the model $\mathcal{M}$. For a given training data set $[X]$ and a given model M, maximising Equation (11.10) is equivalent to maximising the likelihood function $P_{X|\mathcal{M}}(X|\mathcal{M})$. The likelihood of an observation vector sequence X given a model $\mathcal{M}$ can be expressed as

$$f_{X|\mathcal{M}}(X|\mathcal{M}) = \sum_{s} f_{X|S,\mathcal{M}}(X|s,\mathcal{M})\,P_{s|\mathcal{M}}(s|\mathcal{M}) \tag{11.11}$$

where $f_{X|S,M}(X(t)|s(t),M)$, the pdf of the signal sequence X along the state sequence $\mathbf{s} = [s(0), s(1), \ldots, s(T-1)]$ of the model $\mathcal{M}$, is given by

$$f_{X|S,\mathcal{M}}(X|\mathbf{s},\mathcal{M}) = f_{X|S}(\boldsymbol{x}(0)|s(0))\,f_{X|S}(\boldsymbol{x}(1)|s(1))\ldots f_{X|S}(\boldsymbol{x}(T-1)|s(T-1)) \tag{11.12}$$

where $s(t)$, the state at time t, can be one of N states, and $f_{X|S}(X(t)|s(t))$, a shorthand for $f_{X|S,M}(X(t)|s(t),\mathcal{M})$, is the pdf of $\boldsymbol{x}(t)$ given the state $s(t)$ of the model $\mathcal{M}$. The Markovian probability of the state sequence s is given by

$$P_{S|\mathcal{M}}(s|\mathcal{M}) = \pi_{s(0)}a_{s(0)s(1)}a_{s(1)s(2)}\cdots a_{s(T-2)s(T-1)} \tag{11.13}$$

Substituting Equations (11.12) and (11.13) in Equation (11.11) yields

$$\begin{aligned} f_{X|\mathcal{M}}(X|\mathcal{M}) &= \sum_{\mathbf{s}} f_{X|S,\mathcal{M}}(X|s,\mathcal{M})\,P_{s|\mathcal{M}}(s|\mathcal{M}) \\ &= \sum_{\mathbf{s}} \pi_{s(0)} f_{X|S}(\boldsymbol{x}(0)|s(0))\,a_{s(0)s(1)} f_{X|S}(\boldsymbol{x}(1)|s(1))\cdots a_{s(T-2)s(T-1)} f_{X|S}(\boldsymbol{x}(T-1)|s(T-1)) \end{aligned} \tag{11.14}$$

where the summation is taken over all state sequences s. In the training process, the transition probabilities and the parameters of the observation pdfs are estimated to maximise the model likelihood of Equation (11.14). Direct maximisation of Equation (11.14) with respect to the model parameters is a non-trivial task. Furthermore, for an observation sequence of length T vectors, the computational load of Equation (11.14) is $O(N^T)$. This is an impractically large load, even for such modest values as $N=6$ and $T=30$. However, the repetitive structure of the trellis state–time diagram of an HMM implies that there is a large amount of repeated, redundant, computation in Equation (11.14) that can be avoided in an efficient implementation. In the next section we consider the forward–backward method of model likelihood calculation, and then proceed to describe an iterative maximum-likelihood model optimisation method.

11.3.1 Forward–Backward Probability Computation

An efficient recursive algorithm for the computation of the likelihood function $f_{X|M}(X|\mathcal{M})$ is the forward–backward algorithm. The forward–backward computation method exploits the highly regular and repetitive structure of the state–time trellis diagram of Figure 11.8.

In this method, a forward probability variable $\alpha_t(i)$ is defined as the joint probability of the partial observation sequence $X = [\boldsymbol{x}(0), \boldsymbol{x}(1), \ldots, \boldsymbol{x}(t)]$ and the state i at time t, of the model $\mathcal{M}$:

$$\alpha_t(i) = f_{X,S|\mathcal{M}}(\boldsymbol{x}(0), \boldsymbol{x}(1), \cdots, \boldsymbol{x}(t), s(t) = i|\mathcal{M}) \tag{11.15}$$

The forward probability variable $\alpha_t(i)$ of Equation (11.15) can be expressed in a recursive form in terms of the forward probabilities at time $t-1$, $\alpha_{t-1}(i)$:

$$\begin{aligned}\alpha_t(i) &= f_{X,S|\mathcal{M}}(\boldsymbol{x}(0),\boldsymbol{x}(1),\ldots,\boldsymbol{x}(t),s(t)=i\,|\,\mathcal{M})\\ &= \left(\sum_{\mathrm{j}=1}^{N} f_{X,S|\mathcal{M}}(\boldsymbol{x}(0),\boldsymbol{x}(1),\ldots,\boldsymbol{x}(t-1),s(t-1)=\mathrm{j}\,|\,\mathcal{M})a_{\mathrm{ji}}\right) f_{X|S,\mathcal{M}}(\boldsymbol{x}(t)\,|s(t)=i,\mathcal{M})\\ &= \sum_{\mathrm{j}=1}^{N} \left(\alpha_{t-1}(\mathrm{j})a_{\mathrm{ji}}\right) f_{X|S,\mathcal{M}}(\boldsymbol{x}(t)\,|s(t)=i,\mathcal{M})\end{aligned} \tag{11.16}$$

Figure 11.9 illustrates a network for computation of the forward probabilities for the four-state left–right HMM of Figure 11.8. The likelihood of an observation sequence $\boldsymbol{X}=[\boldsymbol{x}(0),\boldsymbol{x}(1),\ldots,\boldsymbol{x}(T-1)]$ given a model $\mathcal{M}$ can be expressed in terms of the forward probabilities as

$$\begin{aligned}f_{X|\mathcal{M}}(\boldsymbol{x}(0),\boldsymbol{x}(1),\ldots,\boldsymbol{x}(T-1)\,|\,\mathcal{M}) &= \sum_{i=1}^{N} f_{X,S|\mathcal{M}}(\boldsymbol{x}(0),\boldsymbol{x}(1),\ldots,\boldsymbol{x}(T-1),s(T-1)=i\,|\,\mathcal{M})\\ &= \sum_{i=1}^{N}\alpha_{T-1}(i)\end{aligned} \tag{11.17}$$

Similar to the definition of the forward probability concept, a backward probability is defined as the probability of the state i at time t followed by the partial observation sequence $[\boldsymbol{x}(t+1), \boldsymbol{x}(t+2),\ldots,\boldsymbol{x}(T-1)]$ as

$$\begin{aligned}\beta_t(i) &= f_{X,S|\mathcal{M}}(s(t)=i,\boldsymbol{x}(t+1),\boldsymbol{x}(t+2),\ldots,\boldsymbol{x}(T-1)\,|\,\mathcal{M})\\ &= \sum_{\mathrm{j}=1}^{N} a_{i\mathrm{j}} f_{X,S|\mathcal{M}}(s(t+1)=\mathrm{j},\boldsymbol{x}(t+2),\boldsymbol{x}(t+3),\ldots,\boldsymbol{x}(T-1))\\ &\quad\times f_{X|S}(\boldsymbol{x}(t+1)\,|s(t+1)=\mathrm{j},\mathcal{M})\\ &= \sum_{\mathrm{j}=1}^{N} a_{i\mathrm{j}}\beta_{t+1}(\mathrm{j}) f_{X|S,\mathcal{M}}(\boldsymbol{x}(t+1)\,|s(t+1)=\mathrm{j},\mathcal{M})\end{aligned} \tag{11.18}$$

In the next section, forward and backward probabilities are used to develop a method for training HMM parameters.

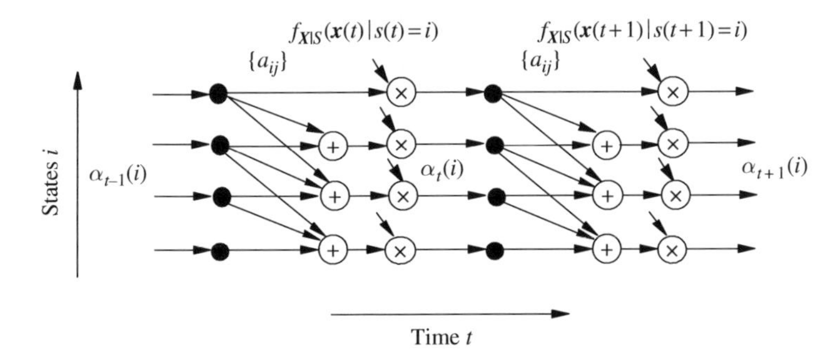

Figure 11.9 A network for computation of forward probabilities for a left–right HMM.

11.3.2 Baum–Welch Model Re-estimation

The HMM training problem is the estimation of the model parameters $\mathcal{M} = (\pi, \boldsymbol{A}, \boldsymbol{F})$ for a given data set $\boldsymbol{X}$. These parameters are the initial state probabilities π, the state transition probability matrix $\boldsymbol{A}$ and the continuous (or discrete) density state observation pdfs. The HMM parameters are estimated from a set of training examples $\{\boldsymbol{X} = [\boldsymbol{x}(0), \ldots, \boldsymbol{x}(T-1)]\}$, with the objective of maximising $f_{X|\mathcal{M}}(\boldsymbol{X}|\mathcal{M})$, the likelihood of the model and the training data. The Baum–Welch method of training HMMs is an iterative likelihood maximisation method based on the forward–backward probabilities defined in the preceding section. The Baum–Welch method is an instance of the EM algorithm described in Chapter 7. For an HMM $\mathcal{M}$, the posterior probability of a transition at time t from state i to state j of the model $\mathcal{M}$, given an observation sequence $\boldsymbol{X}$, can be expressed as

$$\begin{aligned}\gamma_t(i, j) &= P_{S|X,\mathcal{M}}(s(t) = i, s(t+1) = j\,|\boldsymbol{X}, \mathcal{M}) \\ &= \frac{f_{S,X|\mathcal{M}}(s(t) = i, s(t+1) = j, \boldsymbol{X}\,|\mathcal{M})}{f_{X|\mathcal{M}}(\boldsymbol{X}\,|\mathcal{M})} \\ &= \frac{\alpha_t(i)a_{ij}f_{X|S}(\boldsymbol{x}(t+1)\,|s(t+1) = j, \mathcal{M})\,\beta_{t+1}(j)}{\sum\limits_{i=1}^{N}\alpha_{T-1}(i)}\end{aligned} \tag{11.19}$$

where $f_{S,X|\mathcal{M}}(s(t) = i, s(t+1) = j, \boldsymbol{X}\,|\mathcal{M})$ is the joint pdf of the states $s(t)$ and $s(t+1)$ and the observation sequence $\boldsymbol{X}$, and $f_{X|S}(\boldsymbol{x}(t+1)\,|s(t+1) = i)$ is the state observation pdf for the state i. Note that for a discrete observation density HMM the state observation pdf in Equation (11.19) is replaced with the discrete state observation pmf $P_{X|S}(\boldsymbol{x}(t+1)\,|s(t+1) = i)$. The posterior probability of state i at time t given the model $\mathcal{M}$ and the observation $\boldsymbol{X}$ is

$$\begin{aligned}\gamma_t(i) &= P_{S|X,\mathcal{M}}(s(t) = i\,|\boldsymbol{X}, \mathcal{M}) \\ &= \frac{f_{S,X|\mathcal{M}}(s(t) = i, \boldsymbol{X}\,|\mathcal{M})}{f_{X|\mathcal{M}}(\boldsymbol{X}\,|\mathcal{M})} \\ &= \frac{\alpha_t(i)\beta_t(i)}{\sum\limits_{j=1}^{N}\alpha_{T-1}(j)}\end{aligned} \tag{11.20}$$

Now the state transition probability a_{ij} can be interpreted as

$$a_{ij} = \frac{\text{expected number of transitions from state } i \text{ to state } j}{\text{expected number of transitions from state } i} \tag{11.21}$$

From Equations (11.19)–(11.21), the state transition probability can be re-estimated as the ratio

$$\bar{a}_{ij} = \frac{\sum\limits_{t=0}^{T-2}\gamma_t(i, j)}{\sum\limits_{t=0}^{T-2}\gamma_t(i)} \tag{11.22}$$

Note that for an observation sequence $[\boldsymbol{x}(0), \ldots, \boldsymbol{x}(T-1)]$ of length T, the last transition occurs at time $T-2$ as indicated in the upper limits of the summations in Equation (11.22). The initial-state probabilities are estimated as

$$\bar{\pi}_i = \gamma_0(i) \tag{11.23}$$

11.3.3 Training HMMs with Discrete Density Observation Models

In a discrete density HMM, the observation signal space for each state is modelled by a set of discrete symbols or vectors. Assume that a set of M vectors $[\boldsymbol{\mu}_{i1}, \boldsymbol{\mu}_{i2}, \ldots, \boldsymbol{\mu}_{iM}]$ model the space of the signal associated with the ith state. These vectors may be obtained from a clustering process as the centroids of the clusters of the training signals associated with each state. The objective in training discrete density HMMs is to compute the state transition probabilities and the state observation probabilities. The forward–backward equations for discrete density HMMs are the same as those for continuous density HMMs, derived in the previous sections, with the difference that the probability density functions such as $f_{X|S}(\boldsymbol{x}(t)\,|s(t)=i)$ are substituted with probability mass functions $P_{X|S}(\boldsymbol{x}(t)\,|s(t)=i)$ defined as

$$P_{X|S}(\boldsymbol{x}(t)\,|s(t)=i) = P_{X|S}(Q[\boldsymbol{x}(t)]\,|s(t)=i) \tag{11.24}$$

where the function vector $Q[\boldsymbol{x}(t)]$ quantises the observation vector $\boldsymbol{x}(t)$ to the nearest discrete vector in the set $[\boldsymbol{\mu}_{i1}, \boldsymbol{\mu}_{i2}, \ldots, \boldsymbol{\mu}_{iM}]$. For discrete density HMMs, the probability of a state vector μ_{ik} can be defined as the ratio of the number of occurrences of μ_{ik} (or vectors quantised to μ_{ik}) in the state i, divided by the total number of occurrences of all other vectors in the state i:

$$\bar{P}_{ik}(\boldsymbol{\mu}_{ik}) = \frac{\text{expected number of times in state } i \text{ and observing } \boldsymbol{\mu}_{ik}}{\text{expected number of times in state } i} = \frac{\sum_{t\in \boldsymbol{x}(t)\rightarrow \boldsymbol{\mu}_{ik}}^{T-1} \gamma_t(i)}{\sum_{t=0}^{T-1} \gamma_t(i)} \tag{11.25}$$

In Equation (11.25) the summation in the numerator is taken over those time instants $t \in x(t) \rightarrow \mu_{ik}$ where the kth symbol $\boldsymbol{\mu}_{ik}$ is observed in the state i.

For statistically reliable results, an HMM must be trained on a large data set $\boldsymbol{X}$ consisting of a sufficient number of independent realisations of the process to be modelled. Assume that the training data set consists of L realisations $\boldsymbol{X} = [\boldsymbol{X}(0), \boldsymbol{X}(1), \ldots, \boldsymbol{X}(L-1)]$, where $\boldsymbol{X}(k) = [\boldsymbol{x}(0), \boldsymbol{x}(1), \ldots, \boldsymbol{x}(T_k-1)]$. The re-estimation formula can be averaged over the entire data set as

$$\hat{\pi}_i = \frac{1}{L}\sum_{l=0}^{L-1} \gamma_0^l(i) \tag{11.26}$$

$$\hat{a}_{ij} = \frac{\sum_{l=0}^{L-1}\sum_{t=0}^{T_l-2} \gamma_t^l(i,j)}{\sum_{l=0}^{L-1}\sum_{t=0}^{T_l-2} \gamma_t^l(i)} \tag{11.27}$$

and

$$\hat{P}_i(\boldsymbol{\mu}_{ik}) = \frac{\sum_{l=0}^{L-1}\sum_{t\in \boldsymbol{x}(t)\rightarrow \boldsymbol{\mu}_{ik}}^{T_l-1} \gamma_t^l(i)}{\sum_{l=0}^{L-1}\sum_{t=0}^{T_l-1} \gamma_t^l(i)} \tag{11.28}$$

In Equation (11.28) the inner summation in the numerator is taken over those time instants $t \in x(t) \rightarrow \mu_{ik}$ where the kth symbol μ_{ik} is observed in the state i. The parameter estimates of Equations (11.26)–(11.28) can be used in further iterations of the estimation process until the model converges.

11.3.4 HMMs with Continuous Density Observation Models

In continuous density HMMs, continuous probability density functions (pdfs) are used to model the space of the observation signals associated with each state. Baum *et al.* (1970) generalised the parameter re-estimation method to HMMs with concave continuous pdfs such as a Gaussian pdf. A continuous P-variate Gaussian pdf for the state i of an HMM can be defined as

$$f_{X|S}\left(\boldsymbol{x}(t)\,|s(t)=i\right)=\frac{1}{(2\pi)^{P/2}\,|\boldsymbol{\Sigma}_i|^{1/2}}\exp\left\{[\boldsymbol{x}(t)-\boldsymbol{\mu}_i]^T\,\boldsymbol{\Sigma}_i^{-1}\,[\boldsymbol{x}(t)-\boldsymbol{\mu}_i]\right\} \tag{11.29}$$

where $\boldsymbol{\mu}_i$ and $\boldsymbol{\Sigma}_i$ are the mean vector and the covariance matrix associated with the state i. The re-estimation formula for the mean vector of the state Gaussian pdf can be derived as

$$\bar{\boldsymbol{\mu}}_i=\frac{\sum\limits_{t=0}^{T-1}\gamma_t(i)\boldsymbol{x}(t)}{\sum\limits_{t=0}^{T-1}\gamma_t(i)} \tag{11.30}$$

Similarly, the covariance matrix is estimated as

$$\bar{\boldsymbol{\Sigma}}_i=\frac{\sum\limits_{t=0}^{T-1}\gamma_t(i)\,(\boldsymbol{x}(t)-\bar{\boldsymbol{\mu}}_i)\,(\boldsymbol{x}(t)-\bar{\boldsymbol{\mu}}_i)^{\mathrm{T}}}{\sum\limits_{t=0}^{T-1}\gamma_t(i)} \tag{11.31}$$

The proof that the Baum–Welch re-estimation algorithm leads to maximisation of the likelihood function $f_{X|M}(X|\mathcal{M})$ can be found in Baum *et al.* (1970).

11.3.5 HMMs with Gaussian Mixture pdfs

The modelling of the space of a signal process with a mixture of Gaussian pdfs is considered in Sections 6.7.3, 7.7 and 11.2.4. In HMMs with Gaussian mixture pdf for state observation model, the signal space associated with the ith state is modelled with a mixture of M Gaussian densities as

$$f_{X|S}\left(\boldsymbol{x}(t)\,|s(t)=i\right)=\sum_{k=1}^{M}P_{ik}\mathcal{N}\left(\boldsymbol{x}(t),\boldsymbol{\mu}_{ik},\boldsymbol{\Sigma}_{ik}\right) \tag{11.32}$$

where P_{ik} is the prior probability of the kth component of the mixture. The posterior probability of state i at time t and state j at time $t+1$ of the model $\mathcal{M}$, given an observation sequence $\boldsymbol{X}=[\boldsymbol{x}(0),\ldots,\boldsymbol{x}(T-1)]$, can be expressed as

$$\begin{aligned}\gamma_t(i,j)&=P_{S|X,\mathcal{M}}\left(s(t)=i,s(t+1)=j|\boldsymbol{X},\mathcal{M}\right)\\&=\frac{\alpha_t(i)a_{ij}\left[\sum\limits_{k=1}^{M}P_{jk}\mathcal{N}\left(\boldsymbol{x}(t+1),\boldsymbol{\mu}_{jk},\boldsymbol{\Sigma}_{jk}\right)\right]\beta_{t+1}(j)}{\sum\limits_{i=1}^{N}\alpha_{T-1}(i)}\end{aligned} \tag{11.33}$$

and the posterior probability of state i at time t given the model $\mathcal{M}$ and the observation $\boldsymbol{X}$ is given by

$$\gamma_t(i) = P_{S|X,\mathcal{M}}(s(t) = i\,|\boldsymbol{X}, \mathcal{M}) = \frac{\alpha_t(i)\beta_t(i)}{\sum_{j=1}^{N} \alpha_{T-1}(j)} \tag{11.34}$$

Now we define the joint posterior probability of the state i and the kth Gaussian mixture component pdf model of the state i at time t as

$$\zeta_t(i,k) = P_{S,K|X,\mathcal{M}}(s(t) = i, m(t) = k\,|\boldsymbol{X}, \mathcal{M}) = \frac{\sum_{j=1}^{N} \alpha_{t-1}(j) a_{ji} P_{ik} \mathcal{N}(\boldsymbol{x}(t), \boldsymbol{\mu}_{ik}, \boldsymbol{\Sigma}_{ik})\, \beta_t(i)}{\sum_{j=1}^{N} \alpha_{T-1}(j)} \tag{11.35}$$

where $m(t)$ is the Gaussian mixture component at time t. Equations (11.33) to (11.35) are used to derive the re-estimation formula for the mixture coefficients, the mean vectors and the covariance matrices of the state mixture Gaussian pdfs as

$$\bar{P}_{ik} = \frac{\text{expected number of times in state } i \text{ and observing mixture } k}{\text{expected number of times in state } i} = \frac{\sum_{t=0}^{T-1} \xi_t(i,k)}{\sum_{t=0}^{T-1} \gamma_t(i)} \tag{11.36}$$

and

$$\bar{\boldsymbol{\mu}}_{ik} = \frac{\sum_{t=0}^{T-1} \xi_t(i,k)\boldsymbol{x}(t)}{\sum_{t=0}^{T-1} \xi_t(i,k)} \tag{11.37}$$

Similarly the covariance matrix is estimated as

$$\bar{\boldsymbol{\Sigma}}_{ik} = \frac{\sum_{t=0}^{T-1} \xi_t(i,k)\,[\boldsymbol{x}(t) - \bar{\boldsymbol{\mu}}_{ik}]\,[\boldsymbol{x}(t) - \bar{\boldsymbol{\mu}}_{ik}]^{\mathrm{T}}}{\sum_{t=0}^{T-1} \xi_t(i,k)} \tag{11.38}$$

11.4 Decoding Signals Using Hidden Markov Models

Hidden Markov models are used in applications such as speech recognition, image recognition and signal restoration, and for decoding the underlying states of a signal. For example, in speech recognition, HMMs are trained to model the statistical variations of the acoustic realisations of the words in a vocabulary of say size V words. In the word-recognition phase, an utterance is classified and labelled

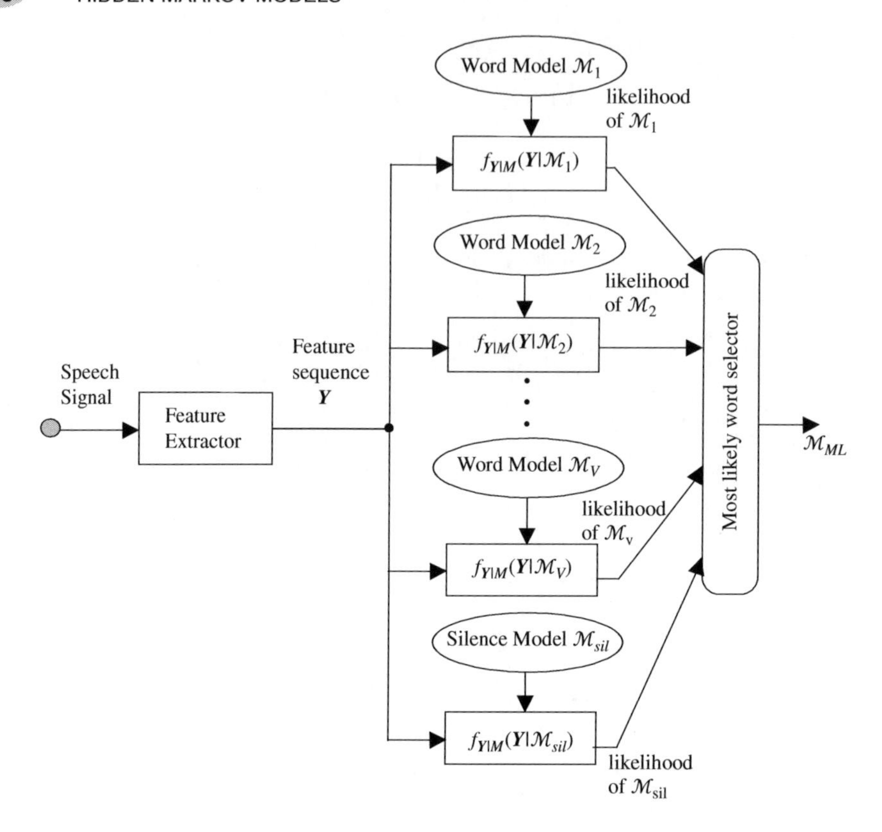

Figure 11.10 Illustration of the use of HMMs in speech recognition.

with the most likely of the $V+1$ candidate HMMs (including an HMM for silence) as illustrated in Figure 11.10.

Consider the decoding of an unlabelled sequence of T signal vectors $\boldsymbol{X} = [\boldsymbol{x}(0), \boldsymbol{x}(1), \ldots, \boldsymbol{x}(T-1)]$ given a set of V candidate HMMs $[\mathcal{M}_1, \ldots, \mathcal{M}_V]$. The probability score for the observation vector sequence $\boldsymbol{X}$ and the model $\mathcal{M}_k$ can be calculated as the likelihood

$$f_{X|\mathcal{M}}(\boldsymbol{X}|\mathcal{M}_k) = \sum_{\mathbf{s}} \pi_{s(0)} f_{X|S}(\boldsymbol{x}(0)|s(0))\, a_{s(0)s(1)} \\ f_{X|S}(\boldsymbol{x}(1)|s(1)) \cdots a_{s(T-2)s(T-1)} f_{X|S}(\boldsymbol{x}(T-1)|s(T-1)) \tag{11.39}$$

where the likelihood of the observation sequence $\boldsymbol{X}$ is summed over all possible state sequences of the model $\mathcal{M}$. Equation (11.39) can be efficiently calculated using the forward–backward method described in Section 11.3.1. The observation sequence $\boldsymbol{X}$ is labelled with the HMM that scores the highest likelihood as

$$\text{Label}(\boldsymbol{X}) = \arg\max_k \left(f_{X|\mathcal{M}}(\boldsymbol{X}|\mathcal{M}_k) \right), \quad k = 1, \ldots, V+1 \tag{11.40}$$

In decoding applications often the likelihood of an observation sequence $\boldsymbol{X}$ and a model $\mathcal{M}_k$ is obtained along the *single* most likely state sequence of model $\mathcal{M}_k$, instead of being summed over all sequences, so Equation (11.40) becomes

$$\text{Label}(X) = \arg\max_k \left[\max_s f_{X,S|\mathcal{M}}(X, s | \mathcal{M}_k) \right] \tag{11.41}$$

In Section 11.6, on the use of HMMs for noise reduction, the most likely state sequence is used to obtain the maximum-likelihood estimate of the underlying statistics of the signal process.

11.4.1 Viterbi Decoding Algorithm

In this section, we consider the decoding of a signal to obtain the maximum a posterior (MAP) estimate of the underlying state sequence. The MAP state sequence s^{MAP} of a model $\mathcal{M}$ given an observation signal sequence $X = [x(0), \ldots, x(T-1)]$ is obtained as

$$\begin{aligned} s^{\text{MAP}} &= \arg\max_s f_{X,S|\mathcal{M}}(X, s | \mathcal{M}) \\ &= \arg\max_s \left(f_{X|S,\mathcal{M}}(X | s, \mathcal{M}) P_{S|\mathcal{M}}(s | \mathcal{M}) \right) \end{aligned} \tag{11.42}$$

The MAP state sequence estimate is used in such applications as the calculation of a similarity score between a signal sequence X and an HMM $\mathcal{M}$, segmentation of a non-stationary signal into a number of distinct quasi-stationary segments, and implementation of state-based Wiener filters for restoration of noisy signals as described in Section 11.6.

For an N-state HMM and an observation sequence of length T, there are altogether N^T state sequences. Even for moderate values of N and T say ($N = 6$ and $T = 30$), an exhaustive search of the state–time trellis for the best state sequence is a computationally prohibitive exercise. The Viterbi algorithm is an efficient method for the estimation of the most likely state sequence of an HMM. In a state–time trellis diagram, such as Figure 11.8, the number of paths diverging from each state of a trellis can grow exponentially by a factor of N at successive time instants. The Viterbi method prunes the trellis by selecting the most likely path to each state. At each time instant t, for each state i, the algorithm selects the most probable path to state i and prunes out the less likely branches. This procedure ensures that at any time instant, only a single path *survives* into each state of the trellis.

For each time instant t and for each state i, the algorithm keeps a record of the state j from which the maximum-likelihood path branched into i, and also records the cumulative probability of the most likely path into state i at time t. The Viterbi algorithm is given next, and Figure 11.11 gives a network illustration of the algorithm.

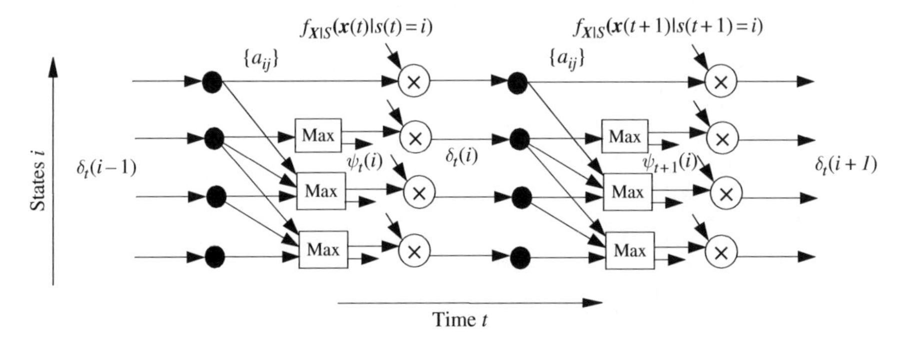

Figure 11.11 A network illustration of the Viterbi algorithm.

Viterbi Algorithm

$\delta_t(i)$ records the cumulative probability of the best path to state i at time t.
$\psi_t(i)$ records the best state sequence to state i at time t.

Step 1: *Initialisation*, at time $t = 0$, for states $i = 1, \ldots, N$

$$\delta_0(i) = \pi_i f_i(\boldsymbol{x}(0))$$

$$\psi_0(i) = 0$$

Step 2: *Recursive calculation* of the ML state sequences and their probabilities
For time $t = 1, \ldots, T-1$
For states $i = 1, \ldots, N$

$$\delta_t(i) = \max_j \left[\delta_{t-1}(j)a_{ji}\right] f_i(\boldsymbol{x}(t))$$

$$\psi_t(i) = \arg\max_j \left[\delta_{t-1}(j)a_{ji}\right]$$

Step 3: *Termination*, retrieve the most likely final state

$$s^{\text{MAP}}(T-1) = \arg\max_i \left[\delta_{T-1}(i)\right]$$

$$\text{Prob}_{\max} = \max_i \left[\delta_{T-1}(i)\right]$$

Step 4: *Backtracking* through the most likely state sequence:
For $t = T-2, \ldots, 0$

$$s^{\text{MAP}}(t) = \psi_{t+1}\left[s^{\text{MAP}}(t+1)\right].$$

The backtracking routine retrieves the most likely state sequence of the model $\mathcal{M}$. Note that the variable $\text{Prob}_{\max}$, which is the probability of the observation sequence $\boldsymbol{X} = [\boldsymbol{x}(0), \ldots, \boldsymbol{x}(T-1)]$ and the most likely state sequence of the model $\mathcal{M}$, can be used as the probability score for the model $\mathcal{M}$ and the observation $\boldsymbol{X}$. For example, in speech recognition, for each candidate word model the probability of the observation and the most likely state sequence is calculated, and then the observation is labelled with the word that achieves the highest probability score.

Matlab function HiddenMarkovProcess( )
Demonstrates the generation of a Markovian random signal from an underlying M-state hidden Markov model. At each sampling interval and at each state the state transition probabilities are used to generate a random number; this number is the identity of the next state. At each state the state mixture Gaussian density is used to generate a HMM-GMM signal.

Matlab function markov( )
Demonstrates the realisation of an M-state discrete Markov process, where the state probabilities are defined by a state transition matrix. The Markov output process is the state sequence.

11.5 HMM in DNA and Protein Sequences

A major application of hidden Markov models is in biosignal processing and computational molecular biology in applications including multiple alignment and functional classification of proteins, prediction of protein folding, recognition of genes in bacterial and human genomes, analysis and prediction of DNA functional sites, and identification of nucleosomal DNA periodical patterns.

Hidden Markov models (HMMs) are powerful probabilistic models for detecting homology among evolutionarily related sequences. Homology is concerned with likeness in structures between parts of different organisms due to evolutionary differentiation from the same or a corresponding part of a remote ancestor.

HMMs are statistical models that consider all possible combinations of matches, mismatches and gaps to generate an alignment of a set of sequences. Figure 11.12 shows a simple example of statistical modelling of DNA observations. In this case the observations are nucleotides and the aim of modelling is to align and estimate the sequential probabilities of observation sequences composed of DNA labels ACGT. Each row shows a DNA sequence. Each column is a state for which the probabilities of occurrence of ACTG are calculated as the normalised number of occurrences of each letter in the column.

Figure 11.13 shows a widely used profile-HMM structure for DNA and protein sequencing. HMMs that represent a sequence profile of a group of related proteins or DNAs are called *profile HMMs*. Again, squares represent main states, diamonds are insertion states and circles are deletion states. There are three possible 'states' for each amino acid position in a particular sequence alignment: a 'main' state where an amino acid can match or mismatch, an 'insert' state where a new amino acid can be added to one of the sequences to generate an alignment, or a 'delete' state where an amino acid can be deleted from one of the sequences to generate the alignment. Probabilities are assigned to each

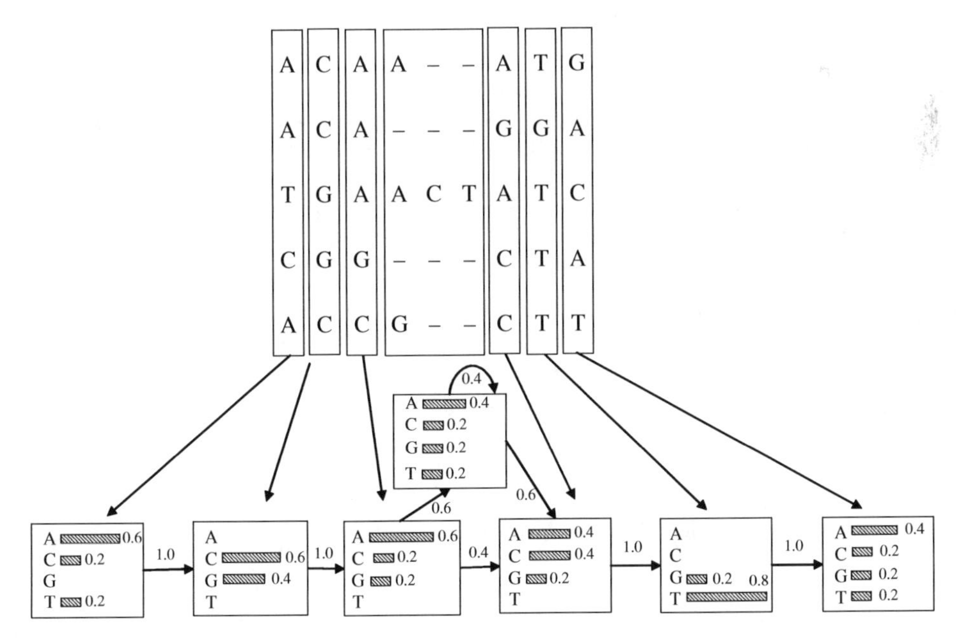

Figure 11.12 A Markov model for a dataset of DNA sequences. The discrete probabilities are histograms of the occurrence of each symbol in a column.

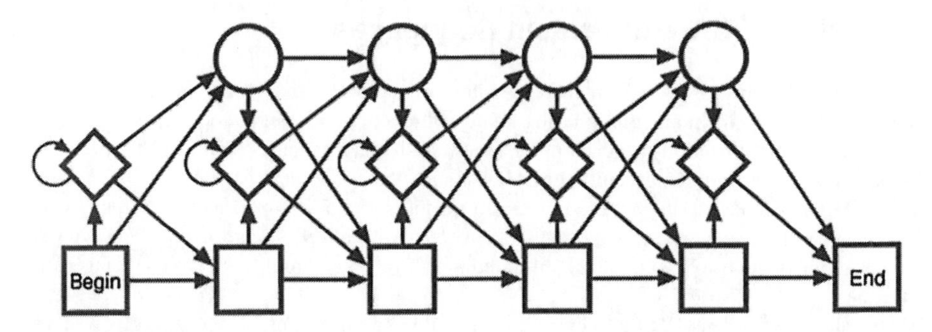

Figure 11.13 A DNA profile HMM: squares represent base states, diamonds are insertion states and circles are deletion states.

of these states based on the number of each of these events encountered in the sequence alignment. An arrow in the model represents a transition from one state to another and is also associated with a transition probability. The greater the number and diversity of sequences included in the training alignment, the better the model will be at identifying related sequences.

An adequately 'trained' profile HMM has many uses. It can align a group of related sequences, search databases for distantly related sequences, and identify subfamily-specific signatures within large proteins or DNA super families.

11.6 HMMs for Modelling Speech and Noise

11.6.1 Modelling Speech

HMMs are the main statistical modelling framework for speech recognition. Normally a three- to five-state HMM, with 10–20 Gaussian mixture pdfs per state, is used to model the statistical variations of the spectral and temporal features of a phonemic unit of speech. Each state of an HMM of a phoneme models a sub-phonemic segment with the first state modelling the first segment of the phoneme and the second state modelling the second segment and so on. For implementation of HMMs of speech the hidden Markov model toolkit (HTK) provides a good platform.

11.6.2 HMM-Based Estimation of Signals in Noise

In this and the following two sections, we consider the use of HMMs for estimation of a signal $\boldsymbol{x}(t)$ observed in an additive noise $\boldsymbol{n}(t)$, and modelled as

$$\boldsymbol{y}(t) = \boldsymbol{x}(t) + \boldsymbol{n}(t) \tag{11.43}$$

From Bayes' rule, the posterior pdf of the signal $\boldsymbol{x}(t)$ given the noisy observation $\boldsymbol{y}(t)$ is defined as

$$\begin{aligned} f_{X|Y}\left(\boldsymbol{x}(t)\,|\boldsymbol{y}(t)\,\right) &= \frac{f_{Y|X}\left(\boldsymbol{y}(t)\,|\boldsymbol{x}(t)\,\right) f_X(\boldsymbol{x}(t))}{f_Y(\boldsymbol{y}(t))} \\ &= \frac{1}{f_Y(\boldsymbol{y}(t))} f_N\left(\boldsymbol{y}(t) - \boldsymbol{x}(t)\right) f_X\left(\boldsymbol{x}(t)\right) \end{aligned} \tag{11.44}$$

For a given observation, $f_Y(\mathbf{y}(t))$ is a constant, and the maximum a posteriori (MAP) estimate is obtained as

$$\hat{\mathbf{x}}^{\text{MAP}}(t) = \arg\max_{\mathbf{x}(t)} f_N(\mathbf{y}(t) - \mathbf{x}(t))\, f_X(\mathbf{x}(t)) \tag{11.45}$$

The computation of the posterior pdf, Equation (11.44), or the MAP estimate, Equation (11.45), requires the pdf models of the signal and the noise processes. Stationary, continuous-valued processes are often modelled by a Gaussian or a mixture Gaussian pdf that is equivalent to a single-state HMM. For a non-stationary process an N-state HMM can model the time-varying pdf of the process as a Markovian chain of N stationary Gaussian sub-processes. Now assume that we have an N_s-state HMM $\mathcal{M}$ for the signal, and another N_n-state HMM η for the noise. For signal estimation, we need estimates of the underlying state sequences of the signal and the noise processes. For an observation sequence of length T, there are N_s^T possible signal state sequences and N_n^T possible noise state sequences that could have generated the noisy signal. Since it is assumed that the signal and the noise are uncorrelated, each signal state may be observed in any noisy state; therefore the number of noisy signal states is of the order of $N_s^T \times N_n^T$.

Given an observation sequence $\mathbf{Y} = [\mathbf{y}(0), \mathbf{y}(1), \ldots, \mathbf{y}(T-1)]$, the most probable state sequences of the signal and the noise HMMs maybe expressed as

$$\mathbf{s}_{\text{signal}}^{\text{MAP}} = \arg\max_{s_{\text{signal}}} \left(\max_{s_{\text{noise}}} f_Y\left(\mathbf{Y}, \mathbf{s}_{\text{signal}}, \mathbf{s}_{\text{noise}} \,|\mathcal{M}, \eta\right) \right) \tag{11.46}$$

and

$$\mathbf{s}_{\text{noise}}^{\text{MAP}} = \arg\max_{s_{\text{noise}}} \left(\max_{s_{\text{signal}}} f_Y\left(\mathbf{Y}, \mathbf{s}_{\text{signal}}, \mathbf{s}_{\text{noise}} \,|\mathcal{M}, \eta\right) \right) \tag{11.47}$$

Given the state sequence estimates for the signal and the noise models, the MAP estimation equation (11.45) becomes

$$\hat{\mathbf{x}}^{\text{MAP}}(t) = \arg\max_{\mathbf{x}} \left(f_{N|S,\eta}\left(\mathbf{y}(t) - \mathbf{x}(t) \,|\mathbf{s}_{\text{noise}}^{\text{MAP}}, \eta\right) f_{X|S,\mathcal{M}}\left(\mathbf{x}(t) \,|\mathbf{s}_{\text{signal}}^{\text{MAP}}, \mathcal{M}\right)\right) \tag{11.48}$$

Implementation of Equations (11.46)–(11.48) is computationally prohibitive. In Sections 11.6.4 and 11.6.5, we consider some practical methods for the estimation of signal in noise.

Example 11.1 Assume a signal, modelled by a binary-state HMM, is observed in an additive stationary Gaussian noise. Let the noisy observation be modelled as

$$\mathbf{y}(t) = \bar{s}(t)\mathbf{x}_0(t) + s(t)\mathbf{x}_1(t) + \mathbf{n}(t) \tag{11.49}$$

where $s(t)$ is a hidden binary-state process such that $s(t) = 0$ indicates that the signal is from the state S_0 with a Gaussian pdf of $\mathcal{N}(\mathbf{x}(t), \boldsymbol{\mu}_{x_0}, \boldsymbol{\Sigma}_{x_0x_0})$, and $s(t) = 1$ indicates that the signal is from the state S_1 with a Gaussian pdf of $\mathcal{N}(\mathbf{x}(t), \boldsymbol{\mu}_{x_1}, \boldsymbol{\Sigma}_{x_1x_1})$. Assume that a stationary Gaussian process $\mathcal{N}(\mathbf{n}(t), \boldsymbol{\mu}_n, \boldsymbol{\Sigma}_{nn})$, equivalent to a single-state HMM, can model the noise. Using the Viterbi algorithm the maximum a posteriori (MAP) state sequence of the signal model can be estimated as

$$\mathbf{s}_{\text{signal}}^{\text{MAP}} = \arg\max_{s} \left[f_{Y|S,\mathcal{M}}\left(\mathbf{Y} \,|\mathbf{s}, \mathcal{M}\right) P_{S|\mathcal{M}}\left(\mathbf{s} \,|\mathcal{M}\right) \right] \tag{11.50}$$

For a Gaussian-distributed signal and additive Gaussian noise, the observation pdf of the noisy signal is also Gaussian. Hence, the state observation pdfs of the signal model can be modified to account for the additive noise as

$$f_{Y|s_0}(\boldsymbol{y}(t)\,|s_0) = \mathcal{N}\left(\boldsymbol{y}(t), (\boldsymbol{\mu}_{x_0} + \boldsymbol{\mu}_n), (\boldsymbol{\Sigma}_{x_0x_0} + \boldsymbol{\Sigma}_{nn})\right) \tag{11.51}$$

and

$$f_{Y|s_1}(\boldsymbol{y}(t)\,|s_1) = \mathcal{N}\left(\boldsymbol{y}(t), (\boldsymbol{\mu}_{x_1} + \boldsymbol{\mu}_n), (\boldsymbol{\Sigma}_{x_1x_1} + \boldsymbol{\Sigma}_{nn})\right) \tag{11.52}$$

where $\mathcal{N}(\boldsymbol{y}(t), \boldsymbol{\mu}, \boldsymbol{\Sigma})$ denotes a Gaussian pdf with mean vector $\boldsymbol{\mu}$ and covariance matrix $\boldsymbol{\Sigma}$. The MAP signal estimate, given a state sequence estimate $\boldsymbol{s}^{\text{MAP}}$, is obtained from

$$\hat{\boldsymbol{x}}^{\text{MAP}}(t) = \arg\max_{\boldsymbol{x}} \left[f_{X|S,\mathcal{M}}\left(\boldsymbol{x}(t)\,|\boldsymbol{s}^{\text{MAP}}, \mathcal{M}\right) f_N(\boldsymbol{y}(t) - \boldsymbol{x}(t)) \right] \tag{11.53}$$

Substitution of the Gaussian pdf of the signal from the most likely state sequence, and the pdf of noise, in Equation (11.53) results in the following MAP estimate:

$$\hat{\boldsymbol{x}}^{\text{MAP}}(t) = \left(\boldsymbol{\Sigma}_{xx,s(t)} + \boldsymbol{\Sigma}_{nn}\right)^{-1} \boldsymbol{\Sigma}_{xx,s(t)}(\boldsymbol{y}(t) - \boldsymbol{\mu}_n) + \left(\boldsymbol{\Sigma}_{xx,s(t)} + \boldsymbol{\Sigma}_{nn}\right)^{-1} \boldsymbol{\Sigma}_{nn}\boldsymbol{\mu}_{x,s(t)} \tag{11.54}$$

where $\boldsymbol{\mu}_{x,s(t)}$ and $\boldsymbol{\Sigma}_{xx,s(t)}$ are the mean vector and covariance matrix of the signal $\boldsymbol{x}(t)$ obtained from the most likely state sequence $[s(t)]$.

11.6.3 Signal and Noise Model Combination and Decomposition

For Bayesian estimation of a signal observed in additive noise, we need to have an estimate of the underlying statistical state sequences of the signal and the noise processes. Figure 11.14 illustrates the outline of an HMM-based noisy speech recognition and enhancement system. The system performs the following functions:

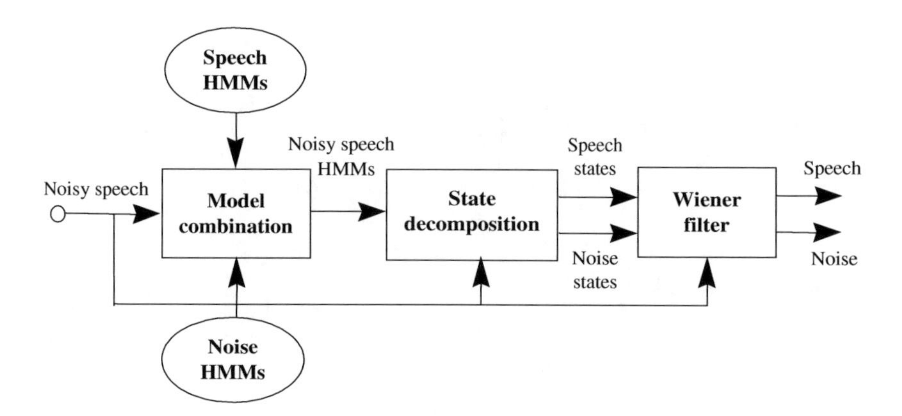

Figure 11.14 Outline configuration of HMM-based noisy speech recognition and enhancement.

(1) combination of the speech and noise HMMs to form the noisy speech HMMs;
(2) estimation of the best combined noisy speech model given the current noisy speech input;
(3) state decomposition, i.e. the separation of speech and noise states given noisy speech states;
(4) state-based Wiener filtering using the estimates of speech and noise states.

11.6.4 Hidden Markov Model Combination

The performance of HMMs trained on clean signals deteriorates rapidly in the presence of noise, since noise causes a mismatch between the clean HMMs and the noisy signals. The noise-induced mismatch can be reduced: either by filtering the noise from the signal (for example using the Wiener filtering and the spectral subtraction methods described in Chapter 15) or by combining the noise and the signal models to model the noisy signal.

The model combination method, illustrated in Figure 11.15, was developed by Gales and Young. In this method HMMs of speech are combined with an HMM of noise to form HMMs of noisy speech signals. In the power-spectral domain, the mean vector and the covariance matrix of the noisy speech can be approximated by adding the mean vectors and the covariance matrices of speech and noise models:

$$\boldsymbol{\mu}_y = \boldsymbol{\mu}_x + g\boldsymbol{\mu}_n \tag{11.55}$$

$$\boldsymbol{\Sigma}_{yy} = \boldsymbol{\Sigma}_{xx} + g^2\boldsymbol{\Sigma}_{nn} \tag{11.56}$$

Model combination also requires an estimate of the current signal-to-noise ratio for calculation of the scaling factor g in Equations (11.55) and (11.56). In cases such as speech recognition, where the models are trained on cepstral features, the model parameters are first transformed from cepstral features into power spectral features before using the additive linear combination equations (11.55) and (11.56). Figure 11.15 illustrates the combination of a four-state left–right HMM of a speech signal with a two-state ergodic HMM of noise. Assuming that speech and noise are independent processes, each speech state must be combined with every possible noise state to give the noisy speech model.

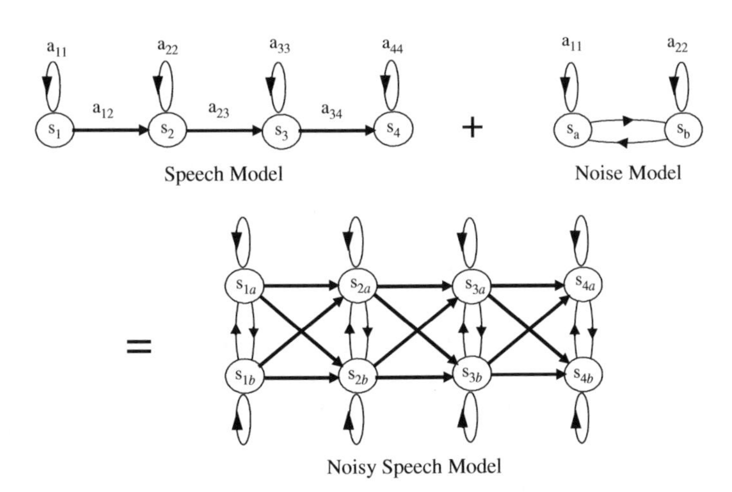

Figure 11.15 Outline configuration of HMM-based noisy speech recognition and enhancement. S_{ij} is a combination of the state i of speech with the state j of noise.

Note that in the combined model of Figure 11.15 as the number of states and the number of transitions from each state increases, compared with the individual constituent models, the state transition probability will also be affected. For example, the probability of a transition from state s_{1a} to s_{2b} is the probability that the speech will move from state 1 to state 2 and the probability that independently the noise will move from state a to state b.

11.6.5 Decomposition of State Sequences of Signal and Noise

The HMM-based state decomposition problem can be stated as follows: given a noisy signal and the HMMs of the signal and the noise processes, estimate the underlying states of the signal and the noise. HMM state decomposition can be obtained using the following method:

(a) Given the noisy signal and a set of combined signal and noise models, estimate the maximum-likelihood (ML) combined noisy HMM for the noisy signal.
(b) Obtain the ML state sequence of the noisy speech from the ML combined model.
(c) Extract the signal and noise states from the ML state sequence of the ML combined noisy signal model obtained in (b).

The ML state sequences provide the probability density functions for the signal and noise processes. The ML estimates of the speech and noise pdfs may then be used in Equation (11.45) to obtain a MAP estimate of the speech signal. Alternatively the mean spectral vectors of the speech and noise from the ML state sequences can be used to program a state-dependent Wiener filter as described in the next section.

11.6.6 HMM-Based Wiener Filters

The least mean square error Wiener filter is derived in Chapter 8. For a stationary signal $x(m)$, observed in an additive noise $n(m)$, the Wiener filter equations in the time and the frequency domains are derived as

$$\boldsymbol{w} = (\boldsymbol{R}_{xx} + \boldsymbol{R}_{nn})^{-1}\boldsymbol{r}_{xx} \tag{11.57}$$

and

$$W(f) = \frac{P_{XX}(f)}{P_{XX}(f) + P_{NN}(f)} \tag{11.58}$$

where $\boldsymbol{R}_{xx}$, $\boldsymbol{r}_{xx}$ and $P_{XX}(f)$ denote the autocorrelation matrix, the autocorrelation vector and the power-spectral functions respectively. The implementation of the Wiener filter, Equation (11.58), requires the signal and the noise power spectra. The power-spectral variables may be obtained from the ML states of the HMMs trained to model the power spectra of the signal and the noise. Figure 11.16 illustrates an implementation of HMM-based state-dependent Wiener filters. To implement the state-dependent Wiener filter, we need an estimate of the state sequences for the signal and the noise. In practice, for signals such as speech there are a number of HMMs: one HMM per word, phoneme, or any other elementary unit of the signal. In such cases it is necessary to classify the signal, so that the state-based Wiener filters are derived from the most likely HMM. Furthermore the noise process can also be modelled by an HMM. Assuming that there are V HMMs $\{\mathcal{M}_1, \ldots, \mathcal{M}_V\}$ for the signal process, and one HMM for the noise, the state-based Wiener filter can be implemented as follows:

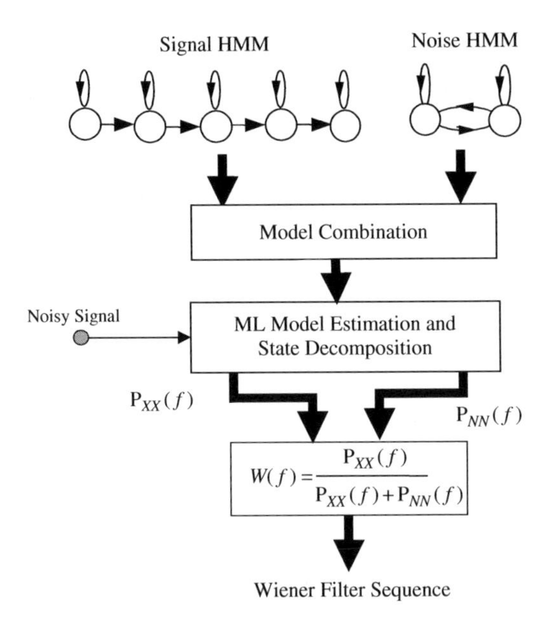

Figure 11.16 Illustration of HMMs with state-dependent Wiener filters.

Step 1: Combine the signal and noise models to form the noisy signal models.
Step 2: Given the noisy signal, and the set of combined noisy signal models, obtain the ML combined noisy signal model.
Step 3: From the ML combined model, obtain the ML state sequence of speech and noise.
Step 4: Use the ML estimate of the power spectra of the signal and the noise to program the Wiener filter equation (11.56).
Step 5: Use the state-dependent Wiener filters to filter the signal.

11.6.7 Modelling Noise Characteristics

The implicit assumption in using an HMM for noise is that noise statistics can be modelled by a Markovian chain of N different stationary processes. A stationary noise process can be modelled by a single-state HMM. For a non-stationary noise, a multi-state HMM can model the time variations of the noise process with a finite number of quasi-stationary states. In general, the number of states required to accurately model the noise depends on the non-stationary character of the noise.

An example of a non-stationary noise process is the impulsive noise of Figure 11.17. Figure 11.18 shows a two-state HMM of the impulsive noise sequence where the state S_0 models the 'off' periods between the impulses and the state S_1 models an impulse. In cases where each impulse has a well-defined temporal structure, it may be beneficial to use a multi-state HMM to model the pulse itself. HMMs are used for modelling impulsive noise and for channel equalisation, see Chapter 17.

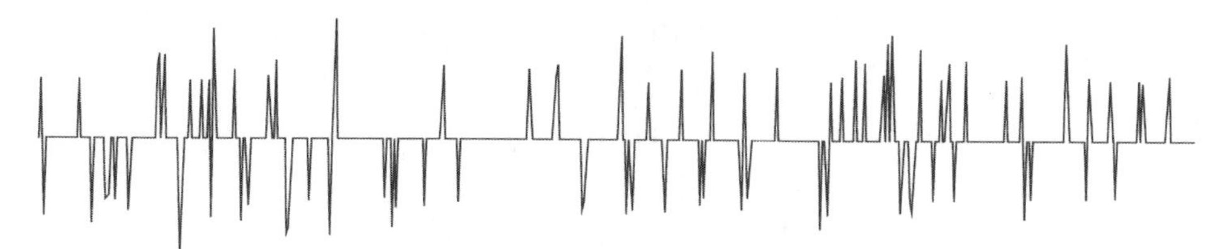

Figure 11.17 Impulsive noise.

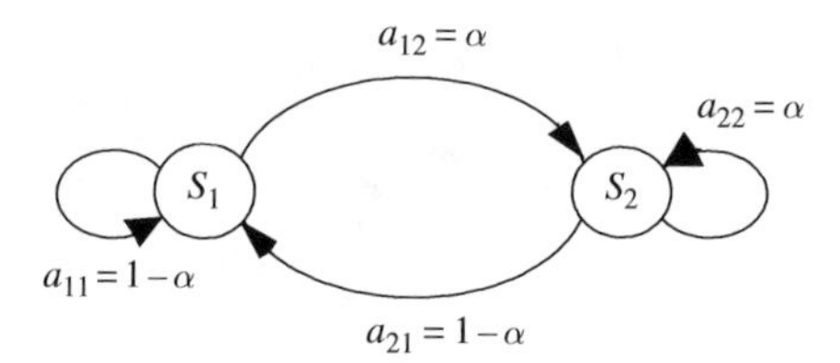

Figure 11.18 A binary-state model of an impulsive noise process.

11.7 Summary

HMMs provide a powerful method for the modelling of non-stationary processes such as speech, noise and time-varying channels. An HMM is a Bayesian finite-state process, with a Markovian state prior, and a state likelihood function that can be either a discrete density model or a continuous Gaussian pdf model. The Markovian prior models the time evolution of a non-stationary process with a chain of stationary sub-processes. The state observation likelihood models the space of the process within each state of the HMM.

In Section 11.3 we studied the Baum–Welch method for training the parameters of an HMM to model a given data set, and derived the forward–backward method for efficient calculation of the likelihood of an HMM given an observation signal. In Section 11.4 we considered the use of HMMs in signal classification and in decoding the underlying state sequence of a signal. The Viterbi algorithm is a computationally efficient method for estimation of the most likely sequence of an HMM. Given an unlabelled observation signal, decoding the underlying state sequence and labelling the observation with one of a number of candidate HMMs are accomplished using the Viterbi method. In Section 11.6 we considered the use of HMMs for MAP estimation of a signal observed in noise, and considered the use of HMMs in implementation of state-based Wiener filter sequences.

Further Reading

Bahl L.R., Jelinek F. and Mercer R.L. (1983) A Maximum Likelihood Approach to Continuous Speech Recognition. IEEE Trans. Pattern Analysis and Machine Intelligence, 5, pp. 179–190.

Bahl L.R., Brown P.F., de Souza P.V. and Mercer R.L. (1986) Maximum Mutual Information Estimation of Hidden Markov Model Parameters for Speech Recognition. IEEE Proc. Acoustics, Speech and Signal Processing, ICASSP-86, Tokyo, pp. 40–43.

Baum L.E. and Eagon J.E. (1967) An Inequality with Applications to Statistical Estimation for Probabilistic Functions of a Markov Process and to Models for Ecology. Bull. AMS, 73, pp. 360–363.

Baum L.E. and Petrie T. (1966) Statistical Inference for Probabilistic Functions of Finite State Markov Chains. Ann. Math. Stat. 37, pp. 1554–1563.

Baum L.E., Petrie T., Soules G. and Weiss N. (1970) A Maximisation Technique Occurring in the Statistical Analysis of Probabilistic Functions of Markov Chains. Ann. Math. Stat., 41, pp. 164–171.

Conner P.N. (1993) Hidden Markov Model with Improved Observation and Duration Modelling. PhD Thesis, University of East Anglia, England.

Epharaim Y., Malah D. and Juang B.H.(1989) On Application of Hidden Markov Models for Enhancing Noisy Speech. IEEE Trans. Acoustics Speech and Signal Processing, 37(12), pp. 1846–1856, Dec.

Forney G.D. (1973) The Viterbi Algorithm. Proc. IEEE, 61, pp. 268–278.

Gales M.J.F. and Young S.J. (1992) An Improved Approach to the Hidden Markov Model Decomposition of Speech and Noise. Proc. IEEE, Int. Conf. on Acoust., Speech, Signal Processing, ICASSP-92, pp. 233–236.

Gales M.J.F. and Young S.J. (1993) HMM Recognition in Noise using Parallel Model Combination. Eurospeech-93, pp. 837–840.

Huang X.D. and Jack M.A. (1989) Unified Techniques for Vector Quantisation and Hidden Markov Modelling using Semi-Continuous Models. IEEE Proc. Acoustics, Speech and Signal Processing, ICASSP-89, Glasgow, pp. 639–642.

Huang X.D., Ariki Y. and Jack M.A. (1990) Hidden Markov Models for Speech Recognition. Edinburgh University Press, Edinburgh.

Jelinek F. and Mercer R. (1980) Interpolated Estimation of Markov Source Parameters from Sparse Data. Proc. of the Workshop on Pattern Recognition in Practice. North-Holland, Amsterdam.

Jelinek F. (1976) Continuous Speech Recognition by Statistical Methods. Proc. of IEEE, 64, pp. 532–556.

Juang B.H. (1984) On the Hidden Markov Model and Dynamic Time Warping for Speech Recognition – A unified Overview. AT&T Technical J., 63, pp. 1213–1243.

Juang B.H. (1985) Maximum-Likelihood Estimation for Mixture Multi-Variate Stochastic Observations of Markov Chain. AT&T Bell Laboratories Tech J., 64, pp. 1235–1249.

Kullback S. and Leibler R.A. (1951) On Information and Sufficiency. Ann. Math. Stat., 22, pp. 79–86.

Lee K.F. (1989) Automatic Speech Recognition: the Development of SPHINX System. Kluwer Academic Publishers, Boston, MA.

Lee K.F. (1989) Hidden Markov Model: Past, Present and Future. Eurospeech-89, Paris.

Liporace L.R. (1982) Maximum Likelihood Estimation for Multi-Variate Observations of Markov Sources. IEEE Trans. IT, IT-28, pp. 729–734.

Markov A.A. (1913) An Example of Statistical Investigation in the Text of *Eugen Onyegin* Illustrating Coupling of Tests in Chains. Proc. Acad. Sci. St Petersburg VI Ser., 7, pp. 153–162.

Milner B.P. (1995) Speech Recognition in Adverse Environments, PhD Thesis, University of East Anglia, England.

Peterie T. (1969) Probabilistic Functions of Finite State Markov Chains. Ann. Math. Stat., 40, pp. 97–115.

Rabiner L.R. and Juang B.H. (1986) An Introduction to Hidden Markov Models. IEEE ASSP. Magazine, pp. 4–16.

Rabiner L.R. and Juang B.H. (1993) Fundamentals of Speech Recognition. Prentice-Hall, Englewood Cliffs, NJ.

Rabiner L.R., Juang B.H., Levinson S.E. and Sondhi M.M. (1985) Recognition of Isolated Digits using Hidden Markov Models with Continuous Mixture Densities. AT&T Technical Journal, 64, pp. 1211–1234.

Varga A. and Moore R.K. (1990) Hidden Markov Model Decomposition of Speech and Noise. Proc. IEEE Int. Conf. on Acoust., Speech, Signal Processing, pp. 845–848.

Viterbi A.J. (1967) Error Bounds for Convolutional Codes and an Asymptotically Optimum Decoding Algorithm. IEEE Trans. on Information Theory, IT-13, pp. 260–226.

Young S., Evermann G., Kershaw D., Moore G., Odell J., Ollason D., Dan P., Valtchev V., and Woodland P. The Hidden Markov Model (HTK) Book. Cambridge University Engineering Department.

12 Eigenvector Analysis, Principal Component Analysis and Independent Component Analysis

Eigen is a German word that translates as 'own', 'peculiar to', 'characteristic' or 'individual'. To understand eigenvectors we need to understand the basic operational function of a matrix. In general, linear matrices change the directions and the magnitudes of the vectors on which they operate. Eigenvectors of a matrix are those vectors whose directions are unaffected by a transformation through the matrix. It is an interesting property of a matrix that if a vector is repeatedly passed through a matrix it eventually aligns itself with the most significant eigenvector of the matrix.

The eigenanalysis of a matrix finds a set of characteristic orthonormal eigenvectors and their magnitudes called eigenvalues. A matrix can be decomposed and expressed in terms of its eigenvectors and eigenvalues. Similarly, matrix operations can be expressed in terms of operations on the eigenvectors and eigenvalues.

Eigenanalysis is useful in applications such as the diagonalisation of correlation matrices, adaptive filtering, radar signal processing, feature extraction, pattern recognition, signal coding, model order estimation, noise estimation, and separation of mixed biomedical or communication signals.

A major application of eigenanalysis is in principal component analysis (PCA). PCA is widely used for feature extraction and dimension reduction by disposing of those orthogonal dimensions or features that have insignificant variance or very low signal-to-noise ratio. PCA is essentially eigenanalysis of the covariance matrix of a random process. PCA is the optimal feature transform for Gaussian processes which do not possess higher than second-order statistics (i.e. covariance statistic). PCA allows the transformation and representation of a signal in terms of the coefficients of a set of orthonormal eigenvectors, such that the principal components (i.e. the most significant components) correspond to the principal dimensions that have the largest eigenvalues (or variances) and are most significant.

Independent component analysis (ICA) is an extension of the PCA for signals that have higher (than second) order statistics. ICA usually involves the determination of a set of parameters that diagonalise

Multimedia Signal Processing: Theory and Applications in Speech, Music and Communications Saeed V. Vaseghi

the second-order (covariance) and fourth-order (kurtosis) statistics of the signal. ICA is often used as an add-on to PCA for the processing of non-Gaussian signals such as speech, image and biomedical signals. ICA is particularly useful for separation of mixed signals in multi-source multi-sensor medical systems and multi-input multi-output (MIMO) communication systems.

12.1 Introduction – Linear Systems and Eigenanalysis

A common requirement in signal processing is the linear transformation of an M-dimensional signal vector $\boldsymbol{x} = [x_1, x_2, x_M]^{\mathrm{T}}$ to an N-dimensional output signal vector $\boldsymbol{y} = [y_1, y_2, y_N]^{\mathrm{T}}$, via an $N \times M$ dimensional matrix system $\boldsymbol{A}$, as

$$\boldsymbol{y} = \boldsymbol{A}\boldsymbol{x} \tag{12.1}$$

The output vector $\boldsymbol{y}$ can be expressed as a weighted combination of the column vectors of $\boldsymbol{A} = [\boldsymbol{a}_1, \boldsymbol{a}_2, \cdots, \boldsymbol{a}_M]$ as

$$\boldsymbol{y} = \boldsymbol{a}_1 x_1 + \boldsymbol{a}_2 x_2 + \cdots + \mathrm{A}_M x_M \tag{12.2}$$

where $\boldsymbol{a}_i = [a_{i1}, \ldots, a_{iN}]^{\mathrm{T}}$ is the N-dimensional i^{th} column of $\boldsymbol{A}$.

Note that all linear filtering and signal transformation methods can be cast in the form of Equation (12.1). The matrix $\boldsymbol{A}$ is the system that transforms the input signal vector $\boldsymbol{x}$ to the output signal vector $\boldsymbol{y}$, hence the analysis of the structure of $\boldsymbol{A}$ and development of methods for efficient representation and application of $\boldsymbol{A}$ are of particular interest in signal processing and system analysis.

Eigenanalysis is a method of representation of a linear system, such as a covariance matrix, in terms of a set of orthonormal eigenvectors and the corresponding eigenvalues. The eigenvalues represent the variance or power of the process along the corresponding eigenvectors. The bigger an eigenvalue the more significant the corresponding eigenvector. The core idea is that the orthonormal eigen components lend themselves to independent processing and easier manipulation and interpretation.

The main advantages of eigenanalysis can be listed as follows:

(1) The eigenvectors of a matrix form a set of orthonormal basis vectors in terms of which the matrix can be expressed and applied.
(2) Eigenanalysis is used in principal component analysis for feature extraction in coding and pattern recognition in applications such as speech, image processing and radar signal processing.
(3) The eigenvectors of the covariance matrix of a process can be used to transform and pre-whiten the process and diagonalise its covariance matrix. Pre-whitening is useful for many applications such as for faster convergence of adaptive filters in echo cancellation and for ease of calculation of the inverse of covariance matrix often required in probability estimation.
(4) Eigenvalues can be used for estimation of the dimensionality of a linear system. It is possible to reduce the dimensions of a matrix or a process to its most significant eigenvalues and eigenvectors.
(5) For white noise, when the dimensions of the covariance matrix of a random process exceeds the required dimension for modelling the signal (e.g. speech or image), then the smallest eigenvalues correspond to the noise power.

Example 12.1 *Eigenvectors of a 'flip' matrix*
Figure 12.1 illustrates a very simple matrix transformation that flips a vector $[x\, y]$ about the y-axis: the transformation changes the sign of the x-component of the vector. It is clear that the y-axis, $[0y]^{\mathrm{T}}$, is an eigenvector of this transformation as the direction of the vector $[0y]$ is not affected by the transformation.

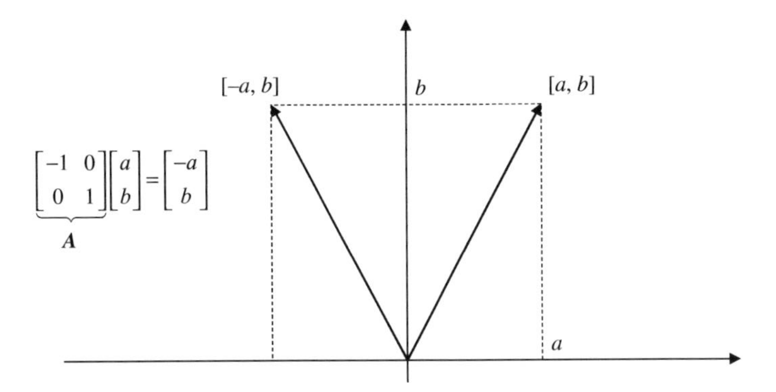

Figure 12.1 Illustration of transformation of a vector. In this simple example the vertical component of the vector is unaffected hence the vector $[0 \quad 1]^T$ is an eigenvector of the matrix $\boldsymbol{A}$.

Example 12.2 *Eigenvectors of* $\mathbf{A} = \begin{bmatrix} 1 & 0.5 \\ 0.5 & 1 \end{bmatrix}$ *and transformation of a circle by* $\mathbf{A}$

Figure 12.2 shows the result of transformation of a circle by a matrix $\boldsymbol{A}$. The transformation changes the shape of the circle and also rotates the points on the circle. For example, after transformation, the point $[x = 0 y = 1]$ on the circle becomes $[x = 0.5 y = 1]$. The eigenvectors of the matrix are superimposed on the diagram. Note that after the transformation the rotated ellipse-shaped curve is positioned with its principal directions in the direction of the eigenvectors of the transformation matrix $\boldsymbol{A}$ as illustrated.

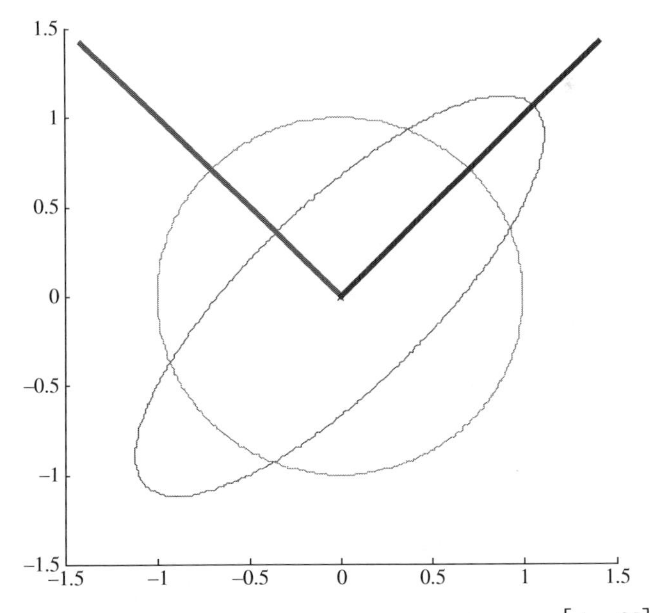

Figure 12.2 The circle is transformed (into an ellipse shape) by the matrix $A = \begin{bmatrix} 1 & 0.5 \\ 0.5 & 1 \end{bmatrix}$. The eigenvectors of the matrix are the principal dimensions of the transformed curve. The eigenvectors are $[0.7071 \quad 0.7071]^T$ and $[-0.7071 \quad 0.7071]^T$.

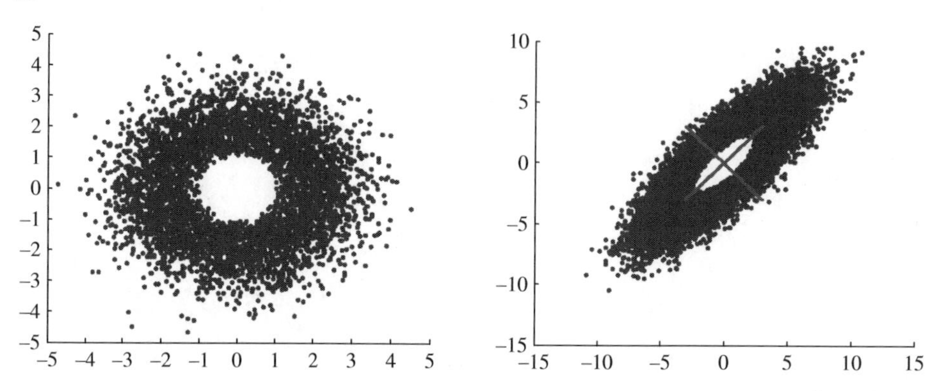

Figure 12.3 (a) A two-dimensional random process, (b) after transformation by the matrix $A = \begin{bmatrix} 1 & 2 \\ 2 & 1 \end{bmatrix}$. Note in (b) that the eigenvectors of the matrix A form the new orthogonal dimensions of the transformed process.

Example 12.3 *Eigenvectors of* $A = \begin{bmatrix} 1 & 2 \\ 2 & 1 \end{bmatrix}$ *and transformation of a random process by matrix* $\boldsymbol{A}$

Figure 12.3(a) shows the scatter diagram for a two-dimensional random process. Figure 12.3(b) shows the process after a linear transformation by a matrix $\boldsymbol{A}$. The values of the individual coefficients, a_{ij}, of the matrix $\boldsymbol{A}$ are also shown. In Figure 12.3(b) the eigenvectors of the transformation matrix $\boldsymbol{A}$ are superimposed on the scatter diagram of transformed process. Note that the eigenvectors of the matrix form the new orthogonal principal dimensions of the process after transformation.

Example 12.4 *Eigenvectors of a rotation matrix* $\boldsymbol{R}$

As a further example consider a rotation matrix $\boldsymbol{R}$ defined as

$$\boldsymbol{R} = \begin{bmatrix} \cos(\theta) & \sin(\theta) \\ -\sin(\theta) & \cos(\theta) \end{bmatrix} \tag{12.3}$$

A rotation matrix $\boldsymbol{R}$ can be used to rotate a vector by an angle of θ. The rotation angle can be selected such that at each rotation one vector element is set to zero; this property is used in Givens transformation methods to diagonalise the covariance matrix of a process. Figure 12.4 shows an ellipse rotated anti-clockwise by a rotation matrix with an angle of $\theta = -\pi/12$. Note that since the rotation matrix rotates the direction of all real-valued vectors therefore it does not possess a real-valued eigenvector whose direction would be unaffected by the transformation. The eigenvectors of the rotation matrix are complex valued.

Matlab function Circle( )

Generates a circle and linearly transforms it. Eigenanalysis is used to orthogonalise the transformed circle and change its shape back to a circle! The results before and after transformation are displayed. It is shown that the eigenvectors of the transform form the orthogonal principal dimensions of the transformed distribution.

Matlab function GaussianDoughnut( )
Generates a random signal with a two-dimensional doughnut-shaped distribution. The signal is then linearly transformed. Eigenvalue analysis shows that the eigenvectors of the transform form the orthogonal principal dimensions of the transformed distribution.

Matlab function MatrixRotation( )
Generates an ellipse and rotates it by an angle of theta using a rotation matrix. Eigen-analysis is used to show that the eigenvectors are aligned with the principle orthogonal dimensions of the rotated signal. The eigenvectors are used to turn the rotated ellipse back.

A Note on the Terms Orthogonal, Orthonormal and Independent

It's useful here to define the terms orthogonal, orthonormal and independent. *Orthogonal* implies that two vectors are geometrically perpendicular or statistically uncorrelated in terms of the second-order statistics. *Orthonormal* vectors are orthogonal and have magnitudes of one. It should be noted that there is a difference between the terms *uncorrelated* and *independent*. Independence is a more strict concept and requires the signals to be probabilistically independent and hence uncorrelated not only in the second-order statistics but also in *all* the higher-order statistics, whereas in contrast the terms uncorrelated and orthogonal commonly refer to second-order statistics and essentially imply that two vectors have zero correlation or that a vector process has a diagonal covariance matrix. This is elaborated later in this chapter in the study of independent component analysis.

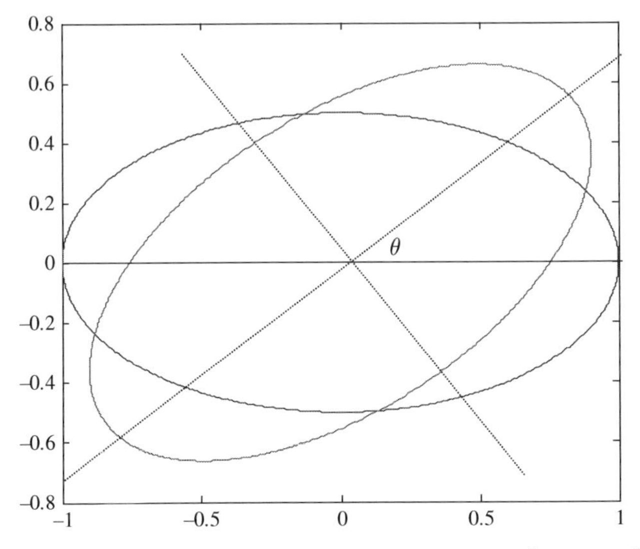

Figure 12.4 An ellipse is rotated by the rotation matrix $A = \begin{bmatrix} \cos(\theta) & -\sin(\theta) \\ \sin(\theta) & \cos(\theta) \end{bmatrix}$.

12.2 Eigenvectors and Eigenvalues

Eigenvectors and eigenvalues are used for analysis of square ($N \times N$) matrices such as covariance matrices. For the more general case of an $N \times M$ dimensional matrix the singular value decomposition (SVD) method is used.

In general, two sets of eigenvectors are defined for a matrix; the right eigenvectors and the left eigenvectors. The set of right eigenvectors of a matrix are defined as

$$\boldsymbol{A}\boldsymbol{v}_i = \lambda_i \boldsymbol{v}_i \quad i = 1, \ldots, N \tag{12.4}$$

where each eigenvector has a magnitude or norm of 1 (i.e. $\sum_{j=1}^{N} v_{ij}^2 = 1$).
Equation (12.4) reveals the main defining characteristic of an eigenvector: when an eigenvector $\boldsymbol{v}_i$ of a matrix $\boldsymbol{A}$ is transformed through the matrix, the direction of the eigenvector is unchanged and only the magnitude of the eigenvector is affected, this explains the name *eigen* meaning 'peculiar to', or 'characteristic' or 'own' (of a matrix).

As a simple example of an eigenvector, note that the differential operation only affects the magnitude of an exponential, $\frac{d}{dt}e^{\lambda t} = \lambda e^{\lambda t}$, hence the exponential function is an eigenvector of the differential operator.

From Equation (12.4) we can write (by adding the equations for different eigenvectors and values)

$$\boldsymbol{A}\boldsymbol{v}_1 + \cdots + \boldsymbol{A}\boldsymbol{v}_N = \lambda_1 \boldsymbol{v}_1 + \cdots + \lambda_N \boldsymbol{v}_N \tag{12.5}$$

Hence we have

$$\boldsymbol{A}[\boldsymbol{v}_1, \boldsymbol{v}_2 \cdots \boldsymbol{v}_N] = [\boldsymbol{v}_1, \boldsymbol{v}_2 \cdots \boldsymbol{v}_N] \begin{bmatrix} \lambda_1 & 0 & \cdots & 0 \\ 0 & \lambda_2 & \cdots & 0 \\ \vdots & \vdots & \ddots & \vdots \\ 0 & 0 & \cdots & \lambda_N \end{bmatrix} \tag{12.6}$$

or in compact matrix notation

$$\boldsymbol{A}\boldsymbol{V}_{\mathrm{R}} = \boldsymbol{V}_{\mathrm{R}}\boldsymbol{\Lambda} \tag{12.7}$$

where the matrix $\boldsymbol{V}_{\mathrm{R}}$ is the set of *right* eigenvectors and $\boldsymbol{\Lambda}$ is the diagonal eigenvalue matrix. From Equation (12.7) it follows that (by post-multiplying both sides by $\boldsymbol{V}_{\mathrm{R}}^{-1}$)

$$\boldsymbol{A} = \boldsymbol{V}_{\mathrm{R}}\boldsymbol{\Lambda}\boldsymbol{V}_{\mathrm{R}}^{-1} \tag{12.8}$$

Similarly, we can define a set of *left* eigenvectors as

$$\boldsymbol{V}_{\mathrm{L}}\,\boldsymbol{A} = \boldsymbol{\Lambda}\,\boldsymbol{V}_{\mathrm{L}} \tag{12.9}$$

Pre-multiplying Equation (12.7) by $\boldsymbol{V}_{\mathrm{L}}$ and post-multiplying Equation (12.9) by $\boldsymbol{V}_{\mathrm{R}}$ we obtain

$$\boldsymbol{V}_{\mathrm{L}}\,\boldsymbol{A}\boldsymbol{V}_{\mathrm{R}} = \boldsymbol{V}_{\mathrm{L}}\,\boldsymbol{V}_{\mathrm{R}}\boldsymbol{\Lambda} \tag{12.10}$$

$$\boldsymbol{V}_{\mathrm{L}}\,\boldsymbol{A}\boldsymbol{V}_{\mathrm{R}} = \boldsymbol{\Lambda}\,\boldsymbol{V}_{\mathrm{L}}\,\boldsymbol{V}_{\mathrm{R}} \tag{12.11}$$

By equating the right-hand sides of Equations (12.10) and (12.11), it follows that

$$\boldsymbol{\Lambda}\,\boldsymbol{V}_{\mathrm{L}}\,\boldsymbol{V}_{\mathrm{R}} = \boldsymbol{V}_{\mathrm{L}}\,\boldsymbol{V}_{\mathrm{R}}\boldsymbol{\Lambda} \tag{12.12}$$

Equation (12.12) is in the form of $\Lambda\, \boldsymbol{C} = \boldsymbol{C}\Lambda$ where $\boldsymbol{C} = \boldsymbol{V}_{\mathrm{L}}\ \boldsymbol{V}_{\mathrm{R}}$. Hence it follows that since Λ is diagonal then for Equation (12.12) to be true in the general case, $\boldsymbol{V}_{\mathrm{L}}\ \boldsymbol{V}_{\mathrm{R}}$ must also be the diagonal identity matrix. It follows that $\boldsymbol{V}_{\mathrm{L}}$ and $\boldsymbol{V}_{\mathrm{R}}$ are orthonormal:

$$\boldsymbol{V}_{\mathrm{R}}\ \boldsymbol{V}_{\mathrm{L}} = \mathbf{I} \tag{12.13}$$

Note from Equation (12.13) that $\boldsymbol{V}_{\mathrm{L}} = \boldsymbol{V}_{\mathrm{R}}^{-1}$ and $\boldsymbol{V}_{\mathrm{R}} = \boldsymbol{V}_{\mathrm{L}}^{-1}$. Post-multiplying both sides of Equation (12.7) by $\boldsymbol{V}_{\mathrm{L}}$ we have

$$\boldsymbol{A}\boldsymbol{V}_{\mathrm{R}}\ \boldsymbol{V}_{\mathrm{L}} = \boldsymbol{V}_{\mathrm{R}}\Lambda\boldsymbol{V}_{\mathrm{L}} \tag{12.14}$$

Using Equation (12.13) in (12.14) we have

$$\boldsymbol{A} = \boldsymbol{V}_{\mathrm{R}}\Lambda\ \boldsymbol{V}_{\mathrm{L}} \tag{12.15}$$

Note that for a symmetric matrix, such as an autocovariance matrix, $\boldsymbol{V}_{\mathrm{L}} = \boldsymbol{V}_{\mathrm{R}}^{\mathrm{T}} = \boldsymbol{V}^{\mathrm{T}}$. Hence for a symmetric matrix we have

$$\boldsymbol{A} = \boldsymbol{V}\ \Lambda\ \boldsymbol{V}^{\mathrm{T}} \tag{12.16}$$

Note that the matrix $\boldsymbol{A}$ can be described in terms of a summation of a set of orthogonal sub-matrices formed from multiplication of the left and right eigenvectors, $\nu_i\nu_i^{\mathrm{T}}$, as

$$\boldsymbol{A} = \lambda_1\nu_1\nu_1^{\mathrm{T}} + \lambda_2\nu_2\nu_2^{\mathrm{T}} + \cdots + \lambda_N\nu_N\nu_N^{\mathrm{T}} \tag{12.17}$$

where $\left(\nu_i\nu_i^{\mathrm{T}}\right)\left(\nu_{\mathrm{j}}\nu_{\mathrm{j}}^{\mathrm{T}}\right) = 0$ for $i \neq \mathrm{j}$. Note that

$$\boldsymbol{A}\ \nu_1\nu_1^{\mathrm{T}} = \lambda_1\nu_1\nu_1^{\mathrm{T}} \tag{12.18}$$

Hence $\nu_1\nu_1^{\mathrm{T}}$ may be called an *eigenmatrix* of $\boldsymbol{A}$. Hence in Equation (12.17) the matrix $\boldsymbol{A}$ is expressed in terms of its eigen matrices.

To summarise so far: the determination of the eigenvectors and eigenvalues of a matrix is important in signal processing for matrix diagonalisation. Each eigenvector is paired with a corresponding eigenvalue. Mathematically, two different kinds of eigenvectors need to be distinguished: left eigenvectors and right eigenvectors. However, for many problems it is sufficient to consider only right eigenvectors and their transpose.

Figure 12.5 shows two eigenvectors of a 2×2 matrix. Note that as illustrated the eigenvectors have unit magnitude and are orthogonal (perpendicular) to each other.

12.2.1 Matrix Spectral Theorem

In Equation (12.17) a matrix $\boldsymbol{A}$ is expressed in terms of a set of orthonormal 'eigen matrices'. The spectral theorem states that a linear transformation of a vector $\boldsymbol{x}$ by a matrix $\boldsymbol{A}$ can be expressed in terms of the sum of a set of orthogonal (subspace) transformations by the eigen matrices as

$$\boldsymbol{A}\ \boldsymbol{x} = \lambda_1\nu_1\nu_1{}^{\mathrm{T}}\boldsymbol{x} + \lambda_2\nu_2\nu_2{}^{\mathrm{T}}\boldsymbol{x} + \cdots + \lambda_N\nu_N\nu_N{}^{\mathrm{T}}\boldsymbol{x} \tag{12.19}$$

In Equation (12.19) the vector $\boldsymbol{x}$ is decomposed in terms of its projection onto orthogonal subspaces (eigen matrices) of $\boldsymbol{A}$.

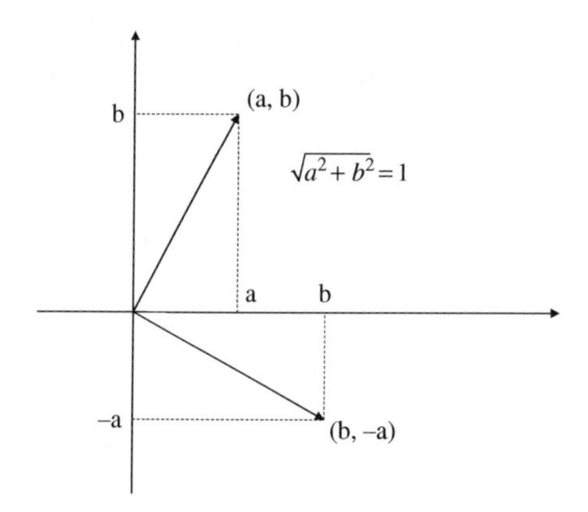

Figure 12.5 A depiction of two orthogonal eigenvectors of a 2×2 matrix.

Example 12.5 *Transforming an input vector x into an eigenvector of a matrix* **A**
Consider a full rank $N \times N$ matrix $\boldsymbol{A}$. The N eigenvectors ν_i of this matrix form the N orthogonal dimensions of an N-dimensional space. Hence any N-dimensional vector $\boldsymbol{x}$ can be expressed as a linear combination of the eigenvectors of $\boldsymbol{A}$ as

$$\boldsymbol{x} = b_1 \nu_1 + b_2 \nu_2 + \cdots + b_N \nu_N \tag{12.20}$$

where b_i are the combination weights. Now the transformation of vector $\boldsymbol{x}$ by matrix $\boldsymbol{A}$ can be expressed as

$$\begin{aligned} \boldsymbol{A}\boldsymbol{x} &= b_1 \boldsymbol{A}\boldsymbol{\nu}_1 + b_2 \boldsymbol{A}\boldsymbol{\nu}_2 + \cdots + b_N \boldsymbol{A}\boldsymbol{\nu}_N \\ &= b_1 \lambda_1 \boldsymbol{\nu}_1 + b_2 \lambda_2 \boldsymbol{\nu}_2 + \cdots + b_N \lambda_N \boldsymbol{\nu}_N \\ &= \lambda_1 \left(b_1 \boldsymbol{\nu}_1 + \frac{b_2 \lambda_2}{\lambda_1} \boldsymbol{\nu}_2 + \cdots + \frac{b_N \lambda_N}{\lambda_1} \boldsymbol{\nu}_N \right) \end{aligned} \tag{12.21}$$

Passing the vector $\boldsymbol{x}$ through the matrix $\boldsymbol{A}$ a large number n of times results in

$$\boldsymbol{A}^n \boldsymbol{x} = \lambda_1^n \left(b_1 \boldsymbol{\nu}_1 + b_2 \frac{\lambda_2^n}{\lambda_1^n} \boldsymbol{\nu}_2 + \cdots + b_N \frac{\lambda_N^n}{\lambda_1^n} \boldsymbol{\nu}_N \right) \approx \lambda_1^n b_1 \boldsymbol{\nu}_1 \tag{12.22}$$

where it is assumed that λ_1 is the largest eigenvalue. Note that as the vector $\boldsymbol{x}$ is passed through the matrix $\boldsymbol{A}$ for a large number of times it turns into the eigenvector of $\boldsymbol{A}$ associated with the largest eigenvalue. Note that the approximation in Equation (12.21) does not hold if the signal has two or more eigenvalues with equal magnitudes.

12.2.2 Computation of Eigenvalues and Eigenvectors

An eigenvalue λ_i of a matrix $\boldsymbol{A}$ can be obtained from the characteristic polynomial of $\boldsymbol{A}$ as explained in the following. Rearranging Equation (12.4) we obtain

$$(\mathbf{A} - \lambda_i \mathbf{I})\, \boldsymbol{\nu}_i = 0 \tag{12.23}$$

Hence we have

$$\det(\boldsymbol{A} - \lambda_i \mathbf{I}) = 0 \tag{12.24}$$

The determinant, $\det(\cdot)$ of the matrix equation gives the characteristic polynomial in terms of the powers of the eigenvalues. For example consider the following small 2×2 matrix

$$\boldsymbol{A} = \begin{pmatrix} a & b \\ c & d \end{pmatrix} \tag{12.25}$$

An eigenvalue of this matrix is obtained from

$$\det \begin{pmatrix} a - \lambda & b \\ c & d - \lambda \end{pmatrix} = 0 \tag{12.26}$$

or

$$(a - \lambda)(d - \lambda) - bc = 0 \tag{12.27}$$

The characteristic polynomial equation of the eigenmatrix is given by

$$\lambda^2 - (a + d)\lambda + (ad - bc) = 0 \tag{12.28}$$

For large matrices the eigenvalues are obtained using an iterative root-finding numerical solution.

Once the eigenvalues of the matrix are obtained, the corresponding eigenvectors can be obtained by solving the following sets of linear equations

$$(\boldsymbol{A} - \lambda_i \mathbf{I})\boldsymbol{\nu}_i = 0 \quad i = 1, \ldots, N \tag{12.29}$$

where for each eigenvalue of λ_i we have a set of linear equations with ν_i as the vector that contains the unknown values to be determined.

12.3 Principal Component Analysis (PCA)

Principal component analysis (PCA), Figure 12.6, is a popular data analysis method used for such purposes as signal compression, feature extraction, model order estimation and signal and noise

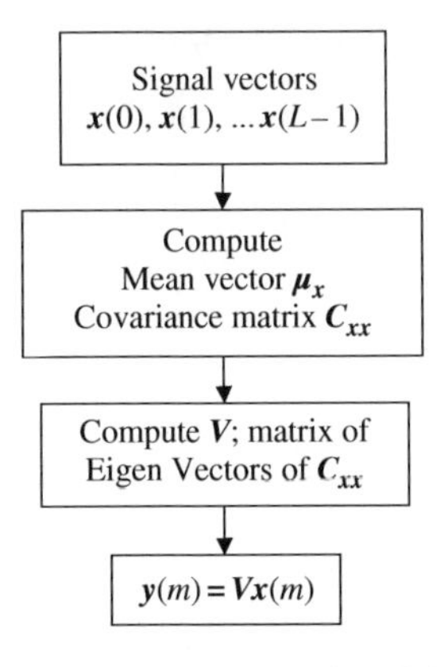

Figure 12.6 A block diagram illustration of the PCA process.

separation. PCA is also called the *Karhunen–Loève transform* (or KLT, named after Kari Karhunen and Michael Loève) or the *Hotelling transform* (after Harold Hotelling).

The principal components (PCs) of a signal process are a set of orthogonal components obtained from an eigenanalysis of its covariance matrix. The purpose of PCA is to transform a signal, such as speech or image, and represent it in terms of an orthogonal set of principal components.

The PCA process will reveal the significant uncorrelated structures of the signal process, in the sense that the most important (principal) components of the signal will be along the dimensions with the largest covariance values. Furthermore, as the PC dimensions are orthogonal, the signal along each dimension can be processed independently. The PCA achieves the following:

(1) PCA transforms a process such that the data are represented along a new set of orthogonal dimensions with a diagonal covariance matrix.
(2) The PC coefficient with the largest variance is the first principal component, the PC coefficient with the second largest variance is the second most important and so on.
(3) The transformed PCs are uncorrelated and can be processed independently.
(4) For an over-modelled process, i.e. where the dimension of the correlation matrix (M) is bigger than the actual number of significant components of the signal (N), the last $N-M$ principal components represent the noise variances.

12.3.1 Computation of PCA

Assume we have L samples of an N-dimensional vector process $\boldsymbol{x}(0), \boldsymbol{x}(1), \ldots, \boldsymbol{x}(L-1)$. These vectors may be speech frames or sub-blocks of an image. The first step in PCA analysis is to obtain an estimate of the mean of the signal vector as

$$\mu = \frac{1}{L} \sum_{m=0}^{L-1} \boldsymbol{x}(m) \tag{12.30}$$

An estimate of the covariance matrix of the signal vector is then obtained as

$$\boldsymbol{C}_{xx} = \frac{1}{L} \sum_{m=0}^{L-1} (\boldsymbol{x}(m) - \boldsymbol{\mu})(\boldsymbol{x}(m) - \boldsymbol{\mu})^T \tag{12.31}$$

The next step in PCA is an eigenanalysis of the covariance matrix $\boldsymbol{C}_{xx}$. The covariance matrix is expressed in terms of its eigenvectors and eigenvalues as

$$\boldsymbol{C}_{xx} = \boldsymbol{V}\,\Lambda \boldsymbol{V}^{\mathrm{T}} = \boldsymbol{V}^{\mathrm{T}} \Lambda \boldsymbol{V} \tag{12.32}$$

Note that since the covariance matrix is real and symmetric its eigenvectors are real and orthonormal. From Equation (12.32) it can be shown that the transformation of the vector process $\boldsymbol{x}$ by the matrix of eigenvectors $\boldsymbol{V}$ diagonalises its covariance matrix as

$$\boldsymbol{y} = \boldsymbol{V}\boldsymbol{x} \tag{12.33}$$

$$\boldsymbol{C}_{yy} = E\left(\boldsymbol{V}\boldsymbol{x}\boldsymbol{x}^{\mathrm{T}}\boldsymbol{V}^{\mathrm{T}}\right) = \boldsymbol{V}\boldsymbol{C}_{xx}\boldsymbol{V}^{\mathrm{T}} = \boldsymbol{V}\boldsymbol{V}^{\mathrm{T}}\Lambda\boldsymbol{V}\boldsymbol{V}^{\mathrm{T}} = \Lambda \tag{12.34}$$

where the operator $E(\cdot)$ represents the expectation operation and $\boldsymbol{V}^{\mathrm{T}}\boldsymbol{V} = \boldsymbol{V}\boldsymbol{V}^{\mathrm{T}} = \mathbf{I}$.

If we wish to '*sphere a process*', i.e. to diagonalise *and* also normalise the covariance matrix of the process, then we need to transform the signal as

$$\boldsymbol{y} = \boldsymbol{C}_{xx}^{-1/2}\boldsymbol{x} = \left(\Lambda^{-1/2}\boldsymbol{V}\right)\boldsymbol{x} \tag{12.35}$$

$$\begin{aligned} C_{yy} &= E\left(\Lambda^{-1/2} V x x^{\mathrm{T}} V^{\mathrm{T}} \Lambda^{-1/2}\right) = \Lambda^{-1/2} V C_{xx} V^{\mathrm{T}} \Lambda^{-1/2} \\ &= \Lambda^{-1/2} V V^{\mathrm{T}} \ \Lambda V V^{\mathrm{T}} \Lambda^{-1/2} = \Lambda^{-1/2} \Lambda \Lambda^{-1/2} = \mathbf{I} \end{aligned} \tag{12.36}$$

As mentioned, the process of diagonalising and equalising the covariance matrix of a process is known as sphereing.

> **Matlab function PCA( )**
> Demonstrates PCA analysis of the sum of two independent signals. The signals are generated independently and then mixed. Eigenanalysis is used to calculate their eigenvalues and eigenvectors and to orthogonalise the signals. The scatter diagrams of the signal before and after orthogonalisation are plotted as are the eigenvalues, original signals, the mixed signals and demixed signals.

12.3.2 PCA Analysis of Images: Eigen-Image Representation

PCA analysis can be used to decompose an image into a set of orthogonal principal component images also known as eigen-images. Eigen-images can be used for image coding, image denoising, or as features for image classification. For example an image sub-block can be reconstructed as a weighted function of the principal eigen-images as

To obtain the set of eigen-images $[E_{\mathrm{ij}}]$ for a given image A of size $r_0 \times c_0$ pixels, the image is first divided into L sub-images (sub-blocks) A_k of size $r \times c$ (typically 8×8 or 16×16). The mean of the sub-images is obtained as

$$A_{\mathrm{mean}} = \frac{1}{L} \sum_{k=0}^{L-1} A_k \tag{12.37}$$

The mean image is then removed from each sub-image as

$$\bar{A}_k = A_k - A_{\mathrm{mean}} \tag{12.38}$$

An $r \times r$ covariance matrix for the rows of the sub-images is obtained

$$C_r = \frac{1}{L} \sum_{k=0}^{L-1} \bar{A}_k \bar{A}_k^{\mathrm{T}} \tag{12.39}$$

Similarly, a $c \times c$ covariance matrix for the columns of the sub-images is obtained

$$C_c = \frac{1}{L} \sum_{k=0}^{L-1} \bar{A}_k^{\mathrm{T}} \bar{A}_k \tag{12.40}$$

The row and column covariance matrices are then subjected to eigenanalysis to yield r row eigenvectors $[r_i;\ 1 \le i \le r]$ and c column eigenvectors $[c_{\mathrm{j}};\ 1 \le \mathrm{j} \le c]$.

An $r \times c$ eigen-image E_{ij} is defined as the product of each row eigenvector with each column eigenvector as

$$E_{\mathrm{ij}} = r_i c_{\mathrm{j}}^{\mathrm{T}} \quad 1 \le i \le r,\ 1 \le \mathrm{j} \le c \tag{12.41}$$

Figure 12.7 shows an example of eigen-images for an image of brain. Figure 12.8 shows the original brain scan image and the image reconstructed from the principal eigen-images.

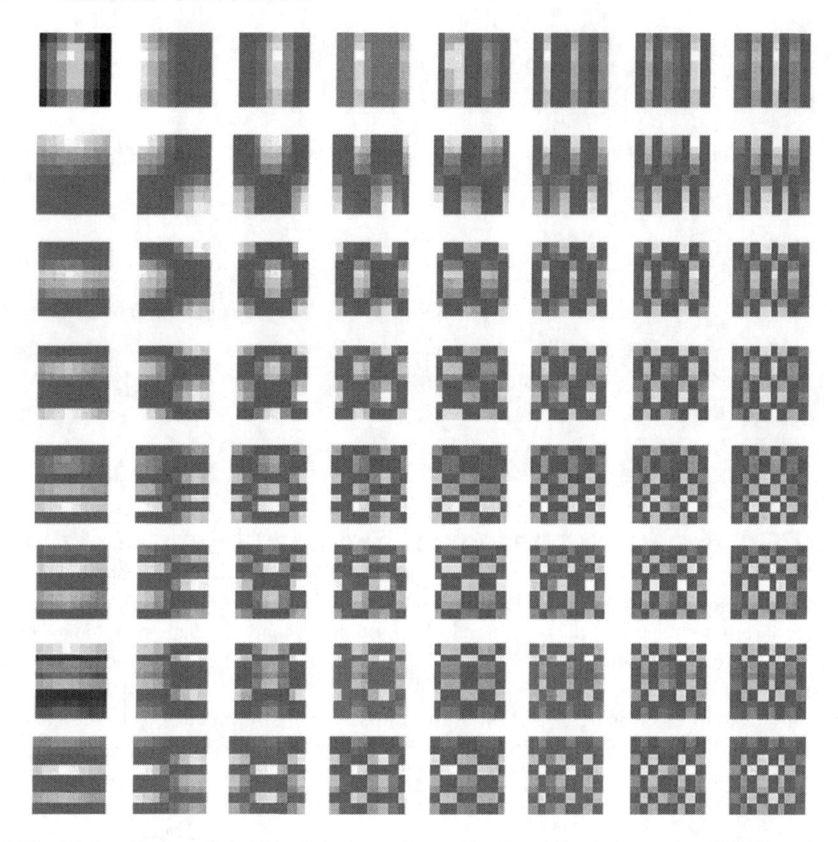

Figure 12.7 64 eigen-images of size 8 by 8 in descending order of associated eigen values. The eigen images are obtained from brain scan image.

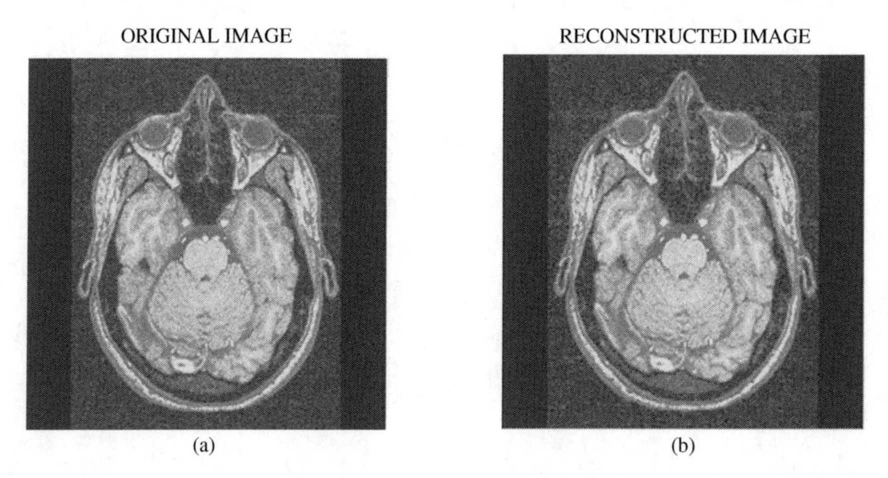

Figure 12.8 (a) An original image scan of brain, (b) reconstructed from the four most significant eigen-images associated with sub-blocks of 8 by 8. Note that images blocks of size 8 by 8 have 64 eigen-images. Used with permission.

MATLAB function PCA_NoisySpeech( )
Analyses noisy speech and reconstructs speech using the most significant principle components with higher SNRs. Speech is segmented into frames of length N samples. Using the signal within each frame a covariance matrix is calculated for sub-frames of length M samples. The covariance matrix is eigenanalysed and the P most significant eigenvalues and the associated eigenvectors are used to reconstruct speech.

12.3.3 PCA Analysis of Speech in White Noise

PCA analysis can be used to decompose a noisy speech signal, retain the most significant signal components that are 'above' a desired signal-to-noise ratio threshold and then discard the remaining noisy components.

In this method a PCA of the correlation matrix of the noisy speech signal is performed. First speech is segmented into frames of N samples. For each frame a correlation matrix for the signal is obtained. The correlation matrix is then decomposed into a matrix of eigenvectors and their corresponding eigenvalues. The most significant components of the signal, corresponding to the largest eigenvalues of the autocorrelation matrix, are retained whereas the noisy components corresponding to the smallest eigenvalues are discarded. The signal is then expressed as a combination of its most significant PC eigenvectors where the combination factors are obtained as normalised cross-correlation of the signal and each PC eigenvector.

Note that for white noise assuming that the dimensions of the correlation matrix are bigger that the dimensions of significant correlation lags in the signal (say < 14) then the smallest eigenvalues correspond to the variance of noise.

Figure 12.9 shows the use of PCA in reconstructing speech from its most significant components. In this case, speech was segmented into segments of 50 samples long and correlation matrices of size 50×50 were calculated as average values over segments of 400 speech samples. Then segments of speech of 50 samples long were expressed in terms of the 16 most significant eigenvectors. Note that the choices of segment size can be varied so long as it conforms to the stationary assumption of PCA.

Figure 12.10(a) shows a speech signal contaminated with additive white noise. Figure 12.10(b) shows the same signal reconstructed from 16 principal components after a PCA was applied to a 50×50 covariance matrix of the speech signal. Note that the noise has been significantly reduced by retaining the PCs with relatively high SNRs and discarding the PCs with lower SNRs.

12.4 Independent Component Analysis

Independent component analysis (ICA) is a powerful method for analysis and extraction of non-Gaussian features and for demixing of mixed signals. ICA builds on principal component analysis (PCA). However, whereas PCA works on the assumed Gaussian structure of the signal and de-correlates the second-order statistics (i.e. covariance matrix) of the signal, ICA works on the non-Gaussian structures of the signal and de-correlates the second-order and the higher-order statistics (usually the fourth order, i.e. kurtosis) of the signal process. Since ICA assumes the signals are non-Gaussian the solutions involve the optimisation of non-linear contrast (objective) functions.

ICA is also known as 'blind' signal separation (BSS) since it requires no knowledge of the signal mixing process other than the assumption that the source signals are statistically independent and

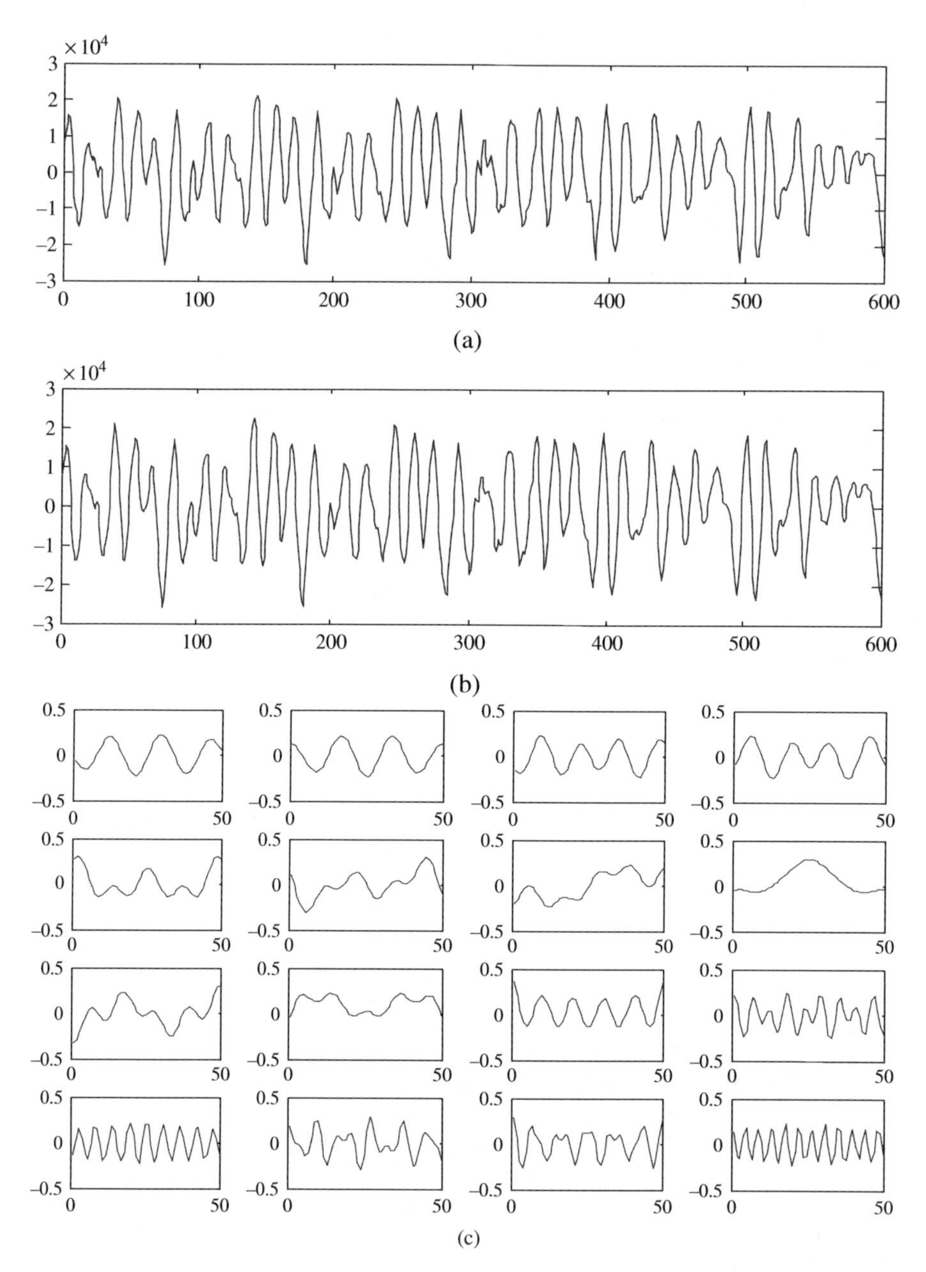

Figure 12.9 Principal component analysis of time domain speech signals: (a) original speech was segmented into segments of 50 samples long and its covariance matrix was obtained for eigenanalysis, (b) shows the speech reconstructed from the 16 principal component eigenvectors, (c) shows 16 principal eigenvectors of speech.

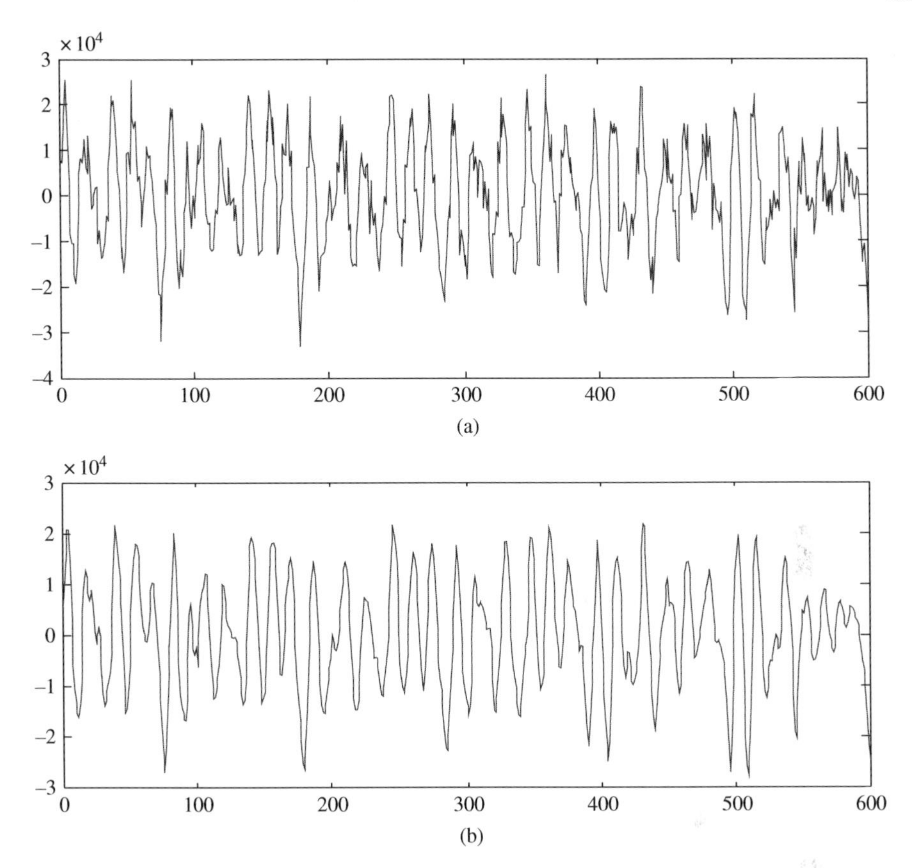

Figure 12.10 Illustration of the use of PCA in removing white noise: (a) noisy speech, (b) segments of 50 sample-long of speech reconstructed from 16 most significant principal components.

that at most no more than one source signal is Gaussian. Some argue that it's more accurate to describe ICA as a tool or a method for solving the general problem of BSS. ICA works on the principal of calculating a demixing matrix whose output vector would have statistically independent components. Whereas classical signal processing theory often assumes that the signals are Gaussian distributed processes and hence do not possess any higher-order statistics above covariance, ICA works on the assumption that the signals are non-Gaussian and possess higher-order statistics.

ICA is primarily used for separation of mixed signals in multi-source multi-sensor applications such as in electrocardiography (EEC) and electroencephalogram (EEG). ICA is also applied in multiple-input multiple-output (MIMO) mobile communication and multiple sensor signal measurement systems where there are N signal sources or transmitters and M sensors or receivers. For example ICA is used in beam-forming in array signal processing applications such as microphone array for directional reception of audio and speech signals (Figure 12.11) and antenna array for directional reception or demixing of electromagnetic radio waves.

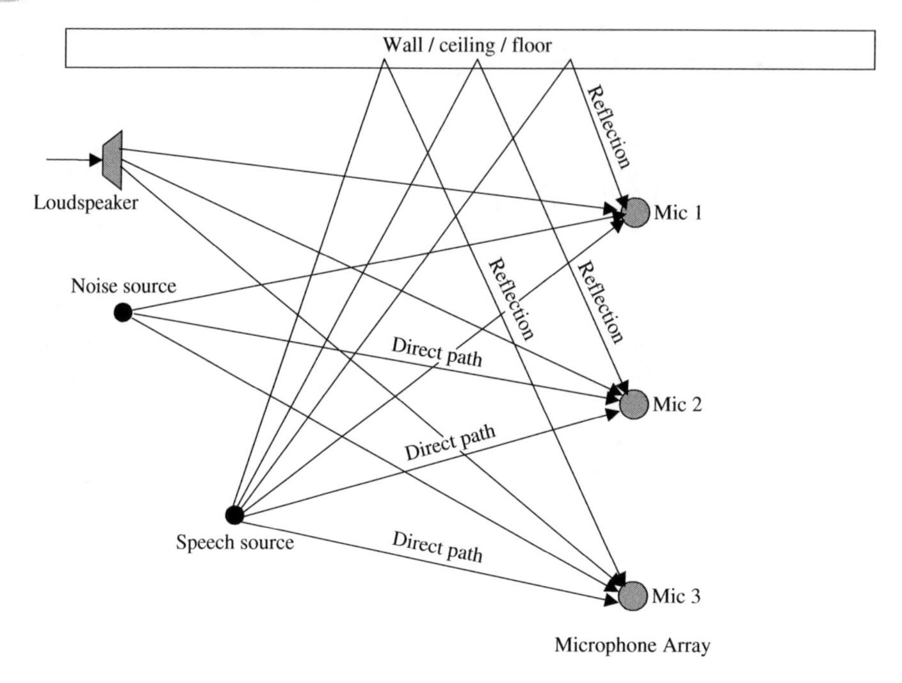

Figure 12.11 Illustration of different sounds and noise arriving at microphones, via direct line-of-sight paths and via reflections from walls/ceiling/floor.

The most common examples of applications of ICA include:

(1) Extraction of N sources from an array of M sensors ($M \geq N$) in biomedical signal processing such as electrocardiograph or electroencephalogram signal measurements.
(2) An array of M microphones in a room or car cabin with N sources of sounds and noise.
(3) An array of N radio transmitter antennas and M receiver antennas in beam-forming smart antennas for mobile phones.
(4) Beam-forming and array signal processing in radar and sonar.
(5) Image/speech feature extraction and image/speech denoising.

12.4.1 Statement of the ICA Problem

Let the signal samples from N sources form an N-dimensional source signal vector $\boldsymbol{s}(m) = [s_1(m), s_2(m), \ldots, s_N(m)]$ and assume that the output of the array of M sensors is represented by vector $\boldsymbol{x}(m) = [x_1(m), x_2(m), \ldots, x_M(m)]$. Assume that each sensor output $x_i(m)$ is a linear combination of the N source signals, that is

$$x_i(m) = a_{i1}s_1(m) + a_{i2}s_2(m) + \cdots + a_{iN}s_N(m) \quad i = 1, \ldots, M \tag{12.42}$$

For each discrete time m, Equation (12.42) can be written in matrix form as

$$\begin{bmatrix} x_1(m) \\ x_2(m) \\ \vdots \\ x_M(m) \end{bmatrix} = \begin{bmatrix} a_{11} & a_{12} & \cdots & a_{1N} \\ a_{21} & a_{22} & \cdots & a_{2N} \\ \vdots & \vdots & \ddots & \vdots \\ a_{M1} & a_{M2} & \cdots & a_{MN} \end{bmatrix} \begin{bmatrix} s_1(m) \\ s_2(m) \\ \vdots \\ s_N(m) \end{bmatrix} \tag{12.43}$$

Note that when we have L observation vectors, with each vector composed of the M samples from the sensors, Equation (12.43) becomes

$$\underbrace{\begin{bmatrix} x_1(0) & x_1(1) & \cdots & x_1(L-1) \\ x_2(0) & x_2(1) & \cdots & x_2(L-1) \\ \vdots & \vdots & \ddots & \vdots \\ x_M(0) & x_M(2) & \cdots & x_M(L-1) \end{bmatrix}}_{M\times K} = \underbrace{\begin{bmatrix} a_{11} & a_{12} & \cdots & a_{1N} \\ a_{21} & a_{22} & \cdots & a_{2N} \\ \vdots & \vdots & \ddots & \vdots \\ a_{M1} & a_{M2} & \cdots & a_{MN} \end{bmatrix}}_{M\times N} \underbrace{\begin{bmatrix} s_1(0) & s_1(1) & \cdots & s_1(L-1) \\ s_2(0) & s_2(1) & \cdots & s_2(L-1) \\ \vdots & \vdots & \ddots & \vdots \\ s_M(0) & s_M(2) & \cdots & s_M(L-1) \end{bmatrix}}_{N\times K} \tag{12.44}$$

In compact notation Equation (12.43) can be rewritten as

$$\boldsymbol{x} = \boldsymbol{A}\boldsymbol{s} \tag{12.45}$$

The matrix $\boldsymbol{A}$ is known as the *mixing matrix* or the observation matrix. In general the N source signals are independent and this implies that the signal vector $\boldsymbol{s}$ has a diagonal covariance matrix and diagonal cumulant matrices and that the elements of $\boldsymbol{s}$ have zero mutual information. However, the observation signal vectors, $\boldsymbol{x}$, are correlated with each other as each observation signal contains a mixture of all the source signals as shown by Equation (12.45). Hence the covariance matrix of the observation signal vector $\boldsymbol{x}$ is a non-diagonal matrix and the elements of $\boldsymbol{x}$ may have non-zero mutual information.

In many practical cases of interest all that we have is the sequence of observation vectors $[\boldsymbol{x}(0), \boldsymbol{x}(1), \ldots, \boldsymbol{x}(L-1)]$, the mixing matrix $\boldsymbol{A}$ is unknown and we wish to estimate a demixing matrix $\boldsymbol{W}$ to obtain an estimate of the original signal $\boldsymbol{s}$. The demixing problem is the estimation of a matrix $\boldsymbol{W}$ such that

$$\hat{\boldsymbol{s}} = \boldsymbol{W}\boldsymbol{x} \tag{12.46}$$

where ideally the demixing matrix $\boldsymbol{W} = \boldsymbol{A}^{-1}$. Note that each element of the source signal is estimated from M sensors as

$$\hat{s}_i = \boldsymbol{w}_i^T \boldsymbol{x} = \sum_{j=1}^{M} w_{ij} x_j \quad i = 1, \ldots, N \tag{12.47}$$

where the vector $\boldsymbol{w}_i$ is the ith row of the matrix $\boldsymbol{W}$. The ICA aims to find a demixing matrix $\boldsymbol{W}$ such that the elements of $\hat{\boldsymbol{s}} = \boldsymbol{W}\boldsymbol{x}$ are probabilistically independent and hence have zero mutual information.

12.4.2 Basic Assumptions in Independent Component Analysis

The identifiability of the source signals in independent component analysis requires that the following assumptions are satisfied:

(1) The N source signals are statistically independent, this implies that their correlation matrix and higher-order cumulant matrices are diagonal. The most complete definition of independence is that the probability of a source vector is the product of the probabilities of the individual sources i.e. $f(\boldsymbol{s}) = f(s_1).f(s_2)\ldots f(s_N)$; this means that the elements of $\boldsymbol{s}$ have zero mutual information.
(2) Only one of the signal sources s_k can be Gaussian. A mix of two Gaussian signals cannot be separated as Gaussian processes are symmetric and do not possess higher than second-order (i.e. covariance) matrix statistics. This issue is explained in detail later.
(3) The number of sensors M, must be at least as large as the number of estimated source components N i.e. the mixing matrix $\boldsymbol{A}$ must be full rank.

12.4.3 Limitations of Independent Component Analysis

(1) ICA cannot identify the scale or the sign of a mixing transformation since the transform $\boldsymbol{A}$ and the source signal $\boldsymbol{s}$ can be multiplied by a sign and scaling factor and their reciprocal as

$$\boldsymbol{x} = \boldsymbol{A}\boldsymbol{s} = (-\alpha\boldsymbol{A})\left(-\frac{1}{\alpha}\boldsymbol{s}\right) \tag{12.48}$$

(2) ICA cannot determine the position of the components in the source vector since any permutation of the source can be cancelled by an inverse permutation of the mixing matrix as

$$\boldsymbol{x} = \boldsymbol{A}\boldsymbol{s} = \boldsymbol{A}\boldsymbol{P}^{-1}\boldsymbol{P}\boldsymbol{s} \tag{12.49}$$

where $\boldsymbol{P}$ is a permutation matrix.
(3) ICA cannot separate the mixed signals if more than one of the mixed signals is Gaussian; this is due to the symmetric distribution of Gaussian processes as explained next.

12.4.4 Why a Mixture of Two Gaussian Signals Cannot Be Separated

A Gaussian process is a second-order process; this means that it is completely defined by its mean vector and covariance matrix and that its higher (than second) order statistics are zero. Here we show that ICA cannot demix a mixture of two Gaussian processes.

Consider a combination of two independent Gaussian signal processes s_1 and s_2

$$x_1 = a_{11}s_1 + a_{12}s_2 \tag{12.50}$$

$$x_2 = a_{21}s_1 + a_{22}s_2 \tag{12.51}$$

Now consider the probability of the two independent signals s_1 and s_2

$$f(s_1, s_2) = f(s_1)f(s_2) = \frac{1}{2\pi}\exp\left(-0.5\left(s_1^2 + s_2^2\right)\right) \tag{12.52}$$

where for simplicity its assumed that s_1 and s_2 have unit variance. The probability distribution of s_1 and s_2 is symmetric and since the argument of the exponential function in Equation (12.52) is the location of a circle, an infinite number of values of s_1 and s_2 on the circle can have the same value of

$s_1^2 + s_2^2$ and hence the same probability. Now assuming that an estimate of $\boldsymbol{s} = [s_1, s_2]^{\mathrm{T}}$ is obtained as $\widehat{s} = \boldsymbol{A}\boldsymbol{x}$, we have

$$\begin{aligned} f(\widehat{s}_1, \widehat{s}_2) &= |\det \boldsymbol{A}|^{-1} \frac{1}{2\pi} \exp\left(-0.5\left(\boldsymbol{x}^{\mathrm{T}}\boldsymbol{A}^{\mathrm{T}}\boldsymbol{A}\boldsymbol{x}\right)\right) \\ &= |\det \boldsymbol{A}|^{-1} \frac{1}{2\pi} \exp\left(-0.5\left(\widehat{s}_1^2 + \widehat{s}_2^2\right)\right) \end{aligned} \quad (12.53)$$

Again there are an infinite number of values of $\widehat{s}_1$ and $\widehat{s}_2$ that yield the same probability in Equation (12.53).

12.4.5 Difference Between Independent and Uncorrelated

ICA may not be achieved if the only optimisation criterion used is that the separated signals should be uncorrelated (in terms of second-order statistics) and hence have a diagonal covariance matrix. This is because two signals may have a common component and at the same time may also have a zero cross-correlation and a diagonal covariance matrix. For a simple example $x_1 = \cos(2\pi f t)$ and $x_2 = \cos(2\pi f t + \pi/2) = \sin(2\pi f t)$ have the same frequency and amplitude but have zero covariance despite the fact that one can be obtained from the other by a simple time-shift. Hence the terms uncorrelated and independent have different meanings and implications.

Generally two variables s_i and s_j are uncorrelated, and have diagonal covariance matrix, if their covariance is zero

$$E[s_i s_j] - E[s_i]E[s_j] = 0 \quad (12.54)$$

For zero-mean random variables Equation (12.54) becomes $E[s_i s_j] = 0$.

Independence is a stronger property than uncorrelated; whereas two independent signals are uncorrelated, two uncorrelated signals are not necessarily independent. The condition for two signals to be independent is that any two functions of them need to have zero covariance, i.e.

$$E[h_1(s_i)h_2(s_j)] - E[h_1(s_i)]E[h_2(s_j)] = 0 \quad (12.55)$$

where $h_1(\cdot)$ and $h_2(\cdot)$ are any two functions, for example $h_1(s_i)h_2(s_j) = s_i^2 s_j^2$ etc.

To achieve ICA the signal vector is required to have independent components in a probabilistic sense such that the probability of the observation vector is the product of the probabilities of its components:

$$f(\hat{s}) = \prod_{i=1}^{M} f(\hat{s}_i) \quad (12.56)$$

As explained in the next section, ICA can be achieved using entropy maximisation based on the idea that the entropy of independent components is greater than the entropy of correlated components. In the context of ICA, the maxent (maximum entropy) is equivalent to minimising the cross entropy between the pdfs of the transformed signal $f(\hat{s})$ and the desired pdf of the source signal with its independent components $f(\boldsymbol{s})$. This cross-entropy minimisation is often expressed in terms of optimisation of an objective function such as the diagonalisation of the higher-order statistics and cumulants.

12.4.6 Independence Measures, Entropy and Mutual Information

The independence property of the components of a multi-variate signal may be quantified by a number of related measures, such as the non-Gaussianity measure, the maximum entropy (maxent) measure, and diagonal cumulant matrices as discussed in this section. First, we introduce the important concepts of differential entropy and mutual information.

Differential Entropy

The entropy of an M-valued discrete random variable $Y = [y_1, \ldots, y_M]$ is defined as

$$H(Y) = -\sum_{i=1}^{M} P_i \log P_i \tag{12.57}$$

where P_i is the probability of the ith symbol.

For continuous-valued random variables the *differential entropy* measure of a random process Y is defined as

$$H(Y) = E(-\log f(y)) = -\int_{-\infty}^{\infty} f(y) \log f(y) dy \tag{12.58}$$

Entropy is the information metric for discrete-valued processes, similarly, differential entropy is often employed as the information metric for continuous-valued processes. Note that entropy and differential entropy can also be interpreted as expectation of the log likelihood of a process.

Maximum Differential Entropy

The entropy of a discrete-valued random variable attains a maximum value for a uniformly distributed variable, whereas the entropy of a continuous-valued random variable, i.e. the differential entropy, is largest for a Gaussian variable compared with other distributions with *the same* covariance matrix. The differential entropy of a variable with a variance of σ^2 is bounded by the inequality

$$H(Y) \leq \log\left(\sqrt{2\pi e}\,\sigma\right) \tag{12.59}$$

where the maximum value of $\log\left(\sqrt{2\pi e}\,\sigma\right)$ is attained for a Gaussian case.

Example 12.6 Find the differential entropy of uniformly distributed random variable. Assume a uniformly distributed random variable y has minimum and maximum limits of $[a, b]$ and hence a pdf of $f(y) = 1/(b-a)$. The differential entropy can be obtained by substituting for the uniform pdf in Equation (12.58) to obtain $H(y) = log(b-a)$. Since the variance of a uniform variable is given by $\sigma^2 = (b-a)^2/12$, the entropy of a uniform variable in terms of its variance becomes $H(y) = \log\left(\sqrt{12}\,\sigma\right)$ which is smaller than the maximum value given by Equation (12.59).

Example 12.7 Show that the differential entropy of two independent variables are additive and hence generalize the result. Consider a two-dimensional vector

$$
\begin{aligned}
H([y_1, y_2]) &= -\int_{-\infty}^{\infty}\int_{-\infty}^{\infty} f(y_1, y_2)\log f(y_1, y_2)dy_1 dy_2 \\
&= -\int_{-\infty}^{\infty}\int_{-\infty}^{\infty} f(y_1)f(y_2)\left(\log f(y_1) + \log f(y_2)\right) dy_1 dy_2 \\
&= -\int_{-\infty}^{\infty} f(y_1)\log f(y_1) \underbrace{\int_{-\infty}^{\infty} f(y_2)dy_2}_{1} dy_1 - \int_{-\infty}^{\infty} f(y_2)\log f(y_2)\underbrace{\int_{-\infty}^{\infty} f(y_1)dy_1}_{1}\, dy_2 \qquad (12.60)\\
&= -\int_{-\infty}^{\infty} f(y_1)\log f(y_1)dy_1 - \int_{-\infty}^{\infty} f(y_2)\log f(y_2)dy_2 \\
&= H(y_1) + H(y_2)
\end{aligned}
$$

Similarly for N independent variables $[y_1, y_2, \ldots, y_N]$ we have

$$H([y_1, y_2, \ldots, y_N]) = \sum_{k=1}^{N} H(y_k) \qquad (12.61)$$

Example 12.8 Find the differential entropy of $\boldsymbol{y} = [y_1 y_2]$ assuming that (1) y_1 and y_2 are the same, (2) y_1 and y_2 are independent identically distributed.
In the first case $f(y_1, y_2) = f(y_1) = f(y_2)$ and hence the differential entropy is equal to $H(\boldsymbol{y}) = H(y_1) = H(y_2)$; in the second case $f(y_1, y_2) = f(y_1)f(y_2)$ and the differential entropy is equal to $H(\boldsymbol{y}) = 2H(y_1) = 2H(y_2)$.

Mutual Information

The mutual information of two or more variables is a measure of their dependence (or independence). The mutual information of discrete-valued elements was defined in Section 6.4.2. The mutual information of the continuous-valued elements of an N-dimensional vector $\boldsymbol{y} = [y_1, y_2, \ldots, y_N]$ can be defined in terms of the differential entropy as

$$I([y_1, y_2, \ldots, y_N]) = \sum_{k=1}^{N} H(y_k) - H(y) \qquad (12.62)$$

Note that the mutual information of independent variables is zero. The limits of the mutual information of N variables is $\sum_{k=1}^{N} H(y_k) \le I([y_1, \ldots, y_N]) \le 0$.

Effect of a Linear Transformation on Mutual Entropy

Consider a linear transformation $\boldsymbol{y} = \boldsymbol{Wx}$. The probability density function of the vector $\boldsymbol{y}$ in terms of that of the vector $\boldsymbol{x}$ (see Section 6.8) is given by

$$f_Y(\mathbf{y} = \boldsymbol{Wx}) = |\det(\boldsymbol{W})|^{-1} f_X(\boldsymbol{W}^{-1}\mathbf{y}) \qquad (12.63)$$

Hence the log-likelihood of the probability of a transformation is

$$\log f_Y(y = \boldsymbol{W}\boldsymbol{x}) = -\log|\det(\boldsymbol{W})| + \log f_X(\boldsymbol{W}^{-1}\boldsymbol{y}) \tag{12.64}$$

Hence, using Equations (12.63) and (12.64) it follows that the differential entropy of $\boldsymbol{x}$ can be expressed in terms of the differential entropy of $\boldsymbol{s}$ as

$$\begin{aligned}
H_Y(y) &= -\int_{-\infty}^{\infty} f_Y(y)\log f_Y(y)dy \\
&= -\int_{-\infty}^{\infty} |\det(\boldsymbol{W})|^{-1} f_X(\boldsymbol{W}^{-1}\boldsymbol{y})\log\left(|\det(\boldsymbol{W})|^{-1} f_X(\boldsymbol{W}^{-1}\boldsymbol{y})\right)d\boldsymbol{y} \\
&= -|\det(\boldsymbol{W})|^{-1}|\det(\boldsymbol{W})|\left(-\log|\det(\boldsymbol{W})|\underbrace{\int_{-\infty}^{\infty} f_X(\boldsymbol{X})d\boldsymbol{x}}_{1} + \underbrace{\int_{-\infty}^{\infty} f_X(\boldsymbol{x})\log(f_X(\boldsymbol{x}))d\boldsymbol{x}}_{-H(\boldsymbol{x})}\right) \\
&= H(\boldsymbol{x}) + \log|\det(\boldsymbol{W})|
\end{aligned} \tag{12.65}$$

Hence, the change in differential entropy due to a linear transformation $\boldsymbol{W}$ is $\log|\det(W)|$. From Equations (12.62) and (12.65) the mutual information of $\boldsymbol{y}$ is given by

$$I([y_1, y_2, \ldots, y_N]) = \sum_{k=1}^{N} H(y_k) - H(\boldsymbol{x}) - \log|\det(\boldsymbol{W})| \tag{12.66}$$

Non-Gaussianity as a Measure of Independence

From the central limit theorem we have that a linear combination of many independent random variables has a Gaussian distribution. A mix of even two non-Gaussian variables is more Gaussian than the original individual source variables. Hence it follows that a measure of independence of signals is their non-Gaussianity; demixed signals are more non-Gaussian than the mixed signals. For signal separation, or principal component/feature analysis, ICA use methods that maximise the non-Gaussian property of the transformed signals. Two main measures of non-Gaussianity used in ICA methods are negentropy and kurtosis, defined as follows.

Negentropy: A measure of NonGaussianity and Independence

The negentropy of a process is a measure of its non-Gaussian structure. The negentropy measure is based on differential entropy.

Whereas the entropy of a discrete-valued variable attains a maximum value for a uniformly distributed variable, the entropy of a continuous-valued variable, the differential entropy, is largest for a Gaussian variable compared with other distributions with *the same* covariance matrix.

Hence differential entropy can be used to define a measure of non-Gaussianity of a process $\boldsymbol{y}$, defined as its negentropy $J(\boldsymbol{y})$, as the difference between the differential entropy of a random process $\boldsymbol{y}$ and that of a Gaussian process $\boldsymbol{y}_{\text{Gauss}}$ with the same covariance matrix, as

$$\mathrm{J}(y) = H(y_{\text{Gauss}}) - H(y) \tag{12.67}$$

Negentropy is a non-negative variable that is zero for a Gaussian process. The advantage of using negentropy is that it is an information-theoretic measure. However, the problem in using negentropy is its high computational complexity. Estimating negentropy using Equation (12.67) would require an estimate of the pdf of the process. Simpler approximations of negentropy will be discussed next.

Fourth Order Moments – Kurtosis

In general the cumulant of four zero-mean variables x_1, x_2, x_3 and x_4 is defined as

$$Q(x_1, x_2, x_3, x_4) = E[x_1x_2x_3x_4] - E[x_1x_2]E[x_3x_4] - E[x_1x_3]E[x_2x_4] - E[x_1x_4]E[x_2x_3] \quad (12.68)$$

The fourth-order cumulant of a zero-mean random variable x is defined as

$$Q(x) = E[x^4] - 3\left(E[x^2]\right)^2 \quad (12.69)$$

where $E[x^2]$ is the variance of x.

Kurtosis is defined as the fourth-order cumulant normalised by the square of the variance as

$$k(x) = \frac{E[x^4]}{\left(E[x^2]\right)^2} - 3 \quad (12.70)$$

For a Gaussian process the fourth-order moment $E[x^4] = 3\left(E[x^2]\right)^2$, hence from Equation (12.70) the fourth-order cumulant and the kurtosis of a Gaussian process are zero. Therefore kurtosis, like negentropy, can be used as a measure of non-Gaussianity. Super-Gaussian signals have positive kurtosis whereas sub-Gaussian signals have negative kurtosis.

Kurtosis-based Contrast Functions – Approximations to Entropic Contrast

Consider two vector-valued random variables $\boldsymbol{s}$ and $\boldsymbol{y}$ with probability densities $p(\boldsymbol{s})$ and $p(\boldsymbol{y})$. Assume $\boldsymbol{s}$ is the target variable and we wish $\boldsymbol{y}$ to have the same probability distribution as $\boldsymbol{s}$. In general the distance between the probability distributions of two random variables can be measured using the Kullback–Leibler criterion (also known as cross entropy) as

$$K\left(p_s|p_y\right) = \int_{-\infty}^{\infty} p_s(x)\log\frac{p_s(x)}{p_y(x)}dx \quad (12.71)$$

A probability distribution can be approximated in terms of the Edgeworth series, that is in terms of its cumulants. Using an Edgeworth expansion it can be shown that the distance between the probability distributions of two random variables can be approximated in terms of the distance between their second-order and fourth-order moments as

$$K_{24}(\boldsymbol{y}|\boldsymbol{s}) = \frac{1}{4}\sum_{ij}\left(R_{ij}^y - R_{ij}^s\right)^2 + \frac{1}{48}\sum_{ijkl}\left(Q_{ijkl}^y - Q_{ijkl}^s\right)^2 \quad (12.72)$$

where the first term on the r.h.s. of Equation (12.72) is the distance between the covariances of the two processes R_{ij} and the second term is the distance between the fourth-order cumulants of the two processes Q_{ij}.

Assuming that the source signals s_i are independent processes their covariance and cumulant matrices would be diagonal and hence we have

$$K_{24}(y|s) = \frac{1}{4}\sum_{ij}\left(R_{ij}^{y} - \sigma^2(s_i)\delta_{ij}\right)^2 + \frac{1}{48}\sum_{ijkl}\left(Q_{ijkl}^{y} - k^2(s_i)\delta_{ijkl}\right)^2 \tag{12.73}$$

where $\sigma^2(s_i)$ and $k(s_i)$ are the variance and the kurtosis of the source signal s_i and the Kronecker delta function $\delta_{ijkl}(\cdot)$ is 1 when all its indices are identical, otherwise it is zero. Note that the minimisation of the distances in Equation (12.73) implies that ideally the diagonal elements of the covariances and cumulants of the estimate $\boldsymbol{y}$ are made identical to those of the source signal $\boldsymbol{s}$ whereas the off-diagonal elements are minimised to zero.

In cases where a prior knowledge of the variance and cumulants of the source signals are not available, the cumulant contrast function can be simplified as

$$K_{24}(y|s) = \frac{1}{4}\sum_{ij\neq ii}\left(R_{ij}^{y}\right)^2 + \frac{1}{48}\sum_{ijkl\neq iiii}\left(Q_{ijkl}^{y}\right)^2 \tag{12.74}$$

Minimisation of the contrast function of Equation (12.74) implies the minimisation of the off-diagonal elements of the covariance and kurtosis functions.

12.4.7 Super-Gaussian and Sub-Gaussian Distributions

A Gaussian random process x has a fourth-order moment related to its variance σ_x^2 as $E\left(x^4\right) = 3\left(E\left(x^2\right)\right)^2 = 3\sigma_x^4$. Hence from Equation (12.69) the kurtosis of a Gaussian process is zero. A zero mean uniform process of variance σ_x^2 has range of variations of $\pm\sqrt{12}\,\sigma/2$ and a probability of $f(x) = 1/\left(\sqrt{12}\,\sigma\right)$ and hence $E\left(x^4\right) = \int_{-1/\left(2\sqrt{12}\sigma\right)}^{1/\left(2\sqrt{12}\,\sigma\right)}\left(1/\left(\sqrt{12}\,\sigma\right)\right)x^4dx = 1.8\sigma^4$. Hence from Equation (12.69) the kurtosis of a uniform process is $-1.2\sigma^4$.

Random processes with a positive kurtosis are called *super-Gaussian* and those with a negative kurtosis are called *sub-Gaussian*. Super-Gaussian processes have a spikier pdf than Gaussian, whereas sub-Gaussian processes have a broader pdf. An example of a super-Gaussian process is a Laplacian process.

Consider a general class of exponential probability densities given by

$$p(x) = k_1 x^{\beta}\exp\left(-k_2|x|^{\alpha}\right) \tag{12.75}$$

for a Gaussian signal $\alpha = 2, k_1 = 1/\sqrt{2\pi}\,\sigma, \beta = 0$ and $k_2 = 1/2\sigma^2$ where σ^2 is the variance. For $0 < a < 2$ the distribution is super-Gaussian and for $a > 2$ the distribution is sub-Gaussian.

Assuming that the signals to be separated are super-Gaussian, independent component analysis may be achieved by finding a transformation that maximises kurtosis. Figure 12.12 shows examples of Gaussian, super-Gaussian and sub-Gaussian pdfs.

12.4.8 Fast-ICA Methods

Prior to application of a fast-ICA method, the data are sphered with PCA as explained in Section 12.3, so that the covariance matrix of the data is an identity matrix.

The fast-ICA methods are the most popular methods of independent component analysis. Fast-ICA methods are iterative optimisation search methods for solving the ICA problem of finding a demixing

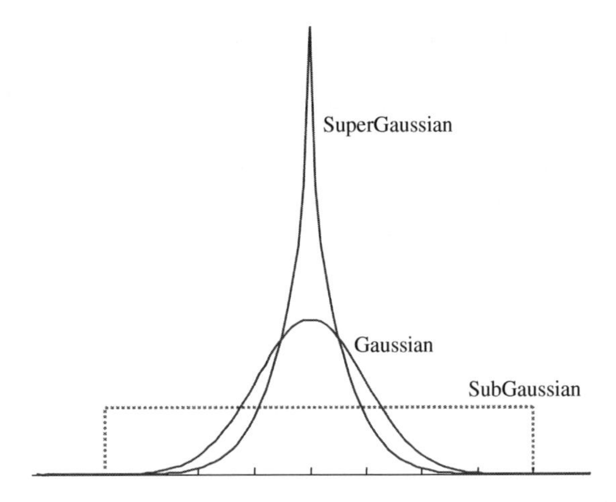

Figure 12.12 Super-Guassian pdfs are more peaky than Gaussian pdfs, whereas sub-Gaussian pdfs are less peaky.

matrix $\boldsymbol{W}$ that is the inverse of the mixing matrix $\boldsymbol{A}$ (within a scalar multiplier and an unknown permutation matrix).

These iterative search methods find a demixing transformation $\boldsymbol{W}$ that optimises a non-linear contrast function. The optimisation methods are typically based on a gradient search or the Newton optimisation method that search for the optimal point of a contrast (objective) function $G(\boldsymbol{Wx})$. At the optimal point of the contrast function the components of $\boldsymbol{y} = \boldsymbol{Wx}$ are expected to be independent.

Gradient Search Optimisation Method

For gradient search optimisation the iterative update methods for estimation of the demixing matrix $\boldsymbol{W}$ is of the general form

$$\boldsymbol{W}_{n+1} = \boldsymbol{W}_n + \mu g(\boldsymbol{W}_n \boldsymbol{x}) \tag{12.76}$$

where $g(\boldsymbol{W}_n \boldsymbol{x}) = \frac{d}{d\boldsymbol{W}_n} G(\boldsymbol{W}_n \boldsymbol{x})$ is the first derivative of the contrast function and μ is an adaptation step size.

Newton Optimisation Method

For Newton optimisation the iterative method for estimation of $\boldsymbol{W}$ is of the form

$$\boldsymbol{W}_{n+1} = \boldsymbol{W}_n - \mu \frac{g(\boldsymbol{W}_n \boldsymbol{x})}{g'(\boldsymbol{W}_n \boldsymbol{x})} \tag{12.77}$$

where $g(\boldsymbol{W}_n \boldsymbol{x})$ is the first derivative of the contrast function and $g'(\boldsymbol{W}_n \boldsymbol{x})$ is the second derivative of the contrast function. The derivative of the contrast function, $g(\boldsymbol{W}_n \boldsymbol{x})$, is also known as the influence function.

12.4.9 Fixed-Point Fast-ICA

In the fixed-point fast-ICA method a batch or block of data (consisting of a large number of samples) are used in each step of the estimation of the demixing matrix. Hence, each step is composed of the iterations on the samples that constitute the batch or block of data for that sample. In this section we consider the one-unit fast-ICA method where at each step one of the sources is estimated or demixed from the observation mixture, i.e. at each step one row vector $\boldsymbol{w}$ of the demixing matrix $\boldsymbol{W}$ is estimated.

A popular version of the fast-ICA is based on a constrained optimisation of the objective function $G(\boldsymbol{w}^T\boldsymbol{x})$ subject to the constraint $E\left((\boldsymbol{w}^T\boldsymbol{x})^2\right) = \|\boldsymbol{w}\|^2 = 1$. The solution is given by

$$E\left(\boldsymbol{x}g(\boldsymbol{w}^T\boldsymbol{x})\right) - \beta\boldsymbol{w} = 0 \tag{12.78}$$

At the optimal value of $\boldsymbol{w}_0$, multiplying both sides of Equation (12.78) by $\boldsymbol{w}_0^T$ and noting that $\boldsymbol{w}_0^T\boldsymbol{w}_0 = \|\boldsymbol{w}_0\| = 1$ yields

$$\beta = E\left(\boldsymbol{w}_0^T\boldsymbol{x}g(\boldsymbol{w}_0^T\boldsymbol{x})\right) \tag{12.79}$$

To obtain a Newton type solution, the Jacobian matrix of Equation (12.78) (the second derivative w.r.t $\boldsymbol{w}$ in this case) is obtained as

$$JF(\boldsymbol{w}) = E\left(\boldsymbol{x}\boldsymbol{x}^T g'(\boldsymbol{w}^T\boldsymbol{x})\right) - \beta\mathbf{I} = 0 \tag{12.80}$$

where $F(\boldsymbol{w})$ denotes the left-hand side of Equation (12.78) and $g'(\boldsymbol{w}^T\boldsymbol{x})$ is the derivative of $g(\boldsymbol{w}^T\boldsymbol{x})$ and the second derivative of $G(\boldsymbol{w}^T\boldsymbol{x})$. Since it is assumed that prior to fast-ICA the data has been sphered so that $E\left(\boldsymbol{x}\boldsymbol{x}^T\right) = \boldsymbol{I}$, the first term of Equation (12.80) can be approximated as

$$E\left(\boldsymbol{x}\boldsymbol{x}^T g'(\boldsymbol{w}^T\boldsymbol{x})\right) \approx E\left(\boldsymbol{x}\boldsymbol{x}^T\right) E\left(g'(\boldsymbol{w}^T\boldsymbol{x})\right) = E\left(g'(\boldsymbol{w}^T\boldsymbol{x})\right) \tag{12.81}$$

Hence the Newton optimisation method, at the nth iteration, can be written as

$$\begin{aligned} \boldsymbol{w}_{n+1} &= \boldsymbol{w}_n - \mu\left[\left[\, E\left(\boldsymbol{x}g(\boldsymbol{w}_n^T\boldsymbol{x})\right) - \beta\boldsymbol{w}_n\right] \Big/ E\left(g'(\boldsymbol{w}_n^T\boldsymbol{x})\right) - \beta\right] \\ \boldsymbol{w}_{n+1} &= \boldsymbol{w}_{n+1} / \|\boldsymbol{w}_{n+1}\| \end{aligned} \tag{12.82}$$

where $\boldsymbol{w}_{n+1}$ on the l.h.s. represents the new value of the estimate at the nth iteration.

12.4.10 Contrast Functions and Influence Functions

The non-linear contract functions $G(\boldsymbol{W}\boldsymbol{x})$ are the objective functions for optimisation of ICA transform $\boldsymbol{W}$: the optimal $\boldsymbol{W}$ is a maxima (or minima) of $G(\boldsymbol{W}\boldsymbol{x})$. The non-linearity of the contrast function exploits the non-Gaussian distribution of the signal and facilitates de-correlation of the higher-order statistics of the process after the second-order statistics are de-correlated by a PCA pre-processing stage.

Maximisation of $G(\boldsymbol{W}\boldsymbol{x})$ is achieved using either the gradient of the contrast function in an iterative optimisation search method such as the gradient ascent (or descent) methods or the Newton optimisation method.

The gradient of the contrast function is given by

$$g(\boldsymbol{W}\boldsymbol{x}) = \frac{\partial}{\partial W} G(\boldsymbol{W}\boldsymbol{x}) \tag{12.83}$$

$g(\boldsymbol{Wx})$ is also known as the *influence function.*

In general for a variable y with density $p(y) \propto \exp(|y|^{\alpha})$ the contrast function would be of the form

$$G(y) = [|y|^{\alpha}] \tag{12.84}$$

Hence for $\alpha = 2$, i.e. a Gaussian process $G(y) = E[|y|^2]$. For a super-Gaussian process $0 < \alpha < 2$. For highly super-Gaussian process when $\alpha < 1$, the contrast function shown in Equation (12.84) is not differentiable at $y = 0$ because in this case the differentiation of y would produce a ratio function with y as the denominator. Hence, for super-Gaussian functions, differentiable approximations to the contrast function are used. One such approximation is $\log(\cosh(y))$ whose derivative is $\tanh(y)$. Another choice of non-linearity for the contrast function is $G(y) = -e^{-y^2/2}$ whose derivative is $g(y) = ye^{-y^2/2}$. For sub-Gaussian functions $G(y) = y^4$ (kurtosis) can be an appropriate choice of a contrast function.

Some of the most popular contrast and influence functions are as follows:

Contrast function	Influence function	Appropriate process	
$G(y) = \log(\cosh(y))$	$g(y) = \tanh(y)$	General purpose	(12.85)
$G(y) = -e^{-y^2/2}$	$g(y) = ye^{-y^2/2}$	Highly super-Gaussian	(12.86)
$G(y) = y^4$	$g(y) = y^3$	Sub-Gaussian	(12.87)

Figure 12.13 illustrates three contrast functions and their respective influence functions. Note that the influence functions are similar to non-linearities used in neural networks.

Note that it may be the case – perhaps it is often the case – that the process subjected to ICA is not identically distributed; that is different elements of the input vector process may have different forms of distributions, hence there is a case for using different contrast functions for different elements of the vector. This can be particularly useful for the fine-tuning stage of ICA, after application of a conventional ICA to the process. Of course one needs to estimate the form of the distribution of each component and hence the appropriate contrast and influence functions.

12.4.11 ICA Based on Kurtosis Maximisation – Projection Pursuit Gradient Ascent

The demixing of a mixed signal may be achieved through iterative estimation of a demixing matrix $\boldsymbol{W}$ using an objective criterion that maximises the kurtosis of the transformed signal vector. Let $\boldsymbol{x}$ be the observation signal vector. In general each element of $\boldsymbol{x}$ is a mixture of the source signals and hence $\boldsymbol{x}$ has a non-diagonal covariance matrix. The signal $\boldsymbol{x}$ is first de-correlated and diagonalised using the eigenvectors of the correlation matrix of $\boldsymbol{x}$. Furthermore, the diagonalised process is sphered (i.e. made to have unit variance for each element of the vector process) by using a normalising eigenvalue matrix. Let z be the sphered (its covariance diagonalised and normalised) version of $\boldsymbol{x}$

$$z = \Delta^{-0.5} \mathrm{U}^{\mathrm{T}} \boldsymbol{x} \tag{12.85}$$

where the matrices $\boldsymbol{U}$ and Δ are the eigenvectors and eigenvalues of the correlation matrix of $\boldsymbol{x}$.

As explained diagonalisation of the covariance matrix alone is not sufficient to achieve demixing. For demixing of a mixed observation we need to diagonalise the higher-order cumulants of the signals.

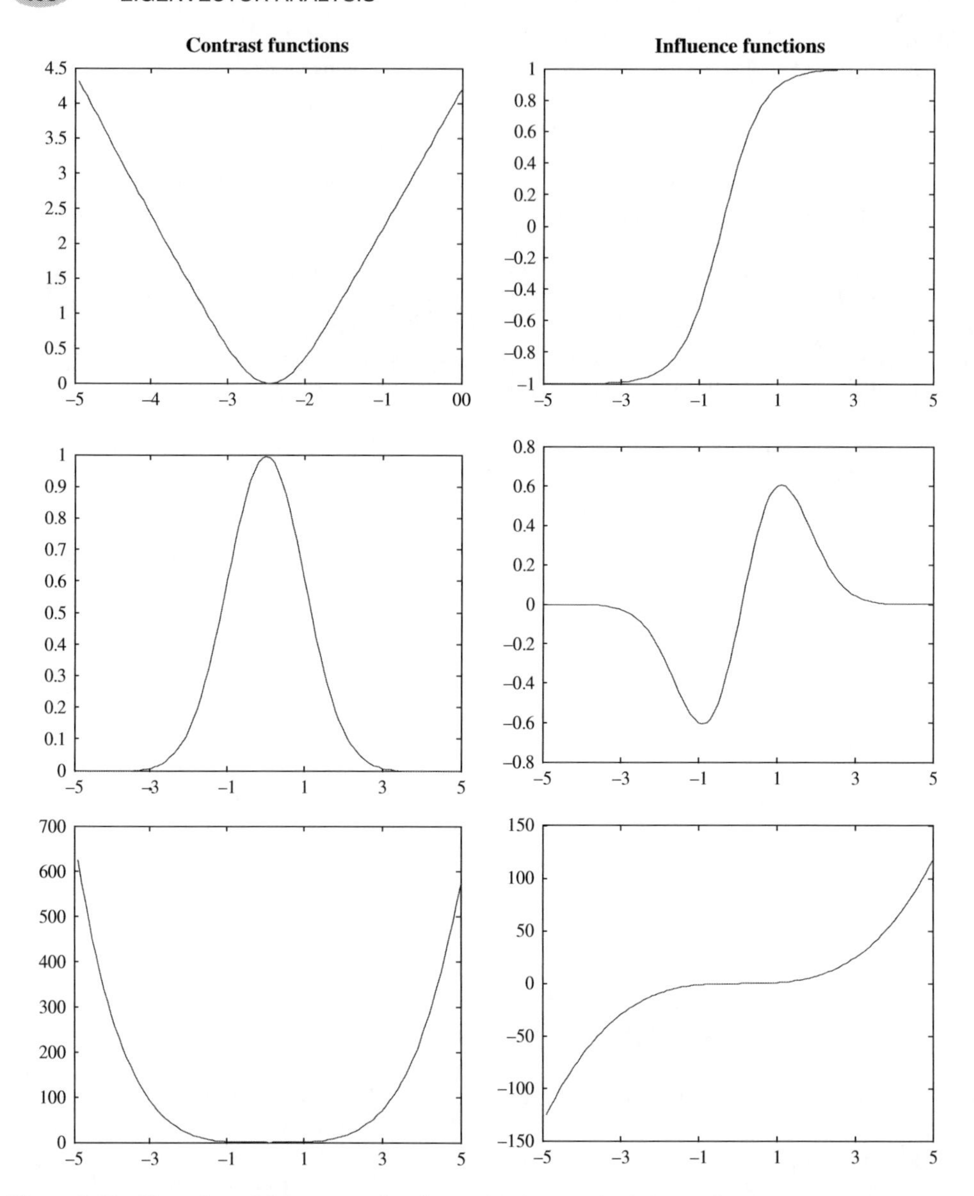

Figure 12.13 Illustrations of three contrast functions and their respective influence functions; the x-axis represents the input to the influence/contrast functions.

One way to achieve this is to search for a transform that maximises kurtosis which in the context of ICA is a measure of non-Gaussianity and independence.

Now assume that $\boldsymbol{w}_i$ is the ith row vector of a demixing matrix $\boldsymbol{W}$ and that it demixes an element of $\boldsymbol{z}$ as

$$y_i = \boldsymbol{w}_i^T \boldsymbol{z} \tag{12.86}$$

The fourth-order cumulant of y_i is given by

$$k(y_i) = E(y_i^4) - 3E(y_i^2) = E\left((\boldsymbol{w}_i^T z)^4\right) - 3E\left((\boldsymbol{w}_i^T z)^2\right) \tag{12.87}$$

The instantaneous rate of change (differential) of kurtosis with $\boldsymbol{w}_i$ is given by

$$\frac{\partial k(y_i)}{\partial \boldsymbol{w}_i} = 4z(\boldsymbol{w}_i^T z)^3 \tag{12.88}$$

An iterative gradient ascent identification method, for the demixing vector $\boldsymbol{w}_i$, based on kurtosis maximisation can be defined as

$$\begin{aligned} \boldsymbol{w}_i(n) &= \boldsymbol{w}_i(n-1) + \mu \frac{\partial k(y_i)}{\partial \boldsymbol{w}_i(n-1)} \\ &= \boldsymbol{w}_i(n-1) + 4\mu z(\boldsymbol{w}_i^T z)^3 \end{aligned} \tag{12.89}$$

where $\partial k(y_i)/\partial \boldsymbol{w}_i(n-1)$ is the rate of change of kurtosis with the transform coefficient vector. At the completion of each update $\boldsymbol{w}_i(n)$ is normalised as $\boldsymbol{w}_i(n)/|\boldsymbol{w}_i(n)|$.

Figure 12.14 illustrates a set of image basis functions obtained from ICA of the brain image of Figure 12.8. Contrasting these with the eigen-images of Figure 12.7 obtained from the PCA method shows that ICA is more efficient as most of the information is packed into fewer independent sub-images.

12.4.12 Jade Algorithm – Iterative Diagonalisation of Cumulant Matrices

The Jade algorithm (cardoso and Saouloumiac 1996) is an ICA method for identification of the demixing matrix. The Jade method is based on a two-stage process: (1) diagonalisation of the covariance matrix,

Figure 12.14 ICA-based independent image basis functions obtained from the brain image of Figure 12.8. Contrast these with the eigen images of Figure 12.7 obtained from the PCA method.

this is the same as PCA, and (2) diagonalisations of the kurtosis matrices of the observation vector sequence.

The Jade method is composed of the following stages:

(1) *Initial PCA stage.* First the covariance matrix of the signal $\boldsymbol{X}$ is formed. Eigenanalysis of the covariance matrix yields a whitening (sphereing) matrix $\boldsymbol{W} = \Lambda^{-0.5}\boldsymbol{U}^{\mathrm{T}}$, where the matrices $\boldsymbol{U}$ and Λ are composed of the eigenvectors and eigenvalues of the covariance of $\boldsymbol{X}$. The signal is sphered by transformation through the matrix $\boldsymbol{W}$ as $\boldsymbol{Y} = \boldsymbol{W}\boldsymbol{X}$. Note that the covariance matrix of $\boldsymbol{Y}$ is $E[\boldsymbol{Y}\boldsymbol{Y}^{\mathrm{T}}] = E[\boldsymbol{W}\boldsymbol{X}\boldsymbol{X}^{\mathrm{T}}\boldsymbol{W}^{\mathrm{T}}] = \Lambda^{-0.5}\boldsymbol{U}^{\mathrm{T}}\boldsymbol{U}\Lambda\boldsymbol{U}^{\mathrm{T}}\boldsymbol{U}\Lambda^{-0.5} = \boldsymbol{I}$.
(2) *Calculation of kurtosis matrices.* The fourth-order (kurtosis) cumulant matrices $\boldsymbol{Q}_i$ of the signal are formed.
(3) *Diagonalisation of kurtosis matrices.* A single transformation matrix $\boldsymbol{V}$ is obtained such that all the cumulant matrices are as diagonal as possible. This is achieved by finding a matrix $\boldsymbol{V}$ that minimises the off-diagonal elements.
(4) *Apply ICA for signal separation.* A separating matrix is formed as $\boldsymbol{W}\boldsymbol{V}^{\mathrm{T}}$ and applied to the original signal.

As mentioned, the diagonalisation and sphereing of the covariance matrix of the observation process is performed using principal component analysis.

The diagonalisation of the cumulant matrices is a more complicated process compared with PCA mainly due to the four-dimensional tensorial nature of the fourth-order cumulant matrices which have in the order of $O(N^4)$ parameters. Instead of using four-dimensional matrices, the cumulant matrices are expressed in terms of a set of two-dimensional matrices with each element of the matrix expressed as

$$Q_{\mathrm{ij}} = \sum_{k,l=1}^{M} \mathrm{cum}\,(X_i, X_{\mathrm{j}}, X_k, X_l) \tag{12.90}$$

Given T samples from each of the M sensors, a set of cumulant matrices can be computed as follows. Assume that the matrix $\boldsymbol{X}$ denotes an $M \times T$ matrix containing the samples from all the M sensors and that vectors $\boldsymbol{x}_i$ and $\boldsymbol{x}_{\mathrm{j}}$ (the ith and j^{th} row of $\boldsymbol{X}$) denote the T samples from sensors i and j respectively. The Jade algorithm calculates a series of $M \times M$ cumulant matrices $\boldsymbol{Q}_{\mathrm{ij}}$ as

$$\boldsymbol{Q}_{\mathrm{ij}} = \left(\boldsymbol{x}_i.\boldsymbol{x}_{\mathrm{j}}.\boldsymbol{X}\right)\boldsymbol{X}^T \quad 1 \le i \le M, \quad 1 \le \mathrm{j} < i \tag{12.91}$$

where $\boldsymbol{x}_i.\boldsymbol{x}_{\mathrm{j}}$ is a vector dot product, that is an element-by-element multiplication of two $T \times 1$ vectors $\boldsymbol{x}_i$ and $\boldsymbol{x}_{\mathrm{j}}$. The $T \times 1$ vector resulting from the product $\boldsymbol{x}_i.\boldsymbol{x}_{\mathrm{j}}$ is then multiplied by every row vector of the $N \times T$ matrix $\boldsymbol{X}$. The result is then multiplied by the $T \times M$ matrix $\boldsymbol{X}^{\mathrm{T}}$ to yield an $M \times M$ *cumulant* matrix $\boldsymbol{Q}_{\mathrm{ij}}$.

The Givens rotation method is used to derive a matrix for the diagonalisation of the cumulant matrices. A Givens rotation matrix $\boldsymbol{R}(i, \mathrm{j}, c, s)$ is equal to the identity matrix but for the following entries

$$\begin{pmatrix} \boldsymbol{r}_{ii} & \boldsymbol{r}_{i\mathrm{j}} \\ \boldsymbol{r}_{\mathrm{j}i} & \boldsymbol{r}_{\mathrm{jj}} \end{pmatrix} = \begin{pmatrix} c & s \\ -s & c \end{pmatrix} \tag{12.92}$$

where $c = \cos\theta$ and $s = \sin\theta$ are calculated to set the off-diagonal elements to zero. For diagonalisation of the cumulant matrix, the angle θ, or equivalently the variables c and s, are calculated to minimise the following objective function over $K = M(M-1)/2$ cumulant matrices

$$O(c, s) = \sum_{k=1}^{K} \mathrm{OffDiag}\left(\boldsymbol{R}\,(i, \mathrm{j}, c, s)\,\boldsymbol{Q}_{\mathrm{ij}}\,(k)\,\boldsymbol{R}^T\,(i, \mathrm{j}, c, s)\right) \tag{12.93}$$

where OffDiag denotes the off-diagonal elements of a matrix. It can be shown that the variables c and s can be derived using the following method. Define a matrix $\boldsymbol{G}$ as

$$\boldsymbol{G} = \sum_{k=1}^{K} [Q_{ii}k - Q_{jj}k \; Q_{ij}k + Q_{ji}k]^T [Q_{ii}k - Q_{jj}k \; Q_{ij}k + Q_{ji}k] \tag{12.94}$$

Now define x, y and r as

$$x = G(1,1) - G(2,2), \; y = G(1,2) + G(2,1), \; r = \sqrt{x^2 + y^2} \tag{12.95}$$

The rotation angles for the ij off-diagonal elements are defined as

$$c = \sqrt{\frac{x+r}{2r}}, \; s = \sqrt{\frac{y+r}{2r(x+r)}} \tag{12.96}$$

The above methods are described in cardoso (1998) and cardoso and Saouloumiac (1996).

Figure 12.15 shows the application of the Jade algorithm to separation of ECGs of a mother and foetus where the measured signals from eight sensors have been used. The results show the efficiency and power of the Jade method.

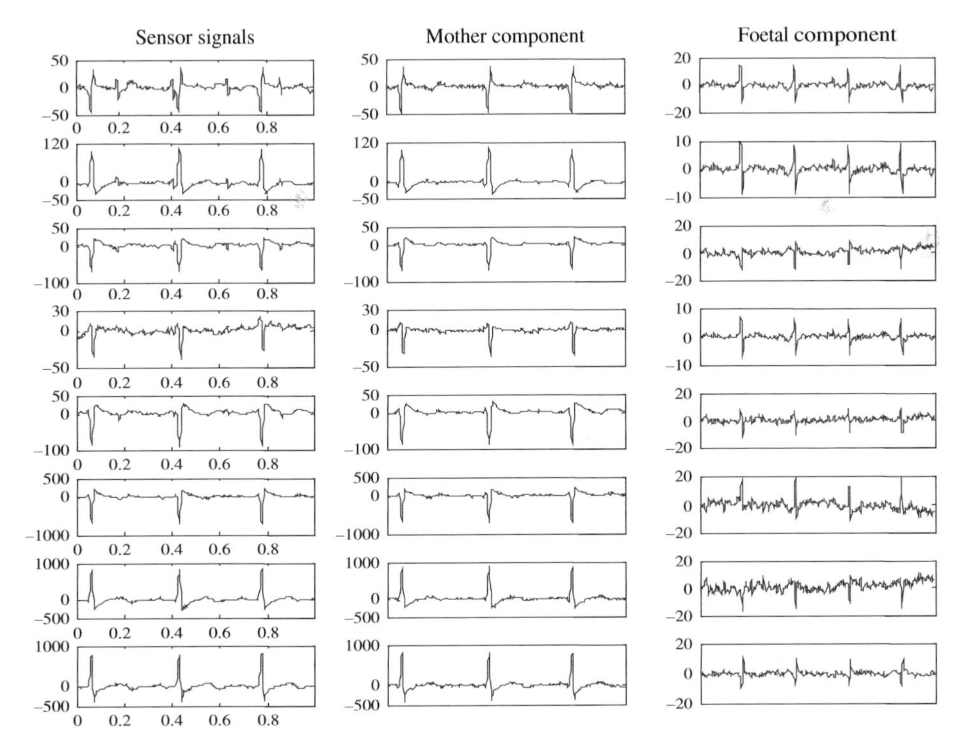

Figure 12.15 Application of Jade algorithm to separation of mother and foetus ECG. Note that signals from eight sensors are used in this example.

12.5 Summary

The in-depth understanding and utilisation of the linear system theory is of immense importance in the study and application of digital signal processing. A central concept in linear system theory is the concept of eigen analysis. Eigen analysis provides the tool for decomposition of a signal in terms of a set of orthogonal basis functions. Eigen vector analysis can be used to white or sphere a signal process, sphereing a vector-valued signal is the process of whitening the signal in such a way that all its components have unit variance. Eigen vector analysis is also used in principal component analysis (PCA) to determine the most significant orthogonal basis components of a signal process. Independent component analysis can be viewed as an extension of eigen analysis or PCA, however, there are substantial differences. Unlike PCA, ICA can be used to whiten signals that are non-Gaussian. One important application of ICA is in blind source separation (BSS), another application is in principle feature extraction for non-Gaussian signals (such as eigen image extraction) as a more efficient alternative to PCA. The main feature of ICA is that it employs non-linear influence/contrast functions in the process of estimation of the optimal linear whitening transformation.

Further Reading

Amari S.-I., Cichocki A. and Yang H.H. (1996) New learning algorithm for blind source separation. In Advances in Neural Information Processing Systems 8, pp. 757–763. MIT Press, Cambridge, MA.

Bell A.J. and Sejnowski T.J. (1995) An Information-Maximization Approach to Blind Separation and Blind Deconvolution. Neural Computation, 7, pp. 1129–1159.

Cardoso J.F. (1997) Infomax and Maximum Likelihood for Source Separation. IEEE Letters on Signal Processing, 4, pp. 112–114.

Cardoso J.F. (1998) Blind Source Separation: Statistical Principles, Proc. IEEE, 9 (10), pp. 2009–2025.

Cardoso J.F. (1999) Higher Order Contrasts for Independent Component Analysis, Neural Computations, 11, 157–192.

Cardoso J.F. and Hvam Laheld B. (1996) Equivariant Adaptive Source Separation. IEEE Trans. on Signal Processing, 44(12), pp. 3017–3030.

Cardoso J.F. and Souloumiac A. (1996) Jacobi Angles for Simultaneous Diagonalization, SIAM Journal on Matrix Analysis and Applications, SIMAX, 17 (1), pp. 161–164.

Cichocki A., Bogner R.E., Moszczynski L. and Pope K. (1997) Modified Herault–Jutten Algorithms for Blind Separation of Sources. Digital Signal Processing, 7, pp. 80–93.

Comon P. (1994) Independent Component Analysis – A New Concept? Signal Processing, 36, pp. 287–314.

Cover T.M. and Thomas J.A. (1991) Elements of Information Theory. John Wiley & Sons.

Delfosse N. and Loubaton P. (1995) Adaptive Blind Separation of Independent Sources: A Deflation Approach. Signal Processing, 45, pp. 59–83.

FastICA MATLAB package. http://www.cis.hut.fi/projects/ica/fastica/.

Friedman J.H. (1987) Exploratory Projection Pursuit. J. American Statistical Association, 82 (397), pp. 249–266.

Friedman J.H. and Tukey J.W. (1974) A Projection Pursuit Algorithm for Exploratory Data Analysis. IEEE Trans. of Computers, c-23 (9), pp. 881–890.

Golub G.H. and Van Loan C.F. (1996) Matrix Computations, 3rd edn. Johns Hopkins University Press.

Huber P.J. (1985) Projection Pursuit. Annals of Statistics, 13 (2), pp. 435–475.

Hyvärinen A., Karhunen J. and Oja E. (2001) Independent Component Analysis. John Wiley & Sons.

Jolliffe I.T. (2002) Principal Component Analysis. Springer-Verlag.

Karhunen J., Oja E., Wang L., Vigário R. and Joutsensalo J. (1997) A Class of Neural Networks for Independent Component Analysis. IEEE Trans. on Neural Networks, 8 (3), pp. 486–504.

Lee T.-W., Girolami M. and Sejnowski T.J. (1999) Independent Component Analysis using an Extended Infomax Algorithm for Mixed Sub-Gaussian and Super-Gaussian Sources. Neural Computation, 11 (2), pp. 417–441.

Makeig S., Bell A.J., Jung T.-P. and Sejnowski T.-J. (1996) Independent Component Analysis of Electroencephalographic Data. Advances in Neural Information Processing Systems, Vol. 8, pp. 145–151. MIT Press.

Marcus M. and Minc H. (1988) Introduction to Linear Algebra. New York: Dover, p. 145.

Pearlmutter B.A. and Parra L.C. (1997) Maximum Likelihood Blind Source Separation: A Context-Sensitive Generalization of ICA. Advances in Neural Information Processing Systems, 9, pp. 613–619.

Pham D.-T., Garrat P. and Jutten C. (1992) Separation of a Mixture of Independent Sources Through a Maximum Likelihood Approach. Proc. EUSIPCO, pp. 771–774.

Press W.H., Flannery B.P., Teukolsky S.A. and Vetterling W.T. (1992) Eigensystems. Chapter 11 in Numerical Recipes in FORTRAN: The Art of Scientific Computing, 2nd edn. Cambridge, England: Cambridge University Press, pp. 449–489.

Ristaniemi T. and Joutsensalo J. (1999) On the performance of blind source separation in CDMA downlink. Proc. Int. Workshop on Independent Component Analysis and Signal Separation (ICA′99), Aussois, France, pp. 437–441.

Vigário R. (1997) Extraction of Ocular Artifacts from EEG using Independent Component Analysis. Electroenceph. Clin. Neurophysiol., 103 (3), pp. 395–404.

Vigário R., Jousmäki V., Hämäläinen M., Hari R. and Oja E. (1998) Independent Component Analysis for Identification of Artifacts in Magnetoencephalographic Recordings. Advances in Neural Information Processing Systems 10, pp. 229–235. MIT Press.

Vigário R., Särelä J. and Oja E. (1998) Independent Component Analysis in Wave Decomposition of Auditory Evoked Fields. Proc. Int. Conf. on Artificial Neural Networks (ICANN'98), Skövde, Sweden, pp. 287–292.

Part III

Applications of Digital Signal Processing to Speech, Music and Telecommunications

13 Music Signal Processing and Auditory Perception

Musical instruments are some of the earliest human inventions and designs that intuitively made use of the relationships between harmonic series well before the mathematics of harmonic series were fully understood and formalised. Many string instruments, such as guitar and piano, can be considered as mechanical forms of harmonic (Fourier series) synthesisers.

Similarly, as explained in this chapter, the musical notes and the layout of the keys of some musical instruments, such as the piano, evolved to 'match' the layout and the frequency resolution of the human auditory system well before the development of a formal scientific knowledge of the anatomical structure and frequency analysis functions of the cochlea of the human ear.

Musical signal processing has a wide range of applications including digital coding of music for efficient storage and transmission on mobile phones and portable music players, modelling and reproduction of the acoustics of music instruments and music halls, digital music synthesisers, digital audio editors, digital audio mixers, spatial-temporal sound effects for home entertainment and cinemas, music content classification and indexing and music search engines for the Internet.

This chapter begins with an introduction to the applications of music signal processing and the methods of classification of different types of musical instruments. The way that musical instruments, such as guitars and violins, produce air vibrations (or sound) is explained. This is followed by a review of the basic physics of vibrations of string and pipe musical instruments, the propagation of sound waves and the frequencies of musical notes.

The human auditory perception system comprising of the outer, middle and inner parts of the ear are studied in this chapter. The factors that affect the perception of audio signals and pitch and the psychoacoustics of hearing, and how these psychoacoustic effects are utilised in audio signal processing methods, are considered.

Digital music processing methods are mostly adaptations and extensions of the signal processing methods developed for speech processing. Harmonic plus noise models of music signals, source-filter models of music instruments, and probability models of the distribution of music signals and their applications to music coding methods such as MP3 and music classification are studied in this chapter.

Multimedia Signal Processing: Theory and Applications in Speech, Music and Communications Saeed V. Vaseghi

13.1 Introduction

Music signal processing methods – used for music coding, electronic/computer music synthesis, music composition, music content recognition and music instrument identification – facilitate some of the essential functional requirements in a modern multimedia communication system.

Widespread digital processing and dissemination of music began with the introduction of CD format digital music in the 1980s, increased with the popularity of MP3 Internet music and the demand for music on multimedia mobile/portable devices such as iPod and continues with the ongoing research in automatic transcription and indexing of music signals and the modelling of music instruments.

Some examples of the applications of music signal processing methods include the following:

- Music coding for efficient storage and transmission of music signals. Examples are MP3, and Sony's adaptive transform acoustic coder.
- Music synthesis, pitch modification, audio mixing, instrument morphing (i.e. the creation of hybrids of different instruments that can be constructed digitally but may not be physically possible), audio morphing (i.e. mixing the characteristics of different vocal or musical sounds), music and vocal signal separation, audio editing and computer music composition.
- Music sound effects as in 3-D (or 5-D etc.) spatial surround music and special effect sounds in cinemas and theatres.
- Music transcription and content classification, music search engines for the Internet.
- Noise reduction and distortion equalisation such as Dolby systems, restoration of old audio records degraded with hiss, crackles etc., and signal processing systems that model and compensate for non-ideal characteristics of loudspeakers and music halls.

Music processing can be divided into two main branches: music signal modelling and music content creation. The music signal modelling approach is based on a well-developed body of signal processing theory using signal analysis and synthesis tools such as filter banks, Fourier transform, cosine transform, harmonic plus noise model, wavelet transform, linear prediction models, probability models, hidden Markov models, hierarchical models, decision-tree clustering, Bayesian inference and perceptual models of hearing. Its typical applications are music coding such as MP3, music synthesis, modelling of music halls and music instruments, music classification and indexing and creation of spatial sound effects.

Music content creation has a different set of objectives concerned with the creative methods of composition of music content. It is driven by the demand for electronic music instruments, computer music software, digital sound editors and mixers and sound effect creation. In this chapter we are mainly concerned with the transformations and modelling of music signals.

13.2 Musical Notes, Intervals and Scales

This section provides an introduction to the system of musical notes, the frequencies of the musical notes, musical intervals and musical scales.

13.2.1 Musical Notes

Musical notes are the symbols or signs that represent the frequencies, durations and timings of the elementary musical sounds. It can be said that musical notes play a similar role to the alphabet of a language; they allow music compositions and scores to be recorded in a symbolic form and read and

Table 13.1 The main musical note frequencies and symbols. Note that on a piano the sharp (or flat) notes are signified by black keys. The full table is given in Table 13.2. In this table the notes are ordered in C-major scale (C, D, E, F, G, A, B).

Notes	Prime		Second		Third	Fourth		Fifth		Sixth		Seventh
Natural	C		D		E	F		G		A		B
Sharp		C#		D#			F#		G#		A#	
Flat		Db		Eb			Gb		Ab		Bb	
Latin	Do		Re		Mi	Fa		Sol		La		Si
Frequency (Hz)	261.63	277.18	293.66	311.13	329.63	349.23	369.99	392.00	415.30	440.00	466.16	493.88
Scale step	$\sqrt[12]{2}$	$\sqrt[12]{2}$	$\sqrt[12]{2}$	$\sqrt[12]{2}$	$\sqrt[12]{2}$	$\sqrt[12]{2}$	$\sqrt[12]{2}$	$\sqrt[12]{2}$	$\sqrt[12]{2}$	$\sqrt[12]{2}$	$\sqrt[12]{2}$	$\sqrt[12]{2}$
Distance in semitones	1	1	1	1	1	1	1	1	1	1	1	1
Distance in cents	100	100	100	100	100	100	100	100	100	100	100	100
	|------------ 1 Octave = 12 Semitones = 1200 Cents ------------|											

played by musicians. The systems of musical notes also allow standardisation of musical instruments and their tuning frequencies.

The Western musical note system, as shown in Table 13.1, is based on seven basic notes, also known as the natural notes, these are:

[C, D, E, F, G, A, B]

There are also five 'sharp' notes:

[C#, D#, F#, G#, A#]

and five 'flat' notes:

[Db, Eb, Gb, Ab, Bb]

The hash sign # denotes a sharp note and the sign 'b' denotes a flat note. The sharp version of a note is a semitone higher than that note, e.g. $C\# = \sqrt[12]{2}C$, whereas the flat version of a note is a semitone lower than that note, e.g. $Db = D/\sqrt[12]{2}$.

Table 13.1 shows the seven natural music notes and their frequencies. Note that under the 12-tone equal-temperament system, C# is equal to Db, D# is equal to Eb, F# is equal to Gb, G# is equal to Ab and A# is equal to Bb.

13.2.2 Fundamental Frequency, Pitch and Intervals

Pitch is the auditory sensation of the fundamental frequency of a musical note or a sound. Whereas the fundamental frequency of a note is the inverse of its vibration period and can be measured by a frequency analyser, its pitch is the sensation felt by the human auditory perception system i.e. the combination of ear and brain.

Table 13.2 Musical sound intervals and their respective frequency ratios.

Interval	Frequency ratio	Number of semitones or cents in an interval	Examples
Octave	2:1	12 semitones (1200 cents)	G3 196 Hz and G4 392 Hz
Major sixth	5:3	9 semitones (900 cents)	G3 196 Hz and E4 329.63 Hz
Minor sixth	8:5	8 semitones (800 cents)	G3 196 Hz and D4# 311.13 Hz
Fifth	3:2	7 semitones (700 cents)	G3 196 Hz and D4 293.66 Hz
Fourth	4:3	5 semitones (500 cents)	G3 196 Hz and C4 261.63 Hz
Major third	5:4	4 semitones (400 cents)	G3 196 Hz and B3 246.94 Hz
Minor third	6:5	3 semitones (300 cents)	G3 196 Hz A3# B3b 233.08 Hz
Unison	1:1	0 semitone (0 cent)	Same note from different instruments or singers

A *music interval* is the relationship between the frequencies of two notes, or the perceived pitches of the notes, at the beginning and the end of the interval. Usually this relationship is expressed as the ratio or the distance of the frequencies (or pitch) of the musical notes at the beginning and the end of the intervals.

The most common music intervals, shown in Table 13.2 are: octave, sixth, fifth, fourth and third. Note that octave is the widest music interval with a doubling of the frequency at the end of each interval whereas the minor third is the shortest interval with a ratio of 6:5. Each octave is divided into 12 semitones and each semitone is 100 cents.

As shown in Figure 13.1, the 88 keys of the keyboard on a piano span 7 octave intervals with 12 piano keys in each interval which are composed of 7 natural notes A, B, C, D, E, F, G and five sharps C#, D#, F#, G# and A#. After each set of 12 keys the frequencies double (go up by an octave) as discussed next.

13.2.3 Consonance and Dissonance Sounds and Music Intervals

The word consonance is derived from a combination of the Latin words ‘con’ and ‘sonare’ meaning sounding ‘well’ together. In music theory sounds that are in consonance are those sounds that together are perceived as stable, harmonious and pleasant. In contrast, sounds that are in dissonance are those sounds that together are perceived as unstable, non-harmonious and unpleasant.

The intervals shown in Table 13.2 result in consonant sounds. Perfect consonances include sounds whose frequency ratios are 1 (unison), an octave, a perfect fifth or a perfect fourth. Imperfect consonances are sounds separated by the intervals of a major third, a minor third, a major sixth and a minor sixth.

13.2.4 Musical Scales

In music theory, a musical scale is a specific pattern of the pitch ratios of successive notes. The pitch difference between successive notes is known as a scale step. The scale step may be a constant or it may vary. Musical scales are usually known by the type of interval and scale step that they contain. Musical scales are typically ordered in terms of the pitch or frequencies of notes. Some of the most important examples of musical scales are the chromatic scale, diatonic scale and Pythagorean scale. Musical materials are written in terms of a musical scale.

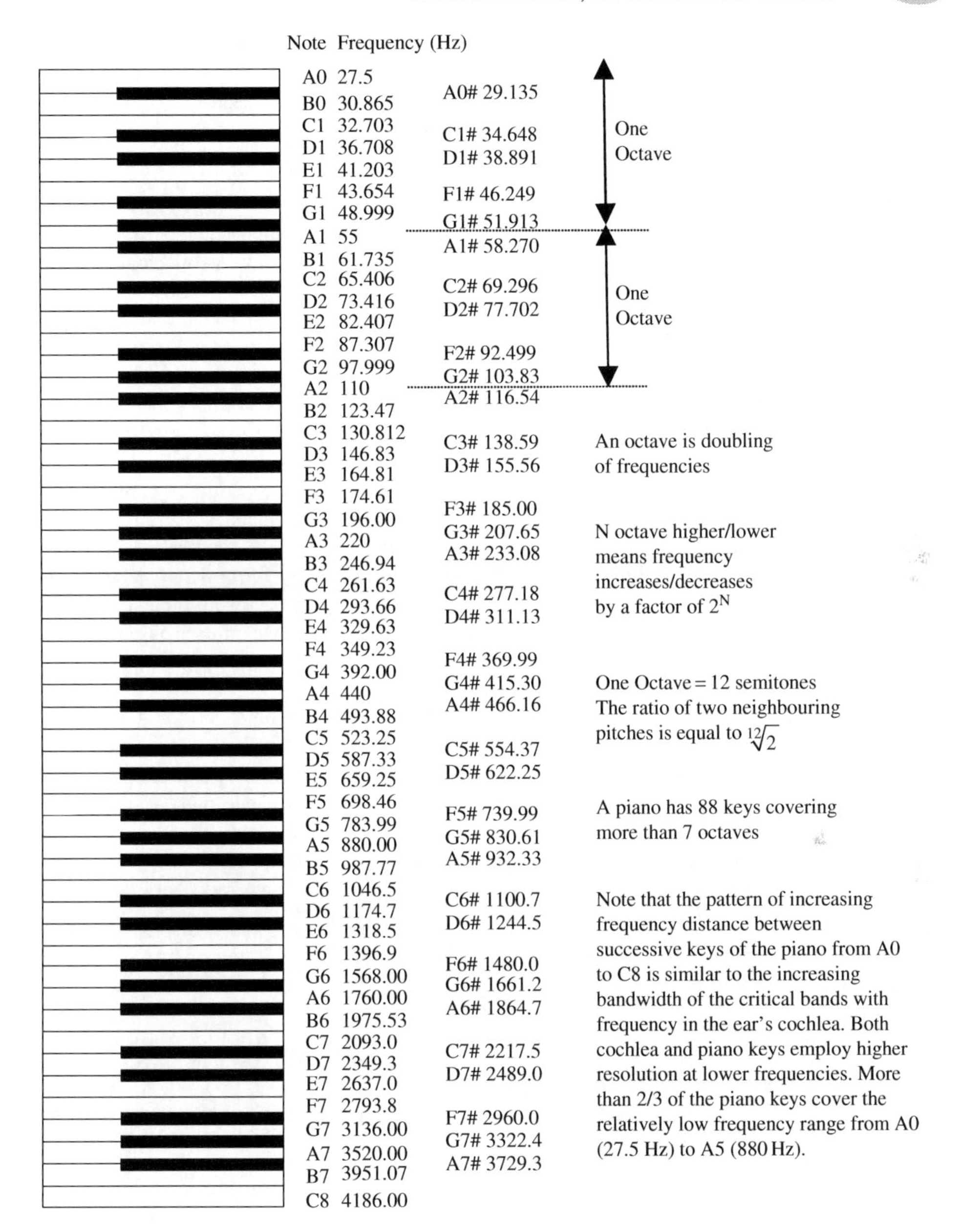

Figure 13.1 The frequencies of the keys on a piano. Note that the piano keys are arranged in groups of 12. Each set of 12 keys spans an octave which is the doubling of frequency. For example the frequency of A_N is $2^N A_0$ or N octaves higher than A_0, e.g. $A_7 = 2^7 \times 27.5 = 3520\,\text{Hz}$. Black keys correspond to sharp notes.

Chromatic Equal Temperament Musical Scale

The chromatic music scale contains all the 12 notes of the Western music scale

$$[A, A\#, B, C, C\#, D, D\#, E, F, F\#, G, G\#]$$

All other music scales can be described in terms of the chromatic scale.

In music theory, the music intervals are characterised by the exact integer number ratios defined in Table 13.2. If these ratios change then the stability and quality of the interval is compromised. A tempered scale is a system of tuning which slightly compromises the exact pitch ratios of the intervals of the musical scales in order to meet other requirements of the system, such as to allow an instrument to be used easily to play different music scales.

In an equal-temperament musical scale an interval is divided into a number of equal steps or equal frequency ratios. For the Western musical system, the most common equal-temperament scale is the 12-tone chromatic equal-temperament scale which divides the octave interval into exactly 12 equal steps known as semitones.

The most commonly accepted pitch standard is the note A4 = 440 Hz, also known as the concert pitch. In *equal-tempered chromatic scales* each successive pitch (e.g. piano key) is related to the previous pitch by a factor of the twelfth root of 2($\sqrt[12]{2} = 1.05946309436$) known as a half-tone or a semi-tone. Hence there are exactly 12 semi-tones (black and white keys on a piano) or steps, in an octave which corresponds to a doubling of pitch as illustrated in Figure 13.1 and Table 13.3.

The frequency of two musical notes separated by N semitones can be expressed in terms of each other by multiplying the starting pitch by $\left(\sqrt[12]{2}\right)^N$. For example, the G above A4 = 440 Hz (that is, G5 which is 10 piano keys above A4 in Figure 13.1) has a frequency of $440\left(\sqrt[12]{2}\right)^{10} = 783.99$ Hz.

As already mentioned, in musical tuning a *temperament* is a system of tuning which slightly compromises the exact intervals (integer pitch ratios) of scales in order to meet other requirements of the musical system such as the ability to play different scales. The chromatic pitch scale is referred to as 'equal-tempered' because a compromise is built into the use of the step size $\sqrt[12]{2}$ as the factor separating each successive pitch. For example, the notes G and C are a fifth ($3/2 = 1.5$) apart. G is seven chromatic steps above C, so, using the 12th root of 2, the pitch ratio between G and C on an equal-tempered chromatic scale standard is $\left(\sqrt[12]{2}\right)^7 = 1.4983$, which is slightly less than the 1.5 required for a perfect fifth. This slight reduction in frequency ratio is referred to as *tempering*. Tempering is necessary so that musical instruments such as a piano can play in different scales.

Matlab function MusicalNotes( )
Generates and plays the equal-temperament musical notes up to the 8th harmonics; the highest frequency played is 30.87*2^8 = 7903 Hz. The program also displays the time domain waveforms and their spectrograms.

Diatonic Scale

A diatonic `musical` scale is a seven-note music scale comprising of five whole-tone notes (notes whose distances are two semitones) and two half-tone notes (notes whose distances are a semitone). The two half-tone notes are maximally separated in that there are two or three whole-tone notes in between them.

Table 13.3 Music frequencies for the equal-tempered scale, $A_4 = 440$.

Note	Frequency (Hz)	Note	Frequency (Hz)	Note	Frequency (Hz)
C_0	16.35	B_2	123.47	$A^{\#}{}_5/B^{b}{}_5$	932.33
$C^{\#}{}_0/D^{b}{}_0$	17.32	C_3	130.81	B_5	987.77
D_0	18.35	$C^{\#}{}_3/D^{b}{}_3$	138.59	C_6	1046.50
$D^{\#}{}_0/E^{b}{}_0$	19.45	D_3	146.83	$C^{\#}{}_6/D^{b}{}_6$	1108.73
E_0	20.60	$D^{\#}{}_3/E^{b}{}_3$	155.56	D_6	1174.66
F_0	21.83	E_3	164.81	$D^{\#}{}_6/E^{b}{}_6$	1244.51
$F^{\#}{}_0/G^{b}{}_0$	23.12	F_3	174.61	E_6	1318.51
G_0	24.50	$F^{\#}{}_3/G^{b}{}_3$	185.00	F_6	1396.91
$G^{\#}{}_0/A^{b}{}_0$	25.96	G_3	196.00	$F^{\#}{}_6/G^{b}{}_6$	1479.98
A_0	27.50	$G^{\#}{}_3/A^{b}{}_3$	207.65	G_6	1567.98
$A^{\#}{}_0/B^{b}{}_0$	29.14	A_3	220.00	$G^{\#}{}_6/A^{b}{}_6$	1661.22
B_0	30.87	$A^{\#}{}_3/B^{b}{}_3$	233.08	A_6	1760.00
C_1	32.70	B_3	246.94	$A^{\#}{}_6/B^{b}{}_6$	1864.66
$C^{\#}{}_1/D^{b}{}_1$	34.65	C_4	261.63	B_6	1975.53
D_1	36.71	$C^{\#}{}_4/D^{b}{}_4$	277.18	C_7	2093.00
$D^{\#}{}_1/E^{b}{}_1$	38.89	D_4	293.66	$C^{\#}{}_7/D^{b}{}_7$	2217.46
E_1	41.20	$D^{\#}{}_4/E^{b}{}_4$	311.13	D_7	2349.32
F_1	43.65	E_4	329.63	$D^{\#}{}_7/E^{b}{}_7$	2489.02
$F^{\#}{}_1/G^{b}{}_1$	46.25	F_4	349.23	E_7	2637.02
G_1	49.00	$F^{\#}{}_4/G^{b}{}_4$	369.99	F_7	2793.83
$G^{\#}{}_1/A^{b}{}_1$	51.91	G_4	392.00	$F^{\#}{}_7/G^{b}{}_7$	2959.96
A_1	55.00	$G^{\#}{}_4/A^{b}{}_4$	415.30	G_7	3135.96
$A^{\#}{}_1/B^{b}{}_1$	58.27	A_4	440.00	$G^{\#}{}_7/A^{b}{}_7$	3322.44
B_1	61.74	$A^{\#}{}_4/B^{b}{}_4$	466.16	A_7	3520.00
C_2	65.41	B_4	493.88	$A^{\#}{}_7/B^{b}{}_7$	3729.31
$C^{\#}{}_2/D^{b}{}_2$	69.30	C_5	523.25	B_7	3951.07
D_2	73.42	$C^{\#}{}_5/D^{b}{}_5$	554.37	C_8	4186.01
$D^{\#}{}_2/E^{b}{}_2$	77.78	D_5	587.33	$C^{\#}{}_8/D^{b}{}_8$	4434.92
E_2	82.41	$D^{\#}{}_5/E^{b}{}_5$	622.25	D_8	4698.64
F_2	87.31	E_5	659.26	$D^{\#}{}_8/E^{b}{}_8$	4978.03
$F^{\#}{}_2/G^{b}{}_2$	92.50	F_5	698.46		
G_2	98.00	$F^{\#}{}_5/G^{b}{}_5$	739.99		
$G^{\#}{}_2/A^{b}{}_2$	103.83	G_5	783.99		
A_2	110.00	$G^{\#}{}_5/A^{b}{}_5$	830.61		
$A^{\#}{}_2/B^{b}{}_2$	116.54	A_5	880.00		

Using a chromatic equal-temperament scale, whose notes are separated by a semitone, a diatonic scale will have notes separated as

tone-tone-semitone-tone-tone-tone-semitone

An example of a diatonic scale is the white keys of a piano as shown in Table 13.1. A diatonic scale can start from any note (the start note is called the tonic note) so long as the separation between the sequence of notes is maintained at *tone-tone-semitone-tone-tone-tone-semitone*. Hence within the 12 steps of a chromatic scale there are 12 diatonic scales with different tonic notes.

Note that the diatonic scale is a very important scale that forms the foundation of European music and includes the modern major and minor scales and the 'church' mode/scale. Out of all the possible seven-note scales it has the highest number of consonant intervals and the greatest number of major and minor triads (a triad is a chord of three tones, in particular one consisting of a given tone with its major or minor third and its perfect, augmented, or diminished, fifth).

> **Matlab function DiatonicScale( )**
> Generates and plays musical notes in the DIATONIC musical scale according to the scale of **tone-tone-semitone-tone-tone-tone-semitone.** The starting note (tonic note) and its harmonic can be varied.

Major Scales

This is one of the diatonic scales consisting of five whole steps (each whole step is two semitones) and two half steps between the third and fourth and seventh and eighth notes. It is often considered to be made up of seven notes (eight if one includes the first note of the next octave of the scale). A prominent example of a major scale is the C-major (C, D, E, F, G, A, B, C) shown in Table 13.1 and reproduced here together with indications of the distances between the notes.

C		D		E	F		G		A		B
Do		Re		Mi	Fa		Sol		La		Si
Semitone	Whole tone		Whole tone	Semi tone		Whole tone		Whole tone		Whole tone	Semitone

Note that using the same scale of tone-tone-semitone-tone-tone-tone-semitone we can construct various other major scales such as

A-Major: A, B, C#, D, E, F#, G#, A
G-Major: G, A, B, C, D, E, F#, G
F-Major: F, G, A, Bb, C, D, E, F

and so on.

Minor Scales

This is a diatonic scale with half steps (semitones) between the second and third, fifth and sixth, and seventh and eighth notes and with whole steps for the other intervals. There are three categories of minor scales: the natural minor, harmonic minor, and melodic minor scales.

A *natural minor scale* has the interval pattern:
tone-semitone-tone-tone-semitone-tone-tone
For example, in the key of A minor, the natural minor scale is A B C D E F G A′. Where A′ is the beginning of next interval. The C-minor scale is C, D, Eb, F, G, Ab, Bb.

A *harmonic minor scale* is the same as the natural minor but with a chromatically raised (sharp) seventh note. For example, in the key of A minor, the harmonic minor scale is A B C D E F G# A′.

In the *melodic minor scale*, the sixth and seventh notes of the scale are each raised by one half step when going up the scale, but return to the natural minor when going down the scale. For example, in the key of A minor, the melodic minor scale is A B C D E F# G# A′.

Matlab function CMajor_CMinor ( )
Generates and plays C-Major and C-Minor musical scales. The program also displays the spectrograms of music signals.

Matlab function CMajor_AMajor ( )
Generates and plays C-Major and A-Major musical scales. The program also displays the spectrograms of music signals.

Pythagorean Scale

This is an important ancient musical scale which employs scale steps (frequency ratios) of 3/2 between successive notes. The Pythagorean scale is credited to the Greek philosopher and mathematician Pythagoras.

The Pythagorean scale is based on the following power series of 3/2 with respect to a reference note or pitch:

$$(3/2)^{-1}, (3/2)^0 = 1, (3/2)^1, (3/2)^2, (3/2)^3, (3/2)^4, (3/2)^5$$

where the reference note has a weight ratio of $(3/2)^0 = 1$. We can 'normalise' this power series by multiplying or dividing each step by a power of 2 in order to ensure that the scale lies within an octave interval of 1 to 2.

For example, normalising the above sequence of ratios via multiplication by weights [2, 1, 1, 1/2, 1/2, 1/4, 1/4] yields the normalised sequence

$$4/3, 1, 3/2, 9/8, 27/16, 81/64, 243/128$$

The above musical scale step ratios can be rearrange in the order of increasing frequency as

$$1, 9/8, 81/64, 4/3, 3/2, 27/16, 243/128$$

This is the ratio of successive notes in the Pythagorean scale.

As an example of a music scale generated by the Pythagorean scale consider a tonic (start) note of C4 = 261.63 Hz. Multiply the tonic note by the set [1, 9/8, 81/64, 4/3, 3/2, 27/16, 243/128] to yield the C-major scale C D E F G A B as shown by the white piano keys of Table 13.1.

13.2.5 Perfect Fifth and the Circle of Fifth

In music theory a perfect fifth interval, also known as a diapente, denotes a ratio of $3:2 = 1.5$. Perfect fifths play an important part in the composition of stable consonant (harmonious sounding) music and is one of the earliest accepted music intervals.

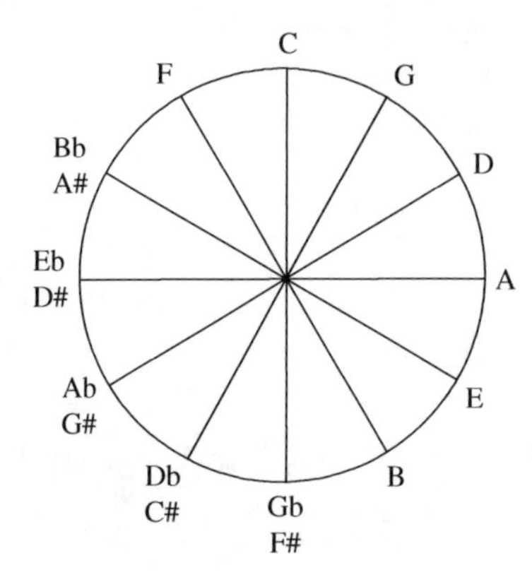

Figure 13.2 The circle of perfect fifths. Note that there is an interval of a fifth (3:2 or 7 semitones) between each two successive notes, say C_k and G_k (e.g. $G3/C3 = 196/130.81 \approx 1.5 = 3/2$) whereas in the 'opposite' direction there is an interval of a fourth (4:3 or 5 semitones) between for example G_{k-1} and C_k (e.g. $C3/G2 = 130.81/98 \approx 1.33 \approx 4/3$).

A perfect fifth interval is approximately equivalent to $\left(\sqrt[12]{2}\right)^7 = 1.4983$, hence it encompasses seven semitones or seven successive keys on a piano. From Figure 13.1 or Table 13.2 we can see that the distances between each pair of successive keys for the so-called natural notes C, G, D, A, E, B are seven piano keys. Hence these notes are a perfect fifth interval apart respectively. The complete circle of musical notes with perfect fifth intervals is shown in Figure 13.2.

Note that from C3 to G3 there are seven semitones constituting a perfect fifth, whereas in the 'opposite' direction from G2 to C3 there are five semitones constituting a perfect fourth. Hence the circle of fifths is also known as the circle of fourths. The perfect fourth interval is also known as the inverse of perfect fifth and vice versa.

13.3 Musical Instruments

There are a number of different systems for the classification of musical instruments. In a popular classification system, musical instruments are divided, according to how they set the air into vibrations, into five types of instruments: (1) string, (2) woodwind, (3) brass, (4) percussion and (5) keyboard instruments. Table 13.4 gives examples of each type of instrument and the excitation form for each instrument.

In an alternative classification system, used in some academic literature, musical instruments are classified according to four major categories, based on the vibrating material that produces the sound. This system of classification named after its inventors as the Sachs–Hornbostel system is shown in Table 13.5. There is an additional class of electronic instruments called electrophones such as the theremin, Hammond organ, and electronic and computer software synthesisers.

13.3.1 Acoustic and Electric Guitars

A guitar, shown in Figure 13.3, has four main parts:

(1) A number of guitar strings, usually six (although there are guitars with seven, eight, ten or twelve strings). The strings have different thickness and density that produce different frequencies and tones.

Table 13.4 A popular classification of music instruments.

Instrument type	Examples	Excitation source	Pitch-changing method
String	Violin, cello, guitar, piano, harp, sitar, banjo, mandolin.	String vibrations produced by plucking, hitting, or bowing the strings.	Using different lengths, thickness, density or tension of strings.
Brass	Trumpet, trombone, horn, tuba, bugle, didgeridoo, conch shell.	Vibration of air inside the tube with the player's lips pressed to the mouthpiece.	Varying the vibration of lips, varying the length of the tube (trombone), playing through different lengths of tubes (instruments with valves).
Woodwind (saxophone and some flutes made of brass)	Saxophone, clarinet, oboe, flute, piccolo, horn, bagpipes, flute, recorder, whistle.	Blowing air: against an edge (flute); between a reed and a surface (saxophone); between two reeds (oboe).	Opening and closing holes along the instrument's length with fingers.
Percussion	Drums, marimba, bells, chimes, cymbals, gong, spoons, log drum, maracas, rhythm sticks, tambourine, triangle.	A vibrating membrane or a vibrating piece of material. The instrument is made to vibrate by hitting, shaking, rubbing.	The pitch (if it exists, as for bell or drum) depends on the material, its thickness and tension.
Keyboards	Piano, pipe organ, accordion, harpsichord, clavichord, celesta.	Strings (piano) or pipes (organ).	Varying string length and tension, varying the pipe length, diameter and density.

Table 13.5 Sachs–Hornbostel classification of musical instruments.

Aerophones	Idiophones
Instruments whose tone is generated by vibration of air within the instrument as for flutes, trumpets and church organ.	Instruments whose sound is produced by the vibration of the material of the instrument. For example bells and cymbals.
Chordophones	**Membranophones**
Instruments with strings stretched between fixed points. The strings vibrate when they are plucked, struck or bowed. For example violin, guitar and harp.	Instruments from which sound is produced mainly by the vibration of a stretched membrane such as the drum.

(2) The head of the guitar containing the tuning pegs for changing the tension of the strings and hence their frequencies and damping.
(3) The neck of the guitar with frets embedded at predetermined intervals on its face for changing the effective length of the strings, and hence their frequency of vibration, by pressing the strings against the frets.
(4) A means of amplifying and shaping the spectrum of the sound emanating from the vibrations of the guitar.

The main difference between an acoustic guitar and an electric guitar is the means by which the vibrations of the strings are 'picked up', amplified and spectrally shaped. In an acoustic guitar the string vibrations are picked up and transmitted via the saddle–bridge mechanism to the guitar's resonating wooden sound box that amplifies and spectrally shapes the vibrations. In an electric guitar a set of electromagnetic pickup coils convert the vibrations of strings into induced electric currents fed to an amplifier.

Main Body of the Acoustic Guitar – The Soundboard

The wooden body of an acoustic guitar, Figure 13.3, amplifies the vibrations of strings and affects the timbre and quality of the sound by 'shaping' the spectrum of amplitude of the harmonics of vibrating strings. Guitar strings are thin and have a small surface area and hence on their own, when they vibrate, they can only displace and vibrate a small amount of air and consequently produce little sound.

To amplify the sound of the strings, the vibrations of the strings are transferred via the saddle and the bridge on which the strings rest, to the larger surface area of the sound board which is the upper part of the body of the wooden box of the guitar with a circular hole cut on it. The circular hole acts as a Helmholtz resonator and affects the overall sound of the guitar. Note that the classical experiment on Helmholtz resonance is to blow air over a bottle which makes the air in the bottle resonate.

The body of an acoustic guitar has a waist, or a narrowing, as shown in Figure 13.3. The two widenings are called bouts. The upper bout is where the neck connects, and the lower bout which is usually larger is where the bridge attaches. The size and shape of the body affect the tone and timbre of guitar. The top plate of a guitar is made so that it can vibrate up and down relatively easily. It is usually made of spruce or some other light, springy wood and is about 2–3 mm thick. On the inside of the plate there is a series of braces that strengthen the plate and keep the plate flat. The braces also affect the way in which the top plate vibrates and hence its 'tone'. The back plate is less important acoustically for most frequencies, partly because it is held against the player's body. The sides of a guitar do not radiate much sound.

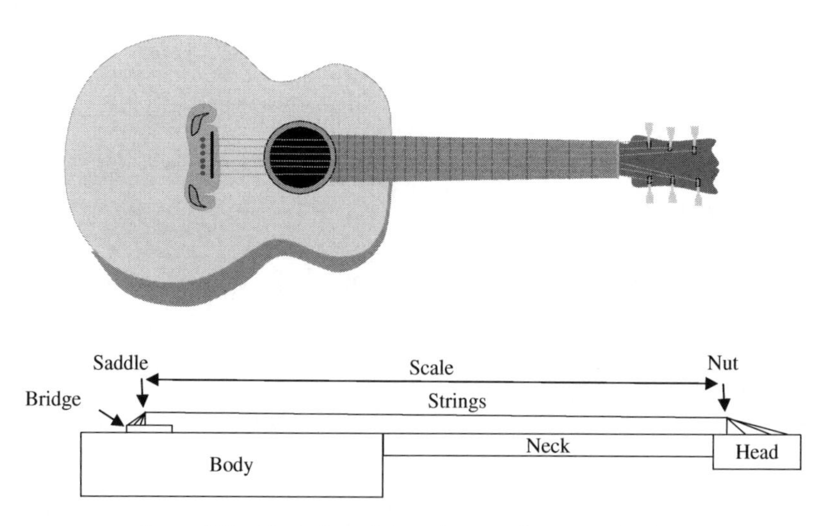

Figure 13.3 Illustration of the main parts of an acoustic guitar.

Table 13.6 The fundamental notes of guitar strings.

String number	Note	Frequency
Sixth string (lowest tone)	E2	82.4 Hz
Fifth string	A2	110 Hz
Fourth string	D3	146.8 Hz
Third string	G3	196.0 Hz
Second string	B3	246 Hz
First string (highest tone)	E4	329.6 Hz

Guitar Strings, Notes and Frequencies

There are six strings on most guitars and they are tuned from the lowest tone string (this is the string closer to the top of the guitar as it rests on a player's lap) to the highest tone string as: E, A, D, G, B, E. The standard tuning of the open strings (meaning strings not pressed on frets) is shown in Table 13.6. The pitch of a vibrating string depends on the following factors:

- *The mass of the string* – heavier strings vibrate with a lower frequency.
- *Tension* – The vibration frequency can be changed by changing the string tension using the tuning pegs; tightening a string gives a higher pitch.
- *Length* – The frequency also depends on the length of the string that is free to vibrate. Shortening the string, by stopping it on a higher fret, gives higher pitch.

Bridge, Saddle and Nut

The bridge, shown in Figure 13.3, is attached to the soundboard of a guitar. It acts as the anchor for one end of the six strings. The bridge has a thin, hard piece embedded in it called the saddle on which the strings rest. The other end of each string rests on the nut which is between the neck and the head of a guitar; the nut is grooved to hold the strings. The vibrations of the strings are transmitted via the saddle and the bridge to the upper part of the body of the guitar which acts as a sound board for amplification of the sound and shapes the frequency spectrum and timbre of the sound. Note that the saddle and the nut hold the two effective 'open' vibrating ends of the string. The distance between these two points is called the scale length of the guitar.

Head, Neck and Frets

The head of a guitar is the part that contains the tuning pegs. The neck is the part that connects the head to the main body of the guitar. The face of the neck, containing the frets, is also called the fingerboard. The frets are metal pieces embedded into the fingerboard at specific intervals. By pressing a string down onto a fret, the effective vibrating length of the string and therefore its fundamental frequency of vibration or tone is changed.

Frequency Intervals Between Frets

The interval between two successive frets corresponds to a change of the effective vibrating length and hence the frequency of vibration of a string by a factor of $\sqrt[12]{2} = 1.059463$. Hence 12 fret intervals correspond to doubling of frequency or an octave (it also corresponds to halving of the string length). Note that the ratio of the vibrating length of the string at the kth and $(k-1)^{\text{th}}$ frets is $L_{k-1}/L_k = \left(\sqrt[12]{2}\right)$. Hence the length of the string at the kth frets is given by

$$L_k = L_{k-1}/\left(\sqrt[12]{2}\right) = L_0/\left(\sqrt[12]{2}\right)^k \tag{13.1}$$

where L_0 is the full length of the vibrating string between the nut and the saddle. The distance between the kth and $(k-1)^{\text{th}}$ fret is given by

$$L_{k-1} - L_k = L_0\left(\left(\sqrt[12]{2}-1\right)/\left(\sqrt[12]{2}\right)^k\right) \tag{13.2}$$

hence the distance between successive frets decreases progressively.

Figure 13.4 illustrates the various notes of guitar strings when the effective length of the strings is changed by pressing them on different frets. Table 13.7 shows the frequencies of the notes of the first string with the string pressed on different frets.

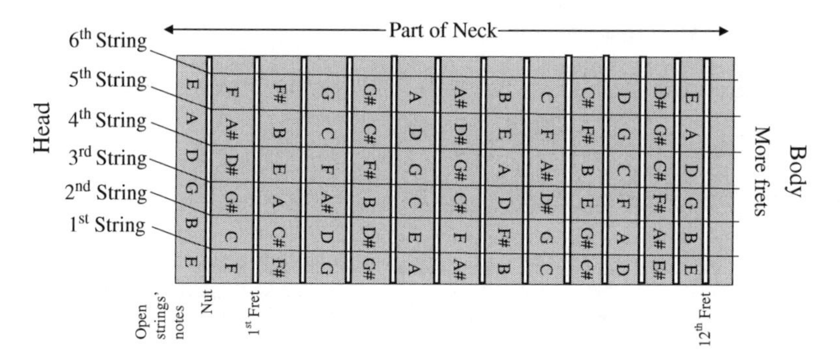

Figure 13.4 Illustration of the first 12 frets and the corresponding notes of guitar strings when the effective length of a string is changed by pressing them on different frets. Note that musical notes at the 12th fret are an octave higher (double) than the open strings' notes.

Table 13.7 The frequencies of the notes of the first string with the string pressed on different frets. It is assumed that the scale length is 26 inches. Note as the length of string halves the frequency of its pitch doubles.

Note	Fret	Frequency (Hz) (1st string)	Fret position from saddle (Hz)	Note	Fret	Frequency (inch) (1st string)	Fret position from saddle (inch)
E4	open	329.6	26.00	B4	7	493.8	17.35
F4	1	349.2	24.54	C4	8	523.2	16.38
F4#	2	370.0	23.16	C4#	9	554.3	15.46
G4	3	392.0	21.86	D4	10	587.3	14.59
G4#	4	415.3	20.64	D4#	11	622.2	13.77
A4	5	440.0	19.48	E5	12	659.2	13.00
A4#	6	466.1	18.38				

Electric Guitars

Electric guitars do not need a hollow vibrating box to amplify the sound. Hence, the body of an electric guitar can be made of a solid of any shape. In an electric guitar the mechanical vibrations of the strings are picked up by a series of electro-magnetic coils placed underneath the strings. The coil-wrapped magnets convert the mechanical vibrations of strings into electric vibrating currents which are then band-pass filtered and amplified by an electronic amplifier. Guitar amplifiers can do more than amplification; they can be operated in their non-linear distortion region to create a variety of rich sounds.

Electromagnetic pickups work on the principle of variable magnetic reluctance. The pickup consists of a permanent magnet wrapped with many turns of fine copper wire. The pickup is mounted on the body of the instrument, close to the strings. When the instrument's metal strings vibrate in the magnetic field of the permanent magnet, they alter the reluctance of the magnetic path. This changes the flux in the magnetic circuit which in turn induces a voltage in the winding. The signal created is then carried for amplification. Electric guitars usually have several rows of pickups, including humbucking pickups, placed at different intervals. A humbucking pickup comprises two standard pickups wired together in series. However, the magnets of the two pickups are reversed in polarity, and the windings are also reversed. Hence, any hum or other common mode electro-magnetic noise that is picked up is cancelled out, while the musical signal is reinforced. The design of the pickup magnetic coils is an important aspect of the design of an electric guitar.

13.3.2 The Violin

The violin evolved from earlier string instruments such as the rebec, a Middle-Eastern bow-string instrument, the lira da braccio and the fiddle. In its modern form the violin, shown in Figure 13.5, emerged in Italy around 1550; the most renowned violins were made by Cremonese violin-makers, like Amati, Stradivari and Guarneri, dating from about 1600 to 1750.

Violin sounds are produced by drawing a bow across one or more of four stretched strings. The string tensions are adjusted by tuning pegs at one end of the string, so that their fundamental frequencies are tuned, at perfect fifth (ratio 3:2) intervals, to about

$$196\,\text{Hz},\ 293.66\,\text{Hz},\ 440\,\text{Hz},\ 659.66\,\text{Hz}$$

corresponding to the notes

$$\text{G3, D4, A4 and E5}$$

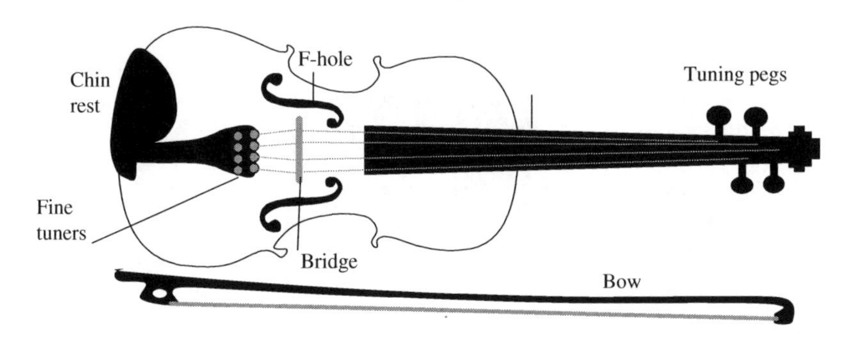

Figure 13.5 Sketch of a violin.

or in Latin notation

Sol, Re, La, Mi

respectively.

The vibrations of strings produce little sound on their own as they can only displace a small amount of air. To amplify the sound and to shape its frequency spectrum, energy from the vibrating strings is transferred to the vibrations of the larger surface area of the wooden sound box. The main plates of the violin's wooden box vibrate, amplify and shape the frequency spectrum of the sound. The strings are supported by the 'bridge', shown in Figure 13.6, which defines the effective vibrating length of the string, and acts as a mechanical transformer. The bridge converts the transverse forces of the strings into the vibrations of the wooden sound box. The bridge has its own frequency response and resonant modes and affects the overall tone and the sound of the instrument.

The front plate of the violin is carved from fine-grained pinewood. The back plate is made from maple and pine is used for the sides.

The two f-shaped holes cut into the front plate influence the timbre of the violin. They allow the vibrations of the air inside the box to be transmitted to the outside air. The size and position of the f-shape holes affect the frequency response of the violin. The resonant frequency is also affected by the area of the f-holes and the volume of the instrument.

The output of the violin is increased by attaching a solid rod, the sound post, between the back and front plates, close to the feet of the bridge. The vibration force exerted by the bowed strings causes the bridge to rock about and the sound box plates to vibrate. This has the effect of increasing the volume of the output sound. The violin has a *bass* bar glued underneath the top plate to dampen its response at higher frequencies and prevent the dissipation of vibration energy into acoustically inefficient higher frequencies.

Vibrations of Violin versus Guitar

The plucking of a guitar's strings produces a set of sinusoidal waves and their harmonics on the strings; these can be modelled by linear system theory. The linearity and superposition principles imply that the sounds produced by plucking two strings of a guitar is the sum of their individual sounds and when a string is plucked harder a sound of greater amplitude with the expected pitch of the string is produced.

In contrast to the plucking of a guitar, the bowing of a violin produces a saw-tooth waveform. Furthermore, the behaviour of a violin and bow system is non-linear in that, for example, a greater amount of force applied via the bow onto a string does not just produce a bigger or longer sound but it may produce a very different, perhaps scratchy and unpleasant, sound. It is this linear behaviour of

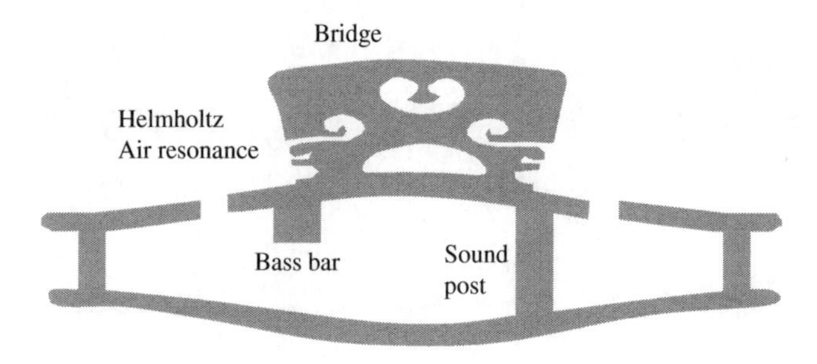

Figure 13.6 Cross-section of violin at the bridge.

a guitar vs. the non-linear behaviour of a violin that underlies the fact that for a beginner it is usually easier to play a guitar in a pleasant musical-sounding way, whereas in contrast it is usually difficult for a beginner to play a violin in a musical and pleasant way.

Helmholtz Waves on Violin, The Slip–Stick Mechanism

Hermann von Helmholtz (a German physician and physicist, 1821–94) investigated the vibrations of bowed violin strings and observed that when a violin string is bowed, it vibrates in a different way from the sinusoidal vibrations set up when the strings of a guitar are plucked.

The bowing of the strings of a violin sets up two sets of string vibrations composed of (i) a set of vibrations parallel to the bowing direction, and (ii) a set of transverse vibrations, made up of V-shaped straight lines.

Hence, the bowing of strings gives rise to a set of V-shaped Helmholtz waves with a vertex separating two straight sections of the strings as shown in Figure 13.7. When the vertex of a V-shaped Helmholtz wave is between the bow and the fingered end of the string, the string moves at the same speed and direction as the bow. In this phase, only a small amount of force is needed to lock the motion of the bow with that of the string. This is known as the *sticking phase*. When the vertex of the Helmholtz waveform moves past underneath the bow on its way to the bridge and back, the string slips past the bow and starts to move in the opposite direction to the bow. This is known as the *slipping phase*.

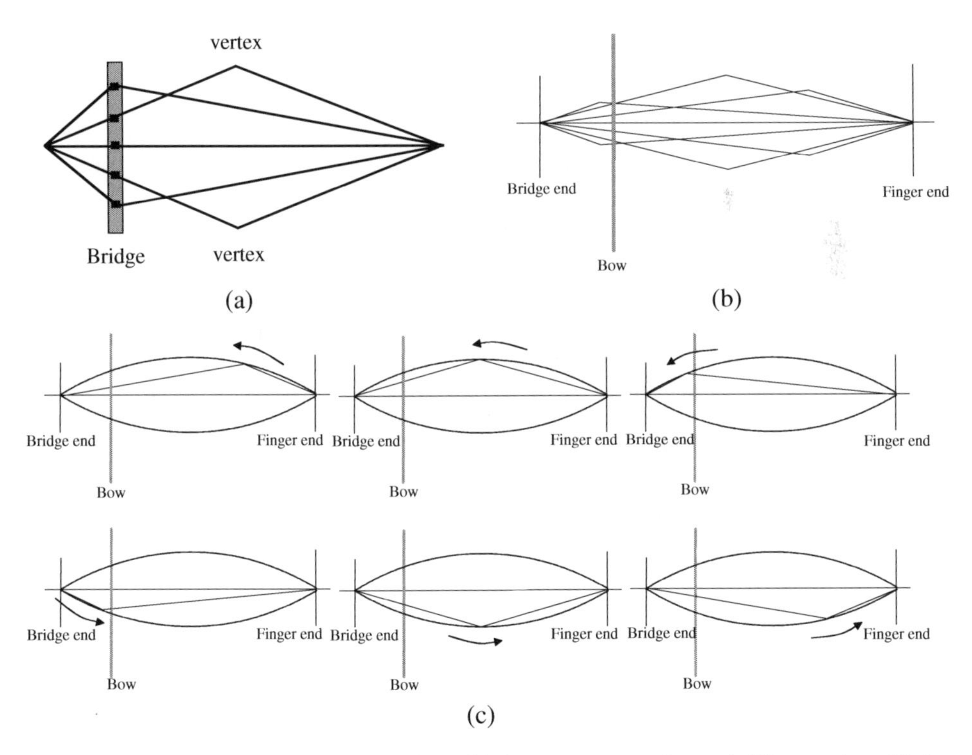

Figure 13.7 Saw-tooth movement of violin strings: (a) shows movement of several different strings, (b) shows how one string may move back and forth, (c) shows 'snap shots' of movements of one string in (b).

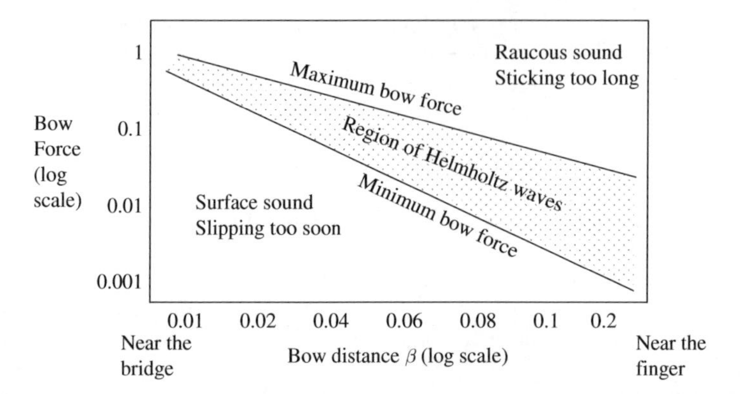

Figure 13.8 The Schelleng diagram of bow force versus position for a long steady bow stroke of a violin. The shaded area is where Helmholtz waves are set up.

For the stick–slip mechanism of the Helmholtz waves of a violin to produce sustained musical sounding waves the bow force exerted on the strings must be within certain maximum and minimum bounds that depend on the distance of the bow from the bridge, as shown in the Schelleng diagram in Figure 13.8.

Assuming that the bow is at a distance of βL from the bridge, where L is the length of the string between the finger and the bridge, then the minimum force is proportional to β^{-2} and the maximum force is proportional to β^{-1}. The formulae for the maximum and minimum forces of the Schelleng diagram are given by

$$f_{\max} = \frac{2Zv_b}{(\mu_s - \mu_d)\beta} \tag{13.3}$$

$$f_{\min} = \frac{Z^2}{2R} \frac{2v_b}{(\mu_s - \mu_d)\beta^2} \tag{13.4}$$

where β is the bow position, v_b is the bow velocity, μ_s is the coefficient of static friction, μ_d is the coefficient of dynamic friction, Z is the characteristic impedance of the string, and R represents losses of the string

The saw-tooth signal generated on the top of the violin bridge by a bowed string has a rich harmonic content. The amplitude of each frequency component of the saw-tooth signal is modified by the frequency response of the instrument, which is determined by the resonance of the bridge and by the vibrations of the body of the violin, Figure 13.9. At low frequencies the bridge acts as a mechanical lever. However, between 2.5 and 3 kHz the bowing action excites a strong resonance of the bridge. This boosts the signal amplitude in this frequency range, where the ear is most sensitive. Another resonance occurs at about 4.5 kHz. Between these two resonances there is a dip in the frequency response.

13.3.3 Wind Instruments

Examples of wind instruments include different forms, shapes and arrangements of brass and wooden cylindrical tubes such as saxophone, clarinet, flute, recorder, whistle, trumpet, trombone, horn, tuba, bugle and didgeridoo.

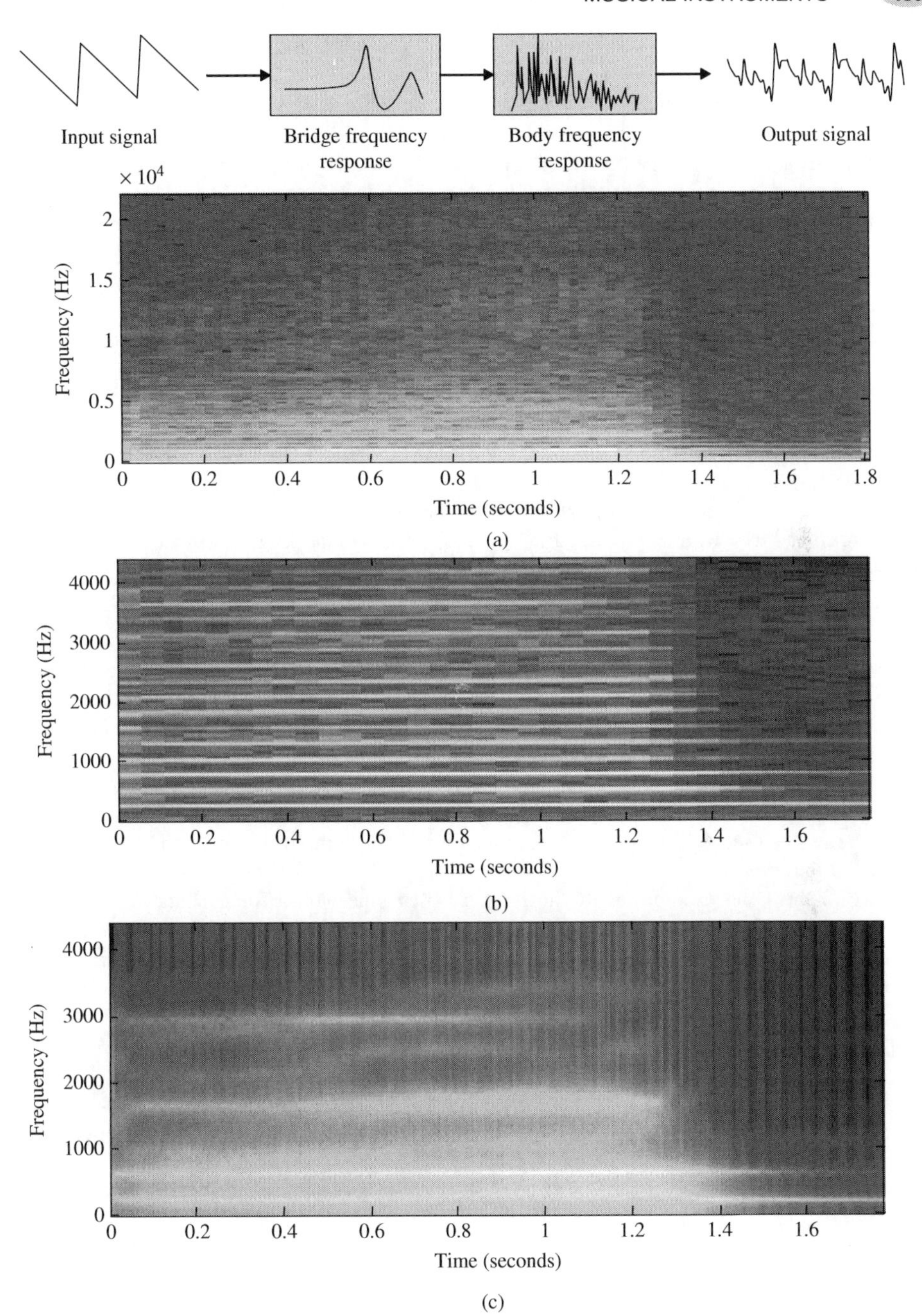

Figure 13.9 (a) An illustration of the input and output waveforms of a violin together with the frequency responses of the bridge and the violin body; (b) the spectrogram of a violin note sampled at 44,100 Hz; (c) the spectrogram of the note in (b) down-sampled by a factor of 4 to 11,050 Hz; (d) the spectrogram of the frequency response of a linear predictor model of the violin sound box obtained from the down-sampled signal. Note lighter shades indicate higher power levels.

Open-End Tubes

To study how wind instruments work we consider the simplest example of a wind instrument, a penny whistle also called tin whistle or Irish whistle. This is a cylindrical instrument open at both ends with six holes cut along its length, and with a flat slanted narrow tube at one end as the mouthpiece. The mouthpiece directs an air stream at a slanted hole with a sharp edge that splits the air stream causing air currents to excite the tube. The whistle is tuned diatonically, which allows it to be used in two major keys (such as D major and G major, or C major and F major) and their corresponding minor keys and modes.

To understand how the whistle produces sound vibrations, assume it has all its finger holes covered. Consider the propagation along the tube of a sudden change in air pressure at one end of the tube, such as the lowering of the pressure by taking some air out of one end. The adjacent air molecules will move in to fill the vacuum, leaving behind a new vacuum, which will be filled by the neighbouring air molecules and so on. In this way a pulse of low-pressure air will propagate along the tube. When the pulse arrives at the open end of the tube it will attract air from the room into the tube and will be reflected back with a changed polarity as a high-pressure pulse of air.

A cycle of low- and high-pressure air along the tube forms the fundamental period of the sound from the open-end tube with a wavelength of $\lambda = 2L$ where L is the length of the tube. Assuming the speed of propagation of sound is c, the fundamental frequency f_1 of an open-end tube is

$$f_1 = \frac{c}{2L} \tag{13.5}$$

The quality of resonance of the whistle's sound depends on the reflection and loss properties of the tube. In an open-ended tube there is no effective containment at the ends of the tube other than the room pressure that forms the pressure nodes. For the fundamental note, there are two pressure nodes at the ends and a pressure anti-node in the middle of the tube. The boundary condition of two pressure nodes at the ends of the tube, is also satisfied with all integer multiples of the fundamental frequency, hence integer multiples of the fundamental note exist with different intensities. In addition, the finger holes of a whistle can be used to change its effective length and hence the wavelength and the fundamental frequency of the sound.

Table 13.8 shows an example of the notes of an open pipe with a fundamental note of C3 at a frequency of 261.63 Hz from an open pipe of length $L = c/(2 \times f) = 340/(2 \times 261.63) \approx 65\,\text{cm}$, where $c = 340\,\text{m/s}$ is the speed of sound at room temperature. The fundamental, its first 11 harmonics and their frequencies and nearest musical notes are shown in Table 13.8.

Closed-End Tubes – The Trumpet

For a closed-end tube the pressure of the reflected wave at the closed end is the same as that of the incoming wave, hence the reflected wave has the same polarity as the incoming wave. This implies

Table 13.8 An example of the harmonics of an open-end pipe with a length of 65 cm and a pitch note of C4.

1	C4 261.63 Hz	7	Bb6 1831.4 Hz
2	C5 523.25 Hz	8	C7 2093 Hz
3	G5 784.86 Hz	9	D7 2354.7 Hz
4	C6 1046.5 Hz	10	E7 2616.3 Hz
5	E6 1308.1 Hz	11	F7 2877.9 Hz
6	G6 1569.7 Hz	12	G7 3139.6 Hz

that the wavelength of the fundamental note is four times the length of the tube (the same as two round trips in the tube). Hence, the fundamental frequency of a closed-end tube is one half of a similar open-end tube or equivalently an octave lower. A closed-end tube will generate a harmonic series based on odd integer multiples of the fundamental frequency. Hence, a closed-end cylindrical tube will produce an unusual set of musical of notes. In this section we consider the trumpet as a closed-end pipe.

Effect of The Trumpet's Shape on its Harmonic Series

Figure 13.10 illustrates the different sections of a trumpet. The effect of the conical pipe (the flare) and the bell-shaped open-end in a trumpet is to introduce the missing even harmonics of the closed-end pipe. The bell-shaped end also increases sound radiation efficiency and hence the power of the radiated higher frequencies of sounds which accounts for the brightness of brass instruments such as the trumpet.

Consider a closed-end instrument such as a trumpet; the player's lips closes one end of the pipe. Since closed-end cylindrical pipes only produce odd harmonics, this will exclude octaves, which are powers of two multiples of the fundamental frequency. However, due to the design of its shape and the effect of its non-cylindrical parts (Figure 13.10), a trumpet produces a harmonic series (or over-tones) that include the octaves. The most non-cylindrical parts are the mouthpiece and the bell.

To analyse the effect of adding a mouthpiece and bell to a cylindrical closed-end pipe, consider the assumed closed-end harmonic series in Table 13.9.

The addition of a mouthpiece lowers the top six odd harmonics 9–19 yielding the new harmonic series in Table 13.10.

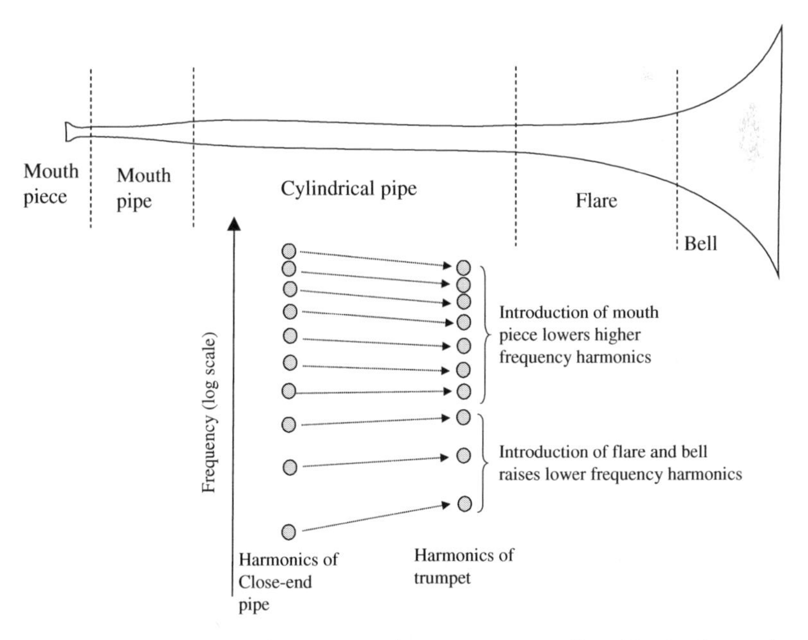

Figure 13.10 The harmonics of a pipe instrument, such as a trumpet, are affected by the shape and construction of the mouth pipe, cylindrical pipe, flare and bell sections.

Table 13.9 The odd harmonics of a closed-end pipe and the nearest notes.

1	70 Hz, C#2	11	770, G5
3	210, G#3	13	910, A5
5	350, F4	15	1050, C6
7	490, B4	17	1190, D6
9	630, D#5	19	1330, E6

Table 13.10 The harmonics of a trumpet pipe with the mouthpiece added.

1	70 Hz, C#2	11	750, F#5
3	210, G#3	13	875, A5
5	350, F4	15	1000, B5
7	490, B4	17	1125, C6
9	625, D#5	19	1250, D#6

Table 13.11 The harmonics of a closed-end pipe with the addition of a mouthpiece and bell section.

1	125 Hz, B2	6	750, A5
2	250, B3	7	875, A5
3	375, F4#	8	1000, B5
4	500, B4,	9	1125, C6
5	625, D#5	10	1250, D#6

The addition of a flare and bell section to the end of the trumpet tube raises the four lower odd harmonics 1–7.

Note that whereas Table 13.9 lists the odd harmonics of a closed-end tube, the harmonic series in Table 13.11 contain all multiples (even and odd) of a pitch note of B2 with a fundamental frequency of 125 Hz. However for a trumpet the fundamental frequency is not part of the series as it does not have appreciable energy; it can however be recreated (as a so-called pedal note) by the vibrations of the lips aided by the vibrations of air in the pipe.

13.3.4 Examples of Spectrograms of Musical Instruments

Figure 13.11 shows some typical spectrograms of examples of string, brass, pipe and percussion instruments, for single note sounds. The figure reveals the harmonic structure and or shaped-noise spectrum of different instruments.

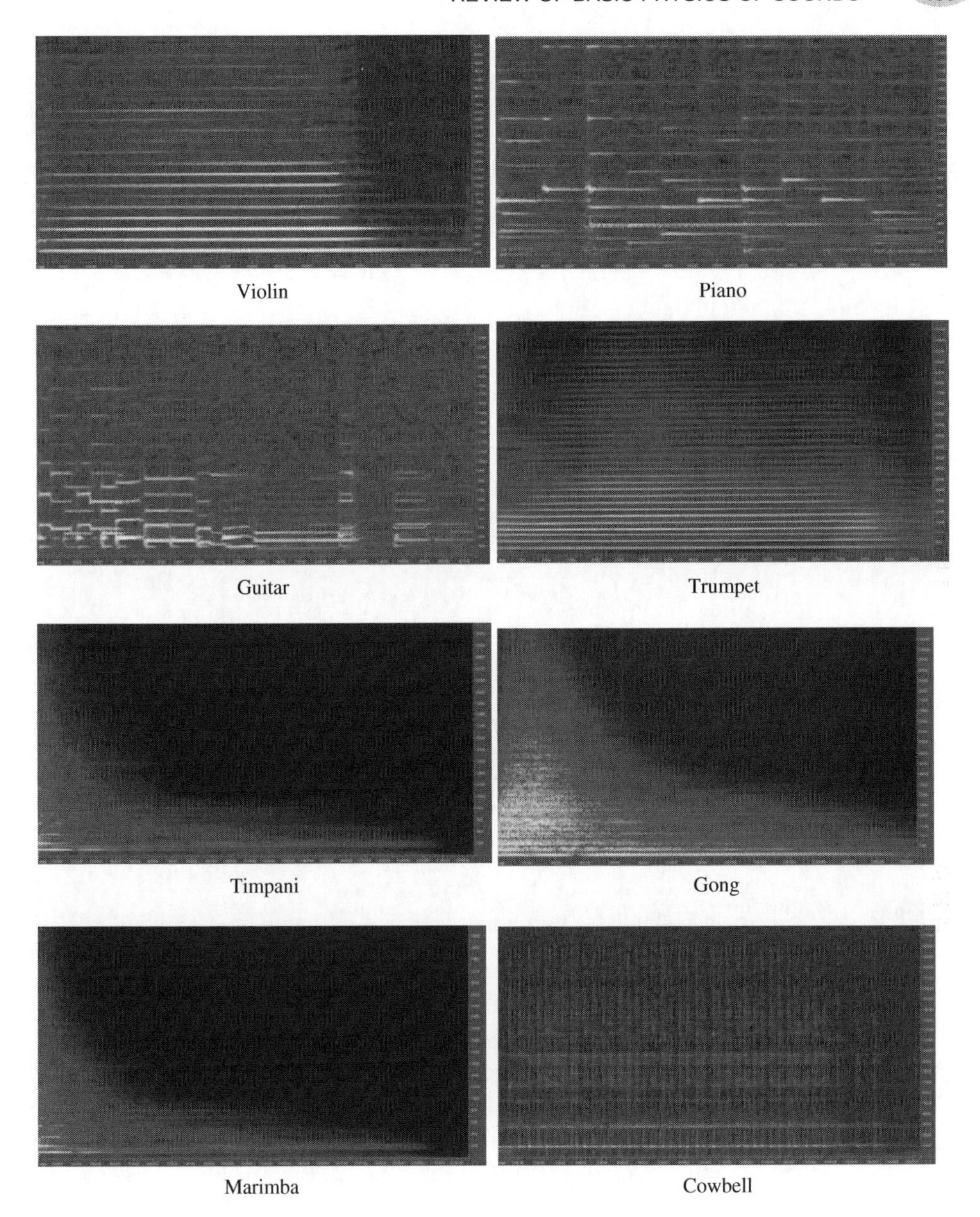

Figure 13.11 Examples of spectrograms of some musical instruments.

13.4 Review of Basic Physics of Sounds

Sound is the audible sensation of air pressure variations caused by the vibrations, movement, friction or collision of objects. In this section we review the basic physics, properties and propagation of sound waves.

13.4.1 Sound Pressure, Power Intensity Levels, and Speed

Sound Pressure Level

The minimum audible air pressure variations (i.e. the threshold of hearing) p_0 is only 10^{-9} of the atmospheric pressure or $2 \times 10^{-5}\,\text{N/m}^2$(Newton/meter^2). Sound pressure is measured relative to p_0 in decibels as

$$\text{SPL}(p) = 20\log_{10}(p/p_0)\ \text{dB} \tag{13.6}$$

From Equation (13.6) the threshold of hearing is 0 dB. The maximum sound pressure level (the threshold of pain) is $10^6 p_0$ (10^{-3} of the atmospheric pressure) or 120 dB. Hence the range of hearing is about 120 dB, although the range of comfortable and safe hearing is less than 120 dB.

Sound Power Level

For an audio tone with a power of w watts, the sound power level is defined in decibels relative to a reference power of $w_0 = 10^{-12}$ watts (or 1 picowatt) as

$$\text{PL} = 10\log_{10}(w/w_0) = 10\log_{10} w + 120\,\text{dB} \tag{13.7}$$

Hence with the power reference level of a picowatt used in Equation ((13.7)), 1 picowatt is 0 dB, 1 watt is 120 dB, 10 watts is 130 dB, 100 watts is 140 dB and so on.

Sound Intensity Level

This is defined as the rate of energy flow across a unit area as

$$\text{IL} = 10\log_{10}(I/I_0) = 10\log_{10} I + 120\,\text{dB} \tag{13.8}$$

where $I_0 = 10^{-12}\,\text{watts/m}^2$.

Loudness of Sounds

The perceived loudness of a sound is measured in units of *phon* and *sone*. One phon is equal to 1 dB of sound pressure level above the nominal threshold of hearing, the sound pressure level SPL of 20 micropascals $= 2 \times 10^{-5}$ Pascal. One sone, defined as the loudness of a 1 kHz tone at 40 dB SPL, is equivalent to 40 phons. The loudness of a sound at SPL (dB) at 1 kHz is found by

$$\text{Loudness} = 2^{(\text{SPL}-40)/10}\ \text{sones} \tag{13.9}$$

Hence every 10 dB increase in SPL corresponds to a doubling of the number of sones and the perceived loudness of sounds.

Speed of Sound Propagation

Sound travels with a speed of

$$c = 331.3 + 0.6t\ \text{m/s} \tag{13.10}$$

where t is the temperature of the air in degrees Celsius. Hence, at 20 °C the speed of sound is about 343.3 metres per second or about 34.3 cm per ms. Sound propagates faster in liquids than in air and faster in solids than in liquids. The speed of propagation of sound in water is 1500 m/s and in metals it can be 5000 m/s.

13.4.2 Frequency, Pitch, Harmonics

Sound waves are produced by vibrating objects and instruments. The frequency of a sound is the same as that of the source and is defined as the number of oscillations per second in the units of hertz. Since a sound wave is a pressure wave, the frequency of the wave is also the number of oscillations per second from a high pressure (compression) to a low pressure (rarefaction) and back to a high pressure.

The human ear is a sensitive detector of the fluctuations of air pressure, and is capable of hearing sound waves in a range of about 20 Hz to 20 kHz. The sensations of prominent sound frequencies are referred to as the *pitch* of a sound. *A high pitch sound corresponds to a high fundamental frequency and a low pitch sound corresponds to a low fundamental frequency*. The harmonics of a fundamental frequency F_0 are its integer multiples kF_0.

Certain sound waves which when played simultaneously produce a pleasant sensation are said to be *consonant*. Such sound waves form the basis of the *intervals* in music. For example, any two sounds whose frequencies make a 2:1 ratio are said to be separated by an *octave* and two sounds with a frequency ratio of 5:4, are said to be separated by an *interval* of a third. Examples of other music sound wave intervals and their respective frequency ratios are listed in Table 13.8.

Wavelength of Sounds

The wavelength λ of a sound wave depends on its speed of propagation c and frequency of vibration f through the equation $\lambda = c/f$. For example at a speed of 344 m/s, a sound wave at a frequency of 10 Hz has a wavelength of 34.4 metres, at 1 kHz it has a wavelength of 34.4 cm and at 10 kHz it has a wavelength of 3.44 cm.

Bandwidths of Music and Voice

The bandwidth of unimpaired hearing is normally from 10 Hz to 20 kHz, although some individuals may have a hearing ability beyond this range of frequencies. Sounds below 10 Hz are called infra-sounds and above 20 kHz are called ultra-sounds. The information in speech (i.e. words, speaker identity, accent, intonation, emotional signals etc.) is mainly in the traditional telephony bandwidth of 300 Hz to 3.5 kHz.

The sound energy above 3.5 kHz mostly conveys quality and sensation essential for high quality applications such as broadcast radio/TV, music and film sound tracks. The singing voice has a wider dynamic range and a wider bandwidth than speech and can have significant energy in the frequencies well above that of normal speech. For music the bandwidth is from 10 Hz to 20 kHz. Standard CD music is sampled at 44.1 kHz or 48 kHz and quantised with the equivalent of 16 bits of uniform quantisation which gives a signal-to-noise ratio of about 100 dB at which the quantisation noise is inaudible and the signal is transparent.

13.4.3 Sound Propagation: Reflection, Diffraction, Refraction and the Doppler Effect

The propagation of sound waves affects the sensation and perception of music. Sound propagates from the source to the receiver through a combination of four main propagation modes: (1) direct propagation path, (2) reflection from walls, (3) diffraction around objects or through openings and (4) refraction due to temperature differences in the layers of air. In general, in propagating through different modes sound is delayed and attenuated by different amounts.

Reflection happens when a sound wave encounters a medium with different impedance from that in which it is travelling, for example when the sound propagating in the air hits the walls of a room as shown in Figure 13.12. Sound reflects from walls, objects, etc. Acoustically, reflection results either in sound reverberation for small round-trip delays (less than 100 ms), or in echo for longer round-trip delays.

Diffraction is the bending of waves around objects and the spreading out of waves beyond openings as shown in Figure 13.13. In order for this effect to be observed the size of the object or gap must be comparable to or smaller than the wavelength of the waves. When sound waves travel through doorways or between buildings they are diffracted, so that the sound is heard around corners.

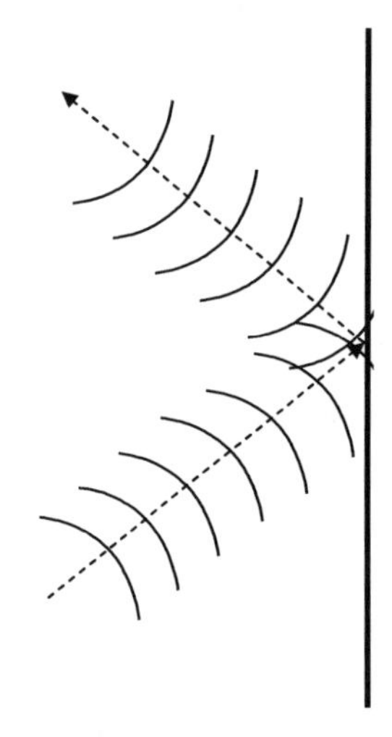

Figure 13.12 Illustration of reflection.

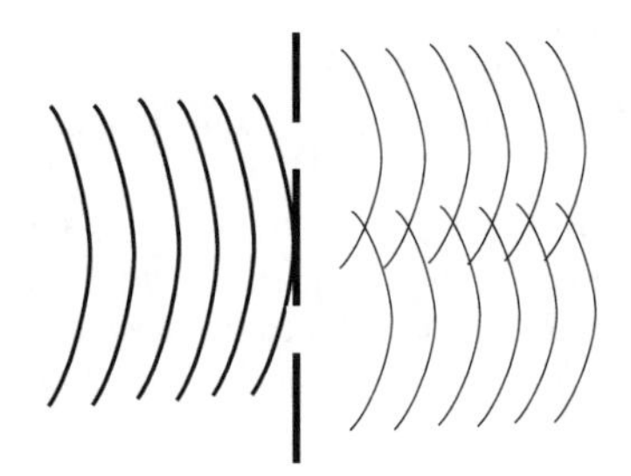

Figure 13.13 Illustration of diffraction.

If we consider two separate 'windows' then each 'window' acts as a new source of sound, and the waves from these secondary sources can act constructively and destructively. When the size of the openings or obstacles is about the same as the wavelength of the sound wave, patterns of maxima and minima are observed. If a single opening is divided into many small sections, each section can be thought of as an emitter of the wave. The waves from each piece of the opening are sent out in phase with each other; at some places they interfere constructively, and at others they interfere destructively.

Refraction, shown in Figure 13.14, is the bending of a wave when it enters a medium where its speed of propagation is different. For sound waves, refraction usually happens due to temperature changes in different layers of air as the speed of sound increases with temperature, so during the day when the higher layers of air are cooler sound is bent upwards (it takes longer for sound to travel in the upper layers) and during the night when a temperature inversion happens the sound is bent downwards.

The *Doppler effect* is the perceived changes in the received frequency of a waveform resulting from relative movements of the source (emitter) towards or away from the receiver. As illustrated in Figure 13.15, the received pitch of the sound increases (sound wave fronts move towards each other) when there is a relative movement of the source towards the receiver and decreases when there is a relative movement of the source away from the receiver. When a sound source approaches a receiver with a relative speed of v, the perceived frequency of the sound is raised by a factor of $c/(c-v)$, where c is the speed of sound. Conversely, when a sound source moves away from the receiver with a relative speed of v, the perceived frequency of the sound is lowered by a factor of $c/(c+v)$. When $v \geq c$ the sound barrier is broken and a sonic boom due to reinforcement of densely packed wave fronts is heard. The sound barrier speed is 761 mph.

Assuming that the sound source is moving with speed $\pm v_{sr}$ (where + is towards receiver and – away from it) and the receiver is moving with speed $\pm v_{rs}$ (where + is towards source and – away from it) then the relationship between the perceived frequency f_r and the source frequency f_s is given by

$$f_r = \frac{c + v_{rs}}{c - v_{sr}} f_s = \frac{1 + \frac{v_{rs}}{c}}{1 - \frac{v_{sr}}{c}} f_s \qquad (13.11)$$

where as explained v_{sr} and v_{rs} can be positive or negative depending on the relative direction of movement.

Figure 13.14 Illustration of refraction.

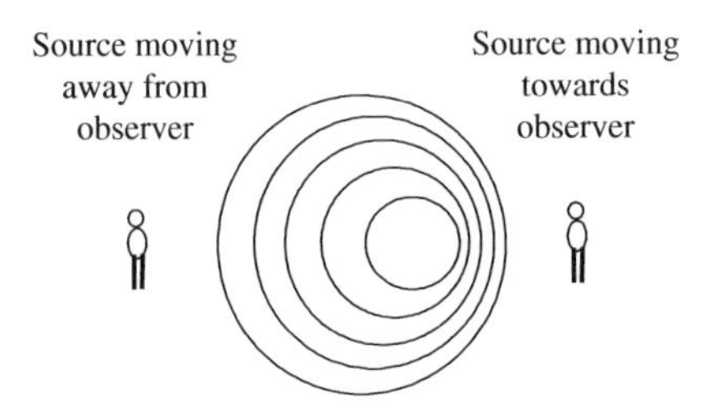

Figure 13.15 Illustration of doppler effect.

Doppler effects also happen with electromagnetic waves such as light waves. For example if a light source is moving towards the observer it seems bluer (shifted to a higher frequency); this is known as *blue shift*. If a light source is moving away from an observer it seems redder (shifted to a lower frequency); this is known as *red shift*. The fact that the light from distant galaxies is red shifted is considered as major evidence that the universe is expanding.

13.4.4 Motion of Sound Waves on Strings

A wave travelling along a string will bounce back at the fixed end and interfere with the part of the wave still moving towards the fixed end. When the wavelength is matched to the length of the string, the result is standing waves. For a string of length L, the wavelength is $\lambda = 2L$, the period, that is time taken to travel one wavelength, is $T = 2L/c$, where c is the speed of the wave and the *fundamental frequency* of the string is

$$f_1 = \frac{c}{2L} \tag{13.12}$$

In general, for a string fixed at both ends the harmonic frequencies are the integer multiples of the fundamental given as:

$$f_n = \frac{nc}{2L} \qquad n = 1, 2, 3, 4, \ldots \tag{13.13}$$

For example, when a guitar string is plucked, waves at different frequencies will bounce back and forth along the string. However, the waves that are not at the harmonic frequencies will have reflections that do not interfere constructively. The waves at the harmonic frequencies will interfere constructively, and the musical tone generated by plucking the string will be a combination of the harmonics.

Example 13.1 The fundamental frequency of a string depends on its mass, length and tension. Assume a string has a length of $L = 63\,\text{cm}$, a mass of $m = 30\,\text{g}$, and a tension of $S = 87\,\text{N}$. Calculate the fundamental frequency of this string.

Solution: The speed of the wave on a string is given by

$$c = \left(\frac{\text{Tension}}{(\text{Mass/Length})}\right)^{1/2} = \left(\frac{87}{(0.03/0.63)}\right)^{1/2} = 42.74\,\text{m/s} \tag{13.14}$$

From Equation (13.12) the fundamental frequency is obtained as

$$f_1 = \frac{c}{2L} = \frac{42.74}{2 \times 0.63} = 33.9\,\text{Hz} \tag{13.15}$$

The harmonic frequencies are given by nf_1.

13.4.5 Longitudinal Waves in Wind instruments and Pipe Organs

The main difference between the sound waves in pipes and on strings is that while strings are fixed at both ends, a tube is either open at both ends or open at one end and fixed at the other. In these cases the harmonic frequencies are given by:

$$\text{Tube open at both ends}: f_n = \frac{nc}{2L} \qquad n = 1, 2, 3, 4, \ldots \tag{13.16}$$

$$\text{Tube open at one end}: f_n = \frac{nc}{4L} n = 1, 2, 3, 4, \ldots \tag{13.17}$$

Hence, the harmonic frequencies of a pipe are changed by varying its effective length. A pipe organ has an array of different pipes of varying lengths, some open-ended and some closed at one end. Each pipe corresponds to a different fundamental frequency. For an instrument like a flute, on the other hand, there is only a single pipe. Holes can be opened along the flute to reduce the effective length, thereby increasing the frequency. In a trumpet, valves are used to make the air travel through different sections of the trumpet, changing its effective length; with a trombone, the change in length is obvious.

Example 13.2 *Calculation of formants: Resonance of vocal tract*
The vocal tract tube can be modelled as a closed-end tube with an average length of 17 cm for a male speaker. Assume that the velocity of sound is $c = 343.3\,\text{m/s}$ at a temperature of 20 °C. The nth harmonic of the resonance of the vocal tract tube is given by

$$f_n = nc/4L = n343.3/(4 \times 0.17) = 505n \tag{13.18}$$

Hence the fundamental resonance frequency of the vocal tract, the first resonance (aka formant) of speech, is 505 Hz. Since this formant varies with temperature it is usually rounded to 500 Hz. The higher formants occur at odd multiples of the frequency of the first formant at 1500, 2500, 3500 and 4500 Hz. Note that this is a simplified model. In reality the shape of the vocal tract is affected by the position of articulators and the formants are a function of the phonetic content of speech and the speaker characteristics.

13.4.6 Wave Equations for Strings

In this section we consider the question of how a force such as plucking or hammering a string sets up a pattern of wave motions on a string. The wave equation for a vibrating string can be derived from Newton's second law of motion which states that force = mass × acceleration.

Consider a short length Δx of an ideal string under tension as illustrated in Figure 13.16. The net vertical force can be expressed in terms of the string tension T and the angular displacements as

$$F_y = T \times (\sin(\phi_1) - \sin(\phi_2)) = T \times (\tan(\phi_1) - \tan(\phi_2)) \tag{13.19}$$

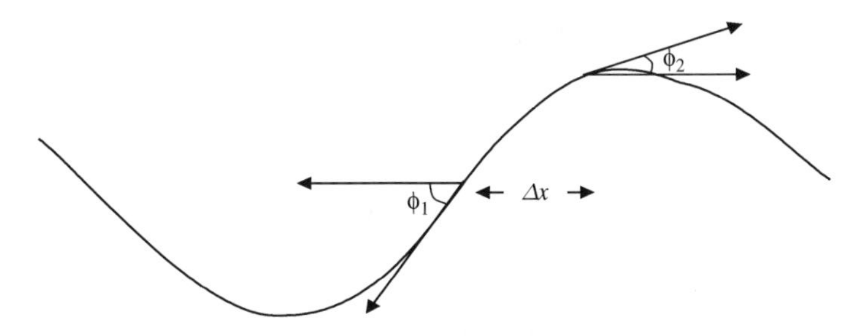

Figure 13.16 Displacement movement of a vibrating string.

where it is assumed that for small values of ϕ, $\sin(\phi) = \tan(\phi) = \frac{\partial y}{\partial x}$. Note $\tan(\phi_1)$ and $\tan(\phi_2)$ are the displacement slopes at x and $x + \Delta x$ respectively given by

$$\tan(\phi_1) = \left.\frac{\partial y}{\partial x}\right|_x \qquad \tan(\phi_2) = \left.\frac{\partial y}{\partial x}\right|_{x+\Delta x} \tag{13.20}$$

and

$$\tan(\phi_2) = \frac{\partial y}{\partial x} + \frac{\partial^2 y}{\partial x^2}\Delta x \tag{13.21}$$

From Equations (13.19)–(13.21) we obtain the tension force as

$$F_y = T\left(\frac{\partial^2 y}{\partial x^2}\right)\Delta x \tag{13.22}$$

Assuming that the string has a uniform mass density of ε per unit length, then the mass of a length Δx is $\varepsilon\Delta x$. Using Newton's second law describing the relationship between force, mass and acceleration we have

$$T\left(\frac{\partial^2 y}{\partial x^2}\right)\Delta x = \varepsilon\Delta x\left(\frac{\partial^2 y}{\partial t^2}\right) \tag{13.23}$$

or

$$c^2\frac{\partial^2 y}{\partial x^2} = \frac{\partial^2 y}{\partial t^2} \tag{13.24}$$

where $c = \sqrt{T/\varepsilon}$ has the dimension of velocity. From Equation (13.24) we can obtain the following types of solutions for waves travelling in time t in positive and negative x directions

$$y^+(x, t) = f(x - ct) \tag{13.25}$$

and

$$y^-(x, t) = f(x + ct) \tag{13.26}$$

The sum of two travelling waves is also a solution and gives a standing wave as

$$y(x, t) = Af(x - ct) + Bf(x + ct) \tag{13.27}$$

A discrete-time version of Equation (13.27) can be obtained as

$$y(n, m) = Af(n - cm) + Bf(n + cm) \tag{13.28}$$

where n and m represent the discrete-space and the discrete-time variables.

13.4.7 Wave Equation for Acoustic Tubes

The wave equations for ideal acoustic tubes are similar to the wave equations for strings with the following differences:

(a) The motion of a vibrating string can be described by a single 2-dimensional variable $y(x, t)$ whereas an acoustic tube has two 2-dimensional variables: the pressure gradient $p(x, t)$ and the volume velocity $u(x, t)$. Note that in reality the motion of string and pressure waves are functions of the 3-dimensional space.
(b) String vibrations are perpendicular, or transverse, to the direction of the wave propagation and string waves are said to be *transversal*, whereas in a tube the motion of waves are in the same direction as the wave oscillations and the waves are said to be *longitudinal*.

In an acoustic tube the pressure gradient and the velocity gradient interact. Using Newton's second law, describing the relationship between force, mass and acceleration, with an analysis similar to that for deriving the wave equation for strings, the equations expressing pressure gradient and velocity gradient functions are

$$\text{Pressure gradient:} \quad c^2\left(\frac{\partial^2 p}{\partial x^2}\right) = \left(\frac{\partial^2 p}{\partial t^2}\right) \tag{13.29}$$

$$\text{Velocity gradient:} \quad c^2\left(\frac{\partial^2 u}{\partial x^2}\right) = \left(\frac{\partial^2 u}{\partial t^2}\right) \tag{13.30}$$

The wave velocity c can be expressed in terms of the mass density of air ρ and the compressibility of air κ as

$$c = \frac{1}{(\rho\kappa)^{0.5}} \tag{13.31}$$

The solutions for pressure and velocity gradient are

$$p(x, t) = Z_0[u^+(x - ct) + u^-(x + ct)] \tag{13.32}$$

$$u(x, t) = u^+(x - ct) - u^-(x + ct) \tag{13.33}$$

where Z_0 is obtained as follows. Using Newton's second law of motion:

$$\frac{\partial p}{\partial x} = -\frac{\rho}{A}\left(\frac{\partial u}{\partial t}\right) \tag{13.34}$$

where A is the cross-section area of the tube. From Equations (13.32) and (13.33) we obtain

$$Z_0 = \frac{\rho c}{A} \tag{13.35}$$

13.5 Music Signal Features and Models

The signal features and models employed for music signal processing are broadly similar to those used for speech processing. The main characteristic differences between music and speech signals are as follows:

(a) The essential features of music signals are pitch (i.e. fundamental frequency, timbre (related to spectral envelope and resonance), slope of attack, slope of sustain, slope of decay and beat.
(b) The slope of the attack at the start of a note or a segment of music, the sustain period, the fall rate and the timings of notes are important acoustic parameters in music. These parameters have a larger dynamic range than those of speech.

(c) Beat and rhythm, absent in normal speech, are important acoustic features of musical signals.
(d) Music signals have a wider bandwidth than speech, extending up to 20 kHz and often have more energy in higher frequencies than speech.
(e) Music signals have a wider spectral dynamic range than speech. Music instruments can have sharper resonance and the excitation can have a sharp harmonic structure (as in string instruments).
(f) Music signals are polyphonic as they often contain multiple notes from a number of sources and instruments played simultaneously. In contrast speech is usually a stream of phonetic events from a single source. Hence, music signals have more diversity and variance in their spectral–temporal composition.
(g) Music signals are mostly stereo signals with a time-varying cross-correlation between the left and right channels.
(h) The pitch and its temporal variations play a central role in conveying sensation in music signals; the pitch is also important in conveying prosody, phrase/word demarcation, emotion and expression in speech.

The signal analysis and modelling methods used for musical signals include:

(a) harmonic plus noise models
(b) linear prediction models
(c) probability models of the distribution of music signals
(d) decision-tree clustering models.

In the following we consider different methods of modelling music signals.

13.5.1 Harmonic Plus Noise Model (HNM)

The harmonic plus noise model describes a signal as the sum of a periodic component and a spectrally-shaped random noise component as

$$x(m) = \underbrace{\sum_{k=1}^{N} [A_k(m)\cos(2\pi k f_0(m)m) + B_k(m)\sin(2\pi k f_0(m)m)]}_{\text{Fourier Series, Harmonics}} + \underbrace{e(m)}_{\text{Noise}} \tag{13.36}$$

where $f_0(m)$ is the time-varying fundamental frequency or pitch, $A_k(m)$ and $B_k(m)$ are the amplitudes of the kth sinusoidal harmonic components and $e(m)$ is the non-harmonic noise-like component at time discrete-time m.

The sinusoids model the main vibrations of the system. The noise models the non-sinusoidal energy produced by the excitation and any non-sinusoidal system response such as breath noise in wind instruments, bow noise in strings, and transients in percussive instruments. For example, for wind instruments, the sinusoids model the oscillations produced inside the pipe and the noise models the turbulence that takes place when the air from the player's mouth passes through the narrow slit.

For bowed strings the sinusoids are the result of the main modes of vibrations of the strings and the sound box, and the noise is generated by the sliding of the bow against the string plus other non-linear behaviour of the bow–string–resonator system.

The amplitudes, and frequencies of sinusoids vary with time and their variations can be modelled by a low-order polynomial. For example $A_k(m)$ can be modelled as a constant, a line, or a quadratic curve as

$$A_k(m) = a_k(m_i) \tag{13.37}$$

$$A_k(m) = a_k(m_i) + b_k(m_i)(m - m_i) \tag{13.38}$$

$$A_k(m) = a_k(m_i) + b_k(m_i)(m - m_i) + c_k(m_i)(m - m_i)^2 \tag{13.39}$$

where m_i is the beginning of the ith segment of music. Similar equations can be written for $B_k(m)$. The rates of variation of a music signal are state-dependent and different sets of polynomial coefficients are required during attack, sustain and fall periods of a musical note. The noise component $e(m)$ is often spectrally shaped and may be modelled by a linear prediction filter as

$$e(m) = \sum_{k=1}^{P} a_k e(m) + \varepsilon(m) \tag{13.40}$$

For the harmonic plus noise model of Equation (13.36) we need to estimate the fundamental frequency f_0, the amplitudes of the harmonics A_k and B_k, the number of harmonics N and the parameters of the noise-shaping filter a_k.

13.5.2 Linear Prediction Models for Music

Linear prediction analysis can be applied to the modelling of music signals in two ways:

(1) to model the music signal within each signal frame,
(2) to model the correlation of the signal across speech frames, e.g. to model the correlation of harmonics and noise across successive frames.

A linear predictor model is described as

$$x(m) = \sum_{k=1}^{P} a_k x(m-k) + e(m) \tag{13.41}$$

where a_k are the predictor coefficients and $e(m)$ is the excitation. For music, signal processing linear prediction model can be combined with harmonic noise model (HNM), so that linear predictor models the spectral envelope of the music whereas HNM models the harmonic plus noise structure of the excitation. Combination of linear predictor and HNM are described in Chapter 15.

13.5.3 Sub-band Linear Prediction for Music Signals

A main assumption of linear prediction theory is that the input signal has a flat spectrum that is shaped as it is filtered through the predictor. Due to the wider spectral dynamic range and the sharper resonance of music signals compared with speech, and due to the non-white harmonically-structured spectrum of the input excitation for string instruments, a linear prediction system has difficulty modelling the entire bandwidth of a music signal and capturing its spectral envelope such that the residue or input is spectrally flat.

The problem can be partly mitigated by a sub-band linear prediction system introduced in Chapter 10. A further reason for using a sub-band based method with music signals is the much larger bandwidth of musical signals. The signal can be divided into N sub-bands and then each sub-band signal can be down-sampled prior to LP modelling. Each sub-band signal having a smaller dynamic range will be better suited to LP modelling. Figures 13.17 and 13.18 show examples of linear prediction analysis of speech and music respectively.

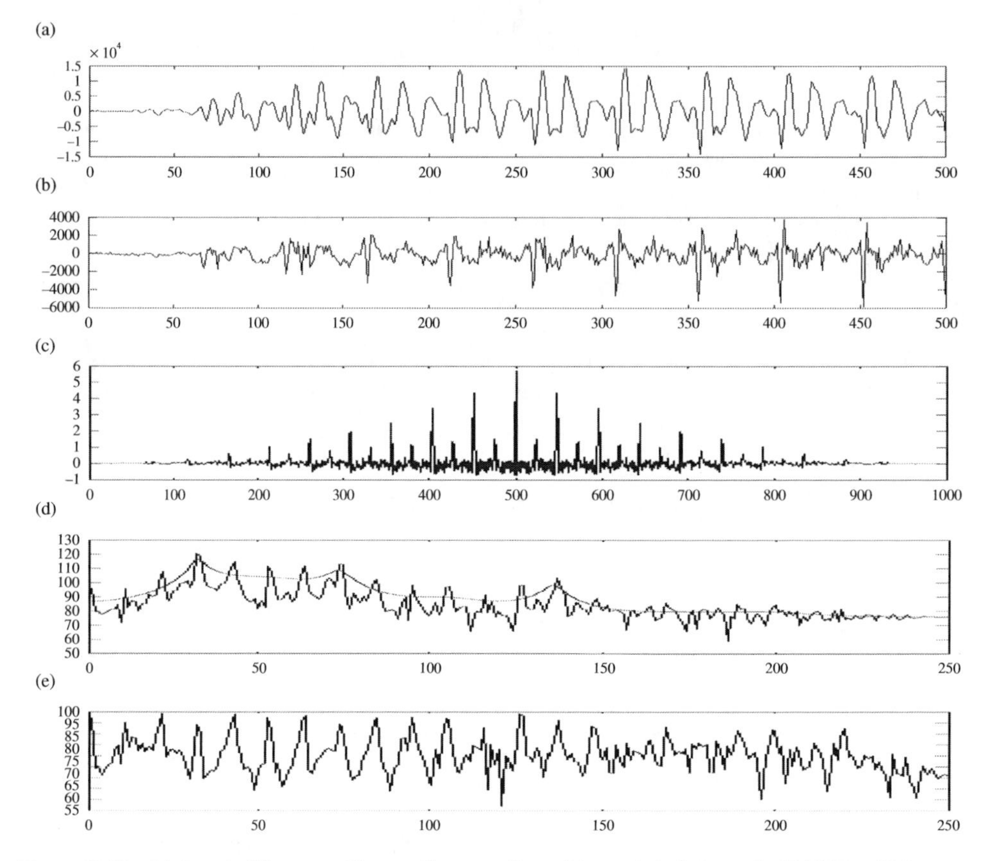

Figure 13.17 (a) Speech, (b) output of inverse linear predictor, (c) correlation of speech, (d) DFT and LP spectra of (a), (e) spectrum of input signal in (b).

13.5.4 Statistical Models of Music

As in speech processing, statistical and probability models are used for music coding and classification. For example, the entropy coding method, where the length of a code assigned to a sample value depends on the probability of occurrence of that sample value (i.e. the more frequent sample values, or symbols, are assigned shorter codes) is used in music coders such as in MP3 music coders described in Section 13.8.

Music compression, music recognition, and computer music composition benefit from probability models of music signals. These probability models describe the signal structures at several different levels:

(a) At the level of signal sample or parameter values, the probability models describe the distribution of different parameters of music such as the pitch, the number of harmonics, the spectral envelope, amplitude variation with time, onset/offset times of music events.

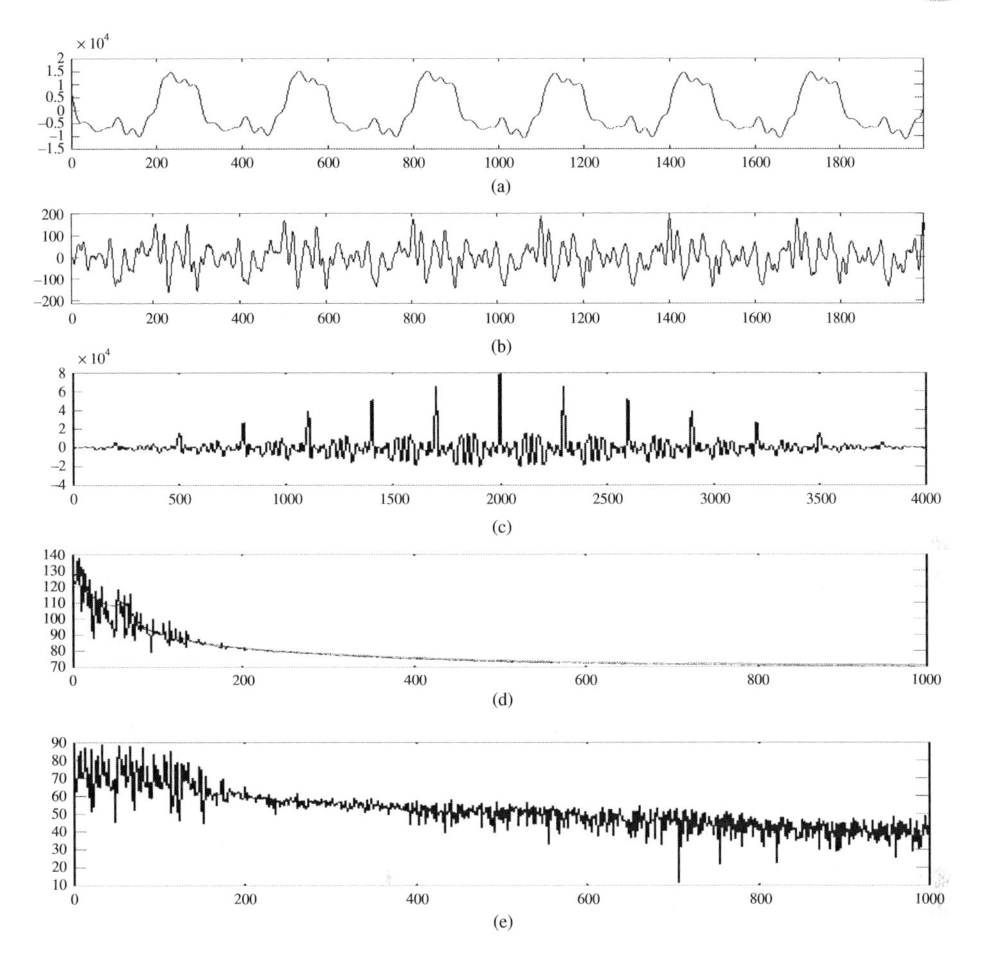

Figure 13.18 (a) A segment of music signal, (b) output of inverse linear predictor, (c) autocorrelation of signal in (a), (d) DFT and LP spectra of (a), (e) spectrum of predictor's input signal in (b).

(b) At the level of grammar, a finite state Markovian probability model describes the concurrency and the sequential dependency of different notes in the contexts of chords, parts and rhythmic and melodic variations.
(c) Hierarchical structures (i.e., signal, notes, sequence of notes, grammar) of music can be modelled using structured multi-level finite state abstractions of the music generation process.

13.6 Anatomy of the Ear and the Hearing Process

Sound is the auditory sensation of air pressure fluctuations picked up by ears. This section introduces the anatomy of the ear and explains how the ears work and act as transducers that transform the variations of air pressure into electrical firings of neurons decoded by brain. We also study aspects

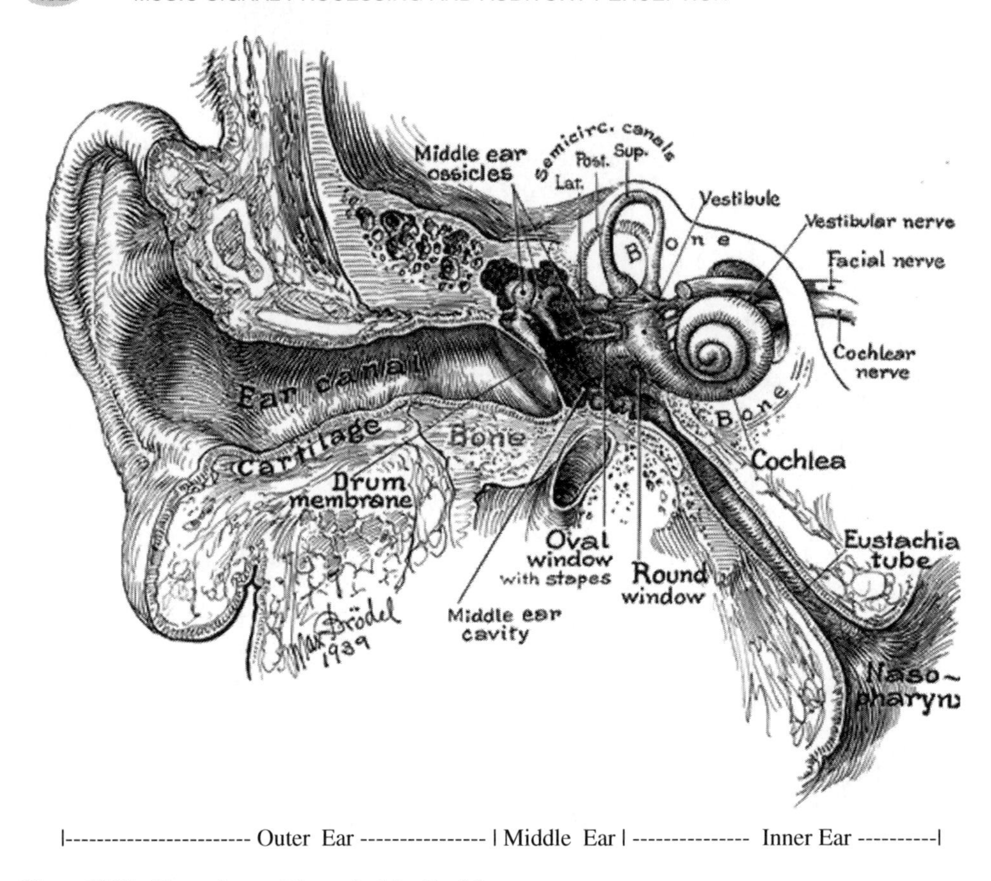

Figure 13.19 The anatomy of the ear by Max Brodel.
Source: Brodel M: The Anatomy of the Human Ear, W.B. Saunders Company © 1946.

of the psychoacoustics of hearing such as the threshold of hearing, critical bandwidth and auditory masking in frequency and time domains.

The ear, illustrated in Figure 13.19, is essentially a transducer that converts the air pressure variations on the eardrums into electrical firings of neurons which are transmitted to brain and decoded as sounds. The presence of an ear on each side of the head allows stereo hearing and the ability to find the direction of arrival of sound from an analysis of the relative intensity and phase (delay) of the sound waves reaching each ear. The ear is composed of three main parts:

(1) The outer ear, composed of the pinna, the ear canal and the outer surface of the eardrum, picks up the air vibrations and directs it to the eardrum.
(2) The middle ear contains a vibration transmission mechanism composed of three small bones. It translates air vibrations that impinge on the eardrum into mechanical vibrations of the bones in the middle ear which impinge on the inner tubes.
(3) The inner ear is the most intricate part of the ear. It transforms mechanical vibrations in the middle ear into hydraulic vibrations of fluid-filled cochlear tubes that set off neural firings of hair cells.

The anatomy of the ear is described in the following.

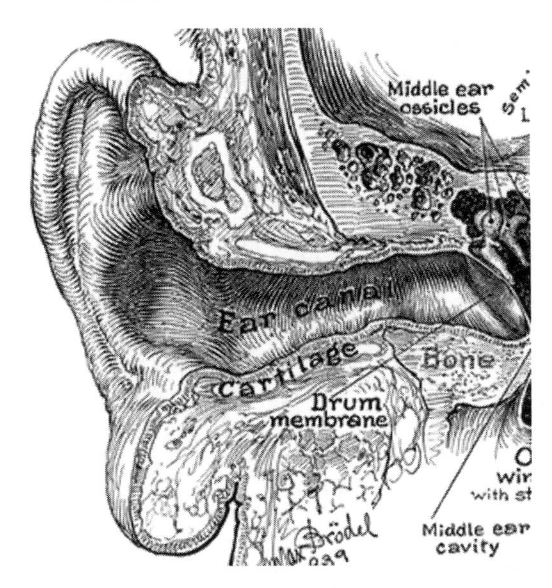

Figure 13.20 The outer ear is composed of the pinna and the ear canal which terminates at the eardrum where the middle ear starts. Source: Brodel M: The Anatomy of the Human Ear, W.B. Saunders Company © 1946.

13.6.1 Outer Ear

The outer ear consists of the pinna, the ear canal, and the outer layer of the eardrum.

The **pinna**, shown in Figure 13.20, is composed of cartilage. The pinna and the ear canal are shaped to facilitate efficient transmission of sound pressure waves to the eardrum. The total length of the ear canal in adults is approximately 2 1/2 cm, which for a closed-end tube gives a resonance frequency of approximately $f = c/4L$ or 3400 Hz, where c is the speed of propagation of sound (assumed to be 340 m/s) and L is the length of the ear canal. Note that this frequency also coincides with the frequency of maximum sensitivity of hearing.

At the end of the ear canal at the **tympanic membrane (eardrum)**, the energy of air vibrations is transformed into the mechanical energy of eardrum vibrations. The tympanic membrane, or eardrum, is approximately 1 cm in diameter and has three layers, with the outer layer continuous with the skin of the outer ear canal. The central portion of the tympanic membrane provides the active vibrating area in response to sound pressure waves.

13.6.2 Middle Ear

The middle ear serves as an impedance-matching transformer and also an amplifier. The middle ear matches the impedance of air in the ear canal to the impedance of the perilymph liquid in the cochlea of the inner ear.

The middle ear, shown in Figure 13.21, is composed of a structure of three bones, known as ossicles, which are the smallest bones in the body. The ossicles transmit the vibrations of the sound pressure waves from the eardrum to the oval window. Due to a narrowing of the contact area of transmission bone structure from the eardrum to the oval window, amplification of pressure is achieved at the oval window. The three bones of the middle ear, in the order of their position from the eardrum, are known as malleus, incus and stapes.

Figure 13.21 The middle ear contains three small bones that transmit eardrum vibrations to the oval window. Source: Brodel M: The Anatomy of the Human Ears, W.B. Saunders Company © 1946.

The **malleus** is the nearest middle ear bone to the eardrum. One end of the malleus is attached to the inner layer of the eardrum and vibrates with it. The other end of the malleus is attached to the incus, see Figure 13.21.

The **incus** is the bone in the middle; it is attached to the malleus at one end, and hence vibrates with it. The other end of the incus is attached to the head of the stapes. As the cross-section of the incus is less than that of the malleus at the eardrum, the incoming sound is given a small boost in energy of about 2.5 dB.

The **stapes** is connected to the incus at one end and at its other end it has a footplate seated in the oval window which separates the middle ear from the inner ear. As the incus vibrates, so does the footplate of the stapes. As the vibrating area of the tympanic membrane is larger than the area of the stapes, the incoming sound is given amplification in energy of over 20 dB.

Other parts of, or connections to, the middle ear include the following.

The **oval window** is an opening at the base of the cochlea, connecting the middle and inner ear, through which sound vibrations of the stapes are transmitted to cochlear fluid where a travelling wave is set up.

The **round window** is at the most basal end of the scala tympani. The round window allows the release of hydraulic pressure of the fluid perilymph that is caused by vibration of the stapes within the oval window.

The **Eustachian tube** connects the middle ear with the nasopharynx of the throat. This tube opens with swallowing or coughing to equalise pressure between the middle ear and the ambient pressure that is found in the throat.

13.6.3 Inner Ear

The inner ear is the main organ of hearing. It transforms the mechanical vibrations of the ossicles of the middle ear into a travelling wave pattern on the basilar membrane and then to the neural firings of the hair cells of the organ of Corti.

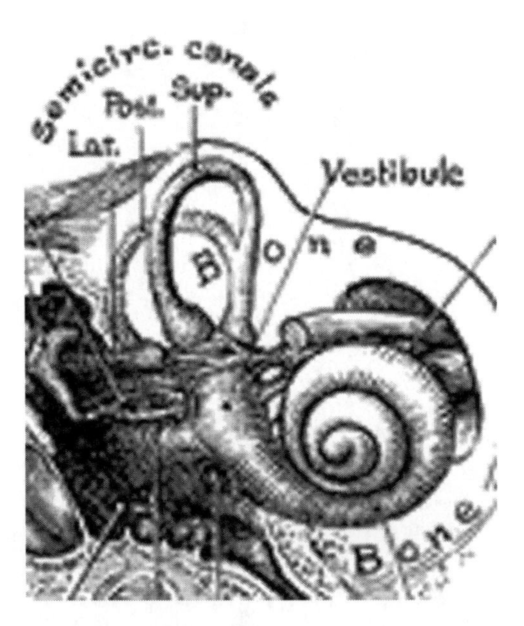

Figure 13.22 The inner ear is composed of the cochlea, a labyrinth of fluids and inner and outer hair cells. Source: Brodel M. The Anatomy of the Human Ear, W.B. Saunders Company ©1946.

The inner ear, Figure 13.22, is composed of two main sections: the vestibular labyrinth where the motions transduced arise from head movements, inertial effects due to gravity, and ground-borne vibrations and the cochlear labyrinth where the motions transduced arise from the vibrations of the air (sound) picked up by the eardrum and transmitted to the cochlea by the ossicles.

The vestibular labyrinth is buried deep in the temporal bone and consists of two organs, (i) the utricle and the sacculus which are specialised to respond to *linear accelerations* of the head and *static head position* and (ii) the semicircular canals which are specialised to respond to *rotational accelerations* of the head.

The **cochlea** (pronounced kok-lee-*uh* or koh-klee-*uh* and derived from the Greek word *kochlias* meaning snail) is a spiral snail-shaped structure that contains three fluid-filled tubes known as the *scala tympani*, *scala media* and *scala vestibule*, as shown in Figures 13.19 and 13.22. The cochlea is the organ that converts sound vibrations into neural signals.

The cochlea converts sounds, delivered to it as the mechanical vibrations of the middle ear at the oval window, into electrical signals of neurons. As explained next, this transduction is performed by specialised sensory cells within the cochlea.

The vibration patterns initiated by movements of the stapes footplate of the middle ear on the oval window sets up a travelling wave pattern within the cochlea's fluid-filled tubes. This wave pattern causes a shearing of the cilia of the outer and inner hair cells. This shearing causes hair cell *depolarisation* resulting in neural impulses that the brain interprets as sound. The neural signals, which code the sound's characteristics, are carried to the brain by the auditory nerve.

The vibrations of the fluids in the cochlea affect a *frequency to place transformation* along the basilar membrane. The higher frequencies excite the part of the cochlea near the oval window and the lower frequencies excite the part of the cochlea further away from the oval window. Hence, as shown in Figure 13.23, distinct regions of the cochlea and their neural receptors respond to different frequencies. Figure 13.24 illustrates how a periodic waveform with uniformly spaced harmonic frequencies may be registered along the length of the basilar membrane.

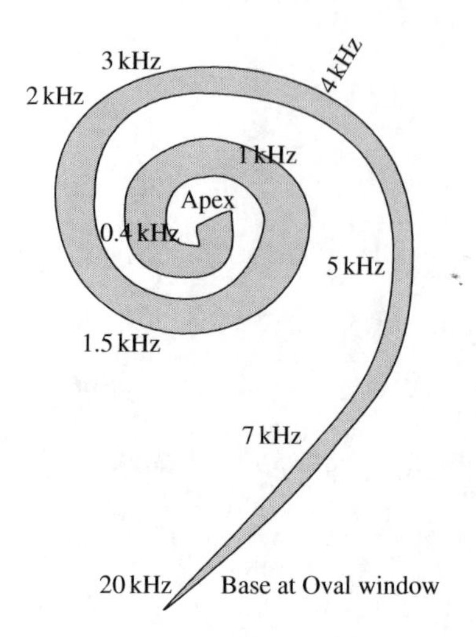

Figure 13.23 Illustration of frequency to place transformation along the length of the cochlea. Width at apex = 0.5 mm Width a base = 0.04 mm Length = 32 mm.

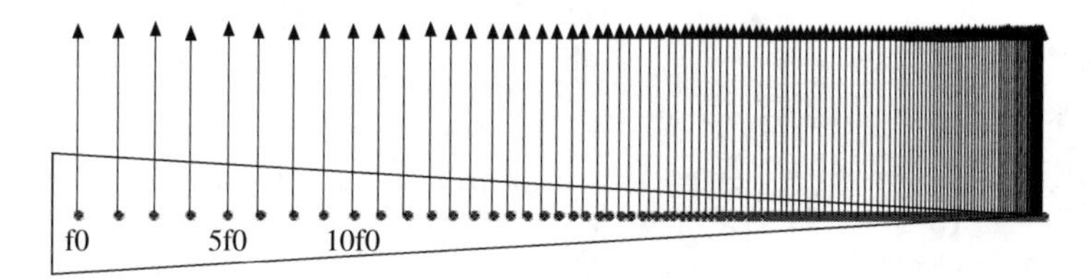

Figure 13.24 Illustration of the frequency to place distribution of the harmonics of a periodic waveform along the length of the basilar membrane of the cochlea. Note that each unit length (e.g. mm) of the basilar membrane at the apex-end where the frequency resolution is higher analyses a smaller bandwidth (is less crowded with frequency components) than at the base end (oval window) of the basilar membrane where the frequency resolution is lower.

Structure of the Cochlea

The cochlea's coiled shell contains a bony core and a thin spiral bony shelf that winds around the core and divides the bony labyrinth of the cochlea into upper (vestibuli) and lower (tympani) chambers. There is another middle membranous tube in between these two, Figure 13.25. These three compartments are filled with fluids that conduct the travelling wave patterns.

The **scala vestibuli** is the upper compartment in the cochlea's coiled shell; it leads from the oval window to the apex of the spiral. Hence the mechanical vibrations of the stapes on the oval window are converted to travelling pressure waves along the scala vestibuli which at the apex connects to the lower compartment, called the scala tympani.

The **scala tympani** is the lower compartment in the cochlea's coiled shell; it extends from the apex of the cochlea to a membrane-covered opening in the wall of the inner ear called the round window, which acts as a pressure release window. These compartments constitute the bony labyrinth of the cochlea and they are filled with **perilymph** fluid which has a low potassium K+ concentration and a high sodium Na+ concentration. The perilymphatic chamber of the vestibular system has a wide connection to the scala vestibuli, which in turn connects to the scala tympani by an opening called the

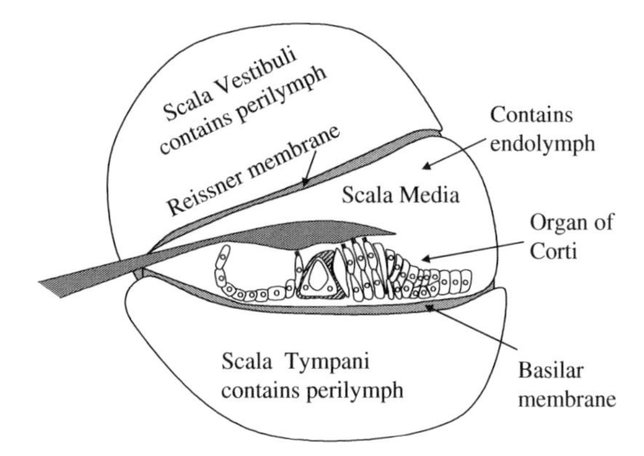

Figure 13.25 A schematic drawing of a cross-section of the cochlear tubes.

helicotrema at the apex of the cochlea. The scala tympani is then connected to the cerebrospinal fluid (CSF) of the subarachnoid space by the *cochlear aqueduct*.

The **scala media**, also known as the cochlear duct, is the membranous labyrinth of the cochlea. It lies between the two bony compartments, the scala vestibuli and scala tympani, and it ends as a closed sac at the apex of the cochlea. The cochlear duct is separated from the scala vestibuli by a vestibular membrane called *Reissner's membrane* and from the scala tympani by a *basilar membrane*. The scala media is filled with **endolymph**. In contrast to perilymph, endolymph has a high potassium K+ concentration and a low sodium Na+ concentration. The endolymphatic system of the cochlear duct (scala media) is connected to the saccule by the *ductus reuniens* and from there connects to the *endolymphatic sac*, which lies in a bony niche within the cranium. The endolymph of the utricle and semicircular canals also connects to the endolymphatic sac.

The **basilar membrane** lies at the *base* of the cochlear duct (scala media), hence the name basilar. It is a ribbon-like structure that extends from the bony shelf of the cochlea and forms the floor of the cochlear duct separating it from the scala tympani. The basilar membrane performs three main functions:

1) It provides a *base* on which rests the organ of Corti and the supporting structure.
2) It *separates* the endolymph fluid of the scala media from the perilymph fluid of the scala tympani; if these fluids are mixed due to an injury the hearing would be severely impaired or destroyed.
3) *Frequency dispersion*. The basilar membrane contains thousands of fibres, whose *lengths* and *stiffness* vary, becoming progressively longer and more compliant (less stiff) from the base of the cochlea to its apex. Because of these two gradients of size and stiffness, high frequencies are coded at the basal end of the basilar membrane with low frequencies progressively coded toward the apical end, Figure 13.23.

As shown in Figure 13.26, sound vibrations enter the perilymph fluid at the oval window, travel along the scala vestibuli and pass through the vestibular membrane to enter the endolymph of the scala media duct, where they cause movements in the basilar membrane. After passing through the basilar membrane, the sound vibrations enter the perilymph of the scala tympani, and their forces are dissipated to the air in the tympanic cavity by movement of the membrane covering the round window.

Due to the incompressibility of fluid and the pressure difference that develops between the fluid ducts as a result of the fluid flux, variations in the longitudinal flow are accompanied by the lateral

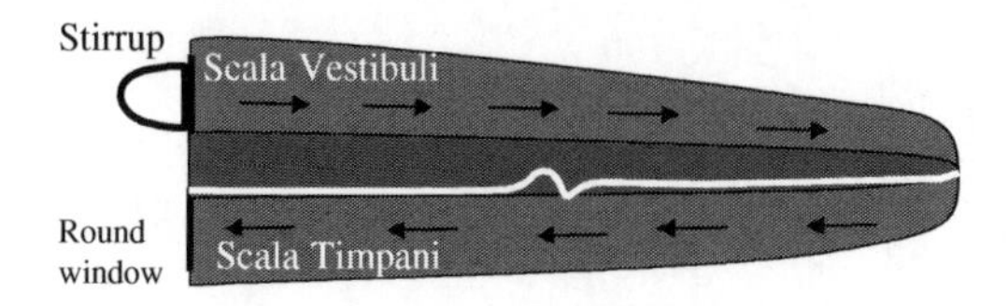

Figure 13.26 Illustration of a travelling wave and the response of the basilar membrane.

motion of the basilar membrane, resulting in a travelling wave that moves from the base towards the apex of the basilar membrane.

The basilar membrane is elastic and has little longitudinal rigidity. However, the membrane's lateral stiffness varies substantially along its length; it decreases by two orders of magnitude from the base to the apex of the cochlea. This varying stiffness results in a frequency-dispersive wave propagation. Hence, the basilar membrane effectively acts as a set of band-pass filters in that different parts of it respond to different frequencies of sounds.

The **organ of Corti**, shown in Figure 13.27, is the main receptor organ of hearing and resides within the scala media. The organ of Corti contains the hearing receptors (hair cells) and is located on the upper surface of the basilar membrane in the scala media and stretches from the base to the apex of the cochlea. Its receptor cells, which are called hair cells, are arranged in rows and they possess numerous hair-like processes that extend into the endolymph of the cochlear duct.

As sound vibrations pass through the inner ear, the hair cells shear back and forth against the tectorial membrane, and the mechanical deformation of the hairs stimulates the receptor cells. Various receptor cells, however, have different sensitivities to such deformation of the hairs. Thus, a sound that produces a particular frequency of vibration will excite receptor cells at a certain position, while a sound involving another frequency will stimulate a set of cells at a different place. The hair cells of the organ of Corti transform vibrations into neural firing transmitted via auditory nerve to the brain.

The **tunnel of Corti** is the space filled with endolymph that is bordered by the pillars of Corti and the basilar membrane.

The **pillars of Corti** are the supporting cells that bound the tunnel of Corti. The tunnel of Corti runs the entire length of the cochlear partition.

The **tectorial membrane** is a flexible, gelatinous membrane, which overlays the sensory receptive inner and outer hair cells as shown in Figure 13.27. The cilia of the outer hair cells are embedded in the tectorial membrane. For inner hair cells, the cilia may or may not be embedded in the tectorial membrane. When the cochlear partition changes position in response to the travelling wave, the shearing of the cilia is thought to be the stimulus that causes depolarisation of the hair cells to produce an action potential.

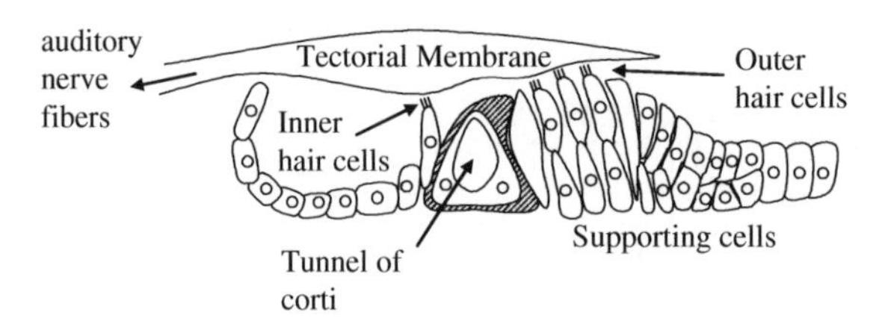

Figure 13.27 Organ of Corti transforms vibrational waves in the fluids of the cochlea to neural firings of hair cells.

Hair cell receptors The auditory receptor cells are called hair cells because they possess stereocilia, which participate in the signal transduction process. The hair cells are located between the basilar (base) membrane and the reticular lamina, a thin membrane that covers the hair cells. The stereocilia extend beyond the reticular lamina into the gelatinous substance of the tectorial (roof) membrane.

Two types of hair cells are present in the cochlea: inner hair cells are located medially, and are separated from the outer hair cells by the inner and outer rods of Corti (Figure 13.27). Hair cells synapse upon dendrites of neurons whose cell bodies are located in the spiral ganglion. Signals detected within the cochlea are relayed via the spiral ganglia to the cochlear nuclei within the brainstem via the VIII auditory nerve.

The outer hair cells (Figure 13.28(a)) consist of three rows of approximately 12,000 hair cells, 36,000 in total. Although they are much greater in number than the inner hair cells, they receive only about 5% of the innervations of the nerve fibres from the acoustic portion of the auditory nerve. These cells contain muscle-like filaments that contract upon stimulation and fine tune the response of the basilar membrane to the movement of the travelling wave. Because of their tuned response, healthy outer hair cells will ring following stimulation.

The inner hair cells (Figure 13.28(b)) are one row of approximately 3500 hair cells (i.e. about 10% of the number of outer hair cells). These cells receive about 95% of the innervations from the nerve fibres from the acoustic portion of the auditory nerve. These cells have primary responsibility for producing our sensation of hearing. When lost or damaged, a severe to profound hearing loss usually occurs.

Synapses of Hair Cells

The stereocilia (hairs) of the hair cells are embedded in the tectorial membrane. The relatively high inertial resistance of the gelatinous tectorial membrane to movements of the hair cells, caused by sound-induced travelling waves in the cochlea, creates shearing forces that bend the cilia hairs. Bending of the cilia causes the hair cells to either depolarise or hyperpolarise, depending on the direction of the bend.

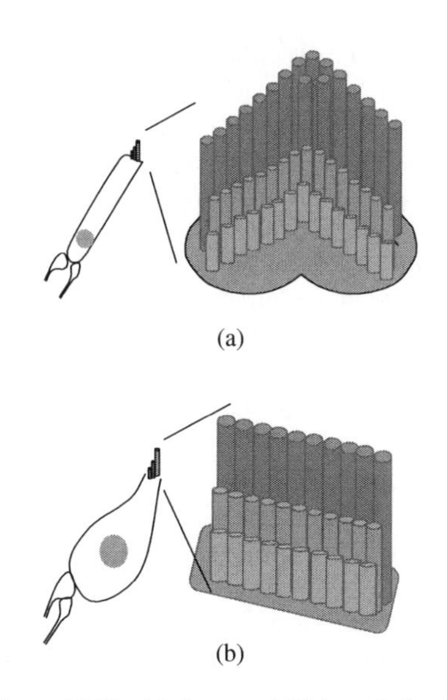

Figure 13.28 (a) Outer and (b) inner hair cells.

The bending of the stereocilia opens mechanically-gated ion channels; this enables positively charged ions, primarily potassium and calcium, to enter the hair cell. The influx of positive ions results in a receptor action potential within the cell, which triggers the cell's signalling activity.

The cilia regulate the flow of potassium from the endolymph into the hair cell. When the hair cell is at rest (i.e. not bent by a sound-induced shearing force), there is a small influx of potassium. As the cilia bend in one direction, the K+ channel is opened and allows potassium to enter the cell; this causes depolarisation of the cell and increases the influx of calcium via voltage-sensitive calcium channels. The influx of calcium causes the release of neurotransmitter onto the spiral ganglion cells. Bending of the cilia hairs in the opposite direction shuts off the potassium influx and hyperpolarises the hair cells.

In the inner hair cells, the receptor potential triggers the release of neurotransmitters chemicals, mainly glutamate, at the basal end of the cell, by opening voltage-gated calcium channels. The neurotransmitters diffuse across the space between the hair cell and a nerve terminal, where they bind to receptors and trigger action potentials in the nerve. In this way, the vibrations of sound is converted into electrical neural signals.

The inner hair cells are responsible for sending auditory neural signals to the brain via the VIII cranial nerve. The outer hair cells play an active mechanical role in regulating the responsivity of the basilar membrane to incoming sound vibrations. Motor proteins within the outer hair cells change the cellular structure to either dampen or amplify the incoming sound vibrations, hence the outer hair cells are referred to as the *cochlear amplifier*.

To summarise, within the cochlea, hair cell sensitivity to frequencies progresses from high frequencies at the base to low frequencies at the apex. The inner hair cells respond passively to deflections of sound-induced pressure waves. The outer hair cells respond actively and can elongate or shorten in response to motion of the basilar membrane to actively produce amplification or attenuation of the response of the inner hair cells. The scala vestibuli and scala tympani are filled with perilymph which has a similar composition to extra-cellular fluid, while the scala media is filled with endolymph which has a similar composition to intra-cellular fluid. The hair cells synapse on neurons of the spiral ganglia, initiating signal transmission into the central nervous system via the VIII cranial nerve.

Cochlear (auditory) nerve fibres are part of vestibulo-cochlear nerve, or VIII cranial nerve; they transmit auditory and balance information to the brain. Cochlear nerve fibres provide synaptic connections between the hair cells of the cochlea and the cochlear nucleus within the brainstem where its fibres make contact with the cochlear nucleus which is the first stage of relaying and processing of auditory neural signals.

In humans the cochlear nerve consists of some 30,000 nerve fibres. Sensory receptor neurons carry nerve impulses from cochlear receptors *toward* the brain, axons coming from cells in the olivary complex carry information in the opposite direction. The sensory neurons are divided into two types: (i) type I which synapse on inner hair cells (composed of some 90–95% of hair cells) and (ii) type II which synapse on outer hair cells (composed of some 5–10% of hair cells). Type I fibres innervate inner hair cells in a many-to-one relationship. Type II fibres typically innervate outer hair cells in a many-to-many relationship. Knowledge of the physiology of the cochlea and associated auditory neural processes has been utilised by researchers to develop neural networks that mimic human auditory processing functions.

13.6.4 Bandwidth, Sensitivity and Dynamic Range of Hearing

The ear is an extremely sensitive miniaturised anatomical structure. At the minimum audible vibrations of the threshold of hearing, the eardrum vibrates with an extremely small displacement of the order of 10^{-12} m (picometre). It is often said that if the ear was an order of magnitude or two more sensitive than it is, it would pick up the sound of the random Brownian motions and collisions of the molecules in the air. The minimum detectable changes in amplitude of audio signals are reported by various researchers to be between 0.25 to 3 dB; this figure varies among individuals and changes with age.

The ear is sensitive to a range of audible frequencies between 20 Hz and 20 KHz. Within this range, the frequency response of the ear varies sharply due to the shape and the physical properties of the ear canal and the middle ear. The ear is least sensitive to sound pressure vibrations below 100 Hz and above 10 kHz. It is most sensitive to sound vibrations in the frequency between 500 Hz and 5 kHz with the maximum sensitivity frequency at somewhere around 3 kHz.

Figure 13.29 shows the *equal loudness contours*. As the name 'equal loudness' implies, along each intensity curve the different frequencies have the same perceived loudness as that of the 1 kHz reference frequency. These perceptual curves were first measured by Fletcher and Munson in 1933. In Fletcher and Munson's experiment, they asked the listening subject to indicate the level of intensity of audio tones at different frequencies that equalled the perceived loudness of a reference 1 kHz tone of a given intensity. The intensity of the 1 kHz reference tone was changed in steps of 10 dB to obtain a family of curves.

A later set of equal loudness contours were obtained by Robinson and Dadson(1956) which were considered to be more accurate than Fletcher and Munson's, and this became the basis for a standard (ISO 226) until 2003 when a new revised standard known as (ISO 226:2003) was produced.

In Figure 13.29, the lowest pressure intensity curve represents the frequency-dependent *threshold of hearing* curve and the highest pressure intensity curve represents the *threshold of pain*. Note that the threshold of hearing is about 70 dB SPL at 10 Hz, 0 dB at 1 kHz (1 kHz is the reference frequency) and about −10 dB around 3 kHz.

The dynamic range of hearing is the difference between the threshold of hearing and the threshold of pain. From Figure 13.29 it is evident that humans have a dynamic range of hearing of approximately 120 dB beyond which the threshold of pain is reached. The dynamic range of human hearing is about 20 dB more than most high-end commercial recording equipment which has a dynamic range of 90 dB.

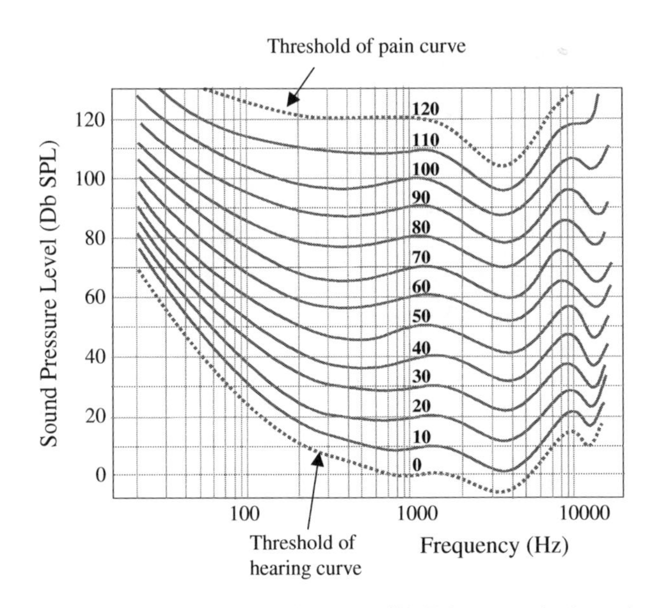

Figure 13.29 Equal loudness curves for different intensities (10 dB increments) of sound pressure level at the 1 kHz reference frequency.

Stereo Hearing and Localisation of Direction of Arrival of Sounds

The ears, positioned on the opposite sides of the head, provide us with binaural or stereo hearing: the ability to hear the fullness of sounds and their reflections from surroundings. Using the differences of the sounds arriving at the two ears we can localise the direction of arrival of sound and approximate the whereabouts of the position of the sound source.

The localisation of a sound source is affected by the following factors:

(1) Differences in the time of arrival (ToA) of the sounds at the left and right ears.
(2) Differences in the intensity levels of the sounds at the left and right ears.
(3) Cross spectral analysis of the signals received by the ears.

The anatomical characteristics that facilitate direction of arrival (DoA) estimation and sound source localisation include:

(1) The distance between the left and right ears and hence the relative propagation delay in a signal reaching the two ears of a person.
(2) The non-spherical shape of the head and the differences between the shape of the front of the head and ears compared with the back of the head and ears.
(3) The shape and orientation of the pinna and its intricate structure and curves.
(4) The effect of the torso in contributing difference in the spectrum of the sound reaching the two ears.

Stereo hearing ability allows us to locate the spatial direction of arrival of the sources of sounds. This is achieved by calculation of the difference in the intensity and in the time of arrival (ToA) of the sounds at each ear. The time difference between the sound arriving at the left and right ears can be as much as a maximum of t_{max} = distance between the ears/ speed of sound. Assuming the average distance between the ears is about 15 cm and the speed of sound is 340 m/s then t_{max} would theoretically be about 0.44 ms which is equivalent to one cycle of a 2273 Hz signal. Note that the time differences that allow the brain to locate the direction of arrival of sounds are minute. In order to estimate the direction of a sound, parts of the hearing pathway in the brainstem (superior olivary complex and cochlear nucleus) can detect delays of the order of 20 microseconds.

The phase-locking principle is also believed to play a role in the localisation of sounds, particularly at low frequencies below 3 kHz. For frequencies below 3 kHz the cochlear neuron firings lock to the audio signal stimulation in the sense that they follow the pattern of sound wave cycles. This is not surprising as phase is the manifestation of time delay in the frequency domain.

13.7 Psychoacoustics of Hearing

The psychoacoustics of hearing explains the relationship between the sensation of sounds and the measurable physical properties of the sounds. Audio signal processing methods utilise models of the psychoacoustics and the sensitivity of hearing, i.e. how the perception of the frequency components of a sound is affected by its frequency–time distribution and by its proximity in time and frequency to other signals. In this relation, three basic concepts were developed from experimental observations. These are:

(a) absolute threshold of hearing
(b) auditory critical bandwidths
(c) spectral and temporal masking.

13.7.1 Absolute Threshold of Hearing

The absolute threshold of hearing (ATH) is the minimum pressure level of a pure tone that can be detected by a person of acute hearing in noiseless conditions. The frequency dependency of the absolute threshold of hearing, shown in Figure 13.30, was obtained in a series of experiments on the determination of equal loudness curves by Fletcher and Munson in the 1930s and then revised by Robinson and Dadson in the 1950s, Figure 13.29. The variations of the absolute threshold of hearing with frequency may be modelled as

$$\mathrm{ATH}(f) = 3.64(f/1000)^{-0.8} - 6.5e^{-0.6(f/1000-3.3)^2} + 10^{-3}(f/1000)^4\,\mathrm{dB\ SPL} \tag{13.42}$$

Note that the maximum sensitivity of hearing (i.e. the minimum ATH) occurs around 3 to 4 kHz and is about −5 dB SPL. Sound pressure levels just detectable at the frequency of the maximum sensitivity of hearing are not detectable at other frequencies.

In general, two frequency tones of equal power but significantly different frequencies will not sound equally loud. The loudness of a sound is the perception of its power and depends on its frequency and the sensitivity of hearing at that frequency. The perceived loudness of a sound may be expressed in *sones* where, as expressed in Equation (13.9), 1 sone is defined as the loudness of a 40 dB SPL tone at 1 kHz. A doubling of sone is perceived as a doubling of the loudness of sound and corresponds to an increase of 10 dB in SPL. Normal talking within a distance of 1 metre has an SPL of 40 to 60 dB and a corresponding loudness of 1 to 4 sones. The threshold of pain happens at about 120 dB which corresponds to 256 sones.

In music and speech coding the absolute threshold of hearing can be interpreted as the maximum allowable coding noise level in frequency (note that this argument does not take into account the

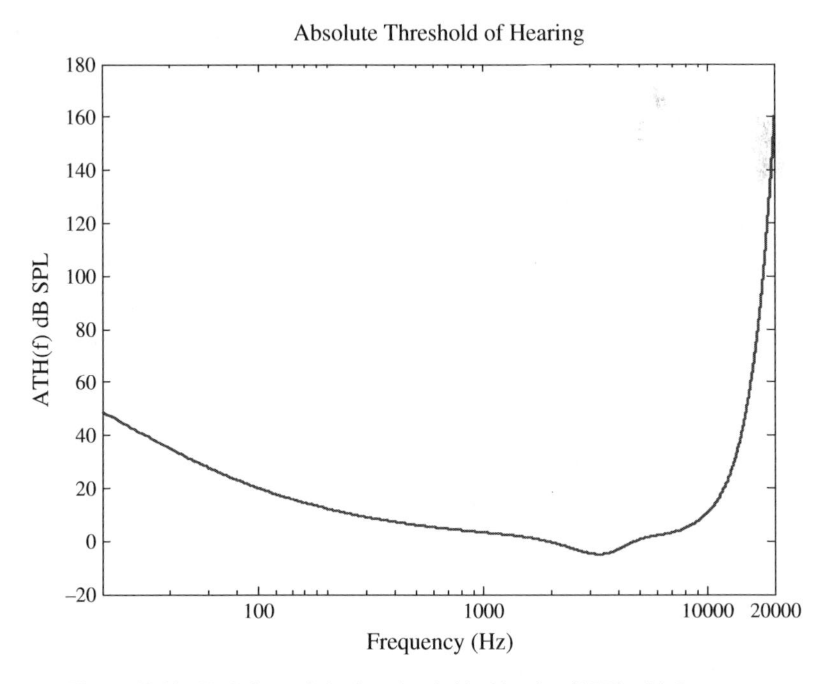

Figure 13.30 Variations of absolute threshold of hearing (ATH) with frequency.

noise-masking process described below). In designing coding system algorithms designers do not know the playback level, hence it is common practice to equate the energy of one bit to that of the sound pressure level at the minimum of the threshold of hearing curve at 4 kHz.

Matlab function ATH( )

Generates, plays back and displays a sine wave of constant magnitude with its frequency sweeping the range 0 to 20,000 Hz.

This program demonstrates the variable loudness of hearing of a constant-amplitude sine wave as a function of frequency. The spectrogram of the wave is also plotted.

13.7.2 Critical Bands of Hearing

Anatomically, critical bands of hearing are segments of cochlea, about 1.3 mm long, which act like band-pass filters. The audible effect of two or more tones residing within a critical band would be different than if the same signals were in different critical bands.

The concept of critical bands of hearing is based on the experimental observations of the perception of audio signals along the basilar membrane of the cochlea where a frequency to place transformation takes place (Figures 13.23 and 13.24). Experiments show that at any frequency along the cochlea the ear behaves like a series of band-pass filters known as critical bands which have frequency-dependent bandwidth. Signals within a critical band affect each other's perception far more than signals in different bands. Critical bandwidth increases with the centre frequency of the band. Below 500 Hz bandwidths are constant at about 100 Hz. Over 500 Hz the bandwidth of each critical band is about 20% larger than the preceding band.

As stated above, the effects of two tones or narrowband noise within a critical band are substantially different than the effect of two tones or narrowband noise in different critical bands. For example the ear is not able to completely resolve two pure tones within a critical band and perceives their combined effect as a beating of the tones, whereas two tones with frequencies in different bands can be resolved and heard as distinct tones.

Critical bandwidth around a frequency can be defined as the bandwidth at which the subjective response of hearing changes abruptly. For example, the perceived loudness of a narrowband noise at a constant sound pressure level remains constant as the bandwidth of noise is increased up to the critical bandwidth. However beyond the critical bandwidth the loudness increases.

Figure 13.31 illustrates two experiments which are often used in the literature to demonstrate the concept of critical bandwidth. In Figure 13.31(a) and (c) the detection threshold of a narrowband noise flanked between two masking tones remains constant as long as the separation between the two masking tones remains within a critical band. In Figure 13.31(b) and (d) the detection level of a tone positioned between two narrowband noise signals remains constant as long as the separation between the two narrowband noise signals remains within a critical band. The detection threshold decreases rapidly as the separation of maskers increases beyond a critical bandwidth.

Critical bandwidth around a frequency f remains constant (at about 100 Hz bandwidth) up to 500 Hz and then increases at about 20% of the centre frequency above 500 Hz. The variation of critical bandwidth BW_c with frequency can be modelled as

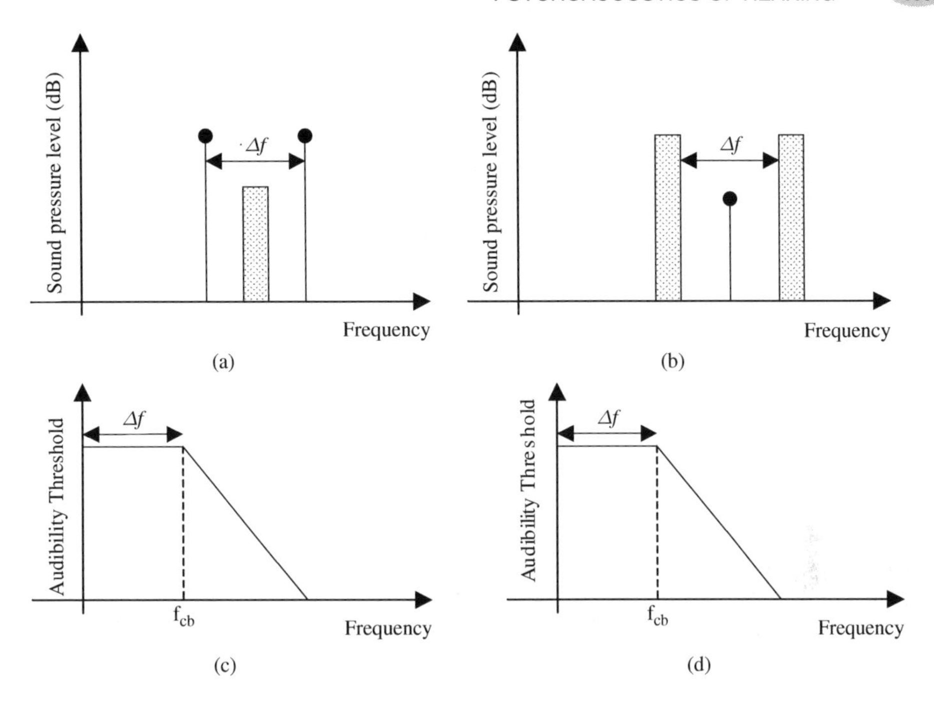

Figure 13.31 Illustration of critical bands: (a) a narrowband noise with two masking tones, (b) a tone with two masking narrowband noises, (c) and (d) show the thresholds of hearing as a function of the distance between the maskers.

$$BW_c(f) = 25 + 75\left[1 + 1.4(f/1000)^2\right]^{0.69} \text{ Hz} \tag{13.43}$$

A distance of one critical band is termed as one *bark* and the formula for mapping linear frequency to bark scale is given by

$$z_c(f) = 13\arctan(0.00076f) + 3.5\arctan\left[(f/7500)^2\right] \quad \text{(bark)} \tag{13.44}$$

Critical bandwidth varies continuously with frequency as shown in Figure 13.32. However in audio processing applications it is useful to define a discrete set of critical bands as shown in Table 13.12.

Matlab function CriticalbandsDemo( )

This is an experiment in critical bands and masking phenomena. The program generates either a tone flanked by two narrowband noises or a narrowband noise flanked by two tones. As the distance between the two maskers is gradually increased the masking signal can be heard with increasing clarity. The spectrograms of the signals are also plotted.

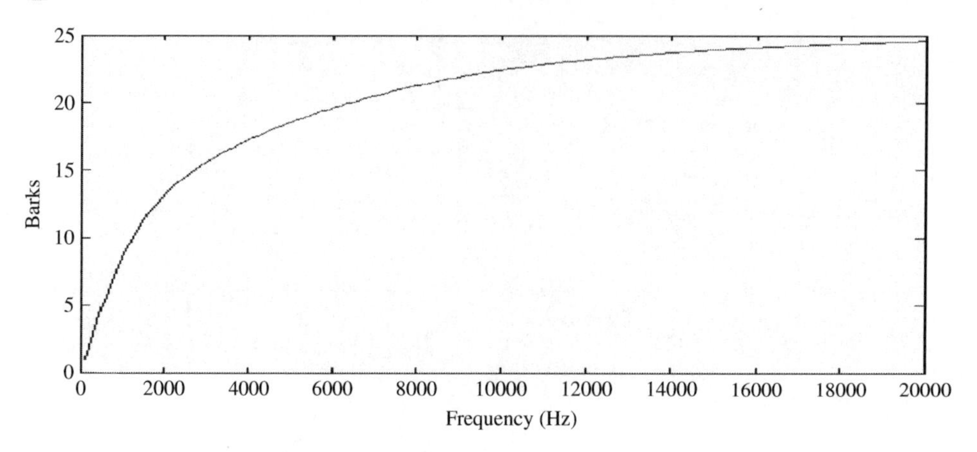

Figure 13.32 Illustration of the relationship between linear frequency units (Hz) and critical band units of bark.

Table 13.12 Discrete critical bands of hearing.

Critical band/centre (Hz)	Frequency (Hz)			Critical band/centre (Hz)	Frequency (Hz)		
	Low	High	Width		Low	High	Width
0/50	0	100	100	13/2150	2000	2320	320
1/150	100	200	100	14/2500	2320	2700	380
2/250	200	300	100	15/2900	2700	3150	450
3/350	300	400	100	16/3400	3150	3700	550
4/450	400	510	110	17/4000	3700	4400	700
5/570	510	630	120	18/4800	4400	5300	900
6/700	630	770	140	19/5800	5300	6400	1100
7/840	770	920	150	20/7000	6400	7700	1300
8/1000	920	1080	160	21/8500	7700	9500	1800
9/1175	1080	1270	190	22/10500	9500	12000	2500
10/1370	1270	1480	210	23/13500	12000	15500	3500
11/1600	1480	1720	240	24/19500	15500	22050	6550
12/1850	1720	2000	280				

13.7.3 Audio Masking

The concept of masking, i.e. a signal made inaudible by a louder signal in its time/space proximity, is a familiar experience. In general a large-amplitude signal that happens to be in time, space and frequency proximity of a smaller-amplitude signal can partially or totally mask the latter.

Simultaneous masking occurs when two sounds occur at the same time, such as when a conversation (the masked signal) is rendered inaudible by a noisy passing vehicle (the masker).

Masking can also happen with non-simultaneous signals. Backward masking occurs when the masked signal ends before the masker begins, i.e. the masker masks a signal preceding it. Forward masking occurs when the masked signal begins after the masker has ended, that is the masker masks a signal succeeding it. Hence a signal at time t, $x(t)$, can be masked by another signal $y(t+\tau)$ or $y(t-\tau)$. Note that maskings of non-simultaneous signals are in part due to delays in the transmission and cognitive processing of audio signals from the ear to the brain.

Masking becomes stronger as the sounds get closer together in time, space and frequency. For example, simultaneous masking is stronger than either forward or backward masking because the sounds occur at the same time. In this section we consider temporal and spectral masking effects.

Spectral Masking

Spectral masking is related to the concept of critical bands of hearing. For music coding two types of spectral masking are defined, tone-masking-noise and noise-masking-tone. A tone-masking-noise masks a noise within a critical bandwidth provided that the noise spectrum is below a threshold that depends on the strength of the masking tone. A noise-masking-tone masks a tone within a critical bandwidth provided that the tone strength is below a threshold dependent on the strength of the noise.

The masking effect of a tone (or noise) is not confined to within the critical bands. Inter-band masking is also observed and depends on the strength of the masking tone or noise. Figure 13.33 illustrates the masking effect of a pure tone. The spread of masking is often modelled by a triangular-shaped filter with the slopes of +25 dB and −10 dB per bark. The shape of the spreading function can be approximated as

$$\mathrm{SF}(x) = 15.81 + 7.5(x + 0.474) - 17.5\sqrt{1 + (x + 0.474)^2}\ \mathrm{dB} \tag{13.45}$$

where x has units of barks.

In psychoacoustic coders, after accounting for the shape of the spread function, the masking thresholds due to noise and tone components are obtained as

$$\mathrm{TH}_N = E_\mathrm{T} - 14.5 - B \quad \mathrm{dB} \tag{13.46}$$

$$\mathrm{TH}_\mathrm{T} = E_N - K \quad \mathrm{dB} \tag{13.47}$$

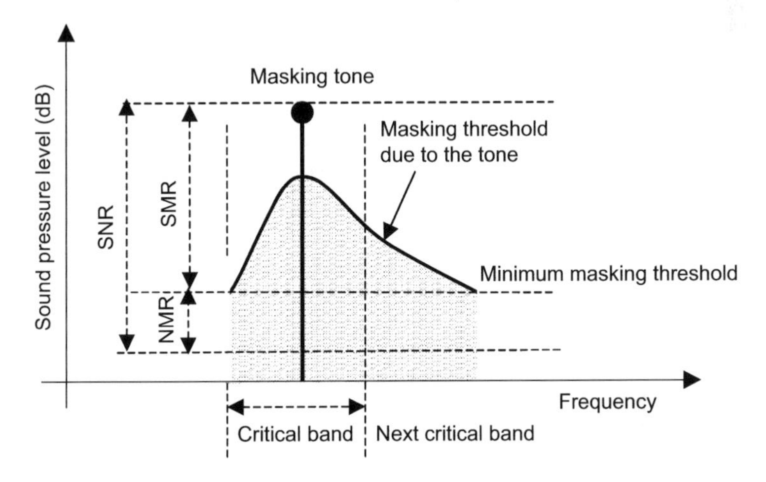

Figure 13.33 Illustration of the masking effects of a tone. SNR = signal-to-noise ratio, SMR = signal-to-mask ratio, NMR = noise-to-(minimum of)mask ratio. Note the minimum masking threshold is the minimum value of the spreading function in the critical band.

where E_T and E_N are energy levels of the tone and noise maskers and B is the critical band number. The parameter K is typically selected between 3 to 5 dB.

Masking thresholds are also referred to as just noticeable distortion (JND) levels. For perceptual coding, masking signals are first classified as either tones or noise maskers and then the JND values are calculated to shape the spectrum and keep the coding noise below audible levels.

Temporal Masking

Temporal masking, due to the simultaneous occurrence of two signals, is a common experience, for example when a passing vehicle's loud engine noise drowns a conversation. However, temporal masking can also be non-simultaneous. A signal can mask the audibility of another signal even before the masker begins; this is a something of a counter-intuitive effect know as *backward masking* or pre-masking. A signal can also mask the audibility of another signal after the masker is ended; this effect is known as *forward masking* or post-masking.

Figure 13.34 shows a sketch of the shape of temporal masking in time. Pre-masking and post-masking clearly point to delays in the processing of sounds in the auditory–brain system. Note that the backward masking effect is much shorter than the forward making effect. Backward masking lasts about 5 ms whereas forward masking can last up to 300 ms.

13.7.4 Example of the Use of a Psychoacoustic Model in Music Coding

In this section we consider an example of calculation of audio masking thresholds in an MPEG-1 layer-1 audio coder. The aim is to calculate the time-varying maximum allowable quantisation noise at each frequency that is masked and rendered inaudible by signal activity in other frequencies. The quantisation noise must be below the so-called just noticeable distortion level. It is assumed that the sampling rate is 44,100 samples per second with 16 bits per sample.

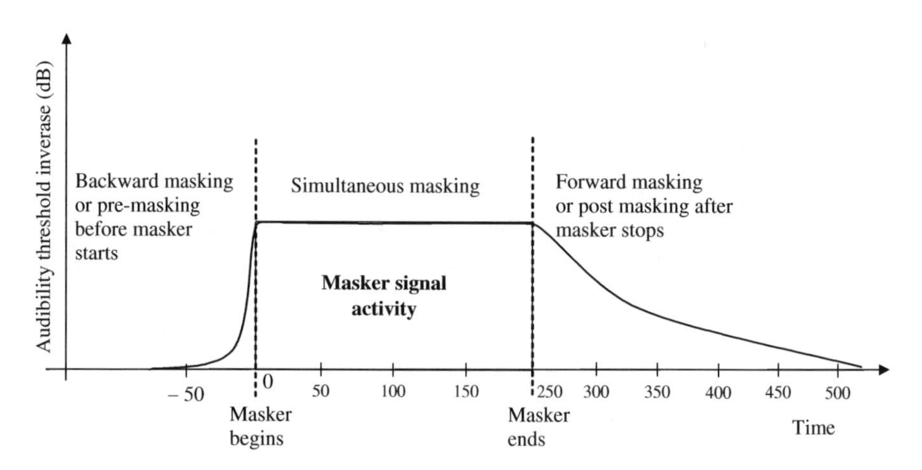

Figure 13.34 Illustration of temporal masking effects.

Step 1 – Spectral Analysis and SPL Normalisation

The psychoacoustic model of hearing uses a DFT of 512 points. The signal is divided into segments of 512 samples (about 11.6 ms) and windowed with an overlap of 32 samples between successive windows, therefore each window contains 480 new samples. At a sampling rate of 44,100 samples/second this gives a DFT spectral resolution of $\Delta f = F_s/N = 44,100/512 = 86.13\,\text{Hz}$.

The audio signal amplitude is normalised as

$$x_n(m) = \frac{x(m)}{N(2^{b-1})} \tag{13.48}$$

where N is the DFT length and b is the number of bits per sample ($b = 16\,\text{bits}$). The adjusted short-time power spectrum of the signal is obtained as

$$P(k) = PN + 10\log\left(|X(k)|^2\right) \tag{13.49}$$

where $PN = 20\log_{10}(2^{b-1}) = 90\,\text{dB}$ which is the full amplitude range of a 16 bit quantiser.

Step 2 – Identification of Tonal and Noise Maskers

Local maxima in the frequency components of the signal spectrum that exceed the neighbouring spectral components within a predefined frequency distance (in bark scale) by more than 7 dB are classified as tonal frequency components. Hence the rule for classification of frequency components as tonal may be expressed as

$$S_\text{T} = \left\{ P(k) \text{ if} \left| \begin{array}{l} P(k) > P(k \pm 1) \text{ and} \\ P(k) > P(k \pm \Delta_k) + 7\,\text{dB} \end{array} \right. \right. \tag{13.50}$$

where

$$\Delta_k \in \begin{cases} 2 & 2 < k < 63 \quad (0.17\text{–}5.5\,\text{kHz}) \\ [2,3] & 63 \le k < 127 \quad (5.5\text{–}11\,\text{kHz}) \\ [2,6] & 127 \le k < 256 \quad (11\text{–}20\,\text{kHz}) \end{cases} \tag{13.51}$$

and S_T denotes the set of total components. Tonal maskers are then computed from the peaks listed in the tonal set S_T as

$$P_\text{TM}(k) = 10\log 10 \sum_{\text{j}=-1}^{1} 10^{0.1P(k+\text{j})}\ \text{dB} \tag{13.52}$$

The noise masker values for each critical band are obtained from the non-tonal signal components as

$$P_\text{NM}(\bar{k}) = 10\log 10 \sum_{\text{j}} 10^{0.1P(\text{j})}\ |\forall P(\text{j}) \notin P_\text{TM}(k \pm \Delta k)\ \ \text{dB} \tag{13.53}$$

where $\forall P(\text{j}) \notin P_\text{TM}(k \pm \Delta k)$ means for all $P(\text{j})$ not part of the identified tonal components P_TM of the signal and $\bar{k}$ is the geometric mean spectral line within a critical band given by

$$\bar{k} = \left(\prod_{\text{j}=l}^{u} \text{j}\right)^{1/(u-l+1)} \quad \text{dB} \tag{13.54}$$

where l and u are the lower and upper frequency (DFT) bin indices of the critical band.

The tonal and noise maskers are then decimated and reorganised in the following manner. To start with the tonal and noise maskers that have a value less than the absolute threshold of hearing are discarded. Then any two tonal components that fall within a half bark distance of each other are replaced by the larger of the two values. The remaining values are sub-sampled without loss of information according to the following formula

$$P_{\mathrm{TM,NM}}(i) = P_{\mathrm{TM,NM}}(k) \tag{13.55}$$

$$i = \begin{cases} k & 1 \le k \le 48 \\ k + (k \bmod 2) & 49 \le k \le 96 \\ k + 3 - ((k-1) \bmod 4) & 97 \le k \le 256 \end{cases} \tag{13.56}$$

The effect of Equation (13.56) is a 2:1 decimation of the masker bins in the critical bands 18–21 and a 4:1 decimation in the critical bands 22–25 with no loss of the masking components.

Calculation of Masking Thresholds

The masking threshold $T_{TM}(i, \mathrm{j})$ at frequency bin i due to a tone masker at frequency bin j, $P_{TM}(\mathrm{j})$, is modelled by

$$T_{TM}(i, \mathrm{j}) = P_{TM}(\mathrm{j}) - 0.275\, z(\mathrm{j}) + SF(i, \mathrm{j}) - 6.025 \quad \text{dB SPL} \tag{13.57}$$

where $z(\mathrm{j})$ is the bark frequency of bin j. $SF(i, \mathrm{j})$ is a spreading function model of the masking effect at bin i due to a masking tone $P_{TM}(\mathrm{j})$ at bin j. $SF(i, \mathrm{j})$ is modelled by a piecewise linear function as

$$SF(i, \mathrm{j}) = \begin{cases} 17\Delta_z - 0.4P_{TM}(\mathrm{j}) + 11 & -3 \le \Delta_z < -1 \\ (0.4P_{TM}(\mathrm{j}) + 6)\Delta_z & -1 \le \Delta_z < 0 \\ -17\Delta_z & 0 \le \Delta_z < 1 \\ (0.15P_{TM}(\mathrm{j}) - 17)\Delta_z - 0.15P_{TM}(\mathrm{j}) & 1 \le \Delta_z < 8 \end{cases} \quad \text{dB SPL} \tag{13.58}$$

where $\Delta_z = z(i) - z(\mathrm{j})$ is the separation in barks between the maskee and the masker. Note that $SF(i, \mathrm{j})$ approximates the basilar masking spread function shown in Figure 13.33.

Similarly, $T_{NM}(i, \mathrm{j})$, the masking threshold at frequency bin i due to a noise masker $P_{NM}(\mathrm{j})$ at frequency bin j is given by

$$T_{NM}(i, \mathrm{j}) = P_{NM}(\mathrm{j}) - 0.175z(\mathrm{j}) + SF(i, \mathrm{j}) - 2.025 \quad \text{dB} \tag{13.59}$$

$SF(i, \mathrm{j})$ for the noise masker is obtained by replacing $P_{TM}(\mathrm{j})$ with $P_{NM}(\mathrm{j})$ in Equation (13.58). The global masking threshold at bin i due to all tone maskers T_{TM} and noise maskers T_{NM} is given by

$$T_g(i) = 10 \log 10 \left(10^{0.1ATH(i)} + \sum_{l=1}^{L} 10^{0.1T_{TM}(i,l)} + \sum_{m=1}^{M} 10^{0.1T_{NM}(i,m)} \right) \quad \text{dB} \tag{13.60}$$

where L and M are the number of tonal and noise maskers.

13.8 Music Coding (Compression)

The transmission bandwidth and the storage capacity requirement for digital music depend on the sampling rate and the number of bits per sample. Stereo music with left and right channels sampled at 44,100 Hz and quantised with 16 bits per sample generates data at a rate of

$$2 \times 44,100 \times 16 = 1,411,200 \text{ bits per second} \tag{13.61}$$

and requires about 5 Gigabits or about 603 Megabytes of storage per hour of music.

The objective of music compression is to reduce the bit-rate as far as possible while maintaining high fidelity. This is usually achieved through decomposition of the music signal into a series of de-correlated time–frequency components or a set of source-filter parameters of a synthesiser model of music. Using a psychoacoustic model, the various components of the decomposed music signal are each allocated the minimum number of bits required to maintain the quantisation noise masked and inaudible and achieve high fidelity of the reconstructed signal.

Signal Structures Used in Music Compression

In general, music coders utilise three aspects of the signals to reduce the bit-rate and simultaneously maintain high fidelity, these are:

(a) The correlation structure of music signals; this means the part of a sample $x(m)$ that is predictable from past samples can be modelled and need not be transmitted.
(b) The psychoacoustics of hearing; this means that the inaudible level of quantisation noise, at each time–frequency component, is time-varying and depends on the amount of noise that can be masked.
(c) The statistical distribution of music signals; this means the probability distribution can be used for efficient non-uniform quantisation and for efficient transmission in variable length coding.

Specifically, the signal structures utilised in music coding are as follows:

- Short-term correlations of successive samples of music, within a frame of say 5 to 30 ms, can be modelled by principle component analysis (e.g. discrete cosine transform (DCT)) or by a linear prediction model.
- Long-term inter-frame correlation and periodic patterns can be modelled by a pitch model and an amplitude envelope model or by modelling the correlation of successive frames of music with a DCT or linear prediction model.
- Non-uniform probability distribution of the signal variables can be utilised in the quantisation process and also in entropy coding to assign variable length codes to different signal values according to their probability, e.g. Huffman code where more frequently occurring values are assigned shorter code lengths.
- Cross-correlation between the left and the right channels of stereo music can be used in joint channel coding to reduce the bit-rate.
- The masking effects of auditory hearing can be used to allocate the minimum number of bits to each part of the signal such that the quantisation noise remains masked below the just noticeable distortion levels.

In this section we consider adaptive transform coding and MPEG (MP3) music coding methods.

13.8.1 Basic Principles of Music Compression

The quantisation noise of a music coder depends on a number of factors that include (a) the number of bits per sample, (b) the efficiency of utilisation of the distributions of the music signal in time and frequency domains and (c) the efficiency of utilisation of the psychoacoustics of hearing. The goal of audio coding is to utilise the time–frequency distribution of the signal and to shape the time–frequency distribution of the quantisation noise such that the quantisation noise is made inaudible and the reconstructed signal is indistinguishable from the original signal.

In general, audio coders operate by decomposing the signal into a set of units; each unit corresponds to a certain range in time and frequency. Using this time–frequency distribution, the signal is analysed according to psychoacoustic principles. This analysis indicates which set of frequency spectral components are critical to hearing and must be coded with higher precision and assigned more bits, and which set of frequency components are less important and can tolerate relatively more quantisation noise without degrading the perceived sound quality. Based on this information, the available bits are distributed and allocated to the time–frequency groups and the spectral coefficients in each frequency group are quantised with the allocated bits. In the decoder, the quantised spectra are reconstructed according to the bit-allocation pattern and then synthesised into an audio signal.

13.8.2 Adaptive Transform Coding

A transform coder, Figure 13.35, consists of the following sections:

(a) Buffer and window, divides the signal into overlapping segments of length N samples. The segment length may be variable as it controls the time and frequency resolutions and affects the severity of pre-echo distortion described in Section 13.7.4.
(b) Frequency analysis transforms the signal into frequency. Discrete cosine transform is often used due to its ability to compress most of the signal energy into a relatively limited number of principal components. Fourier transform may also be used.
(c) A pre-echo detector detects abrupt changes (e.g. attacks) in signal energy which can cause an audible spread of the quantisation noise of the high-energy part of a frame of music into the low-energy part of the frame and hence produce a pre-echo distortion.
(d) Psychoacoustic model calculates the tonal and non-tonal components of the signal and then estimates the just noticeable distortion (JND) levels for each frequency band of each signal frame.
(e) The quantiser represents each frequency component with k bits. One quantiser may be used for all frequencies. However, it may be advantageous to use a set of quantisers to span the frequency range: one quantiser for each group of frequency bins.

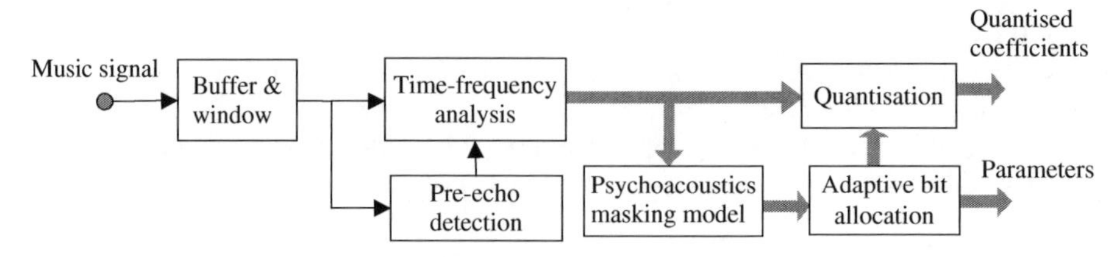

Figure 13.35 Outline of adaptive transform coding.

(f) The bit-allocation modules, known as the rate–distortion loop, allocate bits to each quantiser in order to satisfy two requirements: (i) to keep the total number of bits within the intended bit-rate, (ii) to keep the distortion of each frequency component of the signal below the calculated JND levels.

13.8.3 Time–Frequency Analysis

The time–frequency analysis of music can be performed using a number of different methods:

(a) frequency transformation of frames of music using a DCT or FFT
(b) a bank of band-pass filters to split the signal into sub-bands
(c) a combination of a filter-bank followed by finer frequency analysis using a transform such as DCT as in MPEG and Sony's Atrac.

Time–frequency analysis is a central part of music coders such as MP3 and Atrac. In MP3 the signal is split into 32 equal-width bands of frequency. The sub-band signals are then transformed into finer frequency components using the modified discrete cosine transform (MDCT). The MDCT allows up to 50% overlap between time-domain windows, leading to improved frequency resolution while maintaining critical sampling.

In Sony's Atrac coder the signal is split into three sub-bands which span the frequency ranges 0–5.5 kHz, 5.5–11 kHz and 11–22 kHz. The sub-band decomposition is performed using quadrature mirror filters (QMF). The input signal is divided into upper and lower bands by the first QMF, and the lower band is divided again by a second QMF. The use of QMFs ensures that time-domain aliasing caused by sub-band decomposition will be cancelled during reconstruction.

13.8.4 Variable Length Transform: Pre-Echo Control

A change in the value of a frequency component of a signal segment would have an affect on the values of every time domain sample of that segment and vice versa. When a signal window, which contains a fast rising 'attack' part of a music track, is transformed to the frequency domain, quantised and then transformed back to time (Figure 13.36), the quantisation noise in the high-energy attack

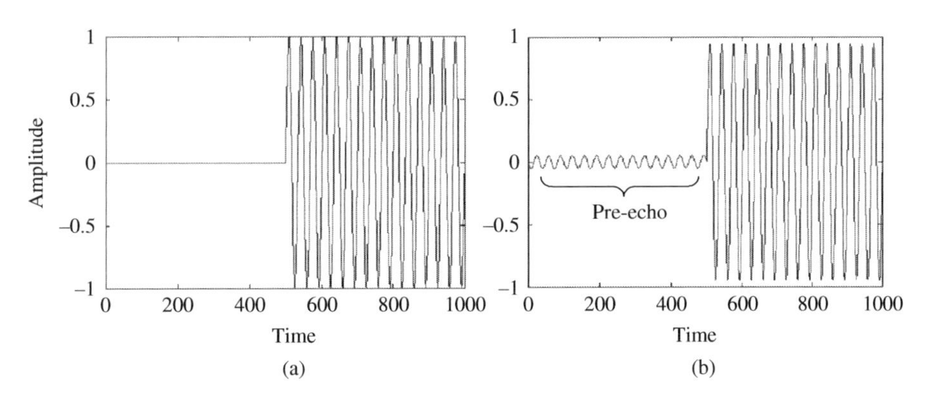

Figure 13.36 Illustration of pre-echo in transform coding of a sine wave: (a) original signal, (b) quantised signal with pre-echo.

part of the signal spreads over the entire length of the window including the relatively lower energy pre-attack part of the signal window. This results in an audible noise in the low-energy part of the window before the attack part of the signal; this noise is known as pre-echo noise.

To prevent pre-echo noise a signal 'attack' detector is used and a shorter duration window is employed during the 'attack' portions of the signal. This would limit the spread of quantisation noise over time. However, as frequency resolution is inversely proportional to the duration of the signal window, a shorter duration window gives a lower frequency resolution. The compromise solution is to use two or more windows of different durations depending on the time-varying characteristics of the music. For example, at a sampling rate of 44,100 Hz, we can switch between three windows: a long window of length 512 samples (11.6 ms), a medium window of length 128 samples (2.9 ms) and a short window of 64 samples (1.45 ms).

Note that a short window (necessary for the fast rising attack part of a signal) is not necessary for the falling or decay part of the signal, because the quantisation noise will be masked by the forward masking process which persists longer than backward masking.

13.8.5 Quantisation of Spectral Values

The spectral coefficients of each frame of a music signal are quantised using two parameters: (1) a word length and (2) a scale factor. The scale factor defines the full-scale range of the quantisation, and the word length defines the precision within that scale. Furthermore, the quantisation steps may be spread non-uniformly to reflect the non-uniform probability distribution of sample values.

Each group of frequencies has the same word length and scale factor, reflecting the psychoacoustic similarity of the frequencies within a group. The scale factor is chosen from a fixed list of possibilities, and reflects the magnitude of the spectral coefficients in each group of frequencies. The word length is determined by the bit-allocation algorithm. For each audio frame the following signal information is stored:

- MDCT block size mode (long or short)
- word length data for each group of frequencies
- scale factor code for each group of frequencies
- quantised spectral coefficients.

Coding of Quantised Spectral Values

The quantised spectral values are encoded with a binary number. Efficient entropy coding of spectral values can be achieved using Huffman coding to assign variable-length binary codes to different quantisation levels. Shorter codewords are assigned to more frequently occurring values and longer codewords to infrequently occurring values.

The Huffman coder and the quantisation process can interact in a rate–distortion loop as increasing the quantisation step size will yield a higher proportion of samples with smaller quantisation levels requiring shorter length codes and a consequent reduction in the bit-rate but at the expense of more quantisation noise and vice versa.

Bit-Allocation Method

The bit-allocation algorithm divides the available bits between the various frequency bands through adjusting the quantisation step size. Frequency bands allocated a large number of bits will have less quantisation noise; bands with fewer bits will have more noise. For high fidelity the bit-allocation algorithm must be adaptively matched to the masking properties of the signal to ensure that sub-bands that are critical to the hearing process have sufficient bits, and that the noise in non-critical bands is not perceptually significant.

Rate–Distortion Loop

An iterative rate–distortion loop is used to adjust the number of quantisation levels, hence the number of bits, for each group of frequency bins until the quantisation noise at each frequency band remains below the just noticeable distortion (JND) threshold level, subject to the constraint that the rate remains within the maximum number of bits available for each frame.

The rate part of the rate–distortion loop employs a 'global' parameter to adjust the overall quantisation parameters to maintain the overall bit-rate within the required level. The distortion part of the loop adjusts the quantisation levels (and hence the bit-rate) within each group of frequencies to maintain the quantisation noise below noticeable distortion level.

The rate control and distortion control form an iterative optimisation loop, the goal of which is to achieve the required target bit-rate and inaudible distortion. There is a maximum number of iterations at which the loop will be terminated in order to prevent the system from iterating infinitely for a frame for which it is not possible to obtain the required audio transparency within the available number of bits.

13.9 High Quality Audio Coding: MPEG Audio Layer-3 (MP3)

MPEG is the acronym for the Moving Picture Experts Group established in 1988 to develop open standards for development of coders for moving pictures and audio. Open standards are specifications that are available to developers interested in implementing the standard. Usually an implementation example is provided to avoid misinterpretation of the text of the standard.

The audio coding standards developed by MPEG is used in many applications including digital audio broadcasting, Internet audio, portable audio, DVD and audio storage. A popular version of MPEG is MP3 developed by a group of engineering laboratories including the Fraunhofer Institute, a German audio research laboratory. MP3 (short for MPEG-1, audio layer 3) is a subset of MPEG compression that can take a two-minute, 22-megabyte, music track on a CD, and reduce it by a factor of 16 to about 1.4 megabytes, the size of a floppy disk.

There are a few versions of the MPEG standard; each version includes a higher level of quality, flexibility and application and offers different choices.

MPEG-1 audio consists of three operating modes, called layers, with increasing complexity, delay and quality from layer-1 to layer-3. MPEG-1 defines audio compression at sampling rates of 32 kHz, 44.1 kHz and 48 kHz. It works with both mono and stereo signals, and a technique called joint stereo coding can be used for efficient coding of the left and right channels. Several bit-rates are specified in the MPEG-1 Layer-3 standard: 32, 40, 48, 56, 64, 80, 96, 112, 128, 160, 192, 224, 256 and 320 kbit/s. The available sampling frequencies are 32, 44.1 and 48 kHz. A sample rate of 44.1 kHz is most common since this is also used for CD audio, the main source used for creating MP3 files. MPEG-1 Layer-3 provides high quality audio at about 128 kbps for stereo signals.

MPEG-2 introduced new concepts for video coding and digital TV. It extends MPEG-1 audio sampling rates to half rates to include 16 kHz, 22.05 kHz and 24 kHz. It also includes the 3/2 channel format comprising right, centre and left channels together with right and left surround channels.

MPEG-3 was to define video coding for high definition television (HDTV) applications. However, as MPEG-2 contains all that is needed for HDTV, MPEG-3 was rolled into MPEG-2.

MPEG-4 is more concerned with new functionalities than better compression efficiency. The major applications of MPEG-4 are mobile and fixed terminals, database access, communications and interactive services. MPEG-4 audio consists of audio coders spanning the range from 2 kbps low bit-rate speech, up to 64 kbps/channel high quality audio.

MPEG-7 is a content representation standard for multimedia information search engines, filtering, management and processing of data.

MPEG Bit-Rates

MPEG audio allows flexible compression ratio. Within the prescribed limits of the open standard, the selection of the bit-rate is left to the implementer and/or the operator of the audio coder. The standard bit-rate is defined in the range from 8 kbps to 320 kbps. The open standard also enables the use of variable bit-rate coding and fixed bit-rate coding within the prescribed limits.

13.9.1 MPEG Structure

Figure 13.37 illustrates a block diagram structure of an MP3 coder. This is a frequency domain coder in that segments of 576 time domain samples are transformed into 576 frequency domain samples. The available bits are then allocated non-uniformly among various frequency components depending on the just noticeable distortion (JND) levels at different frequencies calculated from the psychoacoustic model. It consists of the following subsystems.

The **filter bank** is a uniformly spaced filter bank consisting of 32 equal-bandwidth poly-phase filters of length 512 taps each.

Poly-phase filters divide the bandwidth into N bands (where N is a power of 2) with the signal in each sub-band down-sampled by a factor of N. The filters are said to be critically sampled and have excellent anti-aliasing property in the sense that they can achieve perfect reconstruction of the original signal.

The filters are equally spaced and split a bandwidth of 24 kHz, at a sampling rate of 48 KHz, into 32 bands of width 750 Hz. The output of each filter is down-sampled by a factor of 32:1. After down-

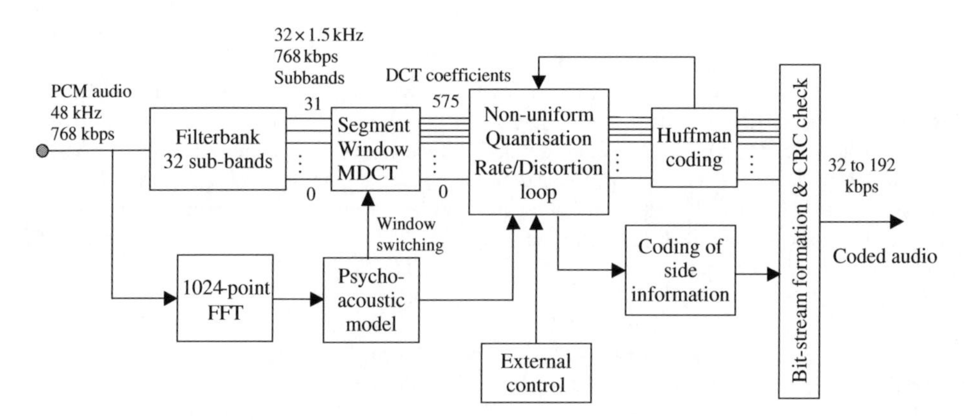

Figure 13.37 Illustration of MPEG-1 layer-3 system.

sampling each sub-band has a sampling rate of 1500 Hz, and the total numbers of samples across 32 sub-bands is 48,000 samples per second: the same as the input sampling rate before band splitting.

Modified discrete cosine transform Each sub-band output is segmented into segments of 18 samples long – corresponding to a segment length of $18 \times 32 = 576$ samples of the original signal before down-sampling or 12 ms duration – and transformed by a modified discrete cosine transform (MDCT). Hence the 750 Hz width of each sub-band is further decomposed into 18 frequency bins with a frequency resolution of $750/18 = 41.7$ Hz.

The **auditory perceptual model** is based on critical bands of hearing and masking thresholds as described in Section 13.6. A 1024-sample FFT of the music signal (with a frequency resolution $F_s/N = 48,000/1024 = 46.875$ Hz) is used to calculate the noise masking thresholds, the so-called just noticeable distortion (JND) levels; this is the amount of quantisation noise in each frequency band that would be masked and made inaudible by the signal energy at and around that band as explained in Section 13.6. The frequency bands for calculation of the masking thresholds are based on the critical bands of hearing. If the quantisation noise energy can be kept below the masking threshold then the compressed signal would have the same transparent perceptual audio quality as the original signal.

Quantisation and coding processes aim to distribute the available bits among the DCT coefficients such that the quantisation noise remains masked. This is achieved through an iterative two-stage optimisation loop. A power law quantiser is used so that large spectral values are coded with a larger quantisation step size, as a higher signal energy masks more quantisation noise. The quantised values are then Huffman coded (Figure 13.38). To adapt the coder to the local statistics of the input audio signal the best Huffman coding table is selected from a number of choices.

The Huffman coder is a probabilistic coding method that achieves coding efficiency through assigning shorter length codewords to more probable (i.e. more frequent) signal values and longer length codewords to less frequent values. Consequently, for audio signals smaller quantised values, which are more frequent, are assigned shorter length codewords and larger values, which are less frequent, are assigned longer length codewords.

Quantisation consists of two loops: (i) an inner loop that adjusts the rate to keep the overall bit-rate within the required limit and (ii) an outer loop that aims to keep the distortion in each critical band masked. The rate-loop interacts with the Huffman coder as explained next.

The **rate loop (inner loop)** adjusts the bit-rate of the Huffman coded signal, and maintains it at or below the target rate, through the use of a global gain variable to control the relative proportion of

Symbol Probability	Huffman Code
$P_8 = 0.05$	00011
$P_7 = 0.1$	0000
$P_6 = 0.15$	001
$P_5 = 0.2$	10
$P_4 = 0.2$	11
$P_3 = 0.15$	010
$P_2 = 0.1$	011
$P_1 = 0.05$	00010

Figure 13.38 Illustration of Huffman coding of the quantisation levels of a coder. In this case the minimum bits/symbol indicated by the entropy is 2.8464. The Huffman code shown achieves 2.9 bits/sample.

samples within each quantisation level. Note that the lower quantisation levels with higher probability are assigned shorter code length. If the number of available bits is not enough to Huffman code a block of data within the pre-specified rate then the global gain is adjusted to result in a larger quantisation step size and hence a higher proportion of smaller-valued quantised samples requiring shorter length codewords in Huffman coding. This is repeated with different values of the quantisation step size until the bit-rate requirement is no more than the target bit-rate for each block.

Distortion loop (outer loop) When the quantisation distortion in a band is noticeable then the number of quantisation levels needs to be increased by adjusting the scale factors. The scale factors are applied to each scale band to shape the quantisation noise so that it remains below the masking threshold. Initially the scale factor for each band is set to 1. If the quantisation noise in a band exceeds the masking threshold then the scale factor for that band is adjusted (increased) to reduce the quantisation step size and keep the noise below the masking threshold. However this adjustment and control of distortion noise can result in a higher bit-rate than allowed since to reduce quantisation noise in each band the number of quantisation levels is increased. So the rate-adjustment loop has to be repeated each time scale factors are adjusted. Therefore the rate loop is nested in the distortion loop. The rate and distortion loops may fail to converge and may go on forever, however several conditions can be checked to stop the iterations in an early stage.

Note on Implementation of the Rate–Distortion Loop

The MDCT values, $R(f)$, which are initially between -1 and 1, are quantised using the equation

$$Q(f) = \text{round}((R(f)2^{N(f)})^{0.75}) \tag{13.62}$$

where

$$N(f) = 0.25(210\text{–global\_gain}) + \text{scale factor}(\text{scf}(f)) \tag{13.63}$$

f is the frequency variable and $\text{scf}(f)$ is the scale factor sub-band index which corresponds more or less to the bark scale. There are 22 scale factor sub-bands. The scale factor for sub-band 22 is always 0. By increasing the scale factor for each sub-band the number of quantisation levels is increased, so less quantisation noise is introduced in that sub-band. This is used for shaping the noise spectrum to fit the masking spectrum. By increasing global gain, the number of quantisation levels for all sub-bands is decreased, hence reducing the number of bits needed to encode them. This can be used in bit-rate control.

After these values have been set, different Huffman tables can be selected, for encoding pairs of frequency lines, to get the greatest amount of compression possible.

13.10 Stereo Music Coding

The relationship between the signals in the left and the right channels of stereo music may be modelled as

$$x_L(m) = fn[x_R(m)] + x_{L\perp R}(m) \tag{13.64}$$

$fn[x_R(m)]$ is the part of the left channel $x_L(m)$ which is correlated with the right channel $x_R(m)$ and can be predicted from it, fn is a mapping function, and $x_{L\perp R}(m)$ is the part of the left channel $x_L(m)$ uncorrelated with $x_R(m)$, $\perp$ denotes orthogonal. The objective of stereo coding is to exploit inter-channel correlation and redundancies so that the part of the music signal that is common to both channels is not coded twice.

Figure 13.39 shows an example of signals in the left and right channels of a segment of stereo music. The spectrograms and the time waveforms show considerable similarity between the two signals. The normalised cross-correlation of the left and right channels, shown in Figure 13.39(d), fluctuates with time. However, there are periods when the normalised cross-correlation approaches the maximum value of one indicating identical signals. In stereo music coding it is assumed that:

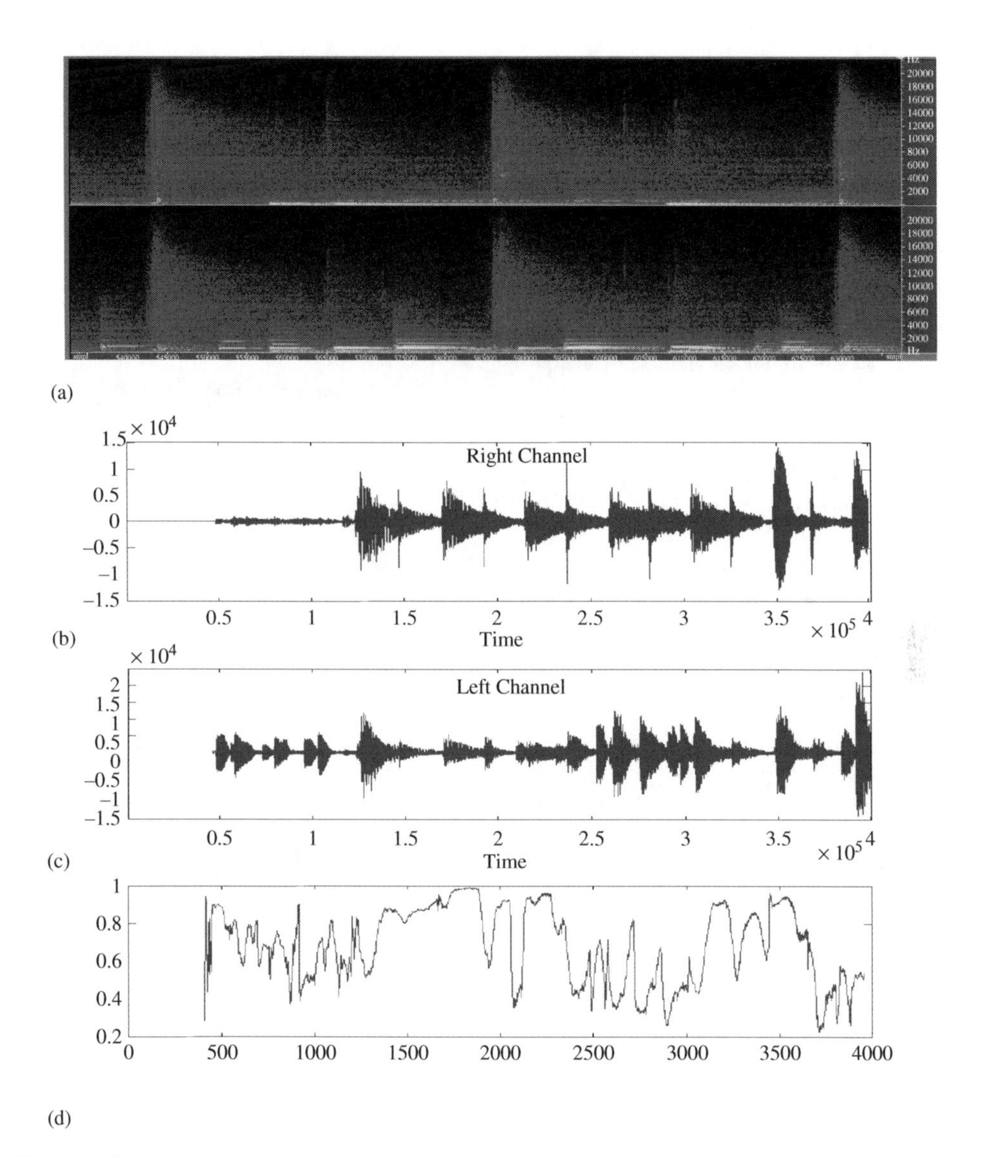

Figure 13.39 (a) The spectrogram of 2 seconds of a segment of stereo Jazz music; (b), (c) and (d) right and left channels of a segment of music of duration 10 seconds and their cross-correlation respectively.

- The *masking threshold* of hearing is the same for both channels. Thus, the mean of the two channels is fed to the masking threshold estimator.
- The audio is centred, i.e. approximately equal in both channels.

In practice, joint stereo coding systems encode the sum (L + R) and the difference (L − R) of the left (L) and right (R) channels. From an analysis of the FFT spectra of L + R and L − R signals a combined set of thresholds of the just noticeable distortion (JND) levels is calculated for both L + R and L − R channels. The sum and difference signals are then coded in the same way as a mono signal using the common set of JNDs for both channels.

More complex stereo coding models may use an inter-channel linear predictor model, or inter-channel orthogonalisation transforms.

13.11 Summary

This chapter began with an introduction to the system of musical notations, pitch intervals and musical scales. The generation of musical sounds by string and pipe instruments was studied. Particular focus was placed on some aspects of the structure, workings and physics of the guitar, violin and trumpet. The basic physical properties of sounds and the methods of propagations of sounds were reviewed as were the methods of modelling of musical signals.

Since the human ear and auditory perception system is the ultimate receiver and decoder of music signals, the anatomy of the ear and the psychoacoustics of hearing were studied in this chapter. An understanding of such important topics as the frequency and neural responses of the cochlea, the critical bands of hearing and the spectral–temporal masking phenomena, is essential to the successful engineering of music processing systems.

In the final part of this chapter we considered the structure and the operation of musical coders and in particular the principles of design and operation of the popular MP3 coder and a stereo coding method were introduced.

Further Reading

Benade A.H. (1976) Fundamentals of Musical Acoustics. Oxford University Press.

Cremer L. (1984) The Physics of the Violin. MIT Press, Cambridge, MA.

Fletcher H. and Munson W.A. (1933) Loudness, its Definition, Measurement, and Calculation, J. Acoust. Soc. Am. 5, 82–108.

Gelfand S. (2004) hearing: an introduction to psychological and physiological acoustics, 4th edn. Marcel Dekker, New York.

Gold B. and Morgan N. (2000) Speech and Audio Signal Processing: Processing and Perception of Speech and Music, John Wiley and Sons.

Helmohltz H. von (1885/1954) On the Sensations of Tone (English edn). Dover, New York.

Klapuri A. and Davy M. (eds) (2006) Signal Processing Methods for Music Transcription, Springer.

ISO/IEC 11172-3:1993 Information Technology – Coding of Moving Pictures and Associated Audio for Digital Storage Media at up to about 1, 5 Mbits/s – Part 3: Audio.

Lawn, R.J. and Hellmer J.L. (1996) Jazz Theory and Practice. Alfred Publishing, New York.

McIntyre M.E. and Woodhouse J. (1981) On the Fundamentals of Bowed String Dynamics, Acustica, 43, p. 93.

Painter T. and Spanias A. (1997) A review of Algorithms for Perceptual Coding of Digital Audio Signals, Digital Signal Processing Proceedings, DSP 97, 13th International Conference Vol. 1, 2–4 July pp. 179–208.

Richardson, B. (1992) The Physics of the Violin. The Cambridge Companion to the Violin (ed. Robin Stowell). Cambridge University Press.

Roads C. (1989) The Music Machine. MIT Press, Cambridge, MA.

Robinson D.W. and Dadson R.S. (1956) A Re-determination of the Equal-Loudness Relations for Pure Tones, Br. J. Appl. Phys. 7, pp. 166–181.

Schelleng J.C. (1973) The Bowed String and the Player. Journal of Acoustic Society of America 53 (1), pp. 26–41.

Tobias J.S. (ed.) (1970) Foundations of Modern Auditory Theory, Vol. 1. Academic Press, New York.

Woodhouse J. and Galluzzo P.M. (2004) Why is the violin so hard to play? http://plus.maths.org/issue31/features/woodhouse/index.html

14 Speech Processing

Speech sounds are auditory sensations of air pressure vibrations produced by air exhaled from the lungs and modulated and spectrally shaped by the vibrations of the glottal cords and the frequency response of the vocal tract and the nasal cavity as the air is pushed out through the lips and nose.

Speech is an immensely information-rich signal exploiting frequency-modulated, amplitude-modulated and time-modulated carriers (e.g. resonance movements, harmonics and noise, pitch intonation, power, duration) to convey information about words, speaker identity, accent, expression, style of speech, emotion and the state of health of the speaker. Most of this information is conveyed primarily within the traditional telephone bandwidth of 4 kHz. The speech energy above 4 kHz mostly conveys audio quality and sensation and some of the information of unvoiced/consonant speech.

In this chapter the fundamentals of speech signals and speech production and perception are studied. We study the mechanisms that produce and convey phonetic speech sounds and examine the acoustic correlates of speaker characteristics such as gender, accent and emotion. The spectral and temporal structures of speech are studied and the most commonly used models and features for capturing speech characteristics in time and frequency are introduced. Speech coding methods for improving bandwidth utilisation and power efficiency in mobile communication are covered. Finally, we study automatic speech recognition and a simple voice-dialling application.

14.1 Speech Communication

Speech is the most natural form of human communication. The temporal–spectral variations of speech signals convey such information as words, intention, expression, intonation, accent, speaker identity, gender, age, style of speaking, state of health of the speaker and emotion.

Speech sounds are produced by air pressure vibrations generated by pushing inhaled air from the lungs through the vibrating vocal cords and vocal tract and out from the lips and nose airways. The air is modulated and shaped by the vibrations of the glottal cords, the resonance of the vocal tract and nasal cavities, the position of the tongue and the openings and closings of the mouth.

Just as the written form of a language is composed of a sequence of elementary symbols English alphabet (A, B, C, . . . , Z), speech is composed of a sequence of elementary acoustic sounds or symbols (e.g. AA, AE, IH, IY, . . . , OW etc), known as phonemes, that convey the spoken form of a

Multimedia Signal Processing: Theory and Applications in Speech, Music and Communications Saeed V. Vaseghi

language. There are about 40–60 phonemes in the spoken English language from which a very large number of words can be constructed. Note that in practice the production of each phonemic sound is affected by the context of the neighbouring phonemes. This phenomenon is called co-articulation.

14.1.1 Information Conveyed by Speech

Speech signals convey much more than spoken words. The information conveyed by speech is multi-layered and includes time–frequency modulation of such carriers of information as formants and pitch intonation. Formants are the resonances of the vocal tract and pitch is the sensation of the fundamental frequency of the opening and closings of the glottal folds. The information conveyed by speech signals includes the following:

(a) Acoustic phonetic symbols. These are the most elementary speech units from which larger speech units such as syllables and words are formed. Some words have only one or two phonemes such as 'I', 'you', 'me', 'to', 'he' etc.
(b) Prosody. This is the rhythms of speech and is composed primarily of the intonation and stress signals carried by changes in the pitch and amplitude trajectories. Prosody helps to signal such information as the boundaries between segments of speech, link sub-phrases and clarify intention and remove ambiguities such as whether a spoken sentence is a statement or a question.
(c) Gender information. Gender is conveyed by the pitch (related to the fundamental frequency of voiced sounds) and the size and physical characteristics of the vocal tract. Due to the differences in vocal anatomy, the female voice has higher resonance frequencies and a higher pitch. Males have an average pitch of about 120 Hz, female pitch is around 220 Hz.
(d) Age, conveyed by the effects of the size and the elasticity of the vocal cords and vocal tract, and the pitch. Children can have a pitch of more than 300 Hz.
(e) Accent, broadly conveyed through (i) changes in the pronunciation vocabulary in the form of substitution, deletion or insertion of phoneme units in the 'standard' transcription of words (e.g. Australian *todie* pronunciation of *today* or US ***Jaan*** pronunciation of ***John***) and (ii) systematic changes in acoustic production of speech, i.e. in resonance frequencies (formants), pitch intonation, duration, emphasis and stress.
(f) Speaker's identity conveyed by the physical characteristics of a person's vocal folds, vocal tract, pitch intonations and stylistics.
(g) Emotion and health, conveyed by changes in vibrations of the vocal folds, vocal tract resonance, duration and stress and by the dynamics of pitch and vocal tract spectrum.

In the remainder of this chapter we will study how various acoustic correlates of speech and speaker can be modelled and used for speech processing applications.

14.2 Acoustic Theory of Speech: The Source–filter Model

An outline of the anatomy of the human speech production system is shown in Figure 14.1. It consists of the lungs, larynx, vocal tract cavity, nasal cavity, teeth, lips, and the connecting tubes. The combined voice production mechanism produces the variety of vibrations and spectral–temporal compositions that form different speech sounds.

The act of production of speech begins with exhaling (inhaled) air from the lungs. Without the subsequent modulations, this air will sound like a random noise from a deflating balloon, with no

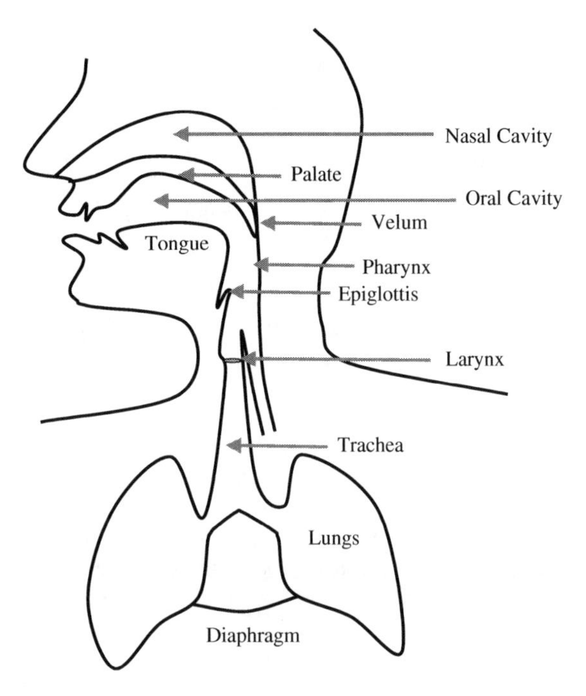

Figure 14.1 Illustration of anatomy of speech production.

information. The information is first modulated onto the passing air by the manner and the frequency of closing and opening of the glottal folds. The output of the glottal folds is the excitation signal to the vocal tract which is further shaped by the resonances of the vocal tract and the effects of the nasal cavities and the teeth and lips.

The vocal tract is bounded by hard and soft tissue structures. These structures are either essentially immobile, such as the hard palate and teeth, or movable. The movable structures associated with speech production are also called *articulators*. The tongue, lips, jaw and velum are the primary articulators; movement of these articulators appear to account for most of the variations in the vocal tract shape associated with making different sounds when speaking. However, additional structures are capable of motion as well. For instance, the glottis can be moved up or down to shorten or lengthen the vocal tract and hence change its frequency response.

Speech sounds result from a combination of a source of sound energy (the larynx) modulated by a time-varying transfer function filter (vocal articulators) determined by the shape and size of the vocal tract and nasal cavity. This results in a shaped spectrum with broadband energy peaks. This model is known as the source–filter model of speech production, shown in Figure 14.2. In this model the

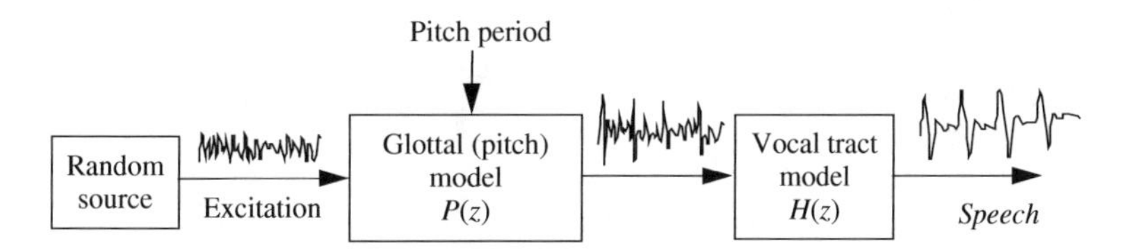

Figure 14.2 A source–filter model of speech production.

source of acoustic energy is at the larynx, and the vocal tract serves as a time-varying filter whose shape determines the phonetic content of the sounds.

14.2.1 The Source Model

The source signal of speech is the noise-like air from the lungs which is temporally and spectrally shaped by the manner and the frequency of the openings and closings of the glottal folds. There are two broad types of speech sounds as shown in Figure 14.3: *voiced* sounds like an '*e*' pronounced as '*iy*', and *unvoiced* sounds like '*s*'.

Voiced sounds are produced by a repeating sequence of opening and closing of glottal folds with a frequency of between 40 (e.g. for a low-frequency male voice) to 600 (e.g. for a female child's voice) cycles per second (Hz) depending on the speaker, the phoneme and the linguistic and emotional/expressional context. First, as the air is pushed out from the lungs the vocal cords are brought together, temporarily blocking the airflow from the lungs and leading to increased sub-glottal pressure. When the sub-glottal pressure becomes greater than the resistance offered by the vocal folds, the folds open and let out a pulse of air. The folds then close rapidly due to a combination of several factors, including their elasticity, laryngeal muscle tension, and the Bernoulli effect of the air stream. If the process is maintained by a steady supply of pressurised air, the vocal cords will continue to open and close in a quasi-periodic fashion. As they open and close, the pulses of air flow through the glottal opening as shown in Figure 14.4.

The periodicity of the glottal pulses determines the fundamental frequency ($F0$) of the source signal and contributes to the perceived pitch of the sound. The time-variations of the glottal pulse period convey the style, the intonation, the stress and emphasis in speech signals. In normal speech the fundamental frequency (pitch) changes constantly, providing linguistic clues and speaker information, as in the different intonation patterns associated with questions or statements, or information about the emotional content, such as differences in speaker mood e.g. calmness, excitement, sadness etc.

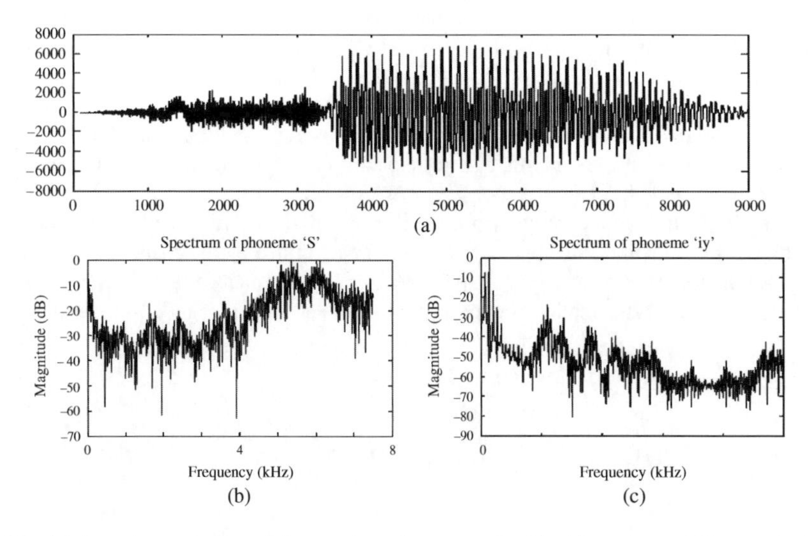

Figure 14.3 (a) Acoustic production of the word ***sea*** (pronounced *s-iy*), (b) spectrum of the unvoiced segment '*s*', and (c) spectrum of the voiced speech segment '*iy*'.

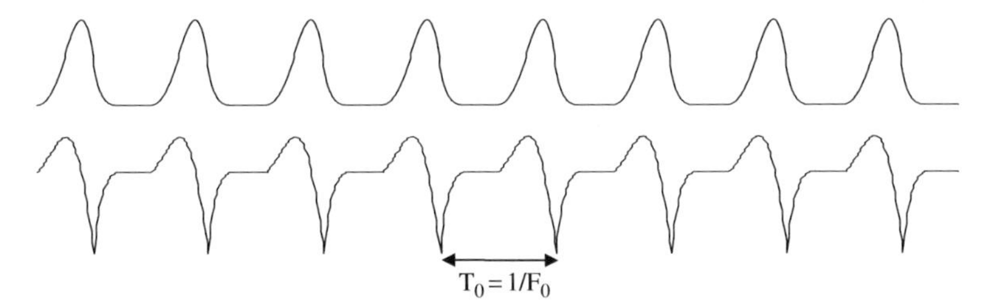

Figure 14.4 A sequence of glottal pulses of air flow (top) and their derivative (bottom).

Figure 14.3 shows an example of a speech segment containing an unvoiced sound '*s*' and a voiced sound '*iy*'. Note that the spectrum of voiced sounds is shaped by the resonance of the vocal tract filter and contains the harmonics of the quasi-periodic glottal excitation, and has most of its power in the lower frequency bands, whereas the spectrum of unvoiced sounds is non-harmonic and usually has more energy in higher frequency bands. The shape of the spectrum of the input to the vocal tract filter is determined by the details of the opening and closing movements of the vocal cords, and by the fundamental frequency of the glottal pulses.

For unvoiced sounds air is passed through some obstacle in the mouth (e.g. when pronouncing 'S'), or is let out with a sudden burst (e.g. when pronouncing 'P'). The position where the obstacle is created depends on which speech sound (i.e. phoneme) is produced. During transitions, and for some mixed-excitation phonemes, the same air stream is used twice: first to make a low-frequency hum with the vocal cords, then to make a high-frequency, noisy hiss in the mouth.

Glottal Pulse Model for Voiced Signals

For voiced sounds, the shape of the glottal pulses of air flow, Figure 14.4, is determined by the manner and the duration of the opening and closing of the glottal folds in each cycle of voice sounds and contributes to the perception of the voice quality and the speaker's identity. The quality of voice and its classification into such types as normal (modal), creaky, breathy, husky, tense etc. depends on the glottal pulse shape.

Figure 14.5 shows the Liljencrants–Fant (LF) model of a single glottal pulse and its derivative. The glottal pulse consists of an open phase, during which a pulse or puff of air is let through, and a closed phase. The open phase of the cycle itself is composed of an opening phase which culminates in the maximum opening of the glottal folds and a closing phase. The maximum negative value of the derivative of the pulse is reached at the point of the fastest rate of closing of the glottal folds.

The LF model of the derivative of the glottal pulse is defined as

$$v_{\mathrm{LF}}(t) = \begin{cases} E_0 e^{\alpha t} \sin \omega t & 0 \le t < T_e \\ E_1 \left(e^{-\beta(t-T_e)} - e^{-\beta(T_c-T_e)} \right) & T_e \le t < T_c \\ 0 & T_c \le t \le T_0 \end{cases} \tag{14.1}$$

where a composition of a segment of less than 3/4 of a period of a sine wave, with a frequency of ω and an exponential envelope $E_0 e^{\alpha t}$, is used to model the derivative of the glottal pulse up to the instance T_e where the derivative of the pulse reaches the most negative value which corresponds to the fastest rate of change of the closing of the glottal folds. The final part of the closing phase of the glottal

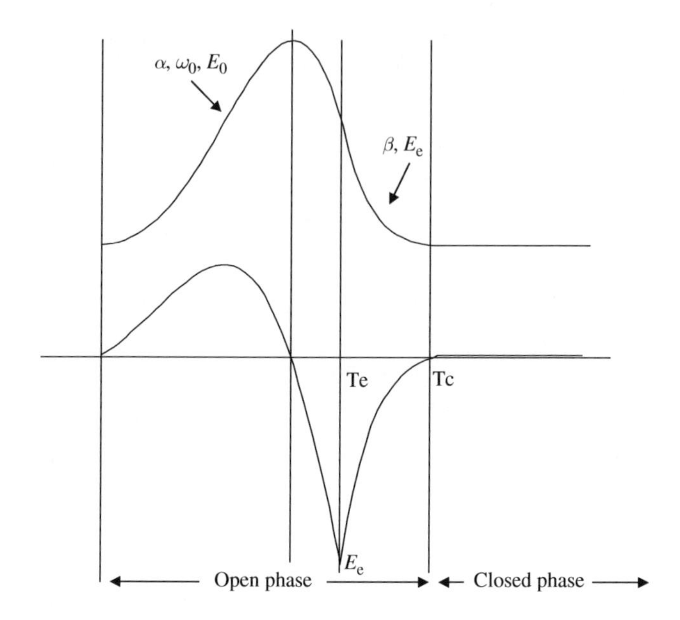

Figure 14.5 The LF model of a glottal pulse and its derivative.

folds, the so-called return phase, is modelled by an exponentially decaying function in the second line of Equation (14.1). In Equation (14.1), T_0 is the period of the glottal waveform, $F_0 = 1/T_0$ is the fundamental frequency (pitch) of speech harmonics, and T_c is the instance of closing of the glottal fold. The parameters E_0 and E_1 can be described in terms of the most negative value of the pulse E_e at the instant T_e; $E_o = E_e/[e^{\alpha T_e} \sin \omega T_e]$ and $E_1 = E_e/[1 - e^{-\beta(T_c - T_e)}]$. The modelling and estimation of the glottal pulse is one of the ongoing challenges of speech processing research.

14.2.2 The Filter Model

Whereas the source model describes the detailed structure of the speech spectrum, the filter model describes the envelope of the speech spectrum. The resonance characteristics of the physical space, such as the vocal tract, through which a sound wave propagates changes the spectrum of sound and its perception.

The vocal tract space, composed of the oral and nasal cavities and the airways, can be viewed as a time-varying acoustic filter that amplifies and filters the sound energy and shapes its frequency spectrum. The resonance frequencies of the vocal tract are called *formants*. The identities of the acoustic realisation of phonemes are conveyed by the resonance frequencies at formants. Depending on the phoneme sound and the speaker characteristics there are about 3 to 5 formants in voiced sounds.

Formants are dependent on the phonemes but are also affected by the overall shape, length, volume and reverberation characteristics of the vocal space and the vocal tract tissues and the associated parts i.e. nasal cavity, tongue, teeth and lips. The detailed shape of the filter transfer function is determined by the entire vocal tract serving as an acoustically resonant system combined with losses including those due to radiations at the lips.

Matlab program function GlottalLF( )
Generates and plots the Liljencrants–Fant (LF) glottal pulse model. The parameters of the model are:

the period of the glottal pulse T_0,
the period of the sine wave component of the pulse ω_0,
the amplitude of envelope of the sine wave E_0,
the rate of rise of the exponential envelope of the sine wave (alpha) α,
the instant of maximum rate of closure of glottal folds T_e and
the instant of actual closure T_0.

Different values of the LF parameters and the pitch will affect the voice quality and produce a variety of different types of voice such as normal, creaky, tense, breathy, gravel etc.

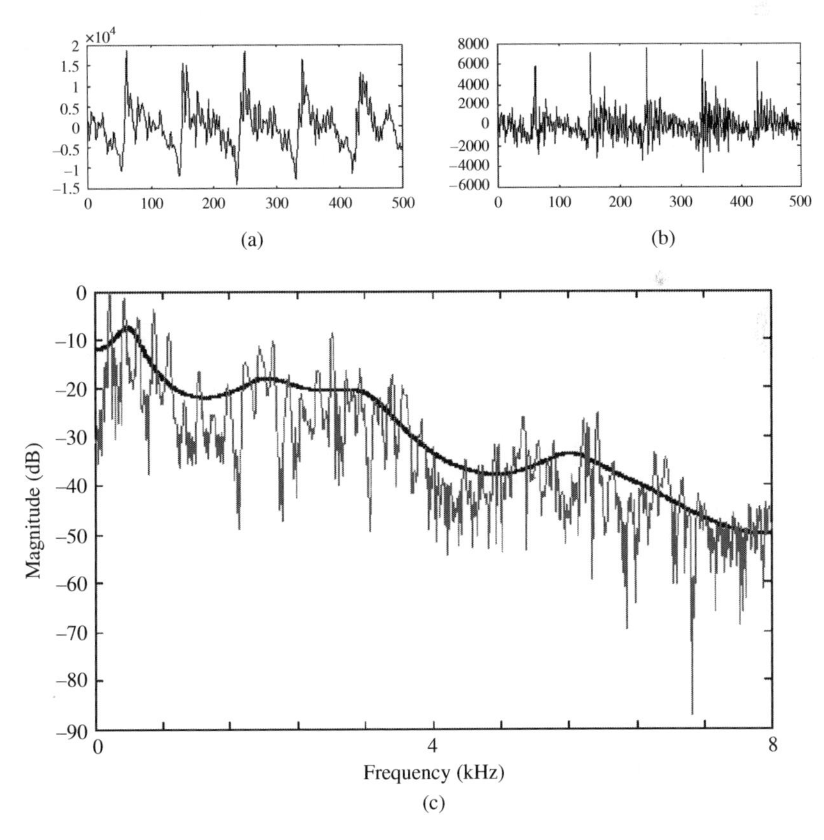

Figure 14.6 (a) A segment of the vowel 'ay', (b) its glottal excitation, and (c) its magnitude Fourier transform and the frequency response of a linear prediction model of the vocal tract.

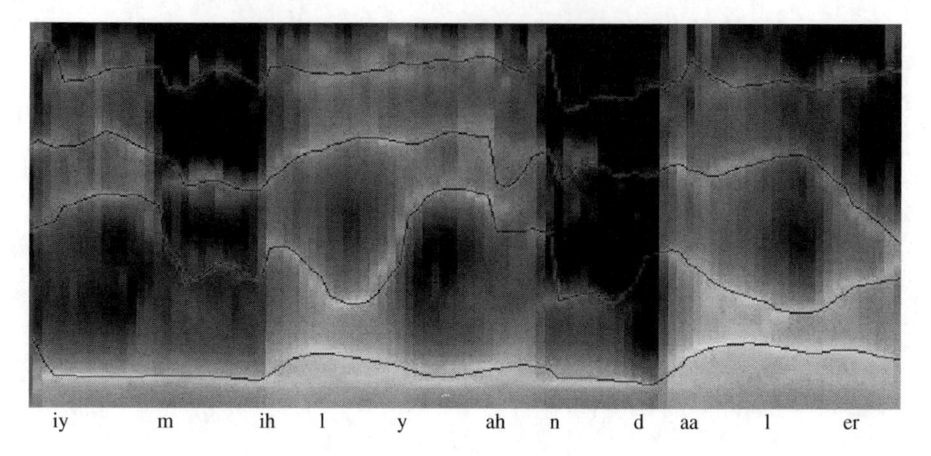

Figure 14.7 An example of formant tracks superimposed on LP spectrogram.

Figure 14.6 shows a segment of a speech signal and its frequency spectrum for the vowel '*ay*'. The formants, which correspond to those frequencies at which the peaks of the frequency response occur, represent the centre points of the main bands of energy that are passed by a particular shape of the vocal tract. In this case they are 500, 2000 and 3000 Hz with bandwidths of 60 to 100 Hz. Figure 14.7 shows formant tracks superimposed on a spectrogram of spectral envelopes obtained from the frequency response of linear prediction models of speech. The flexibility of the human vocal tract, in which the articulators can easily adjust to form a variety of shapes, results in the potential to produce a wide range of sounds. For example, the particular vowel quality of a sound is determined mainly by the shape of the vocal tract, and is reflected in the filter impulse response or frequency response function.

14.3 Speech Models and Features

Speech models and features are used for efficient speech coding, recognition, synthesis, and enhancement. For speech coding, most commercial coders are based on linear prediction models. For text-to-speech synthesis often a harmonic plus noise model of speech is used. For speech recognition the most popular features, derived from the spectral envelope, are the variants of cepstral features (including FFT-cepstral and LPC-cepstral) and their dynamics in time.

The purpose of speech modelling and parameterisation is two-fold:

(a) To represent speech in terms of a compact and efficient set of parameters for speech coding, recognition, synthesis, or enhancement.
(b) To separate speech parameters, such as spectral envelope and pitch intonation curves, which perform different functions or convey different signals such as accent, emotion, gender etc.

The different varieties of speech features and parameters can be listed as follows:

(a) The spectral envelope of speech is modelled by the frequency response of a linear prediction model of the vocal tract, or by the envelope of the DFT spectrum of speech or by the output of a set of filter-banks.
(b) Speech formants, including formant frequencies and bandwidth, and their trajectories in time. Formants are the resonance frequencies of the vocal tract cavity, where the spectral envelope's

peaks occur. Note that the formant information is contained in the spectral envelope or in the linear prediction model of speech.

(c) The fundamental frequency of the opening and closing of the glottal cords, i.e. the pitch.
(d) The temporal dynamics of speech parameters, namely the time variation of the spectral envelope, the formants, and the pitch.
(e) Intonation signals. Intonation signals are conveyed by the temporal dynamics of pitch across a segment of speech.
(f) Stress/emphasis patterns which are functions of pitch intonation, duration and energy.

The most commonly used models of speech production, namely linear prediction models and harmonic noise models are introduced next. The cepstrum features of speech are discussed in Section 14.8.3 on speech recognition.

14.4 Linear Prediction Models of Speech

A widely used source–filter model of speech is the linear prediction (LP) model introduced in detail in Chapter 10. LP models are used for speech coding, recognition and enhancement. A LP model is expressed as

$$x(m) = \sum_{k=1}^{P} a_k x(m-k) + e(m) \tag{14.2}$$

where $x(m)$ is speech signal, a_k are the LP parameters and $e(m)$ is speech excitation. Note that the coefficients a_k model the correlation of each sample with the previous P samples whereas $e(m)$ models the part of speech that cannot be predicted from the past P samples.

In the frequency domain Equation (14.2) becomes

$$X(f) = \frac{E(f)}{1 - \sum_{k=1}^{P} a_k e^{-j2\pi fk}} = \frac{E(f)}{A(f)} = \frac{G \cdot U(f)}{A(f)} \tag{14.3}$$

where $X(f)$ is the speech spectrum, $E(f)$ is the spectrum of excitation, $U(f)$ is the same as $E(f)$ but with normalised power, G is a gain factor and $G/A(f)$ is the spectrum of the LP model of the combination of vocal tract, nasal cavities and lips as well as the spectral slope due to the glottal pulse. In a source–filter LP model of speech the spectral envelope of speech is modelled by the frequency response of the LP model $G/A(f)$ whereas the finer harmonic and random noise-like structure of the speech spectrum is modelled by the excitation (source) signal $E(f)$.

The model parameters $\{a_k, k = 1, \ldots, P\}$ can be factorised and described in terms of a set of complex conjugate and real roots, the so-called poles of the model $(\rho_k, k = 1, \ldots, P\}$. The poles are related to the resonance or formants of speech. The model parameters can also be expressed in terms of the reflection coefficients of a lattice model of speech as described in Section 10.2.3. Figure 14.6(c) shows the frequency response of a linear prediction model of a speech sound.

Matlab function LPCSpeechDemo( )

Demonstrates linear prediction modelling and analysis of speech signals. The program analyses a segment of a speech signal and displays: the signal, its FFT-spectrum,

(Continued)
LP-frequency response, pole-zero diagram, inverse filter output i.e. the estimate of input/excitation signal to the LP model and the autocorrelation function of speech from which a pitch estimation function and the pitch may be obtained.

14.4.1 Line Spectral Frequencies

Line spectral frequencies (LSFs), also known as line spectral pairs (LSP), are an alternative representation of linear prediction parameters. LSFs are used in speech coding, and in the interpolation and extrapolation of LP model parameters, for their good interpolation and quantisation properties. LSFs are derived as the roots of the following two polynomials:

$$\begin{aligned} P(z) &= A(z) + z^{-(P+1)}A(z^{-1}) \\ &= 1 - (a_1 + a_p)z^{-1} - (a_2 + a_{P-1})z^{-2} - \cdots - (a_P + a_1)z^{-P} + z^{-P+1} \end{aligned} \tag{14.4}$$

$$\begin{aligned} Q(z) &= A(z) - z^{-(P+1)}A(z^{-1}) \\ &= 1 - (a_1 - a_P)z^{-1} - (a_2 - a_{P-1})z^{-2} - \cdots - (a_P - a_1)z^{-P} - z^{-P+1} \end{aligned} \tag{14.5}$$

where $A(z) = 1 - a_1z^{-1} - a_2z^{-2} - \cdots - a_Pz^{-P}$ is the inverse linear predictor filter. Clearly $A(z) = [P(z) + Q(z)]/2$. The polynomial equations (14.4) and (14.5) can be written in factorised form as

$$P(z) = \prod_{i=1,3,5,\cdots} (1 - 2\cos\omega_i z^{-1} + z^{-2}) \tag{14.6}$$

$$Q(z) = \prod_{i=2,4,6,\cdots} (1 - 2\cos\omega_i z^{-1} + z^{-2}) \tag{14.7}$$

where ω_i are the LSF parameters. It can be shown that all the roots of the two polynomials have a magnitude of one and they are located on the unit circle and alternate with each other. Hence in LSF representation the parameter vector $[a_1, a_2, \ldots, a_P]$ is converted to the LSF vector $[\omega_1, \omega_2, \ldots, \omega_P]$.

Figure 14.8 shows a segment of voiced speech together with poles of its linear predictor model and the LSF parameters.

14.5 Harmonic Plus Noise Model of Speech

An alternative to a source–filter model of speech is the harmonic plus noise model (HNM). As the name implies speech signals for each frame can be represented by a combination of harmonic and noise-like models as

$$x(m) = \underbrace{\sum_{k=1}^{M} a_k \cos(2\pi k F_0 m) + b_k \sin(2\pi k F_0 m)}_{\text{Harmonic Model (Fourier Series)}} + \underbrace{v(m)}_{\text{Noise Model}} \tag{14.8}$$

where F_0 is the fundamental frequency, a_k, b_k are the amplitudes of the sine and cosine components of the kth harmonic, M is the number of harmonics and $v(m)$ is the noise-like random component that models the fricative and aspiration noise contents of speech.

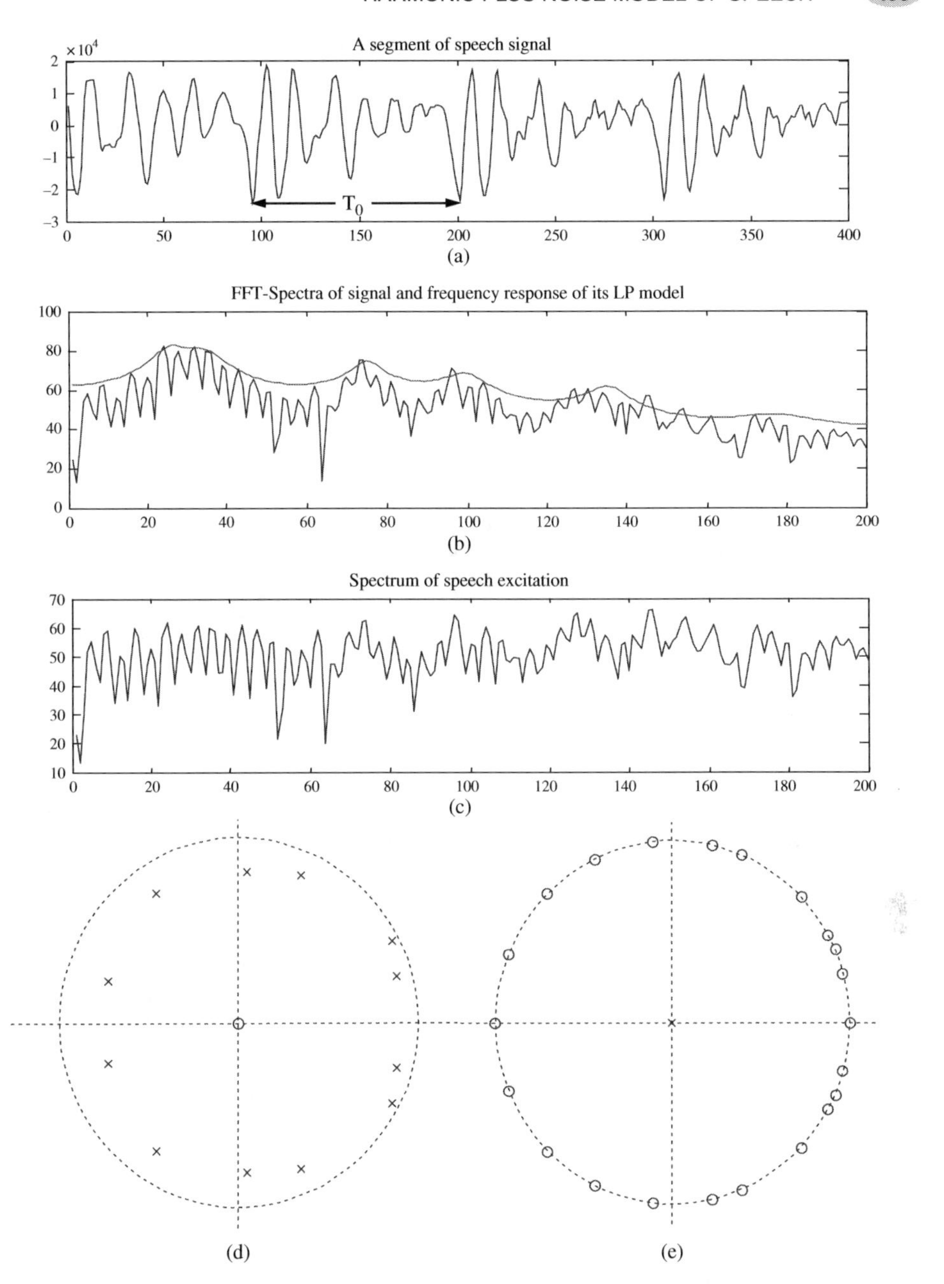

Figure 14.8 (a) A segment of speech signal, (b) its FFT and LP spectra, (c) the spectrum of its excitation, (d) the poles of its LP model, (e) the roots of LSF polynomials $P(z)$ and $Q(z)$.

The spectral shape of the noise-like signal component of speech $v(m)$ is often modelled by a linear prediction model as

$$v(m)=\sum_{k=1}^{P}c_k x(m-k)+g\,e(m) \tag{14.9}$$

where c_k are the LP model coefficients, $e(m)$ is a unit variance random process and g is a gain factor. Hence the parameter vector of the HNM is $[\boldsymbol{a}, \boldsymbol{b}, F_0, \mathbf{c}, g]$. Note that the harmonic part of the model is effectively a Fourier series representation of the periodic component of the signal.

In the frequency domain Equation (14.8) can be expressed as

$$X(f,t)=\underbrace{\sum_{k=1}^{M}A_k(t)\delta(f-kF_0)}_{\text{Harmonic Model (Fourier Series)}}+\underbrace{V(f,t)}_{\text{Noise Model}} \tag{14.10}$$

where t is the frame index, $\delta(f)$ is the Kronecker delta function. Since speech is composed of relatively short duration phonemes and it is divided into frames with typical frame duration of 20 ms, it is more realistic to replace the Kronecker delta function with a harmonic shape-function such as a Gaussian-shaped function as

$$X(f,t)=\underbrace{\sum_{k=1}^{M}A_k(t)M(f-kF_0)}_{\text{Harmonic Model (Fourier Series)}}+\underbrace{V(f,t)}_{\text{Noise Model}} \tag{14.11}$$

where $M(f)$ is a Gaussian-shaped function with a main support of $-F_0$ to F_0.

14.5.1 A Harmonicity Model of Excitation

In general, for voiced speech the signal is composed of a mixture of both harmonics and noise. The proportion of harmonics and noise in voiced speech depends on a number of factors including the speaker characteristics (e.g. to what extent a speaker's voice is normal or breathy), the speech segment character (e.g. to what extent a speech segment is periodic) and on the frequency; the higher frequencies of voiced speech have a higher proportion of noise-like components. Unvoiced speech is mostly composed of a spectrally-shaped noise-like component.

Rather than using a hard decision for the classification of each speech frame or the sub-band around each harmonic as voice/unvoiced, the ratio of the harmonic energy to the noise energy in each sub-band can be calculated as the level of *harmonicity* of that sub-band defined as

$$H_k=1-\frac{\int_{-F_0/2}^{F_0/2}\left[|X(kF_0)|\,M(f-kF_0)-|X(f-kF_0)|\right]^2 df}{\int_{-F_0/2}^{F_0/2}|X(f)|^2 df} \tag{14.12}$$

where H_k is the harmonicity of the speech signal in the kth band, $X(f)$ is the DFT of the speech signal and $M(f)$ is a Gaussian-shaped function shown in Figure 14.9 and defined as

$$M(f)=\beta\exp\left(-(f/\alpha)^2\right) \tag{14.13}$$

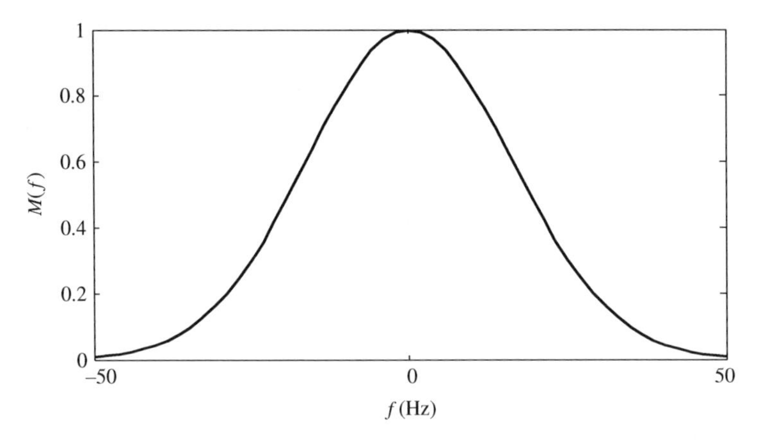

Figure 14.9 Gaussian-shaped $M(f)$ is used for modelling harmonics.

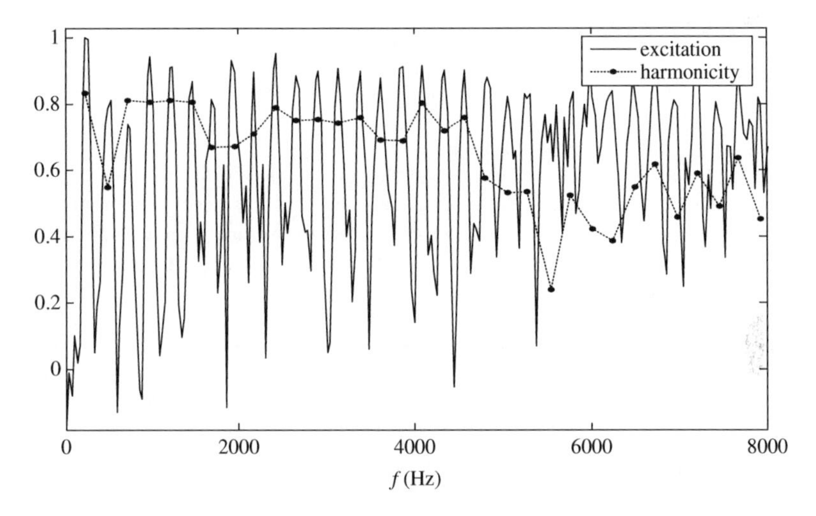

Figure 14.10 Harmonicity of the excitation sub-bands superimposed on the normalised excitation.

where α and β are chosen to produce a unit energy Gaussian pulse of appropriate width. The signal around each sub-band frequency of each frame is then reconstructed as

$$\left|\hat{X}(f)\right| = X(kF_0)\left(\frac{H_k M(f - kF_0)}{\sqrt{\int M^2(f)\,df}} + \frac{(1 - H_k)\,N(f)}{\sqrt{\int N^2(f)\,df}}\right) \quad \text{for } kF_0 - \frac{F_0}{2} < f < kF_0 + \frac{F_0}{2} \tag{14.14}$$

where $N(f)$ is the noise component of the excitation. $N(f)$ is a Rayleigh distributed random variable to comply with the assumption of the Gaussian distribution model of the speech DFT. Figure 14.10 illustrates the excitation of a sample frame together with the harmonicity values of each band.

Estimation of Harmonic Amplitudes

The harmonic amplitudes can be obtained from the tracking of the peak amplitudes of the DFT of speech at the frequency neighbourhoods around the integer multiples of the fundamental frequency F_0. Alternatively, given an estimate of F_0, the following least square error estimation method can be used to obtain the harmonic amplitudes.

A segment of N samples of speech can be expressed in a vector-matrix notation as

$$\begin{bmatrix} x(m) \\ x(m-1) \\ x(m-2) \\ \vdots \\ x(m-N-1) \end{bmatrix} = \begin{bmatrix} \cos 2\pi F_0 m & \cdots & \cos 2\pi M F_0 m & \sin 2\pi F_0 m & \cdots & \sin 2\pi M F_0 m \\ \cos 2\pi F_0 (m-1) & \cdots & \cos 2\pi M F_0 (m-1) & \sin 2\pi F_0 (m-1) & \cdots & \sin 2\pi M F_0 (m-1) \\ \cos 2\pi F_0 (m-2) & \cdots & \cos 2\pi M F_0 (m-2) & \sin 2\pi F_0 (m-2) & \cdots & \sin 2\pi M F_0 (m-2) \\ \vdots & \ddots & \vdots & \vdots & \ddots & \vdots \\ \cos 2\pi F_0 (m-N-1) & \cdots & \cos 2\pi M F_0 (m-N-1) & \sin 2\pi F_0 (m-N-1) & \cdots & \sin 2\pi M F_0 (m-N-1) \end{bmatrix} \begin{bmatrix} a_1 \\ \vdots \\ a_M \\ b_1 \\ \vdots \\ b_M \end{bmatrix} + \begin{bmatrix} v(m) \\ v(m-1) \\ v(m-2) \\ \vdots \\ v(m-N-1) \end{bmatrix} \tag{14.15}$$

In compact notation Equation (14.15) can be written as

$$\boldsymbol{x} = \boldsymbol{S}\boldsymbol{c} + \boldsymbol{v} \tag{14.16}$$

where $\boldsymbol{x}$ is the vector of discrete-time speech samples, $\boldsymbol{S}$ is a matrix of sine and cosine functions, $\boldsymbol{c} = [\boldsymbol{a}\ \boldsymbol{b}]$ is the vector of amplitudes of the harmonics and $\boldsymbol{v}$ is the noise component of the speech model. The harmonics amplitude vector $\boldsymbol{c}$ can be obtained from a least squared error minimisation process. Define an error vector as the difference between speech and its harmonic model as

$$\boldsymbol{e} = \boldsymbol{x} - \boldsymbol{S}\boldsymbol{c} \tag{14.17}$$

The squared error function is given by

$$\boldsymbol{e}\boldsymbol{e}^T = (\boldsymbol{x} - \boldsymbol{S}\boldsymbol{c})\,(\boldsymbol{x} - \boldsymbol{S}\boldsymbol{c})^T \tag{14.18}$$

Minimisation of Equation (14.18) with respect to the amplitudes vector $\boldsymbol{c}$ yields

$$\boldsymbol{c} = \left[\boldsymbol{S}^T\boldsymbol{S}\right]^{-1}\boldsymbol{S}^T\boldsymbol{x} \tag{14.19}$$

14.6 Fundamental Frequency (Pitch) Information

The periodic opening and closing of the vocal folds results in the harmonic structure in voiced speech signals. The inverse of the period is the fundamental frequency of speech. Pitch is the sensation of the fundamental frequency of the pulses of airflow from the glottal folds. The terms pitch and fundamental frequency of speech are used interchangeably.

The pitch of the voice is determined by four main factors. These include the length, tension, and mass of the vocal cords and the pressure of the forced expiration also called the sub-glottal pressure. The pitch variations carry most of the intonation signals associated with prosody (rhythms of speech), speaking manner, emotion, and accent. Figure 14.11 illustrates an example of the variations of the trajectory of pitch over time.

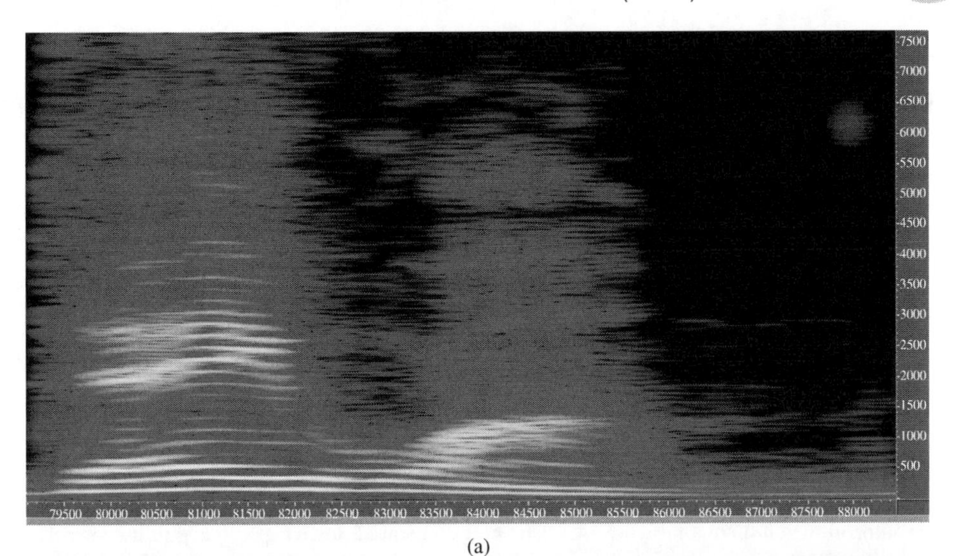

(a)

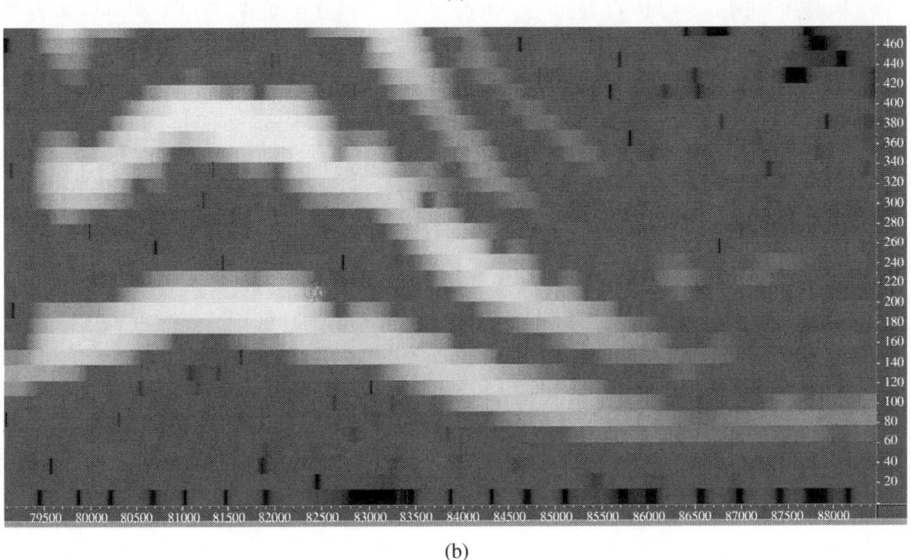

(b)

Figure 14.11 (a) The spectrogram of the utterance '*day one*' showing the pitch and the harmonic structure of speech, (b) a zoomed spectrogram of the fundamental and the second harmonic of pitch.

The following information is contained in the pitch signal:

(a) Gender is conveyed in part by the vocal tract characteristics and in part by the pitch value. The average pitch for females is about 200 Hz whereas the average pitch for males is about 110 Hz. Hence pitch is a main indicator of gender.
(b) Emotion signals in voice, such as excitement and stress, are also in part carried by pitch variations. Shouting as a means of stressing a point or in expression of anger has more to do with raising the

pitch than loudness. Pitch variation is often correlated with loudness variation. Happiness, distress and extreme fear in voice are signalled by fluctuations of pitch.

(c) Accent is in part conveyed by changes in the pitch and rhythm of speech. For example in some accents, such as the Northern Ireland accent, at the end of a sentence the pitch signal is raised instead of being lowered.

(d) Prosody is the rhythmical or intonation signals in speech carried mostly by the time variations of pitch. The functions of prosody are many. Prosody can indicate syntax, demarcation and linking of phrases and sentences, turn-taking in conversational interactions, types of utterance such as questions and statements, and people's attitudes and feelings.

(e) Age and state of health. Pitch can also signal age, weight and state of health. For example children have a high pitched signal of 300–600 Hz.

14.6.1 Fundamental Frequency (Pitch) Estimation

Traditionally the fundamental frequency (whose sensation is known as pitch) is derived from the autocorrelation function as the inverse of the autocorrelation lag corresponding to the second largest peak of the autocorrelation function. Figure 14.12 shows a segment of voiced speech and its autocorrelation function. Note that the largest peak happens at the lag zero and corresponds to the signal energy. For a segment of periodic voiced speech signal the second largest peak occurs at the lag T_0 corresponding to the period of speech.

The autocorrelation of a periodic signal is periodic with a period equal to that of the signal. Hence all the periodic peaks of the autocorrelation function can be usefully employed in the pitch estimation process as in Griffin's methods where the pitch period is found by searching for the value of period T that maximises the energy function

$$E(T) = T \sum_{k=0}^{N(T)} r(kT) \tag{14.20}$$

where $r(kT)$ is the autocorrelation at lag kT and $N(T)$ is the number of autocorrelation values in the summation. Note that the multiplication of the summation in Equation (14.20) by T compensates for the fact that as T increases the number of autocorrelation values in the summation, $N(T)$, decreases. The pitch is obtained as the maximum of Equation (14.20) as

$$T_0 = \arg\max_T E(T) \tag{14.21}$$

Figure 14.13 shows an example of the variation of the $E(T)$ curve with a range of values of periods. For each speech frame N pitch candidates are obtained as the N minimum values of $E(T)$ calculated on a grid of values of $T_{0\min} < T_0 < T_{0\max}$. The Viterbi algorithm is subsequently used to obtain the best pitch trajectory estimate through the given N candidates. Figure 14.14 shows an example of speech and pitch ($F_0 = 1/T_0$) and harmonic tracks.

The pitch estimation method can be formulated in the frequency domain to search for maximum signal-to-noise ratio at the harmonics.

A pitch estimation error criterion over the speech harmonics can be defined as

$$E(F_0) = F_0 \sum_{k=1}^{\mathrm{Max}F} \sum_{l=kF_0-M}^{kF_0+M} W(l) \log |X(l)| \tag{14.22}$$

where $X(l)$ is the DFT of speech, F_0 is a proposed value of the fundamental frequency (pitch) variable and $2M+1$ is a band of values about each harmonic frequency. The use of logarithmic compression

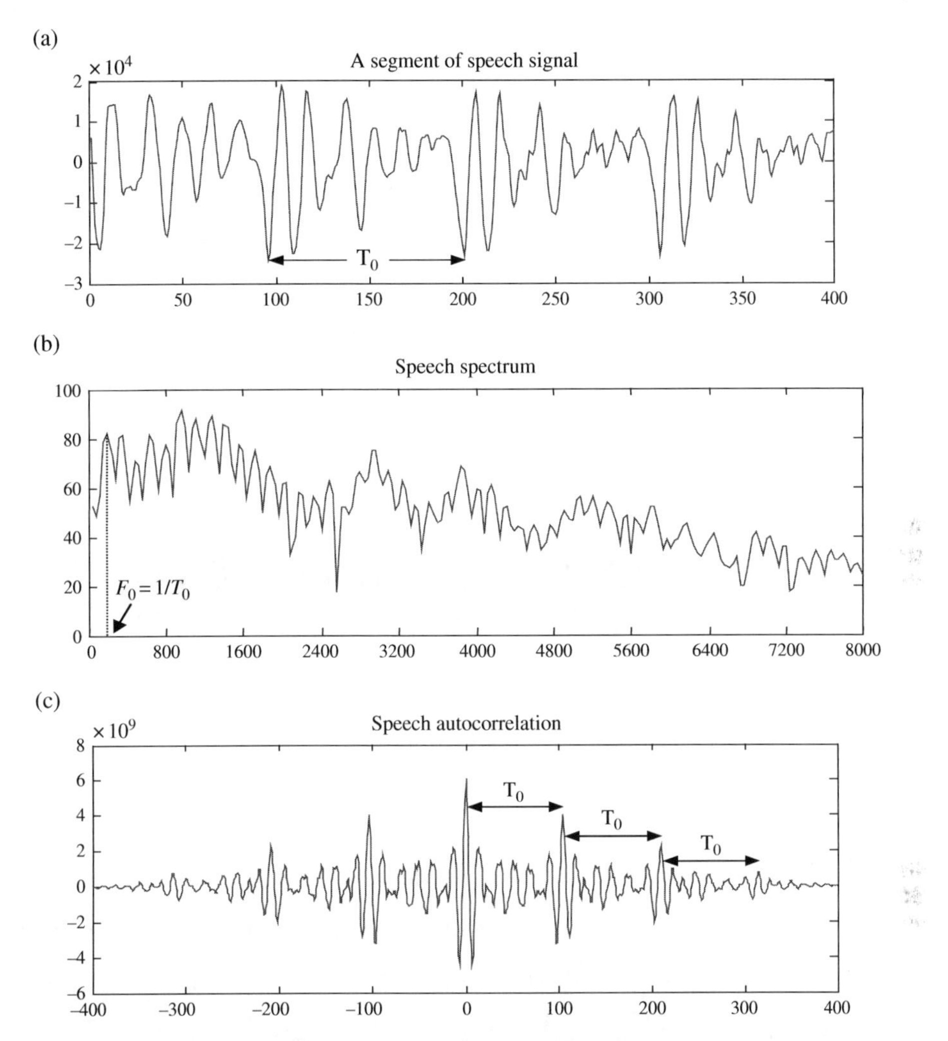

Figure 14.12 (a) A voiced speech segment, (b) its frequency spectrum showing the harmonic structure (note the sampling frequency was 16 kHz) and (c) its autocorrelation function with peaks at integer multiples of the period of voiced speech.

in Equation (14.22) provides for a more balanced influence, on pitch estimation, of the high-energy low-frequency harmonics and the low-energy high-frequency harmonics. The weighting function $W(l)$ is an SNR-dependent Wiener-type weight given by

$$W(l) = \frac{\mathrm{SNR}(l)}{1 + \mathrm{SNR}(l)} \tag{14.23}$$

where $\mathrm{SNR}(l)$ is the signal-to-noise ratio at the proposed l^{th} harmonic.

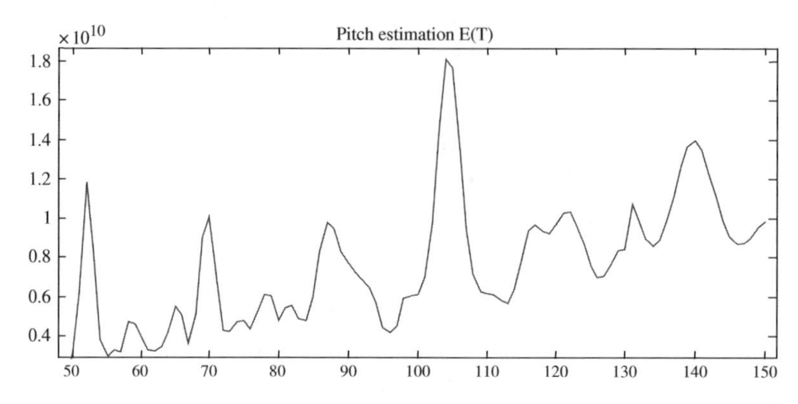

Figure 14.13 An illustration of the variation of the $E(T)$ curve with the proposed values of pitch frequency T.

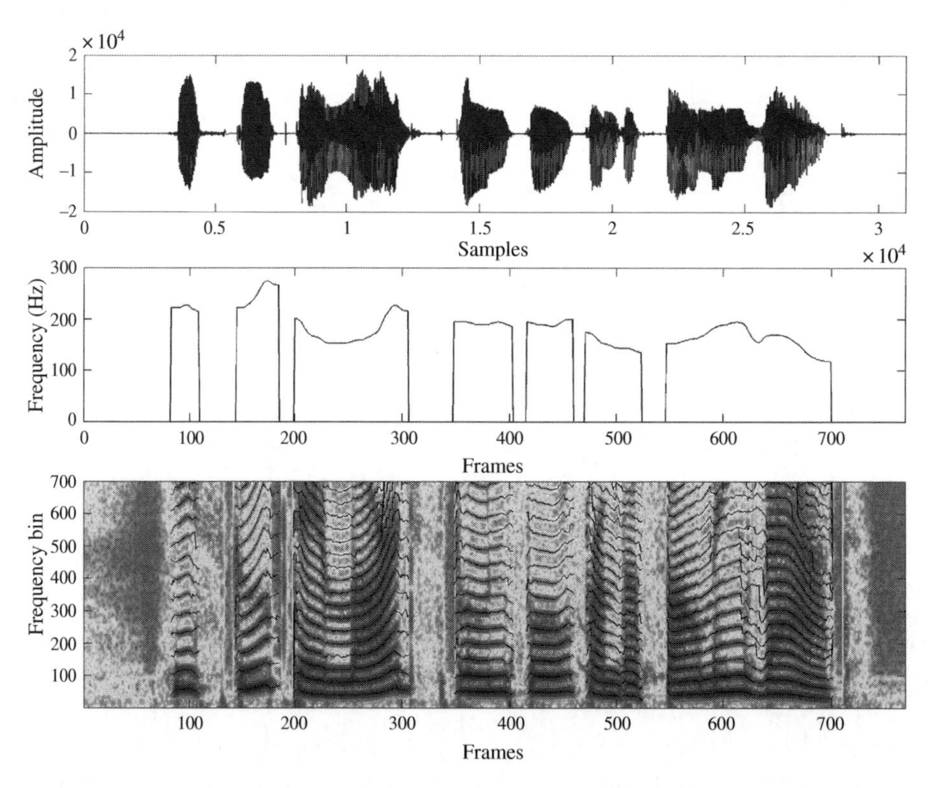

Figure 14.14 An illustration of a speech segment (top panel), its fundamental frequency track (middle panel) and its harmonic tracks.

14.7 Speech Coding

The objective of speech coding is to compress speech and achieve savings in the required transmission bandwidth, transmission power and memory storage space by reducing the number of bits per sample in such a way that the decoded (decompressed) speech is perceptually indistinguishable from the original speech.

High quality speech coders can achieve a big reduction in the bit-rate by a factor of 13 or more (e.g. from 13 bits per sample to 1 bit per sample or less) with no perceptible loss in quality or intelligibility. Specifically speech coding methods achieve the following gains:

- A reduction in speech bit-rate r_b, or equivalently the same reduction in the bandwidth ($BW = k.r_b$) and the memory storage requirement both of which are directly proportional to the bit-rate.
- A reduction in the transmission power requirement because after compression there are less bits (hence less energy) per second to transmit.
- Immunity to noise, as error control coding methods can be used to reintroduce some of the saved bits per sample in order to protect speech parameters from channel noise and distortion.

Speech coding methods achieve a reduction in the bit-rate by utilising the physiology of speech production and the psychoacoustics of audio perception, namely:

- Speech is a correlated signal; from an information theory point of view successive speech samples contain a high level of common or redundant information. The 'redundancy' in natural speech provides pleasant 'smooth' sounds and robustness to background noise, speaker variations and accents. However, speech redundancy can be modelled and removed before transmission and then reintroduced into speech at the receiver.
- Speech is generated by a relatively slowly-varying articulatory system. Therefore, speech model parameters vary slowly and can be efficiently coded.
- Psychoacoustics of hearing. The threshold of hearing and the spectral and temporal noise masking thresholds can be used to allocate *just* sufficient numbers of bits to each sub-band in each speech frame to keep the coding noise masked below the thresholds of hearing.

14.7.1 Linear Prediction and Harmonic Plus Noise Models in Speech Coding

Linear prediction model and harmonic noise model are the two main methods for modelling and coding of speech signals. Linear prediction model is particularly suited to modelling the spectral envelope of speech whereas harmonic noise model is better suited for modelling the fine spectral structure of speech. The two methods can be combined to take advantage of their relative strengths. In general, the main structures used for speech coding are:

(a) The spectral envelope of speech modelled by a LP model and represented by a set of LSF coefficients.
(b) The speech excitation modelled by a harmonic and noise model.

Figure 14.15 shows the outline of a model-based speech coder. The speech signal is analysed and its LP model, excitation signals and pitch are extracted. The LP model represents the spectral envelope of speech. It is converted to a set of LSF coefficients which have good quantisation properties. The LSF coefficients can be scalar quantised or more efficiently they can be grouped into subsets and vector quantised using previously trained LSF vector codebooks.

The most challenging part of speech coding is the coding of the harmonic and noise contents of excitation. To start with, unlike the extraction of LP model parameters, the extraction of the fundamental

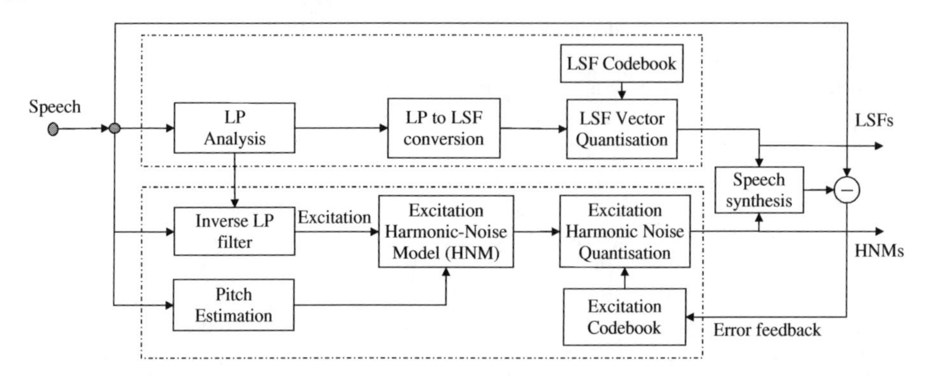

Figure 14.15 An illustration of the outline of a speech coding system composed of a linear prediction model of vocal tract and harmonic noise model of excitation.

frequency is not a straightforward matter; whereas there is a closed-form solution for extraction of LP model parameters (see Chapter 10), the calculation of pitch and harmonics usually requires a search process. The main issues on modelling the harmonic and noise parts of speech excitation are the following:

(1) voiced/unvoiced classification
(2) pitch estimation
(3) estimation of harmonic amplitudes
(4) estimation of noise variance
(5) quantisation of harmonic and noise parameters.

In its simplest form one can model speech excitation as a mixture of harmonics plus noise. More sophisticated methods employ different mixtures of harmonic and noise for each speech sub-band centred on a harmonic.

14.7.2 Modelling Excitation for Voiced Speech

The excitation signal for voiced speech is composed of both harmonic and noise. The harmonic plus noise model described in Section 14.5 can be used for modelling of the excitation of the linear prediction model.

For applications such as mobile phones, speech coders utilise a form of code-excited linear prediction (CELP) model of speech. Figure 14.16 shows an outline of a CELP coder. As shown, CELP models speech as the output of a filter excited by a combination of a periodic component and a non-periodic component.

14.7.3 Principles of Operation of CELP Coders

The principles of the operation of code-excited linear prediction (CELP) coders are as follows:

(1) *Anti-aliasing filter, sampling, PCM quantisation and segmentation.* Before CELP coding, the input speech signal to the handset's microphone is filtered to confine its highest frequency to below 4 kHz and then sampled at a rate of 8000 samples per second. Each sample is quantised with 13 bits per sample. The digitised speech is segmented into segments of 20 ms (160 samples)long.

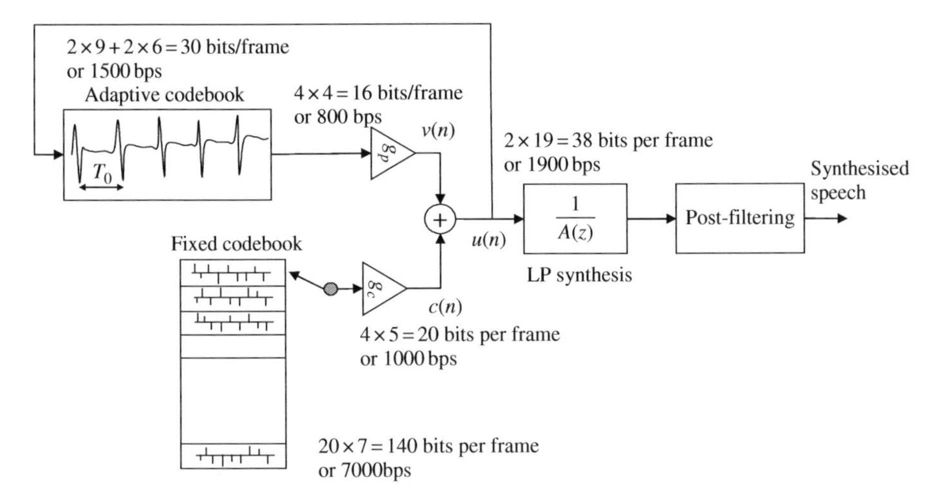

Figure 14.16 The outline of a code excited linear prediction (CELP) decoder. The numbers on top of each block indicate the number of times each parameter is transmitted per frame × the number of bits per parameter, for a 12,200 kbps GSM coder, see Table 14.2.

(2) *Linear prediction analysis.* Each speech segment is windowed and modelled by a tenth-order linear prediction model. The z-transfer function of the linear prediction filter is given as

$$H(z) = \frac{1}{1 - \sum_{k=1}^{P} a_k z^{-k}} \tag{14.24}$$

The linear predictor coefficients are calculated using the Levinson–Durbin method. The predictor coefficients are transformed into line spectral frequencies before quantisation and transmission.

(3) *Calculation of the periodic part of speech.* The periodic part of speech is synthesised by exciting the linear predictor filter with a periodic input obtained from an adaptive codebook. A pitch filter is used to shape the periodic component of excitation. The pitch synthesis filter model has the transfer function

$$\frac{1}{B(z)} = \frac{1}{1 - g_p z^{-T}} \tag{14.25}$$

where T is the pitch period and g is the pitch filter coefficient. For each speech segment the pitch period T is calculated from the autocorrelation function of speech.

(4) *Calculation of the non-periodic part of speech.* The non-periodic part of speech is obtained by exciting the linear prediction model with the 'best' noise-like excitation selected from a number of available 'noise codes' using an analysis-by-synthesis search method. The analysis-by-synthesis approach tests all possible input forms available and selects the one that minimises the perceptually weighted mean squared difference between the original speech and the synthesised speech. This analysis-by-synthesis part of the coding can also compensate for the deficiencies in the earlier part of the coding process.

(5) The reconstructed speech is passed through an adaptive post-filter.

14.7.4 Details of Operation of CELP Coders

Figure 14.17 shows a more detailed block diagram of a code excited linear prediction (CELP) coder used in mobile phones. The analogue speech input is sampled at 8 kHz with an 8-bit A-law device. The 8-bit A-law format is converted to a 13-bit linear format using the method and tables defined by the International Telecommunication Union Telecommunication Standardization Sector Recommendation ITU-T Rec. G.714. The incoming samples are divided into frames of 160 samples (i.e. 20 ms at 8 kHz) during which speech parameters are assumed to be time-invariant. The frame length of 20 ms plus the signal processing time on the handset and the delay at the network determine the 'trans-coding delay' of the communication system. The encoder compresses an input frame of 160 13-bit PCM samples (a total of 2080 bits/frame) to a frame of 260 bits. This is a compression ratio of 8:1. The speech coder of Figure 14.17 consists of the following sections.

Pre-processing High-Pass Filter

This is a second-order high-pass filter for removing unwanted low-frequency components below 80 Hz. The filter transfer function shown in Figure 14.18 is given by

$$H(z) = \frac{0.92727435 - 1.8544941z^{-1} + 0.92727435z^{-2}}{1 - 1.9059465z^{-1} + 0.9114024z^{-2}} \tag{14.26}$$

Linear Prediction Model Analysis and Quantisation

The steps taken in the calculation and quantisation of the short-term linear prediction coefficients, shown in Figure 14.17, are as follows.

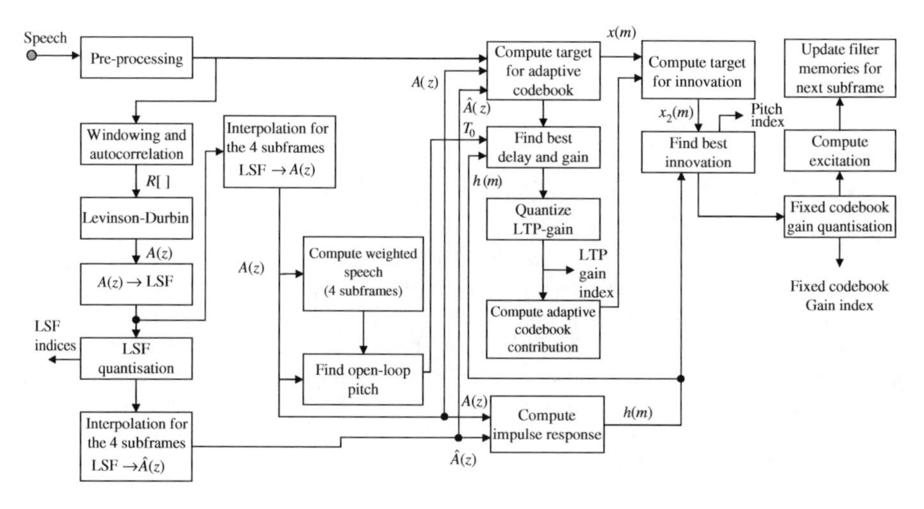

Figure 14.17 Block diagram of a code excited linear prediction (CELP) coder.

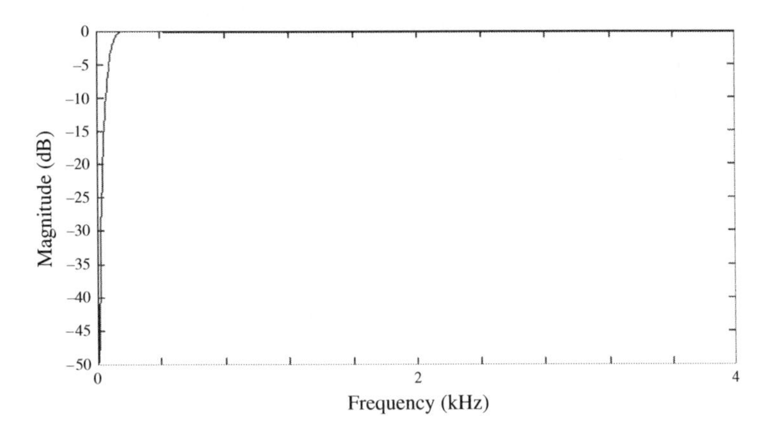

Figure 14.18 The frequency response of the pre-processing filter.

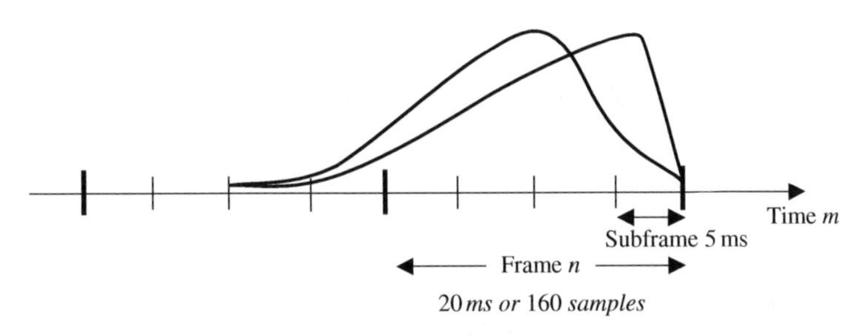

Figure 14.19 Linear prediction analysis windows.

Windowing and autocorrelation: Linear prediction analysis is performed twice per frame with two asymmetric windows. Each window, shown in Figure 14.19, covers one and a half frames of speech to provide some overlap of the signal between the current and the previous frames. One window has its weight concentrated on the second speech subframe and is defined as

$$w_I(m) = \begin{cases} 0.54 - 0.46\cos\left(\pi m \Big/ \left(L_1^{(I)} - 1\right)\right) & m = 0, \ldots, L_1^{(I)} - 1 \\ 0.54 + 0.46\cos\left(\pi(m - L_1^{(I)}) \Big/ \left(L_2^{(I)} - 1\right)\right) & m = L_1^{(I)}, \ldots, L_1^{(I)} + L_2^{(I)} - 1 \end{cases} \tag{14.27}$$

where $L_1^{(I)} = 160$ and $L_2^{(I)} = 80$. The second window has its weight concentrated in the fourth subframe and is given by

$$w_{II}(m) = \begin{cases} 0.54 - 0.46\cos\left(2\pi m \Big/ \left(2L_1^{(II)} - 1\right)\right) & m = 0, \ldots, L_1^{(II)} - 1 \\ \cos\left(2\pi(m - L_1^{(II)}) \Big/ \left(4L_2^{(II)} - 1\right)\right) & m = L_1^{(II)}, \ldots, L_1^{(II)} + L_2^{(II)} - 1 \end{cases} \tag{14.28}$$

where $L_1^{(II)} = 232$ and $L_2^{(II)} = 8$. The windows span 240 samples including 160 samples from the current frame and 80 samples from the previous frame. From each windowed speech segment $P + 1$ autocorrelation coefficients are calculated and used to obtain P LP coefficients.

The windows are given in the following Matlab code.

```
MATLAB function [wI]=GSM_WindowI( )
L1 = 160; L2 = 80;
m = 1 : L1; wI(m) = 0.54 – 0.46* cos(pi*(m – 1)/(L1 – 1));
m = L1 + 1 : L1 + L2; wI(m) = 0.54 + 0.46* cos(pi*(m – L1 – 1)/(L2 – 1));

MATLAB function [wII]=GSM_WindowII( )
L1 = 232; L2 = 8;
m = 1 : L1; wII(m) = 0.54 – 0.46* cos(2*pi*(m – 1)/(2*L1 – 1));
m = L1 + 1 : L1 + L2; wII(m) = cos(2*pi*(m – L1 – 1)/(4*L2 – 1));
```

LP coefficients calculation: The Levinson–Durbin algorithm, described in Section 10.2.2, is used to obtain P LP coefficients $[a_1, a_2, \dots a_P]$ from $P+1$ autocorrelation coefficients $[r(0), r(1), \dots r(P)]$. Alternatively the Schur algorithm can be used.

LP to LSF conversion and quantisation: The LP coefficients are converted to LSF coefficients, described in Section 14.4.1, LSFs have good quantisation and interpolation properties. The LSF quantisation is achieved as follows. First the two sets of LP coefficient vectors from the two windows per speech frame are quantified using the LSF representation in the frequency domain as

$$f_i = \frac{F_s}{2\pi}\arccos(\omega_i) \tag{14.29}$$

where f_i is the line spectral frequency (LSF) in Hz, F_s is the sampling frequency (8 kHz for telephones) and ω_i are the line spectral angular frequencies. The LSF vector is $\boldsymbol{f} = [f_1, f_2, \dots f_P]$. The prediction and quantisation of the LSF vectors are performed as follows. Let $z^{(1)}(m)$ and $z^{(2)}(m)$ denote the mean-removed LSF vectors at frame m. The prediction LSF residual vector is defined by

$$\boldsymbol{r}^{(1)}(m) = \boldsymbol{z}^{(1)}(m) - \boldsymbol{p}(m) \tag{14.30}$$

$$\boldsymbol{r}^{(2)}(m) = \boldsymbol{z}^{(2)}(m) - \boldsymbol{p}(m) \tag{14.31}$$

where $\boldsymbol{p}(m)$ is the predicted LSF vector at frame m given by a first-order moving-average predictor as

$$\boldsymbol{p}(m) = \alpha \boldsymbol{r}^{(2)}(m-1) \tag{14.32}$$

where $\alpha = 0.65$. The two LSF vectors $r^{(1)}(m)$ and $\boldsymbol{r}^{(2)}(m)$ are quantised using split matrix quantisation (SMQ). For a tenth-order linear prediction model the LSF residual matrix $[\boldsymbol{r}_1^{(1)}(m)\boldsymbol{r}_2^{(2)}(m)]$ is split into five 2×2 matrices. For example, the first sub-matrix will be $[r_1^{(1)}(m) r_1^{(2)}(m) r_2^{(1)}(m) r_2^{(2)}(m)]$. The five sub-matrices are then vector quantised using a bit allocation pattern of [7, 8, 9, 8, 6]. Note the total number of bits per frame allocated to LSF coefficients is 38 bits per 20 ms frame. In general an input LSF vector $\boldsymbol{f}$ is quantised using the VQ error relation

$$E = \sum_{i=1}^{10}\left[w_i\left(f_i - \hat{f}_i\right)\right]^2 \tag{14.33}$$

where w_i are weighting coefficients.

Interpolation of LSF Parameters

The LSF parameters are interpolated in the angular frequency domain ω. The LSF coefficients calculated for the windows centred on the second and the fourth subframes are used for those subframes respectively. The LSF coefficients for the first and the third subframes are linearly interpolated from the adjacent subframes as

$$\omega^{(1)}(m) = 0.5\omega^{(4)}(m-1) + 0.5\omega^{(2)}(m) \tag{14.34}$$

$$\omega^{(3)}(m) = 0.5\omega^{(2)}(m) + 0.5\omega^{(4)}(m) \tag{14.35}$$

where $\omega^{(1)}(m)$ is the LSF coefficients vector of the ith subframe of the mth speech frame.

Open-Loop Pitch Analysis

The period T of a periodic speech signal can be obtained from an analysis of the maxima of its autocorrelation function. Open-loop pitch estimation is performed every 10 ms, i.e. twice per frame, from the autocorrelation function calculated as

$$r(k) = \sum_{m=0}^{79} x(m)x(m-k) \tag{14.36}$$

At each stage the three maxima of the autocorrelation function, $r(k)$, are obtained in the following ranges: $k = 18-35, k = 36-71$ and $k = 72-143$. The retained maxima are normalised by $\sqrt{\sum_m x^2(m-t_i)}$, where t_i are the time maxima. The normalised maxima and corresponding delays are denoted as $(M_i, t_i), i = 1, 2, 3$. The following method is then used to select the pitch from the three candidates.

$T_{op} = t_1$
$M(T_{op}) = M_1$
if $M_2 > 0.85M(T_{op})$
  $M(T_{op}) = M_2$
  $T_{op} = t_2$
end
if $M_3 > 0.85M(T_{op})$
  $M(T_{op}) = M_3$
  $T_{op} = t_3$
end

The above procedure of computing the maxima of the correlation function in three ranges, and its bias in favouring the lower range, is designed to avoid choosing pitch multiples.

Adaptive Codebook Search

The adaptive codebook search is performed every 5 ms for each speech subframe to obtain the pitch parameters i.e. the pitch delay and the pitch gain. The pitch values are optimised in a closed-loop pitch analysis performed around the open-loop pitch estimates by minimising the difference between the input speech and the synthesised speech. For the first and the third subframes the range $T_{op} \pm 3$

bounded by [18, 143] is searched. A fractional pitch delay is used with a resolution of 1/6 in the range $\left[17\frac{3}{6}, 94\frac{3}{6}\right]$ and integers only in the range [95, 143]. For the second and fourth subframes, a pitch resolution of 1/6 is always used in the range $\left[T_1 - 5\frac{3}{6}, T_1 + 4\frac{3}{6}\right]$, where T_1 is the nearest integer to the fractional pitch lag of the previous (first or third) subframe. The pitch delay is coded with 9 bits in the first and third subframes and the relative delay of the other frames is coded with 6 bits.

Algebraic Codebook Structure and Search

The algebraic code structure is based on an interleaved single-pulse permutation (ISPP) method. Each codebook vector of size 40 samples contains 10 non-zero pulses with amplitudes ± 1. Each subframe of 40 samples is subdivided into five tracks, where each track contains two pulses as shown in Table 14.1. Each two pulse positions in an eight-position track is coded with 3 bits per pulse (a total of 6 bits/track) and the sign of the first pulse in each track is coded with one bit (a total of 5 bits per subframe). The sign of the second pulse is the opposite of the first pulse if its position is smaller than the first pulse, otherwise it has the same sign as the first pulse. This gives a total of 35 bits for each 40-sample subframe which are Grey coded for robustness. The algebraic codebook is searched for the best vector by minimising the difference between the input speech and the synthesised speech.

Use of 'Excitation Noise' to Compensate for Modelling Errors

Note that in speech coders with a closed-loop error minimisation system the choice of the noise compensates for the inadequacies and errors in the modelling of LP parameters, pitch and harmonic values. Hence, a main reason that much processing power and bit-rate (upto 70 % of the total bit rate) is used to obtain and transmit the best noise sequence is because the 'best noise' sequence also compensates for other errors by trying to match the coded speech with the actual speech in the error minimisation feedback loop.

Calculation of Periodic and Non-Periodic Excitation Gain Factors

The gain factor for the synthesised periodic part of speech $y(m)$ is calculated as the normalised correlation of the synthesised periodic part of speech and the actual speech $x(m)$ as

$$g_p = \frac{\sum_{m=0}^{N-1} \boldsymbol{x}(m)y(m)}{\sum_{m=0}^{N-1} y(m)y(m)} \tag{14.37}$$

Table 14.1 Potential positions of the individual pulses in the algebraic code.

Track	Pulse	Positions
1	i_0, i_5	0, 5, 10, 15, 20, 25, 30, 35
2	i_1, i_6	1, 6, 11, 16, 21, 26, 31, 36
3	i_2, i_7	2, 7, 12, 17, 22, 27, 32, 37
4	i_3, i_8	3, 8, 13, 18, 23, 28, 33, 38
5	i_4, i_9	4, 9, 14, 19, 24, 29, 34, 39

Table 14.2 Bit allocation pattern per frame in a 12.2 kbps GSM coder. Each frame is 20 ms (160 samples) and there are 50 frames per second.

Parameters	1st & 3rd subframes	2nd & 4th subframes	Total bits per frame
2 LSF sets	2×19		38
Pitch delay	2×9	2×6	30
Pitch gain	2×4	2×4	16
Algebraic codebook	2×35	2×35	140
Codebook gain	2×5	2×5	20
Total			**244**

Similarly the gain for the non-periodic synthesised part of speech can be calculated as

$$g_c = \frac{\sum_{m=0}^{N-1} x_2(m)z(m)}{\sum_{m=0}^{N-1} z(m)z(m)} \tag{14.38}$$

where $x_2(m)$ is the difference between the actual speech and its synthesised periodic part and $z(m)$ is the synthesised non-periodic part of speech. That is $z(m)$ is the output of the linear prediction filter in response to the selected noise from the codebook.

Table 14.2 shows the bit allocation pattern in a 12.2 kbps GSM coder.

14.7.5 Decoding and Re-synthesis of Coded Speech

At the receiver the decoding and re-synthesis of speech is performed as follows.

Decoding of LP Filter Coefficients

The received codebook indices for LSF quantisation are used to reconstruct the LSF coefficients for the second and fourth subframes. The LSF coefficients for the first and the third subframes are derived from interpolation of the adjacent LSF vectors as described in Equations (14.26) and (14.27). The LSFs are then converted to LP vectors. Then the following steps are repeated for each subframe:

(6) Decoding of the adaptive codebook vector. The received pitch index is used to find the integer and fractional parts of the pitch lag. The adaptive codebook vector $v(m)$ is obtained by interpolating the past excitation $u(m)$ at the pitch delay.
(7) Decoding of the adaptive codebook gain. The received index is used to find the quantised adaptive codebook gain g_p from the quantisation table.
(8) The received algebraic codebook index is used to extract the positions and amplitude signs of the excitation pulses and to find the algebraic code vector $c(m)$.
(9) Decoding of the fixed codebook gain. The received index is used to compute the quantised fixed codebook gain g_c.

Reconstructing Speech

The input excitation to linear prediction filter is computed as in Figure 14.16:

$$u(n) = g_p v(n) + g_c c(n) \tag{14.39}$$

The input excitation $u(m)$ is filtered by the LP filter $1/A(z)$ and then processed by an adaptive post-filtering filter as described next.

Adaptive Post-Filtering

The adaptive post-filter is the cascade of two filters: a formant post-filter and a tilt compensation filter. The post-filter is updated every subframe of 5 ms. The formant post-filter is given by

$$H_f(z) = \frac{\hat{A}(z/\gamma_n)}{\hat{A}(z/\gamma_d)} \tag{14.40}$$

where $\hat{A}(z)$ is the received quantised and interpolated LP inverse filter and the factors γ_n and γ_d control the amount of the formant post-filtering. Finally the filter $H_t(z)$ compensates for the tilt in the formant post-filter and is given by

$$H_t(z) = (1 - \mu z^{-1}) \tag{14.41}$$

where $\mu = \gamma_t k_1$ is a tilt factor, with k_1 being the first reflection coefficient calculated on the truncated ($L_h = 22$) impulse response, $h_f(m)$ of the filter $H_f(z) = \hat{A}(z/\gamma_n)/\hat{A}(z/\gamma_d)$ of equation (14.40). The reflection coefficient k_1 is given by

$$k_1 = \frac{r_h(1)}{r_h(0)}; \qquad r_h(i) = \sum_{j=0}^{L_h-i-1} h_f(j)h_f(j+i) \tag{14.42}$$

The post-filtering is performed by first passing the synthesised signal through $\hat{A}(z/\gamma_n)$ then through $1/\hat{A}(z/\gamma_d)$ and finally through the tilt filter $H_t(z)$.

14.8 Speech Recognition

Speech recognition systems have a wide range of applications from the relatively simple isolated-word recognition systems for name-dialling, automated costumer service and voice-control of cars and machines to continuous speech recognition as in auto-dictation or broadcast-news transcription. Figure 14.20 shows the outline of a typical speech recognition system. It consists of three main sections:

(a) A front-end section for extraction of a set of spectral–temporal speech features from the time-domain speech samples. The speech features are derived from a bank of filters inspired by a knowledge of how the cochlea of the inner ear performs spectral analysis of audio signals. The most commonly used features are cepstrum features described later in this section.
(b) A middle section that consists of a network of speech models incorporating: pre-trained statistical models of the distributions of speech features, a language model and speaker adaptation. In its simplest form, e.g. for a name-dialling system, a speech model consists of a simple spectral–temporal template for each word. In more high performance systems a lattice network incorporates hidden Markov models of speech with language models.
(c) A speech decoder, usually based on a Viterbi decoding method, takes as input a stream of speech feature vectors and, using a network of acoustic word models, outputs the most likely word sequence.

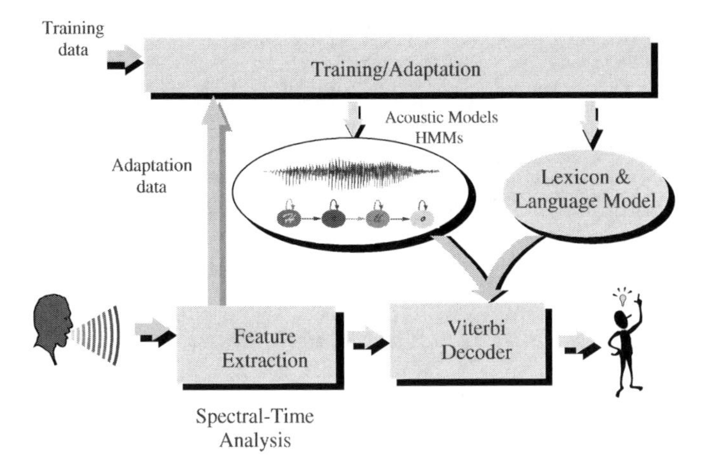

Figure 14.20 The outline of a speech recognition system.

The speech recognition problem can be stated as follows: *given a stream of speech features* $\boldsymbol{X}$ *extracted from the acoustic realisation of a spoken sentence, and a network of pre-trained speech models* $\mathcal{M}$, *decode the most likely spoken word sequence* $\boldsymbol{W} = [w_1, w_2, \ldots, w_N]$. Note that speech model network Λ contains models of acoustic feature representations of words and can also include a model of language grammar. Formally, the problem of recognition of words from a sequence of acoustic speech features can be expressed in terms of maximisation of a probability function as

$$[\hat{w}_1, \hat{w}_2, \ldots, \hat{w}_N] = \max_{[w_1, w_2, w_3, \ldots, w_N]} f\left([w_1, w_2, \ldots, w_N] | \boldsymbol{X}, \mathcal{M}\right) \tag{14.43}$$

where $f(\cdot)$ is the conditional probability of a sequence of words $\boldsymbol{W}$ given a sequence of speech features $\boldsymbol{X}$. Note that in practice speech is processed sequentially and the occurrence of a word, a phoneme or a speech frame is conditioned on the previous words, phonemes and speech frames.

Fundamental Problems of Speech Recognition

Like any pattern-recognition problem, the fundamental problem in speech recognition is the speech pattern variability. Speech recognition errors are caused by the overlap of the distributions of the acoustic realisations of different speech units. In general the sources of speech variability are as follows:

(a) **Duration variability.** No two spoken realisations of a word, even by the same person, have the same duration. Furthermore, often the variation in the duration of a word is non-uniform in that in different realisations of a word parts of a word may be more elongated/shrunk compared with other parts of the same word.
(b) **Spectral variability.** No two spoken realisations of a word, even by the same person, have the same spectral–temporal trajectory.
(c) **Speaker variability.** Speech is affected by the anatomical characteristics, gender, health, and emotional state of the speaker.
(d) **Accent.** Speaker accent can have a major effect on speech characteristics and on speech recognition performance.

(e) **Contextual variability.** The characteristic of a speech unit is affected by the acoustic and phonetic context of the units preceding or succeeding it.
(f) **Co-articulation.** This is similar to contextual variability but is also affected by the speech rate, accent and psychological factors.
(g) **Noise.** Speech recognition is affected by noise, echo, channel distortion and adverse environment.

Speech recognition methods aim to model, and where possible reduce, the effects of the sources of speech variability. The most challenging sources of variations in speech are speaker characteristics including accent, co-articulation and background noise.

Isolated-Word and Continuous-Speech Recognition

In terms of the 'fluency' or continuity of the input speech that a speech recognition system can handle, there are broadly two types of speech recognition systems:

(a) *Isolated-word recognition systems*, with short pauses between spoken words, are primarily used in small vocabulary command-control applications such as name-dialling, Internet navigation, and voice-control of computer menus or accessories in a car. Isolated-word recognition systems may use models trained on whole-word examples or constructed from concatenation of sub-word models.
(b) *Continuous speech recognition* is the recognition and transcription of naturally spoken speech. Continuous speech recognition systems are usually based on word model networks constructed from compilation of phoneme models using a phonetic transcription vocabulary. Usually the word model networks also include an N-gram language model.

Speaker-Dependent and Speaker-Independent Speech Recognition

In any communication system the decoding errors at the receiver increase with the increasing overlap of the probability distributions of the signals that carry the communication symbols. In a speaker-independent speech database, the effect of variations of different speakers' characteristics results in higher variances and hence broader probability distributions of speech features. Broader distributions of different acoustic speech sounds result in higher overlaps between those distributions and hence a higher speech recognition error rate.

Unlike speaker-independent systems, speaker-dependent systems do not have to deal with the extra variance due to the variations of speakers' characteristics and the consequent additional overlap of the probability distributions of different speech sounds. Hence, speaker-dependent systems can be more accurate than speaker-independent systems. However, the performance of speaker-independent systems can be improved using speaker adaptation methods, such as maximum likelihood linear regression (MLLR) method, to modify the parameters of a speaker-independent system towards those of a speaker-dependent system.

14.8.1 Speech Units

Every communication system has a set of elementary symbols (or alphabet) from which larger units such as words and sentences are constructed. For example in digital communication the basic alphabet is '1' and '0', and in written English the basic units are A to Z. The elementary linguistic unit of spoken speech is called a *phoneme* and its acoustic realisation is called a *phone*. There are 60 to 80 phonemes in spoken English; the exact number of phonemes depends on the dialect. For the purpose of automatic speech processing the number of phonemes is clustered and reduced to 40 to 60 phonemes depending on the dialect. The phonemes in American English and British English are listed in Tables 14.3 and 14.4 respectively. The British English phonemes are grouped into ten categories. The US English can be similarly classified.

Note that phonetic units are not produced in isolation and that their articulation and temporal–spectral 'shape' is affected by the context of the preceding and succeeding phones as well as the linguistic, expressional and tonal context in which they are produced. For speech recognition, context-dependent triphone units are used. A triphone is a phone in the context of a preceding and a following phone, for example the triphones for the word *imagination* are $iy+miy-m+aem-ae+gae-g+iyg-iy+niy-n+ay$ where – denotes preceding and + denotes succeeding.

Assuming that there are about 40 phonemes, theoretically there will be about $40\times 40=1600$ context-dependent variations for each phone, and hence a total of $40\times 1600=64,000$ triphones. However, due to linguistic constraints some triphones do not occur in practice.

Table 14.3 A list of American English phonemes.

Phoneme	Example	Phoneme	Example	Phoneme	Example	Phoneme	Example
IY	b**ea**t	IX	ros**e**s	NX	si**ng**	V	**v**at
IH	b**i**t	ER	b**ir**d	P	**p**et	DH	**th**at
EY	b**ai**t	AXR	butt**er**	T	**t**en	Z	**z**oo
EH	b**e**t	AW	d**ow**n	K	**k**it	ZH	a**z**ure
AE	b**a**t	AY	b**uy**	B	**b**et	CH	**ch**urch
AA	b**o**b	OY	b**oy**	D	**d**ebt	JH	**j**u**dg**e
AH	b**u**t	Y	**y**ou	H	**g**et	WH	**wh**ich
AO	b**ough**t	W	**w**it	HH	**h**at	EL	batt**le**
OW	b**oa**t	R	**r**ent	F	**f**at	EM	bott**om**
UH	b**oo**k	L	**l**et	TH	**th**ing	EN	butt**on**
UW	b**oo**t	M	**m**et	S	**s**at	DX	ba**tt**er
AX	**a**bout	N	**n**et	SH	**sh**ut	Q	(glottal stop)

Table 14.4 A categorised list of British English phonemes. Note the prefixes V = Voiced and UnV = Unvoiced.

Vowels	Semivowel	Nasal	V-fricative	Whisper
aa	l	m	v	hh
ae	r	n	dh	sil
ah	w	ng	z	
ao	y	**Affricative**	zh	
ax	**Diphthong**	ch	**UnV-stop**	
eh	aw	jh	p	
er	ay	**UnV-fricative**	t	
ih	ea	f	k	
iy	ey	s	**V-stop**	
oh	ia	sh	b	
ow	oy	th	d	
uh	ua		g	
uw				

Syllables are also subword units, but they are larger than phonemes. A word may be composed of one or more syllables and a syllable may be composed of one or more phonemes. For example, the word '*imagination*' can be deconstructed into the following sub-word units:

Word	*imagination*
Phonetic transcription	*iy m ae g iy n ay sh e n*
Triphone transcription	$iy+m$ $iy-m+ae$ $m-ae+g$ $ae-g$ $+iy$ $g-iy+n$ $iy-n+ay$ $n-ay+sh$ $ay-sh+ay$ $sh-e+sh$ $e-n+sh$ $e-n$
Syllable transcription	*iyma giy nay shen*

14.8.2 Entropy of Speech

In information theory, entropy is defined as a measure of the randomness or information content of a process. Entropy quantifies the information content or the capacity of an information source, as described in Section 6.4.1. The information in speech – due to such random variables as words, speaker, intonation and accent – can be quantified in terms of entropy.

The entropy of an information source gives the theoretical lower bound for the number of binary bits required to encode the source. The entropy of a set of communication symbols is obtained as the probabilistic average of $\log_2$ of the probabilities of the symbols. The entropy of a random variable X with M states or symbols $X = [x_1, \ldots, x_M]$ and the state or symbol probabilities $[p_1, \ldots, p_M]$, where $P_X(x_i) = p_i$, is given by

$$H(X) = -\sum_{i=1}^{M} P_X(x_i) \log P_X(x_i) \tag{14.44}$$

We can consider the entropy of speech at several levels, such as the entropy of words contained in a sequence of speech or the entropy of intonation or the entropy of the speech signal features. The calculation of the entropy of speech is complicated as speech signals simultaneously carry various forms of information such as phonemes, topic, intonation signals, accent, speaker voice, speaker stylistics.

Example 14.1 *Entropy of phonemic symbols*
Speech is made up of about 40 basic acoustic symbols, known as phonemes, which are used to construct words, sentences etc. Assuming that all phonetic units are equi-probable, and that the average speaking rate is 10 phonemes/second, calculate the minimum number of bits per second required to encode speech at the average speaking rate. For speech $M = 40$ assume $P(x_i) = 1/40$, the entropy of the symbolic labels speech is given by

$$H(X) = \sum_{i=1}^{40} -\frac{1}{40} \log_2 \frac{1}{40} = 5.3 \text{ bits} \tag{14.45}$$

Number of bits/sec = Number of symbols/sec × Entropy = $5.3 \times 10 = 53$ bps.

Note the above over-simplified calculation ignores the fact that phonemes, like letters of a written language, have a non-uniform probability distribution. More importantly it also ignores the constraints imposed on the occurrence of phonemes by the vocabulary, the rules of grammar and the context of conversation, and it ignores the speaker characteristics and intonation information.

Effects of Vocabulary, Grammar and Topic on Entropy

The occurrence of a speech unit in a correctly constructed speech sentence depends on the previous speech units and is constrained by the vocabulary, the grammar, the context and the topic of conversation. The probability of a word in a stream of words should therefore be conditioned on the previous words. This is usually achieved using a language probability in the form of a bi-gram in which the probability of a word is conditioned on the previous word, or a tri-gram in which the probability of a word is conditioned on the previous two words. The topic of a conversation can be used to influence the likelihood of a word.

Using a conditional probability function, based on an N-gram grammar and a topic T, the conditional entropy of words is expressed as

$$H(w_m|w_{m-1},\ldots,w_{m-N+1},T)=-\sum_{i=1}^{M}P_X(w_i|w_{m-1},\ldots,w_{m-N+1},T)\log P_X(w_i|w_{m-1},\ldots,w_{m-N+1},T) \tag{14.46}$$

The net effect of conditioning the probability of an observation on the previous observations and on the topic, is to constrain and reduce the randomness and hence the entropy of the observation.

Effect of Speaker Characteristics on Speech Entropy

One of the main sources of variability and information in speech is the variations of speaker characteristics. Speaker variability is due to variations in the anatomical characteristics as well as the speaking manner. The variables that convey speaker characteristics are the vocal tract model parameters, pitch value, pitch range, and prosody parameters. These variables naturally increase the dimensionality, the randomness and the entropy of speech.

14.8.3 Probabilistic Speech and Language Models

The probability that a sequence of speech feature vectors $\boldsymbol{X}$ is an acoustic production of a word w may be expressed, using Bayes' rule, as

$$P(w|\boldsymbol{X})=\frac{f(\boldsymbol{X}|w)P(w)}{f(\boldsymbol{X})} \tag{14.47}$$

where $P(w|\boldsymbol{X})$ is the probability of a word w given a sequence of speech observation feature vectors $\boldsymbol{X}$, $f(\boldsymbol{X}|w)$ is the pdf of the speech observation $\boldsymbol{X}$ conditioned on the word w and $P(w)$ is the prior probability of w obtained from a language model.

The maximum a posteriori (MAP) estimate of the word w conveyed by the feature sequence $\boldsymbol{X}$ is obtained as the word that maximises $P(w|\boldsymbol{X})$. Since for a given observation feature sequence $\boldsymbol{X}$, its probability $P(\boldsymbol{X})$ is a constant, the MAP estimate may be expressed as

$$\hat{w}=\arg\max_{w} f(\boldsymbol{X}|w)P(w) \tag{14.48}$$

A 'language grammar' model can be incorporated in the Bayesian probability model of Equation (14.47) by conditioning the probability of a word w_m on the sequence of speech observation feature vectors $\boldsymbol{X}$ and on the previous $N-1$ words $w_{m-1}, w_{m-2}, \ldots, w_{m-N+1}$ (N-gram grammar) as

$$P(w_m|\boldsymbol{X}, w_{m-1}, \ldots w_{m-N+1}) = \frac{f(\boldsymbol{X}|w_m)P(w_m|w_{m-1}, \ldots, w_{m-N+1})}{f(\boldsymbol{X})} \tag{14.49}$$

where $P(w_m|w_{m-1}, w_{m-2}, \ldots w_{m-N+1})$, the probability of word w_m conditioned on the previous $N-1$ words $w_{m-1}, w_{m-2}, \ldots w_{m-N+1}$, is an N-gram language model. The MAP estimate is then given by

$$\hat{w}_m = \arg \max_{w} f(\boldsymbol{X}|w_m)P(w_m|w_{m-1}, w_{m-2}, \ldots, w_{m-N+1}) \tag{14.50}$$

N-gram language models can be trained using a large database of transcriptions of speech or simply using texts of books or newspapers. Usually a bi-gram language model $P(w_m|w_{m-1})$ in which the occurrence of a word w_m is conditioned on the previous word w_{m-1}, or a tri-gram model $P(w_m|w_{m-1}, w_{m-2})$ in which the occurrence of a word w_m is conditioned on the preceding two words w_{m-1} and w_{m-2}, is used.

Front-End Feature Extraction

The feature extraction subsystem converts time-domain raw speech samples into a compact and efficient sequence of spectral–temporal feature vectors that retain the phonemic information but discard some of the variations due to speaker variability and noise. The most widely used features for speech recognition are cepstral feature vectors which are obtained from a discrete cosine transform function of the logarithm of magnitude spectrum of speech. The temporal dynamics of speech parameters, i.e. the direction and the rate of change of time-variation of speech features, play an important role in improving the accuracy of speech recognition. Temporal dynamics are often modelled by the first- and second-order differences of cepstral coefficients as explained below.

Cepstral Features

For speech recognition, the most commonly used features are cepstral coefficients. As shown in Figures 14.21 and 14.22, cepstral coefficients are derived from an inverse discrete Fourier transform (IDFT) of logarithm of short-term power spectrum of a speech segment (with a typical segment length of 20–25ms) as

$$c(m) = \sum_{k=0}^{N-1} \ln[|X(k)|] e^{\frac{j2\pi mk}{N}} \tag{14.51}$$

where $X(k)$ is the FFT-spectrum of speech $x(m)$. As the spectrum of real-valued speech is symmetric, the DFT in Equation (14.51) can be replaced by a discrete cosine transform (DCT) as

$$c\,(m) = \mathrm{DCT}\,\{\ln\,[|X\,(k)\,|]\} \tag{14.52}$$

As shown in Figure 14.23, cepstral parameters encode the shape of the log-spectrum. For example the coefficient $c(0)$ is given by

$$c(0) = \log|X(0)| + \cdots + \log|X(N-1)| = \log[|X(0)| \times \cdots \times |X(N-1)|] \tag{14.53}$$

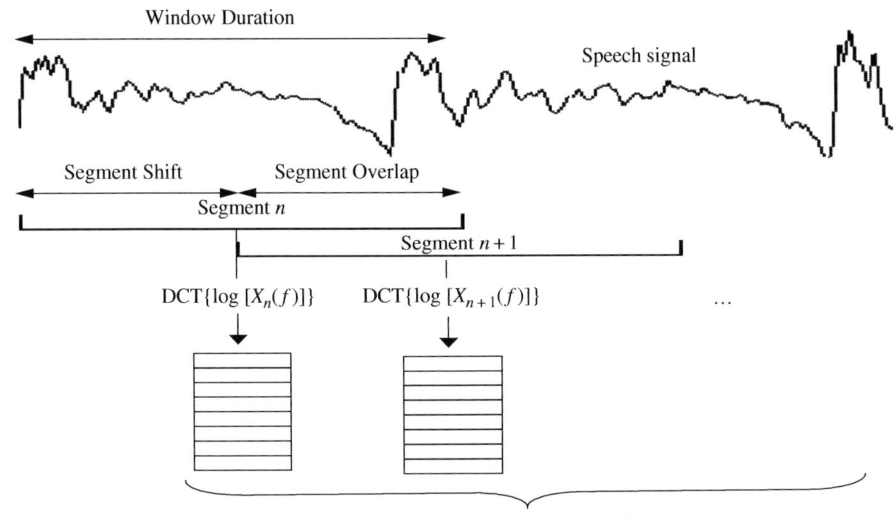

Figure 14.21 Illustration of speech feature extraction.

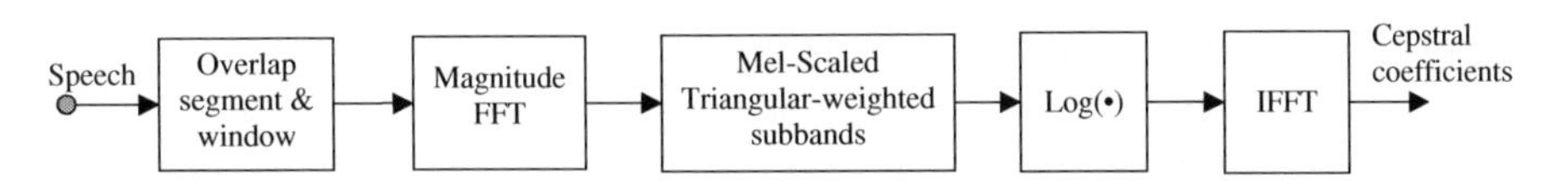

Figure 14.22 Block diagram of a typical cepstral feature extraction system.

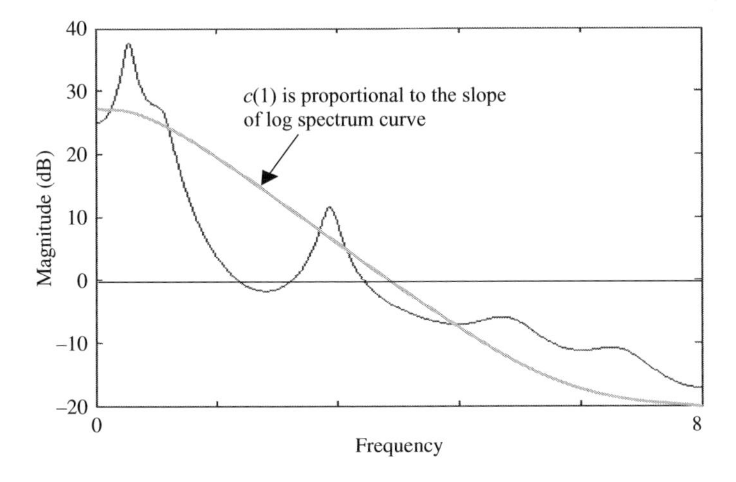

Figure 14.23 Cepstral coefficients encode the shape of the spectrum. For example $c(1)$, corresponding to the DCT basis function shown, encodes the tilt of the spectrum.

Note that $c(0)$ is the average of the log magnitude spectrum, or equivalently the geometric mean of the magnitude spectrum. Similarly the coefficient $c(1)$ describes the tilt or the slope of the log spectrum which is usually negative for vowels (e.g. '*e*' pronounced as '*iy*') with their energy mostly in low frequencies and positive for consonants (e.g. '*s*') with their energies mostly in higher frequencies. Some useful properties of cepstrum features are as follows:

(a) The lower index cepstral coefficients represent the spectral envelope of speech, whereas the higher indexed coefficients represent fine details (i.e. excitation) of the speech spectrum.
(b) Logarithmic compression of the dynamic range of spectrum, benefiting lower power higher frequency speech components.
(c) Insensitivity to loudness variations of speech if the coefficient $c(0)$ is discarded.
(d) As the distribution of the power spectrum is approximately log-normal, the logarithm of the power spectrum, and hence the cepstral coefficients, are approximately Gaussian.
(e) The cepstral coefficients are relatively de-correlated allowing simplified modelling assumptions such as the use of diagonal covariance in modelling the speech feature vectors distribution.

There are a number of variants of cepstral coefficients. The two most common choices for extraction of cepstral coefficients are based on a filter bank model and a linear predictive (LP) model of speech.

Mel Frequency Cepstral Coefficients

A widely used form of cepstrum is mel frequency cepstral coefficients (MFCC). To obtain MFCC features, the spectral magnitude of FFT frequency bins are averaged within frequency bands spaced according to the mel scale which is based on a model of human auditory perception. The scale is approximately linear up to about 1000 Hz and approximates the sensitivity of the human ear as

$$f_{\text{mel}} = 1125 \log(0.0016f + 1) \tag{14.54}$$

where f_{mel} is the mel-scaled frequency of the original frequency f in Hz. An example of the mapping process and also the spacing of bands for a 19-channel filter-bank is shown in Figure 14.24.

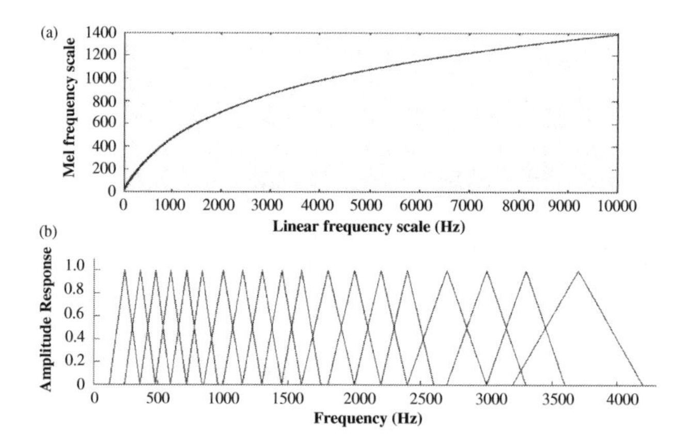

Figure 14.24 Illustration of (a) linear to mel frequency scale mapping, (b) mel frequency scale bands for a filter bank analysis of speech.

LP-Cepstrum

Cepstrum coefficients can also be derived from linear prediction model parameters. The LP-cepstrum (LPC) coefficients are given by (Furui 1989)

$$c(m) = -a(m) + \sum_{k=1}^{m} \left(1 - \frac{k}{m}\right) a(k)c(m-k) \quad 1 \le m \le P \tag{14.55}$$

where P is the order of the LPC analysis.

Temporal Dynamic Features of Speech

Although over a short time segment of say 20 ms speech parameters are assumed to be time-invariant, over a longer segment the parameters of speech vary with time. Furthermore, the time variations of speech convey essential linguistic, prosodic and speaker-characteristic information. Hence, speech parameters, such as linear prediction model coefficients, pitch and excitation gain, are time-varying variables. Temporal difference features are a simple and effective means for description of the trajectory of speech parameters in time. The modelling of difference features is essential in improving the accuracy in speech recognition or the perceptual quality in speech synthesis. For speech recognition relatively simple cepstral difference features are defined by

$$\partial c(m) = c(m+1) - c(m-1) \tag{14.56}$$

$$\partial\partial c(m) = \partial c(m+1) - \partial c(m-1) \tag{14.57}$$

where $\partial c(m)$ is the first-order time-difference of cepstral features; in speech processing it is also referred as the velocity features. The second-order time-difference of cepstral features $\partial\partial c(m)$ is also referred to as the acceleration. Equation (14.56) can be derived from the slope of the minimum mean squared line model of three cepstral samples $c(m-1)$, $c(m)$ and $c(m+1)$.

Dynamic or difference features are effective in improving speech recognition. The use of difference features is also important for improving quality in speech coding, synthesis and enhancement applications.

14.8.4 Statistical Models of Acoustic Features of Speech

For speech recognition, an efficient set of acoustic models is needed to capture the mean and variance of the spectral–temporal trajectory of speech sounds and to discriminate between different speech sounds. In selecting a speech model we have the following broad options:

(a) Templates of averaged speech feature vector sequences. This is mostly used for isolated-word applications. Time alignment of speech examples of different length and time alignment of a speech example with speech templates is achieved using dynamic time warping methods (DTW).
(b) Hidden Markov models (HMMs) or artificial neural networks (ANNs). There are also hybrid models that combine the complimentary power of HMMs and ANNs.
(c) Context-independent models or context-dependent models where several models are used for each word (or phoneme) to capture the variations in the acoustic production of the words caused by the variations of different acoustic context within which words occurs.

For small-vocabulary isolated-word recognition, as in name-dialling, a template of cepstral feature vectors can be used to model the acoustic realisation of each word. The template for each word is obtained from one or more examples of the word. For large-vocabulary and continuous speech recognition, sub-word models are used due to their efficiency for training and adaptation of models, and for the ease of expansion of the vocabulary size. Hidden Markov models, described in Chapter 11, are commonly used for medium- to large-vocabulary speech recognition systems.

14.8.5 Resolutions of Speech Features and Models

The resolution of the features and models employed in a speech recognition system depends on the following factors:

(a) acoustic feature resolution, i.e. spectral–temporal resolution
(b) acoustic model resolution, i.e. the smallest segment modelled
(c) statistical model resolution
(d) contextual resolution.

Spectral–Temporal Feature Resolution

The spectral and temporal resolution of speech features are determined by the following factors: (i) the speech signal window size for feature extraction (typically 25 ms), (ii) the rate at which speech features are sampled (usually every 5 to 10 ms), and (iii) speech feature vector dimensions; typically 13 cepstral features plus 13 first difference and 13 second difference cepstral features.

Statistical Model Resolution

Model resolution is determined by (i) the number of models, (ii) the number of states per model, and (iii) the number of sub-state models per state. For example, when using hidden Markov models, each HMM has N states (typically $N = 3 - 5$), and the distribution of feature vectors within each state is modelled by a mixture of M multi-variate Gaussian densities. Therefore the model for each phoneme has $N \times M$ Gaussian distributions with each Gaussian density parameterised by a mean feature vector and a covariance matrix. Model resolution also depends on whether full-covariance or diagonal-covariance matrices are used for the Gaussian distribution of speech feature vectors. Using full-covariance matrix Gaussian pdfs, the number of Gaussian model parameters for N states, with M mixtures per state and P-dimensional features is $N \times M \times (P^2 + P)$, with typical numbers used (i.e. $N = 3$, $M = 15$ and $P = 3 \times 13 = 39$ including the first and second difference features) we have $3 \times 15 \times (39^2 + 39) = 70,200$ parameters. The use of diagonal covariance matrices reduces this to about 3510 parameters.

Model Context Resolution

The acoustic production of a speech unit is affected by its acoustic context, that is by the preceding and succeeding speech units. In phoneme-based systems, context-dependent triphones (see Section 14.8.1) are used. Since there are about 40 phonemes, the total number of triphones is $40^3 = 64,000$, although many of these cannot occur due to linguistic constraints. The states of triphone models are often clustered and tied to reduce the total number of parameters and hence obtain a compromise between contextual resolution and the number of model parameters used.

14.8.6 Voice-Activated Name Dialling

This section describes a voice-activated name dialling method for mobile phones. The dialling method is developed with the objective of optimising the requirements for high accuracy, simplicity of use and minimum memory, computational and power requirements. The system stores an acoustic template for each name along with its text form and telephone number. The texts for each name and telephone number are input manually. The acoustic template for each name is formed from a single spoken example of the name. The recognition system is an isolated-word recognition, Figure 14.25, that makes use of the LPC parameters and the voice activity detector (VAD) of the mobile phone system. LPC-cepstrum parameters are used as speech features, and the VAD output is used for endpoint detection. The distance metric used for selection of the most likely spoken name is the minimum mean squared distance. In order to minimise the cost of making wrong calls due to speech recognition errors, the best score is compared with a confidence threshold level; if the score is less than the threshold the system prompts: 'Name not recognised'. The system was evaluated on a Laptop PC. Initial experiments obtained very a high accuracy and reliability of 97%. This is not surprising for a speaker-dependent small-vocabulary isolated-word recognition task.

Name-Dialling Vocabulary

For a name-dialling task the vocabulary is simple and consists of N names

$$\text{Names} = \text{sil} < \text{Name1}|\text{Name2}|\text{Name3}|\ldots|\text{Name}N > \text{sil}$$

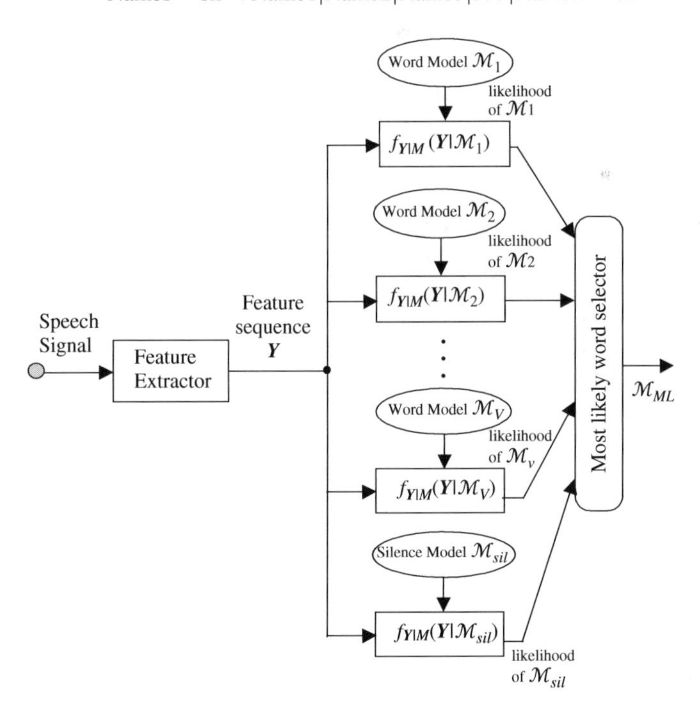

Figure 14.25 Illustration of a speech recognition system. Note that for simple voice-activated dialling, each word model is a single spoken example of a name; the feature extraction includes duration normalisation; and instead of a probability metric a mean squared distance metric is used.

Table 14.5 Illustration of stored codebook of name, number and feature template for name-dialling.

Name	Number	Acoustic template
Luke	08890552077	
William	01763 242 383	....
Colin	....	....
Kate	....	....

where N is typically 8, and sil indicates silence/noise i.e. no speech activity and the bar '|' indicates 'or' (i.e. Name1 or Name2 or ...). The names, numbers and the acoustic templates for each phonebook entry are stored as shown. Table 14.5 shows a name-dialling structure for a mobile phone.

Speech Features and Models for Name Dialling

Since mobile phones use LPC models for speech coding, LPC-based cepstrum coefficients and differential cepstral coefficients are used as speech features. An important point is the dimensionality of cepstral features and the feature vector sampling rate for representation of each word. The feature template size is chosen as a compromise between the requirements for high accuracy and low memory storage and computational complexity.

Hidden Markov models are normally used for speech recognition. However for this relatively simple voice-dialling application we simply store and use the cepstral vector sequence from one example of each word as the template for that word.

Dealing with Speech Duration Variability

For continuous speech recognition the variability in speech duration is usually dealt with using the model flexibility afforded by the state transitions of a Hidden Markov model together with the Viterbi time-warping algorithm. However, for the isolated-word voice-dialling system, there are two relatively simple methods that can be used:

(a) Time-normalisation can be achieved by adjusting the frame shift rate for calculation of cepstrum feature vector sequences of each input name, to yield a pre-specified number of uniformly-spaced cepstrum feature vectors across the duration of the spoken name. For small-vocabulary isolated-word recognition this method works well.

(b) Time normalisation between two feature vector sequences of different lengths can be achieved using dynamic time warping as described next and illustrated in Figure 14.26.

Time-Alignment: Dynamic Time Warping (DTW)

Speech is a time-varying process in which the duration of a word and its subwords varies randomly. Hence a method is required to find the best time-alignment between a sequence of vector features representing a spoken word and the model candidates. The best time-alignment between two sequences

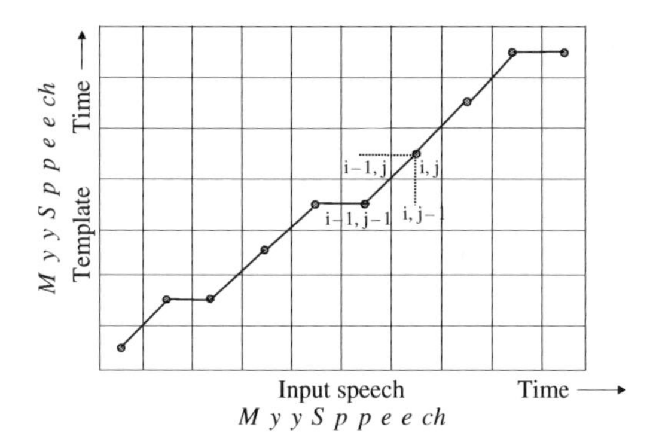

Figure 14.26 Illustration of DTW of input speech and a template model.

of vectors may be defined as the alignment with the minimum Euclidean distance. For isolated-word recognition the time-alignment method used is dynamic time warping (DTW) illustrated in Figure 14.26. The best matching template is the one with the lowest distance path aligning the input pattern to the template.

To illustrate DTW, consider a point (i, j) in the time–time matrix (where i indexes the input pattern frame, and j the template frame) of Figure 14.26, then the previous point must have been $(i-1, j-1)$, $(i-1, j)$ or $(i, j-1)$. The key idea in dynamic programming is that at point (i, j) we continue with the lowest accumulated distance path from $(i-1, j-1)$, $(i-1, j)$ or $(i, j-1)$. The DTW algorithm operates in a time-synchronous manner: each column of the time-time matrix is considered in succession (equivalent to processing the input frame-by-frame) so that, for a template of length N, the maximum number of paths considered at any time is N. If $D(i, j)$ is the global distance up to (i, j) and the local distance at (i, j) is $d(i, j)$ then we have the recursive relation

$$D(i, j) = \min[D(i-1, j), D(i, j-1), D(i-1, j-1)] + d(i, j) \tag{14.58}$$

Given that $D(1, 1) = d(1, 1)$, we have the basis for an efficient recursive algorithm for computing $D(i, j)$. The final global distance $D(n, N)$ gives us the overall matching score of the template with the input. The input word is then recognised as the word corresponding to the template with the lowest matching score.

Distance Metric Score for Word Recognition

The distance metric used for the selection of the nearest name in the memory $\{\mathcal{M}_1, \mathcal{M}_2, \ldots \mathcal{M}_k\}$ to the spoken input name $\boldsymbol{X}$ can be the mean squared error or alternatively the mean absolute value of error. The labelling of the input feature matrix $\boldsymbol{X}$ is achieved as

$$\text{Label}(\boldsymbol{X}) = \arg\min_k \left\{|\boldsymbol{X} - \mathcal{M}_k|^2\right\} k = 1, \ldots, N \tag{14.59}$$

where the squared magnitude function is given by

$$|\boldsymbol{X} - \mathcal{M}_k|^2 = \sum_{t=1}^{T} \sum_{i=1}^{P} |X(i, t) - \mathcal{M}_k(i, t)|^2 \tag{14.60}$$

Minimising Dialling Error: Calculating a Confidence Threshold

Recognition errors in voice-activated dialling can result in wrong phone calls. To minimise recognition errors the speech recognition scores are analysed and a confidence threshold is required so that when the metric distances between the spoken word and the best and second-best candidates are too close then the system will output a message such as 'Name not recognised'. The confidence threshold depends on the metric and can be set experimentally to achieve a desired risk/cost ratio.

Automatic Activation of Name Dialling

Name dialling can be activated by either pressing a button on the phone or by automatic speech/noise detection. If automatic speech/noise detection is used then a fairly high threshold level should be chosen to minimise the risk of inadvertent name dialling. Name dialling for mobile phones has a real practical value in a hands-free environment, for example when the user is driving etc.

Figure 14.27 illustrates a signal flow diagram for a voice-dialling system. Figure 14.28 shows a diagram illustrating the outline of automatic activation of a voice dialling system.

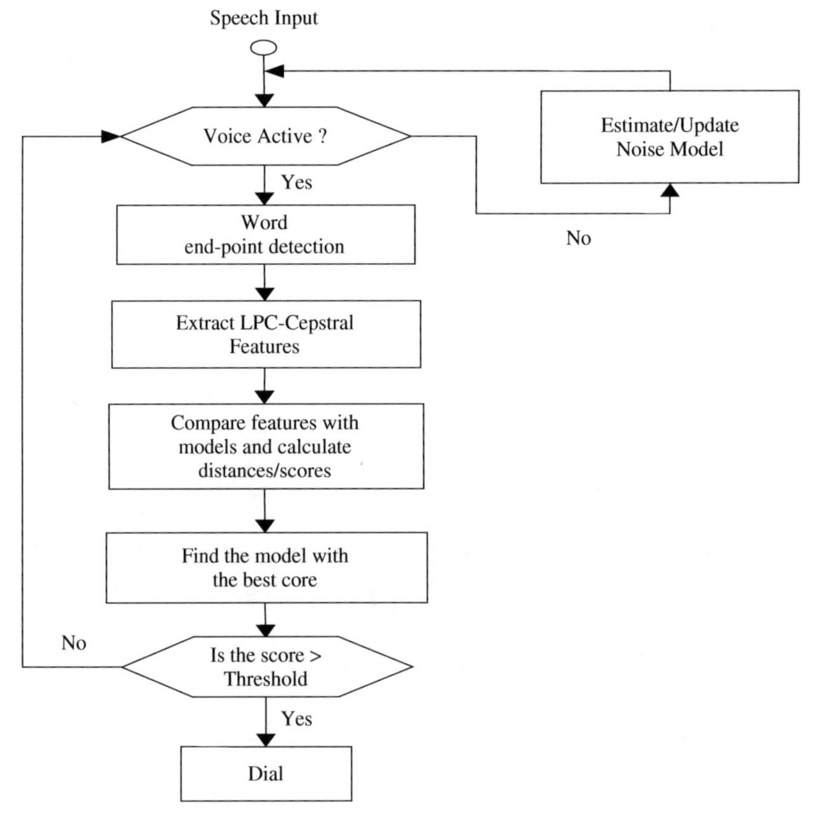

Figure 14.27 A block flow diagram for voice-activated dialling.

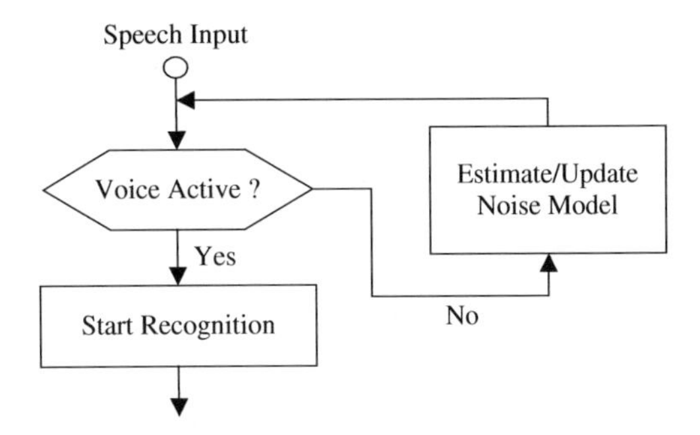

Figure 14.28 Illustration of automatic activation of name dialling.

14.9 Summary

This chapter provided an introduction to speech signal processing methods, specifically the speech production process, speech models and their applications to speech coding for mobile phones and automatic speech recognition. The physical mechanism of speech production composed of a mixture of glottal pulses and fricative noise shaped by the frequency responses of vocal tract and nasal cavity was introduced. The speech models considered here were linear prediction model and harmonic plus noise model and their combination. For speech coding the popular code excited linear prediction (CELP) model was introduced. For speech recognition, issues such as the choices of speech units, speech features and speech models were discussed. The design of a simple voice dialling method was described in detail.

Further Reading

Berouti M., Schwartz R. and Makhoul J. (1979) Enhancement of speech corrupted by acoustic noise. Proc. ICASSP, pp. 208–211.

Deller J.R., Proakis J.G. and Hansen J.H.L. (1993) Discrete-Time Processing of Speech Signals. Maxwell Macmillan International.

Furui S. (1989) Digital Speech Processing, Synthesis and Recognition, Marcel Dekker.

Griffin D.W. and Lim J.S. (1988) Multiband Excitation Vocoder. IEEE Trans. Acoust. Speech Signal Process. 36 (8), pp. 1223–1235.

Hermansky H. (1990) Perceptual Linear Predictive (PLP) Analysis of Speech. Journal of the Acoustical Society of America, 87, 1738–1752.

Ifeachor E.C. and Jervis B.W. (1993) Digital Signal Processing: A Practical Approach. Addison-Wesley.

Makhoul J. (1975) Linear Prediction: A Tutorial Review. Proceedings of the IEEE, 63(4), pp. 561–580.

Moore B.C.J. (1982) An Introduction to the Psychology of Hearing, 2nd edn. Academic Press, London.

Rabiner L.R. and Juang B.H. (1993) Fundamentals of Speech Recognition. Prentice-Hall, Englewood Cliffs, NJ.

Rabiner L.R. and Schafer R.W. (1978) Digital Processing of Speech Signals. Prentice Hall, Englewood Cliffs, NJ.

Young S., Evermann G., Kershaw D., Moore G., Odell J., Ollason D., Dan P., Valtchev V., and Woodland P. The Hidden Markov Model (HTK) Book. Cambridge University Engineering Deparment.

15 Speech Enhancement

Speech enhancement in noisy environments, such as in streets, cars, trains, aircrafts, and noisy public venues, improves the quality and intelligibility of speech and reduces communication fatigue.

Noise reduction benefits a wide range of applications such as mobile phones, hands-free phones, teleconferencing, in-car cabin communication, hearing aids, automated voice services based on speech recognition and synthesis, forensics and reconstruction of old archived records. This chapter provides an overview of the main methods for single-input and multiple-input speech enhancement in noise.

Single-input noise reduction systems strive to suppress audibility of the noise by utilising the temporal–spectral structures of signal and noise processes. Often the signal frequency components are multiplied by a gain factor derived from estimates of the prior and the posterior signal-to-noise ratios.

Multiple-input systems, on the other hand, strive to cancel out the noise as in adaptive noise cancellation or screen out the noise as in microphone array beam formers.

The simplest form of noise reduction is spectral subtraction which subtracts an estimate of the magnitude spectrum of the noise from that of the noisy signal. Spectral subtraction only requires an estimate of the average noise spectrum, usually obtained from the speech inactive noise-only periods, but it often introduces artifacts and distortions.

The classical noise reduction method is the Wiener filter which modifies the magnitude frequency spectrum of the noisy input signal in proportion to an estimate of the signal-to-noise ratio at each frequency. Wiener filter requires estimates of the power spectra, or equivalently the correlation matrices, of speech and noise.

High performance speech enhancement methods are based on Bayesian estimation methods requiring estimates of the parameters of the functions that describe the likelihood and the prior distributions of the signal and noise processes. The Bayesian speech enhancement methods include Kalman filters, minimum mean squared error (MMSE), maximum a posteriori (MAP) and ultimately the signal restoration methods that are based on hidden Markov models (HMMs) of speech and noise.

We will study signal processing methods for extension of the bandwidth of narrowband conventional telephony speech (300 Hz to 3400 Hz) to wideband broadcast speech (20 Hz to 10,000 Hz).

This chapter concludes with multiple-input multiple-output (i.e. microphone arrays) noise reduction methods which are particularly useful for teleconference and in-car communication and include beam-forming arrays, adaptive noise cancellation and stereophonic echo cancellation.

Multimedia Signal Processing: Theory and Applications in Speech, Music and Communications Saeed V. Vaseghi

15.1 Introduction

Figure 15.1 illustrates a classification of the main signal processing methods for enhancement of noisy speech into the following two broad types:

(a) Single-input speech enhancement systems where the only available signal is the noise-contaminated speech picked up by a single microphone. Single-input systems do not *cancel* noise, rather they *suppress* the noise using estimates of the signal-to-noise ratios of the frequency spectrum of the input signal. Single-input systems rely on the statistical models of speech and noise, which may be estimated on-line from the speech-inactive periods or decoded from a set of pre-trained models of speech and noise. An example of a useful application of a single-microphone enhancement system is a mobile phone system used in noisy environments.

(b) Multiple-input speech enhancement systems where a number of signals containing speech and noise are picked up by several microphones. Examples of multiple-input systems are adaptive noise cancellation, adaptive beam-forming microphone arrays and multiple-input multiple-output (MIMO) acoustic echo cancellation systems. In multiple-input systems the microphones can be spatially configured and adapted for optimum performance. Multiple-input noise reduction systems are useful for teleconference systems and for in-car cabin communication systems.

Note that in order to achieve the best noise reduction performance, where possible, the advantages of the signal processing methods developed for single-input noise suppression and multiple-input noise cancellation systems are combined.

15.2 Single-Input Speech Enhancement Methods

In single-input systems the only available signal is the noisy speech; in addition statistical models of speech or noise may be available. In applications where speech enhancement and recognition are performed in the same system, the results of speech recognition can provide the speech enhancement method with such statistical information as the power spectra or correlation matrices obtained from decoding the most-likely speech and noise models. Single-input noise reduction methods include Wiener filter, spectral subtraction, Kalman filter, the MMSE noise suppression method and speech restoration via model-based analysis and synthesis methods as described in this section.

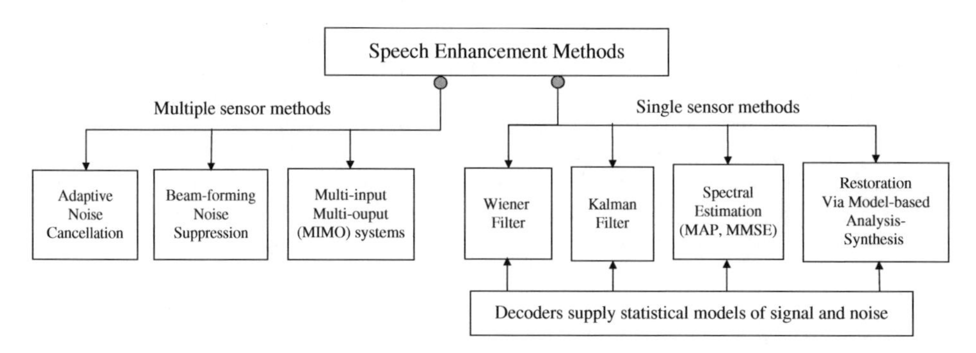

Figure 15.1 A categorisation of speech enhancement methods. Note that statistical models can optionally provide single-input noise reduction methods with the additional information needed for improved performance.

15.2.1 An Overview of Single-Input Speech Enhancement

Assuming that the speech signal $x(m)$ and the noise $n(m)$ are additive the noisy speech $y(m)$ is modelled as

$$y(m) = x(m) + n(m) \tag{15.1}$$

where the integer variable m denotes the discrete-time index. It is generally assumed that the speech signal is uncorrelated with noise; this is a reasonable assumption in most cases when the signal and noise are generated by independent sources.

The general form of a typical speech enhancement method is shown in Figure 15.2. The speech enhancement system is composed of a combination of the following modules:

1. Speech segmentation into a sequence of overlapping frames (of about 20–30 ms) followed by windowing of each segment with a popular window such as the Hamming, Hanning or Hann window.
2. Discrete Fourier transformation (DFT) of the speech samples within each frame to a set of short-time spectral samples.
3. Estimation of the spectral amplitudes of clean speech. This involves a modification of the magnitude spectrum of noisy speech according to an estimate of the signal-to-noise ratio at each frequency.
4. An inter-frame signal-smoothing method to utilise the temporal correlations of the spectral values across successive frames of speech.
5. Speech and noise models, and a speech and noise decoder, supply the speech estimator with the required statistics (power spectra, correlation matrices etc.) of speech and noise.
6. Voice activity detection (integrated with, or substituting, speech and noise recognition) is used to estimated and adapt noise models from the noise-only periods and also for applying extra attenuation to noise-only periods.

In the following the elements of a speech enhancement system are described in more detail.

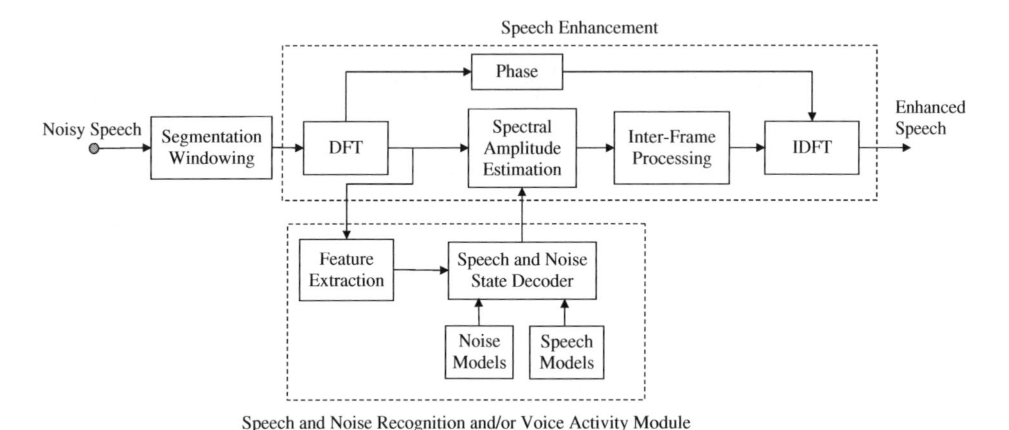

Figure 15.2 Block diagram illustration of a speech enhancement system.

Segmentation and Windowing of Speech Signals

Speech processing systems divide the sampled speech signal into overlapping frames of about 20–30 ms duration. The N speech samples within each frame are processed and represented by a set of spectral features or by a linear prediction model of speech production. The signal within each frame is assumed to be a stationary process. The choice of the length of speech frames (typically set to 20–30 ms) is constrained by the stationarity assumption of linear time-invariant signal processing methods, such as Fourier transform or linear prediction model, and by the maximum allowable delay for real-time communication systems such as voice coders. Note that with a window length of N samples and a sampling rate of F_s Hz the frequency resolution of DFT is F_s/N Hz.

Spectral Representation of Speech and Noise

Speech is segmented into overlapping frames of N samples and transformed to the frequency domain via discrete Fourier transform (DFT). In the frequency domain the noisy speech samples can be represented as

$$Y(k) = X(k) + N(k) \qquad k = 0, \ldots, N-1 \tag{15.2}$$

where $X(k)$, $N(k)$ and $Y(k)$ are the short-time discrete Fourier transforms of speech, noise and noisy speech respectively. The integer k represents the discrete frequency variable; it corresponds to an actual frequency of $2k\pi/N$ (rad/sec) or kF_s/N (Hz) where F_s are the sampling frequencies.

Equation (15.2) can be written in complex polar form in terms of the magnitudes and the phases of the signal and noise at discrete frequency k as

$$Y_k e^{j\theta_{Y_k}} = X_k e^{j\theta_{X_k}} + N_k e^{j\theta_{N_k}} \qquad k = 0, \ldots, N-1 \tag{15.3}$$

where $Y_k = |Y(k)|$ and $\theta_{Y_k} = \tan^{-1}(\mathrm{Im}(Y(k))/\mathrm{Re}(Y(k)))$ are the magnitude and phase of the frequency spectrum of $Y(k)$ respectively. Note that the Fourier transform models the correlation of speech samples with sinusoidal basis functions. The DFT bins can then be processed individually or in groups of frequencies, taking into account the psychoacoustics of hearing in critical bands of the auditory spectral analysis systems.

Linear Prediction Model Representation of Speech and Noise

The correlation of speech (or noise) samples can be modelled with a linear prediction (LP) model, as introduced in Chapter 10. Using a linear prediction model of speech and noise, the noisy speech is expressed as

$$y(m) = \underbrace{\sum_{k=1}^{P} a_k x(m-k) + e(m)}_{\text{Speech LP model}} + \underbrace{\sum_{k=1}^{Q} b_k n(m-k) + v(m)}_{\text{Noise LP model}} \tag{15.4}$$

where a_k and b_k are the coefficients of LP models of speech and noise, $e(m)$ and $v(m)$ are the random inputs of the LP models and P and Q are the model orders respectively. Linear prediction models can be used in a variety of speech enhancement methods including Wiener filters, Kalman filters and speech restoration via decomposition and re-synthesis.

Inter-Frame and Intra-Frame Correlations

The two main issues in modelling noisy speech are the following:

(a) Modelling and utilisation of the probability distributions and the intra-frame correlations of speech and noise samples *within* each noisy speech frame of N samples.
(b) Modelling and utilisation of the probability distributions and the inter-frame correlations of speech and noise features *across* successive frames of noisy speech.

Most speech enhancement systems are based on estimates of the short-time amplitude spectrum or the linear prediction model of speech. The phase distortion of speech is ignored. In the case of DFT-based features, each spectral sample $X(k)$ at a discrete frequency k is the correlation of the speech samples $x(m)$ with a sinusoidal basis function $e^{-j2\pi km/N}$. The intra-frame spectral correlation, that is the correlation of spectral samples within a frame of speech, is often ignored, as is the inter-frame temporal correlation of spectral samples across successive speech frames.

In the case of speech enhancement methods based on linear prediction models (LP) of speech, the LP model's poles model the spectral correlations within each frame. However, the de-noising of linear prediction model is achieved through de-noising the discrete samples of the frequency response of noisy speech and that process ignores the correlation of spectral samples. The optimal utilisation of the inter-frame and intra-frame correlations of speech samples is a continuing research issue.

Speech Estimation Module

At the heart of a speech enhancement system is the speech estimation module. For speech enhancement usually the spectral amplitude, or a linear prediction model, of speech is estimated and this estimate is subsequently used to reconstruct speech samples.

A variety of methods have been proposed for estimation of clean speech including Wiener filter, spectral subtraction, Kalman filter, the minimum mean squared error (MMSE) and the maximum a posterior (MAP) methods. For the proper functioning of the speech estimation module knowledge of the statistics of speech and noise is required and this can be estimated from the noisy speech or it can be obtained from pre-trained models of speech and noise.

Probability Models of Speech and Noise

The implementation of a noise reduction method, such as the Wiener filter, Kalman filter, spectral subtraction or a Bayesian estimation method, requires estimates of the time-varying statistical parameters (and in particular the power spectra or equivalently the correlation matrices) of the speech and noise processes. An estimate of the noise statistics can be obtained from the speech-inactive periods, however for the best results the speech and noise statistical parameters are obtained from a network of probability models of speech and noise and this essentially implies that in an optimal speech processing system speech recognition and speech enhancement need to be integrated.

The most commonly used probability models for speech are hidden Markov models (HMMs). Hidden Markov models, or alternatively Gaussian mixture models (GMMs), can also be used for modelling non-stationary noise. To model different types of noise a number of HMMs need to be trained, one HMM for each type of noise. Alternatively, one can use a GMM of noise with a large number of components, with each component effectively modelling a different type of noise.

Cost of Error Functions in Speech Estimation

The Bayesian cost function, introduced in Chapter 7, provides a general framework for speech estimation and for calculation and minimisation of the cost of estimation error function.

Most speech enhancement methods use a cost function that involves minimising some function of the average squared difference (error) between the discrete-frequency spectrum of clean speech X_k and its estimate $\hat{X}_k$ i.e. $\left(X_k - \hat{X}_k\right)^2$. However, non-linear estimation error functions such as $\left(X_k^{\beta_k} - \hat{X}_k^{\beta_k}\right)^\alpha$ are also used, where varying the parameters β and α provides a family of estimators; note that for the special case of $\beta = 1$ and $\alpha = 2$ this function is the squared error function. Some cost of error functions attempt to utilise the knowledge of the non-linear transformations of frequency and amplitude of speech that take place within the cochlea as explained in the description of the working of the ear in Chapter 13.

15.2.2 Wiener Filter for De-noising Speech

Wiener filter theory, introduced in Chapter 8, forms the foundation of speech de-noising systems. The output of a discrete-time Wiener filter is given by

$$\hat{x}(m) = \sum_{i=0}^{P} w(i)y(m-i) \tag{15.5}$$

where $w(i)$ are the filter coefficients for de-noising the input speech $y(m)$ and $\hat{x}(m)$ is the Wiener estimate of the clean speech $x(m)$. The Wiener filter coefficient vector $\boldsymbol{w} = [w(0), w(1), \ldots, w(P)]^{\mathrm{T}}$ was derived in Chapter 8 as

$$\boldsymbol{w} = \boldsymbol{R}_{yy}{}^{-1}\boldsymbol{r}_{xy} \tag{15.6}$$

where $\boldsymbol{R}_{yy}$ is the autocorrelation matrix of the noisy speech signal $\boldsymbol{y}$ and $\boldsymbol{r}_{yx}$ is the cross-correlation vector of the clean speech $\boldsymbol{x}$ and noisy speech $\boldsymbol{y}$.

For uncorrelated additive speech and noise the Wiener filter equation (15.6) can be written as

$$\boldsymbol{w} = [\boldsymbol{R}_{xx} + \boldsymbol{R}_{nn}]^{-1}\boldsymbol{r}_{xx} \tag{15.7}$$

where $\boldsymbol{R}_{xx}$ and $\boldsymbol{R}_{nn}$ are the autocorrelation matrices of the speech and noise respectively and $\boldsymbol{r}_{xx}$ is the autocorrelation vector of the speech. In the frequency domain, for additive noise uncorrelated with speech, the Wiener filter equation was derived in Chapter 8 as

$$W(k) = \frac{P_{XX}(k)}{P_{XX}(k) + P_{NN}(k)} \tag{15.8}$$

where $W(k)$ is the frequency response of the Wiener filter and $P_{XX}(k)$ and $P_{NN}(k)$ are the power spectra of speech and noise respectively and k is the discrete frequency variable. Figure 15.3 shows a block diagram implementation of a frequency domain Wiener filter.

By dividing the numerator and the denominator of Equation (15.8) by $P_{NN}(k)$, the Wiener filter can be expressed in terms of the signal-to-noise ratio as

$$W(k) = \frac{\mathrm{SNR}(k)}{\mathrm{SNR}(k) + 1} \tag{15.9}$$

Equation (15.9) reveals an important aspect of the general workings of signal-input noise reduction system: *noise suppression methods effectively use a function of the estimates of the signal-to-noise ratios to modify the spectral amplitudes of the noisy signal.*

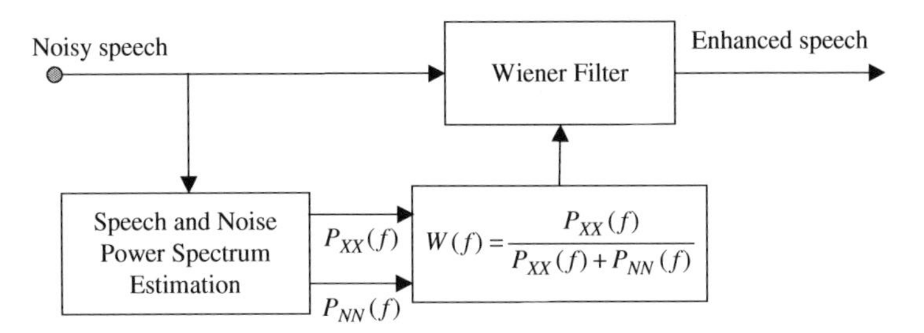

Figure 15.3 Block diagram overview of implementation of a Wiener filter for a speech enhancement system.

Wiener Filter Based on Linear Prediction Models

Wiener filters employing linear prediction models of speech and noise may be used for speech enhancement. The frequency response of Wiener filter can be expressed in terms of the ratio of power spectra of autoregressive (i.e. linear prediction) models of speech and noise as

$$W(f) = \frac{P_{XX}(f)}{P_{YY}(f)} = \frac{G_X^2/A_X^2(f)}{G_Y^2/A_Y^2(f)} = \frac{G_X^2}{G_Y^2}\frac{A_Y^2(f)}{A_X^2(f)} \tag{15.10}$$

where $G_X(f)/A_X(f)$ and $G_Y(f)/A_Y(f)$ are the frequency responses of linear prediction models of speech and noisy speech respectively. In time domain a square root Wiener filter equation (15.10) can be implemented as

$$\hat{x}(m) = \sum_{k=1}^{P} a_x(k)\hat{x}(m-k) + \frac{G_X}{G_Y}\sum_{k=0}^{Q} a_y(k)y(m-k) \tag{15.11}$$

where $a_x(k)$ and $a_y(k)$ are the coefficients of autoregressive models of clean speech and noisy speech respectively.

HMM-Based Wiener Filters

The key to the successful implementation of a Wiener filter is the accurate estimation of the power spectra of speech and noise, $P_{XX}(k)$ and $P_{NN}(k)$, for the frequency domain Wiener filter of Equation (15.8) or equivalently the estimation of the correlation matrices of speech and noise, $\boldsymbol{R}_{xx}$ and $\boldsymbol{R}_{nn}$, for the time domain Wiener filter of Equation (15.7). This is not a trivial task as speech and most noise processes are non-stationary processes.

Given the noisy speech signal, the time-varying power spectra of speech and noise may be estimated from a set of pre-trained hidden Markov models (HMMs), or Gaussian mixture models (GMMs), of speech and noise using a Viterbi decoder as illustrated in Figure 15.4. HMM state-based Wiener filter involves the following signal processing steps:

(a) Speech and noise decomposition – This involves the estimation of the most likely combination of speech and noise HMMs given the noisy speech signal. Using Viterbi state decoders the most likely combination of speech and noise yield the pdfs of the spectra of the most likely estimates of the clean speech and noise.
(b) The speech and noise power spectra from (a) are used to implement state-based Wiener filters.

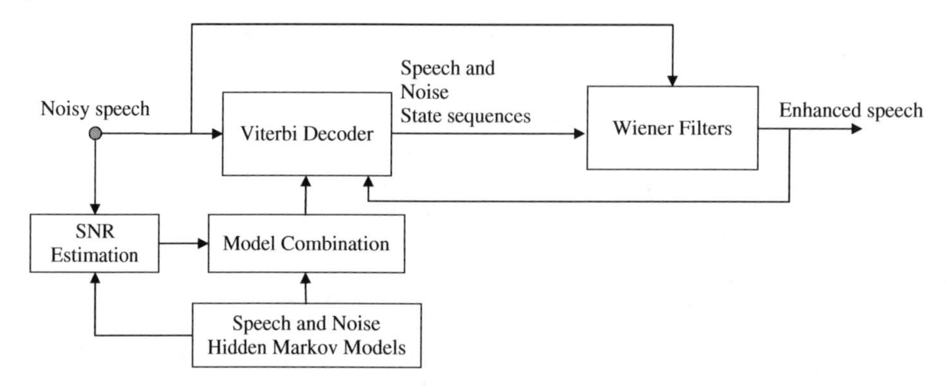

Figure 15.4 Block diagram illustration of a speech enhancement system based on Wiener filters and hidden Markov models (HMMs).

In HMM-based Wiener filtering the choice of the speech features needs to be appropriate for both speech recognition and enhancement. Linear prediction-based cepstrum features provide a suitable choice as the cepstrum coefficients obtained from the HMM's states can be mapped to the linear prediction model coefficients and thereafter to the linear prediction model spectrum for use in the implementation of the Wiener filter equations (15.8) and (15.10).

Assuming that for a noisy speech signal spectrum, $Y(k)$, the Viterbi decoder returns M different most likely state sequences, and that in each state the probability density function of the speech spectrum is represented by a mixture of L Gaussian pdfs, the Wiener filter is given by

$$\hat{X}(k) = \left[\sum_{\beta=1}^{M} \sum_{\gamma=1}^{L} p(\beta, \gamma) W_{\beta,\gamma}(k) \right] Y(k) \tag{15.12}$$

where $p(\beta, \gamma)$ is the estimated probability of speech and noise spectra from mixture γ of HMM state β.

15.2.3 Spectral Subtraction of Noise

A simple and widely studied speech enhancement method is the spectral subtraction method illustrated in Figure 15.5. In spectral subtraction an estimate of the average magnitude spectrum of the noise is subtracted from the magnitude spectrum of noisy speech. The spectral subtraction filter can be expressed as the product of the noisy speech spectrum $Y(k)$ and a spectral gain function $W_{SS}(k)$

$$\hat{X}(k) = W_{SS}(k) Y(k) \tag{15.13}$$

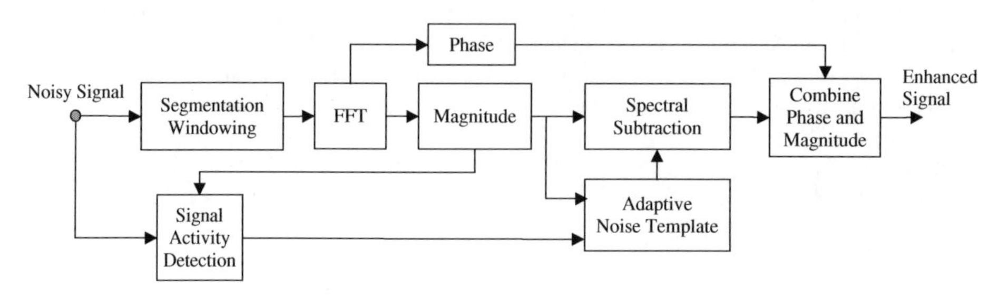

Figure 15.5 Block diagram illustration of an FFT-based spectral subtraction system for de-noising speech.

where the frequency response of the spectral subtraction filter $W_{SS}(k)$ is

$$W_{SS}(k) = fn\left(1 - \frac{\alpha(k)\hat{N}(k)}{Y(k)}\right) \tag{15.14}$$

where $\hat{N}(k)$ is an estimate of the noise average amplitude spectrum, $\alpha(k)$ is a frequency-dependent subtraction factor and the function $fn(\cdot)$ is chosen to avoid negative values of $W_{SS}(k)$ and provide a smoother frequency response when the signal-to-noise ratio drops to relatively lower values. The form of the function $fn(\cdot)$ can be chosen as

$$W_{SS}(k) = \begin{cases} 1 - \alpha(k)\hat{N}(k)/Y(k) & \text{if } \mathrm{SNR}(k) < \mathrm{SNR}_{\mathrm{Thresh}} \\ \gamma \exp\left(-\beta\left(\mathrm{SNR}_{\mathrm{Thresh}} - \mathrm{SNR}(k)\right)\right) & \text{else} \end{cases} \tag{15.15}$$

where SNR(k) is an estimate of the signal-to-noise ratio at the discrete frequency k and $\mathrm{SNR}_{\mathrm{Thresh}}$ is a threshold SNR below which spectral subtraction switches to a form of exponential attenuation, γ is a parameter that provides continuity at the switching point and β is an attenuation control factor.

The problem with spectral subtraction is that it often distorts the speech and results in the appearance of annoying short bursts of noise known as musical noise. The shortcomings of the spectral subtraction method can be summarised as follows:

(1) The only statistic used in spectral subtraction is the mean of the magnitude spectrum of the noise. The mean and variance of the clean speech and the variance of the noise are not employed in the estimation process. Consequently noise variations about the mean are not suppressed and this results in more distortions than would be the case if the variance information was also used.
(2) A hard decision needs to be employed to avoid the values of the estimates of the magnitude spectrum after subtraction going negative or below a noise-floor value.
(3) The spectral subtraction method is not speech-specific; the spectral trajectories of speech across time are not modelled and used in the de-noising process.

Spectral Subtraction Using LP Model Frequency Response

Spectral subtraction can be applied either on the short-time spectral amplitude (STSA) of noisy speech obtained from DFT or on the magnitude of the frequency response of a linear prediction (LP) model of noisy speech, as illustrated in Figure 15.6.

For LP-based spectral subtraction (LPSS), the filter response $W_{LPSS}(k)$ is obtained from equations similar to (15.14) and (15.15) with the main difference that instead of the DFT-based power-spectrum,

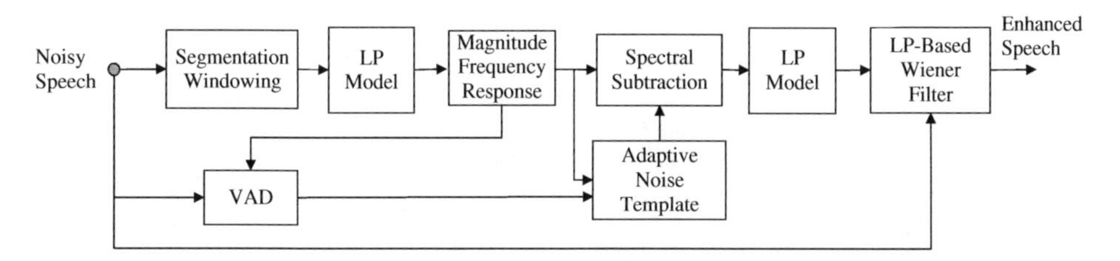

Figure 15.6 Block diagram illustration of LP-based spectral subtraction system for de-noising speech. Note that the LP model following the spectral subtraction module is intended to smooth the output of spectral subtraction.

the LP power-spectrum of noisy speech and the average LP power-spectrum of noise are used. LP-spectral subtraction involves the following steps:

(a) Obtain the coefficient vectors of the LP models of noisy speech and noise, and hence the magnitude frequency responses of the LP models of noisy speech $Y_{LP}(k)$ and noise $\hat{N}_{LP}(k)$.
(b) Using the magnitude frequency responses of the LP models of speech and noise, obtain the frequency response of the spectral subtraction filter $W_{LPSS}(k)$ from Equation (15.14).
(c) Restore the speech signal through application of the smoothed LP-spectral subtraction filter to the noisy speech.

15.2.4 Bayesian MMSE Speech Enhancement

The probabilistic minimum mean squared error estimation (MMSE) of the short-time spectral amplitude (STSA) of speech is a Bayesian estimation method with a mean squared error cost function. The Bayesian MMSE estimation of spectral amplitude is obtained as the posterior mean of the signal as

$$\hat{X}_k = \frac{\int_{-\infty}^{\infty}\int_{0}^{2\pi} X_k\, p\left(Y(k)|X_k, \theta_{X_k}\right) p\left(X_k, \theta_{X_k}\right) d\theta_{X_k}\, dX_k}{\int_{-\infty}^{\infty}\int_{0}^{2\pi} p\left(Y(k)|X_k, \theta_{X_k}\right) d\theta_{X_k}\, dX_k} \tag{15.16}$$

where $p\left(X(k)|Y_k, \theta_{Xk}\right) = p\left(Y(k)|X_k, \theta_{X_k}\right) p\left(X_k, \theta_{X_k}\right)/p\left(Y_k, \theta_{X_k}\right)$ is the posterior probability of clean speech $X(k)$ given noisy observation $Y(k)$. The MMSE equation (15.16) requires the likelihood of the noisy speech $p\left(Y(k)|X_k, \theta_{X_k}\right)$ and the prior probability density function of the clean speech $p\left(X_k, \theta_{X_k}\right)$.

Ephraim and Malah (1985) derived a MMSE spectral amplitude estimation algorithm known as the Ephraim–Malah suppression rule. They used a Gaussian distribution for clean speech (which results in a Rayleigh distribution for the magnitude spectrum of speech), a uniform distribution for the phase of the clean speech and a complex Gaussian distribution for noisy speech. The resulting estimator is of the form:

$$\hat{X}_k = W_{\mathrm{MMSE}}(k) Y_k \tag{15.17}$$

where the gain factor $W_{\mathrm{MMSE}}(k)$ is given by

$$W_{\mathrm{MMSE}}(k) = \Gamma(1.5)\frac{\sqrt{v_k}}{\gamma_k}\exp\left(-\frac{v_k}{2}\right)\left[(1+v_k)I_0\left(\frac{v_k}{2}\right) + v_k I_1\left(\frac{v_k}{2}\right)\right] \tag{15.18}$$

where $\Gamma(\cdot)$ is the gamma function, $I_n(\cdot)$ is Bessel function of order n and v_k and γ_k are defined as

$$v_k = \frac{\xi(k)}{1+\xi(k)}\gamma_k, \quad \xi_k = \frac{\sigma_X^2(k)}{\sigma_N^2(k)}, \quad \gamma_k = \frac{Y^2(k)}{\sigma_N^2(k)} \tag{15.19}$$

where $\sigma_X^2(k)$ and $\sigma_N^2(k)$ are the variances of speech and noise spectra, ξ_k is known as the *prior signal-to-noise ratio* and γ_k is known as the *posterior signal-to-noise ratio*.

Figure 15.7 shows a comparison of the performance of spectral subtraction based on linear prediction model of speech (LPSS) and the MMSE method for de-noising of speech observed in train noise (Yan and Vaseghi 2003).

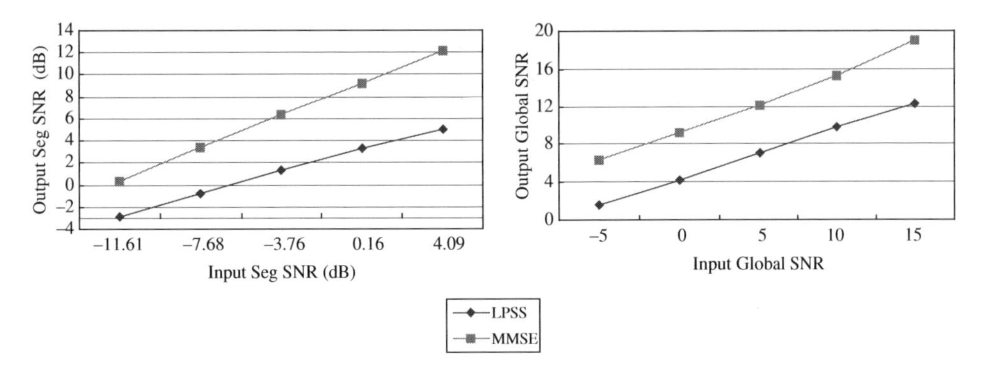

Figure 15.7 Performance comparison of LPSS and MMSE: output SNR vs input SNR.

15.2.5 Kalman Filter for Speech Enhancement

Kalman filter, described in Chapter 9, differs from Wiener filter in several ways, the most important of which is that the Kalman formulation permits explicit inclusion of time-varying equations that model the dynamics of speech and noise process. For de-noising speech, Kalman filter can be implemented in the time or frequency domain. The conventional method is time-domain Kalman filter.

Kalman State-Space Equations of Signal and Noise Models

The speech signal $x(m)$ and the noise $n(m)$ are modelled by autoregressive processes as

$$x(m) = \sum_{k=1}^{P} a_k x(m-k) + e(m) \tag{15.20}$$

$$n(m) = \sum_{k=1}^{Q} b_k n(m-k) + v(m) \tag{15.21}$$

where a_k and b_k are the coefficients of Pth order and Q^{th} order autoregressive models of speech and noise respectively. Equations (15.20) and (15.21) can be expressed in a state-space form for Kalman filtering as

$$\begin{cases} \boldsymbol{x}(m) = \boldsymbol{A}_x \boldsymbol{x}(m-1) + \boldsymbol{g}_x e(m) & (15.22) \\ x(m) = \boldsymbol{h}_x \boldsymbol{x}(m) & (15.23) \end{cases}$$

$$\begin{cases} \boldsymbol{n}(m) = \boldsymbol{A}_n \boldsymbol{n}(m-1) + \boldsymbol{g}_n v(m) & (15.24) \\ n(m) = \boldsymbol{h}_n \boldsymbol{n}(m) & (15.25) \end{cases}$$

where the AR model coefficient matrices of speech and noise are defined as

$$\boldsymbol{A}_x = \begin{bmatrix} a_1 & a_2 & \cdots & a_{P-1} & a_P \\ 1 & 0 & \cdots & 0 & 0 \\ 0 & 1 & \cdots & 0 & 0 \\ \vdots & \vdots & \ddots & \vdots & \vdots \\ 0 & 0 & 0 & 1 & 0 \end{bmatrix} \quad \boldsymbol{A}_n = \begin{bmatrix} b_1 & b_2 & \cdots & b_{Q-1} & b_Q \\ 1 & 0 & \cdots & 0 & 0 \\ 0 & 1 & \cdots & 0 & 0 \\ \vdots & \vdots & \ddots & \vdots & \vdots \\ 0 & 0 & 0 & 1 & 0 \end{bmatrix} \tag{15.26}$$

The signal and noise vectors are $\boldsymbol{x}(m) = [x(m), x(m-1), \ldots, x(m-P)]^{\mathrm{T}}, \boldsymbol{n}(m) = [n(m), n(m-1), \ldots, n(m-Q)]^{\mathrm{T}}$ and $\boldsymbol{h}_x = \boldsymbol{g}_x = [1, 0, \ldots, 0]^{\mathrm{T}}$ are P-dimensional vectors and $\boldsymbol{h}_n = \boldsymbol{g}_n = [1, 0, \ldots, 0]^{\mathrm{T}}$ are Q-dimensional vectors.

Process Equations

The signal and noise processes can be arranged into an augmented vector process equation as

$$\underbrace{\begin{bmatrix} \boldsymbol{x}(m) \\ \boldsymbol{n}(m) \end{bmatrix}}_{\substack{z(m) \\ (P+Q)\times 1}} = \underbrace{\begin{bmatrix} \boldsymbol{A}_x & \boldsymbol{0} \\ \boldsymbol{0} & \boldsymbol{A}_n \end{bmatrix}}_{\substack{A \\ (P+Q)\times(P+Q)}} \underbrace{\begin{bmatrix} \boldsymbol{x}(m-1) \\ \boldsymbol{n}(m-1) \end{bmatrix}}_{\substack{z(m-1) \\ (P+Q)\times 1}} + \underbrace{\begin{bmatrix} \boldsymbol{g}_x & \boldsymbol{0} \\ \boldsymbol{0} & \boldsymbol{g}_n \end{bmatrix}}_{\substack{G \\ (P+Q)\times 2}} \underbrace{\begin{bmatrix} e(m) \\ v(m) \end{bmatrix}}_{\substack{w(m) \\ 2\times 1}} \tag{15.27}$$

Equation (15.27) shows the dimensions of each vector and matrix. As indicated underneath the terms in Equation (15.27), in a compact notation the state process equation can be written as

$$z(m) = \boldsymbol{A}z(m-1) + \boldsymbol{G}\boldsymbol{w}(m) \tag{15.28}$$

Note that within the Kalman methodology an estimate of the state equation (15.28) provides a decomposition of the signal and noise processes.

Noisy Observation (Measurement) Equation

The noisy observation, i.e. the sum of speech and noise signals, may be expressed as

$$y(m) = \boldsymbol{h}^T z(m) = x(m) + n(m) \tag{15.29}$$

where

$$\boldsymbol{h} = \begin{bmatrix} \boldsymbol{h}_x \\ \boldsymbol{h}_n \end{bmatrix} \tag{15.30}$$

and $z(m) = [\boldsymbol{x}(m)\boldsymbol{n}(m)]^{\mathrm{T}}$ as defined in Equation (15.27). Note that in this formulation of the Kalman filter, the signal and noise form the state process vector in Equation (15.27) and the noisy observation, given by Equation (15.29), is a linear transformation of the state vector that simply adds signal and noise. The Kalman filter equations given in Chapter 9 and adapted here are as follows.

State vector prediction equation

$$\hat{z}(m|m-1) = \boldsymbol{A}(m-1)\hat{z}(m-1) \tag{15.31}$$

Covariance matrix of prediction error

$$\boldsymbol{P}(m|m-1) = \boldsymbol{A}(m-1)\boldsymbol{P}(m-1)\boldsymbol{A}^T(m-1) + \boldsymbol{G}\boldsymbol{W}(m-1)\boldsymbol{G} \tag{15.32}$$

$\boldsymbol{W} = \text{diag}[\sigma_e^2, \sigma_v^2]$ is the covariance matrix of the vector $\boldsymbol{w}$ in Equations (15.27)–(15.28).

Kalman gain vector

$$\boldsymbol{K}(m) = \boldsymbol{P}(m|m-1)\boldsymbol{h}\left(\boldsymbol{h}^{\mathrm{T}}\boldsymbol{P}(m|m-1)\boldsymbol{h}\right)^{-1} \tag{15.33}$$

State update estimate equation

$$\hat{z}(m) = \hat{z}(m|m-1) + \boldsymbol{K}(m)\left(y(m) - \boldsymbol{h}^T\hat{z}(m|m-1)\right) \tag{15.34}$$

Note that the innovation signal $y(m) - \boldsymbol{h}^{\mathrm{T}}\hat{z}(m|m-1)$ is a mixture of the unpredictable parts of the signal and the noise. The covariance matrix of estimation error is given by

$$\boldsymbol{P}(m) = \left(\boldsymbol{I} - \boldsymbol{K}\boldsymbol{h}^T\right)\boldsymbol{P}(m|m-1) \tag{15.35}$$

The application of Kalman filter requires estimates of the AR models of speech and noise $\boldsymbol{a} = [a_1, a_2, \ldots, a_P]$ and $\boldsymbol{b} = [b_1, b_2, \ldots, b_Q]$, Figure 15.8. These are obtained from an application of the estimate–maximize (EM) algorithm, described in Chapter 7, which effectively yields the following normal equation (Yule–Walker equation) as

$$\hat{\boldsymbol{a}} = \boldsymbol{R}_{\hat{x}\hat{x}}^{-1}\boldsymbol{r}_{\hat{x}\hat{x}} \tag{15.36}$$

$$\hat{\boldsymbol{b}} = \boldsymbol{R}_{\hat{n}\hat{n}}^{-1}\boldsymbol{r}_{\hat{n}\hat{n}} \tag{15.37}$$

where the autocorrelation matrices of speech and noise, $\boldsymbol{R}_{\hat{x}\hat{x}}$ and $\boldsymbol{R}_{\hat{n}\hat{n}}$, are obtained from Kalman estimates as

$$\boldsymbol{R}_{\hat{x}\hat{x}} = E\left([\hat{x}(m), \ldots, \hat{x}(m-P)]^T[\hat{x}(m), \ldots, \hat{x}(m-P)]\right) \tag{15.38}$$

$$\boldsymbol{R}_{\hat{n}\hat{n}} = E\left([\hat{n}(m), \ldots, \hat{n}(m-Q)]^T[\hat{n}(m), \ldots, \hat{n}(m-Q)]\right) \tag{15.39}$$

Figure 15.8 is an outline illustration of the Kalman filter method of enhancement of noisy speech.

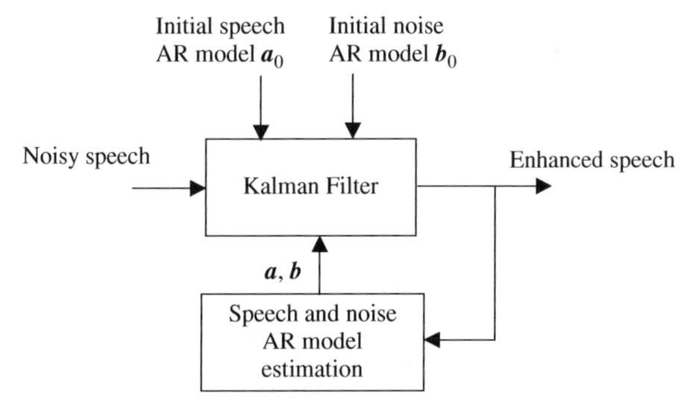

Figure 15.8 Illustration of Kalman filter method of enhancement of noisy speech.

15.2.6 Speech Enhancement Using LP-HNM Model

Linear prediction (LP) and harmonic noise models (HNM), described in Chapter 14, are the two main methods for modelling and synthesis of speech waveforms. LP and HNM offer complementary advantages; an LP model provides a good fit for the spectral envelope of speech whereas an HNM is good at modelling the details of the harmonic plus noise structures of speech. These two methods can be combined for speech enhancement.

The temporal trajectory of the estimates of the parameters of LP and HNM models can be tracked with Viterbi classifiers, described in Chapter 11, and smoothed with Kalman filters. For noisy speech processing this is a different approach to spectral amplitude estimation methods which generally model each individual spectral sample in isolation without fully utilising the information on the wider spectral–temporal structures that may be used to good effect in the de-noising process to obtain improved results.

In this subsection a speech enhancement method is described based on a formant-tracking LP (FTLP) model of the spectral envelope and an HNM model of the excitation. The FTLP model obtains enhanced estimates of the LP parameters of speech along the formant trajectories. Formants are the resonances of the vocal tract and their trajectories describe the contours of energy concentrations in time and frequency. Although formants are mainly defined for voiced speech, characteristic energy concentration contours also exist for unvoiced speech at relatively higher frequencies.

Overview of LP-HNM Enhancement System

The LP-HNM speech enhancement method is illustrated in Figure 15.9 and consists of the following sections:

(1) A pre-cleaning module, composed of a spectral amplitude estimator, for de-noising speech spectrum prior to estimation of the LP model and formant parameters.
(2) A formant-tracking method incorporating Viterbi decoders and Kalman filters for tracking and smoothing the temporal trajectories of the estimates of the formants and poles of the LP model.
(3) A pitch extraction method incorporating Viterbi decoders and Kalman filters for pitch smoothing.
(4) A method for estimation of a harmonic noise model of clean excitation with Kalman filters used for modelling and de-noising the temporal trajectory of the noisy excitation.

The LP model of speech $X(z, m)$ may be expressed as

$$X(z, m) = E(z, m)\, V(z, m) \tag{15.40}$$

where $E(z, m)$ is the z-transform of the excitation signal and $V(z, m)$, the z-transform of an LP model of the spectral envelope of speech, can be expressed as

$$V(z, m) = G(m)\frac{1}{1 + r_0(m)z^{-1}} \prod_{k=1}^{P/2} \frac{1}{1 - 2r_k(m)\cos\left(\varphi_k(m)\right) z^{-1} + r_k^2(m)z^{-2}} \tag{15.41}$$

where $r_k(m)$ and $\varphi_k(m)$ are the time-varying radii and the angular frequencies of the poles of the LP model respectively, $P+1$ is the LP model order and $G(m)$ is the gain.

The speech excitation can be modelled as a combination of the harmonic and the noise contents of the excitation as

$$E(f, m) = \sum_{k=1}^{L(m)} A_k(m) G\left(f - kF_0(m) + \Delta_k(f, m)\right) + V(f, m) \tag{15.42}$$

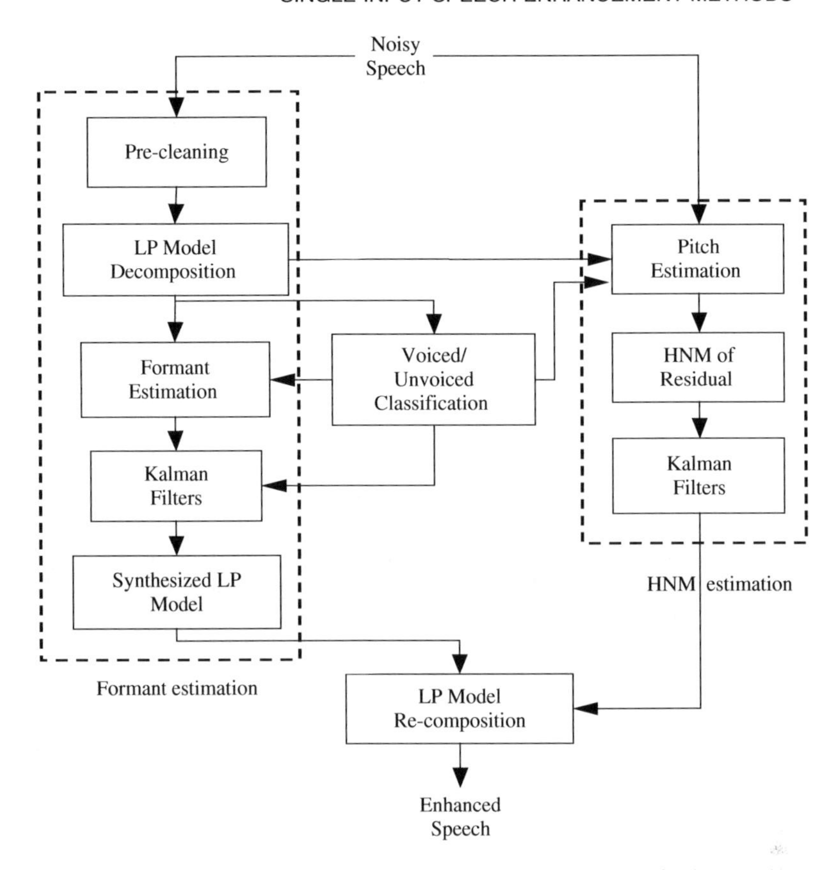

Figure 15.9 Overview of the FTLP-HNM model for enhancement of noisy speech.

where f is the frequency variable, $L(m)$ denotes the number of harmonics, $F_0(m)$ is the time-varying fundamental frequency, $\Delta_k(f, m)$ is the deviation of the kth harmonic from the nominal value of kF_0, $A_k(m)$ are the complex amplitudes of excitation harmonics, $G(f)$ is a Gaussian-shaped model of harmonic shape, and $V(f, m)$ is the noise part of the excitation. The harmonic shape function $G(f)$ has a frequency support equal to F_0 and is selected to model the shape of the harmonics of speech in the frequency domain.

Formant Estimation from Noisy Speech

In this section a robust formant-tracking LP model is described composed of pre-cleaning of noisy speech spectrum followed by formant track estimation and Kalman smoothing of formant tracks.

Initial-Cleaning of Noisy Speech

Before formant estimation, the noisy speech spectrum is pre-cleaned using the MMSE spectral amplitude estimation method described in Section 15.2.4. After pre-cleaning, the spectral amplitude of speech is converted to a correlation function from which an initial estimate of the LP model of speech is obtained using the Levinson–Durbin method. A formant tracker is then used to process the poles, or equivalently the line spectral frequency (LSF) parameters, of the LP model and obtain an improved estimate of the LP model parameters as described next.

Formant Tracking

The poles of the LP model of the pre-cleaned speech are the formant candidates represented by formant feature vectors ν_k comprising the frequency F_k, bandwidth B_k and magnitude M_k of the resonance at formants together with their velocity derivatives as

$$\nu_k = [F_k, B_k, M_k, \Delta F_k, \Delta B_k, \Delta M_k] \; k = 1, \ldots, N \tag{15.43}$$

where the number of formants is typically set to $N = 5$. Velocity derivatives, denoted by Δ, are computed as the slopes of the formant features over time. The probability distributions of formants can be modelled by Gaussian mixture models (GMMs) or HMMs, as described in Chapter 11. A Viterbi classifier is used to classify and track the poles of the LP model and label the poles with different formants. Kalman filters are subsequently employed to smooth formant trajectories. Note that instead of formants one can employ the line spectral frequencies.

For assessment of the performance of the formant tracker, the Wall Street Journal speech database is used to investigate the effect of noise on the estimates of formants (Yan and Vaseghi 2003). The speech examples are degraded by moving car noise or moving train noise with an average SNR in the range 0 to 20 dB. To quantify the contamination of formants by noise a local formant signal-to-noise ratio measure (FSNR) is defined as

$$\text{FSNR}(k) = 10\log\left[\sum_{l\in(F_k\pm B_k/2)} X_l^2 \Bigg/ \sum_{l\in(F_k\pm B_k/2)} N_l^2\right] \tag{15.44}$$

where X_l and N_l are the magnitude spectra of speech and noise respectively and F_k and B_k are the frequency and bandwidth of the resonance at the kth formant. Figure 15.10 displays the FSNRs of noisy speech in moving car and moving train environments. It is evident that the signal-to-noise ratio at formants, the FSNRs, are higher than the average SNR.

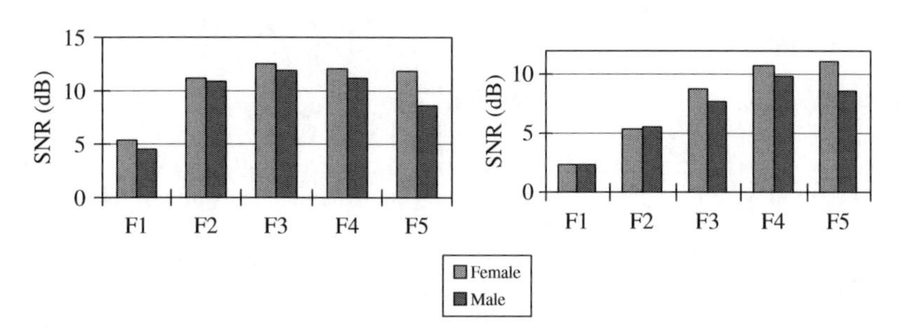

Figure 15.10 Variation of speech SNR at different formants in car noise (left) train noise (right) at average SNR = 0 dB.

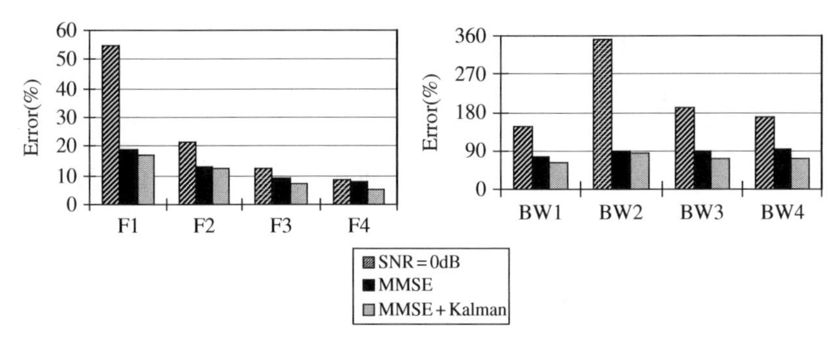

Figure 15.11 Average % error of formant tracks (frequency F_k and bandwidth BW_k) in train noise and cleaned speech using MMSE and Kalman filters; the results were averaged over five males.

To quantify the effects of the noise on formant estimation, an average formant track error measure is defined as

$$E_k = \frac{1}{L}\sum_{m=1}^{L}\left[\left|F_k(m) - \hat{F}_k(m)\right|/F_k(m)\right] \times 100\% \quad k = 1, \ldots, N \tag{15.45}$$

where $F_k(m)$ and $\hat{F}_k(m)$ are the formant tracks of clean and noisy speech respectively, m is frame index and L is the number of speech frames over which the error is measured.

Figure 15.11 shows the improvement in formant estimation resulting from pre-cleaning of speech spectra followed by LP analysis and the subsequent enhancement of the poles of the LP model using a Viterbi classifier and Kalman filters. The reference formant tracks, used for calculation of the estimation errors, are obtained from HMMs of formants of clean speech (Yan and Vaseghi 2003). It can be seen that the application of MMSE noise suppression results in a significant reduction of formant tracking error. Further improvement is obtained through the application of Kalman filtering.

Harmonic Plus Noise Model (HNM) of Speech Excitation

The HNM of speech introduced in Chapter 14 can be applied for modelling of noisy excitation. The procedure includes the following steps:

(a) Voiced/unvoiced classification of speech.
(b) Estimation and smoothing of the trajectories of the fundamental frequency and harmonic tracks.
(c) Estimation and smoothing of the trajectories of the amplitudes of harmonics.
(d) Estimation of the harmonicity (i.e. proportion of harmonic energy to fricative or aspiration noise) values.
(e) Estimation of the noise component of the excitation.

The estimation of HNM parameters is discussed next.

Fundamental Frequency Estimation

Traditionally pitch is derived as the inverse of the time interval τ corresponding to the second largest peak of the autocorrelation of speech. Since autocorrelation of a periodic signal is itself periodic, all

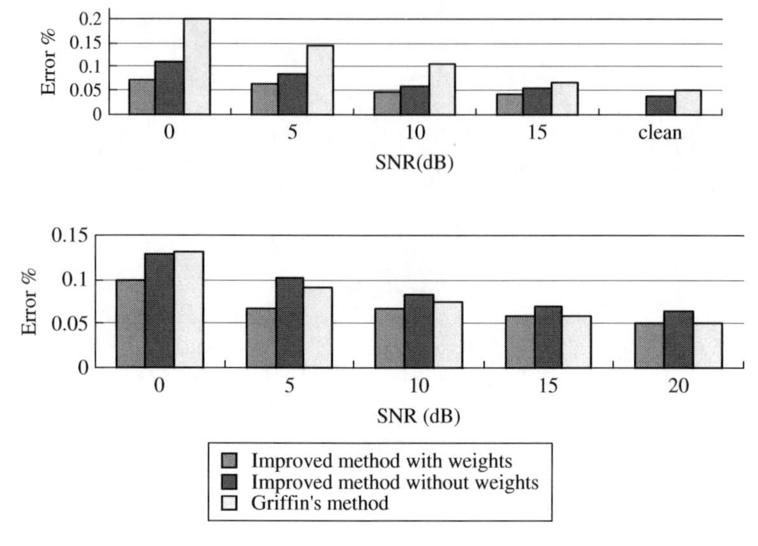

Figure 15.12 Comparison of different pitch track methods for speech in train noise (top) and car noise (bottom) from 0 dB SNR to clean.

the periodic peaks of the autocorrelation can be used in the pitch estimation process. The proposed pitch estimation method is an extension of the autocorrelation-based method to the frequency domain. A pitch estimation error is defined as

$$E(F_0) = E - F_0 \sum_{k=1}^{\text{Max}F} \sum_{l=kF_0-M}^{kF_0+M} W(l) \log |X(l)| \tag{15.46}$$

where $X(l)$ is the DFT of speech, F_0 is a proposed value of the fundamental frequency (pitch) variable, E is sum of log spectral energy, and $2M+1$ is a band of values about each harmonic frequency. The weighting function $W(l)$ is a SNR-dependent Wiener-type weight. Figure 15.12 provides a comparative illustration of the performance of the proposed pitch estimation method with Griffin's method (Griffin and Lim 1988), at different SNRs for car noise and train noise. It can be seen that the proposed frequency method with SNR weighting provides improved performances in all cases evaluated.

Estimation of Harmonic Amplitudes

The harmonic content of speech excitation is modelled as

$$e_h(m) = \sum_{k=1}^{L(m)} A_k(m) \cos(2\pi(kF_0(m) + \Delta_k(m)m' + \varphi_k(m)) = \boldsymbol{A}^T\boldsymbol{S} \tag{15.47}$$

where $L(m)$ denotes the number of harmonics and $F_0(m)$ is the pitch, $\boldsymbol{A}$ and $\boldsymbol{S}$ are the harmonic amplitude vector and the harmonically related sinusoids vector respectively. Given the harmonics frequencies, the amplitudes $\boldsymbol{A}$ can be obtained either from searching for the peaks of the speech

DFT spectrum or through a least square error estimation. The maximum significant harmonic number is obtained from the ability of the harmonic model to synthesise speech locally at the higher harmonics of the pitch. Note that the deviation of the harmonic frequencies, $\Delta_k(m)$, from the nominal value can be obtained from a search for the peak amplitudes about the nominal harmonic frequencies.

The estimate of the amplitudes of clean excitation harmonics is obtained from a set of Kalman filters, one for each harmonic. The Kalman filter is the preferred method here as it models the trajectory of the successive samples of each harmonic.

Estimation of Noise Component of HNM

For unvoiced speech the excitation is a noise-like signal across the entire speech bandwidth. For voiced speech the excitation is a mostly noise-like signal above some variable maximum harmonic frequency.

The main effect of the background noises on the estimate of the excitation of the LP model is an increase in its variance. In fact, perceptually good results can be obtained by replacing the noise part of the excitation to the LP model with a Gaussian noise with the appropriate variance estimated as the difference between the variance of the noisy signal and that of the noise. Finally the time-varying spectrum of synthetic HNM of excitation signal is obtained as

$$\hat{E}(f, m) = E_{\mathrm{H}}(f, m) + E_{\mathrm{V}}(f, m) \tag{15.48}$$

where $E_{\mathrm{H}}(f, m)$ and $E_{\mathrm{V}}(f, m)$ are the harmonic and noise parts of the excitation respectively.

Kalman Smoothing of Trajectories of Formants and Harmonics

The Kalman filter equations for de-noising or smoothing of all the parameters of speech are essentially the same; for this reason we describe the Kalman smoothing of formant tracks. The formant trajectory is modelled by an AR process as

$$\hat{F}_k(m) = \sum_{i=1}^{P} c_{ki} \hat{F}_k(m-i) + e_k(m) \tag{15.49}$$

where c_{ki} are the coefficients of a low-order (3 to 5) AR model of the kth formant track and $e_k(m) = N(0, Q_k)$ is a zero mean Gaussian random process. The variance of $e_k(m)$, Q_k is estimated from the previous estimates of e_k. The algorithm for Kalman filter adapted for formant track estimation is as follows.

Time updates (prediction) equations

$$\hat{\boldsymbol{F}}_k(m|m-1) = \boldsymbol{C}\hat{\boldsymbol{F}}_k(m-1) \tag{15.50}$$

$$\boldsymbol{P}(m|m-1) = \boldsymbol{P}(m-1) + \boldsymbol{Q} \tag{15.51}$$

Measurement updates (estimation) equations

$$\boldsymbol{K}(m) = \boldsymbol{P}(m|m-1)\left(\boldsymbol{P}(m|m-1) + \boldsymbol{R}\right)^{-1} \tag{15.52}$$

$$\hat{\boldsymbol{F}}_k(m) = \hat{\boldsymbol{F}}_k(m|m-1) + \boldsymbol{K}(m)\left(\boldsymbol{p}_k(m) - \hat{\boldsymbol{F}}_k(m|m-1)\right) \tag{15.53}$$

$$\boldsymbol{P}(m) = (\boldsymbol{I} - \boldsymbol{K}(m))\boldsymbol{P}(m|m-1) \tag{15.54}$$

where $\hat{F}_k(m|m-1)$ denotes a prediction of $F_k(m)$ from estimates of the formant track up to time $m-1$, $\boldsymbol{P}(m)$ is the formant estimation error covariance matrix, $\boldsymbol{P}(m|m-1)$ is the formant prediction error covariance matrix, $\boldsymbol{K}(m)$ is the Kalman filter gain, $\boldsymbol{R}$ is the measurement noise covariance matrix, estimated from the variance of the differences between the noisy formant observation and estimated tracks. The covariance matrix $\boldsymbol{Q}$ of the process noise is obtained from the prediction error of formant tracks.

Kalman theory assumes the signal and noise can be described by linear systems with random Gaussian excitation. Kalman filter is unable to deal with relatively sharp changes in the signal process, for example when speech changes from a voiced to a non-voiced segment. However, state-dependent Kalman filters can be used to solve this problem. For example a two-state voiced/unvoiced classification of speech can be used to employ two separate sets of Kalman filters: one set of Kalman filters for voiced speech and another set for unvoiced speech. In HMM-based speech models in each state of HMM the signal trajectory can be modelled as a Kalman filter.

Figure 15.13–15.15 show an evaluation of the performance of LP-HNM speech enhancement method. The databases used for the evaluation of the performance of the speech enhancement systems are a subset of five male speakers and five female speakers from *Wall Street Journal* (WSJ). For each speaker, there are over 120 sentences. The speech signal is down-sampled to 10 kHz from an original sampling rate of 16 kHz. The speech signal is segmented into overlapping frames of length 250 samples (25 ms) with an overlap of 150 samples (15 ms) between successive frames. Each speech frame is windowed with a Hann window.

Figure 15.13 shows the significant improvement in the harmonicity measure resulting from a FTLP-HNM model. Figure 15.14 illustrates the results of perceptual evaluation of speech quality (PESQ) of noisy speech and speech restored with MMSE and FTLP-HNM methods. It is evident that the LP-HNM method achieves improved results. Figure 15.15 shows the improvement in ISD measure compared with the MMSE system. It is evident that the new speech processing system achieves a better ISD score.

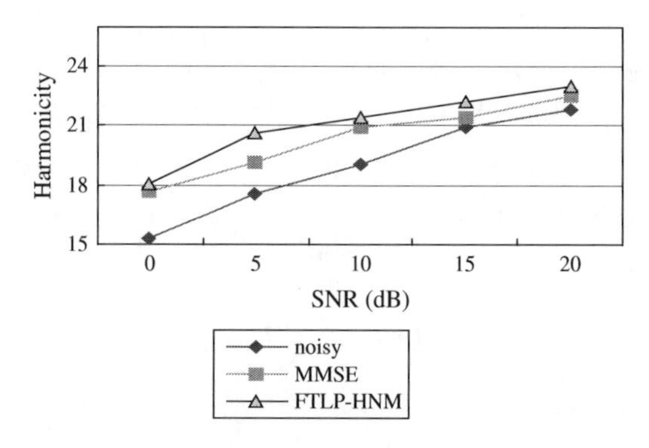

Figure 15.13 Comparison of harmonicity of MMSE and FTLP-HNM systems on train noisy speech at different SNRs.

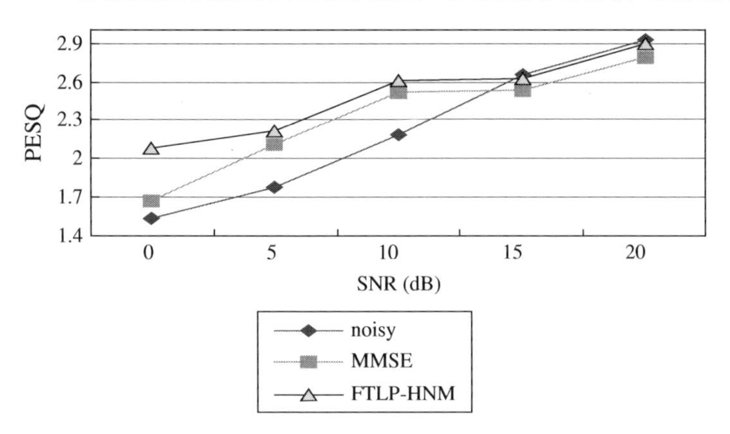

Figure 15.14 Performance of MMSE and FTLP-HNM on train noisy speech at different SNRs.

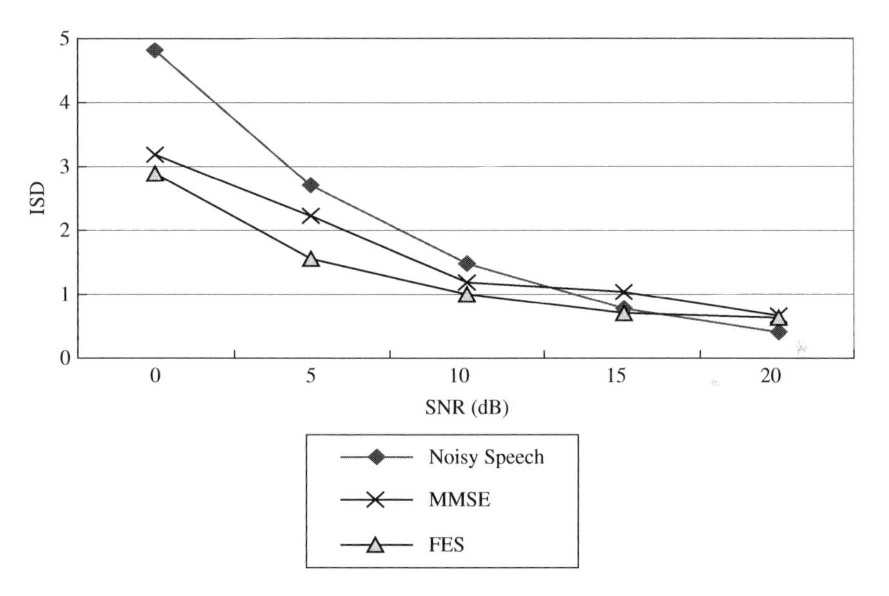

Figure 15.15 Comparison of ISD of noisy speech in train noise pre-cleaned with MMSE and improved with formant-based enhancement system (FES) at SNR = 0, 5, 10, 15 dB.

15.3 Speech Bandwidth Extension – Spectral Extrapolation

Conventional telephony speech is limited to a narrow bandwidth of less than 4 KHz; normally 300–3400 Hz. In contrast, wideband broadcast quality speech can be as wide as 20–20,000 Hz. Although most of the energy and the information content of speech are contained within the telephony bandwidth of 300–3400 Hz, there exists significant energy in other frequencies that conveys quality and sensation. In particular the upper frequency bands above 3400 Hz convey significant information regarding the identity of the consonants and helps to differentiate for example an '*s*' from an '*f*'.

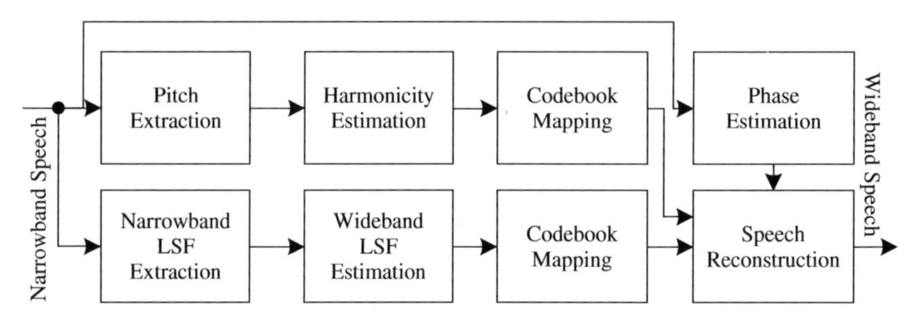

Figure 15.16 Block diagram of a bandwidth extension system based on LP-HNM model of speech. The parameters of the LP-HNM for missing speech are extracted from codebooks trained on joint feature vectors of narrowband and wideband speech.

The aim of bandwidth extension methods is to reconstruct the missing upper (and lower) band spectrum of speech signals using an extrapolation of the available spectrum in order to regain the sensation of wide bandwidth and higher quality speech. Two major remaining research challenges in development of bandwidth extension systems are (1) the recovery of information for the class of speech signals where valuable information resides in the (missing) upper bands rather than in the available lower bands (e.g. fricatives) and (2) language dependency of current systems i.e. the lack of language portability.

This section describes an example of a bandwidth extension system developed by Zavarehei and Vaseghi. Most bandwidth extension techniques, such as that shown in Figure 15.16, use codebook mapping methods for extraction of the missing bands from the available information. The parameters of the missing spectral envelope and excitation in the higher bands are obtained from codebooks trained on joint feature vectors of narrowband and wideband speech. The spectral envelope representation for codebook mapping is often based on the line spectral frequency (LSF) parameters derived from a linear prediction model of speech (see Chapter 14).

An effective method for extrapolation of excitation is based on a harmonic plus noise (HNM) model of speech excitation, described in Section 14.5. For voiced speech the fundamental frequency and harmonics of excitation are tracked and extrapolated. An essential aspect of the reproduction of harmonics is the idea of the extrapolation of the ‘quality’ of harmonics of speech. For this purpose a measure of harmonicity of speech excitation is defined in Section 14.5.1. Harmonicity of each harmonic is a measure of the harmonic to fricative (aspiration noise) energy ratio at and around that harmonic. This measure is used to characterise both the harmonic structure of voiced speech and the non-harmonic structure of unvoiced speech. This approach has an added advantage as it circumvents the need for hard decisions regarding classification of speech into voiced/unvoiced parts. Figure 15.17 shows an example of a speech segment synthesised using a harmonic plus noise model where the proportion of the harmonic and noise components at each harmonic frequency is obtained from the harmonicity of speech.

The estimates of the spectral envelope are then combined with an estimate of the excitation signal and converted to time domain to yield the wideband output speech signal.

15.3.1 LP-HNM Model of Speech

The speech model used here for bandwidth extension, shown in Figure 15.16, is a linear prediction (LP) model of the spectral envelope and a harmonic noise model (HNM) of speech excitation, described in detail in Sections 14.4–14.6. The parameters of the LP-HNM model are the fundamental frequency

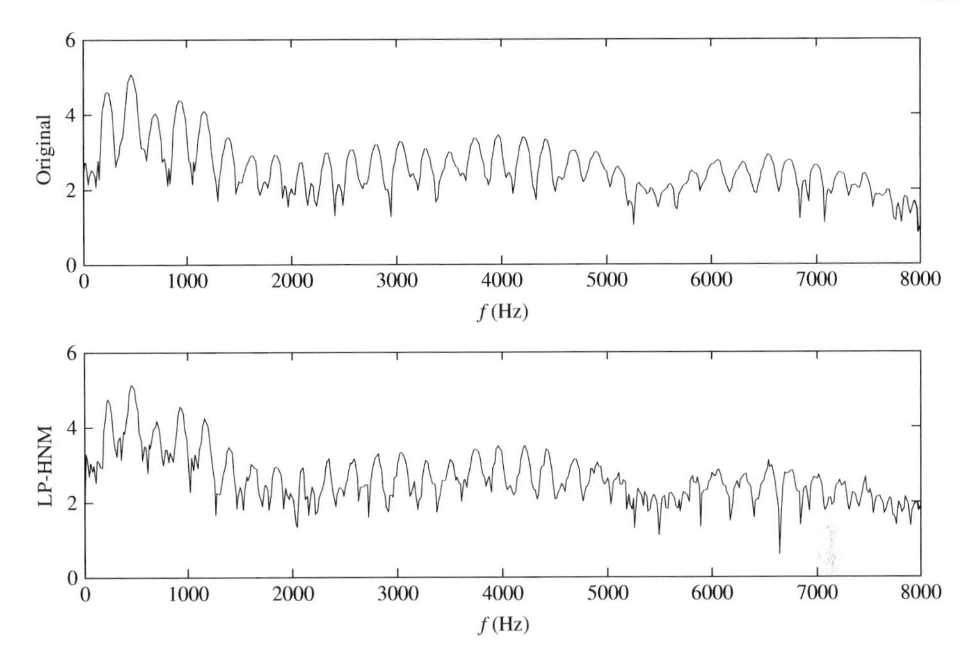

Figure 15.17 Harmonicity-based reconstruction of speech segment: original (top) and LP-HNM reconstructed (bottom).

F_0, the adjusted harmonic frequencies F_k, the harmonic amplitudes A_k, the harmonicity values V_k, the LSF vector representation of the LP model coefficients $\boldsymbol{Q}$ and the LP model gain C.

The use of the degree of harmonicity (i.e. the ratio of the harmonic to aspiration noise-like energy around each harmonic) for this application, results in a soft decision during codebook mapping rather than a hard decision. Moreover, during temporal fluctuations of the spectrogram, the harmonics gradually gain or lose strength which is modelled using the harmonicity degree.

During the reconstruction of the high-frequency part of the spectrum, the excitation amplitudes in each sub-band, at around the harmonics $E(kf_0)$, is assumed to be 1 which is reasonable assuming that the order of the LP model is high enough to have whitened the excitation. The gain of the frame is modelled using the LP gain value. The excitation reconstructed from this model is multiplied by the LP spectrum and the phase component to form the speech spectrum of each frame.

For the excitation codebook to have a fixed size, the harmonicity of a maximum of 32 bands is calculated for each frame. This is a design parameter that may vary in different systems. Sub-bands above this range are assumed to be entirely unvoiced. Experiments indicate that the efficiency of the system seems to saturate for a larger number of harmonics. A codebook of harmonicity feature vectors is trained on wideband speech signals from a training database such as the Wall Street Journal (WSJ) speech database using the K-means algorithm. The distortion measure is the Euclidean distance measure which is applied only to the lower 4 kHz of the signal. The upper 4 kHz harmonicity values act as a shadow codebook for retrieving the harmonicity of the upper sub-bands.

The initial fundamental frequency of excitation F_0 can be obtained using the method described in Section 14.6, which may be refined by pick-peaking and finding the harmonic frequencies and then adjusting the fundamental frequency according to harmonic frequencies. Furthermore, at each stage a number of pitch candidates corresponding to several peak values (local peaks and 'global' peaks) of the correlation or power-spectrum curves are selected and the Viterbi algorithm is used to obtain the best candidate and track the fundamental frequency.

15.3.2 LP Spectral Envelope Extrapolation

This section describes a method for extrapolation of the spectral envelope using codebooks trained on narrowband and wideband speech. For extrapolation of the spectral envelope of speech, the LP model coefficients are usually converted to line spectral frequency (LSF) coefficients as these have good quantisation and interpolation/extrapolation qualities. A 12th order speaker-independent LSF codebook is trained based on narrowband speech signals derived from a database such as the Wall Street Journal (WSJ) speech database down-sampled to 8 kHz sampling rate (4 kHz bandwidth). A shadow codebook, of 24th order LSF, for wideband speech sampled at 16 kHz rate is trained in conjunction with the 8 kHz sampled codebook. (Note that the bandwidth of the wideband signal may be chosen higher depending on the application.) The distortion measure used both in the training stage and in the estimation stage is the mean square spectral distortion (error) of the corresponding LP spectra:

$$D(X_{\mathrm{LSF}}, Y_{\mathrm{LSF}}) = \int_{-0.5}^{0.5} |X_{\mathrm{LP}}(f) - Y_{\mathrm{LP}}(f)|^2 \, df \tag{15.55}$$

where X_{LSF} and Y_{LSF} are the set of narrowband LSF vectors and $X_{\mathrm{LP}}(f)$ and $Y_{\mathrm{LP}}(f)$ are the corresponding LP spectra.

Estimation of the gain of the LP model, G, is crucial in bandwidth extension. Two different approaches for gain estimation were implemented: (i) estimation of the gain using energy normalisation and (ii) codebook mapping.

In the first approach the gain of the wideband LP model is calculated to result in the same amount of energy in the low-band portion as that of the narrowband speech signal. While this approach works well in the voiced segments, experiments show that it does not result in good estimates for the frames where most of the energy is concentrated in the high frequency (e.g. fricatives).

In the second approach, the ratio of the gain of the narrowband LP model of each frame, $G_{\mathrm{narrowband}}$, divided by that of the wideband signal for the same frame, G_{wideband}, is calculated:

$$R_{\mathrm{G}} = G_{\mathrm{narrowband}} / G_{\mathrm{wideband}} \tag{15.56}$$

The gain ratio R_{G} is calculated for each frame and a shadow codebook is trained in conjunction with the LSF codebook, similar to the shadow codebook for the wideband LSF.

During the estimation stage, the narrowband LSF values are calculated for each frame and the closest (measured in terms of mean squared error spectral distortion) codeword is chosen from the narrowband LSF codebook. The corresponding wideband codeword and gain ratio are obtained from the shadow codebooks. The narrowband LP model gain is divided by the gain ratio and the result is used as the wideband LP model gain.

The LSF, gain and harmonicity codebooks are trained using speech signals obtained from the WSJ speech database, spoken by several speakers. The WSJ speech is originally sampled at 16 kHz and has a bandwidth of 8 kHz. Speech is segmented to frames of 25 ms duration with a frame overlap of 15 ms i.e. a frame rate of 10 ms. To produce narrowband speech, wideband speech is filtered to a bandwidth of 4 kHz and down-sampled to a sampling rate of 8 kHz.

For the purpose of system evaluation, 40 test sentences are randomly chosen from the WSJ speech database, the test sentences were not among those used for training the codebooks. The mean log spectral distance (LSD) between the wideband and bandwidth-extended signals are calculated and averaged over all frames and sentences for each modification. The average LSD is calculated using the fast Fourier transform (FFT) of the frames (FFT-LSD) and LP spectrum of the frames (LP-LSD).

Phase Estimation

There are several different methods for coding and estimation/prediction of the phase of the missing spectra such as phase codebooks and phase predictors. As most of the signal in the higher bands of the wideband speech is not harmonically structured, the issue of phase estimation is not extensively explored in the literature of bandwidth extension. In this example work the phase of the upper band is estimated from the lower band so that the unwrapped phase of each frame is linear. Some random phase, proportional to the inverse of the harmonicity of each sub-band is then added to the phase to account for the non-harmonic random phase. In experimental evaluations, absolutely no perceptible difference is audible when the upper band phase of a wideband signal is replaced by its predicted value.

Codebook Mapping of the Gain

The estimation of the LP gain of the wideband signal is crucial in bandwidth extension of narrowband signals. The use of the codebook mapping method for estimation of the LP gain is compared with the energy normalisation method. It is observed that using energy normalisation results in suppression of the signals, especially during fricative segments where most of the energy of the signal is concentrated in the higher bands of the signal. Estimation of the LP gain value through codebook mapping results in superior quality of the wideband signals. A comparison of the LSD values of these two methods is presented in Table 15.1 which shows that codebook mapping results in lower averaged LSD distances in every case.

15.3.3 LP Spectral Excitation Extrapolation

The use of the harmonicity model for reconstruction of the excitation of the speech signal in missing frequency bands is compared with the alternative method of using a band-pass envelope modulated Gaussian noise (BP-MGN). While the rest of the system is similar for evaluations carried out here,

Table 15.1 Average log spectral distance of the wideband signal for energy normalization and codebook mapping of the gain factor.

Distortion measure (dB)	Energy normalisation	Codebook mapping
Overall FFT-LSD	5.83	5.10
High band FFT-LSD	8.25	7.17
Overall LP-LSD	4.96	4.27
High band LP-LSD	6.83	6.09

Table 15.2 Average log spectral distance of the wideband signal for the band-pass envelope modulated Gaussian noise (BP-MGN) model and the harmonicity model.

Distortion measure (dB)	BP-MGN model	Harmonicity model
Overall FFT-LSD	5.85	5.10
High band FFT-LSD	8.79	7.17
Overall LP-LSD	4.51	4.27
High band LP-LSD	6.63	6.09

only the excitation is estimated using the two different methods. It is observed that both methods result in reasonably good quality output speech. However, from the study of the spectrograms of the wideband signals produced by these systems it was observed that the extended bands which had more harmonically structured patterns were better reconstructed using the harmonicity model. Table 15.2 summarises the averaged LSD values calculated for these two cases. While the difference linear prediction based LSDs are large, those of the FFT-LSD seem to be higher. This is due to the more detailed modelling of the harmonics of the higher bands in the proposed method.

Sensitivity to Pitch

To accurately estimate the harmonicity levels, it is crucial that reasonably accurate estimates of the fundamental frequency are extracted from the narrowband speech signal. Inaccurate fundamental frequency estimates normally do not happen in harmonically well-structured speech frames, which are also more likely to have higher frequency harmonics. It is observed that inaccurate fundamental frequencies during harmonically structured (voiced) frames results in inaccurate extrapolation of harmonicity. However, this is not the case during unvoiced frames. During unvoiced frames the excitation signal is normally reconstructed using only noise as, even if a fundamental frequency is assigned to it, the harmonicity values calculated for sub-bands will be very low which will result in domination of the noise component.

Figure 15.18 shows an example of the original wideband speech with a bandwidth of 8 kHz, the filtered speech with a bandwidth of 4 kHz and the restored extended bandwidth speech. LP gain and harmonicity values are estimated using the proposed method of codebook mapping. The figure clearly shows both the harmonic and noise structures in the upper band have been well restored.

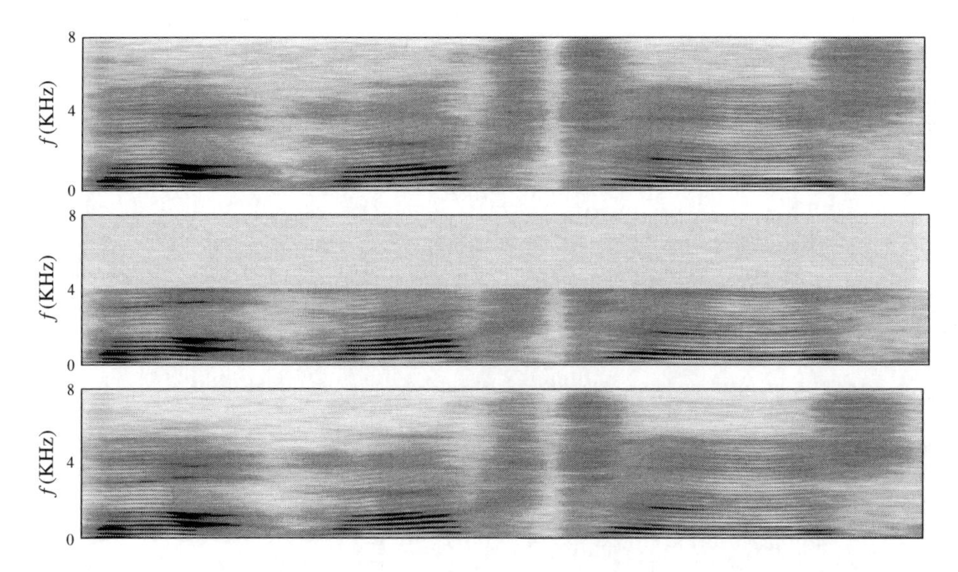

Figure 15.18 Spectrograms of (top) wideband signal, (middle) narrowband signal and (bottom) bandwidth extended signal.

15.4 Interpolation of Lost Speech Segments – Packet Loss Concealment

Interpolation of lost speech segments is an essential requirement in a number of speech applications for the reconstruction of speech segments that are missing or lost to noise or dropouts such as for packet loss concealment (PLC) in speech communication over mobile phones or voice over IP (VoIP), restoration of archived speech recordings and for general purpose speech interpolation.

Algorithms specifically designed for speech gap restoration can be categorised into two classes:

i) Predictive or extrapolative methods, where only the past samples are available.
ii) Estimative or interpolative methods, where some future samples are also available.

In general, speech interpolation methods utilise signal models that capture the correlations of speech parameters on both sides of the missing speech segment. The speech signal gap either can be interpolated in the time domain or one can convert speech in its spectrogram and then interpolate the missing spectral-time components. The latter method, which utilises the spectral temporal structure, is the focus of this section.

Autoregressive (AR) models are of particular interest in restoration of lost samples of speech. Janssen *et al.* (1986) applied AR models for estimation of non-recoverable errors in compact disk systems. Least square error AR (LSAR) and maximum a posterior (MAP) AR are among the most common criteria used for estimation of the gaps. Vaseghi and Rayner (1990) proposed a pitch-based AR interpolator to take into account the long-term correlation structure of the speech and near periodic signals. Kauppinen and Roth (2002) proposed a method that replaces the missing portion of the excitation with zeros which effectively uses the LP model for estimation of the missing samples from past samples only. Estimation of the excitation of the AR source filter is crucial in AR-based interpolators. Esquef and Biscainho (2006) proposed a time-reversed excitation substitution algorithm with a multi-rate post-processing module for audio gap restoration.

Lindblom and Hedelin (2002b) proposed methods for extrapolation (prediction) of lost speech packets using linear prediction; they used LP analysis for modelling the spectral envelope of the signal and sinusoidal models for modelling the excitation. They also considered the interpolation problem. Wang and Gibson (2000) investigated the robustness of the inter-frame and intra-frame line spectral frequency (LSF) coders to packet loss.

Another strategy that has proved to be popular for PLC is waveform substitution. The International Telecommunications Union (ITU) has standardised a waveform substitution algorithm for extrapolation of speech signals up to 60 ms after first frame loss. Waveform substitution techniques search the past few received frames to find appropriate substitutions for the lost packet. This is an improvement on repeating the exact previous frame which can produce annoying artefacts. Goodman *et al.* (1986) reported a predictive waveform substitution technique for application in PLC. Valenzuela and Animalu (1989) proposed a pitch synchronous overlap-add interpolation method using a similar technique.

In other works combinations of the AR-based algorithms and waveform substitution are used (Gündüzhan and Momtahan, 2001; Elsabrouty *et al.*, 2003). Wang and Gibson (2001) proposed the use of linearly interpolated LSFs throughout the gap and optimised selection of the coded excitation from previous frames for restoration of gaps in a CELP coded speech signal.

Sinusoidal models, and their extended version HNM, which have found extensive applications in text-to-speech (TTS) synthesis (Stylianou, 2001), coding (Griffin and Lim, 1988; Kondoz, 1999), and many other speech processing applications including bandwidth extension (Vaseghi *et al.*, 2006; Raza and Cheung-Fat, 2002). Rødbro *et al.* (2003) proposed a linear interpolation technique for estimation of the sinusoidal model parameters of missing frames. The time-domain signal is then, synthesised using a combination of symmetric and asymmetric windows and an overlap-add technique.

In this section an interpolation method is described based on a LP-HNM model of speech where the spectral envelope of speech is modelled using a LSF representation of a linear prediction (LP) model and the excitation is modelled with a harmonic plus noise (HNM), whose parameters are the harmonic frequencies, amplitudes and harmonicities (voicing levels). These parameters are interpolated throughout the gap resulting in natural sounding synthesised speech. The advantage of using LP-HNMs for modelling the spectral envelope and excitation is that the time-varying contours of the formants and the harmonic energies across the signal gap are modelled and tracked.

Furthermore, a codebook-mapping technique is used as a post-processing module for fitting the interpolated parameters to a prior speech model which is trained beforehand. Interpolation of LSF values of speech may sometimes result in estimation of some unusually sharp poles, giving rise to tonal artefacts. The conventional method of mitigating such effects is bandwidth expansion through damping of the poles of the LP model at the cost of smearing the signal in the frequency domain (broadening the poles' bandwidth). Application of the codebook-mapping technique to interpolated parameters of the LP-HNM of the signal mitigates the effects of unwanted tonal artefacts resulting from the interpolation of LSF parameters without an undesirable broadening of the poles' bandwidths. Codebook-mapping results in significant improvement in the performance of the algorithm in restoration of longer gaps (e.g. $> 50\,\text{ms}$).

The LP-HNM based speech interpolation method combines the advantages of the different methods described above and has the following distinct features:

(1) The speech structure is modelled by a time-varying LP-HNM model; i.e. a combination of an LP model of the spectral envelope and a HNM of the excitation.
(2) Interpolation is performed in spectral-time domain, where the non-stationary dynamics of the prominent contours of energies of the spectral envelope (LSF) and the excitation tracks (pitch, amplitudes, phase and harmonicities) of speech on both sides of the missing signal gap are modelled and interpolated.
(3) The temporal dynamics of the LP-HNM parameter tracks are modelled by a combination of linear and AR models.
(4) A pre-trained post-processing codebook is employed to mitigate artefacts and enhance the interpolation process; in this way prior statistical model information is brought into the interpolation process.

Figure 15.19 shows the interpolation method where speech interpolation is transformed into interpolation of the LP-HNM frequency-time tracks. The interpolation of each LP-HNM track is achieved using two methods: (1) a simple linear interpolation, (2) a combination of linear and autoregressive interpolation methods.

Assume that T_G consecutive frames of speech are missing where each speech frame has W samples including S new (non-overlapping) samples; and let T_A and T_B be the number of available speech frames after and before the speech gap respectively. Our goal is to estimate the LP-HNM parameters of T_G missing frames using the T_B frames before and T_A frames after the gap as shown in Figure 15.20. It is assumed that the interpolation delay is less than the maximum acceptable system delay.

Assuming $T_B = T_A = 1$, i.e. only one speech frame is available before and after the gap, the LP-HNM parameters can be linearly interpolated. We define the general linear interpolating function as

$$I_L(x_1, x_2, T_G, t) = \frac{tx_2 + (T_G - t + 1)\,x_1}{T_G + 1} \quad 1 \le t \le T_G \tag{15.57}$$

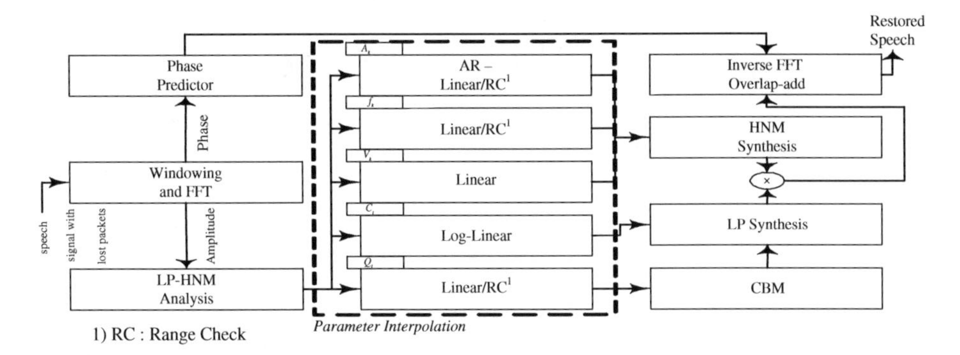

Figure 15.19 LP-HNM + CBM interpolation systems. LP-HNM: linear prediction harmonic noise model, CBM: codebook mapping.

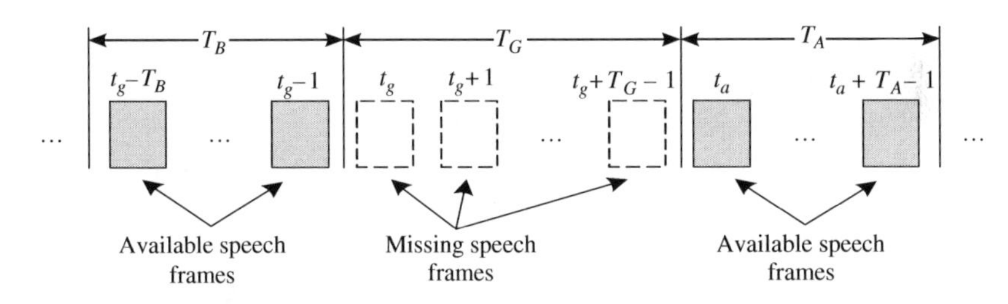

Figure 15.20 Illustration of speech frame interpolation where T_G frames of speech signal are missing.

where x_1, x_2 are the known values at the two ends of the gap. Each LP-HNM parameter track composed of the harmonic amplitudes A_{kt}, the harmonicity V_{kt}, the harmonic frequencies f_{kt}, the LSF coefficients Q_{it} and the LP model gain C_t, can be linearly interpolated using Equation (15.57). Note the subscripts k, i and t represent the harmonic, LSF and frame indices respectively.

The linear interpolation method joins the LP-HNM parameters of speech across the gap with a straight line. As expected the quality of the interpolated speech is sensitive to estimation error of the excitation harmonic amplitudes and LSF values. Furthermore experiments show that classical high-order polynomial interpolators result in artefacts in the output.

The response of a stable AR model with non-zero initial conditions to a zero excitation decays with time towards zero. The proposed interpolation method exploits this fact in order to obtain an estimate of the parameter sequence which has a smooth transition at each side of the gap and is modelled by the mean values of the LP-HNM parameters in the middle.

Assume the values of the time series x_t are missing from the time instance t_g to $t_a - 1$. One solution would be the LSAR interpolator which incorporates information from both sides of the gap simultaneously. However, LSAR assumes that the signals on both sides of the gap are from a stationary process. Furthermore, a large number of samples are required for a reliable estimate of the LSAR models of the time series before and after the gap.

In the method shown in Figure 15.21 two low-order AR models are used for the estimation of the zero-mean trend of the time-series parameter from each side of the gap. The predicted values from each side are overlap-added. The mean value of the time series is estimated by linear interpolation of the mean values of both sides of the gap. Let the number of available frames at each side of the gap

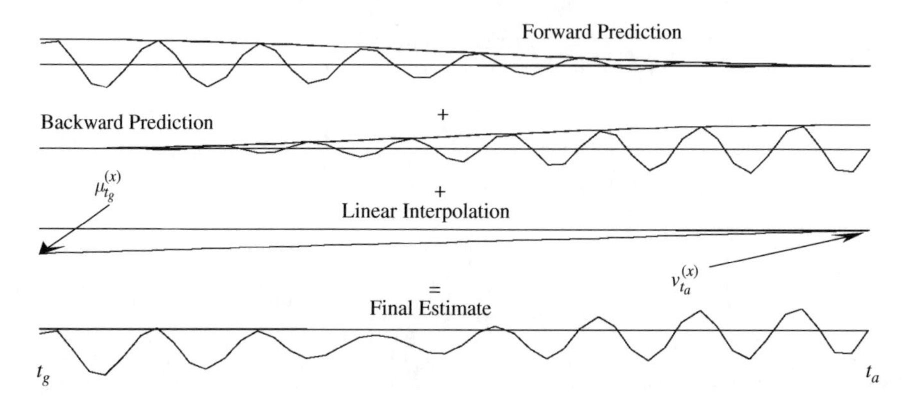

Figure 15.21 The mean-subtracted time-series is linearly predicted from both sides, weighted-averaged and added to the linearly interpolated mean.

$T_A = T_B \geq 3$. The missing parameter values are estimated as

$$x_t^{(AR)} = W_t \sum_{i=1}^{P} a_{t_g}^{(i)} \left(x_{t-i} - \mu_{t_g}^{(x)} \right) + (1 - W_t) \sum_{i=1}^{P} b_{t_a}^{(i)} \left(x_{t+i} - \nu_{t_a}^{(x)} \right) + I_L \left(\mu_{t_g}^{(x)}, \nu_{t_a}^{(x)}, T_G, t - t_g + 1 \right) \quad (15.58)$$

where P is the LP order of the AR model and

$$\mu_{t_g}^{(x)} = \frac{1}{T_B} \sum_{i=1}^{T_B} x_{t_g} - i \text{ and } \nu_{t_a}^{(x)} = \frac{1}{T_A} \sum_{i=1}^{T_A} x_{t_a+i-1} \quad (15.59)$$

are the means and $a_{t_g}^{(i)}$ and $b_{t_a}^{(i)}$ are the ith LP coefficients of the series $\left[x_{t_g - T_B}, \ldots, x_{t_g - 1} \right] - \mu_{t_g}^{(x)}$ and $\left[x_{t_a + T_A}, \ldots, x_{t_a + 1} \right] - \nu_{t_a}^{(x)}$, respectively. The weights, W_t, are chosen from half of a Hanning window of length $2T_G$. Figure 15.22 shows an example of the AR interpolation used for interpolation of the fifth harmonic of a sample signal. The gap, T_G, is rather long (i.e. equal to 70 ms).

15.4.1 Phase Prediction

Short-time phase plays an important role in the perceived quality and naturalness of the HNM-synthesised speech. In order to obtain an acceptable quality of reconstructed speech, the phase estimation method must be based on a model that exploits the continuity of the harmonic parts of speech and maintains the randomness of the non-harmonics. In this section a method is introduced for one-frame forward prediction of speech phase, based on the HNM model of speech. In order to maintain inter-frame and intra-frame continuity of the reconstructed signal, the phase is estimated so that:

1) speech harmonics across successive frames have phase continuity (inter-frame continuity, Figure 15.23 (left);
2) the adjacent frequency bins around each harmonic are in phase halfway through the frame (intra-frame continuity, Figure 15.23 (right); and
3) the amount of randomness added to the phase of each frequency channel increases with its frequency and its distance from the nearest harmonic centre (unvoiced sub-bands, Figure 15.24).

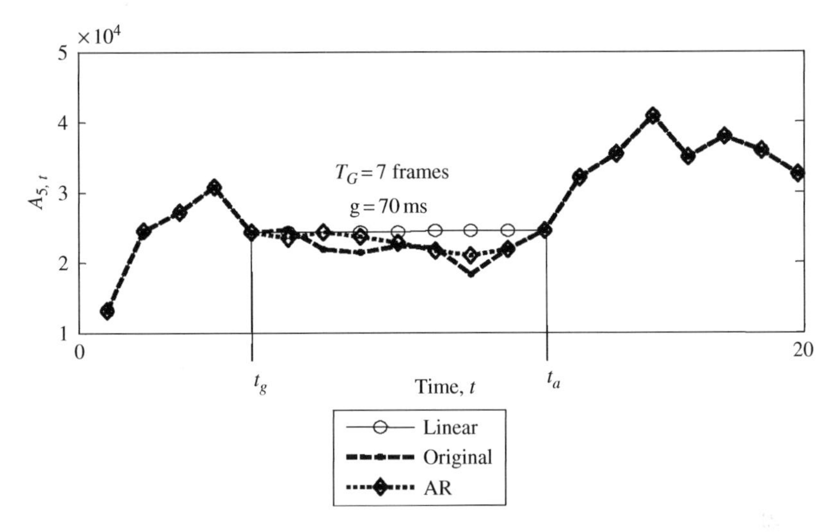

Figure 15.22 AR interpolation of the fifth harmonic of a sample signal.

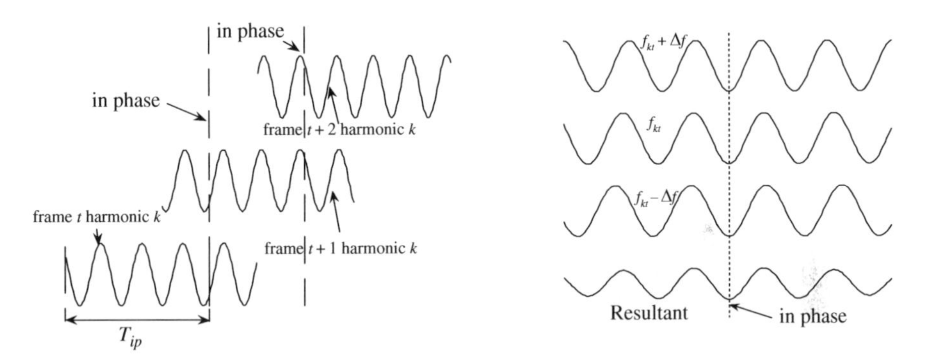

Figure 15.23 Harmonics of successive frames are in phase (left); adjacent bins of harmonics are in phase with them half-way through the frame (right).

Condition 1 guarantees the continuity between successive frames. Condition 2 determines the slope of the linear phase around each harmonic. This ensures positive reinforcement of the energy of signal components around each harmonic. Condition 3 is useful in reconstruction of unvoiced sub-bands and helps to avoid tonality and unwanted artefacts. In order to satisfy the first condition, as shown in Figure 15.23 (left), the equation for phase at harmonics is obtained as

$$\Phi(f_{kt}) = \Phi(f_{k,t-1}) + \frac{2\pi}{F_s}\left[T_{\text{ip}}(f_{k,t-1} - f_{k,t}) + Sf_{k,t}\right] \tag{15.60}$$

where $\Phi(f_{kt})$ is the phase of the kth harmonic at frame t (i.e. with frequency f_{kt}), T_{ip} is the 'in-phase' sample index that is where harmonic f_{kt} and $f_{k,t-1}$ are in phase, S is the shift size and F_s is the sampling frequency. Note that the subscript of the harmonic frequency has been extended to include the frame index as well. This subscript might be used for other LP-HNM parameters as well wherever

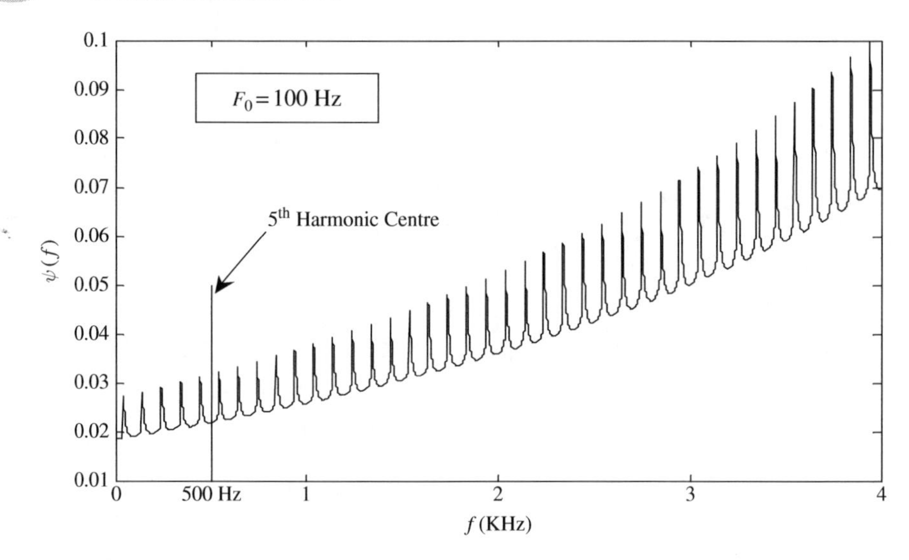

Figure 15.24 Variations of the amount of noise added to phase across frequency.

necessary. The in-phase sample is chosen to be halfway through the overlap as shown by vertical lines in Figure 15.23 (left), i.e.

$$T_{\mathrm{ip}} = (W + S)/2 \tag{15.61}$$

where W is the window size. In order to obtain positive energy reinforcement at the middle of each window, the frequency channels in the vicinity of each harmonic should be in phase at that point (Figure 15.23 (right)). Furthermore a level of randomness needs to be added to the phase for unvoiced (noise) synthesis, i.e. for each harmonic sub-band of an arbitrary frame:

$$\Phi(f) = \Phi(f_k) + (f - f_k)\,a + \frac{\Phi_R(f)}{V_k}\psi(f) \quad \text{for} \quad 0 < |f - f_k| < F_0/2 \tag{15.62}$$

$$a = -\pi W/F_S \tag{15.63}$$

where a is the slope of the phase, $\Phi_R(f)$ is a random variable uniformly distributed in the range $[-\pi, \pi]$, $\psi(f)$ is a weighting factor that increases with the frequency and distance from the centre of the harmonic. The latter determines the level of randomness of the phase. We propose the use of the following function for this purpose:

$$\psi(f) = \sum_{k=1}^{N}\left[\left(\frac{1 - h(f - f_k)}{h(f - f_k)}\right)0.002 + 0.05\right]\exp\left(\frac{f - 3000}{3000}\right) \tag{15.64}$$

where $h(f)$ is a Hamming window in the range $[-F_0/2, F_0/2]$ and f_k are harmonic frequencies. An example of $\psi(f)$ is shown in Figure 15.24. Note that the local minimums correspond to harmonic frequency channels. This figure is further shaped according to the harmonicity of each sub-band.

15.4.2 Codebook Mapping

Codebook-mapping (CBM) is a heuristic technique normally used for partial estimation of a set of parameters, e.g. estimation of upper band parameters based on lower band for bandwidth extension, or correction of over-suppressed harmonics in a noise reduction system. Codebook mapping forces a model upon the parameters through use of pre-trained codebooks.

In estimation of LSF parameters, using linear interpolators, we note that the resulting spectral envelope may have sharp peaks or sound unnatural. These artifacts can be even more annoying than the original packet loss introduced. One technique that can be particularly useful is damping the poles of the LP model, perhaps proportional to the distance from the two ends of the gap. This would mitigate the problem of perceiving sharp peaks in the spectrum at the cost of a de-shaped spectrum.

The codebook mapping technique can be used to fit the estimated values into a pre-trained speech model through the use of codebooks. A codebook is trained on LSF parameters of various speech utterances. The utterances were taken from the Wall Street Journal (WSJ) database of spoken language. Each interpolated LSF vector is then compared with the vectors in the codebook and the K nearest codewords, $\left[\overline{\mathbf{Q}}_{k_1}, \cdots, \overline{\mathbf{Q}}_{k_K}\right]$, are selected according to the Euclidean distance:

$$D_k = \left\| \mathbf{Q}^{(\mathrm{L})} - \overline{\mathbf{Q}}_k \right\| \tag{15.65}$$

where $\mathbf{Q}^{(\mathrm{L})}$ is a linearly-interpolated LSF vector, $\overline{\mathbf{Q}}_k$ is the kth codeword of the LSF codebook, D_k is the Euclidean distance between the two and $k_1, k_2, \ldots, k_K$ are the indices of the nearest codewords to $\mathbf{Q}^{(\mathrm{L})}$. These codewords are weighted averaged where the weights are inversely related to their distances from the original LSF vector. The resulting vector replaces the interpolated LSF vector.

$$\mathbf{Q}^{(\mathrm{CBM})} = \left[\sum_{i=1}^{K} \frac{1}{D_{k_i}} \right]^{-1} \times \sum_{i=1}^{K} \frac{\overline{\mathbf{Q}}_{k_i}}{D_{k_i}} \tag{15.66}$$

where the superscript (CBM) shows the codebook-mapped estimate of the LSF vector. Three different versions of the proposed algorithm are evaluated and compared with some alternative methods in this section. Besides parametric LP-HNM interpolation with and without codebook mapping, a different method which interpolates the HNM parameters extracted from the speech spectrum itself (and not the excitation) is also evaluated.

The multirate gap restoration algorithm, introduced by Esquef and Biscainho (2006), is chosen for comparison purposes. This algorithm is composed of two modules: (i) a *core* module which uses an AR model for each side of the gap and estimates the signal using an estimated excitation signal, and (ii) a *multirate* post-processing module, which further enhances the interpolated signal in two low-frequency sub-bands. In addition to the complete algorithm, the performance of the core method (i.e. without the multirate post-processing) is also evaluated and compared with the proposed algorithms.

Many PLC algorithms proposed in the literature are compared with the standard ITU-T G.711 PLC algorithm. Even though the G.711 PLC algorithm is based on a different set of assumptions than the proposed algorithm, its performance is evaluated compared with the proposed algorithm as a reference point.

A two-state Markov model is used to model the frame loss introduced in the speech signal. The probability of a 'lost' frame after a 'good' frame is p and that of a good frame after a bad frame is q. This model emphasises the burst errors that might occur in some applications.

Evaluation of LP-HNM Interpolation

After introducing the gaps in the signals, each signal is restored using different algorithms, e.g. ITU G.711 PLC algorithm (G.711), multirate gap restoration (Multirate), the core AR-based algorithm of Janssen *et al.* (1986) (AR) and the proposed algorithms (HNM, LP-HNM, LP-HNM + CBM).

The performance of these algorithms is evaluated using perceptual evaluation of speech quality (PESQ) scores and log spectral distance (LSD) measure. The results are calculated and averaged for 100 sentences randomly selected from the WSJ database. The performances of different algorithms in restoration of the gaps generated by a two-state Markov model are illustrated in Table 15.3. Note that the probability of a 'bad' or lost frame after a 'good' or available frame is p and that of a good frame after a bad frame is q.

A set of five utterances are selected randomly from the WSJ database. Three different sets of packet loss patterns are generated, using the two-state Markov model described in the previous section, with a fixed loss rate of 40% and different average gap lengths of 2, 5 and 7 frames. An experiment similar to ITU's Comparison Category Rating (CCR) is conducted. After introduction of the gaps, each signal is restored using the three proposed methods and the G.711 method. Ten listeners were asked to listen to the resulting signals, each played after its G.711 restored counterpart and compare the second utterance with the first one and rate it from −3 to 3 representing 'much worse' and 'much better' respectively. The results are summarised in Table 15.3.

Figure 15.25 shows the spectrograms of a part of a speech signal: its distorted (with missing samples) and restored versions. It is evident that the restored consonant in the middle of the sample (before 0.4 ms) suffers from different artefacts in different methods. The upper bands of the speech signal, after restoration with the proposed algorithms, have a higher level of harmonicity compared with the interpolation method used in G.711. This is due to the more harmonic start of the consonant, available to the algorithms. A freezing effect can be seen throughout the restored gaps of the G.711 algorithm which is a known problem of this method. Furthermore, it is observed that the formant trajectories are best recovered using LP-HNM based algorithms.

It is rather difficult to evaluate and compare the performance of speech gap restoration algorithms. Not only is each designed for a particular application and uses specific resources available, but also they perform differently in reconstruction of different parts of speech signals. Through exhaustive experiments it is concluded that gap restoration algorithms, in general, are less successful in restoration of vowel–consonant and consonant–vowel transition and even less successful in restoration of vowel–consonant–vowel in which the restored quality is reduced to that of mumbled speech.

Table 15.3 Performance of different algorithms for restoration of two-state Markov generated gaps.

	PESQ				LSD			
q	0.85	0.7	0.5	0.4	0.85	0.7	0.5	0.4
p	0.1	0.2	0.3	0.6	0.1	0.2	0.3	0.6
Loss rate %	11	22	38	60	11	22	38	60
Av. gap length	1.18	1.43	2.00	2.50	1.18	1.43	2.00	2.50
HNM	3.15	2.73	2.43	2.12	0.52	0.72	0.85	1.06
LP-HNM	3.15	2.74	2.44	2.13	0.52	0.74	0.86	1.00
LP-HNM + CBM	3.14	2.74	2.48	2.21	0.52	0.70	0.81	0.96
G.711 – A1	3.14	2.59	2.07	1.51	1.59	1.99	2.09	2.19
AR	3.00	2.60	2.25	1.77	0.42	0.56	0.64	0.80
Multirate	2.54	2.07	1.73	1.24	0.58	0.77	0.88	1.11
Distorted	2.76	2.01	1.18	0.44	–	–	–	–

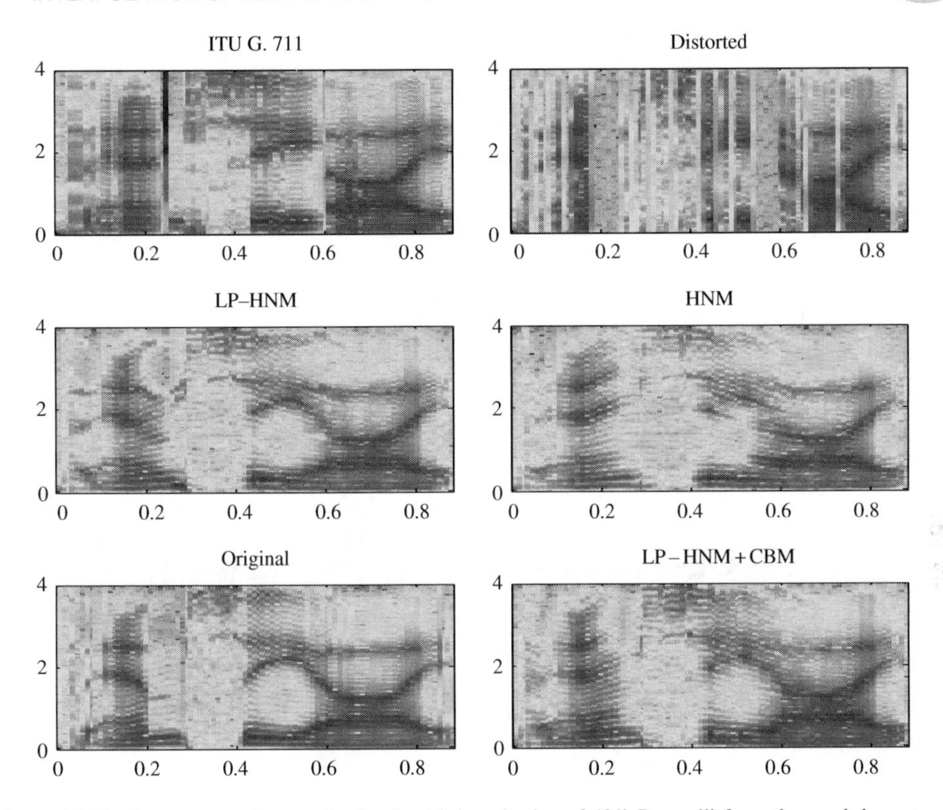

Figure 15.25 Spectrogram of a sample signal, with introduction of 40% Bernoulli frame loss and the restored versions.

The objective results represented in the previous section shows that the LP-HNM method outperforms other algorithms discussed here in most cases. While a very similar output quality is gained in restoration of short gaps, the proposed algorithms are particularly powerful in restoration of longer gaps. The CBM technique results in a level of noise which is believed to be the result of the quantisation of the LSF vectors. While at shorter gap lengths this reduces the quality in comparison with some other methods, it makes the algorithm more robust to increases in the gap length particularly for gap lengths greater than 5 frames, as evident in Table 15.4. Figure 15.25 shows the spectrogram of a sample signal, with introduction of 40% Bernoulli frame loss and the restored versions. It is clearly evident that LP-HNM + BCM produces the better results.

Table 15.4 Comparative subjective results of proposed methods with a loss rate of 40%.

Restoration method	HNM			LP-HNM			LP-HNM + CBM		
q	0.5	0.2	0.14	0.5	0.2	0.14	0.5	0.2	0.14
p	0.3	0.13	0.95	0.3	0.13	0.95	0.3	0.13	0.95
Av. gap length	2	5	7	2	5	7	2	5	7
Subjective score	1.64	1.12	0.61	2.36	1.88	0.73	1.68	1.60	1.29

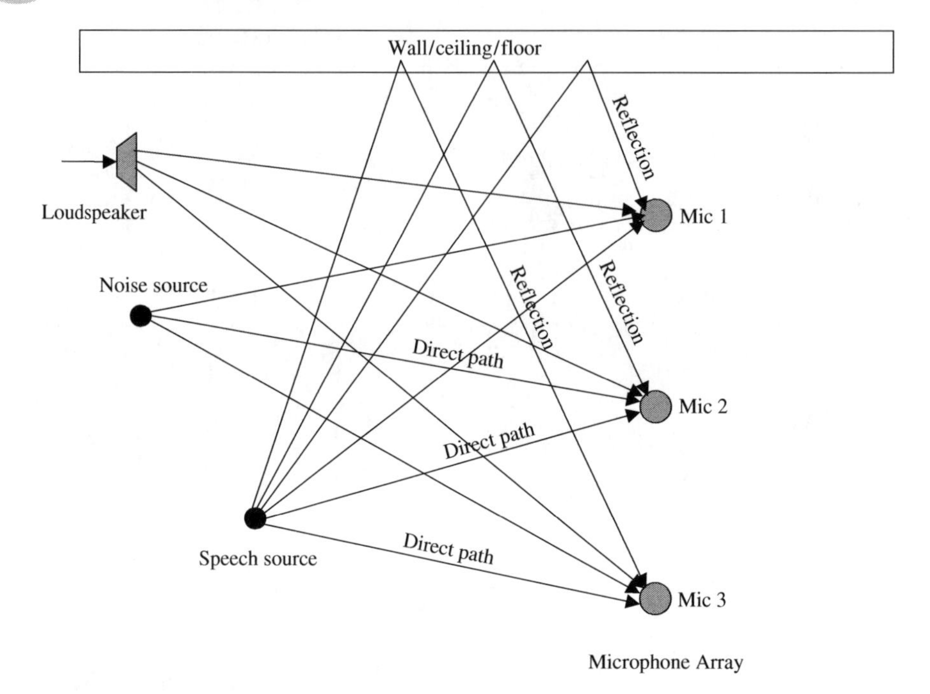

Figure 15.26 Illustration of different sounds and noise arriving at microphones, via direct line of sight paths and via reflections from walls/ceiling/floor.

15.5 Multi-Input Speech Enhancement Methods

In multiple input noise reduction systems several noisy input signals, picked up by an array of microphones, are filtered and combined to reduce noise, echo, interfering speech, room reverberations and distortions.

Multi-input speech enhancement systems include adaptive beam forming, adaptive noise cancellation, multi-input multi-output (MIMO) teleconferencing systems, stereophonic echo cancellation and in-car cabin communication systems.

In a typical multi-input system, Figure 15.26, there are several microphones. The output of each microphone is a mixture of the speech signal, feedback from loudspeakers, speech reflections from walls and noise.

Assuming that there are M microphones and N sets of signal and noise sources, then there are $N \times M$ different acoustic channels between the sources of signals and the microphones. We can write a system of linear equations to describe the relationship between the signals emitted from different sources $x_i(m)$, and the signals picked up by the microphones $y_j(m)$ as

$$y_j(m) = \sum_{i=1}^{N} \sum_{k=0}^{P} h_{ij}(k) x_i(m-k) \quad j = 1, \ldots, M \tag{15.67}$$

where $h_{ij}(k)$ denotes the impulse response of the channel from source i to microphone j modelled by a finite impulse response (FIR) filter. Note that for simplicity each source of signal, noise or interference is denoted with the same letter x and different index as $x_i(m)$; m is the discrete-time index.

In the simplest MIMO model, the response of an acoustic channel from sound source i to microphone j via a direct or reflected path can be represented by two parameters, an attenuation factor $\alpha_{ij}(m)$ and a propagation time delay $\tau_{ij}(m)$, as

$$h_{ij}(m) = \alpha_{ij}(m)\delta\left(m - \tau_{ij}(m)\right) \tag{15.68}$$

Note that each source of sound may reach a microphone via a direct path and via a number of indirect paths after reflections in which case the response from source i to microphone j needs to be expressed as

$$h_{ij}(m) = \sum_{k=1}^{L} \alpha_{ijk}(m)\delta\left(m - \tau_{ijk}(m)\right) \tag{15.69}$$

where $\alpha_{ijk}(m)$ and $\tau_{ijk}(m)$ are the attenuation factor and the propagation time delay along the kth path from source i to microphone j.

Multi-input multi-output (MIMO) speech enhancement systems continue to be the subject of much research. In general the main issues in MIMO noisy speech processing systems are as follows:

(a) Identification of the channel responses or the room transfer functions, $\{h_{ij}(m)\}$, from the speech and/or noise sources to microphones.
(b) The problem of non-uniqueness of the solutions for room channel responses, when the noise sources are correlated. This is explained in the discussion on stereo acoustic echo cancellation in Chapter 16.
(c) The speed of convergence of adaptation of the filter coefficients at the output of each microphone to achieve cancellation of noise and acoustic feedback echo.
(d) Estimation of the time-varying speech and noise characteristics for noise suppression.

The outputs of the MIMO systems can be processed in many different ways. One of the most promising applications of MIMO systems is beam-forming microphone arrays, described next.

15.5.1 Beam-Forming with Microphone Arrays

Beam-forming is a noise reduction, or noise screening, method in which an array of microphones and adaptive filters are employed to selectively seek a 'beam' of signal from a particular direction (usually the direction towards a desired source and where the signal from that source is strongest) and suppress noise, reverberations and acoustic feedbacks coming from other directions.

Beam-forming uses temporal and spatial filtering to create directional spatial selectivity. In this way it is possible to adaptively adjust the filters to selectively pick up a sound wave from the direction of the source where the sound energy is strongest and screen out noise, feedback and reflections of sounds from other directions. Beam-forming has application in hands-free communication such as in-car communication, personal computer voice communication, teleconferencing and robust speech recognition. Beam-forming can also be combined with acoustic feedback cancellation.

Principles of Microphone Array Beam-Forming

Beam-forming relies on the use of delays (or equivalently phase) for constructive and destructive combination of the incoming signals arriving at the microphone from different directions. Figure 15.27(a) shows the sound field of an omni-directional microphone that picks sound equally from all directions.

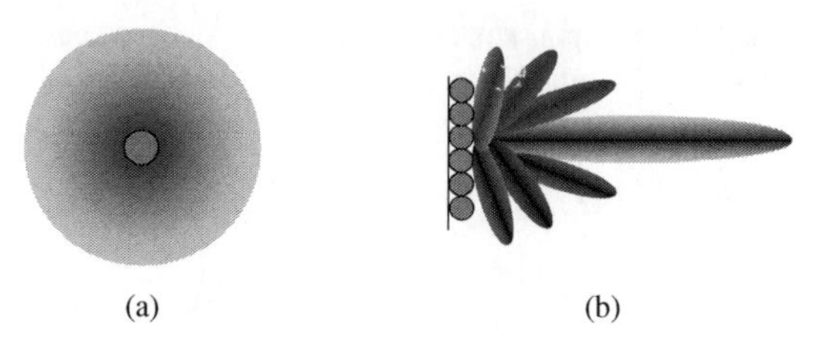

Figure 15.27 Illustration of the reception fields (aperture) of (a) an omnidirectional microphone, (b) an array of microphones.

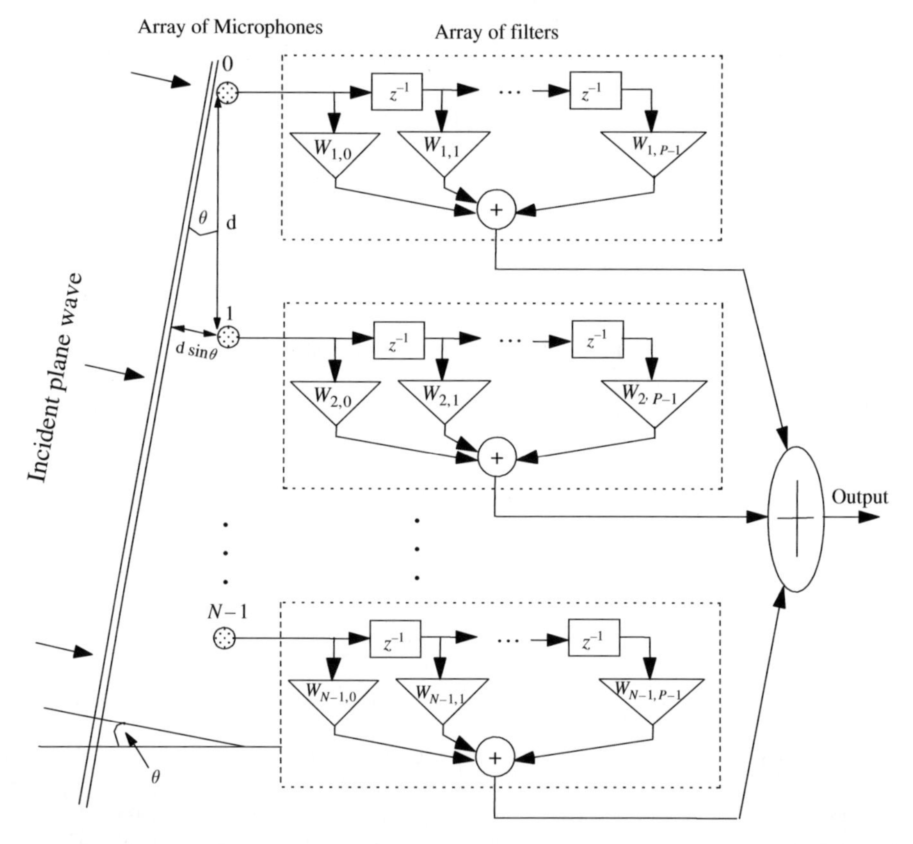

Figure 15.28 Illustration of beam-forming. The array of filters can be adjusted to change the 'look' direction of the beam.

The combined output of several microphones placed in an array exhibits directional selectivity as shown in Figure 15.27(b).

The principle of beam-forming can be illustrated by considering a linear microphone array in the far field of a sound source as illustrated in Figure 15.28. The assumption that the microphone is in the far field of the sound source has the useful implication that the sound waves reaching the microphone array are planar (as opposed to spherical) waves.

Sound waves arriving at the microphone array, at an incident angle of θ reach successive microphones at different times due to the different distances that the waves have to travel. The additional distance that a wave has travelled to reach an adjacent microphone is $d\sin(\theta)$. Note that if θ is zero then the wave is perpendicular to the microphone array and in that case all microphones receive each plane wave at the same time.

Sound propagates with a speed of $v = 342$ metres per second at a room temperature of 25° C. The additional time taken for sound to travel a distance of $d\sin(\theta)$ metres to reach an adjacent microphone is given by

$$\tau = \frac{d\sin(\theta)}{v} \tag{15.70}$$

For a sine wave it is easy to see that if the distance between two adjacent microphones and the angle of incidence of the wave θ is such that it results in a time delay of arrival of sound of say half a wavelength then the sum of the outputs of the microphones will cancel out. Note that this illustration is oversimplified and in practice a combination of temporal and spatial filtering provides spatial selectivity. In general signals arriving at a microphone array at the same time (in phase) add up, whereas signals arriving at microphone array at different times cancel out partially or completely.

The combined output of N microphones can be expressed as

$$y(m) = \sum_{i=0}^{N-1} x(m - \tau_i) \tag{15.71}$$

where for simplicity we assume the channel attenuation factors $\alpha_i = 1$. The response of the microphone array to a frequency f and direction of arrival of signal θ can be expressed as

$$\begin{aligned} H(f,\theta) &= \sum_{i=0}^{N-1} e^{-j2\pi f i \tau} \\ &= \sum_{i=0}^{N-1} e^{-j2\pi f i\left(\frac{d\sin(\theta)}{v}\right)} \end{aligned} \tag{15.72}$$

Equation (15.72) is used to plot the response of a microphone array for a selected frequency and varying direction of arrival, Figure 15.29. The angle of arrival can be changed or adapted using the adaptive filters shown in Figure 15.28.

Note that the bandwidth of the beam decreases with the increasing number of microphones but it is also a function of the distance between the microphones and the frequency of the sound.

15.6 Speech Distortion Measurements

This section describes methods for measuring the quality of speech degraded by noise. Speech quality measures have a number of important uses including:

(i) When evaluated off-line they can be used to compare an algorithm with alternatives or to evaluate the use of different speech models within an algorithm,
(ii) When evaluated in real time they can be used to select the best parameter configuration, codebooks or posterior probability from a number of choices.
(iii) They can provide the user/operator with an online report of the communication system performance.

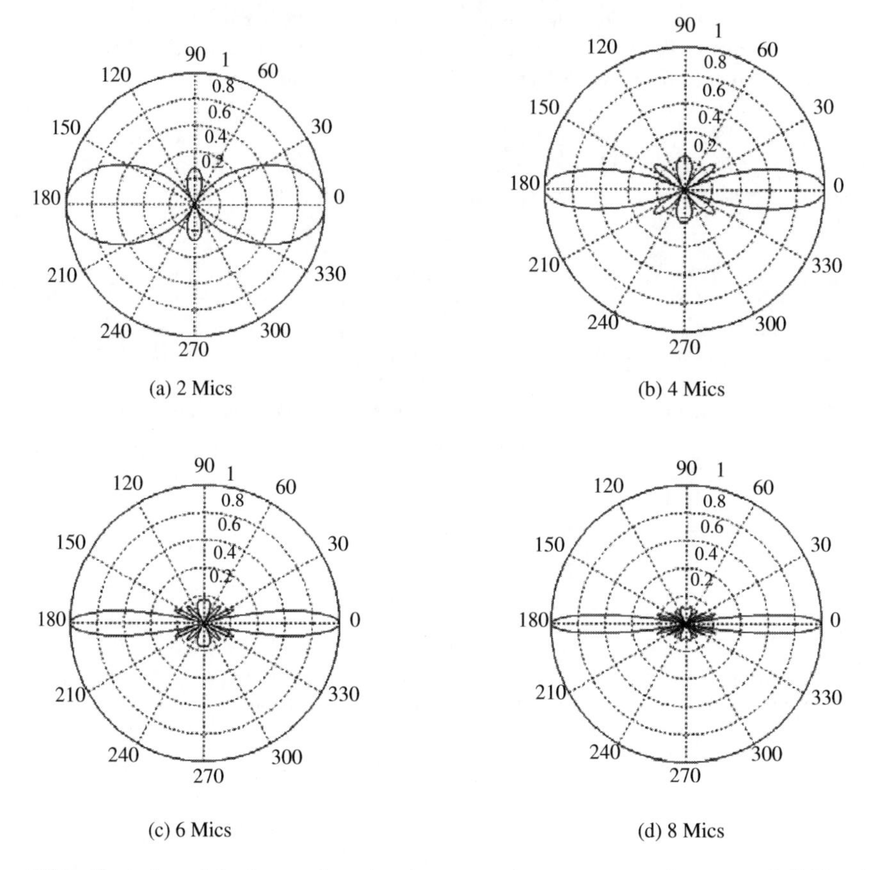

Figure 15.29 Illustration of the beam of a microphone array in response to a frequency of 2 kHz with the microphones spaced at $d = 10\,\text{cm}$.

There are two types of distortion and quality measures: objective measures such as signal-to-noise ratio and subjective measures such as mean opinion score. The most commonly used distortion measures for speech signals are the following.

15.6.1 Signal-to-Noise Ratio – SNR

The most commonly used measure for quality of speech is signal-to-noise ratio (SNR). An average SNR measure is defined as

$$\text{SNR} = 10\log_{10}\left(\frac{P_{\text{Signal}}}{P_{\text{Noise}}}\right)\text{dB} \tag{15.73}$$

where P_{Signal} and P_{Noise} are the power of signal and noise respectively.

15.6.2 Segmental Signal to Noise Ratio – $\mathrm{SNR}_{\mathrm{seg}}$

The segmental SNR is defined as

$$\mathrm{SNR}_{\mathrm{seg}} = \frac{1}{K}\sum_{k=0}^{K-1} 10^{*}\log_{10}\left(\sum_{m=0}^{N-1} x_k^2(m) \Bigg/ \sum_{m=0}^{N-1} (x_k(m) - \hat{x}_k(m))^2\right) \mathrm{dB} \tag{15.74}$$

where $x_k(m)$ and $\hat{x}_m(m)$ are the clean signal and restored signal at frame m, K is the total number of frames and N is the number of samples in each frame. The segmental SNR of speech signals can fluctuate widely as illustrated in Figure 15.15 which shows the variation of segmental SNR at average SNRs of 0, 5 and 10 dB.

The signal-to-noise ratio is not the best measure of speech quality as it does not take into account the structure of the speech or the psychoacoustics of hearing.

15.6.3 Itakura–Saito Distance – ISD

The Itakura–Saito Distance (ISD) measure is defined as

$$\mathrm{ISD}_{12} = \frac{1}{N}\sum_{\mathrm{j}=1}^{N} \frac{(\boldsymbol{a}_1(j) - \boldsymbol{a}_2(j))^T \boldsymbol{R}_1(j)\,(\boldsymbol{a}_1(j) - \boldsymbol{a}_2(j))}{\boldsymbol{a}_1(j)\boldsymbol{R}_1(j)\boldsymbol{a}_1(j)^T} \tag{15.75}$$

where $\boldsymbol{a}_1(j)$ and $\boldsymbol{a}_2(j)$ are the linear predication model coefficient vectors calculated from clean and transformed speech at frame j and $\boldsymbol{R}_1(j)$ is an autocorrelation matrix derived from the clean speech. Due to the asymmetry of ISD measure (i.e. $\mathrm{ISD}_{21} \neq \mathrm{ISD}_{12}$) the following segmental ISD measure is used

$$\mathrm{ISD}_{\mathrm{sym}} = (\mathrm{ISD}_{12} + \mathrm{ISD}_{21})/2 \tag{15.76}$$

The ISD criterion is a more balanced measure of the distance between an original clean speech signal and a distorted speech signal compared with the SNR measures of Equations (15.73)–(15.74).

15.6.4 Harmonicity Distance – HD

To measure the distortions of the harmonic structure of speech, a harmonic contrast function is defined as

$$\mathrm{HD} = \frac{1}{\mathrm{NH} \times N_{\mathrm{frames}}} \sum_{N_{\mathrm{frames}}} \sum_{k=1}^{\mathrm{NH}} 10\log\frac{P_k + P_{k+1}}{2P_{k,k+1}} \tag{15.77}$$

where P_k is the power at harmonic k, $P_{k,k+1}$ is the power at the trough between harmonics k and $k+1$, NH is the number of harmonics and N_{frames} is the number of speech frames.

15.6.5 Diagnostic Rhyme Test – DRT

The diagnostic rhyme test consists of 96 word pairs which differ by a single acoustic feature in the initial consonant. Word pairs are chosen to evaluate the six phonetic characteristics listed in

Table 15.5 DRT characteristics.

Characteristics	Description	Examples
Voicing	voiced – unvoiced	veal – feel, dense – tense
Nasality	nasal – oral	reed – deed
Sustention	sustained – interrupted	vee – bee, sheat – cheat
Sibilation	sibilated – unsibilated	sing – thing
Graveness	grave – acute	weed – reed
Compactness	compact – diffuse	key – tea, show – sow

Table 15.5. The listener hears one word at a time and marks the answering sheet to indicate which of the two words she or he thinks is the correct one. The results are summarised by averaging the error rates from all the answers. Usually, only the total error rate percentage is given, but also single consonants and how they are confused with each other can be investigated with confusion matrices.

DRT provides useful diagnostic information on recognition of the initial consonant. However, it does not test vowels or prosodic features and hence it is not suitable for evaluation of the overall quality of speech. A further deficiency is that the test material is rather limited and the test items do not occur with equal probability, hence the method does not test all possible confusions between various consonants.

15.6.6 Mean Opinion Score – MOS

The mean opinion score provides a numerical evaluation of the perceived quality of speech. The MOS, as shown in Table 15.6, is expressed as a number in the range 1 to 5, where 1 is the lowest perceived quality, and 5 is the highest perceived quality. MOS can be used to assess the perceived qualities of the inputs and outputs of speech a communication channel or a speech enhancement method.

MOS tests for voice are specified by ITU-T recommendation P. 800. The MOS is obtained by averaging the results of a set of standard, subjective tests where a number of listeners rate the quality of test speech sentences read by male and female speakers. Each listener gives each sentence a rating using the rating scheme in Table 15.6.

The MOS is the mean of all the individual test scores. Note that one could ask the listening subjects to rate the quality or impairment (degradation) of speech.

Table 15.6 Mean Opinion Score (MOS).

MOS	Quality	Distortion/Impairment
5	Excellent	Imperceptible
4	Good	Perceptible but not annoying
3	Fair	Slightly annoying
2	Poor	Annoying
1	Bad	Very annoying

15.6.7 Perceptual Evaluation of Speech Quality – PESQ

PESQ stands for 'perceptual evaluation of speech quality'. It is an enhanced perceptual quality measurement for voice quality in telecommunications. PESQ was specifically developed to be applicable to end-to-end voice quality testing under real network conditions, such as VoIP, POTS, ISDN, GSM etc. The PESQ algorithm is designed to predict subjective opinion scores of a degraded speech sample. PESQ returns a score from 4.5 to −0.5, with higher scores indicating better quality.

PESQ is officially approved as ITU-T recommendation P. 862. PESQ was developed by KPN Research in the Netherlands and British Telecommunications (BT). PESQ analyses specific parameters of the speech signal, including time warping, variable delays, transcoding, and noise. It is primarily intended for applications in codec evaluation and network testing but is now increasingly used in speech enhancement. It is best to use PESQ as an addition to other methods such as mean opinion scores (MOS), where an average is taken of the subjective opinion of human listeners.

15.7 Summary

This chapter began with an introduction to the applications and methodologies of speech enhancement. The main speech enhancement topics considered here are background noise reduction, bandwidth extension for conversion of narrowband telephone quality speech to wideband broadcast quality speech, packet loss concealment and multiple-input beam-forming method. For noise reduction the spectral subtraction method, the Bayesian MMSE method, state-space Kalman filters, and a model-based LP-HNM method were described.

For bandwidth extension, various existing spectral extrapolation methods were reviewed and a state of art method is described in detail and its performance is evaluated. For concealment of lost speech packets various existing interpolation and extrapolation methods are reviewed and a state of art method is described in detail and evaluated. Finally beam-forming for screening out noise is covered and the different criteria for measurement of speech quality are described.

Further Reading

Appendix I. (1999) A High Quality Low-Complexity Algorithm for Packet Loss Concealment with G.711, ITU-T Recommend. G.711.

Bakamidis S., Dendrinos M. and Carayannis G. (1991) Speech Enhancement From Noise: A Regenerative Approach, Speech Communication, 10, pp. 44–57.

Berouti M., Schwartz R. and Makhoul J. (1979) Enhancement of Speech Corrupted by Acoustic Noise, Proc. IEEE Int. Conf. Acoust., Speech, Signal Process, pp. 208–211, Apr.

Boll S.F. (1979) Suppression of Acoustic Noise in Speech Using Spectral Subtraction. IEEE Trans. Acoust., Speech and Signal Proc., ASSP-27, pp. 113–120.

Cohen I. (2004) Speech Enhancement Using a Noncausal A Priori SNR Estimator, IEEE Signal Processing Letters, 11 (9), pp. 725–728.

De Moor B. (1993) The Singular Value Decomposition and Long and Short Spaces of Noisy Matrices, IEEE Trans. Signal Processing, 41 (9) pp. 2826–2838.

Deller J.R., Proakis J.G. and Hansen J.H. (1993) Discrete-Time Processing of Speech Signals, New York: Macmillan Publishing Company.

Elsabrouty M., Bouchard M. and Aboulnasr T. (2003) A new hybrid long-term and short-term prediction algorithm for packet loss erasure over IP-networks, Proc. 7th Int. Symp. Signal Processing and its Applications, vol. 1, pp. 361–364.

Ephraim Y. (1992) Statistical-Model Based Speech Enhancement Systems, Proceedings of the IEEE, 80 (10), pp. 1526–1554.

Ephraim Y. and Malah D. (1985) Speech Enhancement Using a Minimum Mean Square Error Log-Spectral Amplitude Estimator, IEEE Trans. Acoust., Speech, Signal Processing, ASSP-33, pp. 443–445.
Ephraim Y. and Van Trees H.L. (1995) A Signal Subspace Approach for Speech Enhancement, IEEE Trans. Speech and Audio Processing 3 (4), pp. 251–266.
Esquef P.A. and Biscainho L.W.P. (2006) An Efficient Model-Based Multirate Method for Reconstruction of Audio Signals Across Long Gaps, IEEE Transactions on Audio, Speech and Language Processing, 14, (4), pp. 1391–1400.
Gannot S., Burshtein D. and Weinstein E. (1998) Iterative and Sequential Kalman Filter-based Speech Enhancement, IEEE Transactions on Speech and Audio Processing, 6, (4), pp. 373–385.
Goodman D.J., Lockhart G.B., Wasem O.J. and Wang W.C. (1986) Waveform substitution techniques for recovering missing speech segments in packet voice communications, IEEE Trans. Acoustics, Speech, Signal Processing, ASSP-34 (5), pp. 1440–1448.
Griffin D.W. and Lim J.S. (1988) Multiband-excitation vocoder, IEEE Trans. Acoust., Speech, Signal Processing, ASSP-36 (2) pp. 236–243.
Gündüzhan E. and Momtahan K. (2001) A linear prediction based packet loss concealment algorithm for PCM coded speech, IEEE Trans. Speech Audio Process., 9 (6), pp. 778–785.
Hansen J.H.L. and Clements M.A. (1987) Iterative Speech Enhancement with Spectral Constraints, Proc. of ICASSP, pp. 189–192.
Janssen A.J.E.M., Veldhuis R. and Vries L.B. (1986), Adaptive interpolation of discrete-time signals that can be modelled as ar processes. IEEE Trans. Acoustics, Speech and Signal Processing, ASSP-34(2), pp. 317–330.
Kalman R. (1960) A New Approach to Linear Filtering and Prediction Problems, Transactions of the ASME, Journal of Basing Engineering, 82, pp. 34–35.
Kauppinen I. and Roth K. (2002) Audio signal restoration – theory and applications, Proc. 5th Int. Conf. on Digital Audio Effects (Hamburg, Germany), pp. 105–110.
Kondoz A.M. (1999) Digital Speech: Coding for Low Bit Rate Communication Systems, John Wiley & Sons.
Lim J.S. and Oppenheim A.V. (1978) All-pole Modelling of Degrade Speech, IEEE Trans. Acoust. Speech, Signal Processing ASSP-26, (3), pp. 197–210.
Lim J.S. and Oppenheim A.V. (1979) Enhancement and Bandwidth Compression of Noisy Speech, Proc. IEEE 67, pp. 1586–1604, Dec.
Lindblom J. and Hedelin P. (2002a) Packet Loss Concealment Based on Sinusoidal Modelling, Proc. IEEE Workshop on Speech Coding, Ibaraki, Japan, October, pp. 65–67.
Lindblom J. and Hedelin P. (2002b) Packet loss concealment based on sinusoidal extrapolation, Proc. IEEE International Conference on Acoustics, Speech, and Signal Processing 2002, Orlando, FL, vol. 1, pp. 173–176.
Mack G.A. and Jain V.K. (1985) A Compensated-Kalman Speech Parameter Estimator, ICASSP pp. 1129–1132.
Martin R. (2002) Speech Enhancement Using MMSE Short Time Spectral Estimation with Gamma Distributed Speech Priors, IEEE ICASSP'02, Orlando, FL, May.
Martin R. and Breithaupt C. (2003) Speech Enhancement in the DFT Domain Using Laplacian Speech Priors, Proc. Int. Workshop Acoustic Echo and Noise Control (IWAENC), pp. 87–90.
McAulay R.J. and Malpass M.L. (1980) Speech Enhancement Using a Soft-Decision Noise Suppression Filter, IEEE Trans. Acoust., Speech, Signal Processing, ASSP-28, pp. 137–145.
McCandles S.S. (1974) An Algorithm for Automatic Formant Extraction Using Linear Prediction Spectra, IEEE Trans. Acoustics, Speech and Signal Processing 22, pp. 135–141.
Milner B.P. and James A.B. (2004) An Analysis of Packet Loss Models for Distributed Speech Recognition, Proc. ICSLP 2004, pp. 1549–1552.
Murthi M.N., Rødbro C.A., Andersen S.V. and Jensen S.H. (2006) Packet Loss Concealment with Natural Variations Using HMM, ICASSP 2006, 1, pp. I-21–24.
Paliwal K.K. and Alsteris L.D. (2003) Usefulness of Phase Spectrum in Human Speech Perception, Proc. European Conf. Speech Communication and Technology, EUROSPEECH-03, Geneva, Switzerland, pp. 2117–2120.
Paliwal K.K. and Basu A. (1987) A Speech Enhancement Method Based on Kalman Filtering, Proc. of ICASSP, pp. 177–180.
Rabiner L.R. (1989) A Tutorial on Hidden Markov Models and Selected Application in Speech Recognition, IEEE Proceedings, 77 (2), pp. 257–286.
Rabiner L.R. and Juang B.H. (1993) Fundamentals of Speech Recognition, Prentice Hall, Englewood Cliffs, NJ.
Rabiner L.R. and Schafer R.W. (1978) Digital Processing of Speech Signals, Prentice Hall, Englewood Cliffs, NJ.

Raza D.G. and Cheung-Fat C. (2002) Enhancing quality of CELP coded speech via wideband extension by using voicing GMM interpolation and HNM re-synthesis, IEEE Acoustics, Speech, and Signal Processing, Proc. vol. 1, pp. I-241–244.

Rentzos D., Vaseghi S., Yan Q., Ho C. and Turajlic E. (2003) Probability Models of Formant Parameters for Voice Conversion, Proc. of Eurospeech, pp. 2405–2408.

Rigoll G. (1986) A New Algorithm for Estimation of Formant Trajectories Directly from the Speech Signal Based on an Extended Kalman-filter, Proc. of ICASSP, pp. 1229–1232.

Rødbro C.A., Christensen M.G., Andersen S.V. and Jensen S.H. (2003) Compressed Domain Packet Loss Concealment of Sinusoidally Coded Speech, Proc. IEEE Int. Conf. Acoustics, Speech, Signal Proc., 1, pp. 104–107.

Rødbro C.A., Murthi M.N., Andersen S.V. and Jensen S.H. (2005) Hidden Markov Model-based Packet Loss Concealment for Voice Over IP, IEEE Transactions on Audio, Speech, and Language Processing, PP (99), pp. 1–15.

Sameti H., Sheikhzadeh H., Deng L. and Brennan R.L. (1998) HMM-Based Strategies for Enhancement of Speech Signals Embedded in Non-Stationary Noise. IEEE Trans. Speech and Audio Processing, 6 (5) pp. 445–455.

Schroeder M.R. (1999) Computer Speech: Recognition, Compression, Synthesis, Springer.

Shafer R.W. and Rabiner L.R. (1970) System for Automatic Formant Analysis of Voiced Speech, J. Acoust. Soc. Am. 47 (2), pp. 634–650.

Stylianou Y. (1996) A Pitch and Maximum Voiced Frequency Estimation Technique Adapted to Harmonic Models of Speech, IEEE Nordic Signal Processing Symp., Helsinki, Finland, Sept.

Stylianou Y. (2001) Applying the harmonic plus noise model in concatenative speech synthesis, IEEE Trans. Speech and Audio Processing, 9 (1), pp. 21–29.

Valenzuela R.A. and Animalu C.N. (1989) A new voice-packet reconstruction technique, Proc. IEEE Int. Conf. Acoustics, Speech, Signal Processing, pp. 1334–1336.

Vaseghi S. (2005) Advanced Digital Signal Processing and Noise Reduction, 3rd edn. John Wiley & Sons.

Vaseghi S. and Rayner P.J.W. (1990) Detection and suppression of impulsive noise in speech communication systems, IEE Proceedings, Part 1, 137(1), pp. 38–46.

Vaseghi S., Zavarehei E. and Yan Q. (2006) Speech Bandwidth Extension: Extrapolations of Spectral Envelope and Harmonicity Quality of Excitation, ICASSP 2006, 3, pp. III-844–847.

Wang J. and Gibson J.D. (2000) Performance comparison of intraframe and interframe LSF quantization in packet networks, IEEE Workshop on Speech Coding, pp. 126–128.

Wang J. and Gibson J.D. (2001) Parameter interpolation to enhance the frame erasure robustness of CELP coders in packet networks, Proc. IEEE Int. Conf. Acoustics, Speech, Signal Processing, vol. 2, pp. 745–748.

Wang, D.L. and Lim, J.S. (1982) The Unimportance of Phase in Speech Enhancements, IEEE Trans. Acoust., Speech and Signal Processing, 30, pp. 679–681.

Weber K., Bengio S. and Bourlard H. (2001) HMM2-Extraction of Formant Structures and Their use for Robust ASR, Proc. of Eurospeech. pp. 607–610.

Yan Q, Vaseghi S., Zavarehei E. and Milner B. (2005) Formant-Tracking Linear Prediction Model for Speech Processing in Noisy Environmen,t Eurospeech.

Yan Q. and Vaseghi S. (2003) Analysis, Modelling and Synthesis of Formants of British, American and Australian Accents, IEEE International Conference on Acoustics, Speech, and Signal Processing, ICASSP vol. 1 (6–10), pp. I-712–715.

Zavarehei E., Vaseghi S. and Yan Q. (2006) Weighted Codebook Mapping for Noisy Speech Enhancement using Harmonic-Noise model, ICSLP 2006.

Zavarehei E., Vaseghi S. and Yan Q., Interframe Modelling of DFT Trajectories of Speech and Noise for Speech Enhancement using Kalman filters, Speech Communication, *in press*.

16 Echo Cancellation

Echo is the repetition of a waveform either due to reflections at points where the characteristics of the medium through which the wave propagates changes or due to the acoustic feedback between the speaker and the microphone of a communication system. In telecommunication, echo degrades the quality of service, and echo cancellation is an essential part of communication systems. The development of echo reduction began in the late 1950s and continues today as new integration of landline, wireless cellular networks, Internet, multiple-input multiple-output communication devices and teleconferencing systems place additional requirements on the performance of echo cancellers.

There are two types of echo in communication systems: acoustic echo and telephone line echo. Acoustic echo results from a direct and/or indirect acoustic feedback path set up between the loudspeaker and the microphone in a mobile phone, hands-free phone, teleconference or hearing aid system. Acoustic echo may be reflected from a multitude of different surfaces, such as walls, ceilings and floors, and travels through multiple paths.

Telephone line echoes result from an impedance mismatch at the telephone exchange hybrids where the subscriber's two-wire line is connected to a four-wire line. The perceptual effects of an echo depend on the time delay between the incident and reflected waves, the strength of the reflected waves, and the number of paths through which the waves are reflected. Telephone line echoes and acoustic feedback echoes in teleconference and hearing aid systems are undesirable and annoying and can be severely disruptive. In this chapter we study some methods for removing line echo from telephone and data telecommunication systems, and acoustic feedback echoes from microphone–loudspeaker systems.

16.1 Introduction: Acoustic and Hybrid Echo

Echo is the reflection of a signal from the points or surfaces where the characteristics of the medium through which the signal propagates changes. Echo is usefully employed for detection, exploration and navigation purposes in electronic detection and imaging instruments such as in sonar, ultrasonic imaging, infrared imaging and radar, and by some animals such as bats and dolphins.

In telecommunications, echo can severely affect the quality and intelligibility of voice conversation in telephone, teleconference or cabin communication systems. The perceived effect of an echo depends on its amplitude and time delay. In general, echoes with appreciable amplitudes and a delay of more

Multimedia Signal Processing: Theory and Applications in Speech, Music and Communications Saeed V. Vaseghi

than 1 msec can be noticeable. Provided the round-trip delay of the echo signal is of the order of a few milliseconds, echo may give a telephone call a perceived sense of 'liveliness'. However, echoes become increasingly annoying and objectionable with the increasing round-trip delay and amplitude, in particular for delays of more than 20 ms. Above a delay of 200 ms echoes can be disruptive for voice communication.

Echo cancellation is an important aspect of the design of modern telecommunication systems such as conventional wire-line telephones, hands-free phones, cellular mobile (wireless) phones, teleconference systems and in-car cabin communication systems. There are two types of echo in a telephone system (Figure 16.1):

(a) Acoustic echo due to acoustic coupling between the speaker and the microphone in hands-free phones, mobile phones and teleconference systems;
(b) Electrical line echo due to mismatch at the hybrid circuit connecting a two-wire subscriber line to a four-wire trunk line in the public switched telephone network.

In telephone landlines echoes are mostly due to the impedance mismatch at the point of connection of the two-wire local line to the four-wire trunk line. In the early days of expansion of telephone networks, the cost of running a four-wire line from the local exchange to subscribers' premises was considered uneconomical. Hence, at the exchange the four-wire trunk lines are converted to two-wire subscriber local lines using a two/four-wire hybrid bridge circuit. At the receiver due to any imbalance between the four/two-wire bridge circuit, some of the signal energy of the four-wire circuit is bounced back towards the transmitter, constituting an echo signal. If the echo is more than a few milliseconds long then it becomes noticeable, and can be annoying and disruptive.

In digital mobile phone systems, the echo is often due to the acoustic feedback coupling between the speaker and the microphone on the handset. In mobile phones the voice signals are processed at two points in the network: first at the voice coder on the hand set, the speech signals are digitised, divided into frames, compressed, coded and modulated; then the signal is processed at the radio frequency interface of the network. The total delay introduced by the various stages of digital signal processing range from 80 ms to 100 ms, resulting in a total round-trip delay of 160–200 ms for any echo. A delay of this magnitude will make any appreciable echo disruptive to the communication process. Owing to the inherent processing delay in digital mobile communication systems, it is essential and mandatory to employ echo cancellers in mobile phone switching centres.

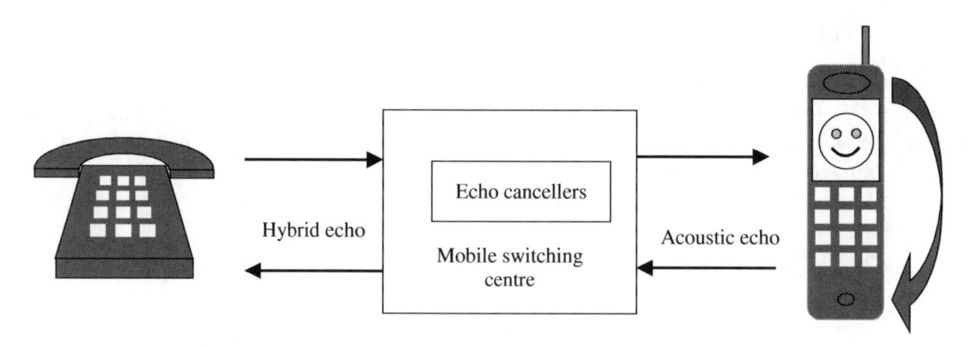

Figure 16.1 Illustration of sources of echo in a mobile to landline system.

16.2 Telephone Line Hybrid Echo

Hybrid echo is the main source of echo generated from the public-switched telephone network (PSTN). Echoes on a telephone line are due to the reflection of speech signals at the points of impedance mismatch on the connecting circuits. Conventionally, telephones in a given geographical area are connected to an exchange by a two-wire twisted line, called the subscriber's landline, which serves to receive and transmit signals. Both transmit and receive signals are present on the two-wire lines of the subscriber loop. In a conventional system a local call is set up by establishing a direct connection, at the telephone exchange, between two subscribers' loops. For a local call, there is usually no noticeable echo either because there is not a significant impedance mismatch on the connecting two-wire local lines or because the distances are relatively small and the resulting low-delay echoes (less than 30 msec) are perceived as a slight amplification and 'livening' effect. For long-distance communication between two exchanges, it is necessary to use repeaters to amplify the speech signals; therefore a separate two-wire telephone line is required for each direction of transmission.

To establish a long-distance call, at each end, a two-wire subscriber's line must be connected to a four-wire line at the exchange, as illustrated in Figure 16.2. The device that connects the two-wire subscriber's loop to the four-wire line is called a hybrid, and is shown in Figure 16.3. As shown the hybrid is basically a three-port bridge circuit that separates the transmit and receive signals into two separate pairs of wires. If the hybrid bridge circuit were perfectly balanced then there would be no reflection of signal or echo. However, each hybrid circuit serves a number of subscribers' lines. The subscribers' lines do not all have the same length and impedance characteristics; therefore it is not possible to achieve perfect balance for all subscribers at the hybrids. When the bridge is not perfectly balanced, some of the signal energy on the receiving four-wire lines becomes coupled back onto itself and produces an echo.

Echo Return Time: Sources of Delay in Telephone Networks

The delay of an echo is the round-trip time taken for the signal to arrive back at the source. End-to-end propagation delay is the sum of delays required for the voice of the speaker to propagate through different network devices and network links to reach the listener. Round-trip delay of an echo is the propagation time from the transmitter to the receiver and then back to the transmitter.

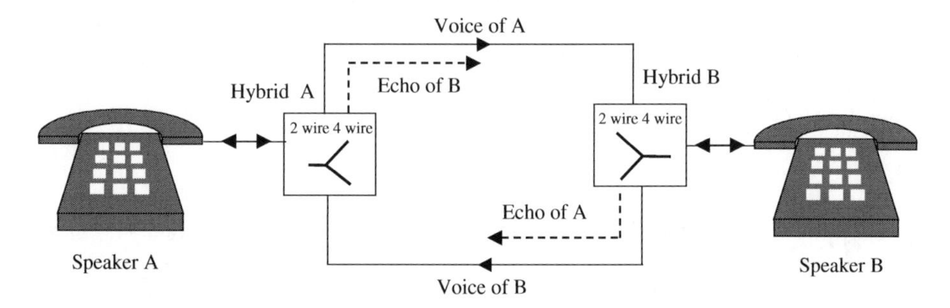

Figure 16.2 Illustration of a telephone call set up by connection of two-wire subscriber's phone via hybrids to four-wire lines at the exchange.

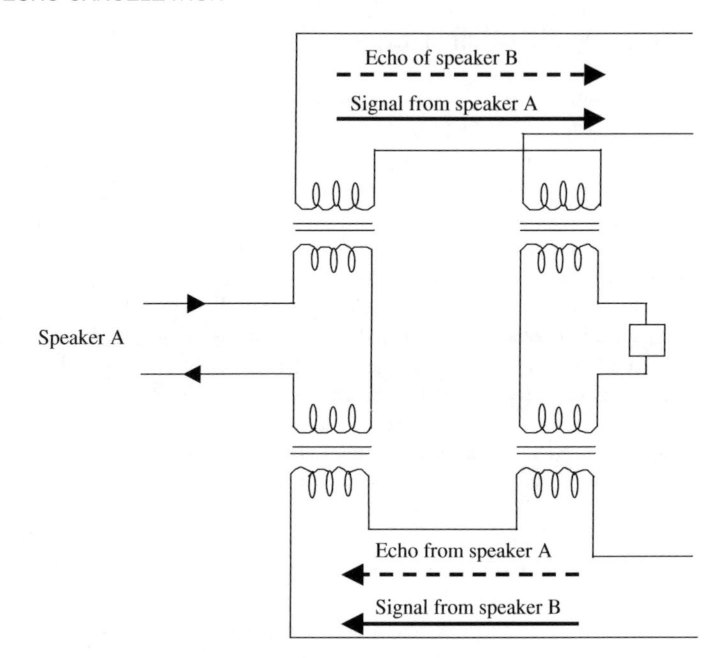

Figure 16.3 A two-wire to four-wire hybrid connection circuit.

For example when a long-distance call is made via a mobile phone and a satellite, the round-trip echo delay can be as long as, or sometimes even more than, 600 ms and echoes can become disruptive. For this reason the employment of echo cancellers in mobile switching centres and satellite networks is mandatory.

The sources of delay in PSTN telephone networks are as follows:

- Speech coder delay: 2.5 msec – 40 msec
- Inter-process hand-offs delay: about 10 msec at each end
- Transmission line delay: about 1 msec per 100 miles of cable
- Satellite links delay: 250–300 msec; multiple hops can yield longer delays
- Packet delay: from 0.5 msecs for a 128 bytes packet on a 2 Mbps line, to 128.6 msec for a 1024 bytes packet on a 64 Kbps line
- Voice over IP (VoIP) gateway node: 50–100 msec
- Decompression delay: typically 10 msec or less.

Acoustic echo can be longer that line echo. The duration of acoustic echo depends on the dimensions of the room and the number of reflections off the walls that the echo goes through. For example, sound travels at a speed of 340 metres/sec at a room temperature of 25 °C. Hence the time taken for sound to travel one metre will be about 2.94 msec. A distance of 10 metres from speaker to microphone will take about 29.4 msec and to this must be added the delay for coding and transmission through the communication network as described above.

Echo Return Loss

The intensity of echo is measured in terms of the echo return loss (ERL) defined as the power ratio, in dB, of the transmitted signal to that of the returned echo as

$$\text{ERL} = 10\log_{10}\left(\frac{\text{Transmitted Signal Power}}{\text{Echo Return Signal Power}}\right) \quad \text{dB} \tag{16.1}$$

The higher the echo return loss the lower will be the power of the echo.

The echo return loss enhancement (ERLE) is the difference in ERL before and after application of echo cancellation.

16.3 Hybrid (Telephone Line) Echo Suppression

The development of echo reduction began in the late 1950s with the advent of echo suppression systems. Echo suppressors were first employed to manage the echo generated primarily in satellite circuits. An echo suppresser (Figure 16.4) is primarily a switch that lets the speech signal through during the speech-active periods and attenuates the line echo during the speech-inactive periods. A line echo suppresser is controlled by a speech/echo detection device. The echo detector monitors the signal levels on the incoming and outgoing lines, and decides if the signal on a line from, say, speaker B to speaker A is the speech from speaker B to speaker A, or the echo of speaker A. If the echo detector decides that the signal is an echo then the signal is heavily attenuated. There is a similar echo suppression unit from speaker A to speaker B.

The performance of an echo suppresser depends on the accuracy of the echo/speech classification subsystem. The echo of speech often has a smaller amplitude level than the speech signal, but otherwise it has mainly the same spectral characteristics and statistics as those of the speech. Therefore the only basis for discrimination of speech from echo is the signal level. As a result, the speech/echo classifier may wrongly classify and let through high-level echoes as speech, or attenuate low-level speech as echo. For terrestrial circuits, echo suppressers have been well designed, with an acceptable level of false decisions and a good performance. The performance of an echo suppresser depends on the time delay of the echo. In general, echo suppressers perform well when the round-trip delay of the echo is less than 100 ms. For a conversation routed via a geostationary satellite the round-trip delay may be

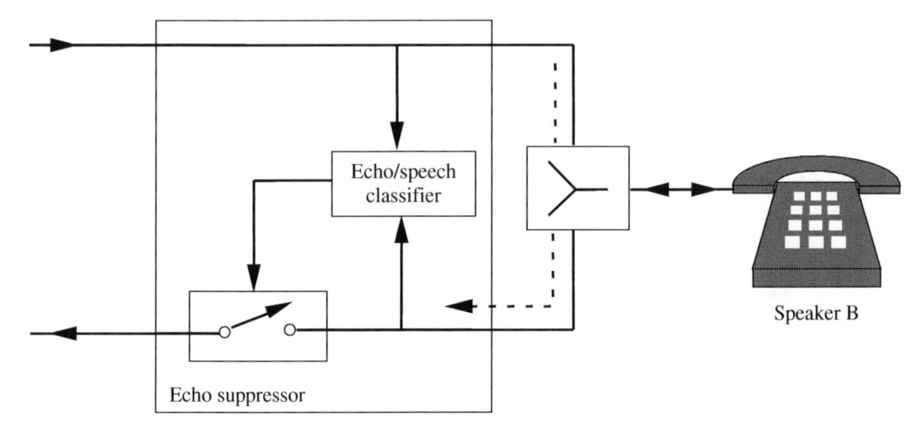

Figure 16.4 Block diagram illustration of an echo suppression system.

as much as 600 ms. Such long delays can change the pattern of conversation and result in a significant increase in speech/echo classification errors. When the delay is long, echo suppressers fail to perform satisfactorily, and this results in choppy first syllables and artificial volume adjustment. A system that is effective with both short and long time delays is the adaptive echo canceller introduced next.

16.4 Adaptive Echo Cancellation

Echo cancellation was developed in the early 1960s by AT&T Bell Labs and later by COMSAT TeleSystems. The first echo cancellation systems were experimentally implemented across satellite communication networks to demonstrate network performance for long-distance calls.

Figure 16.5 illustrates the operation of an adaptive line echo canceller. The speech signal on the line from speaker A to speaker B is input to the 4/2 wire hybrid B and to the echo canceller. The echo canceller monitors the signal on line from B to A and attempts to model the echo path and synthesise a replica of the echo of speaker A. This replica is used to subtract and cancel out the echo of speaker A on the line from B to A. The echo canceller is basically an adaptive linear filter. The coefficients of the filter are adapted so that the energy of the signal on the line is minimised. The echo canceller can be an infinite impulse response (IIR) or a finite impulse response (FIR) filter. The main advantage of an IIR filter is that a long-delay echo can be synthesised by a relatively small number of filter coefficients. In practice, echo cancellers are based on FIR filters. This is mainly due to the practical difficulties associated with the adaptation and stable operation of adaptive IIR filters.

Assuming that the signal on the line from speaker B to speaker A, $y_{\mathrm{B}}(m)$, is composed of the speech of speaker B, $x_{\mathrm{B}}(m)$, plus the echo of speaker A, $x_{\mathrm{A}}^{\mathrm{echo}}(m)$, we have

$$y_{\mathrm{B}}(m) = x_{\mathrm{B}}(m) + x_{\mathrm{A}}^{\mathrm{echo}}(m) \tag{16.2}$$

In practice, speech and echo signals are not simultaneously present on a phone line unless both speakers are speaking simultaneously. This, as pointed out shortly, can be used to simplify the adaptation process. Assuming that the truncated impulse response of the echo path can be modelled by an FIR filter, the output estimate of the synthesised echo signal can be expressed as

$$\hat{x}_{\mathrm{A}}^{\mathrm{echo}}(m) = \sum_{k=0}^{P} h_k(m) x_{\mathrm{A}}(m-k) \tag{16.3}$$

where $h_k(m)$ are the time-varying coefficients of an adaptive FIR filter model of the echo path and $\hat{x}_{\mathrm{A}}^{\mathrm{echo}}(m)$ is an estimate of the echo of speaker A on the line from speaker B to speaker A. The residual

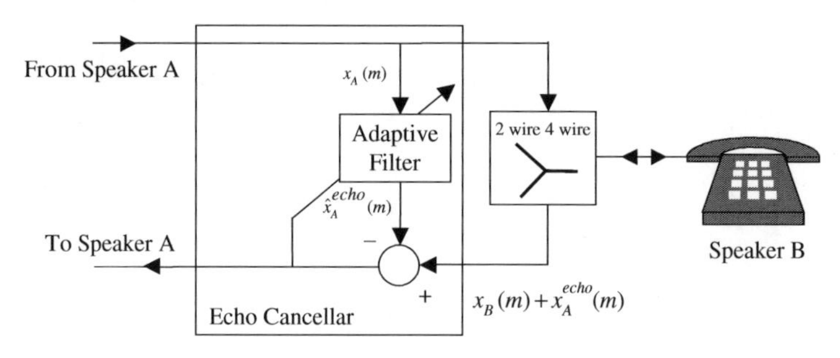

Figure 16.5 Block diagram illustration of an adaptive echo cancellation system.

echo signal, or the error signal, after echo subtraction is given by

$$e(m) = y_B(m) - \hat{x}_A^{echo}(m)$$
$$= x_B(m) + x_A^{echo}(m) - \sum_{k=0}^{P} h_k(m) x_A(m-k) \tag{16.4}$$

For those time instants when speaker A is talking, and speaker B is listening and silent, and only echo is present from line B to A, we have

$$e(m) = \tilde{x}_A^{echo}(m) = x_A^{echo}(m) - \hat{x}_A^{echo}(m)$$
$$= x_A^{echo}(m) - \sum_{k=0}^{P} h_k(m) x_A(m-k) \tag{16.5}$$

where $\tilde{x}_A^{echo}(m)$ is the residual echo. An echo canceller using an adaptive FIR filter is illustrated in Figure 16.6. The magnitude of the residual echo depends on the ability of the echo canceller to synthesise a replica of the echo, and this in turn depends on the adaptation algorithm discussed next.

16.4.1 Echo Canceller Adaptation Methods

The echo canceller coefficients $h_k(m)$ are adapted to minimise the energy of the echo signal on a telephone line, say from speaker B to speaker A. Assuming that the speech signals $x_A(m)$ and $x_B(m)$ are uncorrelated, the energy on the telephone line from B to A is minimised when the echo canceller output $\hat{x}_A^{echo}(m)$ is equal to the echo $x_A^{echo}(m)$ on the line. The echo canceller coefficients may be adapted using one of the variants of the recursive least square error (RLS) or the least mean squared error (LMS) adaptation methods. One of the most widely used algorithms for adaptation of the coefficients of an echo canceller is the normalised least mean square error (NLMS) method. The time-update equation describing the adaptation of the filter coefficient vector is

$$\boldsymbol{h}(m) = \boldsymbol{h}(m-1) + \mu \frac{e(m)}{\boldsymbol{x}_A(m)^T \boldsymbol{x}_A(m)} \boldsymbol{x}_A(m) \tag{16.6}$$

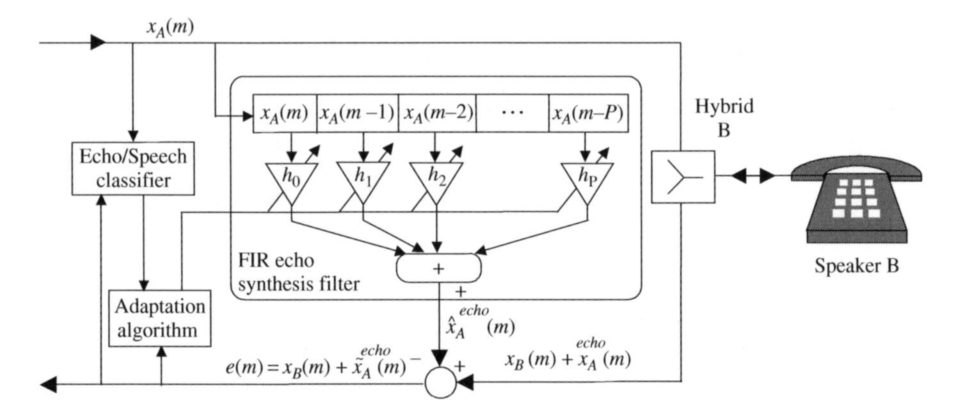

Figure 16.6 Illustration of an echo canceller using an adaptive FIR filter and incorporating an echo/speech classifier.

where $\boldsymbol{x}_A(m) = [x_A(m), \ldots, x_A(m-P)]$ and $\boldsymbol{h}(m) = [h_0(m), \ldots, h_P(m)]$ are the input signal vector and the coefficient vector of the echo canceller, and $e(m)$ is the error signal that is the difference between the signal on the echo line and the output of the echo synthesiser. Note that the normalising quantity $\boldsymbol{x}(m)_A^T \boldsymbol{x}_A(m)$ is the energy of the input speech to the adaptive filter. The scalar μ is the adaptation step size, and controls the speed of convergence, the steady-state error and the stability of the adaptation process.

16.4.2 Convergence of Line Echo Canceller

For satisfactory performance, the echo canceller should have a fast convergence rate, so that it can adequately track changes in the communication link and the signal characteristics. The convergence of an echo canceller is affected by the following factors:

(a) *Non-stationary characteristics of telephone line and speech.* The echo characteristics depend on the impedance mismatch between the subscribers' loop and the hybrids. Any changes in the connecting paths affect the echo characteristics and the convergence process. Also as explained in Chapter 9, the non-stationary character and the eigenvalue spread of the input speech signal of an LMS adaptive filter affect the convergence rates of the filter coefficients.
(b) *Simultaneous conversations – double talk.* In a telephone conversation, usually the speakers do not talk simultaneously, and hence speech and echo are seldom present on a line at the same time. This observation simplifies the echo cancellation problem and substantially aids the correct functioning of adaptive echo cancellers. Problems arise during the periods when both speakers talk at the same time. This is because speech and its echo have similar characteristics and occupy basically the same bandwidth. When the reference signal contains both echo and speech, the adaptation process can lose track, and the echo cancellation process can attempt to cancel out and distort the speech signal. One method of avoiding this problem is to use a voice activity detector (VAD), and freeze the adaptation process during periods when speech and echo are simultaneously present on a line, as shown in Figure 16.6. In this system, the effect of a speech/echo misclassification is that the echo may not be optimally cancelled out. This is more acceptable than is the case in echo suppressors, where the effect of a misclassification is the suppression and loss of part of the speech.

The adaptation algorithm. Most echo cancellers use variants of the LMS adaptation algorithm. The attractions of the LMS algorithm are its relatively low memory and computational requirements and its ease of implementation and monitoring. The main drawback of the LMS algorithm is that it can be sensitive to the eigenvalue spread of the input signal and is not particularly fast in its convergence rate. However, in practice, LMS adaptation has produced effective line echo cancellation systems. The recursive least square (RLS) error methods have a faster convergence rate, are less sensitive to the eigenvalue spread of the signal and have a better minimum mean square error performance. With the increasing availability of low-cost high-speed dedicated DSP processors, implementation of higher performance and computationally intensive echo cancellers based on RLS are now feasible.

16.4.3 Echo Cancellation for Digital Data Transmission

Echo cancellation becomes more complex with the increasing integration of wireline telephone systems and mobile cellular systems, and the use of digital transmission methods such as asynchronous transfer mode (ATM) for integrated transmission of data, image and voice. For example, in ATM based systems, the voice transmission delay varies depending on the route taken by the cells that carry the

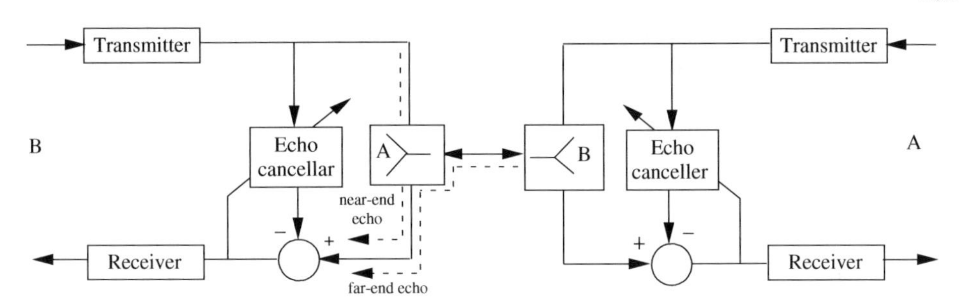

Figure 16.7 Echo cancellation in digital modems using two-wire subscriber's loop.

voice signals. This variable delay added to the delay inherent in digital voice coding complicates the echo cancellation process.

The two-wire subscriber telephone lines that were originally intended to carry relatively low-bandwidth voice signals are now used to provide telephone users with high-speed digital data links and digital services such as video-on-demand and Internet services using digital transmission methods such as the asynchronous digital subscriber line (ADSL). Traditionally, the bandwidth of the subscriber's line is limited by low-pass filters at the core network to 3.4 kHz. Within this bandwidth, voice-band modems can provide data rates of around 30 kilobits per second (kbps). However the copper wire itself has a much higher usable bandwidth extending into megahertz regions, although attenuation and interference increase with both the frequency and the length of the wire. Using advanced signal processing and modulation schemes methods such as ADSL can achieve a 10 megabits per second data rate over 240 MHz bandwidth of subscriber's twisted wire line.

Figure 16.7 shows a system for providing a full-duplex digital service over a two-wire subscriber's loop. To provide simultaneous transmission of data in both directions within the same bandwidth over the subscriber's line, echo cancellation is needed. The echoes on a line consist of the near-end echo which loops back at the first or the near-end hybrid, and the far-end echo which is the signal that loops back at a hybrid some distance away. The main purpose of the echo canceller is to cancel the near-end echo. Since the digital signal coming from a far-end echo may be attenuated by 40–50 dB, the near echo on a high speed data transmission line can be as much as 40–50 dB above the desired signal level. For reliable data communication the echo canceller must provide 50–60 dB attenuation of the echo signal so that the signal power remains at 10 dB above the echo.

16.5 Acoustic Echo

Acoustic echo results from a feedback path set up between the speaker and the microphone in a mobile phone, hands-free phone, teleconference or hearing aid system. Acoustic echo is usually reflected from a multitude of different surfaces, such as walls, ceilings and floors, and travels through different paths. If the time delay is not too long then the acoustic echo may be perceived as a soft reverberation, and may add to the artistic quality of the sound. Concert halls and church halls with desirable reverberation characteristics can enhance the quality of a musical performance. However, acoustic echo is a well-known problem with hands-free telephones, teleconference systems, public address systems, mobile phones, and hearing aids, and is due to acoustic feedback coupling of sound waves between the loudspeakers and microphones. Acoustic echo can result from a combination of direct acoustic coupling and a multi-path effect where the sound wave is reflected from various surfaces and then picked up by the microphone. In its worst case, acoustic feedback can result in howling if a significant proportion of the sound energy transmitted by the loudspeaker is received back at the microphone and circulated in the feedback loop. The overall round gain of an acoustic feedback loop depends on the frequency responses of the electrical and the acoustic signal paths. The undesirable effects of the electrical sections on the acoustic feedback can be reduced by designing systems that have a flat frequency response. The main problem is in the acoustic feedback path and the reverberating characteristics of the room. If the microphone–speaker–room system is excited at a frequency whose loop gain is greater than unity then the signal is amplified each time it circulates round the loop, and feedback howling results. In practice, the howling is limited by the non-linearity of the electronic system.

There are a number of methods for removing acoustic feedback. One method for alleviating the effects of acoustic feedback and the room reverberations is to place a frequency shifter (or a phase shifter) in the electrical path of the feedback loop. Each time a signal travels round the feedback loop it is shifted by a few hertz before being re-transmitted by the loudspeaker. This method has some effect in reducing howling but it is not effective for removal of the overall echo of the acoustic feedback. Another approach is to reduce the feedback loop-gain at those frequencies where the acoustic feedback energy is concentrated. This may be achieved by using adaptive notch filters to reduce the system gain at frequencies where acoustic oscillations occur. The drawback of this method is that in addition to reducing the feedback the notch filters also result in distortion of the desired signal frequencies.

The most effective method of acoustic feedback removal is the use of an adaptive feedback cancellation system. Figure 16.8 illustrates a model of an acoustic feedback environment, comprising a microphone, a loudspeaker and the reverberating space of a room. The z-transfer function of a linear model of the acoustic feedback environment may be expressed as

$$H(z) = \frac{G(z)}{1 - G(z)A(z)} \tag{16.7}$$

where $G(z)$ is the z-transfer function model for the microphone–loudspeaker system and $A(z)$ is the z-transfer function model of reverberations and multi-path reflections of a room environment. Assuming that the microphone–loudspeaker combination has a flat frequency response with a gain of G, Equation (16.7) can be simplified to

$$H(z) = \frac{G}{1 - GA(z)} \tag{16.8}$$

Note that in Equations (16.7) and (16.8), owing to the reverberating character of the room, the acoustic feedback path $A(z)$ is itself a feedback system. The reverberating characteristics of the acoustic environment may be modelled by an all-pole linear predictive model, or alternatively a relatively long FIR model.

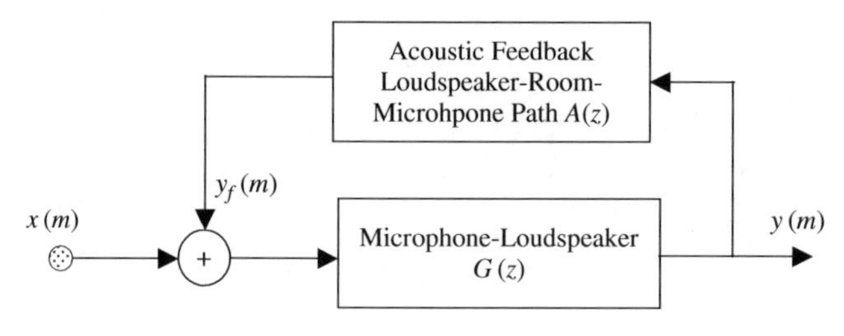

Figure 16.8 Configuration of a feedback model for a microphone–loudspeaker–room system.

The equivalent time-domain input/output relation for the linear filter model of Equation (16.8) is given by the following difference equation:

$$y(m) = G\sum_{k=0}^{P} a_k(m)y(m-k) + Gx(m) \tag{16.9}$$

where $a_k(m)$ are the coefficients of an all-pole linear feedback model of the reverberating room environment, G is the microphone–loudspeaker amplitude gain factor, and $x(m)$ and $y(m)$ are the time domain input and output signals of the microphone–loudspeaker system.

Figure 16.9 is an illustration of an acoustic feedback cancellation system. In an acoustic feedback environment, the total input signal to the microphone is given as the sum of any new input to the microphone $x(m)$ plus the unwanted acoustic feedback signal $y_f(m)$:

$$y(m) = x(m) + y_f(m) \tag{16.10}$$

The most successful acoustic feedback control systems are based on adaptive estimation and cancellation of the feedback signal. As in a line echo canceller, an adaptive acoustic feedback canceller attempts to synthesise a replica of the acoustic feedback at its output as

$$\hat{y}_f(m) = \sum_{k=0}^{P} a_k(m)y(m-k) \tag{16.11}$$

The filter coefficients are adapted to minimise the energy of an error signal defined as

$$e(m) = x(m) + y_f(m) - \hat{y}_f(m) \tag{16.12}$$

The adaptation criterion is usually the minimum mean square error criterion and the adaptation algorithm is the NLMS or the RLS method. The problem of acoustic echo cancellation is more complex than line echo cancellation for a number of reasons. First, acoustic echo is usually much longer (up to a second) than terrestrial telephone line echoes. In fact, the delay of an acoustic echo is similar to or more than a line echo routed via a geostationary satellite system.

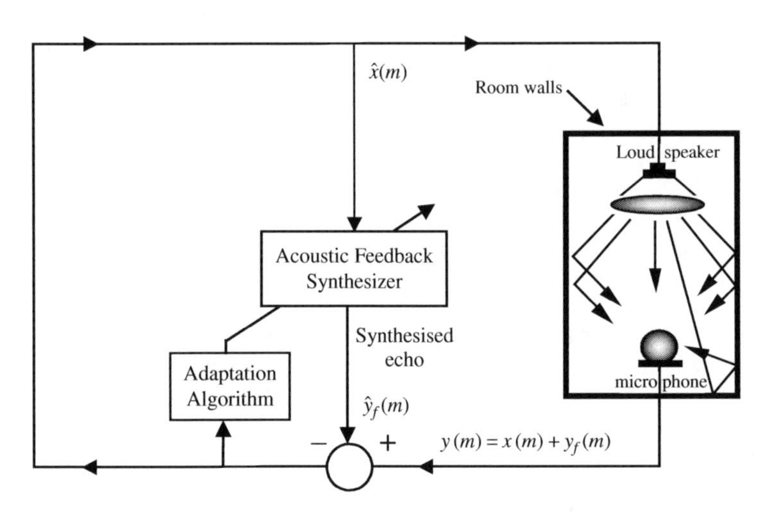

Figure 16.9 Illustration of adaptive acoustic feedback cancellation in a conference room environment.

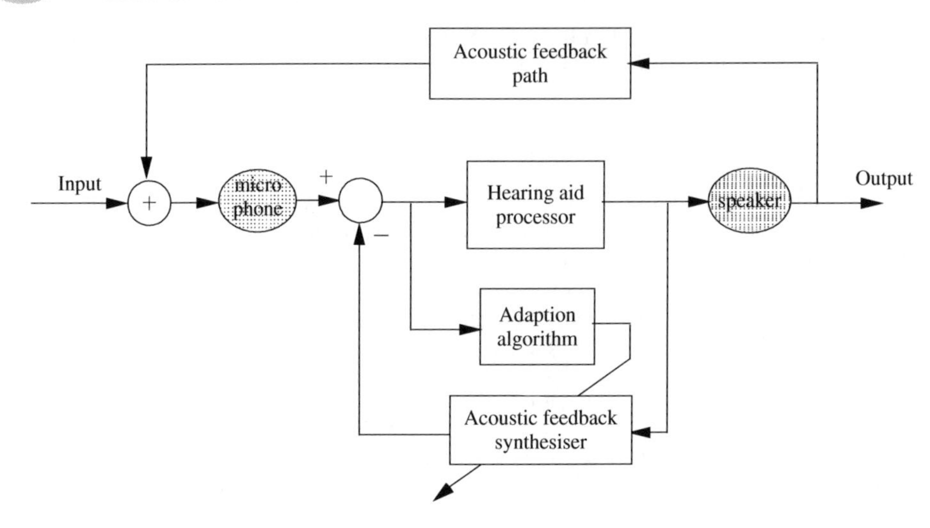

Figure 16.10 Configuration of an acoustic feedback canceller incorporated in a hearing aid system.

The large delay of an acoustic echo path implies that impractically large filters of the order of a few thousand coefficients may be required. The stable and speedy adaptation of filters of such length presents a difficult problem. Secondly, the characteristics of an acoustic echo path are more non-stationary compared with those of a telephone line echo. For example, the opening or closing of a door, or people moving in or out of a room, can suddenly change the acoustic character of a conference room. Thirdly, acoustic echoes are due to signals reflected back from a multitude of different paths, off the walls, the floor, the ceiling, the windows etc. Finally, the propagation and diffusion characteristics of the acoustic space of a room are a non-linear process, and are not well approximated by a lumped FIR (or IIR) linear filter. In comparison, it is more reasonable to model the characteristics of a telephone line echo with a linear filter. In any case, for acoustic echo cancellation, the filter must have a large impulse response and should be able to quickly track fast changes in echo path characteristics.

An important application of acoustic feedback cancellation is in hearing aid systems, Figure 16.10. The maximum usable gain of a hearing aid system is limited by the acoustic feedback between the microphone and the speaker. Figure 16.10 illustrates the configuration of a feedback canceller in a hearing aid system. The acoustic feedback synthesiser has the same input as the acoustic feedback path. An adaptation algorithm adjusts the coefficients of the synthesiser to cancel out the feedback signals picked up by the microphone, before the microphone output is fed into the speaker.

16.6 Sub-Band Acoustic Echo Cancellation

In addition to the complex and varying nature of room acoustics, there are two main problems in acoustic echo cancellation. First, the echo delay is relatively long, and therefore the FIR echo synthesiser must have a large number of coefficients, say 2000 or more. Secondly, the long impulse response of the FIR filter and the large eigenvalue spread of the speech signals result in a slow and uneven rate of convergence of the adaptation process.

A sub-band-based echo canceller alleviates the problems associated with the required filter length and the speed of convergence. The sub-band-based system is shown in Figure 16.11. The sub-band analyser splits the input signal into N sub-bands. Assuming that the sub-bands have equal bandwidth, each sub-band occupies only $1/N$ of the baseband frequency, and can therefore be decimated (down-sampled)

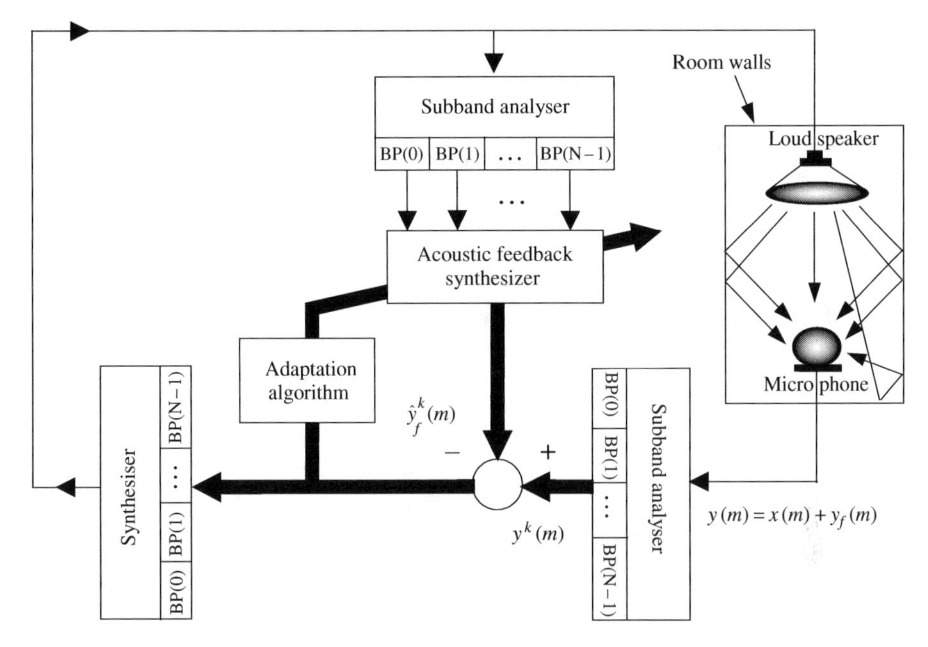

Figure 16.11 Configuration of a sub-band acoustic echo cancellation system.

without loss of information. For simplicity, assume that all sub-bands are down-sampled by the same factor R. The main advantages of a sub-band echo canceller are a reduction in filter length and a gain in the speed of convergence as explained below:

(a) *Reduction in filter length.* Assuming that the impulse response of each sub-band filter has the same duration as the impulse response of the full band FIR filter, the length of the FIR filter for each down-sampled sub-band is $1/R$ of the full band filter.
(b) *Reduction in computational complexity.* The computational complexity of an LMS-type adaptive filter depends directly on the product of the filter length and the sampling rate. As for each sub-band, the number of samples per second and the filter length decrease with $1/R$, it follows that the computational complexity of each sub-band filter is $1/R^2$ of that of the full band filter. Hence the overall gain in computational complexity of a sub-band system is R^2/N of the full band system.
(c) *Speed of convergence.* The speed of convergence depends on both the filter length and the eigenvalue spread of the signal. The speed of convergence increases with the decrease in the length of the FIR filter for each sub-band. A more important factor affecting the convergence of adaptive filter is the eigenvalue spread of the autocorrelation matrix of the input signal. As the spectrum of a signal becomes flatter, the spread of its eigenvalues decreases, and the speed of convergence of the adaptive filter increases. In general, the signal within each sub-band is expected to have a flatter spectrum than the full band signal. This aids the speed of convergence. However, it must be noted that the attenuation of sub-band filters at the edges of the spectrum of each band creates some very small eigenvalues.

16.7 Echo Cancellation with Linear Prediction Pre-whitening

Adaptive echo cancellation systems work better (i.e. converge faster) if the input and the reference signals are uncorrelated white noise processes. Speech signals are highly correlated but can be

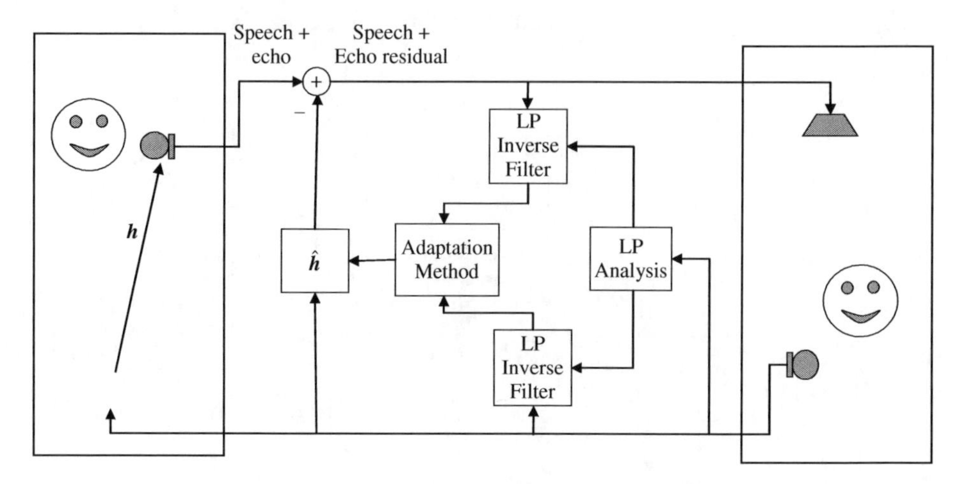

Figure 16.12 An acoustic echo cancellation system incorporating inverse linear prediction filters as pre-whitening filters.

pre-whitened by first modelling the speech with a linear prediction model (introduced in Chapter 10) and then using an inverse linear predictor for whitening the signals as illustrated in Figure 16.12. Linear prediction models and pre-whitening filters are described in detail in Chapter 10. The pre-whitened input to adaptive filter, i.e. pre-whitened incoming signal, is given by

$$e(m) = x(m) - \sum_{k=1}^{P} a_k x(m-k) \tag{16.13}$$

A similar equation can be used to pre-whiten the adaptive echo canceller's reference signal as shown in Figure 16.12.

Note that for the purpose of synthesis of the echo the input to the filter $\widehat{\boldsymbol{h}}$ is the non-whitened speech signal, whereas for the purpose of the adaptation of the filter coefficients $\widehat{\boldsymbol{h}}$ the whitened speech and whitened reference signals are used.

The process of pre-whitening the input and reference signals of the adaptive filter can substantially improve the performance of echo cancellation systems.

16.8 Multi-Input Multi-Output Echo Cancellation

Multiple-input multiple-output (MIMO) echo-cancellation systems have applications in car-cabin communications systems, stereophonic teleconferencing systems and conference halls. Stereophonic echo cancellation systems have been developed relatively recently and MIMO systems are still the subject of ongoing research and development.

In a typical MIMO system there are P speakers and Q microphones in the room. As an acoustic feedback path is set up between each speaker and each microphone, there are altogether $P \times Q$ such acoustic feedback paths that need to be modelled and estimated. The truncated impulse response of each acoustic path from loudspeaker i to microphone j may be modelled by an FIR filter $\boldsymbol{h}_{ij}$. The truncated impulse response of each acoustic path from human speaker i to microphone j is modelled by an FIR filter $\boldsymbol{g}_{ij}$.

For a large number of speakers and microphones, the modelling and identification of the numerous acoustic channels becomes a major problem due to the correlations of the echo signals, from a common number of sources, propagating through different channels as discussed later.

16.8.1 Stereophonic Echo Cancellation Systems

Figure 16.13 shows the configuration of an echo cancellation for a stereophonic communication system. There are two microphones and two loudspeakers at each end of the communication link. Each microphone receives the feedback echo from two different speakers through two different paths. In addition there are usually multi-path reflections of sounds from walls. Let the speech signal $s(m)$ after reaching the right and left microphones be denoted as $x_r(m)$ and $x_l(m)$ respectively. We may write

$$x_r(m) = \boldsymbol{g}_r^T(m)\boldsymbol{s}(m) \text{ and } x_l(m) = \boldsymbol{g}_l^T(m)\boldsymbol{s}(m) \tag{16.14}$$

where $\boldsymbol{g}_r$ and $\boldsymbol{g}_l$ are the truncated room impulse responses from the source speaker to the right and left microphones respectively.

In Figure 16.13, the truncated impulse responses of the acoustic feedback paths from the right and left loudspeakers to the right microphone are denoted as to $\boldsymbol{h}_{rr}(m)$ and $\boldsymbol{h}_{lr}(m)$ respectively and combined as $\boldsymbol{h}_r(m) = [\boldsymbol{h}_{rr}(m), \boldsymbol{h}_{lr}(m)]$. The signals from the right and left loudspeakers, $\boldsymbol{x}_r(m)$ and $\boldsymbol{x}_l(m)$, may be combined as $\boldsymbol{x}(m) = [\boldsymbol{x}_r(m), \boldsymbol{x}_l(m)]$. There exist similar paths from loudspeakers to each microphone which are not shown here in order to avoid overcrowding the figure. The synthesised replication of the echo signal in the right microphone is given by

$$\hat{x}_{\text{echo},r}(m) = \underbrace{\hat{\boldsymbol{h}}_{rr}^T(m-1)\boldsymbol{x}_r(m)}_{\substack{\text{Synthesised echo from right} \\ \text{loudspeaker to right microphone}}} + \underbrace{\hat{\boldsymbol{h}}_{lr}^T(m-1)\boldsymbol{x}_l(m)}_{\substack{\text{Synthesised echo from left} \\ \text{loudspeaker to right microphone}}} = \hat{\boldsymbol{h}}_r^T(m-1)\boldsymbol{x}(m) \tag{16.15}$$

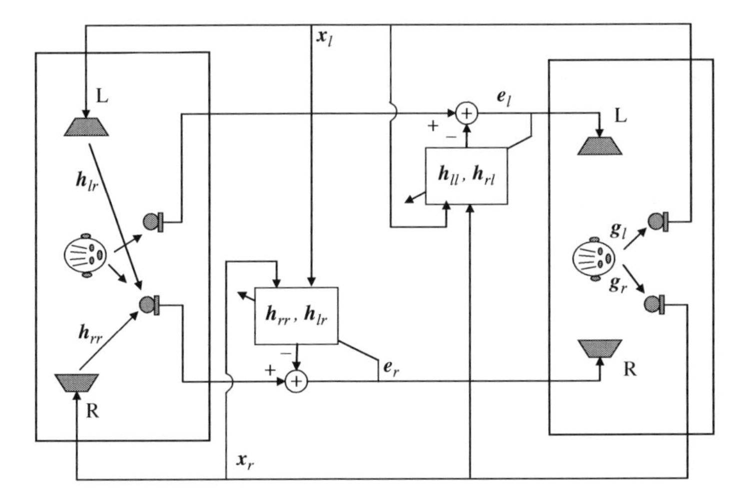

Figure 16.13 Illustration of the feedback signals and adaptive cancellation of acoustic feedbacks in a stereophonic echo-cancellation system.

The error signal composed of speech and the residue echo is given by

$$e_r(m) = y(m) - \hat{x}_{\text{echo},r}(m) \tag{16.16}$$

The NLMS adaptation method for FIR model of stereophonic echo paths is

$$\hat{\boldsymbol{h}}_r(m) = \hat{\boldsymbol{h}}_r(m-1) + \mu \frac{e(m)}{\boldsymbol{x}(m)^T \boldsymbol{x}(m)} \boldsymbol{x}(m) \tag{16.17}$$

Similar equations describing the echo and adaptive echo cancellation can be written for the right microphones.

Non-uniqueness Problem in MIMO Echo Channel Identification

A problem in MIMO echo cancellation systems is that the speech signals from different loudspeakers reaching a microphone are highly correlated. For the stereo echo cancellation, the loudspeakers' signals are $\boldsymbol{x}(m) = [\boldsymbol{x}_r(m)\boldsymbol{x}_l(m)]$ and the channels to be identified, for example from right and left loudspeakers to say the right microphone, are $\boldsymbol{h}_r(m) = [\boldsymbol{h}_{rr}(m), \boldsymbol{h}_{lr}(m)]$. The Wiener solution to this echo path estimation problem is given by

$$\hat{\boldsymbol{h}}_r = \boldsymbol{R}_{xx}^{-1} \boldsymbol{r}_{xy} \tag{16.18}$$

where $\boldsymbol{R}_{xx}$ is the autocorrelation matrix of $\boldsymbol{x}(m) = [\boldsymbol{x}_r(m), \boldsymbol{x}_l(m)]$. The problem is that due to the high correlation of $\boldsymbol{x}_r(m)$ and $\boldsymbol{x}_l(m)$ the autocorrelation matrix $\boldsymbol{R}_{xx}$ is not a full rank matrix and hence the solution is not unique.

The non-uniqueness problem can also be explained in the frequency domain by considering the sum of the feedbacks from the loudspeakers into say the right microphone $X_{\text{echo},r}(f)$:

$$\begin{aligned} X_{\text{echo},r}(f) &= H_{lr}(f)G_l(f)S(f) + H_{rr}(f)G_r(f)S(f) \\ &= [H_{lr}(f)G_l(f) + H_{rr}(f)G_r(f)]S(f) \end{aligned} \tag{16.19}$$

where $G_l(f)$, $G_r(f)$ are the frequency responses of the paths from the source to the transmitter microphones and $H_{lr}(f)$ and $H_{rr}(f)$ are the frequency responses of the loudspeakers feedback paths to the receiver's right microphone. Note there are many combinations of different values of $G_l(f)$, $G_r(f)$, $H_{lr}(f)$ and $H_{rr}(f)$ that would satisfy Equation (16.19). A solution to this problem is to de-correlate the stereo signals. The problem of stereophonic echo cancellation is the subject of ongoing research. A good paper on the problems of stereophonic echo cancellation is by Benesty et al. (1998).

MIMO In-Cabin Communication Systems

MIMO systems have application for in-car cabin communication systems (CCS) and for multipurpose and large vehicles. The problem in CCS systems is background noise reduction and acoustic feedback cancellation. Figure 16.14 illustrates the configuration of a two-loudspeaker two-microphone system and the configurations of the FIR adaptive filters used for modelling the truncated impulse response of the acoustic feedback path from each loudspeaker via the cabin to reach the microphone. The synthesised feedback is subtracted from the signal received by each microphone in order to cancel feedback echo signals.

An interesting system solution to mitigate the effect of feedback in MIMO systems is to combine beam-forming microphone arrays with echo cancellation. Microphone arrays, introduced in Chapters 1 and 15, form a beam where any signal arriving from directions outside the beam is screened out and attenuated. This would help to reduce the feedback and can be used in combination with echo cancellers.

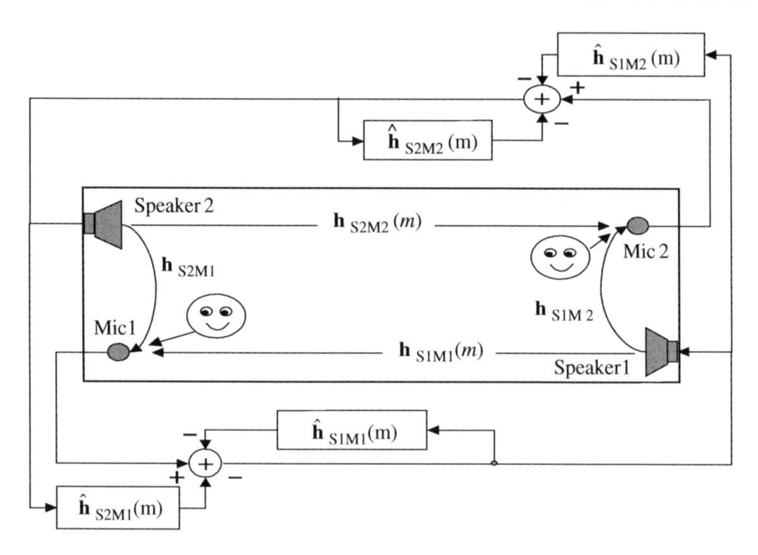

Figure 16.14 A block diagram illustration of a MIMO echo-cancellation in car communication system.

16.9 Summary

Telephone line echo and acoustic feedback echo affect the functioning of telecommunication and teleconferencing systems. In general, line echo cancellation is a relatively less complex problem than acoustic echo cancellation because acoustic cancellers need to model the more complex environment of the space of a room.

We began this chapter with a study of the telephone line echoes arising from the mismatch at the two/four-wire hybrid bridge. In Section 16.2, line echo suppression and adaptive line echo cancellation were considered. For adaptation of an echo canceller, the LMS or the RLS adaptation methods can be used. The RLS methods provide a faster convergence rate and better overall performance at the cost of higher computational complexity.

In Section 16.3, we considered the acoustic coupling between a loudspeaker and a microphone system. Acoustic feedback echo can result in howling, and can disrupt the performance of teleconference, hands-free telephones, and hearing aid systems. The main problems in implementation of acoustic echo cancellation systems are the requirement for a large filter to model the relatively long echo, and the adaptation problems associated with the eigenvalue spread of the signal. The sub-band echo canceller introduced in Section 16.4 alleviates these problems.

For stereophonic and MIMO systems the problem of acoustic echo cancellation remains an important and challenging research issue.

Further Reading

Allen J., Berkley D. and Blauret J. (1977) Multi-Microphone Signal Processing Technique to Remove Room Reverberation from Speech Signals. J. Acoust. Soc. Am., 62, 4.

Armbruster W. (1992) Wideband Acoustic Echo Canceller with Two Filter Structure. Proc. Eusipco-92, 3, pp. 1611–1617.

Benesty J., Morgan D.R. and Sondhi M.M. (1998) A Better Understanding of an Improved Solution to the Problem of Stereophonic Acoustic Echo Cancellation, IEEE Trans. Speech and Audio Processing, 6(2), pp. 156–165.

Carter G. (1987) Coherence and Time Delay Estimation. Proc. IEEE. 75, 2, pp. 236–255.
Flanagan J.L, Johnston J.D., Zahn R. and Elko G. (1985) Computer-steered Microphone Arrays for Sound Transduction in Large Rooms, J. Acoust. Soc. Am.,. 78 (5), pp. 1508–1518.
Flanagan J.L. *et al.* (1991) Autodirective Microphone systems. Acoustica 73, pp. 58–71.
Flanagan J.L., Berkley D., Elko G., West J. and Sondhi M. (1991) Autodirective Microphone Systems. Acoustica 73, pp. 58–71.
Gao X.Y. and Snelgrove W.M. (1991) Adaptive Linearisation of a Loudspeaker, IEEE. Proc. Int. Conf. Acoustics, Speech and Signal Processing, ICASSP–91, 3, pp. 3589–3592.
Gilloire A. and Vetterli M. (1994) Adaptive Filtering in Sub-bands with Critical Sampling: Analysis, Experiments and Applications to Acoustic Echo Cancellation, IEEE. Trans. Signal Processing, 40, pp. 320–328.
Gritton C.W. and Lin D.W. (1984) Echo Cancellation Algorithms, IEEE ASSP Mag., 1 (2), pp. 30–37.
Gustafsson S. and Martin R. (1997) Combined Acoustic Echo Control and Noise Reduction for Mobile Communications, Proc. EuroSpeech-97, pp. 1403–1406.
Hansler E. (1992) The Hands-Free Telephone Problem – An Annotated Bibliography. Signal Processing, 27, pp. 259–271.
Hart J.E., Naylor P.A. and Tanrikulu O. (1993) Polyphase Allpass IIR Structures for Sub-band Acoustic Echo Cancellation. EuroSpeech-93, 3, pp. 1813–1816.
Hua Ye and Bo-Xia Wu (1991) A New Double-Talk Detection Algorithm Based on the Orthogonality Theorem. IEEE Trans on Communications, 39 (11), pp. 1542–1545.
Kellermann W. (1988) Analysis and Design of Multirate Systems for Cancellation of Acoustical Echoes. IEEE. Proc. Int. Conf. Acoustics, Speech and Signal Processing, ICASSP-88, pp. 2570-2573.
Knappe M.E. (1992) Acoustic Echo Cancellation: Performance and Structures. M. Eng. Thesis, Carleton University, Ottawa, Canada.
Martin R. and Altenhoner J. (1995) Coupled Adaptive Filters for Acoustic Echo Control and Noise Reduction. IEEE. Proc. Int. Conf. Acoustics, Speech and Signal Processing, ICASSP-95, 5, pp. 3043-3046.
McCaslin S.R., Hemkumar N. and Redheendran B. (1997) Double-Talk Detector for Echo Canceller. US Patent No. 5631900, May 20.
Oslen H.F. (1964) Acoustical Engineering. D. Van Nostrand, Toronto.
Schroeder M.R. (1964) Improvement of Acoustic-Feedback Stability by Frequency Shifting. J. Acoust. Soc. Amer., 36, pp. 1718–1724.
Silverman H. and Kirtman E. (1992) A Two-stage Algorithm for Determining Talker Location from Linear Microphone Array Data, Comput. Speech Lang., 6, pp. 129–152.
Sondhi M.M. (1967) An Adaptive Echo Canceller. Bell Syst. Tech. J., 46, pp. 497–511.
Sondhi M.M. and Berkley D.A. (1980) Silencing Echoes on the Telephone Network. Proc. IEEE, 68, pp. 948–963.
Sondhi M.M. and Morgan D.R. (1991) Acoustic Echo Cancellation for Stereophonic Teleconferencing. IEEE Workshop on Applications of Signal Processing to Audio and Acoustics.
Tanrikulu O., Baykal B., Constantinides A.G. and Chambers J.A. (1995) Finite-Precision Design and Implementation of All-Pass Polyphase Networks for Echo Cancellation in Sub-bands. IEEE. Proc. Int. Conf. Acoustics, Speech and Signal Processing, ICASSP-95, 5, pp. 3039–3042.
Vaidyanathan P.P. (1993) Multirate Systems and Filter Banks. Prentice-Hall.
Widrow B., McCool J.M., Larimore M.G. and Johnson C.R. (1976) Stationary and Nonstationary Learning Characteristics of the LMS Adaptive Filters. Proceedings of the IEEE, 64 (8), pp. 1151–1162.

17 Channel Equalisation and Blind Deconvolution

Blind deconvolution is the process of unravelling two unknown signals that have been convolved. An important application of blind deconvolution is in blind equalisation for restoration of a signal distorted in transmission through a communication channel. Blind equalisation has a wide range of applications, for example in digital telecommunications for removal of inter-symbol interference, in speech recognition for removal of the effects of microphones and channels, in deblurring of distorted images, in dereverberation of acoustic recordings, in seismic data analysis, etc.

In practice, blind equalisation is only feasible if some useful statistics of the channel input, and perhaps also of the channel itself, are available. The success of a blind equalisation method depends on how much is known about the statistics of the channel input, and how useful this knowledge is in the channel identification and equalisation process. This chapter begins with an introduction to the basic ideas of deconvolution and channel equalisation. We study blind equalisation based on the channel input power spectrum, equalisation through separation of the input signal and channel response models, Bayesian equalisation, non-linear adaptive equalisation for digital communication channels, and equalisation of maximum-phase channels using higher-order statistics.

17.1 Introduction

In this chapter we consider the recovery of a signal distorted, in transmission through a channel, by a convolutional process and observed in additive noise. The process of recovery of a signal convolved with the impulse response of a communication channel, or a recording medium, is known as deconvolution or equalisation. Figure 17.1 illustrates a typical model for a distorted and noisy signal, followed by an equaliser. Let $x(m)$, $n(m)$ and $y(m)$ denote the channel input, the channel noise and the observed channel output respectively. The channel input/output relation can be expressed as

$$y(m) = h[x(m)] + n(m) \tag{17.1}$$

where the function $h[\cdot]$ is the channel distortion. In general, the channel response may be time-varying and non-linear. In this chapter, it is assumed that the effects of a channel can be modelled using a

Multimedia Signal Processing: Theory and Applications in Speech, Music and Communications Saeed V. Vaseghi

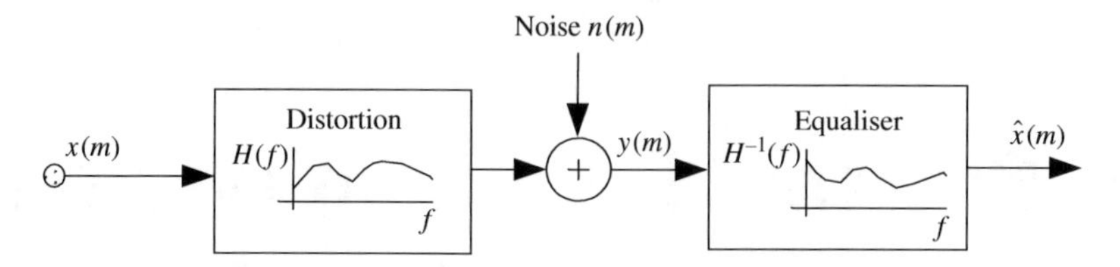

Figure 17.1 Illustration of a channel distortion model followed by an equaliser.

stationary, or a slowly time-varying, linear transversal filter. For a linear transversal filter model of the channel, Equation (17.1) becomes

$$y(m) = \sum_{k=0}^{P-1} h_k(m)x(m-k) + n(m) \tag{17.2}$$

where $h_k(m)$ are the coefficients of a Pth order linear FIR filter model of the channel. For a time-invariant channel model, $h_k(m) = h_k$.

In the frequency domain, Equation (17.2) becomes

$$Y(f) = X(f)H(f) + N(f) \tag{17.3}$$

where $Y(f)$, $X(f)$, $H(f)$ and $N(f)$ are the frequency spectra of the channel output, the channel input, the channel response and the additive noise respectively. Ignoring the noise term and taking the logarithm of Equation (17.3) yields

$$\ln|Y(f)| = \ln|X(f)| + \ln|H(f)| \tag{17.4}$$

From Equation (17.4), in the log-frequency domain the effect of channel distortion is the addition of a 'tilt' term $\ln|H(f)|$ to the signal spectrum.

17.1.1 Ideal Inverse Channel Filter

The ideal inverse-channel filter, or the ideal equaliser, recovers the original input from the channel output signal. In the frequency domain, the ideal non-causal inverse channel filter can be expressed as

$$H(f)H^{\text{inv}}(f) = 1 \tag{17.5}$$

In Equation (17.5) $H^{\text{inv}}(f)$ is used to denote the inverse channel filter. For the ideal equaliser we have $H^{\text{inv}}(f) = H^{-1}(f)$, or, expressed in the log-frequency domain $\ln H^{\text{inv}}(f) = -\ln H(f)$. The general form of Equation (17.5) is given by the z-transform relation

$$H(z)H^{\text{inv}}(z) = z^{-N} \tag{17.6}$$

for some value of the delay N that makes the channel inversion process causal. Taking the inverse Fourier transform of Equation (17.5), we have the following convolutional relation between the impulse responses of the channel $\{h_k\}$ and the ideal inverse channel response $\{h_k^{\text{inv}}\}$:

$$\sum_k h_k^{\text{inv}} h_{i-k} = \delta(i) \tag{17.7}$$

where $\delta(i)$ is the Kronecker delta function. Assuming the channel output is noise-free and the channel is invertible, the ideal inverse channel filter can be used to reproduce the channel input signal with zero error, as follows. The inverse filter output $\hat{x}(m)$, with the distorted signal $y(m)$ as the input, is given as

$$\begin{aligned}\hat{x}(m) &= \sum_k h_k^{\text{inv}} y(m-k) \\ &= \sum_k h_k^{\text{inv}} \sum_{\text{j}} h_{\text{j}} x(m-k-\text{j}) \\ &= \sum_i x(m-i) \sum_k h_k^{\text{inv}} h_{i-k}\end{aligned} \tag{17.8}$$

The last line of Equation (17.8) is derived by a change of variables $i = k + \text{j}$ in the second line and rearrangement of the terms. For the ideal inverse channel filter, substitution of Equation (17.7) in Equation (17.8) yields

$$\hat{x}(m) = \sum_i \delta(i) x(m-i) = x(m) \tag{17.9}$$

which is the desired result. In practice, it is not advisable to implement $H^{\text{inv}}(f)$ simply as $H^{-1}(f)$ because, in general, a channel response may be non-invertible. Even for invertible channels, a straightforward implementation of the inverse channel filter $H^{-1}(f)$ can cause problems. For example, at frequencies where $H(f)$ is small, its inverse $H^{-1}(f)$ is large, and this can lead to noise amplification if the signal-to-noise ratio is low.

17.1.2 Equalisation Error, Convolutional Noise

The equalisation error signal, also called the convolutional noise, is defined as the difference between the channel equaliser output and the desired signal:

$$\begin{aligned}v(m) &= x(m) - \hat{x}(m) \\ &= x(m) - \sum_{k=0}^{P-1} \hat{h}_k^{\text{inv}} y(m-k)\end{aligned} \tag{17.10}$$

where $\hat{h}_k^{\text{inv}}$ is an estimate of the inverse channel filter. Assuming that there is an ideal equaliser h_k^{inv} that can recover the channel input signal $x(m)$ from the channel output $y(m)$, we have

$$x(m) = \sum_{k=0}^{P-1} h_k^{\text{inv}} y(m-k) \tag{17.11}$$

Substitution of Equation (17.11) in Equation (17.10) yields

$$\begin{aligned}v(m) &= \sum_{k=0}^{P-1} h_k^{\text{inv}} y(m-k) - \sum_{k=0}^{P-1} \hat{h}_k^{\text{inv}} y(m-k) \\ &= \sum_{k=0}^{P-1} \tilde{h}_k^{\text{inv}} y(m-k)\end{aligned} \tag{17.12}$$

where $\tilde{h}_k^{\text{inv}} = h_k^{\text{inv}} - \hat{h}_k^{\text{inv}}$. The equalisation error signal $v(m)$ may be viewed as the output of an error filter $\tilde{h}_k^{\text{inv}}$ in response to the input $y(m-k)$, hence the name 'convolutional noise' for $v(m)$. When the equalisation process is proceeding well, such that $\hat{x}(m)$ is a good estimate of the channel input $x(m)$, then the convolutional noise is relatively small and de-correlated and can be modelled as a zero mean Gaussian random process.

17.1.3 Blind Equalisation

The equalisation problem is relatively simple when the channel response is known and invertible, and when the channel output is not noisy. However, in most practical cases, the channel response is unknown, time-varying, non-linear, and may also be non-invertible. Furthermore, the channel output is often observed in additive noise.

Digital communication systems provide equaliser-training periods, during which a *training* pseudo-noise (PN) sequence, also available at the receiver, is transmitted. A synchronised version of the PN sequence is generated at the receiver, where the channel input and output signals are used for the identification of the channel equaliser as illustrated in Figure 17.2(a). The obvious drawback of using training periods for channel equalisation is that power, time and bandwidth are consumed for the equalisation process.

It is preferable to have a 'blind' equalisation scheme that can operate without access to the channel input, as illustrated in Figure 17.2(b). Furthermore, in some applications, such as the restoration of acoustic recordings, or blurred images, all that is available is the distorted signal and the only restoration method applicable is blind equalisation.

Blind equalisation is feasible only if some statistical knowledge of the channel input, and perhaps that of the channel, is available. Blind equalisation involves two stages of channel identification, and deconvolution of the input signal and the channel response, as follows:

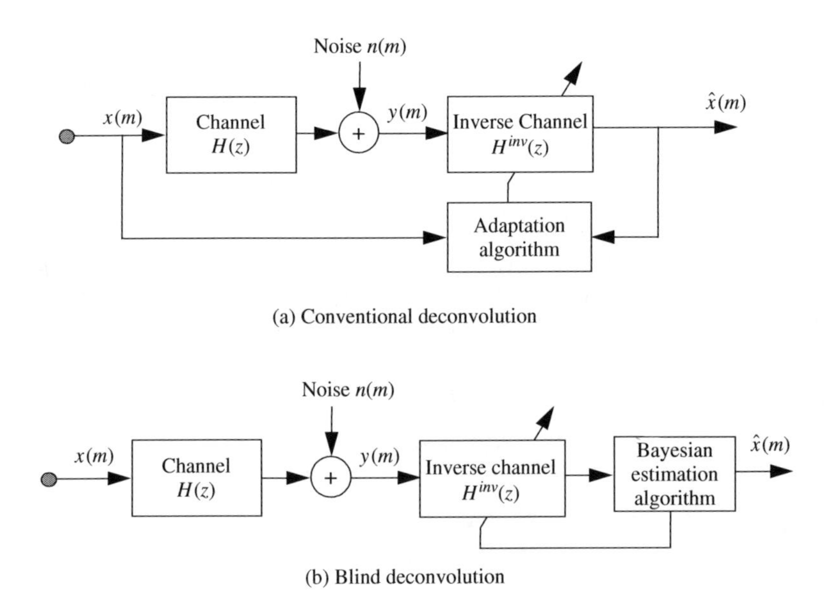

Figure 17.2 A comparative illustration of (a) a conventional equaliser with access to channel input and output, and (b) a blind equaliser.

(a) *Channel identification.* The general form of a channel estimator can be expressed as

$$\hat{\boldsymbol{h}} = \psi(\boldsymbol{y}, \mathcal{M}_x, \mathcal{M}_h) \tag{17.13}$$

where ψ is the channel estimator, the vector $\hat{\boldsymbol{h}}$ is an estimate of the channel response, $\boldsymbol{y}$ is the channel output, and M_x and M_h are statistical models of the channel input and the channel response respectively.

Channel identification methods rely on utilisation of a knowledge of the following characteristics of the input signal and the channel:

(i) The distribution of the channel input signal: for example, in decision-directed channel equalisation, described in Section 17.5, the knowledge that the input is a binary signal is used in a binary decision device to estimate the channel input and to 'direct' the equaliser adaptation process.
(ii) The relative durations of the channel input and the channel impulse response: the duration of a channel impulse response is usually orders of magnitude smaller than that of the channel input. This observation is used in Section 17.3.1 to estimate a stationary channel from the long-time averages of the channel output.
(iii) The stationary, versus time-varying characteristics of the input signal process and the channel: in Section 17.3.1, a method is described for the recovery of a non-stationary signal convolved with the impulse response of a stationary channel.

(b) *Channel equalisation.* Assuming that the channel is invertible, the channel input signal $x(m)$ can be recovered using an inverse channel filter as

$$\hat{x}(m) = \sum_{k=0}^{P-1} \hat{h}_k^{\text{inv}} y(m-k) \tag{17.14}$$

In the frequency domain, Equation (17.14) becomes

$$\hat{X}(f) = \hat{H}^{\text{inv}}(f)Y(f) \tag{17.15}$$

In practice, perfect recovery of the channel input may not be possible, either because the channel is non-invertible or because the output is observed in noise. A channel is non-invertible if:

(i) The channel transfer function is maximum-phase: the transfer function of a maximum-phase channel has zeros outside the unit circle, and hence the inverse channel has unstable poles. Maximum-phase channels are considered in the next section.
(ii) The channel transfer function maps many inputs to the same output: in these situations, a stable closed-form equation for the inverse channel does not exist, and instead an iterative deconvolution method is used. Figure 17.3 illustrates the frequency response of a channel that has one invertible and two non-invertible regions. In the non-invertible regions, the signal frequencies are heavily attenuated and lost to channel noise. In the invertible region, the signal is distorted but recoverable. This example illustrates that the inverse filter must be implemented with care in order to avoid undesirable results such as noise amplification at frequencies with low SNR.

17.1.4 Minimum- and Maximum-Phase Channels

For stability, all the poles of the transfer function of a channel must lie inside the unit circle. If all the zeros of the transfer function are also inside the unit circle then the channel is said to be a

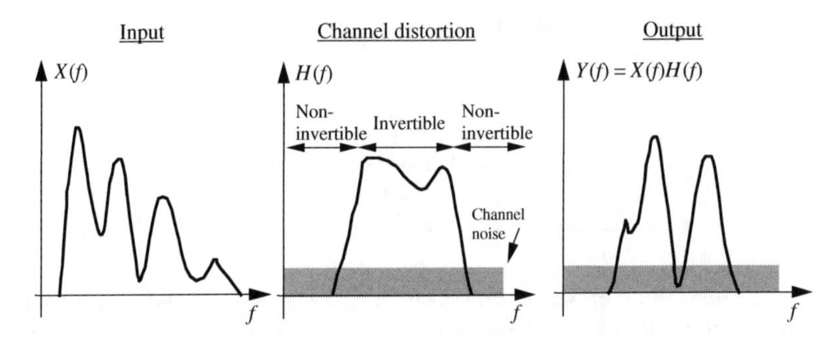

Figure 17.3 Illustration of the invertible and non-invertible regions of a channel.

minimum-phase channel. If some of the zeros are outside the unit circle then the channel is said to be a maximum-phase channel. The inverse of a minimum-phase channel has all its poles inside the unit circle, and is therefore stable. The inverse of a maximum-phase channel has some of its poles outside the unit circle; therefore it has an exponentially growing impulse response and is unstable. However, a stable approximation of the inverse of a maximum-phase channel may be obtained by truncating the impulse response of the inverse filter. Figure 17.4 illustrates examples of maximum-phase and minimum-phase fourth-order FIR filters.

When both the channel input and output signals are available, in the correct synchrony, it is possible to estimate the channel magnitude and phase response using the conventional least square error criterion. In blind deconvolution, there is no access to the exact instantaneous value or the timing of the channel input signal. The only information available is the channel output and some statistics of the channel input. The second-order statistics of a signal (i.e. correlation, covariance or power spectrum) do not include the phase information; hence it is not possible to estimate the channel phase from the second-order statistics. Furthermore, the channel phase cannot be recovered if the input signal is Gaussian, because a Gaussian process of known mean is entirely specified by the autocovariance

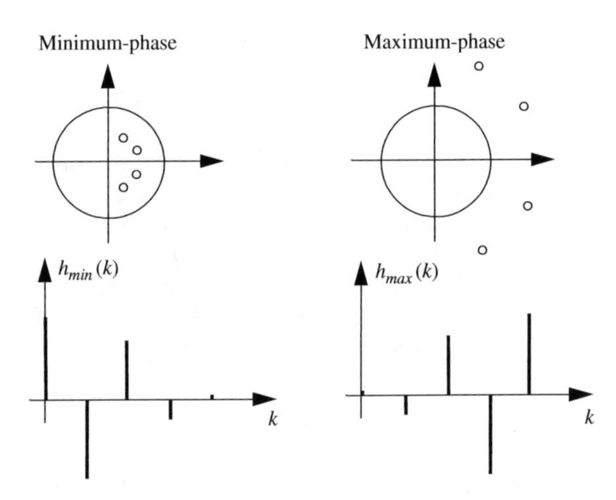

Figure 17.4 Illustration of the zero diagram and impulse response of fourth-order maximum-phase and minimum-phase FIR filters.

matrix, and autocovariance matrices do not include phase information. For estimation of the phase of a channel, we can either use a non-linear estimate of the desired signal to direct the adaptation of a channel equaliser as in Section 17.5, or we can use the higher-order statistics as in Section 17.6.

17.1.5 Wiener Equaliser

In this section, we consider least squared error Wiener equalisation. Note that, in its conventional form, Wiener equalisation is not a form of blind equalisation, because the implementation of a Wiener equaliser requires the cross-correlation of the channel input and output signals, which are not available in a blind equalisation application. The Wiener filter estimate of the channel input signal is given by

$$\hat{x}(m) = \sum_{k=0}^{P-1} \hat{h}_k^{\text{inv}} y(m-k) \tag{17.16}$$

where $\hat{h}_k^{\text{inv}}$ is an FIR Wiener filter estimate of the inverse channel impulse response. The equalisation error signal $v(m)$ is defined as

$$v(m) = x(m) - \sum_{k=0}^{P-1} \hat{h}_k^{\text{inv}} y(m-k) \tag{17.17}$$

The Wiener equaliser with input $y(m)$ and desired output $x(m)$ is obtained from Equation (8.10) as

$$\hat{\boldsymbol{h}}^{\text{inv}} = \boldsymbol{R}_{yy}^{-1} \boldsymbol{r}_{xy} \tag{17.18}$$

where $\boldsymbol{R}_{yy}$ is the $P \times P$ autocorrelation matrix of the channel output, and $\boldsymbol{r}_{xy}$ is the P-dimensional cross-correlation vector of the channel input and output signals. A more expressive form of Equation (17.18) can be obtained by writing the noisy channel output signal in vector equation form as

$$\boldsymbol{y} = \boldsymbol{H}\boldsymbol{x} + \boldsymbol{n} \tag{17.19}$$

where $\boldsymbol{y}$ is an N-sample channel output vector, $\boldsymbol{x}$ is an $(N+P)$-sample channel input vector including the P initial samples, $\boldsymbol{H}$ is an $N \times (N+P)$ channel distortion matrix whose elements are composed of the coefficients of the channel filter, and $\boldsymbol{n}$ is a noise vector. The autocorrelation matrix of the channel output can be obtained from Equation (17.19) as

$$\boldsymbol{R}_{yy} = \mathcal{E}[\boldsymbol{y}\boldsymbol{y}^{\text{T}}] = \boldsymbol{H}\boldsymbol{R}_{xx}\boldsymbol{H}^{\text{T}} + \boldsymbol{R}_{nn} \tag{17.20}$$

where $E[\cdot]$ is the expectation operator. The cross-correlation vector $\boldsymbol{r}_{xy}$ of the channel input and output signals becomes

$$\boldsymbol{r}_{xy} = \mathcal{E}[x\boldsymbol{y}] = \boldsymbol{H}\boldsymbol{r}_{xx} \tag{17.21}$$

Substitution of Equations (17.20) and (17.21) in (17.18) yields the Wiener equaliser as

$$\hat{\boldsymbol{h}}^{\text{inv}} = \left(\boldsymbol{H}\boldsymbol{R}_{xx}\boldsymbol{H}^{\text{T}} + \boldsymbol{R}_{nn}\right)^{-1} \boldsymbol{H}\boldsymbol{r}_{xx} \tag{17.22}$$

The derivation of the Wiener equaliser in the frequency domain is as follows. The Fourier transform of the equaliser output is given by

$$\hat{X}(f) = \hat{H}^{\text{inv}}(f)Y(f) \tag{17.23}$$

where $Y(f)$ is the channel output and $\hat{H}^{\text{inv}}(f)$ is the frequency response of the Wiener equaliser. The error signal $V(f)$ is defined as

$$\begin{aligned} V(f) &= X(f) - \hat{X}(f) \\ &= X(f) - \hat{H}^{\text{inv}}(f)Y(f) \end{aligned} \tag{17.24}$$

As in Section 8.5, minimisation of the expectation of the squared magnitude of $V(f)$ results in the frequency Wiener equaliser given by

$$\begin{aligned} \hat{H}^{\text{inv}}(f) &= \frac{P_{XY}(f)}{P_{YY}(f)} \\ &= \frac{P_{XX}(f)H^*(f)}{P_{XX}(f)\,|H(f)|^2 + P_{NN}(f)} \end{aligned} \tag{17.25}$$

where $P_{XX}(f)$ is the channel input power spectrum, $P_{NN}(f)$ is the noise power spectrum, $P_{XY}(f)$ is the cross-power spectrum of the channel input and output signals, and $H(f)$ is the frequency response of the channel. Note that in the absence of noise, $P_{NN}(f) = 0$ and the Wiener inverse filter becomes $H^{\text{inv}}(f) = H^{-1}(f)$.

17.2 Blind Equalisation Using Channel Input Power Spectrum

One of the early papers on blind deconvolution was by Stockham *et al.* (1975) on dereverberation of old acoustic recordings. Early acoustic recorders, as illustrated in Figure 17.5, had a bandwidth of about 200 Hz to 4 kHz. However, the limited bandwidth, or even the additive noise or scratch noise pulses, are not considered to be the major causes of distortions of acoustic recordings. The main distortion on acoustic recordings is due to reverberations of the recording horn instrument. An acoustic recording can be modelled as the convolution of the input audio signal $x(m)$ and the impulse response

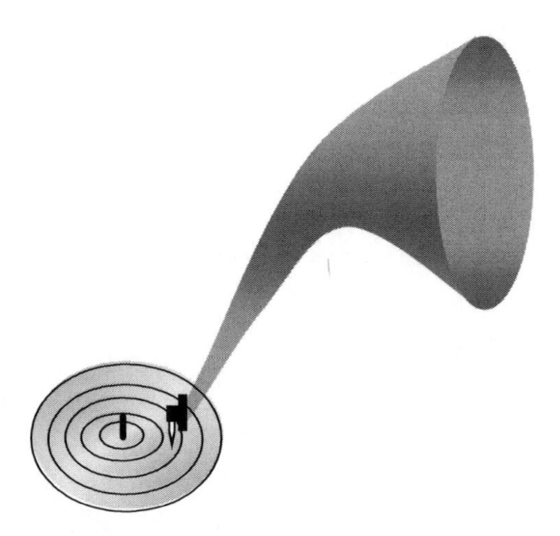

Figure 17.5 Illustration of the early acoustic recording process on a wax disc. Acoustic recordings were made by focusing the sound energy through a horn via a sound box, diaphragm and stylus mechanism, onto a wax disc. The sound was distorted by reverberations of the horn.

of a linear filter model of the recording instrument $\{h_k\}$, as in Equation (17.2), reproduced here for convenience

$$y(m) = \sum_{k=0}^{P-1} h_k x(m-k) + n(m) \quad (17.26)$$

or in the frequency domain as

$$Y(f) = X(f)H(f) + N(f) \quad (17.27)$$

where $H(f)$ is the frequency response of a linear time-invariant model of the acoustic recording instrument, and $N(f)$ is an additive noise. Multiplying both sides of Equation (17.27) with their complex conjugates, and taking the expectation, we obtain

$$\mathcal{E}[Y(f)Y^*(f)] = \mathcal{E}\left[(X(f)H(f) + N(f))(X(f)H(f) + N(f))^* \right] \quad (17.28)$$

Assuming the signal $X(f)$ and the noise $N(f)$ are uncorrelated Equation (17.28) becomes

$$P_{YY}(f) = P_{XX}(f)\,|H(f)|^2 + P_{NN}(f) \quad (17.29)$$

where $P_{YY}(f)$, $P_{XX}(f)$ and $P_{NN}(f)$ are the power spectra of the distorted signal, the original signal and the noise respectively. From Equation (17.29) an estimate of the spectrum of the channel response can be obtained as

$$|H(f)|^2 = \frac{P_{YY}(f) - P_{NN}(f)}{P_{XX}(f)} \quad (17.30)$$

In practice, Equation (17.30) is implemented using time-averaged estimates of the power spectra.

17.2.1 Homomorphic Equalisation

In homomorphic equalisation, the convolutional distortion is transformed, first into a multiplicative distortion through a Fourier transform of the distorted signal, and then into an additive distortion by taking the logarithm of the spectrum of the distorted signal. A further inverse Fourier transform operation converts the log-frequency variables into cepstral variables as illustrated in Figure 17.6. Through homomorphic transformation convolution becomes addition, and equalisation becomes subtraction.

Ignoring the additive noise term and transforming both sides of Equation (17.27) into log-spectral variables yields

$$\ln Y(f) = \ln X(f) + \ln H(f) \quad (17.31)$$

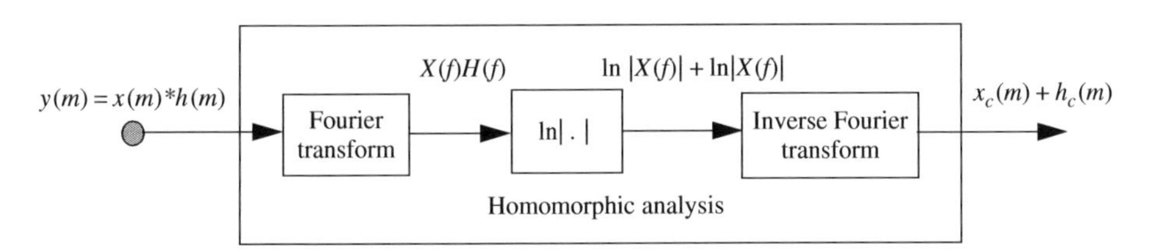

Figure 17.6 Illustration of homomorphic analysis in deconvolution.

Note that in the log-frequency domain, the effect of channel distortion is the addition of a tilt to the spectrum of the channel input. Taking the expectation of Equation (17.31) yields

$$\mathcal{E}[\ln Y(f)] = \mathcal{E}[\ln X(f)] + \ln H(f) \tag{17.32}$$

In Equation (17.32), it is assumed that the channel is time-invariant; hence $\mathcal{E}[\ln H(f)] = \ln H(f)$. Using the relation $\ln z = \ln |z| + j\angle z$, the term $\mathcal{E}[\ln X(f)]$ can be expressed as

$$\mathcal{E}[\ln X(f)] = \mathcal{E}[\ln |X(f)|] + j\mathcal{E}[\angle X(f)] \tag{17.33}$$

The first term on the right-hand side of Equation (17.33), $\mathcal{E}[\ln |X(f)|]$, is non-zero, and represents the frequency distribution of the signal power in decibels, whereas the second term $\mathcal{E}[\angle X(f)]$ is the expectation of the phase, and can be assumed to be zero. From Equation (17.32), the log-frequency spectrum of the channel can be estimated as

$$\ln H(f) = \mathcal{E}[\ln Y(f)] - \mathcal{E}[\ln X(f)] \tag{17.34}$$

In practice, when only a single record of a signal is available, the signal is divided into a number of segments, and the average signal spectrum is obtained over time across the segments. Assuming that the length of each segment is long compared with the duration of the channel impulse response, we can write an approximate convolutional relation for the ith signal segment as

$$y_i(m) \approx x_i(m) * h_i(m) \tag{17.35}$$

The segments are windowed, using a window such as a Hamming or a Hanning window, to reduce the spectral leakage due to end effects at the edges of the segment. Taking the complex logarithm of the Fourier transform of Equation (17.35) yields

$$\ln Y_i(f) = \ln X_i(f) + \ln H_i(f) \tag{17.36}$$

Taking the time averages over N segments of the distorted signal record yields

$$\frac{1}{N}\sum_{i=0}^{N-1} \ln Y_i(f) = \frac{1}{N}\sum_{i=0}^{N-1} \ln X_i(f) + \frac{1}{N}\sum_{i=0}^{N-1} \ln H_i(f) \tag{17.37}$$

Estimation of the channel response from Equation (17.37) requires the average log spectrum of the undistorted signal $X(f)$. In Stockham's method for restoration of acoustic records, the expectation of the signal spectrum is obtained from a modern recording of the same musical material as that of the acoustic recording. From Equation (17.37), the estimate of the logarithm of the channel is given by

$$\ln \hat{H}(f) = \frac{1}{N}\sum_{i=0}^{N-1} \ln Y_i(f) - \frac{1}{N}\sum_{i=0}^{N-1} \ln X_i^{\mathcal{M}}(f) \tag{17.38}$$

where $X^{\mathcal{M}}(f)$ is the spectrum of a modern recording. The equaliser can then be defined as

$$\ln H^{\text{inv}}(f) = \begin{cases} -\ln \hat{H}(f), & 200\,\text{Hz} \le f \le 4000\,\text{Hz} \\ -40\,\text{dB}, & \text{otherwise} \end{cases} \tag{17.39}$$

In Equation (17.39), the inverse acoustic channel is implemented in the range between 200 and 4000 Hz, where the channel is assumed to be invertible. Outside this range, the signal is dominated by noise, and the inverse filter is designed to attenuate the noisy signal.

17.2.2 Homomorphic Equalisation Using a Bank of High-Pass Filters

In the log-frequency domain, channel distortion may be eliminated using a bank of high-pass filters. Consider a time sequence of log-spectra of the output of a channel described as

$$\ln Y_t(f) = \ln X_t(f) + \ln H_t(f) \tag{17.40}$$

where $Y_t(f)$ and $X_t(f)$ are the channel input and output derived from a Fourier transform of the tth signal segment. From Equation (17.40), the effect of a time-invariant channel is to add a constant term $\ln H(f)$ to each frequency component of the channel input $X_t(f)$, and the overall result is a time-invariant tilt of the log-frequency spectrum of the original signal. This observation suggests the use of a bank of narrowband high-pass notch filters for the removal of the additive distortion term $\ln H(f)$. A simple first-order recursive digital filter with its notch at zero frequency is given by

$$\ln \hat{X}_t(f) = \alpha \ln \hat{X}_{t-1}(f) + \ln Y_t(f) - \ln Y_{t-1}(f) \tag{17.41}$$

where the parameter α controls the bandwidth of the notch at zero frequency. Note that the filter bank also removes any dc component of the signal $\ln X(f)$; for some applications, such as speech recognition, this is acceptable.

17.3 Equalisation Based on Linear Prediction Models

Linear prediction models, described in Chapter 10, are routinely used in applications such as seismic signal analysis and speech processing, for the modelling and identification of a minimum-phase channel. Linear prediction theory is based on two basic assumptions: that the channel is minimum-phase and that the channel input is a random signal. Standard linear prediction analysis can be viewed as a blind deconvolution method, because both the channel response and the channel input are unknown, and the only information is the channel output and the assumption that the channel input is random and hence has a flat power spectrum. In this section, we consider blind deconvolution using linear predictive models for the channel and its input. The channel input signal is modelled as

$$X(z) = E(z)A(z) \tag{17.42}$$

where $X(z)$ is the z-transform of the channel input signal, $A(z)$ is the z-transfer function of a linear predictive model of the channel input and $E(z)$ is the z-transform of a random excitation signal. Similarly, the channel output can be modelled by a linear predictive model $H(z)$ with input $X(z)$ and output $Y(z)$ as

$$Y(z) = X(z)H(z) \tag{17.43}$$

Figure 17.7 illustrates a cascade linear prediction model for a channel input process $X(z)$ and a channel response $H(z)$. The channel output can be expressed as

$$\begin{aligned} Y(z) &= E(z)A(z)H(z) \\ &= E(z)D(z) \end{aligned} \tag{17.44}$$

where

$$D(z) = A(z)H(z) \tag{17.45}$$

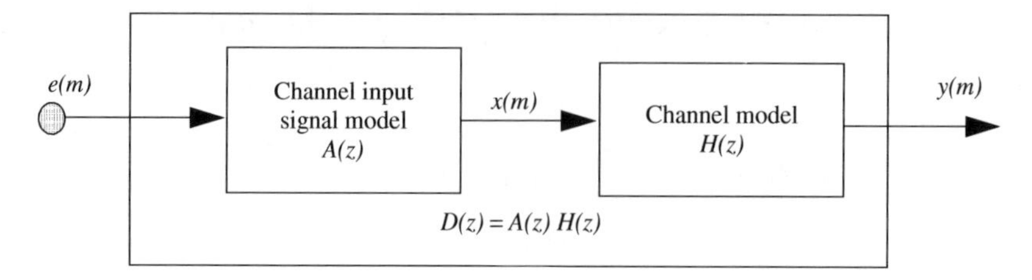

Figure 17.7 A distorted signal modelled as cascade of a signal model and a channel model.

The z-transfer function of the linear prediction models of the channel input signal and the channel can be expanded as

$$A(z) = \frac{G_1}{1 - \sum_{k=1}^{P} a_k z^{-k}} = \frac{G_1}{\prod_{k=1}^{P} (1 - \alpha_k z^{-1})} \tag{17.46}$$

$$H(z) = \frac{G_2}{1 - \sum_{k=1}^{Q} b_k z^{-k}} = \frac{G_2}{\prod_{k=1}^{Q} (1 - \beta_k z^{-1})} \tag{17.47}$$

where $\{a_k, \alpha_k\}$ and $\{b_k, \beta_k\}$ are the coefficients and the poles of the linear prediction models for the channel input signal and the channel respectively. Substitution of Equations (17.46) and (17.47) in Equation (17.45) yields the combined input-channel model as

$$D(z) = \frac{G}{1 - \sum_{k=1}^{P+Q} d_k z^{-k}} = \frac{G}{\prod_{k=1}^{P+Q} (1 - \gamma_k z^{-1})} \tag{17.48}$$

The total number of poles of the combined model for the input signal and the channel is modelled as the sum of the poles of the input signal model and the channel model.

17.3.1 Blind Equalisation Through Model Factorisation

A model-based approach to blind equalisation is to factorise the channel output model $D(z) = A(z)H(z)$ into a channel input signal model $A(z)$ and a channel model $H(z)$. If the channel input model $A(z)$ and the channel model $H(z)$ are non-factorable then the only factors of $D(z)$ are $A(z)$ and $H(z)$. However, z-transfer functions are factorable into the roots, the so-called poles and zeros, of the models. One approach to model-based deconvolution is to factorise the model for the convolved signal into its poles and zeros, and classify the poles and zeros as either belonging to the signal or belonging to the channel.

Spencer and Rayner (1990) developed a method for blind deconvolution through factorisation of linear prediction models, based on the assumption that the channel is stationary with time-invariant poles whereas the input signal is non-stationary with time-varying poles. As an application, they considered the restoration of old acoustic recordings where a time-varying audio signal is distorted by the time-invariant frequency response of the recording equipment. For a simple example, consider the case when the signal and the channel are each modelled by a second-order linear predictive model. Let the time-varying second-order linear predictive model for the channel input signal $x(m)$ be

$$x(m) = a_1(m)x(m-1) + a_2(m)x(m-2) + G_1(m)e(m) \tag{17.49}$$

where $a_1(m)$ and $a_2(m)$ are the time-varying coefficients of the linear predictor model, $G_1(m)$ is the input gain factor and $e(m)$ is a zero-mean, unit variance, random signal. Now let $\alpha_1(m)$ and $\alpha_2(m)$ denote the time-varying poles of the predictor model of Equation (17.49); these poles are the roots of the polynomial

$$1 - a_1(m)z^{-1} - a_2(m)z^{-2} = \left[1 - z^{-1}\alpha_1(m)\right]\left[1 - z^{-1}\alpha_2(m)\right] = 0 \tag{17.50}$$

Similarly, assume that the channel can be modelled by a second-order stationary linear predictive model as

$$y(m) = h_1 y(m-1) + h_2 y(m-2) + G_2 x(m) \tag{17.51}$$

where h_1 and h_2 are the time-invariant predictor < /h_1> coefficients and G_2 is the channel gain. Let β_1 and β_2 denote the poles of the channel model; these are the roots of the polynomial

$$1 - h_1 z^{-1} - h_2 z^{-2} = (1 - z^{-1}\beta_1)(1 - z^{-1}\beta_2) = 0 \tag{17.52}$$

The combined cascade of the two second-order models of Equations (17.49) and (17.51) can be written as a fourth-order linear predictive model with input $e(m)$ and output $y(m)$:

$$y(m) = d_1(m)y(m-1) + d_2(m)y(m-2) + d_3(m)y(m-3) + d_4(m)y(m-4) + Ge(m) \tag{17.53}$$

where the combined gain $G = G_1 G_2$. The poles of the fourth-order predictor model of Equation (17.53) are the roots of the polynomial

$$\begin{aligned} &1 - d_1(m)z^{-1} - d_2(m)z^{-2} - d_3(m)z^{-3} - d_4(m)z^{-4} \\ &= \left[1 - z^{-1}\alpha_1(m)\right]\left[1 - z^{-1}\alpha_2(m)\right]\left[1 - z^{-1}\beta_1\right]\left[1 - z^{-1}\beta_2\right) = 0 \end{aligned} \tag{17.54}$$

In Equation (17.54) the poles of the fourth-order predictor are $\alpha_1(m)$, $\alpha_2(m)$, β_1 and β_2. The above argument on factorisation of the poles of time-varying and stationary models can be generalised to a signal model of order P and a channel model of order Q.

In Spencer and Rayner (1990), the separation of the stationary poles of the channel from the time-varying poles of the channel input is achieved through a clustering process. The signal record is divided into N segments and each segment is modelled by an all-pole model of order $P+Q$ where P and Q are the assumed model orders for the channel input and the channel respectively. In all, there are $N(P+Q)$ values which are clustered to form $P+Q$ clusters. Even if both the signal and the channel were stationary, the poles extracted from different segments would have variations due to the random character of the signals from which the poles are extracted. Assuming that the variances of the estimates of the stationary poles are small compared with the variations of the time-varying poles, it is expected that, for each stationary pole of the channel, the N values extracted from N segments will form an N-point cluster of a relatively small variance. These clusters can be identified and the centre of each cluster taken as a pole of the channel model. This method assumes that the poles of the time-varying signal are well separated in space from the poles of the time-invariant signal.

17.4 Bayesian Blind Deconvolution and Equalisation

The Bayesian inference method, described in Chapter 7, provides a general framework for inclusion of statistical models of the channel input and the channel response. In this section we consider the

Bayesian equalisation method, and study the case where the channel input is modelled by a set of hidden Markov models. The Bayesian risk for a channel estimate $\hat{h}$ is defined as

$$\begin{aligned}\mathcal{R}(\hat{\boldsymbol{h}}|\boldsymbol{y}) &= \int_{H}\int_{X} C(\hat{\boldsymbol{h}},\boldsymbol{h}) f_{X,H|Y}(\boldsymbol{x},\boldsymbol{h}|\boldsymbol{y})\,d\boldsymbol{x}d\boldsymbol{h} \\ &= \frac{1}{f_Y(\boldsymbol{y})}\int_{H} C(\hat{\boldsymbol{h}},\boldsymbol{h}) f_{Y|H}(\boldsymbol{y}|\boldsymbol{h}) f_H(\boldsymbol{h})d\boldsymbol{h}\end{aligned} \tag{17.55}$$

where $C(\hat{\boldsymbol{h}},\boldsymbol{h})$ is the cost of estimating the channel $\boldsymbol{h}$ as $\hat{\boldsymbol{h}}$, $f_{X,H|Y}(\boldsymbol{x},\boldsymbol{h}|\boldsymbol{y})$ is the joint posterior density of the channel $\boldsymbol{h}$ and the channel input $\boldsymbol{x}$, $f_{Y|H}(\boldsymbol{y}|\boldsymbol{h})$ is the observation likelihood, and $f_H(\boldsymbol{h})$ is the prior pdf of the channel. The Bayesian estimate is obtained by minimisation of the risk function $\mathcal{R}(\hat{\boldsymbol{h}}|\boldsymbol{y})$. There are a variety of Bayesian-type solutions depending on the choice of the cost function and the prior knowledge, as described in Chapter 7.

In this section, it is assumed that the convolutional channel distortion is transformed into an additive distortion through transformation of the channel output into log-spectral or cepstral variables. Ignoring the channel noise, the relation between the cepstra of the channel input and output signals is given by

$$\boldsymbol{y}(m) = \boldsymbol{x}(m) + \boldsymbol{h} \tag{17.56}$$

where the cepstral vectors $\boldsymbol{x}(m)$, $\boldsymbol{y}(m)$ and $\boldsymbol{h}$ are the channel input, the channel output and the channel respectively.

17.4.1 Conditional Mean Channel Estimation

A commonly used cost function in the Bayesian risk of Equation (17.55) is the mean square error $C(\boldsymbol{h}-\hat{\boldsymbol{h}}) = |\boldsymbol{h}-\hat{\boldsymbol{h}}|^2$, which results in the conditional mean (CM) estimate defined as

$$\hat{\boldsymbol{h}}^{\mathrm{CM}} = \int_{H} \boldsymbol{h} f_{H|Y}(\boldsymbol{h}|\boldsymbol{y})d\boldsymbol{h} \tag{17.57}$$

The posterior density of the channel input signal may be conditioned on an estimate of the channel vector $\hat{\boldsymbol{h}}$ and expressed as $f_{X|Y,H}(\boldsymbol{x}|\boldsymbol{y},\hat{\boldsymbol{h}})$. The conditional mean of the channel input signal given the channel output $\boldsymbol{y}$ and an estimate of the channel $\hat{\boldsymbol{h}}$ is

$$\begin{aligned}\hat{\boldsymbol{x}}^{\mathrm{CM}} &= \mathcal{E}[\boldsymbol{x}|\boldsymbol{y},\hat{\boldsymbol{h}}] \\ &= \int_{X} \boldsymbol{x} f_{X|Y,H}(\boldsymbol{x}|\boldsymbol{y},\hat{\boldsymbol{h}})d\boldsymbol{x}\end{aligned} \tag{17.58}$$

Equations (17.57) and (17.58) suggest a two-stage iterative method for channel estimation and the recovery of the channel input signal.

17.4.2 Maximum-Likelihood Channel Estimation

The ML channel estimate is equivalent to the case when the Bayes cost function and the channel prior are uniform. Assuming that the channel input signal has a Gaussian distribution with mean vector $\boldsymbol{\mu}_x$

and covariance matrix Σ_{xx}, the likelihood of a sequence of N P-dimensional channel output vectors $\{y(m)\}$ given a channel input vector h is

$$\begin{aligned} f_{Y|H}(y(0),\ldots,y(N-1)\,|h) &= \prod_{m=0}^{N-1} f_X(y(m)-h) \\ &= \prod_{m=0}^{N-1} \frac{1}{(2\pi)^{P/2}\,|\Sigma_{xx}|^{1/2}} \exp\left\{[y(m)-h-\mu_x]^{\mathrm{T}}\Sigma_{xx}^{-1}[y(m)-h-\mu_x]\right\} \end{aligned} \tag{17.59}$$

It is assumed that the channel output is transformed into cepstral variables so that the channel distortion is additive, it is further assumed that the channel output vectors are independent. To obtain the ML estimate of the channel h, the derivative of the log likelihood function $\ln f_Y(y|h)$ with respect to h is set to zero to yield

$$\hat{h}^{\mathrm{ML}} = \frac{1}{N}\sum_{m=0}^{N-1}(y(m)-\mu_x) \tag{17.60}$$

17.4.3 Maximum A Posteriori Channel Estimation

The MAP estimate, like the ML estimate, is equivalent to a Bayesian estimator with a uniform cost function. However, the MAP estimate includes the prior pdf of the channel. The prior pdf can be used to confine the channel estimate within a desired subspace of the parameter space. Assuming that the channel input vectors are statistically independent, the posterior pdf of the channel given the observation sequence $Y = \{y(0),\ldots,y(N-1)\}$ is

$$\begin{aligned} f_{H|Y}(h\,|y(0),\ldots,y(N-1)) &= \prod_{m=0}^{N-1} \frac{1}{f_Y(y(m))} f_{Y|H}(y(m)\,|h)\,f_H(h) \\ &= \prod_{m=0}^{N-1} \frac{1}{f_Y(y(m))} f_X(y(m)-h)\,f_H(h) \end{aligned} \tag{17.61}$$

Assuming that the channel input $x(m)$ is Gaussian, $f_X(x(m)) = N(x, \mu_x, \Sigma_{xx})$, with mean vector μ_x and covariance matrix Σ_{xx}, and that the channel h is also Gaussian, $f_H(h) = N(h, \mu_h, \Sigma_{hh})$, with mean vector μ_h and covariance matrix Σ_{hh}, the logarithm of the posterior pdf is

$$\begin{aligned} \ln f_{H|Y}(h\,|y(0),\ldots,y(N-1)) &= -\sum_{m=0}^{N-1}\ln f\,(y\,(m)) - NP\ln(2\pi) - \frac{1}{2}\ln\left(|\Sigma_{xx}|\,|\Sigma_{hh}|\right) \\ &-\sum_{m=0}^{N-1}\frac{1}{2}\left\{[y\,(m)-h-\mu_x]^{\mathrm{T}}\Sigma_{xx}^{-1}(y\,(m)-h-\mu_x)+(h-\mu_h)^{\mathrm{T}}\Sigma_{hh}^{-1}(h-\mu_h)\right\} \end{aligned} \tag{17.62}$$

The MAP channel estimate, obtained by setting the derivative of the log posterior function $\ln f_{H|Y}(h|Y)$ to zero, is

$$\hat{h}^{\mathrm{MAP}} = (\Sigma_{xx}+\Sigma_{hh})^{-1}\,\Sigma_{hh}\,(\bar{y}-\mu_x) + (\Sigma_{xx}+\Sigma_{hh})^{-1}\,\Sigma_{xx}\mu_h \tag{17.63}$$

where

$$\bar{y} = \frac{1}{N}\sum_{m=0}^{N-1} y(m) \tag{17.64}$$

is the time-averaged estimate of the mean of observation vector. Note that for a Gaussian process the MAP and conditional mean estimates are identical.

17.4.4 Channel Equalisation Based on Hidden Markov Models

This section considers blind deconvolution in applications where the statistics of the channel input are modelled by a set of hidden Markov models. An application of this method, illustrated in Figure 17.8, is in recognition of speech distorted by a communication channel or a microphone. A hidden Markov model (HMM) is a finite-state Bayesian model, with a Markovian state prior and a Gaussian observation likelihood (see Chapter 11). An N-state HMM can be used to model a non-stationary process, such as speech, as a chain of N stationary states connected by a set of Markovian state transitions. The likelihood of an HMM $\mathcal{M}_i$ and a sequence of N P-dimensional channel input vectors $\boldsymbol{X}=[\boldsymbol{x}(0),\ldots,\boldsymbol{x}(N-1)]$ can be expressed in terms of the state transition and the observation pdfs of $\mathcal{M}_i$ as

$$f_{X|\mathcal{M}}(\boldsymbol{X}|\mathcal{M}_i)=\sum_{\boldsymbol{s}} f_{X|\mathcal{M},S}(\boldsymbol{X}|\mathcal{M}_i,\boldsymbol{s})P_{S|\mathcal{M}}(\boldsymbol{s}|\mathcal{M}_i) \tag{17.65}$$

where $f_{X|\mathcal{M},S}(\boldsymbol{X}|\mathcal{M}_i,\boldsymbol{s})$ is the likelihood that the sequence $\boldsymbol{X}=[\boldsymbol{x}(0),\ldots,\boldsymbol{x}(N-1)]$ was generated by the state sequence $\boldsymbol{s}=[s(0),\ldots,s(N-1)]$ of the model $\mathcal{M}_i$, and $P_{s|\mathcal{M}}(\boldsymbol{s}|\mathcal{M}_i)$ is the Markovian prior pmf of the state sequence $\boldsymbol{s}$. The Markovian prior entails that the probability of a transition to the state i at time m depends only on the state at time $m-1$ and is independent of the previous states. The transition probability of a Markov process is defined as

$$a_{ij}=P\left(s(m)=\mathrm{j}|s(m-1)=i\right) \tag{17.66}$$

where a_{ij} is the probability of making a transition from state i to state j. The HMM state observation probability is often modelled by a multivariate Gaussian pdf as

$$f_{X|\mathcal{M},S}(\boldsymbol{x}|\mathcal{M}_i,s)=\frac{1}{(2\pi)^{P/2}|\Sigma_{\boldsymbol{xx},s}|^{1/2}}\exp\left\{-\frac{1}{2}[\boldsymbol{x}-\boldsymbol{\mu}_{x,s}]^{\mathrm{T}}\,\Sigma_{\boldsymbol{xx},s}^{-1}[\boldsymbol{x}-\boldsymbol{\mu}_{x,s}]\right\} \tag{17.67}$$

where $\boldsymbol{\mu}_{x,s}$ and $\Sigma_{\boldsymbol{xx},s}$ are the mean vector and the covariance matrix of the Gaussian observation pdf of the HMM state s of the model $\mathcal{M}_i$.

The HMM-based channel equalisation problem can be stated as follows. Given a sequence of NP-dimensional channel output vectors $\boldsymbol{Y}=[\boldsymbol{y}(0),\ldots,\boldsymbol{y}(N-1)]$, and the prior knowledge that the channel input sequence is drawn from a set of V HMMs $M=\{\mathcal{M}_i, i=1,\ldots,V\}$, estimate the channel response and the channel input.

The joint posterior pdf of an input word $\mathcal{M}_i$ and the channel vector $\boldsymbol{h}$ can be expressed as

$$f_{\mathcal{M},H|Y}(\mathcal{M}_i,\boldsymbol{h}|\boldsymbol{Y})=P_{\mathcal{M}|H,Y}(\mathcal{M}_i|\boldsymbol{h},\boldsymbol{Y})f_{H|Y}(\boldsymbol{h}|\boldsymbol{Y}) \tag{17.68}$$

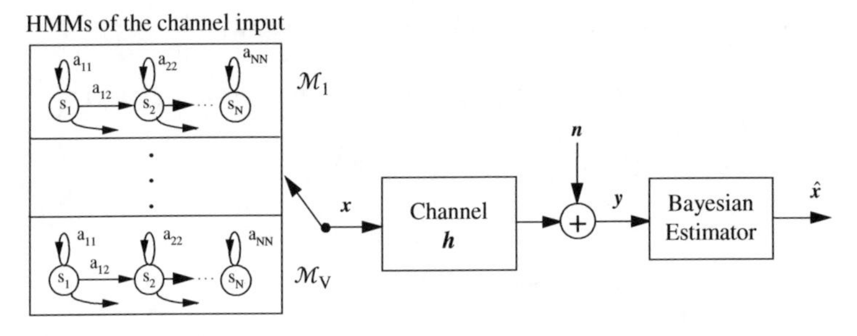

Figure 17.8 Illustration of a channel with the input modelled by a set of HMMs.

Simultaneous joint estimation of the channel vector $\boldsymbol{h}$ and classification of the unknown input word $\mathcal{M}_i$ is a non-trivial exercise. The problem is usually approached iteratively by making an estimate of the channel response, and then using this estimate to obtain the channel input as follows. From Bayes' rule, the posterior pdf of the channel $\boldsymbol{h}$ conditioned on the assumption that the input model is $\mathcal{M}_i$ and given the observation sequence $\boldsymbol{Y}$ can be expressed as

$$f_{H|\mathcal{M},Y}(\boldsymbol{h}\,|\mathcal{M}_i,\boldsymbol{Y}) = \frac{1}{f_{Y|\mathcal{M}}(\boldsymbol{Y}|\mathcal{M}_i)} f_{Y|\mathcal{M},H}(\boldsymbol{Y}\,|\mathcal{M}_i,\boldsymbol{h}) f_{H|\mathcal{M}}(\boldsymbol{h}|\mathcal{M}_i) \tag{17.69}$$

The likelihood of the observation sequence, given the channel and the input word model, can be expressed as

$$f_{Y|\mathcal{M},H}(\boldsymbol{Y}|\mathcal{M}_i,\boldsymbol{h}) = f_{X|\mathcal{M}}(\boldsymbol{Y}-\boldsymbol{h}|\mathcal{M}_i) \tag{17.70}$$

where it is assumed that the channel output is transformed into cepstral variables so that the channel distortion is additive. For a given input model $\mathcal{M}_i$, and state sequence $\boldsymbol{s} = [s(0), s(1), \ldots, s(N-1)]$, the pdf of a sequence of N independent observation vectors $\boldsymbol{Y} = [\boldsymbol{y}(0), \boldsymbol{y}(1), \ldots, \boldsymbol{y}(N-1)]$ is

$$\begin{aligned} f_{Y|H,\mathrm{S},\mathcal{M}}(\boldsymbol{Y}|\boldsymbol{h},\mathbf{s},\mathcal{M}_i) &= \prod_{m=0}^{N-1} f_{X|\mathrm{S,M}}(\boldsymbol{y}(m)-\boldsymbol{h}|s(m),\mathcal{M}_i) \\ &= \prod_{m=0}^{N-1} \frac{1}{(2\pi)^{\mathrm{P}/2}|\Sigma_{\boldsymbol{xx},s(m)}|^{1/2}} \exp\left\{-\frac{1}{2}\left[\boldsymbol{y}(m)-\boldsymbol{h}-\boldsymbol{\mu}_{\boldsymbol{x},s(m)}\right]^{\mathrm{T}} \Sigma^{-1}_{\boldsymbol{xx},s(m)}\left[\boldsymbol{y}(m)-\boldsymbol{h}-\boldsymbol{\mu}_{\boldsymbol{x},s(m)}\right]\right\} \end{aligned} \tag{17.71}$$

Taking the derivative of the log-likelihood of Equation (17.71) with respect to the channel vector $\boldsymbol{h}$ yields a maximum likelihood channel estimate as

$$\hat{\boldsymbol{h}}^{ML}(\boldsymbol{Y},\boldsymbol{s}) = \sum_{m=0}^{N-1}\left(\sum_{k=0}^{N-1}\Sigma^{-1}_{xx,s(k)}\right)^{-1}\Sigma^{-1}_{xx,s(m)}\left(\boldsymbol{y}(m)-\boldsymbol{\mu}_{x,s(m)}\right) \tag{17.72}$$

Note that when all the state observation covariance matrices are identical the channel estimate becomes

$$\hat{\boldsymbol{h}}^{ML}(\boldsymbol{Y},\boldsymbol{s}) = \frac{1}{N}\sum_{m=0}^{N-1}\left(\boldsymbol{y}(m)-\boldsymbol{\mu}_{\boldsymbol{x},s(m)}\right) \tag{17.73}$$

The ML estimate of Equation (17.73) is based on the ML state sequence $\boldsymbol{s}$ of $\mathcal{M}_i$. In the following section we consider the conditional mean estimate over all state sequences of a model.

17.4.5 MAP Channel Estimate Based on HMMs

The conditional pdf of a channel $\boldsymbol{h}$ averaged over all HMMs can be expressed as

$$f_{H|Y}(\boldsymbol{h}|\boldsymbol{Y}) = \sum_{i=1}^{V}\sum_{s} f_{H|Y,S,\mathcal{M}}(\boldsymbol{h}\,|\boldsymbol{Y},\mathbf{s},\mathcal{M}_i)\,P_{S|\mathcal{M}_i}(\mathbf{s}\,|\mathcal{M}_i)P_{\mathcal{M}}(\mathcal{M}_i) \tag{17.74}$$

where $P_{\mathcal{M}}(\mathcal{M}_i)$ is the prior pmf of the input words. Given a sequence of N P-dimensional observation vectors $\boldsymbol{Y} = [\boldsymbol{y}(0), \ldots, \boldsymbol{y}(N-1)]$, the posterior pdf of the channel $\boldsymbol{h}$ along a state sequence $\boldsymbol{s}$ of an HMM $\mathcal{M}_i$ is defined as

$$f_{Y|H,\mathrm{S,M}}(\boldsymbol{h}\,|\boldsymbol{Y},\mathbf{s},\mathcal{M}_i) = \frac{1}{f_Y(\boldsymbol{Y})} f_{Y|H,\mathrm{S},\mathcal{M}}(\boldsymbol{Y}\,|\boldsymbol{h},\mathbf{s},\mathcal{M}_i)\,f_H(\boldsymbol{h})$$

$$= \frac{1}{f_Y(\boldsymbol{Y})} \prod_{m=0}^{N-1} \frac{1}{(2\pi)^P \left|\Sigma_{\boldsymbol{xx},s(m)}\right|^{1/2} \left|\Sigma_{\boldsymbol{hh}}\right|^{1/2}} \exp\left\{-\frac{1}{2}\left[\boldsymbol{y}(m)-\boldsymbol{h}-\boldsymbol{\mu}_{\boldsymbol{x},s(m)}\right]^{\mathrm{T}} \Sigma_{\boldsymbol{xx},s(m)}^{-1} \left[\boldsymbol{y}(m)-\boldsymbol{h}-\boldsymbol{\mu}_{\boldsymbol{x},s(m)}\right]\right\}$$

$$\times \exp\left[-\frac{1}{2}(\boldsymbol{h}-\boldsymbol{\mu}_{\boldsymbol{h}})^{\mathrm{T}} \Sigma_{\boldsymbol{hh}}^{-1} (\boldsymbol{h}-\boldsymbol{\mu}_{\boldsymbol{h}})\right] \tag{17.75}$$

where it is assumed that each state of the HMM has a Gaussian distribution with mean vector $\boldsymbol{\mu}_{\boldsymbol{x},s(m)}$ and covariance matrix $\Sigma_{\boldsymbol{xx},s(m)}$, and that the channel $\boldsymbol{h}$ is also Gaussian-distributed, with mean vector $\mu_{\boldsymbol{h}}$ and covariance matrix $\Sigma_{\boldsymbol{hh}}$. The MAP estimate along state $\boldsymbol{s}$, on the left-hand side of Equation (17.75), can be obtained as

$$\begin{aligned} \hat{\boldsymbol{h}}^{MAP}(\boldsymbol{Y},\mathbf{s},\mathcal{M}_i) = {} & \sum_{m=0}^{N-1}\left[\sum_{k=0}^{N-1}\left(\Sigma_{\boldsymbol{xx},s(k)}^{-1}+\Sigma_{\boldsymbol{hh}}^{-1}\right)\right]^{-1} \Sigma_{\boldsymbol{xx},s(m)}^{-1}\left[\boldsymbol{y}(m)-\boldsymbol{\mu}_{\boldsymbol{x},s(m)}\right] \\ & + \left[\sum_{k=0}^{N-1}\left(\Sigma_{\boldsymbol{xx},s(k)}^{-1}+\Sigma_{\boldsymbol{hh}}^{-1}\right)\right]^{-1} \Sigma_{\boldsymbol{hh}}^{-1}\mu_{\boldsymbol{h}} \end{aligned} \tag{17.76}$$

The MAP estimate of the channel over all state sequences of all HMMs can be obtained as

$$\hat{\boldsymbol{h}}(\boldsymbol{Y}) = \sum_{i=1}^{V}\sum_{\mathbf{S}} \hat{\boldsymbol{h}}^{MAP}(\boldsymbol{Y},\mathbf{s},\mathcal{M}_i) P_{\mathbf{S}|\mathcal{M}}(\mathbf{s}|\mathcal{M}_i) P_{\mathcal{M}}(\mathcal{M}_i) \tag{17.77}$$

17.4.6 Implementations of HMM-Based Deconvolution

In this section, we consider three implementation methods for HMM-based channel equalisation.

Method I: Use of the Statistical Averages Taken Over All HMMs

A simple approach to blind equalisation, similar to that proposed by Stockham, is to use as the channel input statistics the average of the mean vectors and the covariance matrices, taken over all the states of all the HMMs as

$$\boldsymbol{\mu}_{\boldsymbol{x}} = \frac{1}{VN_s}\sum_{i=1}^{V}\sum_{j=1}^{N_s} \boldsymbol{\mu}_{\mathcal{M}_i,j}, \quad \Sigma_{\boldsymbol{xx}} = \frac{1}{VN_s}\sum_{i=1}^{V}\sum_{j=1}^{N_s} \Sigma_{\mathcal{M}_i,j} \tag{17.78}$$

where $\boldsymbol{\mu}_{\mathcal{M}_i,j}$ and $\Sigma_{\mathcal{M}_i,j}$ are the mean and the covariance of the jth state of the ith HMM, V and N_s denote the number of models and number of states per model respectively. The maximum likelihood estimate of the channel, $\hat{\boldsymbol{h}}^{ML}$, is defined as

$$\hat{\boldsymbol{h}}^{ML} = (\bar{\boldsymbol{y}} - \boldsymbol{\mu}_{\mathbf{x}}) \tag{17.79}$$

where $\bar{y}$ is the time-averaged channel output. The estimate of the channel input is

$$\hat{\boldsymbol{x}}(m) = \boldsymbol{y}(m) - \hat{\boldsymbol{h}}^{ML} \tag{17.80}$$

Using the averages over all states and models, the MAP channel estimate becomes

$$\hat{\boldsymbol{h}}^{MAP}(\boldsymbol{Y}) = \sum_{m=0}^{N-1} (\Sigma_{\boldsymbol{xx}}+\Sigma_{\boldsymbol{hh}})^{-1}\Sigma_{\boldsymbol{hh}}(\boldsymbol{y}(m)-\boldsymbol{\mu}_{\boldsymbol{x}}) + (\Sigma_{\boldsymbol{xx}}+\Sigma_{\boldsymbol{hh}})^{-1}\Sigma_{\boldsymbol{xx}}\boldsymbol{\mu}_{\boldsymbol{h}} \tag{17.81}$$

Method II: Hypothesised-Input HMM Equalisation

In this method, for each candidate HMM in the input vocabulary, a channel estimate is obtained and then used to equalise the channel output, prior to the computation of a likelihood score for the HMM. Thus a channel estimate $\hat{\boldsymbol{h}}_w$ is based on the hypothesis that the input word is w. It is expected that a better channel estimate is obtained from the correctly hypothesised HMM, and a poorer estimate from an incorrectly hypothesised HMM. The hypothesised-input HMM algorithm is as follows (Figure 17.9):

For $i = 1$ to number of words V {
step 1 Using each HMM, $\mathcal{M}_i$, make an estimate of the channel, $\hat{\boldsymbol{h}}_i$,
step 2 Using the channel estimate, $\hat{\boldsymbol{h}}_i$, estimate the channel input $\hat{x}(m) = \boldsymbol{y}(m) - \hat{\boldsymbol{h}}_i$
step 3 Compute a probability score for model $\mathcal{M}_i$, given the estimate $[\hat{x}(m)]$.}

Select the channel estimate associated with the most probable word.

Figure 17.10 shows the ML channel estimates of two channels using unweighted average and hypothesised-input methods.

Method III: Decision-Directed Equalisation

Blind adaptive equalisers are often composed of two distinct sections: an adaptive linear equaliser followed by a non-linear estimator to improve the equaliser output. The output of the non-linear estimator is the final estimate of the channel input, and is used as the desired signal *to direct* the equaliser adaptation. The use of the output of the non-linear estimator as the desired signal assumes that the linear equalisation filter removes a large part of the channel distortion, thereby enabling the non-linear estimator to produce an accurate estimate of the channel input. A method of ensuring that the equaliser locks into and cancels a large part of the channel distortion is to use a startup, equaliser training period, during which a known signal is transmitted.

Figure 17.11 illustrates a blind equaliser incorporating an adaptive linear filter followed by a hidden Markov model classifier/estimator. The HMM classifies the output of the filter as one of a number of likely signals and provides an enhanced output, which is also used for adaptation of the linear filter. The output of the equaliser $z(m)$ is expressed as the sum of the input to the channel $x(m)$ and a so-called convolutional noise term $v(m)$ as

$$z(m) = x(m) + v(m) \tag{17.82}$$

The HMM may incorporate state-based Wiener filters for suppression of the convolutional noise $v(m)$ as described in Section 11.6. Assuming that the LMS adaptation method is employed, the adaptation

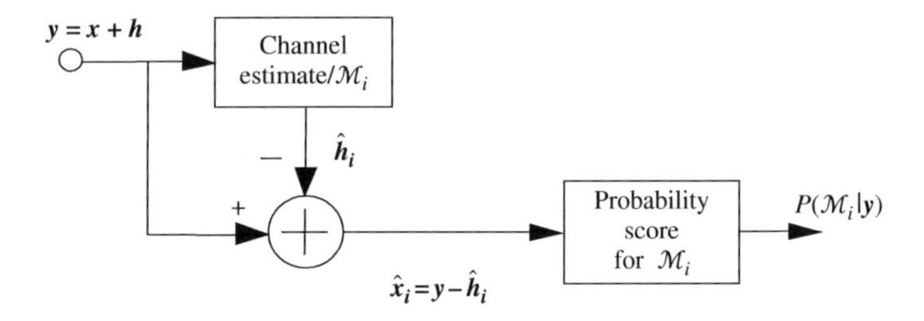

Figure 17.9 Hypothesised channel estimation procedure.

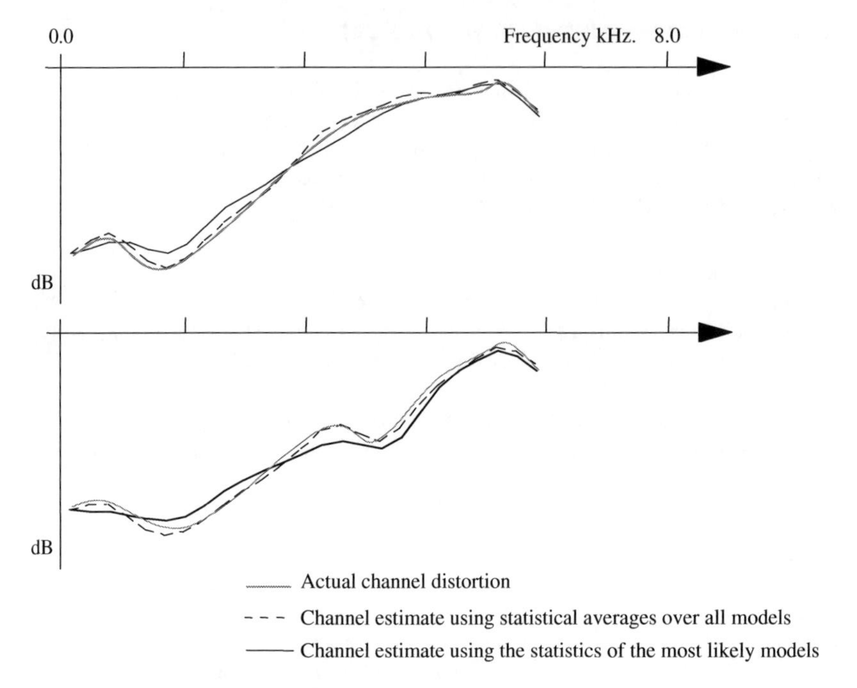

Figure 17.10 Illustration of actual and estimated channel response for two channels.

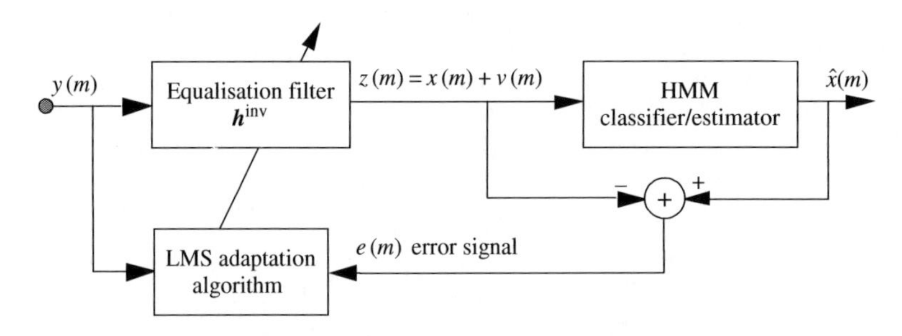

Figure 17.11 A decision-directed equaliser.

of the equaliser coefficient vector is governed by the recursive equation

$$\hat{h}^{\text{inv}}(m) = \hat{h}^{\text{inv}}(m-1) + \mu e(m)y(m) \tag{17.83}$$

where $\hat{h}^{\text{inv}}(m)$ is an estimate of the optimal inverse channel filter, μ is an adaptation step size and the error signal $e(m)$ is defined as

$$e(m) = \hat{x}^{\text{HMM}}(m) - z(m) \tag{17.84}$$

where $\hat{x}^{\text{HMM}}(m)$ is the output of the HMM-based estimator and is used as the correct estimate of the desired signal to direct the adaptation process.

17.5 Blind Equalisation for Digital Communication Channels

High speed transmission of digital data over analogue channels, such as telephone lines or a radio channels, requires adaptive equalisation to reduce decoding errors caused by channel distortions. In telephone lines, the channel distortions are due to the non-ideal magnitude response and the non-linear phase response of the lines. In radio channel environments, the distortions are due to non-ideal channel response as well as the effects of multi-path propagation of the radio waves via a multitude of different routes with different attenuations and delays. In general, the main types of distortions suffered by transmitted symbols are amplitude distortion, time dispersion and fading. Of these, time dispersion is perhaps the most important, and has received a great deal of attention. Time dispersion has the effect of smearing and elongating the duration of each symbol. In high speed communication systems, where the data symbols closely follow each other, time dispersion results in an overlap of successive symbols, an effect known as inter-symbol interference (ISI), illustrated in Figure 17.12.

In a digital communication system, the transmitter modem takes N bits of binary data at a time, and encodes them into one of 2^N analogue symbols for transmission, at the signalling rate, over an analogue channel. At the receiver the analogue signal is sampled and decoded into the required digital format. Most digital modems are based on multilevel phase-shift keying, or combined amplitude and phase shift keying schemes. In this section we consider multi-level pulse amplitude modulation (M-ary PAM) as a convenient scheme for the study of adaptive channel equalisation.

Assume that at the transmitter modem, the kth set of N binary digits is mapped into a pulse of duration T_s seconds and an amplitude $a(k)$. Thus the modulator output signal, which is the input to the communication channel, is given as

$$x(t) = \sum_k a(k) r(t - kT_s) \tag{17.85}$$

where $r(t)$ is a pulse of duration T_s and with an amplitude $a(k)$ that can assume one of $M = 2^N$ distinct levels. Assuming that the channel is linear, the channel output can be modelled as the convolution of the input signal and channel response:

$$y(t) = \int_{-\infty}^{\infty} h(\tau) x(t - \tau) d\tau \tag{17.86}$$

where $h(t)$ is the channel impulse response. The sampled version of the channel output is given by the discrete-time equation

$$y(m) = \sum_k h_k x(m - k) \tag{17.87}$$

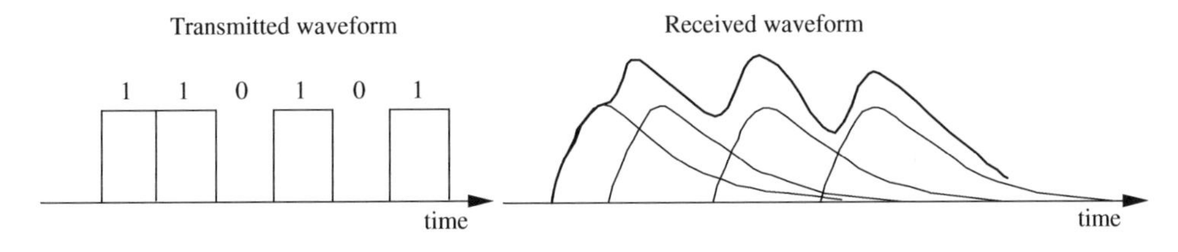

Figure 17.12 Illustration of inter-symbol interference in a binary pulse amplitude modulation system.

To remove the channel distortion, the sampled channel output $y(m)$ is passed to an equaliser with an impulse response $\hat{h}_k^{\mathrm{inv}}$. The equaliser output $z(m)$ is given as

$$\begin{aligned} z(m) &= \sum_k \hat{h}_k^{\mathrm{inv}} y(m-k) \\ &= \sum_j x(m-j) \sum_k \hat{h}_k^{inv} h_{j-k} \end{aligned} \tag{17.88}$$

where Equation (17.87) is used to obtain the second line of Equation (17.88). The ideal equaliser output is $z(m) = x(m-D) = a(m-D)$ for some delay D that depends on the channel response and the length of the equaliser. From Equation (17.88), the channel distortion would be cancelled if

$$h_m^c = h_m * \hat{h}_m^{\mathrm{inv}} = \delta(m-D) \tag{17.89}$$

where h_m^c is the combined impulse response of the cascade of the channel and the equaliser. A particular form of channel equaliser, for the elimination of ISI, is the Nyquist *zero-forcing* filter, where the impulse response of the combined channel and equaliser is defined as

$$h^c(kT_s + D) = \begin{cases} 1, & k = 0 \\ 0, & k \neq 0 \end{cases} \tag{17.90}$$

Note that in Equation (17.90), at the sampling instants the channel distortion is cancelled, and hence there is no ISI at the sampling instants. A function that satisfies Equation (17.90) is the sinc function $h^c(t) = \sin(\pi f_s t)/\pi f_s t$, where $f_s = 1/T_s$. Zero-forcing methods are sensitive to deviations of $h^c(t)$ from the requirement of Equation (17.90), and also to jitters in the synchronisation and the sampling process.

17.5.1 LMS Blind Equalisation

In this section, we consider the more general form of the LMS-based adaptive equaliser followed by a non-linear estimator. In a conventional sample-adaptive filter, the filter coefficients are adjusted to minimise the mean squared distance between the filter output and the desired signal. In blind equalisation, the desired signal (which is the channel input) is not available. The use of an adaptive filter for blind equalisation requires an internally generated desired signal, as illustrated in Figure 17.13. Digital blind equalisers are composed of two distinct sections: an adaptive equaliser that removes a

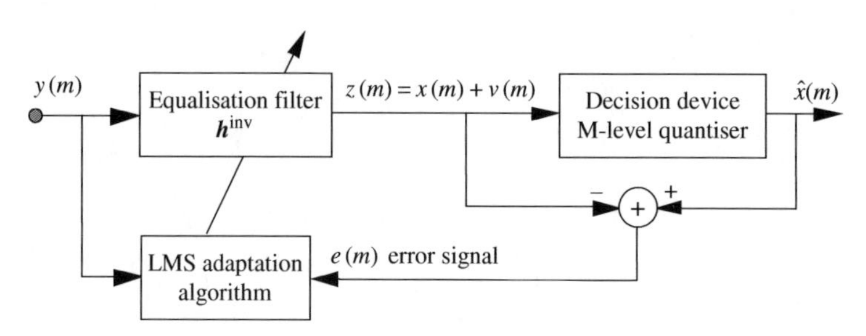

Figure 17.13 Configuration of an adaptive channel equaliser with an estimate of the channel input used as an 'internally' generated desired signal.

large part of the channel distortion, followed by a non-linear estimator for an improved estimate of the channel input. The output of the non-linear estimator is the final estimate of the channel input, and is used as the desired signal *to direct* the equaliser adaptation. A method of ensuring that the equaliser removes a large part of the channel distortion is to use a start-up, equaliser training, period during which a known signal is transmitted.

Assuming that the LMS adaptation method is employed, the adaptation of the equaliser coefficient vector is governed by the recursive equation

$$\hat{\boldsymbol{h}}^{\text{inv}}(m) = \hat{\boldsymbol{h}}^{\text{inv}}(m-1) + \mu e(m)\boldsymbol{y}(m) \tag{17.91}$$

where $\hat{\boldsymbol{h}}^{\text{inv}}(m)$ is an estimate of the optimal inverse channel filter $\boldsymbol{h}^{\text{inv}}$, the scalar μ is the adaptation step size, and the error signal $e(m)$ is defined as

$$\begin{aligned} e(m) &= \psi(z(m)) - z(m) \\ &= \hat{x}(m) - z(m) \end{aligned} \tag{17.92}$$

where $\hat{x}(m) = \psi(z(m))$ is a non-linear estimate of the channel input. For example, in a binary communication system with an input alphabet $\{\pm a\}$ we can use a signum non-linearity such that $\hat{x}(m) = a.\text{sgn}(z(m))$ where the function sgn$(\cdot)$ gives the sign of the argument. In the following, we use a Bayesian framework to formulate the non-linear estimator $\psi()$.

Assuming that the channel input is an uncorrelated process and the equaliser removes a large part of the channel distortion, the equaliser output can be expressed as the sum of the desired signal (the channel input) plus an uncorrelated additive noise term:

$$z(m) = x(m) + v(m) \tag{17.93}$$

where $\nu(m)$ is the so-called convolutional noise defined as

$$\begin{aligned} \nu(m) &= x(m) - \sum_k \hat{h}_k^{\text{inv}} y(m-k) \\ &= \sum_k (h_k^{\text{inv}} - \hat{h}_k^{\text{inv}}) y(m-k) \end{aligned} \tag{17.94}$$

In the following, we assume that the non-linear estimates of the channel input are correct, and hence the error signals $e(m)$ and $\nu(m)$ are identical. Owing to the averaging effect of the channel and the equaliser, each sample of convolutional noise is affected by many samples of the input process. From the central limit theorem, the convolutional noise $e(m)$ can be modelled by a zero-mean Gaussian process as

$$f_E(e(m)) = \frac{1}{\sqrt{2\pi}\sigma_e} \exp\left(-\frac{e^2(m)}{2\sigma_e^2}\right) \tag{17.95}$$

where σ_e^2, the noise variance, can be estimated using the recursive time-update equation

$$\sigma_e^2(m) = \rho\sigma_e^2(m-1) + (1-\rho)e^2(m) \tag{17.96}$$

where $\rho < 1$ is the adaptation factor. The Bayesian estimate of the channel input given the equaliser output can be expressed in a general form as

$$\hat{x}(m) = \arg\min_{\hat{x}(m)} \int_X C(x(m), \hat{x}(m)) f_{X|Z}(x(m)|z(m))\, dx(m) \tag{17.97}$$

where $C(x(m), \hat{x}(m))$ is a cost function and $f_{X|Z}(x(m)|z(m))$ is the posterior pdf of the channel input signal. The choice of the cost function determines the type of the estimator as described in Chapter 7. Using a uniform cost function in Equation (17.97) yields the maximum a posteriori (MAP) estimate

$$\begin{aligned}\hat{x}^{MAP}(m) &= \arg\max_{x(m)} f_{X|Z}(x(m)|z(m)) \\ &= \arg\max_{x(m)} f_E(z(m)-x(m))\, P_X(x(m))\end{aligned} \tag{17.98}$$

Now, as an example consider an M-ary pulse amplitude modulation system, and let $\{a_i i = 1, \ldots, M\}$ denote the set of M pulse amplitudes with a probability mass function

$$P_X(x(m)) = \sum_{i=1}^{M} P_i \delta(x(m) - a_i) \tag{17.99}$$

The pdf of the equaliser output $z(m)$ can be expressed as the mixture pdf

$$f_Z(z(m)) = \sum_{i=1}^{M} P_i f_E(x(m) - a_i) \tag{17.100}$$

The posterior density of the channel input is

$$P_{X|Z}(x(m) = a_i|z(m)) = \frac{1}{f_Z(z(m))} f_E(z(m) - a_i) P_X(x(m) = a_i) \tag{17.101}$$

and the MAP estimate is obtained from

$$\hat{x}^{MAP}(m) = \arg\max_{a_i} \left(f_E(z(m) - a_i) P_X(x(m) = a_i)\right) \tag{17.102}$$

Note that the classification of the continuous-valued equaliser output $z(m)$ into one of M discrete channel input symbols is basically a non-linear process. Substitution of the zero-mean Gaussian model for the convolutional noise $e(m)$ in Equation (17.102) yields

$$\hat{x}^{MAP}(m) = \arg\max_{a_i} \left(P_X(x(m) = a_i) \exp\left(-\frac{(z(m) - a_i)^2}{2\sigma_e^2} \right) \right) \tag{17.103}$$

Note that when the symbols are equi-probable, the MAP estimate reduces to a simple threshold decision device. Figure 17.13 shows a channel equaliser followed by an M-level quantiser. In this system, the output of the equaliser filter is passed to an M-ary decision circuit. The decision device, which is essentially an M-level quantiser, classifies the channel output into one of M valid symbols. The output of the decision device is taken as an internally generated desired signal to direct the equaliser adaptation.

17.5.2 Equalisation of a Binary Digital Channel

Consider a binary PAM communication system with an input symbol alphabet $\{a_0, a_1\}$ and symbol probabilities $P(a_0) = P_0$ and $P(a_1) = P_1 = 1 - P_0$. The pmf of the amplitude of the channel input signal can be expressed as

$$P(x(m)) = P_0 \delta(x(m) - a_0) + P_1 \delta(x(m) - a_1) \tag{17.104}$$

Assume that at the output of the linear adaptive equaliser in Figure 17.13, the convolutional noise $v(m)$ is a zero-mean Gaussian process with variance σ_v^2. Therefore the pdf of the equaliser output $z(m) = x(m) + v(m)$ is a mixture of two Gaussian pdfs and can be described as

$$f_Z(z(m)) = \frac{P_0}{\sqrt{2\pi}\sigma_v} \exp\left\{-\frac{(z(m)-a_0)^2}{2\sigma_v^2}\right\} + \frac{P_1}{\sqrt{2\pi}\sigma_v} \exp\left\{-\frac{(z(m)-a_1)^2}{2\sigma_v^2}\right\} \tag{17.105}$$

The MAP estimate of the channel input signal is

$$\hat{x}(m) = \begin{cases} a_0 & \text{if } \dfrac{P_0}{\sqrt{2\pi}\sigma_v} \exp\left\{-\dfrac{[z(m)-a_0]^2}{2\sigma_v^2}\right\} > \dfrac{P_1}{\sqrt{2\pi}\sigma_v} \exp\left\{-\dfrac{[z(m)-a_1]^2}{2\sigma_v^2}\right\} \\ a_1 & \text{otherwise} \end{cases} \tag{17.106}$$

For the case when the channel alphabet consists of $a_0 = -a$, $a_1 = a$ and $P_0 = P_1$, the MAP estimator is identical to the signum function $\text{sgn}(x(m))$, and the error signal is given by

$$e(m) = z(m) - \text{sgn}\,(z(m))\, a \tag{17.107}$$

Figure 17.14 shows the error signal as a function of $z(m)$. An undesirable property of a hard non-linearity, such as the $\text{sgn}(\cdot)$ function, is that it produces a large error signal at those instances when $z(m)$ is around zero, and a decision based on the sign of $z(m)$ is most likely to be incorrect.

A large error signal based on an incorrect decision would have an unsettling effect on the convergence of the adaptive equaliser. It is desirable to have an error function that produces small error signals when $z(m)$ is around zero. Nowlan *et al.* (1993) proposed a soft non-linearity of the following form

$$e(m) = z(m) - \frac{e^{2az(m)/\sigma^2} - 1}{e^{2az(m)/\sigma^2} + 1} a \tag{17.108}$$

The error $e(m)$ is small when the magnitude of $z(m)$ is small and large when the magnitude of $z(m)$ is large.

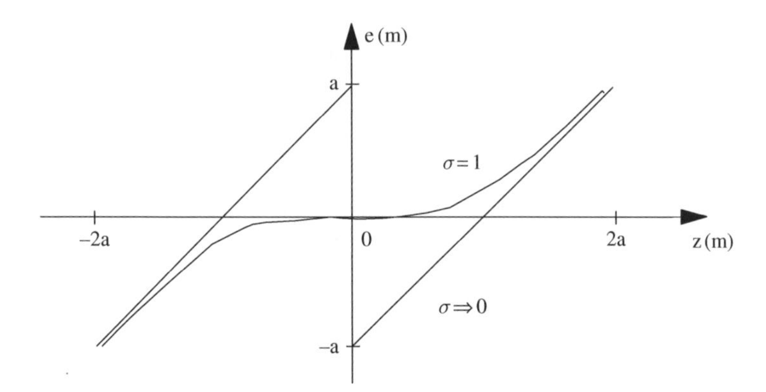

Figure 17.14 Comparison of the error functions produced by the hard nonlinearity of Equation (17.107) and the soft non-linearity of Equation (17.108).

17.6 Equalisation Based on Higher-Order Statistics

The second-order statistics of a random process, namely the autocorrelation or its Fourier transform the power spectrum, are central to the development of linear estimation theory, and form the basis of most statistical signal processing methods such as Wiener filters and linear predictive models. An attraction of the correlation function is that a Gaussian process, of a known mean vector, can be completely described in terms of the covariance matrix, and many random processes can be well characterised by Gaussian or mixture Gaussian models.

A shortcoming of second-order statistics is that they do not always include the correct phase characteristics of the process. That is because the covariance matrix of a minimum phase and its equivalent maximum phase system are identical (for example try an experiment in Matlab with the covariance matrices of the output responses, to the same input, of the following filters: $H_1(z) = G_1\left(1-\alpha z^{-1}\right)$ and $H_2(z) = G_2\left(1-(1/\alpha)z^{-1}\right)$ where $G_1 = 1/(1+\alpha)$ and $G_2 = 1/(1+(1/\alpha))$ for unity gain). Therefore, given the channel output, it is not always possible to estimate the channel phase from the second-order statistics. Furthermore, as a Gaussian process of known mean depends entirely on the autocovariance function, it follows that blind deconvolution, based on a Gaussian model of the channel input, cannot estimate the channel phase.

Higher-order statistics, and the probability models based on them, can model both the magnitude and the phase characteristics of a random process. In this section, we consider blind deconvolution based on higher-order statistics and their Fourier transforms known as the higher-order spectra. The prime motivation in using the higher-order statistics is their ability to model the phase characteristics. Further motivations are the potential of the higher order statistics to model channel non-linearities, and to estimate a non-Gaussian signal in a high level of Gaussian noise.

17.6.1 Higher-Order Moments, Cumulants and Spectra

The kth *order* moment of a random variable X is defined as

$$\begin{aligned} m_k &= \mathcal{E}[x^k] \\ &= (-\mathrm{j})^k \left.\frac{\partial^k \Phi_X(\omega)}{\partial \omega^k}\right|_{\omega=0} \end{aligned} \tag{17.109}$$

where $\Phi_X(\omega)$ is the *characteristic function* of the random variable X defined as

$$\Phi_X(\omega) = \mathcal{E}[\exp(\mathrm{j}\omega x)] \tag{17.110}$$

From Equations (17.109) and (17.110), the first moment of X is $m_1 = E[x]$, the second moment of X is $m_2 = E[x^2]$, and so on. The joint k^{th} order moment ($k = k_1 + k_2$) of two random variables X_1 and X_2 is defined as

$$\mathcal{E}[x_1^{k_1} x_2^{k_2}] = (-\mathrm{j})^{k_1+k_2} \left.\frac{\partial^{k_1}\partial^{k_2}\Phi_{X_1X_2}(\omega_1,\omega_2)}{\partial\omega_1^{k_1}\,\partial\omega_2^{k_2}}\right|_{\omega_1=\omega_2=0} \tag{17.111}$$

and in general the joint k^{th} order moment of N random variables is defined as

$$\begin{aligned} m_k &= \mathcal{E}[x_1^{k_1} x_2^{k_2} \dots x_N^{k_N}] \\ &= (-\mathrm{j})^k \left.\frac{\partial^k \Phi(\omega_1,\omega_2,\dots,\omega_N)}{\partial\omega_1^{k_1}\partial\omega_2^{k_2}\cdots\partial\omega_N^{k_N}}\right|_{\omega_1=\omega_2=\cdots=\omega_N=0} \end{aligned} \tag{17.112}$$

where $k = k_1 + k_2 + \ldots + k_N$ and the joint characteristic function is

$$\Phi(\omega_1, \omega_2, \ldots, \omega_N) = \mathcal{E}\left[\exp\left(j\omega_1 x_1 + j\omega_2 x_2 + \cdots + j\omega_N x_N\right)\right] \tag{17.113}$$

Now the higher-order moments can be applied for characterisation of discrete-time random processes. The kth order moment of a random process $x(m)$ is defined as

$$m_x(\tau_1, \tau_2, \ldots, \tau_{K-1}) = \mathcal{E}[x(m), x(m+\tau_1)x(m+, \tau_2)\cdots x(m+\tau_{k-1})] \tag{17.114}$$

Note that the second-order moment $E[x(m)x(m+\tau)]$ is the autocorrelation function.

Cumulants

Cumulants are similar to moments; the difference is that the moments of a random process are derived from the characteristic function $\Phi_X(\omega)$, whereas the cumulant generating function $C_X(\omega)$ is defined as the logarithm of the characteristic function as

$$C_X(\omega) = \ln \Phi_X(\omega) = \ln \mathcal{E}[\exp(j\omega x)] \tag{17.115}$$

Using a Taylor series expansion of the term $E[\exp(j\omega x)]$ in Equation (17.115) the cumulant generating function can be expanded as

$$C_X(\omega) = \ln\left(1 + m_1(j\omega) + \frac{m_2}{2!}(j\omega)^2 + \frac{m_3}{3!}(j\omega)^3 + \cdots + \frac{m_n}{n!}(j\omega)^n + \cdots\right) \tag{17.116}$$

where $m_k = E[x^k]$ is the kth moment of the random variable x. The kth order cumulant of a random variable is defined as

$$c_k = (-j)^k \left.\frac{\partial^k C_X(\omega)}{\partial \omega^k}\right|_{\omega=0} \tag{17.117}$$

From Equations (17.116) and (17.117), we have

$$c_1 = m_1 \tag{17.118}$$

$$c_2 = m_2 - m_1^2 \tag{17.119}$$

$$c_3 = m_3 - 3m_1 m_2 + 2m_1^2 \tag{17.120}$$

and so on. The general form of the kth order $(k = k_1 + k_2 + \cdots + k_N)$ joint cumulant generating function is

$$c_{k_1 \cdots k_N} = (-j)^{k_1 + \cdots + k_N} \left.\frac{\partial^{k_1 + \cdots + k_N} \ln \Phi_X(\omega_1, \cdots, \omega_N)}{\partial \omega_1^{k_1} \cdots \partial \omega_N^{k_N}}\right|_{\omega_1 = \omega_2 = \cdots = \omega_N = 0} \tag{17.121}$$

The cumulants of a zero mean random process $x(m)$ are given as

$$c_x = \mathcal{E}[x(k)] = m_x = 0 \quad \text{(mean)} \tag{17.122}$$

$$\begin{aligned} c_x(k) &= \mathcal{E}[x(m)x(m+k)] - \mathcal{E}[x(m)]^2 \\ &= m_x(k) - m_x^2 = m_x(k) \quad \text{(covariance)} \end{aligned} \tag{17.123}$$

$$\begin{aligned} c_x(k_1, k_2) &= m_x(k_1, k_2) - m_x[m_x(k_1) + m_x(k_2) + m_x(k_2 - k_1)] + 2\,(m_x)^3 \\ &= m_x(k_1, k_2) \quad \text{(skewness)} \end{aligned} \tag{17.124}$$

$$\begin{aligned} c_x(k_1, k_2, k_3) = {} & m_x(k_1, k_2, k_3) - m_x(k_1)m_x(k_3 - k_2) \\ & - m_x(k_2)m_x(k_3 - k_1) - m_x(k_3)m_x(k_2 - k_1) \end{aligned} \tag{17.125}$$

and so on. Note that $m_x(k_1, k_2, \ldots, k_N) = E[x(m)x(m+k_1), x(m+k_2), \ldots, x(m+k_N)]$. The general formulation of the kth order cumulant of a random process $x(m)$ (Rosenblatt *et al.* (1985)) is defined as

$$c_x(k_1, k_2, \ldots, k_n) = m_x(k_1, k_2, \ldots, k_n) - m_x^G(k_1, k_2, \ldots, k_n) \quad \text{for } n = 3, 4, \ldots \tag{17.126}$$

where $m_x^G(k_1, k_2, \ldots, k_n)$ is the kth order moment of a Gaussian process having the same mean and autocorrelation as the random process $x(m)$. From Equation (17.126), it follows that for a Gaussian process, the cumulants of order greater than 2 are identically zero.

Higher-Order Spectra

The kth order spectrum of a signal $x(m)$ is defined as the $(k-1)$-dimensional Fourier transform of the kth order cumulant sequence as

$$C_X(\omega_1, \ldots, \omega_{k-1}) = \frac{1}{(2\pi)^{k-1}} \sum_{\tau_1=-\infty}^{\infty} \cdots \sum_{\tau_{k-1}=-\infty}^{\infty} c_x(\tau_1, \ldots, \tau_{k-1}) e^{-j(\omega_1\tau_1 + \cdots + \omega_{k-1}\tau_{k-1})} \tag{17.127}$$

For the case $k = 2$, the second-order spectrum is the power spectrum given as

$$C_X(\omega) = \frac{1}{2\pi} \sum_{\tau=-\infty}^{\infty} c_x(\tau) e^{-j\omega\tau} \tag{17.128}$$

The *bi-spectrum* is defined as

$$C_X(\omega_1, \omega_2) = \frac{1}{(2\pi)^2} \sum_{\tau_1=-\infty}^{\infty} \sum_{\tau_2=-\infty}^{\infty} c_x(\tau_1, \tau_2) e^{-j(\omega_1\tau_1 + \omega_2\tau_2)} \tag{17.129}$$

and the *tri-spectrum* is

$$C_X(\omega_1, \omega_2, \omega_3) = \frac{1}{(2\pi)^3} \sum_{\tau_1=-\infty}^{\infty} \sum_{\tau_2=-\infty}^{\infty} \sum_{\tau_3=-\infty}^{\infty} c_x(\tau_1, \tau_2, \tau_3) e^{-j(\omega_1\tau_1 + \omega_2\tau_2 + \omega_3\tau_3)} \tag{17.130}$$

Since the term $e^{j\omega t}$ is periodic with a period of 2π, it follows that higher order spectra are periodic in each ω_k with a period of 2π.

17.6.2 Higher-Order Spectra of Linear Time-Invariant Systems

Consider a linear time-invariant system with an impulse response sequence $\{h_k\}$, input signal $x(m)$ and output signal $y(m)$. The relation between the kth-order cumulant spectra of the input and output signals is given by

$$C_Y(\omega_1, \ldots, \omega_{k-1}) = H(\omega_1) \ldots H(\omega_{k-1}) H^*(\omega_1 + \cdots + \omega_{k-1}) C_X(\omega_1, \ldots, \omega_{k-1}) \tag{17.131}$$

where $H(\omega)$ is the frequency response of the linear system $\{h_k\}$. The magnitude of the kth order spectrum of the output signal is given as

$$|C_Y(\omega_1,\ldots,\omega_{k-1})| = |H(\omega_1)|\cdots|H(\omega_{k-1})|\,|H(\omega_1+\cdots+\omega_{k-1})|\,|C_X(\omega_1,\ldots,\omega_{k-1})| \quad (17.132)$$

and the phase of the kth order spectrum is

$$\Phi_Y(\omega_1,\ldots,\omega_{k-1}) = \Phi_H(\omega_1)+\cdots+\Phi_H(\omega_{k-1}) - \Phi_H(\omega_1+\cdots+\omega_{k-1}) + \Phi_X(\omega_1,\ldots,\omega_{k-1}) \quad (17.133)$$

17.6.3 Blind Equalisation Based on Higher-Order Cepstra

In this section, we consider blind equalisation of a maximum-phase channel, based on higher order cepstra. Assume that the channel can be modelled by an all-zero filter, and that its z-transfer function $H(z)$ can be expressed as the product of a maximum-phase polynomial factor and a minimum-phase factor as

$$H(z) = GH_{\min}(z)H_{\max}(z^{-1})z^{-D} \quad (17.134)$$

$$H_{\min}(z) = \prod_{i=1}^{P_1}(1-\alpha_i z^{-1}) \; |\alpha_i| < 1 \quad (17.135)$$

$$H_{\max}(z^{-1}) = \prod_{i=1}^{P_2}(1-\beta_i z) \; |\beta_i| < 1 \quad (17.136)$$

where G is a gain factor, $H_{\min}(z)$ is a minimum-phase polynomial with all its zeros inside the unit circle, $H_{\max}(z^{-1})$ is a maximum-phase polynomial with all its zeros outside the unit circle, and z^{-D} inserts D unit delays in order to make Equation (17.134) causal. The complex cepstrum of $H(z)$ is defined as

$$h_c(m) = Z^{-1}(\ln H(z)) \quad (17.137)$$

where Z^{-1} denotes the inverse z-transform. At $z = e^{j\omega}$, the z-transform is the discrete Fourier transform (DFT), and the cepstrum of a signal is obtained by taking the inverse DFT of the logarithm of the signal spectrum. In the following we consider cepstra based on the power spectrum and the higher-order spectra, and show that the higher-order cepstra have the ability to retain maximum-phase information. Assuming that the channel input $x(m)$ is a zero-mean uncorrelated process with variance σ_x^2, the power spectrum of the channel output can be expressed as

$$P_Y(\omega) = \frac{\sigma_x^2}{2\pi}H(\omega)H^*(\omega) \quad (17.138)$$

The cepstrum of the power spectrum of $y(m)$ is defined as

$$\begin{aligned} y_c(m) &= \text{IDFT}(\ln P_Y(\omega)) \\ &= \text{IDFT}\left(\ln\left(\sigma_x^2 G^2/2\pi\right) + \ln H_{\min}(\omega) + H_{\max}(-\omega) + \ln H^*_{\min}(\omega) + H^*_{\max}(-\omega)\right) \end{aligned} \quad (17.139)$$

where IDFT is the inverse discrete Fourier transform. Substituting Equations (17.135) and (17.36) in (17.139), the cepstrum can be expressed as

$$y_c(m) = \begin{cases} \ln\left(G^2\sigma_x^2/2\pi\right), & m=0 \\ -\left(A^{(m)}+B^{(m)}\right)/m, & m>0 \\ \left(A^{(-m)}+B^{(-m)}\right)/m, & m<0 \end{cases} \quad (17.140)$$

where $A^{(m)}$ and $B^{(m)}$ are defined as

$$A^{(m)} = \sum_{i=1}^{P_1} \alpha_i^m \tag{17.141}$$

$$B^{(m)} = \sum_{i=1}^{P_2} \beta_i^m \tag{17.142}$$

Note from Equation (17.140) that along the index m, the maximum-phase information $B^{(m)}$ and the minimum-phase information $A^{(m)}$ overlap and cannot be separated.

Bi-cepstrum

The bi-cepstrum of a signal is defined as the inverse Fourier transform of the logarithm of the bi-spectrum:

$$y_c(m_1, m_2) = \text{IDFT}_2[\log C_Y(\omega_1, \omega_2)] \tag{17.143}$$

where $\text{IDFT}_2[.]$ denotes the two-dimensional inverse discrete Fourier transform. The relationship between the bi-spectra of the input and output of a linear system is

$$C_Y(\omega_1, \omega_2) = H(\omega_1)H(\omega_2)H^*(\omega_1+\omega_2)C_X(\omega_1, \omega_2) \tag{17.144}$$

Assuming that the input $x(m)$ of the linear time-invariant system $\{h_k\}$ is an uncorrelated non-Gaussian process, the bi-spectrum of the output can be written as

$$\begin{aligned} C_Y(\omega_1, \omega_2) = & \frac{\gamma_x^{(3)} G^3}{(2\pi)^2} H_{\min}(\omega_1) H_{\max}(-\omega_1) H_{\min}(\omega_2) H_{\max}(-\omega_2) \\ & \times H_{\min}^*(\omega_1+\omega_2) H_{\max}^*(-\omega_1-\omega_2) \end{aligned} \tag{17.145}$$

where $\gamma_x^{(3)}/(2\pi)^2$ is the third-order cumulant of the uncorrelated random input process $x(m)$. Taking the logarithm of Equation (17.145) yields

$$\begin{aligned} \ln C_y(\omega_1, \omega_2) = & \ln|A| + \ln H_{\min}(\omega_1) + \ln H_{\max}(-\omega_1) + \ln H_{\min}(\omega_2) + \ln H_{\max}(-\omega_2) \\ & + \ln H_{\min}^*(\omega_1+\omega_2) + \ln H_{\max}^*(-\omega_1-\omega_2) \end{aligned} \tag{17.146}$$

where $A = \gamma_x^{(3)} G^3/(2\pi)^2$. The bi-cepstrum is obtained through the inverse Discrete Fourier transform of Equation (17.146) as

$$y_c(m_1, m_2) = \begin{cases} \ln|A|, & m_1 = m_2 = 0 \\ -A^{(m_1)}/m_1, & m_1 > 0, m_2 = 0 \\ -A^{(m_2)}/m_2, & m_2 > 0, m_1 = 0 \\ -B^{(-m_1)}/m_1, & m_1 < 0, m_2 = 0 \\ B^{(-m_2)}/m_2, & m_2 < 0, m_1 = 0 \\ -B^{(m_2)}/m_2, & m_1 = m_2 > 0 \\ A^{(-m_2)}/m_2, & m_1 = m_2 < 0 \\ 0, & \text{otherwise} \end{cases} \tag{17.147}$$

Note from Equation (17.147) that the maximum-phase information $B^{(m)}$ and the minimum-phase information $A^{(m)}$ are separated and appear in different regions of the bi-cepstrum indices m_1 and m_2.

The higher-order cepstral coefficients can be obtained either from the IDFT of higher-order spectra as in Equation (17.147) or by using parametric methods as follows. In general, the cepstral and cumulant coefficients can be related by a convolutional equation. Pan *et al.* (1988) have shown that the recursive relation between the bi-cepstrum coefficients and the third-order cumulants of a random process is

$$y_c(m_1, m_2) * \left[-m_1 c_y(m_1, m_2)\right] = -m_1 c_y(m_1, m_2) \tag{17.148}$$

Substituting Equation (17.147) in Equation (17.148) yields

$$\begin{aligned}&\sum_{i=1}^{\infty} A^{(i)}[c_x(m_1 - i, m_2) - c_x(m_1 + i, m_2 + i)] + B^{(i)}[c_x(m_1 - i, m_2 - i) - c_x(m_1 + i, m_2)] \\ &\quad = -m_1 c_x(m_1, m_2)\end{aligned} \tag{17.149}$$

The truncation of the infinite summation in Equation (17.149) provides an approximate equation as

$$\begin{aligned}&\sum_{i=1}^{P} A^{(i)}[c_x(m_1 - i, m_2) - c_x(m_1 + i, m_2 + i)] \\ &+ \sum_{i=1}^{Q} B^{(i)}[c_x(m_1 - i, m_2 - i) - c_x(m_1 + i, m_2)] \approx -m_1 c_x(m_1, m_2)\end{aligned} \tag{17.150}$$

Equation (17.150) can be solved for the cepstral parameters $A^{(m)}$ and $B^{(m)}$.

Tri-cepstrum

The tri-cepstrum of a signal $y(m)$ is defined as the inverse Fourier transform of the tri-spectrum:

$$y_c(m_1, m_2, m_3) = \mathrm{IDFT}_3[\ln C_Y(\omega_1, \omega_2, \omega_3)] \tag{17.151}$$

where $\mathrm{IDFT}_3[\cdot]$ denotes the three-dimensional inverse discrete Fourier transform. The tri-spectra of the input and output of the linear system are related by

$$C_Y(\omega_1, \omega_2, \omega_3) = H(\omega_1)H(\omega_2)H(\omega_3)H^*(\omega_1 + \omega_2 + \omega_3)C_X(\omega_1, \omega_2, \omega_3) \tag{17.152}$$

Assuming that the channel input $x(m)$ is uncorrelated, Equation (17.152) becomes

$$C_Y(\omega_1, \omega_2, \omega_3) = \frac{\gamma_x^{(4)} G^4}{(2\pi)^3} H(\omega_1)H(\omega_2)H(\omega_3)H^*(\omega_1 + \omega_2 + \omega_3) \tag{17.153}$$

where $\gamma_x^{(4)}/(2\pi)^3$ is the fourth-order cumulant of the input signal. Taking the logarithm of the tri-spectrum gives

$$\begin{aligned}\ln C_Y(\omega_1, \omega_2, \omega_3) = &\frac{\gamma_x^{(4)} G^4}{(2\pi)^3} + \ln H_{\min}(\omega_1) + \ln H_{\max}(-\omega_1) + \ln H_{\min}(\omega_2) + \ln H_{\max}(-\omega_2) \\ &+ \ln H_{\min}(\omega_3) + \ln H_{\max}(-\omega_3) + \ln H^*_{\min}(\omega_1 + \omega_2 + \omega_3) \\ &+ \ln H^*_{\max}(-\omega_1 - \omega_2 - \omega_3)\end{aligned} \tag{17.154}$$

From Equations (17.151) and (17.154), we have

$$y_c(m_1, m_2, m_3) = \begin{cases} \ln A, & m_1 = m_2 = m_3 = 0 \\ -A^{(m_1)}/m_1, & m_1 > 0, m_2 = m_3 = 0 \\ -A^{(m_2)}/m_2, & m_2 > 0, m_1 = m_3 = 0 \\ -A^{(m_3)}/m_3, & m_3 > 0, m_1 = m_2 = 0 \\ B^{(-m_1)}/m_1, & m_1 < 0, m_2 = m_3 = 0 \\ B^{(-m_2)}/m_2, & m_2 < 0, m_1 = m_3 = 0 \\ B^{(-m_3)}/m_3, & m_3 < 0, m_1 = m_2 = 0 \\ -B^{(m_2)}/m_2, & m_1 = m_2 = m_3 > 0 \\ A^{(m_2)}/m_2, & m_1 = m_2 = m_3 < 0 \\ 0 & \text{otherwise} \end{cases} \tag{17.155}$$

where $A = \gamma_x^{(4)} G^4/(2\pi)^3$. Note from Equation (17.155) that the maximum-phase information $B^{(m)}$ and the minimum-phase information $A^{(m)}$ are separated and appear in different regions of the tri-cepstrum indices m_1, m_2 and m_3.

Calculation of Equaliser Coefficients from the Tri-cepstrum

Assuming that the channel z-transfer function can be described by Equation (17.134), the inverse channel can be written as

$$H^{\text{inv}}(z) = \frac{1}{H(z)} = \frac{1}{H_{\min}(z)H_{\max}(z^{-1})} = H_{\min}^{\text{inv}}(z)H_{\max}^{\text{inv}}(z^{-1}) \tag{17.156}$$

where it is assumed that the channel gain G is unity. In the time domain Equation (17.156) becomes

$$h^{\text{inv}}(m) = h_{\min}^{\text{inv}}(m) * h_{\max}^{\text{inv}}(m) \tag{17.157}$$

pan *et al.* (1988) describe an iterative algorithm for estimation of the truncated impulse response of the maximum-phase and the minimum-phase factors of the inverse channel transfer function. Let $\hat{h}_{\min}^{\text{inv}}(i, m)$, $\hat{h}_{\max}^{\text{inv}}(i, m)$ denote the estimates of the mth coefficients of the maximum-phase and minimum-phase parts of the inverse channel at the ith iteration. The Pan and Nikias algorithm is the following:

(a) Initialisation

$$\hat{h}_{\min}^{\text{inv}}(i, 0) = \hat{h}_{\max}^{\text{inv}}(i, 0) = 1 \tag{17.158}$$

(b) Calculation of the minimum-phase polynomial

$$\hat{h}_{\min}^{\text{inv}}(i, m) = \frac{1}{m} \sum_{k=2}^{m+1} \hat{A}^{(k-1)} \hat{h}_{\min}^{\text{inv}}(i, m-k+1) \quad i = 1, \ldots, P_1 \tag{17.159}$$

(c) Calculation of the maximum-phase polynomial

$$\hat{h}_{\max}^{\text{inv}}(i, m) = \frac{1}{m} \sum_{k=m+1}^{0} \hat{B}^{(1-k)} \hat{h}_{\max}^{\text{inv}}(i, m-k+1) \quad i = -1, \ldots, -P_2 \tag{17.160}$$

The maximum-phase and minimum-phase components of the inverse channel response are combined in Equation (17.157) to give the inverse channel equaliser.

17.7 Summary

In this chapter, we considered a number of different approaches to channel equalisation. The chapter began with an introduction to models for channel distortions, the definition of an ideal channel equaliser, and the problems that arise in channel equalisation due to noise and possible non-invertibility of the channel. In some problems, such as speech recognition or restoration of distorted audio signals, we are mainly interested in restoring the magnitude spectrum of the signal, and phase restoration is not a primary objective. In other applications, such as digital telecommunication the restoration of both the amplitude and the timing of the transmitted symbols are of interest, and hence we need to equalise for both the magnitude and the phase distortions.

In Section 17.1, we considered the least square error Wiener equaliser. The Wiener equaliser can only be used if we have access to the channel input or the cross-correlation of the channel input and output signals.

For cases where a training signal cannot be employed to identify the channel response, the channel input is recovered through a blind equalisation method. Blind equalisation is feasible only if some statistics of the channel input signal are available. In Section 17.2, we considered blind equalisation using the power spectrum of the input signal. This method was introduced by Stockham for restoration of the magnitude spectrum of distorted acoustic recordings. In Section 17.3, we considered a blind deconvolution method based on the factorisation of a linear predictive model of the convolved signals.

Bayesian inference provides a framework for inclusion of the statistics of the channel input and perhaps also those of the channel environment. In Section 17.4, we considered Bayesian equalisation methods, and studied the case where the channel input is modelled by a set of hidden Markov models. Section 17.5 introduced channel equalisation methods for removal of inter-symbol interference in digital telecommunication systems, and finally in Section 17.6, we considered the use of higher-order spectra for equalisation of non-minimum-phase channels.

Further Reading

Belfiore C.A. and Park J.H. (1979) Decision Feedback Equalization. Proc. IEEE, 67, pp. 1143–1156.

Bellini S. (1986) Bussgang Techniques for Blind Equalization. IEEE GLOBECOM Conf. Rec., pp. 1634–1640.

Bellini S. and Rocca F. (1988) Near Optimal Blind Deconvolution. IEEE Proc. Int. Conf. Acoustics, Speech, and Signal Processing. ICASSP-88, pp. 2236–2239.

Benveniste A., Goursat M. and Ruget G. (1980) Robust Identification of a Non-minimum Phase System: Blind Adjustment of Linear Equalizer in Data Communications. IEEE Trans, Automatic Control, AC-25, pp. 385–399.

Gersho A. (1969) Adaptive Equalization of Highly Dispersive Channels for Data Transmission. Bell System Technical Journal, 48, pp. 55–70.

Godard D.N. (1974) Channel Equalization using a Kalman Filter for Fast Data Transmission. IBM J. Res. Dev., 18, pp. 267–273.

Godard D.N. (1980) Self-recovering Equalization and Carrier Tracking in a Two-Dimensional Data Communication System. IEEE Trans. Comm., COM-28, pp. 1867–1875.

Hanson B.A. and Applebaum T.H. (1993) Sub-band or Cepstral Domain Filtering for Recognition of Lombard and Channel-Distorted Speech. IEEE Int. Conf. Acoustics, Speech and Signal Processing, pp. 79–82.

Hariharan S. and Clark A.P. (1990) HF Channel Estimation using a Fast Transversal Filter Algorithm. IEEE Trans. Acoustics, Speech and Signal Processing, 38, pp. 1353–1362.

Hatzinako S.D. (1990) Blind Equalization Based on Polyspectra. PhD Thesis, Northeastern University, Boston, MA.

Hermansky H. and Morgan N. (1992) Towards Handling the Acoustic Environment in Spoken Language Processing. Int. Conf. on Spoken Language Processing Tu.fPM.1.1, pp. 85–88.

Lucky R.W. (1965) Automatic Equalization of Digital Communications. Bell System Technical Journal, 44, pp. 547–588.

Lucky R.W. (1965) Techniques for Adaptive Equalization of Digital Communication Systems. Bell System Technical Journal, 45, pp. 255–286.

Mendel J.M. (1990) Maximum Likelihood Deconvolution: A Journey into Model Based Signal Processing. Springer-Verlag, New York.

Mendel J.M. (1991) Tutorial on Higher Order Statistics (Spectra) in Signal Processing and System Theory: Theoretical Results and Some Applications. Proc. IEEE, 79, pp. 278–305.

Mokbel C., Monne J. and Jouvet D. (1993) On-Line Adaptation of a Speech Recogniser to Variations in Telephone Line Conditions, Proc. 3rd European Conf. on Speech Communication and Technology. EuroSpeech-93, 2, pp. 1247–1250.

Monsen P. (1971) Feedback Equalization for Fading Dispersive Channels. IEEE Trans. Information Theory, IT-17, pp. 56–64.

Nikias C.L. and Chiang H.H. (1991) Higher-Order Spectrum Estimation via Non-Causal Autoregressive Modeling and Deconvolution. IEEE Trans. Acoustics, Speech and Signal Processing, ASSP-36, pp. 1911–1913.

Nowlan S.J. and Hinton G.E. (1993) A Soft Decision-Directed Algorithm for Blind Equalization. IEEE Transactions on Communications, 41 (2), pp. 275–279.

Pan R. and Nikias C.L. (1988) Complex Cepstrum of Higher Order Cumulants and Non-minimum Phase Identification. IEEE Trans. Acoustics, Speech and Signal Processing, ASSP-36, pp. 186–205.

Picchi G. and Prati G. (1987) Blind Equalization and Carrier Recovery using a Stop-and-Go Decision-Directed Algorithm, IEEE Trans. Commun, COM-35, pp. 877–887.

Qureshi S.U. (1985) Adaptive Equalization. IEEE Proc. 73 (9), pp. 1349–1387.

Raghuveer M.R. and Nikias C.L. (1985) Bispectrum Estimation: A Parameteric Approach. IEEE Trans. Acoustics, Speech, and Signal Processing, ASSP-33, 5, pp. 35–48.

Rosenblatt M. (1985) Stationary Sequences and Random Fields. Birkhauser, Boston, MA.

Spencer P.S. and Rayner P.J.W. (1990) Separation of Stationary and Time-Varying Systems and Its Applications to the Restoration of Gramophone Recordings. PhD Thesis, Cambridge University.

Stockham T.G., Cannon T.M. and Ingebretsen R.B (1975) Blind Deconvolution Through Digital Signal Processing. IEEE Proc., 63 (4), pp. 678–692.

Ungerboeck G. (1972) Theory on the Speed of Convergence in Adaptive Equalizers for Digital Communication. IBM J. Res. Dev., 16, pp. 546–555.

18 Signal Processing in Mobile Communication

Mobile communication systems rely on digital signal processing methods for almost every essential function from locating and tracking the geographical positions of the mobile users to source coding (e.g. code excited linear prediction speech coders, MP3 music coders, JPEG image/video coders) to the division and allocation of time and bandwidth resources among the mobile users (e.g. time, space or code division multiple access systems).

Furthermore, the capacity and performance of mobile communication systems is noise-limited and the use of signal processing algorithms in noise reduction, channel equalisation, echo cancellation, error control coding, and more recently in space–time diversity signal processing using smart adaptive beam-forming antennas, are playing an increasingly pivotal role in the development of efficient and intelligent communication systems. In this chapter we consider how communication signal processing methods are used to improve the speed and capacity of communication systems.

18.1 Introduction to Cellular Communication

Mobile communication systems constitute the largest mass-market application area of digital signal processing hardware and software systems. Whereas the early generations of mobile radio communication systems were relatively simple bulky analogue radio devices connected to base stations, modern cellular communication systems are sophisticated digital signal processing systems to the extent that a modern mobile phone handset is a powerful special-purpose computer in its own right incorporating many more multimedia systems and functions in addition to a mobile phone. Current mobile phones integrate a host of multimedia services including video, speech, music, Internet and teleconferencing.

The increasing integration of multimedia services on mobile phones necessitates the use of broadband high-speed transmission systems for downloading and uploading data and that in turn requires new technology to develop better methods of utilisation of the available radio frequency spectrum. Bandwidth utilisation is an extremely important research and development area of mobile communications in which digital signal process theory and tools are essential. In the rest of this chapter we consider the applications of digital signal processing to mobile communication systems.

Multimedia Signal Processing: Theory and Applications in Speech, Music and Communications Saeed V. Vaseghi

18.1.1 A Brief History of Radio Communication

The age of radio communication began in the 1860s with James Clark Maxwell's development of the theory of electromagnetic waves. Maxwell predicted the existence of electromagnetic radio waves with various frequencies propagating at the speed of light and concluded that light itself was also an electromagnetic wave. It appears that Maxwell did not realise that electromagnetic waves can travel in free space and assumed that some kind of 'ether' mediated the propagation of radio waves just as air mediates the propagation of sound (pressure) waves.

In 1884 Heinrich Rudolf Hertz reformulated Maxwell's equations. Later, between 1885 and 1888, in a series of pioneering experiments Hertz demonstrated that a rapidly oscillating electric current could be launched into space as an electromagnetic wave and detected by a wire loop receiver. For generating and transmitting oscillating radio waves Hertz used a high voltage induction coil and a capacitor (i.e. an LC oscillator circuit) connected to a rod with a gap in the middle and a spark sphere attached to the rod at each end of the gap. This device created oscillatory sparks across the gap as the spheres charged and discharged with opposite polarity electric charges. Like thunderstorms, sparks generate electromagnetic waves.

To detect the electromagnetic radiation, Hertz used a copper wire bent into a loop, with a small brass sphere connected to one end while the other end of the wire loop that pointed to the sphere had a screw mechanism so that the point could be adjusted and moved close to the sphere. The presence of an oscillating charge in the receiver wire loop caused sparks across the very small gap between the end points of the wire loop.

Hertz also demonstrated that radio waves had all the well-known properties of light waves – reflection from obstacles, diffraction from openings and around obstacles, refraction interference and polarisation. It is said that, in response to questions from his students who witnessed his classroom experiments on generation and reception of electromagnetic waves, Hertz replied that he saw no practical use for electromagnetic waves.

In 1895 Guglielmo Marconi, the inventor of radiotelegraph, inspired by Hertz's experiments, demonstrated the feasibility of radio communication by transmitting and receiving his first radio signal in Italy. In 1899 Marconi telegraphed the first wireless signal across the English Channel and in 1901 he sent the first transatlantic radiotelegraph message from England to Newfoundland. Note however that in 1943 the US Supreme Court overturned Marconi's patent in favour of Nicola Tesla who is now credited with the invention of radio communication. Another scientist who is also credited with the invention of radio is Jagadis Chandra Bose who demonstrated his radio in 1896.

From the beginning of the radio age, the quality of radio transmission and robustness of radio signals to noise and fading has been a major challenge in the design of wireless communication systems. Amplitude modulation (AM), where the radio carrier's amplitude is modulated by the signal massage, consumes no more bandwidth than the base band signal but is susceptible to noise and fading. High-quality radio transmission was made possible by Edwin Howard Armstrong's invention of wideband frequency modulation (FM) in 1933. In the FM modulation method, before transmission, amplitude variations of the signal are converted to frequency variations. FM can trade off more bandwidth for more robustness to noise and provides transparent broadcast quality at the cost of using more bandwidth than AM. FM transmission remained the main method for high-quality civilian radio transmission until the recent deployment of digital radio communication.

The era of mobile cellular communication began in the 1970s when AT&T proposed the first high-capacity analogue telephone system called the advanced mobile phone service (AMPS). This system was later extended to a cellular mobile system. The first generation of mobile phone systems had low user capacity and the large 'brick-like' handsets were cumbersome and power inefficient. Over the past three decades cellular mobile phone technology has developed rapidly through the advent of digital mobile phone standards such as the European Global System for Mobile (GSM) communication standard and the American IS-95 and IS-96 standards. In the late 1990s, the increasing demand for mobile phone and multimedia services led to the development of the third generation (3G) and universal mobile telecommunications system (UMTS) standards.

18.1.2 Cellular Mobile Phone Concept

The main limited resources in modern mobile communication systems are bandwidth and on-board battery power. The available radio bandwidth is limited by the need to share the finite radio frequency spectrum among many different mobile users and for many other different applications and purposes. The available power is limited by the capacity, size and weight of the onboard (handset) batteries. In a mobile communication system the radio frequency bandwidth needs to be used efficiently in order to maximise the capacity.

A major solution to the problem of limited radio channel capacity, resulting from finite bandwidth, is cellular systems. In a cellular system the same band of frequencies in suitably distant cells are reused to transmit different data. Figure 18.1 depicts the geometric topology of a cellular mobile radio system. In cellular systems a city or a town is divided into a number of geographical cells. The cells are thought of as having a hexagonal shape.

Note that smaller cell size also requires less power as within smaller cells the base station and the mobile devices operate on less power. This is advantageous in terms of the onboard power requirements and also in terms of the possible electromagnetic radiation hazard.

A key aspect of cellular mobile communication technology is the very large increase in spectral efficiency achieved through the arrangements of cells in clusters and the *reusing* of the same radio frequency channels in non-adjacent cells; this is possible because in a cellular systems the cell phones and base stations operate on low-power transmitters whose electromagnetic wave energy fades away before they reach non-adjacent cells. Note that a cellular system also results in a more efficient use of batteries due to less power used in smaller cells.

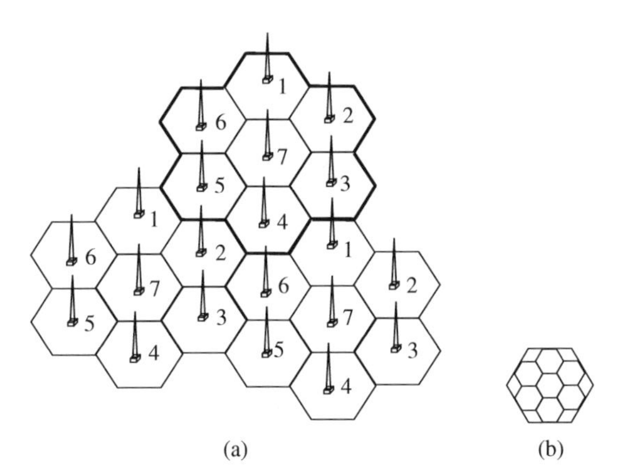

Figure 18.1 (a) A cellular system increases capacity by dividing a geographical area into clusters of cells and reusing the radio spectrum in non-adjacent cells. Each set of radio channel frequencies used in a cell is shown by a different number. (b) In an area with large user demand, a cell can be split into several subcells to further increase the overall capacity of the mobile communication system.

In the cellular configuration of Figure 18.1(a), each cell, within each cluster of seven cells, uses a set of different frequencies (each set of distinct radio channel frequencies used in a cell is shown by a different number).

Different clusters of seven cells reuse the same sets of frequencies as shown by the number codes in each cluster of cells.

The system capacity of a cellular phone system can be increased by reducing the cell size and transmission power. The cost is of course more cell infrastructure. In busy population centres, such as city centres, where there is a much higher than average demand for network service, smaller cells can be used by a process of cell-splitting. Figure 18.1(b) shows the process of cell splitting which is used in high density population areas with high user demand in order to increase the overall capacity of the cellular systems.

18.1.3 Outline of a Cellular Communication System

Figure 18.2 shows the outline of basic network architecture of a GSM cellular mobile phone system. Each cell has a base station that accommodates the transmitter/receiver antennas, the switching networks and the call routing and all the call management equipment of the base station. The basic cellular system consists of the following subsystems.

MS – Mobile Station. The MS is the physical equipment used by a subscriber; usually this is a hand-held cellular telephone handset. It is comprised of two distinct elements, the ME (Mobile Equipment) and the SIM (Subscriber Identity Module). The ME equipment consists of the screen, speaker, microphone, keypad, antenna and all the inside hardware and software elements including signal processing microchips. The SIM contains information that enables the user to access the mobile network.

SIM – Subscriber Identity Module is a smart card which stores the key number identifying a mobile phone service subscriber, as well as subscription information, preferences, network state information such as its current location area identity (LAI) and text messages. If the handset is turned off and back on again it will take data off the SIM and search for the LAI it was in; this avoids searching the whole list of frequencies that the telephone normally would. Each SIM is uniquely identified by its International Circuit Card ID (ICCID).

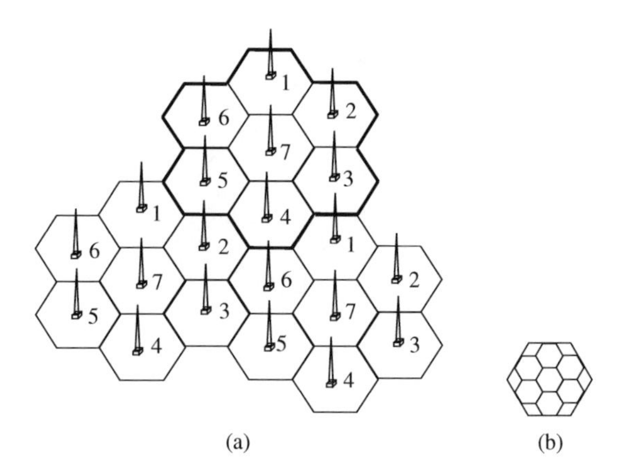

Figure 18.2 Illustration of a basic GSM mobile phone network.

BTS – Base Transceiver Station, usually placed on top of tall buildings or on poles, contains the equipment for transmission and reception of radio signals (transceivers), antennas, and equipment for encrypting and decrypting communication with the Base Station Controller (BSC). Typically a BTS will have several transceivers which allow it to serve several different frequencies and different sectors of the cell in the case of sectorised base stations. A typical BTS site may have from 1 to 12 transceivers in one, two or three sectors, although these numbers may vary widely.

BSC – Base Station Controller controls the BTSs and is responsible for allocation and release of the radio channels, terrestrial channel management, mapping of the radio channels onto wired channels and execution of the hand-over function as mobile users move across different cells. Typically a BSC has 10s or 100s of BTSs under its control. The databases for all the sites, including such information as carrier frequencies, frequency hopping lists, power reduction levels, receiving levels for cell border calculation, are stored in the BSC. This data is obtained directly from radio planning engineering which involves modelling of the signal propagation as well as traffic projections. A main function of the BSC is to act as a concentrator where many different low-capacity connections to BTSs are reduced to a smaller number of connections, with a high level of utilisation, towards the Mobile Switching Center (MSC).

MSC – Mobile Switching Centre is responsible for switching and routing of telephone calls. MSC provides circuit-switched calling, mobility management, and services to the mobile phones roaming within the area that it serves. This includes voice, data and fax services, as well as SMS and call divert. The tasks of the MSC include delivering calls to subscribers as they arrive in a cell based on information from the VLR (Visitor Location Register); connecting outgoing calls to other mobile subscribers or the PSTN; delivering SMSs from subscribers to the SMS centre (SMSC) and vice versa; arranging handovers between BSCs; carrying out *handovers* from and to other MSCs and supporting supplementary services such as conference calls, call hold and billing information.

Gateway MSC is the MSC that determines the location of the visited MSC where the subscriber who is being called is currently located. It also interfaces with the Public Switched Telephone Network. All mobile-to-mobile calls and PSTN-to-mobile calls are routed through a GMSC. The **Visited MSC** is the MSC where a subscriber is currently located. The VLR associated with this MSC will have the subscriber's data in it. The **Anchor MSC** is the MSC from which a handover has been initiated. The **Target MSC** is the MSC towards which a handover should take place.

The MSC connects to the following elements: the home location register (HLR) for obtaining data about the SIM and MSISDN (Mobile Station ISDN Number, the 'directory number' of an MS); the Base Station Subsystem which handles the radio communication with mobile phones; the network which handles the radio communication with mobile phones; the VLR for determining where other mobile subscribers are located.

HLR – Home Location Register is a central database that holds general information about mobile phone subscribers. The main function of the HLR is to track and manage the movements and location area of mobile SIMs and phones. HRL implements the following procedures: manages and tracks the mobility of subscribers by updating their position in administrative areas called 'location areas'; sends the subscriber data to a VLR when a subscriber first roams there; brokers between the mobile switching centre and the subscriber's current VLR in order to allow incoming calls or text messages to be delivered and removes subscriber data from the previous VLR when a subscriber has roamed away.

The HLR database stores details of every SIM card issued by the mobile phone operator. Each SIM has a unique identifier which is one of the primary keys to each HLR record. The SIMs have associated telephone numbers used to make and receive calls to the mobile phone. Other data stored in the HLR in a SIM record is: GSM services that the subscriber has requested or been given, General Packet Radio Service (GPRS) settings to allow the subscriber to access packet services, current location of subscriber and call divert settings.

The HLR connects to the following elements: the Gateway MSC (GMSC) for handling incoming calls; the VLR for handling requests from mobile phones to attach to the network; the SMSC for handling incoming SMS; the voice mail system for delivering notifications to the mobile phone that a message is waiting

VLR – Visitor Location Register is a temporary database that holds visiting subscribers' general information during their visit to a cell. It also holds location area identities of roaming users. Each base station is served by one VLR.

The primary functions of the VLR are: to inform the HLR that a subscriber has arrived in the particular area covered by the VLR; to track where the subscriber is within the VLR area when no call is ongoing; to allow or disallow which services the subscriber may use; to allocate roaming numbers during the processing of incoming calls; to remove the subscriber record if a subscriber becomes inactive whilst in the area of a VLR. The VLR deletes the subscriber's data after a fixed time period of inactivity and informs the HLR (e.g. when the phone has been switched off and left off or when the subscriber has moved to an area with no coverage for a long time).

The data stored in the VLR has either been received from the HLR, or collected from the MS. Data stored in the VLR includes: the subscriber's identity number; authentication data; the subscriber's phone number; the services that the subscriber is allowed to access; access point subscribed and the HLR address of the subscriber.

The VLR connects to the following elements: the visited MSC (VMSC) to pass data needed by the VMSC during its procedures, e.g. authentication or call setup; the HLR to request data for mobile phones attached to its serving area; other VLRs to transfer temporary data concerning the mobile when they roam into new VLR areas.

EIR – Equipment and Identity Register is a database that holds user equipment identities. The EIR is often integrated to the HLR. The EIR keeps a list of mobile phones which are to be banned from the network or monitored. This is designed to allow tracking of stolen mobile phones. The EIR data does not have to change in real time, which means that this function can be less distributed than the function of the HLR.

AUC – When a phone is powered on, the authentication centre (AUC) authenticates the SIM card that attempts to connect to the network. Once the authentication is successful, the HLR is allowed to manage the SIM and services described above. An encryption key is also generated that is subsequently used to encrypt all wireless communications between the mobile phone and the core network. If the authentication fails, then no services is available to the mobile phone. The AUC does not engage directly in the authentication process, but instead generates data known as triplets for the MSC to use during the procedure. The security of the process depends upon a shared secret key, called the Ki, between the AUC and the SIM. The 128-bit secret key Ki is securely burned into the SIM during manufacture and is also securely replicated onto the AUC. This Ki is not transmitted between the AUC and SIM, but is combined with the International Mobile Subscriber Identity number (IMSI) to produce a challenge/response for identification purposes and an encryption key called Kc for use in over-the-air communications.

OMC – Operation and Maintenance Centre is a database that holds relevant information about the overall operation and maintenance of the network. The OMC is connected to all equipment in the switching system and to the BSC. The implementation of OMC is called the operation and support system (OSS). The OSS is the functional entity from which the network operator monitors and controls the system. The purpose of OSS is to offer the customer cost-effective support for centralised, regional, and local operational and maintenance activities that are required for a GSM network. An important function of OSS is to provide a network overview and support the maintenance activities of different operation and maintenance organisations.

GMSC – Gateway Mobile Switching Centre is the point to which a MS terminating call is initially routed, without any knowledge of the MS's location. The GMSC is thus in charge of obtaining the MSRN (Mobile Station Roaming Number) from the HLR based on the MSISDN (Mobile Station ISDN Number, the 'directory number' of an MS) and routing the call to the correct visited MSC.

SMS-G – This is the term used to collectively describe the two Short Message Services Gateways described in the GSM recommendations. The SMS-GMSC (Short Message Service Gateway Mobile Switching Centre) is for mobile terminating short messages, and SMS-IWMSC (Short Message Service Inter-Working Mobile Switching Centre) for mobile originating short messages.

18.2 Communication Signal Processing in Mobile Systems

Modern mobile communication systems rely on advanced signal processing methods for fast, efficient, reliable and low-cost multimedia communication. In fact almost every aspect of the functioning of transmission and reception of information on a mobile phone is dependent on advanced signal processing systems. The signal processing functions in a mobile communication system includes the following modules:

1. **Source Coder/Decoder** – At the transmitters source coders compress the number of bits per sample of the input signal (such as speech, image and music) by removing the correlation and redundancies from the signals; source decoders decompress and reconstruct the signals at the receiver. Source coding achieves a reduction in bit rate and hence a proportional reduction in the bandwidth and the power required to transmit the bits.

   Source coding involves the modelling and utilisation of the correlation structure of the signal. This can be achieved through a combination of (i) the use of signal transforms such as discrete Fourier transform and cosine transform, (ii) the use of signal generation models such as linear prediction models and (iii) signal probability models such as entropy models, for the compression of the source data which may include voice, image, video and text. Source coding methods, such as Huffman coders, MPEG music coders and CELP voice coders, and JPEG image coders can significantly reduce (often by a factor of more than 10 to 20) the required bit rate and bandwidth and hence increase the capacity and speed of transmission of audio, image, text and other data. Speech and music compression are covered in Chapters 13 and 14.
2. **Channel Coder/Decoder** – The purpose of channel coding is to introduce the ability to detect and correct bit errors and hence reduce transmission errors due to noise, fading and loss of data packets. The simplest (and not very efficient) form of a channel coder is a repetition code, for example instead of transmitting the bits '1' or '0' the bit pairs '11' or '00' are transmitted.

   In mobile phones channel coding involves the use of a combination of convolution and block coders for the addition of error-control bits to the source data in order to increase the distance between the allowed sequences of transmitted data and hence improve the error detection and error correction capability of communication systems.
3. **Multiple Access Signalling** – In a communication system time and bandwidth needs to be shared between different users. Multiple access signalling, as the name implies, provides simultaneous access to multiple users on the same shared bandwidth and time resources. Multiple access systems are based on division of time or frequency, code or space among different users leading to time division multiple access (TDMA), frequency division multiple access (FDMA), code division multiple access (CDMA) and space division multiple access (SDMA) methods respectively. In code division multiplexing, orthogonal codes are assigned to different users who can then communicate simultaneously on the same bandwidth.

4. **Cell Handover** – The tracing and determination of the geographical location of a mobile phone user and the issue of which cell (and antenna) at any given time should serve a mobile user, as the user roams across different cells, is accomplished by processing the strengths of the radio signals from a mobile user received by different base stations. Usually the base station, or a number of base stations, that receives the strongest signal serves the mobile user. The signals from a mobile device received by different base stations can be used to locate the user within a resolution of a few metres or less depending on the cell size.
5. **Rake Correlators** – In a broadband mobile environment the multi-path reflections of an electromagnetic signal, from different surfaces and through different paths, arrive at different times as several distorted replicas of the transmitted signal. Rake receivers advantageously use the effect of the multi-path propagation by combining the reflections of a signal received from different propagation paths. This form of the so-called space–time diversity signal processing can reduce fading and add to the strength of the received signal.
6. **Channel Equalisation** – This is used to remove the distortions and time-dispersion of signals that result from the non-ideal characteristics of radio channels. Channel equalisation requires the identification of the channel response and the implementation of the inverse channel response i.e. the equaliser. Channel equalisation reduces the symbol overlaps (ISI) and bit error rate at the receiver. Channel equalisation is described in Chapter 17.
7. **Echo Cancellation** – This is used to reduce both acoustic feedback echo between the speaker and the microphone of a mobile phone and also the telephone exchange hybrid line echo. Echo cancellation is necessary for voice and data communication systems. Acoustic echo cancellation is particularly important for mobile phones, hands-free phones and teleconferencing systems. Echo cancellation is described in detail in Chapter 16.
8. **Smart Antenna Array** is used for a variety of purposes from increasing the signal-to-noise ratio to space division multiple access. Smart antennas are arrays of phased antennas whose beam direction and gain is controlled by adaptive signal processing methods so that the transmitted electromagnetic power is more efficiently beamed and selectively directed towards the mobile users.

18.3 Capacity, Noise, and Spectral Efficiency

A principal challenge in the design and development of mobile communication systems stems from the limitations imposed on data transmission rates due to finite bandwidth and the physical properties of the radio communication channels, which can include noise and distortions such as signal dispersion, fading, impulsive noise, co-channel/multiple-access interference as well as other phenomena most notably multi-path effects.

The rate at which binary data bits (i.e. sequences of ones and zeros) can be transmitted on a communication channel is limited by (a) the available bandwidth and (b) disturbances such as noise, interference, distortion and multi-path effects.

Binary data are signalled using communication symbols. In its simplest form a communication symbol may be a pulse-modulated sinusoidal carrier that takes two different amplitude levels for signalling a binary data bit. The maximum rate at which communication symbols can be transmitted is limited by the bandwidth. The symbol rate r_s, that is the number of symbols per second of transmission, is about half the bandwidth.

However, each communication symbol may simultaneously carry M bits provided that there are 2^M resolvable patterns of variation of the symbol such as the amplitude, or phase or time delay (position) or frequency of the symbol. Assuming each symbol carries M bits, the bit rate is M times the symbol rate, i.e. $r_b = Mr_s$. The maximum number of bits that can be signalled by each symbol is limited by noise and acceptable delay.

In practice the number of bits that each communication symbol can carry in an M-ary signalling scheme is limited by noise, interference, multi-path effect, channel distortions, echo and fading. The constraints imposed on the channel capacity due to noise and bandwidth, limit the rate at which information can be transferred, even when multi-level encoding techniques are used. This is because at low signal-to-noise ratios (SNR) the noise can obliterate the differences that distinguish the various signal levels, limiting in practice the number of detection levels we can use in the decoder.

Therefore, from the above argument it follows that the capacity of a communication link, that is the rate at which data can be transmitted, is proportional to the available bandwidth and the signal-to-noise ratio, the limit of which, in additive white Gaussian noise (AWGN), is expressed by the **Shannon–Hartley theorem** as

$$C = B\log_2\left(1+\frac{S}{N}\right) \tag{18.1}$$

where C is the channel capacity in bits per second; B is the channel bandwidth in Hz; and S/N is the signal-to-noise ratio expressed as a linear (as opposed to logarithmic) power ratio. The theorem, proved by Claude Shannon in 1948, describes the maximum capacity of a communication channel with error-correcting methods versus the levels of noise and interference. Equation (18.1) gives the theoretical limit that the best possible coding and modulation method may achieve. Figure 18.3 shows a plot of the bit-rate per Hz vs signal-to-noise ratio. Note that at an SNR of 0 dB the maximum theoretical bit rate is 1 bit per Hz, at 15 dB it is 5 bits per Hz and this approaches a theoretical rate of 13 bits per Hz at 40 dB SNR.

The challenge in communication signal processing is to increase the capacity in bits per second per Hz of bandwidth through reducing noise, interference and multi-path distortion. Signal processing methods play a central role in removing or compensating for the effect of noise and thereby improving data transmission capacity.

The signal processing methods that are used to improve the capacity of mobile communication systems include source coding, channel coding, channel equalisation, echo cancellation, multi-path models and multiple access methods including space–time signal processing via beam-forming antenna arrays. The use of multiple receiving and transmitting antennas combined with multiple access and noise reduction methods is opening a myriad of possibilities in the use of signal processing for enhanced mobile communication.

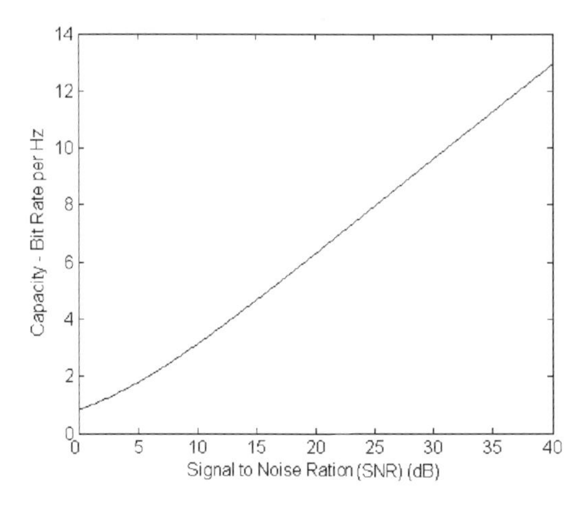

Figure 18.3 Illustration of the variation of the maximum capacity (bits/sec/Hz) of a communication channel with the SNR (dB).

18.3.1 Spectral Efficiency in Mobile Communication Systems

The radio frequency bandwidth available to each operator of a cellular communication service is usually limited to a range of about several hundred kHz to several mega Hz, usually centred between 1 to 4 GHz. The very large number of subscribers using mobile communication devices and the ever-increasing demand on the bandwidth is accommodated through the efficient use of communication resources that results in a large increase in the capacity per Hz, also known as the spectral efficiency defined as the data rate in bits per second per Hz unit of bandwidth:

$$\text{spectral efficiency} = \text{channel throughput/channel bandwidth}$$

Depending on the efficiency of the communication systems and the signal-to-noise ratio, the actual spectral efficiency may vary from 0.1 to 4 bps/Hz.

The main method currently used to increase the capacity of radio channels is based on frequency reuse. As explained earlier this involves the reusing of the same frequencies in non-adjacent cells where the power of the transmitted electromagnetic wave from a cell fades to insignificance by the time it reaches the non-adjacent cells using the same radio channels. In congested urban areas, the frequency reuse factor can be increased through a reduction of the cell size and transmission power at the expense of more base station infrastructures, as shown in Figure 18.1(b). In order to minimise the interference among non-adjacent cells, which reuse the same frequencies, the base-stations and the mobile phones operate on low-power transmitters and receivers. Low-power transmitters/receivers have the following advantages:

(a) Low co-channel interference: Due to the low power of the transmissions from base-stations and mobile phones, the electromagnetic waves from each cell fades away before they reach non-adjacent cells that reuse the same frequencies.
(b) With all mobile devices and base stations operating on low power, the signal-to-noise ratios at the receivers of base stations and mobile phones improve.
(c) With low transmission power, the power consumption of cell phones is relatively low. Low power consumption implies small handset batteries with longer talk time and also less exposure to possibly harmful electromagnetic radiation.

18.4 Multi-path and Fading in Mobile Communication

In classical communication theory it is assumed that the received signal is corrupted by additive white Gaussian noise (AWGN) and the signal distortion due to the channel response is modelled by inter-symbol interference. In reality there are many different sources of noise and interference that may limit the performance of a communication system. The most common sources of distortion and noise in a mobile environment include receiver antenna thermal noise, interference from electromagnetic devices, radiation noise, background noise, echo and most importantly multi-path and fading, described next.

18.4.1 Multi-path Propagation of Electromagnetic Signals

Three main mechanisms impact the propagation of a radio frequency (RF) electromagnetic wave in a mobile communication environment. These are:

(1) *Reflection* occurs when an electromagnetic wave impinges a smooth surface with much larger dimensions than the wavelength λ of the radio frequency (RF) electromagnetic signal. Note that

at the speed of light $c = 0.3 \times 10^9$ m/s a radio frequency wave with a frequency $f = 1$ GHz has a wavelength of $\lambda = c/f = 30$ cm. In comparison light has a wavelength of 700 nm for red to 400 nm for violet.

(2) *Diffraction* occurs when an electromagnetic wave is obstructed by a dense object with dimensions larger than the wavelength λ of the RF electromagnetic signal. The wave then bends and appears as several secondary waves from behind the obstructing object. Diffraction and reflection phenomena account for the propagation of electromagnetic waves in cases where there is no line-of-sight radio connection between the transmitter and the receiver

(3) *Scattering* occurs when a wave impinges from rough objects with large dimensions or from any object with dimensions comparable or smaller that the wavelength λ of the RF electromagnetic signal, causing the wave to scatter in all directions.

In a wireless communication environment the transmitted electromagnetic wave, usually encounters a number of different obstacles, reflectors and diffractors in its propagation path. Hence the transmitted signal and its reflections arrive at the receiver from several different directions over a multiplicity of different paths with each path having a different length and characteristic in terms of fading, phase and time of arrival. This phenomenon is called the *multi-path effect.* For mobile systems the communication environments change with time, space and the speed of movement of the mobile user and these result in time-varying multi-path channel effects.

A simple illustration of multi-path effects in wireless connections is shown in Figure 18.4. The multi-path effect is usually described by:

- Line-of-sight path: this is the direct path between the transmitter and the receiver antennas.
- None-line-of-sight paths: the paths arriving after refraction and reflections from various objects and surfaces.

A multi-path propagation effect can be modelled as the impulse response of a linear channel as

$$h(t) = \sum_{l=1}^{L} \alpha_l \delta(t - \tau_l) \tag{18.2}$$

where L is the number of different propagation paths. The multi-path equation (18.2) has two parameters for each propagation path l; these are the propagation time delay τ_l and the amplitude fading factor α_l.

The multi-path effect results in amplitude and phase fluctuations and time delay in the received signals and this can reduce the transmission capacity and in severe cases create signal outage and loss of connection. The multi-path effect is described by the following characteristics:

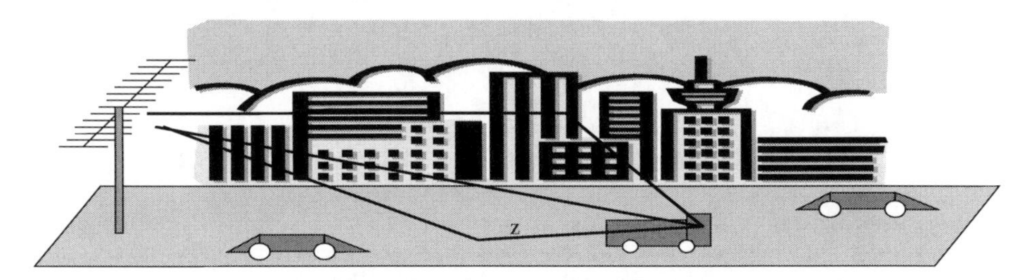

Figure 18.4 A simple illustration of multi-path propagation effects in a mobile phone environment.

- *Multi-path fading characteristics:* When the reflected waves, which arrive from different propagation paths, are out of phase, a reduction of the signal strength, or fading, at the receiver can occur. There are two types of fading: slow fading which occurs due to movements of a mobile user over large areas and fast fading which occurs due to movements over smaller distances comparable to the wavelength.
- *Multi-path delay spread*: As the multiple reflections of the transmitted signal may arrive at the receiver at different times, this can result in inter-symbol interference and time-dispersion and broadening of the signal. This time-dispersion characteristic of the channel is called multi-path delay spread, which is an important parameter for assessment of the performance of wireless systems.

A common measure of multi-path delay spread is the root mean square (rms) delay spread. For reliable communication without using adaptive equalisation or other multi-path modelling techniques, the transmitted data rate should be much smaller than the inverse of the rms delay spread which is called the coherence bandwidth. When the transmitted data rate is much smaller than the coherent bandwidth, the wireless channel is referred to as a flat channel or narrowband channel. When the transmitted data is equal to or larger than the coherent bandwidth, the wireless channel is called a frequency-selective channel or wideband channel.

18.4.2 Rake Receivers for Multi-path Signals

The actual form of the signal distortion due to multi-path reflections depends on the signal bandwidth, the durations of transmitted and reflected signals and on the time delays incurred in propagation of a signal via different reflection paths.

The duration of a signal is inversely proportional to its bandwidth. For very wideband very short duration signals, multi-path reflection does not result in inter-symbol interference, rather it results in the appearance of multiple distorted replicas of the transmitted pulses. For example at a bandwidth of 10 megaHz the duration of a pulse is roughly about 0.1 microseconds. Now the distance travelled by an electromagnetic wave in 0.1 microsecond is only 30 metres. Hence, any two reflections of pulses travelling distances with differences of more than 30 metres would appear as distinct pulses arriving at different times.

Assuming that the noise and fading in different propagation paths are independent, the different versions of a pulse arriving from different paths can be combined, in a Rake receiver, to improve the signal-to-noise ratio. A Rake receiver uses several correlators to individually process several multi-path components of a signal. The correlators' outputs are combined to achieve improved SNR and communications reliability and performance. This is a type of the so-called space diversity gain, i.e. combining the same information arriving from different spatial routes.

Each correlator in a Rake receiver is called a Rake-receiver finger. Two primary methods are used to combine Rake-receiver finger outputs. One method weights each output equally. The second method uses the data to estimate weights which maximise the SNR of the combined output.

18.4.3 Signal Fading in Mobile Communication Systems

Modelling the fading and the propagation loss of electromagnetic wave energy in space is important for the calculation of the required transmitter power in mobile communication systems. In the idealised model of the propagation of an electromagnetic wave in a *free* space there are no obstacles or particles and hence no reflection or loss of energy occurs. The electromagnetic energy radiated by an isotropic source of radio frequency fades in a free space with the square of the distance d as

$$L_s(d) = \left(\frac{4\pi d}{\lambda}\right)^2 = \left(\frac{4\pi df}{c}\right)^2 \tag{18.3}$$

where L_s is the free space path loss (or power loss), $\lambda = c/f$ is the wavelength of an electromagnetic wave with a frequency of f Hz and speed of $c = 0.3 \times 10^9$ m/s, which is the speed of light. The fading of the signal strength in a free space is not due to a loss of energy, as no energy is lost in propagation of a signal through a free space, but it is due to dilution of the energy of the wave, as the same amount of wave energy spreads in propagation through an increasingly larger surface area of a sphere which expands with the increasing distance (radius) d.

There are two types of signal fading in mobile communication systems: large-scale *slow fading* due to the movements of the mobile user over large areas and small-scale *fast fading* due to the movements of the mobile user over small distances of the order of the wavelength λ of the signal (the wavelength of a mobile carrier at 1 GHz frequency is about $\lambda = c/f = 0.3 \times 10^9/10^9 = 30$ cm).

In a mobile communication system, the received signal $r(t)$ can be modelled as the convolution of the transmitted signal $s(t)$ and the impulse response of the radio channel $h_c(t)$ as

$$r(t) = s(t)^* h_c(t) \tag{18.4}$$

where * denotes convolution. For mobile phone systems the received signal $r(t)$ can be further expressed as the product of a slow-fading term $m(t)$ and a fast-fading term $r_0(t)$ as

$$r(t) = m(t) \times r_0(t) \tag{18.5}$$

Figure 18.5 illustrates the variations of large-scale slow fading and small-scale fast fading signals.

18.4.4 Large-Scale Signal Fading

The large-scale path loss due to the movement of a mobile phone user over a distance d, is obtained from experimental measurements in urban/rural environments. The experimental results for path loss, $L_p(d)$,

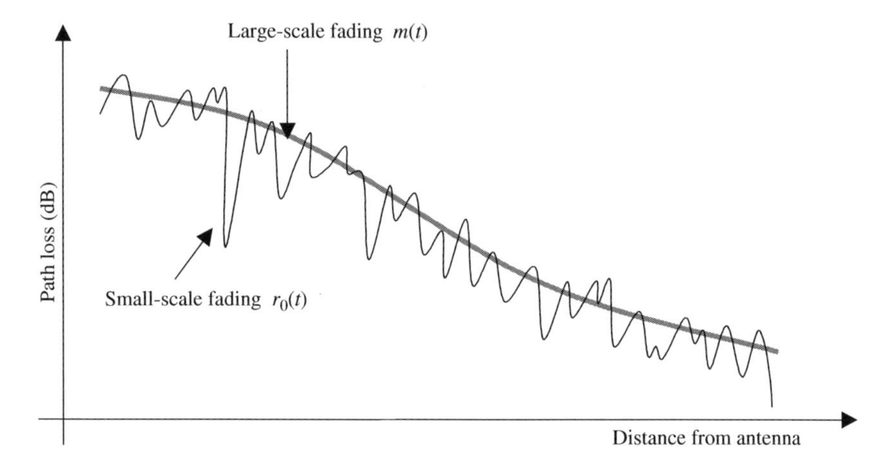

Figure 18.5 Illustration of large-scale and small-scale fading.

can be modelled with an nth power of distance d as

$$L_p(d) \propto \left(\frac{d}{d_0}\right)^n \tag{18.6}$$

where d_0 is a reference distance which is typically about 1 km for large cells, 100 m for micro cells and 1 m for indoor radio channels. The propagation path loss can be written in terms of a logarithmic dB measure as

$$10\log L_p(d) = 10\log L_s(d_0) + 10n\log\left(\frac{d}{d_0}\right) + X_\sigma \text{ dB} \tag{18.7}$$

where $L_s(d_0)$ is the free space loss for the reference distance d_0, $10n\log(d/d_0)$ is the average path loss in dB as a function of the distance d and X_σ is a random variable that models the random fluctuations of large-scale fading due to random changes in different terrains and environments between the base-stations and mobile users.

The path loss exponent n depends on the propagation environment. In a free space $n = 2$. In some urban street environments with large buildings where there may be a strong waveguide effect which contains the wave, n can be less than 2 (i.e. less propagation loss than in a free space). Generally, where there are obstructions to propagation of the energy of an electromagnetic wave, n is greater than 2.

18.4.5 Small-Scale Fast Signal Fading

Small-scale fading is due to the movements of mobile users over small distances comparable to the wavelength of the radio wave. When a radio wave signal is made of multiple reflective rays and a non-faded line-of-sight component then the fluctuations of the amplitude of the signal due to small-scale fading has a distribution that can be modelled by a Rician probability density function and hence it is known as Rician fading.

In the absence of a line-of-sight component the distribution of the amplitude fluctuations of a radio signal caused by small-scale fading has a Rayleigh pdf expressed as

$$p(r) = \begin{cases} \dfrac{r}{\sigma^2}\exp\left(-\dfrac{r^2}{2\sigma^2}\right) & r \geq 0 \\ 0 & r < 0 \end{cases} \tag{18.8}$$

where r is the envelope of the amplitude of the received signal and σ^2 is its variance. Often in mobile communication it is assumed that the received signal has a Rayleigh distribution.

The main effects of small-scale fading are:

(a) Time spreading of the signal pulses. This can result in the received signal having a longer or shorter time span than the transmitted signal. This effect is similar to inter-symbol interference (ISI). When the time span of the received signal is greater than the transmitted signal, and there are multiple reflections, the effect can be mitigated by the use of a Rake receiver as explained earlier. The effect of time spreading can also be mitigated by the use of error control coding and through adding redundancy via diversity gain methods.

(b) Time-varying behaviour of the communication channel due to spatial movements of the mobile user's antenna. Relative movement of transmitter and receiver antennas causes a time-varying change in the received amplitude and phase of the signal due to the extra distance that the signal has to travel. Hence, whenever there is spatial movement of the mobile user the channel becomes time-variant.

18.5 Smart Antennas – Space–Time Signal Processing

Antennas launch signals in the air in the form of electromagnetic waves. An antenna is essentially a wire-to-air impedance matching device: an antenna transmitter converts an alternating RF electric current signal propagating in a cable or wire to an electromagnetic wave propagating in space and an antenna receiver does the reverse function.

The manner in which an electromagnetic energy is distributed onto and collected from the surrounding space has a profound influence on the quality of the reception of radio signals and on the efficient use of the available radio bandwidth and battery power. Development of more efficient and intelligent antennas is central to the efforts in improving the capacity and reliability of mobile communication systems.

Figure 18.6 shows an illustration of the radiation patterns of an omni-directional antenna and a directional antenna. Omni-directional antennas radiate electromagnetic energy equally in all directions. Although this is desirable for radio/TV broadcast applications, for personal mobile communication applications it is inefficient and wasteful of power and spectrum because at any given time the mobile user is present only at one place in space and is receiving only a very small fraction of the total radiated power. Furthermore, each user of a mobile system with an omni-directional antenna contributes to the interference with the communication of other users of mobile systems who may otherwise be accommodated on the same frequency.

To reduce the inefficiency of omni-directional antennas, existing mobile base stations divide each cell into a number of fixed sectors with each sector covered by a fixed directional antenna. The conventional practice is to divide each cell into three sectors with each sector covered by an antenna with a 120° beam width.

Instead of using a single antenna for the transmission and reception of signals in a cell or a sector, an array of antennas, and an adaptive signal processing method, can be employed to selectively direct a beam of electromagnetic energy to the mobile user in order to improve the quality of reception and the spectral efficiency.

In general, an array of N antennas can be used in several different ways:

(a) The output of the transmitter antennas can be adaptively filtered and combined to form a narrow electromagnetic beam directed towards and centred on the mobile user. The beam should follow the movements of the user. This arrangement is known as smart antennas and in addition to improving reception and power utilisation it allows a higher level of frequency reuse and hence better spectral efficiency.

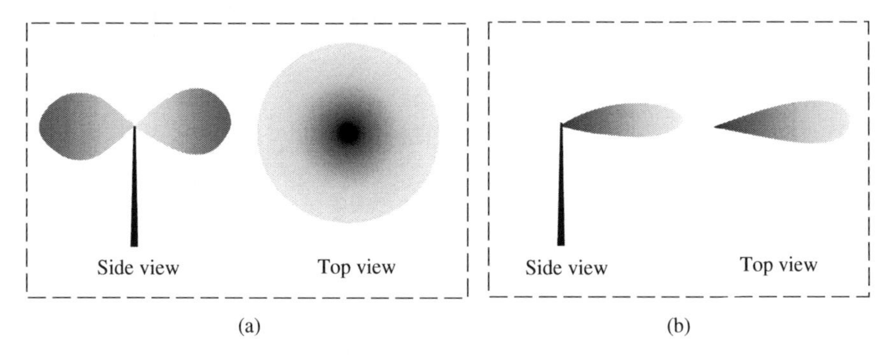

Figure 18.6 Illustration of the radiation pattern of (a) an omni-directional dipole antenna and (b) a directional antenna.

(b) The outputs of the receiver antennas can be added to improve the SNR. Assuming that N antennas receive the same transmitted signal but each with different uncorrelated noise and fading, the output SNR of the combined signals is improved by a factor of N.
(c) The output of the antenna with the maximum SNR can be used. In its simplest form, with two antennas separated by a half wavelength, the selection of the output of the antenna with the higher SNR can improve the SNR by more than 10 dB.

Smart beam-forming antennas in their basic form are effectively adaptive phased array antennas with an adaptive signal processing unit that adjusts the phase of the signals that are fed into (or received by) each element of an antenna array in order to achieve a desired beam width and direction. A similar methodology is also used at audio frequency to form adaptive beam-forming directional microphones for speech processing in noise as explained in Chapter 15.

18.5.1 Switched and Adaptive Smart Antennas

Smart antennas combine inputs from multiple antennas to improve power efficiency and directivity, reduce interference and increase spectral efficiency. There are two main forms of smart beam-forming antennas, namely: switched antenna array and adaptive antenna array.

As illustrated in Figure 18.7, a switched antenna array use a number of pre-selected and programmed RF beams. Switched beam systems combine the outputs of multiple antennas in such a way as to form finely sectorised beams. At any time the signal processing unit estimates the location of the mobile user from its signal strength and then switches the transmitter antenna beam to the one that provides the best signal coverage for the mobile receiver.

Adaptive smart antenna use adaptive signal processing algorithms to continuously adapt the antenna's direction, beam width and gain to follow the movements of the mobile receiver system and to provide optimal reception as the radio environment of the mobile user changes.

18.5.2 Space–Time Signal Processing – Diversity Schemes

Space–time signal processing refers to the signal processing methods that utilise the signal processing possibilities offered by the transmission and or reception of several signals across time and space using multiple transmitter/receiver antennas.

The use of multiple-input multiple-output (MIMO) transmitter/receiver antennas together with adaptive signal processing schemes opens up a myriad of possibilities in the area of space–time signal

Figure 18.7 Illustration of a switched antenna where a beam is selected from a number of pre-programmed options.

processing. Signals from different sources can be arranged, combined and processed in many different ways for optimal transmission and reception in MIMO mobile systems.

There are three main areas of research and development in the application of antenna arrays to MIMO mobile communication systems. These are:

(a) The design of the physical antennas as electromagnetic radiators. This is mainly concerned with such antenna characteristics as the radiation pattern, the main-lobe beamwidth, side-lobe levels and power efficiency.
(b) The estimation of direction of arrival of the electromagnetic pattern in beam-forming antennas using such signal processing methods as multiple signal classification (MUSIC) and estimation of signal parameters via rotational invariant techniques (ESPIRIT).
(c) Development of signal processing methods, and MIMO coders/decoders, for antenna arrays in order to improve the spectral efficiency and hence the capacity of radio channels.

An interesting application of space–time processing is the so-called diversity schemes used for robust and efficient transmission in mobile environments.

Diversity schemes deal with the transmission/reception of the replicas of a signal, or a combination of several signals, transmitted via several independent routes, namely time slots, frequency channels, multi-path reflections, spatial directions or polarisations. In a diversity scheme all signal routes carry the same combination of messages; however the channel characteristics, noise and fading are independent and uncorrelated. Hence replicas of the messages from different routes can be processed and combined to increase the signal-to-noise ratio. The success of diversity schemes depends on the degree to which the noise and fading on the different diversity branches are uncorrelated and how the information from different routes and channels is processed and combined. Diversity schemes can help to overcome noise and fading in wireless communication channels and increase the channel capacity.

A number of diversity schemes are described in the literature, and we will briefly consider some basic diversity schemes here.

- Space diversity: If the receiver has multiple antennas, the distance between the receiving antennas may be made large enough to ensure independent fading. This arrangement is a form of space diversity. Space separation of half of the wavelength is sufficient to obtain two signals with uncorrelated noise and fading. A form of space diversity recently proposed is co-operative communication in which it is proposed that different handsets from different mobile users can co-operate and act as a virtual antenna arrays to receive multiple copies of a signal through a diversity of propagation routes in space.
- Polarisation diversity: Antennas can transmit either a horizontally polarised wave or a vertically polarised wave. When both waves are transmitted simultaneously, received signals will exhibit uncorrelated fading statistics. This scheme can be considered as a special case of space diversity because separate antennas are used. However, only two diversity branches are available, since there are only two orthogonal polarisations.
- Angle diversity: Since the received signal arrives at the antenna via several paths, each with a different angle of arrival, the use of directional antennas can isolate the signal component. Each directional antenna will receive a different angular component. Hence, the noise and fading received by different directional antennas pointing at different angles may be uncorrelated.
- Frequency diversity: Signals with different carrier frequencies far apart from each other could have independent noise and fading. The carrier frequencies must be separated enough so that the fading associated with the different frequencies are uncorrelated. For frequency separations of more than several times the coherence bandwidth the signal fading would be essentially uncorrelated.

- Time diversity: When the same data are sent over the channel at different time instants, the received signals can be uncorrelated if the time separations are large enough. The required time separation is at least as great as the reciprocal of the fading bandwidth, which is two times the speed of the mobile station divided by the wavelength. Hence, the time separation is inversely proportional to the speed of the mobile station. For a stationary device, time diversity is useless. This is in contrast to all of the other diversity types discussed above because they are independent of the speed of the mobile station.

18.6 Summary

The reduction of noise, interference, channel distortion and multi-path effects are some of the most important challenges in mobile communication. Cellular architecture and the use of smart antennas are the main methods for improving the capacity and spectral efficiency in mobile communication systems. This chapter provided an overview of some of the main issues in modelling and reduction of noise and interference in wireless communication systems.

As the demand for multimedia communication on the relatively limited radio spectrum grows, digital array signal processing will play a central part in the development of smart antennas and array noise reduction methods that would take advantage of the opportunities presented by time/space diversity schemes.

Further Reading

Balanis C.A. (2005) Antenna Theory: Analysis and Design, 3rd edn, John Wiley.

Lee Y.W.C. (1989) Mobile Cellular Communications, New York, McGraw-Hill.

Liberti J.C. and Rappaport T.S. (1999) Smart Antennas for Wireless Communications: IS-95 and Third-Generation CDMA Applications, Prentice Hall.

Padgett J.E., Günther C.G. and Hattori T. (1995) Overview of Wireless Personal Communications, IEEE Communications Magazine, 33 (1), pp. 28–41.

Pahlavan K. and Levesque A.H. (1996) Wireless Information Networks, John Wiley.

Paulraj A., Nabar R. and Gore D. (2003) Introduction to Space–Time Signal Processing, Cambridge University Press.

Rappaport T.S. (1996) Wireless Communications: Principles and Practice, 2nd edn, Prentice-Hall.

Schilling D.L. (1994) Wireless Communications Going into the 21st Century, IEEE Transactions on Vehicular Technology, 43 (3), pp. 645–651.

Shannon C.E. (1948) A Mathematical Theory of Communication, Bell System Technical Journal, 27, in two parts, pp. 379–423, 623–656, July, October.

Sklar B. (1997) Rayleigh Fading Channels in Mobile Digital Communication Systems, Part I: Characterization; Part II: Mitigation, IEEE Communication Magazine, 35 (7), pp. 90–109.

Van Trees H.L. (1898) Detection, Estimation and Modulation Theory, Part I, NewYork, Wiley.

Index

Printed and bound by CPI Group (UK) Ltd, Croydon, CR0 4YY
17/04/2025
14658866-0001